17th International Conference on Strangeness in Quark Matter (SQM 2017)

EPJ Web of Conferences Volume 171 (2018)

Utrecht, The Netherlands
10 - 15 July 2017

Editors:

Andre Mischke
Paul Kuijer

ISBN: 978-1-5108-5724-7

Printed from e-media with permission by:

Curran Associates, Inc.
57 Morehouse Lane
Red Hook, NY 12571

proceedings.com

Some format issues inherent in the e-media version may also appear in this print version.

Printed by Curran Associates, Inc. (2018)

For additional information, please contact EDP Sciences – Web of Conferences at the address below.

EDP Sciences – Web of Conferences
17, Avenue du Hoggar
Parc d'Activité de Courtabœuf
BP 112
F-91944 Les Ulis Cedex A
France

Phone: +33 (0) 1 69 18 75 75
Fax: +33 (0) 1 69 28 84 91

contact-edps@webofconferences.org

Additional copies of this publication are available from:

Curran Associates, Inc.
57 Morehouse Lane
Red Hook, NY 12571 USA
Phone: 845-758-0400
Fax: 845-758-2633
Email: curran@proceedings.com
Web: www.proceedings.com

TABLE OF CONTENTS

Proceedings of the
International Conference on Strangeness in Quark Matter 2017

Preface

The 17th edition of the *International Conference on Strangeness in Quark Matter* took place from 10-15 July 2017 in Utrecht in the Netherlands (http://sqm2017.nl). This conference series focuses on new experimental and theoretical developments on the role of strangeness and heavy-flavour production in heavy-ion collisions, and in astrophysical phenomena related to strangeness. This year's conference attracted more than 210 participants from 25 countries, with 20% of female researchers. A two-day long graduate school on the role of strangeness in heavy ion collisions with 40 young participants preceded the conference, which was supported by the Dutch Physical Society (NNV).

The scientific programme consisted of 53 invited plenary talks, 70 contributed parallel talks and a poster session. Three discussion sessions provided scope for the necessary debates on crucial observables to characterise strongly interacting matter at extreme conditions of high baryon density and high temperature and to define possible future directions. One of the discussions centred on hadronic resonance production and their vital interactions in the partonic and hadronic phase that provide evidence for an extended hadronic lifetime even in small collision systems and might affect other observables for the quark-gluon plasma. Moreover, future astrophysical consequences for SQM following the recent detection of gravitational waves were outlined: gravitational waves from relativistic neutron star collisions can serve as cosmic messengers for the phase structure and equation-of-state of dense and strange matter, quite similar to the environment created in relativistic heavy-ion collisions.

Representatives from all major collaborations at the Large Hadron Collider (LHC) and Super Proton Synchrotron at CERN, Brookhaven's Relativistic Heavy Ion Collider (RHIC), and the Heavy Ion Synchrotron SIS at the GSI Helmholtz Centre in Germany made special efforts to release new data at this conference. Thanks to the excellent performance of these accelerator facilities and detectors, a wealth of new data on the production of strangeness and heavy flavour in nuclear collisions has become available.

Among the highlights presented at the conference, the ALICE Collaboration reported new results on strange and multi-strange hyperon production in 5.02 TeV heavy-ion collisions and the first measurement of charm baryons (Λ_c and Ξ_c) in proton-proton and proton-lead collisions at the LHC. Furthermore, ALICE performed the most precise measurement of the (anti-)hypertriton lifetime, an exotic nucleus composed of a proton, a neutron and a lambda particle. The CMS Collaboration reported progress in understanding the energy loss of charm and beauty quarks in the hot QCD medium, while the STAR experiment at RHIC gave an update on global lambda polarisation, which reveals that the curl of the fluid created at RHIC is much higher than that in any fluid ever observed. Enhanced strangeness production in small systems, as reported by the

HADES, NA61/SHINE and ALICE Collaborations, has also reignited the discussion surrounding strangeness production as a signature of the quark-gluon plasma.

Experimentally, the field faces high prospects for future measurements at the Facility for Antiproton and Ion Research in Darmstadt, NICA at JINR Dubna, and at CERN (namely detector upgrades at the LHC during long shutdown 2 and the AFTER programme).
On the theory side, new developments and vigorous research efforts are taking place towards a full understanding of strangeness production and open heavy-flavour dynamics in heavy-ion collisions. Global polarisation in heavy-ion collisions is also a highly debated topic since it allows studying the vorticity of the medium and the initial magnetic field.

A dedicated session was held in memory of Helmut Oeschler who was an active member of the ALICE collaboration and passed away unexpectedly in 2017. Margit Oeschler, Helmut's wife, was present at the memorial session.

Four young scientist prizes, sponsored by the European Physical Journal A, were awarded to the best parallel talk and poster presenters: Heidi Schuldes (Goethe University Frankfurt, Germany), Christian Bierlich (Lund University, Sweden), Yingru Xu (Duke University, US) and Vojtech Pacik (Niels Bohr Institute, Denmark).

André Mischke and Paul Kuijer, Editors

More than 210 participants attended SQM2017 conference in Utrecht (courtesy: Pieter van Dorp van Vliet, Utrecht University).

edp sciences

Web of Conferences

Statement of Peer review

In submitting conference proceedings to *Web of Conferences*, the editors of the proceedings certify to the Publisher that

1. They adhere to its **Policy on Publishing Integrity** in order to safeguard good scientific practice in publishing.
2. All articles have been subjected to peer review administered by the proceedings editors.
3. Reviews have been conducted by expert referees, who have been requested to provide unbiased and constructive comments aimed, whenever possible, at improving the work.
4. Proceedings editors have taken all reasonable steps to ensure the quality of the materials they publish and their decision to accept or reject a paper for publication has been based only on the merits of the work and the relevance to the journal.

Title, date and place of the conference

17th International Conference on Strangeness in Quark Matter, 10-15 July 2017, Utrecht, the Netherlands

Proceedings editor(s):

Andre Mischke (Utrecht University) and Paul Kuijer (Nikhef Amsterdam)

Date and editor's signature

14 December 2017

Studying Strangeness Production with HADES

Heidi Schuldes[1,*] for the HADES Collaboration

[1] *Goethe-University Frankfurt*

Abstract. The High-Acceptance DiElectron Spectrometer (HADES) operates in the 1 - 2*A* GeV energy regime in fixed target experiments to explore baryon-rich strongly interacting matter in heavy-ion collisions at moderate temperatures with rare and penetrating probes. We present results on the production of strange hadrons below their respective NN threshold energy in Au+Au collisions at 1.23*A* GeV ($\sqrt{s_{NN}}$ = 2.4 GeV). Special emphasis is put on the enhanced feed-down contribution of ϕ mesons to the inclusive yield of K^- and its implication on the measured spectral shape of K^-. Furthermore, we investigate global properties of the system, confronting the measured hadron yields and transverse mass spectra with a Statistical Hadronization Model (SHM) and a blast-wave parameterization, respectively. These supplement the world data of the chemical and kinetic freeze-out temperatures.

1 Introduction

The HADES mission is to do detailed studies of the properties of strongly interacting matter at high net-baryon densities with rare and penetrating probes. Whereas at low baryo-chemical potential μ_b ab-inito calculations from lattice QCD are predicting a smooth cross-over from deconfined to hadronic matter, calculations at high μ_b are not reliable when based on first principles due to the fermion determinant sign problem [1]. Therefore, hadron properties have to be extracted from models based on effective Lagrangians.

A phenomenological tool to characterize the system produced in Heavy-Ion Collisions (HIC) at various beam energies is the Statistical Hadronization Model (SHM) [2–4]. It allows to extract freeze-out parameters by fitting particle yields. These parameters show a remarkable regularity of falling on a smooth curve in the temperature T-μ_B plane from lowest SIS18 energies up to highest available energies at LHC [5, 6]. Whereas at highest energies the appearance of chemical equilibrium in the system can be assumed, it is still under discussion in the SIS18 energy regime, where the collision dynamics show long interpenetration times and a large amount of stopping of the colliding nuclei in the reaction zone.

The production of strange hadrons below their respective nucleon-nucleon threshold energies[1] is a particular suitable probe for the high-density phase of the HIC. As particles carrying strangeness can not be produced in binary NN collisions, the energy to produce these particles has to be provided by the system. In the framework of transport model calculations the necessary energy is accumulated

[*] e-mail: h.schuldes@gsi.de

[1] The thresholds are for $NN \to N\Lambda K^+/K^0$: $\sqrt{s_{thr}}$ = 2.55 GeV, for $NN \to NNK^+K^-$: $\sqrt{s_{thr}}$ = 2.86 GeV and for $NN \to NN\phi$: $\sqrt{s_{thr}}$ = 2.89 GeV

in multi-step processes via intermediate resonances. As kaons carry an anti-strange quark, their coupling to the baryons of the medium is suppressed. Therefore, they are expected to have a rather long mean free path in nuclear matter of $\lambda \approx 5$ fm when applying the low-density approximation to measurements of the K^+N cross-section in these models and are supposed to transmit information about the hot and dense stage of the HIC undisturbed [7]. Anti-kaons, on the other hand, show a strong coupling to baryons in these models, leading to a complicated and strongly modified spectral function in the medium. Since the suggestion of the possibility of an $\bar{K}$ condensate in dense nuclear matter from Kaplan and Nelson in the 1980ies [8] and possible implications on astrophysical objects, various attempts to extract information about the strength of the suggested attractive $\bar{K}N$ potential in different approaches have been made (see for example [9–14]).
The first data on sub-threshold K^- production in HIC obtained by the KaoS collaboration revealed that the K^- and K^+ multiplicities show, despite their different NN threshold energies, a similar increase with collision centrality, whereas the transverse mass spectra of K^- are significantly softer compared to the ones of K^+ (for a review of the data see [15]). These two observations found a possible explanation within transport model calculations when assuming the strangeness exchange reaction $\Lambda\pi \rightarrow NK^-$, which was predicted by Ko in the early 1980ies [16] to be the dominant production channel for K^- below threshold [17]. This reaction couples the production of K^- to K^+, thus explaining their similar centrality dependence and the later decoupling of K^-, resulting in the softer spectrum. However, recent data on strangeness production in smaller collision systems indicate that ϕ feed-down decays constitute also a sizeable source for K^- production below the threshold [18–20], which was not taken into account in transport models.
The measurement of a close to complete set of strange hadrons in the same collision system will help to do a detailed characterization of the produced medium, as well as to constrain different model assumptions concerning effective potentials and production mechanisms. In particular the low part of the transverse momentum spectrum of K^0_S is sensitive to KN potentials employed in transport models, as it is in contrast to the one of K^+, not affected by the coulomb potential. The detailed comparison of various observables from K^0_S and Λ to different models is subject of a future publication. In these proceedings we discuss the first measurement of ϕ so far below the NN threshold and its implications for sequential K^+/K^- freeze-out in section 3 and the general freeze-out properties of the system in section 4.

2 Experimental setup and data analysis

The presented data has been recorded with the High Acceptance Di-Electron Spectrometer (HADES), located at the SIS18 accelerator of the GSI Helmholtz center for heavy-ion research in Darmstadt, Germany. The detector is composed of six identical sectors, surrounding the beam axis and covering almost the full azimuthal and polar angles between 18 and 85°. For a detailed description of the spectrometer and its components see [21]. The track and momentum reconstruction are performed with a magnet spectrometer consisting of two Mini Drift Chambers (MDC) in front of and two behind a superconducting magnet. In combination with the time-of-flight determination via a scintillator hodoscope (TOF) and a Resistive Plate Chamber (RPC) at the end of the setup, particles can be identified. In order to reduce the contribution of protons and pions to the signal of the rarely produced kaons, the energy loss signal measured in the drift chambers and the TOF detector can be employed. The background is described using a third order polynomial function and is subtracted in an iterative fitting procedure. Unstable neutral hadrons are identified via their decay into charged particles ($\Lambda \rightarrow p\pi^-$, $K^0_S \rightarrow \pi^+\pi^-$, $\phi \rightarrow K^+K^-$). In order to increase the signal-to-background ratio and to reduce the amount of uncorrelated pairs, cuts on the specific decay topology of the weakly decaying Λ and K^0_S are applied. The remaining background is subtracted using the mixed-event technique.

In total 2.1 $\times 10^9$ Au+Au events at a kinetic energy of 1.23*A* GeV, corresponding to $\sqrt{s_{NN}} = 2.4$ GeV, are used in the analysis, corresponding to the 40% most central events, with $\langle A_{part} \rangle = 191 \pm 11$, as has been estimated with Glauber model calculations [22, 23]. The analysis is performed in small cells in reduced transverse mass $m_t - m_0$ and rapidity y and in four (K^+, K^0 and Λ) or two (K^- and ϕ) centrality bins. The raw signal count rates are corrected in each phase space cell for acceptance and efficiency based on detailed simulations of the detector response. As input for the simulations the particles are generated with a thermal distribution and embedded into UrQMD events [24], which serve as realistic background.
The spectra are parameterized by Boltzmann functions of the following form:

$$\frac{1}{m_t^2}\frac{d^2N}{dm_t dy} = c(y) \cdot \exp - \frac{(m_t - m_0)}{T_B(y)}, \tag{1}$$

with a rapidity dependent normalization $c(y)$ and the slope parameter $T_B(y)$, in order to extrapolate to unmeasured transverse mass regions and to extract the so-called effective temperature of the particles $T_{\text{eff}} = T_B(y_{cm})$. The resulting rapidity density distributions are extrapolated to rapidity regions not covered by the spectrometer using Gaussian functions to extract the total multiplicities of the particles. For a detailed description of the reconstruction of strange hadrons see [27–29].

3 ϕ production and implications for sequential K^+/K^- freeze-out

The measured ϕ/K^- multiplicity ratio for the 0-40% most central collisions [27] is shown on the left side of Fig. 1 in comparison with measurements at similar energies in lighter systems [18–20] as well as data at higher $\sqrt{s_{NN}}$ [30–32]. Whereas this ratio is flat at energies above $\sqrt{s_{NN}} \geq 4$ GeV with values around 0.15, it is increasing for energies below the *NN* threshold. As a consequence ϕ feed-down becomes a sizeable source for K^- production below the threshold in the large Au+Au system.
The estimated multiplicities of all strange hadrons show, despite their different elementary production thresholds, a similar dependence on the centrality of the collision, as observed also at higher energies. Whereas the coupling of K^- to K^+, and to K^0_S and Λ due to strangeness conservation, can be explained in a microscopic picture by strangeness exchange reactions, most transport models are not able to reproduce the large ϕ/K^- multiplicity ratio below the *NN* threshold and can not explain the centrality independence seen for ϕ production.
As observed at slightly higher energies [15], we obtained a systematic lower effective temperature for K^- than for K^+ and the other strange hadrons. However, when taking the 25% feed-down contribution to the K^- production from ϕ decays into account, we find that the K^- originating from these decays are significantly cooler than thermally produced kaons, and therefore result in an overall softer K^- spectrum. The effect can be seen on the right side of Fig. 1, where a simulated two-component cocktail of the K^- transverse mass spectrum around mid-rapidity is shown. To the 75% thermally produced K^- (dotted red), generated with the measured effective temperature of the K^+ in the 0-40% most central collisions [27] ($T_{\text{eff}}^{K+} = (104 \pm 1 \pm 1)$ MeV), we add 25% K^- from ϕ decays (dashed blue), which have been generated according to the measured value of ϕ ($T^{\phi}_{eff} = (108 \pm 7)$ MeV). As a consequence the obtained inverse slope parameter of the cocktail (solid black) is significantly lower for K^- than the one of K^+ ($T_{\text{eff}}^{K-} = 84$ MeV, obtained using Eq. 1), as observed in the experiment (measured $T_{\text{eff}}^{K-} = (84 \pm 6)$ MeV).
In conclusion, we see no indication for a sequential kaon freeze-out scenario, which has been suggested formerly by transport model calculations as an explanation for the significantly lower effective temperature of K^- compared to K^+, when simply taking into account the strong ϕ feed-down contribution. As a consequence, conclusions about an attractive K^-N potential and its implications on production mechanisms and astrophysical objects need to be revisited.

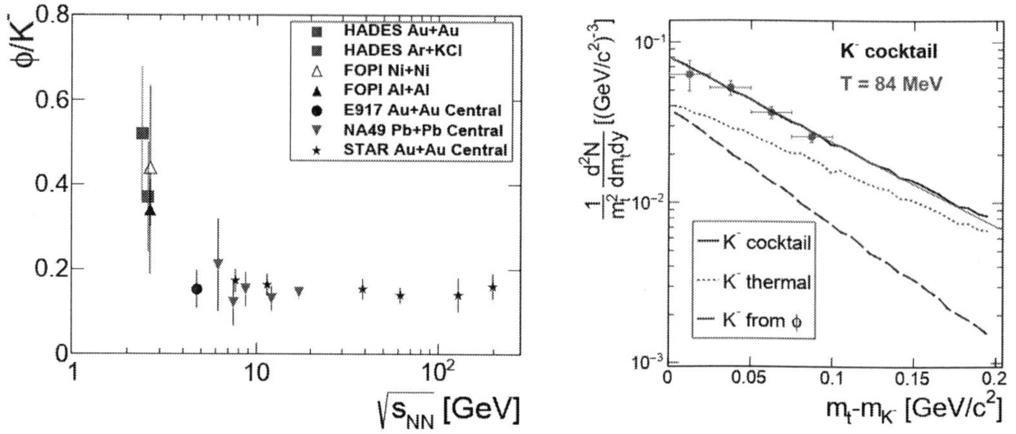

Figure 1. Left: ϕ/K^- multiplicity ratio as a function of $\sqrt{s_{NN}}$ [18–20, 27, 30–32]. Right: Simulated transverse mass spectra of thermally produced K^- (dotted red), K^- from ϕ decays (dashed blue), and their sum, here denoted as cocktail (black), generated according to the effective temperatures measured for 0-40% most central collisions. The spectra can be described by Eq. 1 in order to extract the effective temperature T_{eff}.

4 Freeze-out parameters

The universal centrality dependence of the yields of strange hadrons indicates that the produced system shows some macroscopic behavior. We find a good description of the measured hadron yields in 20% most central collisions with a SHM fit based on a mixed-canonical ensemble. The model assumes particle emission from a homogenous source which is in thermal equilibrium, therefore the quantum numbers, except strangeness, are calculated grand canonically. As strangeness is produced very rarely in the analyzed energy regime (typical one $s\bar{s}$ pair per event), its production is additionally suppressed by introducing a sub-volume, characterized by the canonical suppression radius R_C, in which strangeness has to be conserved exactly and not on average. This suppression is only affecting the open strangeness and not the ϕ production, therefore this model is able to describe the enhanced ϕ/K^- ratio. A chemical freeze-out temperature of $T_{\text{ch}} = (68 \pm 2)$ MeV is extracted.
Furthermore, we find a fair description of the transverse mass spectra at mid-rapidity ($y_{cm}^{Au+Au} = 0.74$) of protons, kaons and the high energy tail of π^-, for the 10% most central collisions, by a blast-wave parameterization according to [34]

$$\frac{dN}{m_t dm_t} \propto \int_0^R r dr m_t I_0\Big(\frac{p_t \sinh\rho(r)}{T_{\text{kin}}}\Big) \times K_1\Big(\frac{m_t \cosh\rho(r)}{T_{\text{kin}}}\Big), \tag{2}$$

where $\rho(r) = \tanh^{-1}\beta$ and I_0 and K_1 are the modified Bessel functions. We employ the radial flow velocity profile of the form

$$\beta = \beta_S (r/R)^n, \tag{3}$$

where β_S is the surface velocity, r/R the relative position in the radial source and $n = 1$ the linear flow velocity profile, shown on the right side of Fig. 2. The averaged radial flow velocity is given by

$$\langle\beta\rangle = \frac{2}{2+n}\beta_S. \tag{4}$$

We obtain the freeze-out parameters $T_{\text{kin}} = (62 \pm 10)$ MeV and $\langle\beta\rangle = 0.36 \pm 0.04$. Λ and ϕ are slightly deviating from the simultaneous fit, indicating that these particles are preferentially emitted

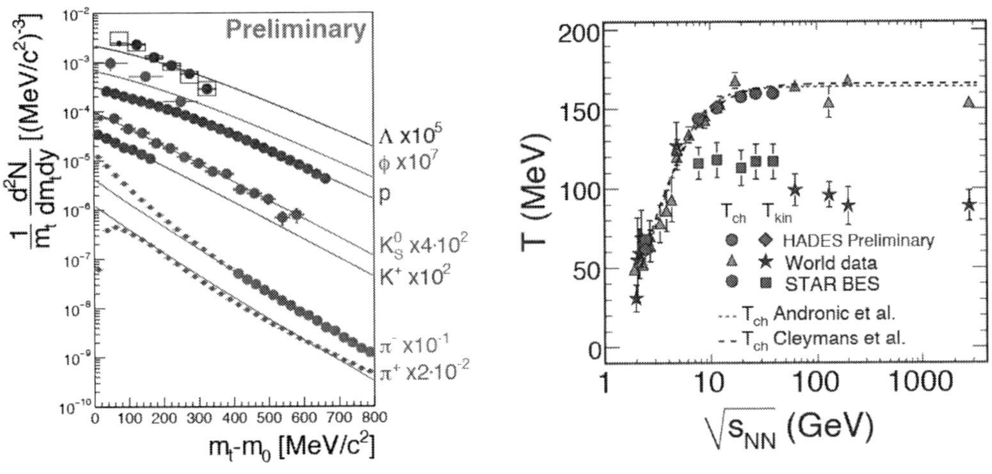

Figure 2. Left: Simultaneous fit to the transverse mass spectra at mid-rapidity of hadrons by a blast-wave parameterization according to Eq. 2. Only fat points are included in the fit. Right: Chemical (T_{ch}) and kinetic (T_{kin}) freeze-out temperatures obtained in Au+Au collisions with HADES (red) compared to world data (summarized in [35]).

from a source with higher temperature and less radial flow.

If we compare the obtained freeze-out temperatures to the world data (summarized in [35]) as shown on the right side of Fig. 2, we find a nice agreement. The measured T_{kin} is smaller than T_{ch}, which was opposite in smaller collision systems in the analyzed energy regime [36].

5 Summary and outlook

The high-quality data on strange hadron production in Au+Au collisions at 1.23*A* GeV measured by HADES allow to characterize the strongly-interacting medium produced in heavy-ion collisions at high net-baryon densities. We find good agreement of the measured hadron yields by a fit with the SHM, assuming a mixed-canonical ensemble in which strangeness is canonically suppressed by R_C. The extracted chemical freeze-out temperature of T_{ch} = (68 ± 2) MeV is in good agreement with the world data, as well as the kinetic freeze-out temperature T_{kin} = (62 ± 10) MeV, extracted from a simultaneous fit to the measured transverse mass spectra of hadrons using a blast-wave approach. Furthermore, we find an enhanced ϕ/K^- multiplicity ratio. Taken this into account, we find a natural explanation of the lower effective temperature measured for K^- compared to K^+ without assuming a sequential freeze-out scenario of the two kaon species. In conclusion, our results on strangeness production are consistent with the assumption of reaching thermal equilibrium also at SIS18 energies.

In 2018/2019 HADES will continue its broad physics program at the SIS18 accelerator in the context of the FAIR phase-0, focusing on the production of multi-strange hadrons and intermediate-mass dileptons in Ag+Ag reactions at 1.65*A* GeV, as well as studying the electromagnetic structure of baryonic resonances in pion-induced reactions. In the later future this research will be continued at higher energies at the SIS100 accelerator.

References

[1] O. Philipsen, EPJ Web Conf. **137** (2017) 03016.
[2] P. Braun-Munzinger, K. Redlich, J. Stachel, arXiv:nucl-th/0304013 (2003).
[3] A. N. Tawfik, Int. J. Mod. Phys. A **29** 17 (2014) 1430021.
[4] M. Floris, Nucl. Phys. A **931** (2014) 103.
[5] J. Cleymans, H. Oeschler, K. Redlich, S. Wheaton, Phys. Rev. C **73** (2006) 034905.
[6] J. Stachel, A. Andronic, P. Braun-Munzinger, K. Redlich, J. Phys. Conf. Ser. **509** (2014) 012019.
[7] C. Hartnack, H. Oeschler, Y. Leifels, E. L. Bratkovskaya, J. Aichelin, Phys. Rept. **510** (2012) 119.
[8] D. B. Kaplan, A. E. Nelson, Phys. Lett. B **175** (1986) 57.
[9] C. H. Lee, G. E. Brown, D. P. Min, M. Rho, Nucl. Phys. A **585** (1995) 401.
[10] J. Schaffner-Bielich, J. Bondorf, A. Mishustin, Nucl. Phys. A **625** (1997) 325.
[11] M. F. M. Lutz, A. Steiner, W. Weise, Nucl. Phys. A **574** (1994) 755.
[12] V. Koch, Phys. Lett. B **337** (1994) 7.
[13] W. Cassing, E. L. Bratkovskaya, U. Mosel, S. Teis, A. Sibirtsev, Nucl. Phys. A **614** (1997) 415.
[14] D. Cabrera, L. Tolos, J. Aichelin, E. Bratkovskaya, J. Phys. Conf. Ser. **668** no.1 (2016) 012048.
[15] A. Förster, F. Uhlig, I. Bottcher, D. Brill, M. Debowski *et al.* [KaoS Collaboration], Phys. Rev. C **75** (2007) 024906.
[16] C. M. Ko, Phys. Lett. B **120** (1983) 294.
[17] C. Hartnack, H. Oeschler, Y. Leifels, E. L. Bratkovskaya, J. Aichelin, Phys. Rept. **510** (2012) 119.
[18] G. Agakishiev *et al.* [HADES Collaboration], Phys. Rev. C **80** (2012) 025209.
[19] K. Piasecki *et al.* [FOPI Collaboration], Phys. Rev. C **91** (2015) 054904.
[20] P. Gasik *et al.* [FOPI Collaboration], Eur. Phys. J. A **52** (2016) 177.
[21] G. Agakishiev *et al.* [HADES Collaboration], Eur. Phys. J. A **41** (2009) 243.
[22] R. J. Glauber, G. Matthiae, Nucl. Phys. B **21** (1970) 135.
[23] B. Kardan, Diploma Thesis, Goethe-University Frankfurt (2016).
[24] S.A. Bass *et al.*, Prog. Part. Nucl. Phys.**41** (1998) 255-369.
[25] W. Cassing, E.L. Bratkovskaya, Phys. Rep. **308** (1999) 65-233.
[26] C. Hartnack, R.K. Rajeev, J. Aichelin, J. Konopka, S.A. Bass *et al.*, Eur. Phys. J. A **1** (1998) 151-169.
[27] J. Adamczewski-Musch *et al.* [HADES Collaboration], arXiv:nucl-ex/1703.08418 (2017).
[28] H. Schuldes, PhD Thesis Goethe-University Frankfurt (2016).
[29] T. Scheib, PhD Thesis Goethe-University Frankfurt (2017).
[30] B. Holzman *et al.* [E917 Collaboration], Nucl. Phys. A **698** (2002) 643.
[31] S. V. Afanasiev *et al.* [NA49 Collaboration], Phys. Lett. B **491** (2000) 59.
[32] J. Adams *et al.* [STAR Collaboration], Phys. Lett. B **612** (2005) 181.
[33] S. Wheaton, J. Cleymans, Comput. Phys. Commun. **180** (2009) 84.
[34] E. Schnedermann, J. Sollfrank, U. W. Heinz, Phys. Rev. C **48** (1993) 2462.
[35] L. Adamczyk *et al.* [STAR Collaboration], arXiv:nucl-ex/1701.07065 (2017).
[36] G. Agakishiev *et al.* [HADES Collaboration], Eur. Phys. J. A **47** (2011) 21.

EPJ Web of Conferences **171**, 01002 (2018) https://doi.org/10.1051/epjconf/201817101002
SQM 2017

News from the NA61/SHINE experiment

Tatjana Šuša for the NA61/SHINE Collaboration[1,*]

[1] *Institute Ruđer Bošković*

Abstract. The main goals of the NA61/SHINE experiment at the CERN SPS are the search for the critical point of strongly interacting matter and the study of the properties of the onset of deconfinement. These aims are pursued by performing a two-dimensional scan of the phase diagram of strongly interacting matter by varying the momentum and size of the colliding nuclei.
This contribution summarises the latest results from the NA61/SHINE experiment, in particular, new results on spectra and yields of ϕ meson in p+p interactions at 40, 80 and 158 GeV/c and K^+ and K^- production in central Be+Be collisions at mid-rapidity. In addition, results on system size dependence of particle yield ratios and fluctuations are presented.

1 Introduction

NA61/SHINE [1] is a fixed-target experiment located in the H2 beam line of the CERN SPS accelerator complex. The main detector system is a set of five Time Projection Chambers (TPCs) which allows a precise measurement of the particle momenta and provides particle identification via the measurement of the specific energy loss, dE/dx. Two time-of-flight walls improve the kaon identification capabilities. The geometrical layout of the TPCs allows particle detection down to $p_T = 0$ GeV/c in a broad interval of forward rapidities. A high resolution hadron calorimeter, the Projectile Spectator Detector, measures forward going energy E_F and is used to select centrality in nuclear collisions.

The main goal of the strong interactions programme of the NA61/SHINE experiment is to discover the critical point of strongly interacting matter and study the properties of the onset of deconfinement. The programme is motivated by the discovery of the onset of deconfinement in Pb+Pb collisions at 30A GeV/c by the NA49 experiment [2, 3]. To achieve this goal a two-dimensional scan of the phase diagram of strongly interactiong matter is performed by varying the beam momentum (13A-150/158A GeV/c) and the size of the colliding nuclei (p+p, p+Pb, Be+Be, Ar+Sc, Xe+La, Pb+Pb).

2 Recent results from NA61

2.1 ϕ meson production in p+p interactions

The production of ϕ meson is measured in p+p interactions at 40, 80 and 158 GeV/c. ϕ mesons were identified by their charged kaon decays $\phi \rightarrow K^+K^-$ using the invariant mass method. The statistics

*e-mail: tatjana.susa@irb.hr

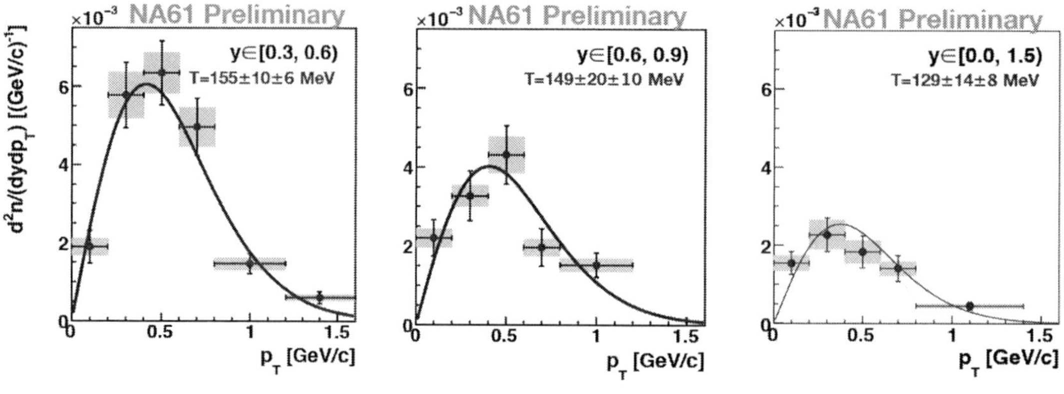

Figure 1. Examples of transverse momentum distributions of ϕ mesons in p+p interactions at beam momenta of 158 (left), 80 (middle) and 40 (right) GeV/c.

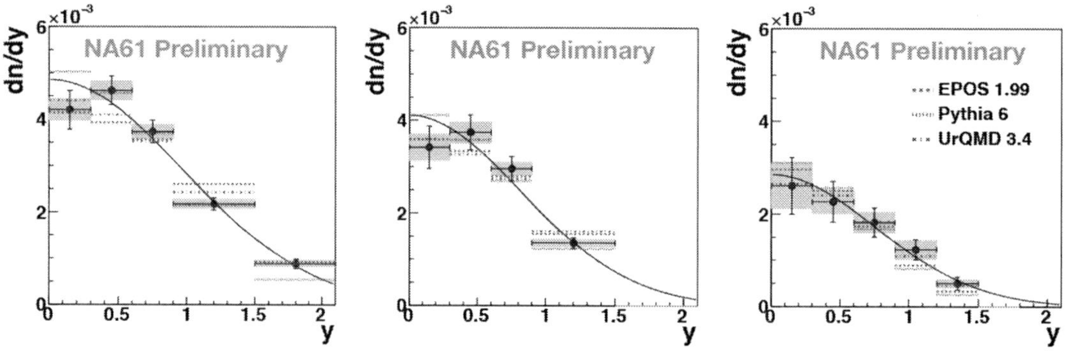

Figure 2. Rapidity distributions of ϕ mesons in p+p interactions at beam momenta od 158 (left), 80 (middle) and 40 (right) GeV/c.

of 40 GeV/c dataset only allowed to extract single differential dn/dy and dn/dp_T spectra in one wide transverse momentum ($p_T \in [0, 1.4]$ GeV/c) and one rapidity ($y \in [0, 1.5]$) bin, respectively, while at 80 and 158 GeV/c a double-differential analysis in bins of rapidity and transverse momentum could be performed. These are the first measurements of ϕ production at 40 and 80 GeV/c and the first double-differential measurements at 158 GeV/c.

Examples of p_T spectra are presented in Fig. 1. The systematic uncertainties are shown by the blue shaded boxes while horizontal error bars indicate the width of the p_T bins. The spectra are fitted with an exponential function in m_T shown by the blue curve. The extracted values of the inverse slope parameter T at 80 and 158 GeV/c are in the range 120 - 160 MeV and decrease with increasing rapidity (see conference slides).

The rapidity spectra of ϕ mesons are presented in Fig. 2. The systematic uncertainties are shown by the blue shaded boxes while horizontal error bars indicate the width of the rapidity bins. Shown are also predictions for the shape of rapidity spectra from the Epos1.99 [4, 5] (version in CRMC1.6.0 package [6]), Pythia6.4.28 [7] and UrQMD3.4 [8, 9] models which are normalised to the integral of the data. All models approximately describe the shape of the spectra. The result of the fit with a single Gaussian to the experimental points is shown by the black curve. The width of the rapidity spectra

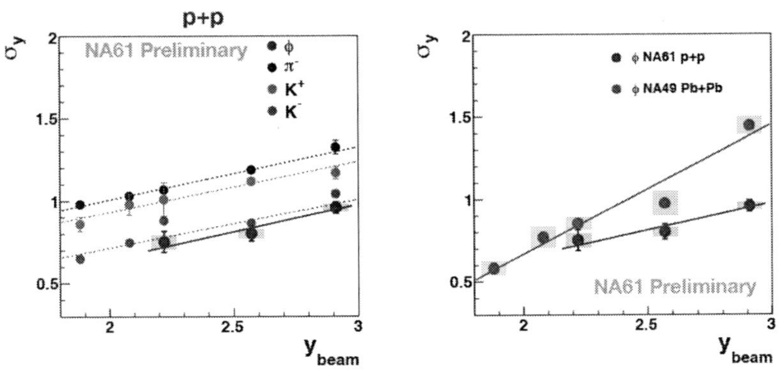

Figure 3. Width of the rapidity distributions of ϕ, π^- and $K^\pm$ mesons in p+p collisions (left), width of the rapidity distributions of ϕ mesons in p+p and Pb+Pb collisions (right) as a function of beam rapidity.

derived from the fit is compared to the width of the rapidity spectra of π^-, K^+ and K^- in p+p [10] collisions in Fig. 3 (left) and to the width of the rapidity spectra of ϕ meson in Pb+Pb [11] collisions in Fig. 3 (right). While in p+p collisions the ϕ follows the trend of the other hadrons, the widths of the rapidity distributions of ϕ mesons in p+p and Pb+Pb exhibit different behavior.

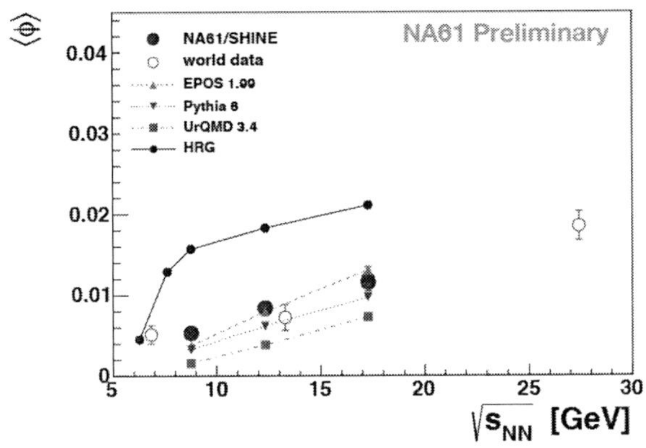

Figure 4. Energy dependence of the mean ϕ multiplicity in p+p collisions, together with the data from other experiments [12–15] and predictions from Epos1.99 [4, 5], Pythia6.4.28 [7], UrQMD3.4 [8, 9] and HRG [16] models.

The mean ϕ multiplicity, $\langle\phi\rangle$, is derived by summing the measured rapidity spectra and by extrapolation to the unmeasured high rapidity region using the fitted single Gaussian. The obtained values are doubled to take into account the contribution from the backward hemisphere. The NA61 results on mean ϕ multiplicity are compared in Fig. 4 to the world data [12–15] and to the predictions of the Epos, Pythia6, UrQMD and hadron resonance gas (HRG [16]) models. The NA61/SHINE results on mean ϕ multiplicity are consistent with the world data and are much more accurate. The Epos model describes the data quite well, Pythia6 underestimates the experimental data, while UrQMD underestimates and HRG overestimates NA61/SHINE results approximately by a factor two.

The ratio of mean multiplicity of ϕ mesons and pions in p+p and central Pb+Pb collisions as a function of center of mass energy per nucleon pair is shown in Fig. 5 (left), while the double ratio (Pb+Pb)/(p+p) of the ratios $\langle\phi\rangle/\langle\pi\rangle$ and $\langle K^\pm\rangle/\langle\pi^\pm\rangle$ is presented in Fig. 5 (right). Pb+Pb data are from NA49 [2, 3, 11], while p+p kaon and pion data are taken from Ref. [10]. As seen from Fig. 5 (left) the $\langle\phi\rangle/\langle\pi\rangle$ ratio increases with $\sqrt{s_{NN}}$. In Pb+Pb collisions it is approximately three times larger than in p+p collisions, independently of the interaction energy. Figure 5 (right) shows that the enhancement between pp and PbPb of the double ratio $\langle\phi\rangle/\langle\pi\rangle$ is close to that of $\langle K^+\rangle/\langle\pi^+\rangle$ and systematically larger than that of $\langle K^-\rangle/\langle\pi^-\rangle$.

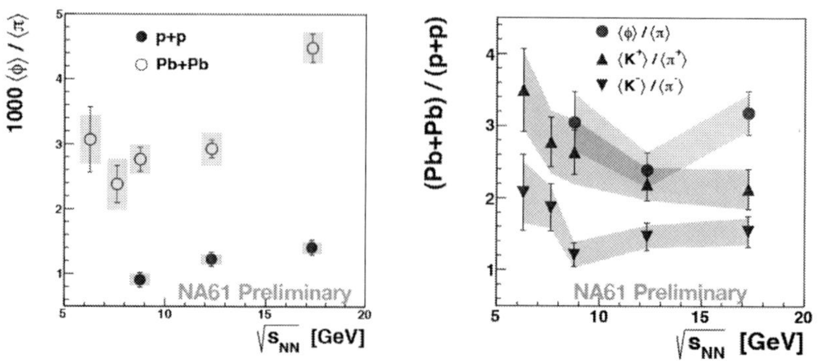

Figure 5. Energy dependence of the $\langle\phi\rangle/\langle\pi\rangle$ ratio in p+p and central Pb+Pb collisions (left) and the double (Pb+Pb)/(p+p) ratio of the $\langle\phi\rangle/\langle\pi\rangle$ and $\langle K^{\pm}\rangle/\langle\pi^{\pm}\rangle$ ratios (see text for references).

2.2 Kaon production in Be+Be interactions

NA61/SHINE performed the first measurement of $K^{\pm}$ production at mid-rapidity in the most violent Be+Be interactions (the 20% of events with the smallest forward energy, E_F) at 30A, 40A, 75A and 150A GeV/c. The $K^{\pm}$ mesons were identified based on the measurement of the energy loss in the TPCs (dE/dx) and time-of-flight in the ToF detectors.

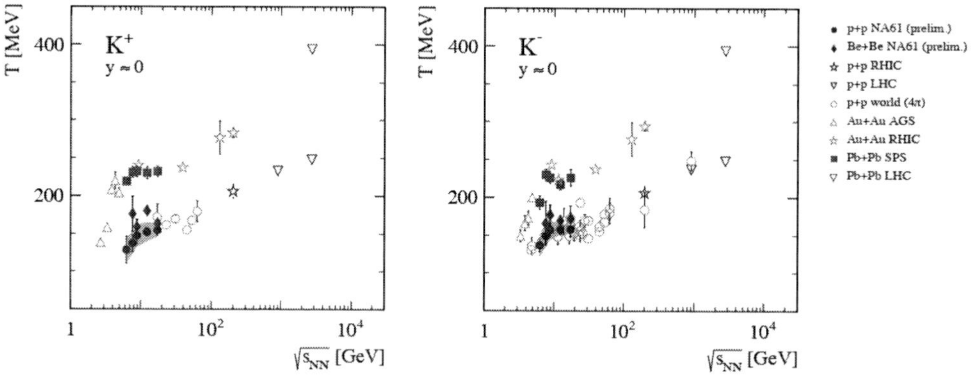

Figure 6. Energy dependence of the inverse slope parameter of m_T spectra at mid-rapidity of positively (left) and negatively (right) charged kaons.

Fig. 6 and Fig. 7 present the energy dependence of the inverse slope parameter of m_T spectra of charged kaons and the multiplicity ratio of charged kaons to pions at mid-rapidity, respectively. Results for Be+Be interactions are compared with NA61/SHINE p+p results and with results from central Pb+Pb collisions from NA49 [2, 3] and other experiments [17–24]. A plateau in the energy dependence of the inverse slope parameter and a peak observed in the energy dependence of the multiplicity ratio of charged kaons to pions in Pb+Pb collisions were predicted by the SMES [25] model as signatures of the onset of deconfinement. The NA61/SHINE results on the energy dependence of the inverse slope parameter in p+p and Be+Be interactions are also consistent with the plateau in the SPS energy range, while a step structure with plateau is observed in the energy dependence of the K^+/π^+ ratio, suggesting that some properties of hadron production previously attributed to onset of deconfinement in heavy ion collisions might be present also in p+p and Be+Be interactions.

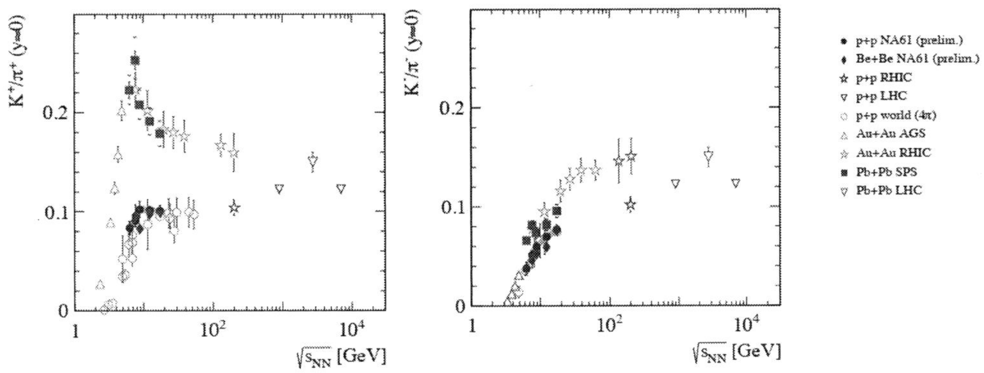

Figure 7. Energy dependence of the ratio of positively (left) and negatively (right) charged kaon to pion yield at mid-rapidity.

2.3 System size dependence of particle yields and fluctuations

NA61/SHINE results on the system size dependence of particle yield ratios and fluctuations are presented in Fig. 8 and Fig. 9, respectively.

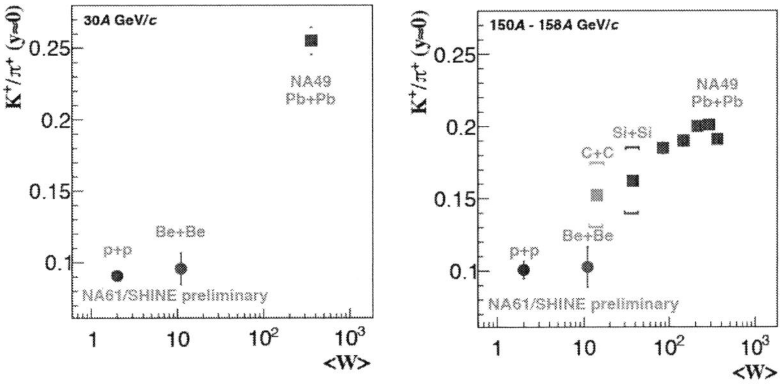

Figure 8. Ratio of positively charged kaon to pion yield at mid-rapidity as a function of mean number of wounded nucleons $\langle W \rangle$ at 30A (left) and 150/158A (right) GeV/c.

In Fig. 8 the ratios of positively charged kaon to pion yield at mid-rapidity in p+p and Be+Be collision at 30A and 150/158A GeV/c are shown, together with the C+C, Si+Si and Pb+Pb data from the NA49 experiment. Mid-rapidity K^+/π^+ ratios in p+p and Be+Be collisions are very close, independently of the collison energy and differ significanty from those in medium and heavy system.

NA61/SHINE uses the scaled variance of the multiplicity distribution $\omega[N] \equiv Var(N)/\langle N \rangle$ to characterize the strength of multiplicity fluctuations. It is an intensive variable, insensitive to the volume of the system, but sensitive to fluctuations of the volume. In Fig. 9 the scaled variance of the multiplicity distribution of negatively charged hadrons in p+p, Be+Be and Ar+Sc collisions at 30A and 150/158 AGeV/c is shown as a function of the mean number of wounded nucleons $\langle W \rangle$.

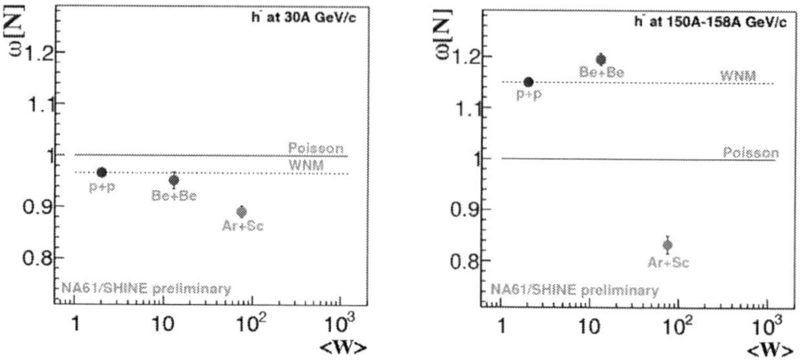

Figure 9. The scaled variance of the multiplicity distribution of negatively charged hadrons in p+p, Be+Be and Ar+Sc collisions at 30A (left) and 150/158A (right) GeV/c as a function of the mean number of wounded nucleons $\langle W \rangle$.

The multiplicity fluctuations in p+p and Be+Be collisions are similar and significantly higher than in Ar+Sc collisions. The observed suppression of the scaled variance in Ar+Sc collisions is not expected for an Ideal Boltzmann gas in the Grand Canonical Ensamble treatment nor in the Wounded Nucleon Model [26].

References

[1] N. Abgrall et al., JINST **9**, P06005 (2014)
[2] C. Alt et al., Phys. Rev. C**77**, 024903 (2008)
[3] S. V. Afanasiev et al., Phys. Rev. **C66**, 054902 (2002)
[4] K. Werner, F. Liu, and T. Pierog, Phys. Rev. C**74**, 044902 (2006)
[5] T. Pierog and K. Werner, Nucl. Phys. B (Proc. Suppl.) **196**, 102 (2009)
[6] C. Baus, T. Pierog, and R. Ulrich, https://web.ikp.kit.edu/rulrich/crmc.html
[7] T. Sjöstrand, S. Mrenna, and P. Skands, J. High Energy Phys. **05**, 026 (2006)
[8] S. Bass et al., Prog. Part. Nucl. Phys. **41**, 255 (1998)
[9] M. Bleicher et al., J. Phys. G**25**, 1859 (1999)
[10] S. Pulawski, PoS CPOD **2014**, 010 (2015)
[11] C. Alt et al., Phys. Rev. C**78**, 044907 (2008)
[12] V. Blobel et al., Phys. Lett. B**59**, 88 (1975)
[13] C. Daum et al., Nucl. Phys. B**186**, 205 (1981)
[14] Aguilar-Benitez et al., Z. Phys. C**50**, 405 (1991)
[15] S. Afanasiev et al., Phys. Lett. B**491**, 59 (2000)
[16] V. Vovchenko, V. V. Begun, M. I. Gorenstein, Phys. Rev. C**93**, 064906 (2016)
[17] M. Gazdzicki, D. Roehrich, Z. Phys. C**65**, 215 (1995)
[18] M. Gazdzicki, D. Roehrich, Z. Phys. C**71**, 55 (1996)
[19] M. Kliemant, B. Lungwitz, M. Gazdzicki, Phys. Rev. C**69**, 044903 (2004)
[20] I. Arsene et al., Phys. Rev. C**72**, 014908 (2005)
[21] K. Aamodt et al., Eur. Phys. J. C**71**, 1655 (2011)
[22] B. Abelev et al., Phys. Rev. C**79**, 034909 (2009)
[23] B. Abelev et al., Phys. Lett. B**736**, 196 (2014)
[24] B. Abelev et al., Phys. Rev. Lett. **109**, (2012)
[25] M. Gazdzicki, M. I. Gorenstein, Acta Phys. Polon. B**30**, 2705 (1999)
[26] A. Bialas, M. Bleszynski and W. Czyz, Nucl. Phys. B**111**, 461 (1976)

Highlights from PHENIX at RHIC

Rachid Nouicer[1,2,*], for the PHENIX Collaboration

[1] *Physics Department, Brookhaven National Laboratory, Upton, New York 11973, United States*
[2] *RIKEN-BNL Research Center, Brookhaven National Laboratory, Upton, New York 11973, United States*

Abstract. Hadrons conveying strange quarks or heavy quarks are essential probes of the hot and dense medium created in relativistic heavy-ion collisions. With hidden strangeness, ϕ meson production and its transport in the nuclear medium have attracted high interest since its discovery. Heavy quark-antiquark pairs, like charmonium and bottomonium mesons, are mainly produced in initial hard scattering processes of partons. While some of the produced pairs form bound quarkonia, the vast majority hadronize into particles carrying open heavy flavor. In this context, the PHENIX collaboration carries out a comprehensive physics program which studies the ϕ meson production, and heavy flavor production in relativistic heavy-ion collisions at RHIC. In recent years, the PHENIX experiment upgraded the detector in installing silicon vertex tracker (VTX) at mid-rapidity region and forward silicon vertex tracker (FVTX) at the forward rapidity region. With these new upgrades, the experiment has collected large data samples, and enhanced the capability of heavy flavor measurements via precision tracking. This paper summarizes the latest PHENIX results concerning ϕ meson, open and closed charm and beauty heavy quark production in relativistic heavy-ion collisions. These results are presented as a function of rapidity, energy and system size, and their interpretation with respect to the current theoretical understanding.

1 Physics Motivation

Until now, Quantum Chromodynamics (QCD) is considered the fundamental theory of the strong interaction between quarks and gluons. Conforming to QCD, at ordinary temperatures or densities this force confines the quarks into composite hadrons. However, when the temperature reaches the QCD energy scale or its density rises to the point where the average inter-quark separation is less than 1 fm, hadronic matter under extremely dense and hot conditions undergoes a phase transition to form a Quark-Gluon Plasma (QGP or QCD matter) in which quarks and gluons no longer are confined to the size of a hadron [1, 2]. Since the discovery of the QCD matter at the Relativistic Heavy Ion Colllider (RHIC) [3–7] at Brookhaven National Laboratory (BNL), the perception of the properties of strongly interacting matter at high temperatures has been a central goal of our research. This task is carried out by making measurements of multiple observables over a large range of energies and collision systems. To achieve this goal, RHIC provided collisions of gold-gold (Au + Au), copper-copper (Cu + Cu), uranium-uranium (U + U), copper-gold (Cu + Au), deuteron-gold (d + Au), helium-gold (^{3}He + Au), proton-gold (p + Au) and proton-Aluminum (p + Al) at center of mass energies from $\sqrt{s_{NN}}$ = 7.7 to

*e-mail: rachid.nouicer@bnl.gov

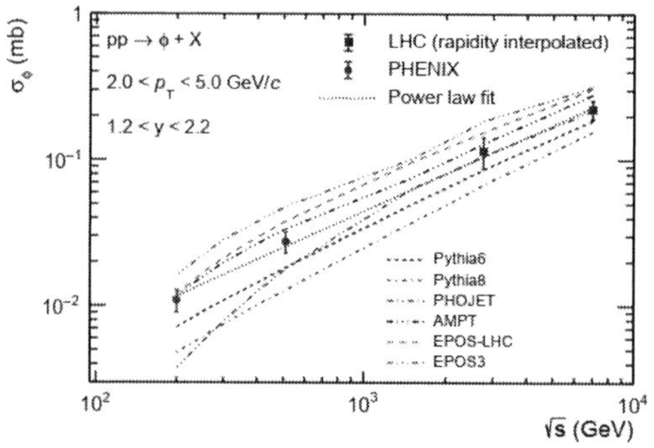

Figure 1. Production cross section of ϕ meson in $1.2 < y < 2.2$ and $2 < p_T < 5$ GeV/c in $p + p$ collisions as a function of the center-of-mass energy $\sqrt{s}$. The experimental results are compared to different model predictions. The LHC data points are interpolated at the PHENIX forward rapidity, for details see Ref. [8]

200 GeV, and proton-proton ($p + p$) from $\sqrt{s} = 62.4$ to 510 GeV [7]. To study the properties of QCD matter created at RHIC, the PHENIX experiment established a comprehensive physics program which studies hadrons carrying strange quarks, open heavy flavor and quarkonium production as a function of rapidity, energy and system size.

2 ϕ Meson Results

The ϕ vector meson (the lightest bound state of s and $\bar{s}$ quarks) is proposed as a good probe to investigate the properties of the QCD matter formed in heavy-ion collisions. A hydrodynamical study of transverse momentum (p_T) distribution of ϕ meson implies that it freezes out early compared to other hadrons [9, 10]. The ϕ meson life time is about 42 fm/c. Due to its relatively longer life time; the ϕ meson will mostly decay outside the fireball. Consequently, its daughter will not have much time to rescatter in the hadronic phase. This implies that properties of ϕ meson are basically controlled by the conditions in the early partonic phase which can be conceded as a good probe to investigate the properties of the QCD matter. Therefore, it is important to study the ϕ meson production as a function of centrality, rapidity, energy, and system size in heavy-ion collisions. Figure 1 shows the energy dependence of the ϕ meson production cross section integrated in $1.2 < y < 2.2$ and $2 < p_T < 5$ GeV/c in $p + p$ collisions [8]. We observe that the experimental measurements of the ϕ meson production increase from RHIC to LHC energies and it follow a power-law versus the colliding energy defined as $\sigma_\phi(s) \propto s^n$, with $n = 0.42 \pm 0.03$ (black dotted line). The results are compared to different model predictions, PYTHIA6, PYTHIA8, PHOJET, AMPT, EPOS3 and EPOS-LHC, for details see Ref. [8]. These models exhibit a similar energy dependence trend to that of the data. To study system size dependence of the ϕ meson production, the PHENIX experiment collected data from ^{3}He+Au collisions in 2014, and p+Au and p+Al collisions in 2015 RHIC runs. Figure 2 presents nuclear modification factors for minimum-bias ^{3}He+Au, p+Au and p+Al collisions at $\sqrt{s_{NN}} = 200$ GeV as a function of transverse momentum (panels (a), (b) and (c)) and rapidity (panels (d), (e) and (f)). These results may provide a constraint on various cold nuclear matter effects included in models.

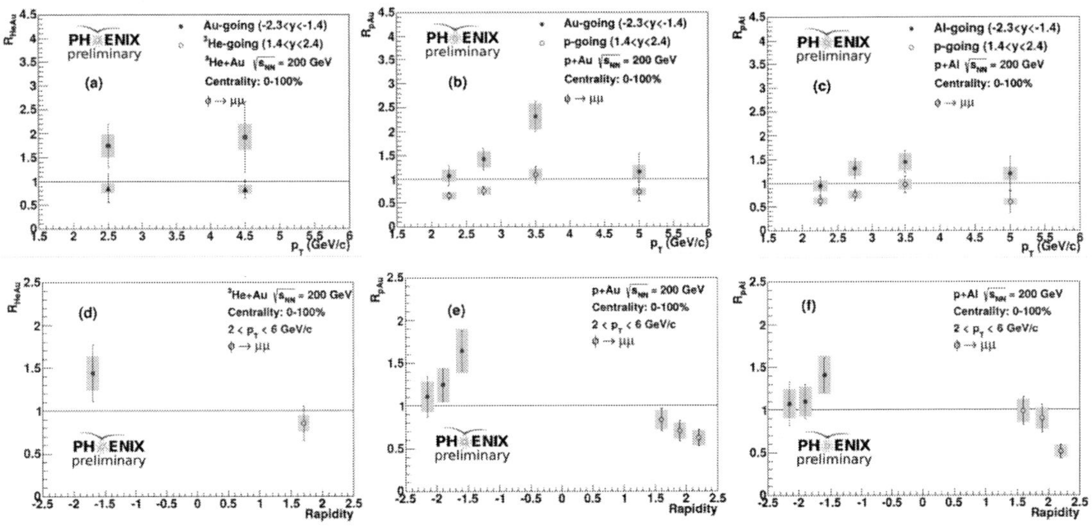

Figure 2. ϕ meson nuclear modification factors in minimum-bias ^{3}He+Au, p+Au and p+Al collisions at $\sqrt{s_{NN}}$ = 200 GeV. Panels (a), (b) and (c), ϕ meson nuclear modification factors versus transverse momentum, p_{T}. Panels (d), (e) and (f), ϕ meson nuclear modification factors versus rapidity.

3 Heavy Flavor Results

The PHENIX experiment studies heavy quark production via measurements of single leptons from semi-leptonic decays in both the electron channel at mid-rapidity and in the muon channel at forward-rapidity. To enhance the capability of heavy flavor measurements via precision tracking, the PHENIX collaboration has installed a silicon vertex tracker both in central rapidity (VTX) and in forward rapidity (FVTX) regions. The VTX [11] and FVTX [12] detectors cover respectively in rapidity $| y |< 1.2$ and $1.2 < y < 2.2$, and 2π in azimuth acceptance. With these two upgrades, the PHENIX experiment has collected large data samples. Recently, the PHENIX collaboration published first results on single electron yields from semileptonic charm and bottom decays in Au + Au collisions at $\sqrt{s_{NN}}$ = 200 GeV using the VTX at mid-rapidity [13], and the FVTX published B-meson production at forward and backward rapidity in $p + p$ collisions at $\sqrt{s}$ = 200 GeV and 510 GeV, and Cu + Au at $\sqrt{s_{NN}}$ = 200 GeV [15, 21]. These results will be discussed in the next sections.

3.1 Separation of Bottom and Charm in Au+Au collisions at $\sqrt{s_{NN}}$ = 200 GeV

The large suppression R_{AA} of $e^{\pm}_{HF}$ at high-p_{T} observed in Au + Au collisions at $\sqrt{s_{NN}}$ = 200 GeV, and similar to that of light mesons π^0, is striking because perturbative QCD calculations indicate a substantial contribution from bottom quark decays for $p_{\mathrm{T}} > 5$ GeV/c [16]. For the specific purpose of separating the contributions of charm and bottom quarks at mid-rapidity, the PHENIX has added micro-vertexing capabilities in the form of a silicon vertex tracker (VTX). The different lifetimes and kinematics for D-meson ($c\tau_{D^0} \approx 123\,\mu$m) and B-meson ($c\tau_{B^0} \approx 455\,\mu$m) hadrons decaying to electrons enable separation of charm and bottom contributions with measurements of displaced tracks.

Figure 3 shows the DCA$_T$ distribution (in transverse plane) for electrons ($e^{\pm}$) in $2.0 < p^e_{\mathrm{T}} < 2.5$ GeV/c range obtained in 0-10% central Au + Au collisions at $\sqrt{s_{NN}}$ = 200 GeV. In

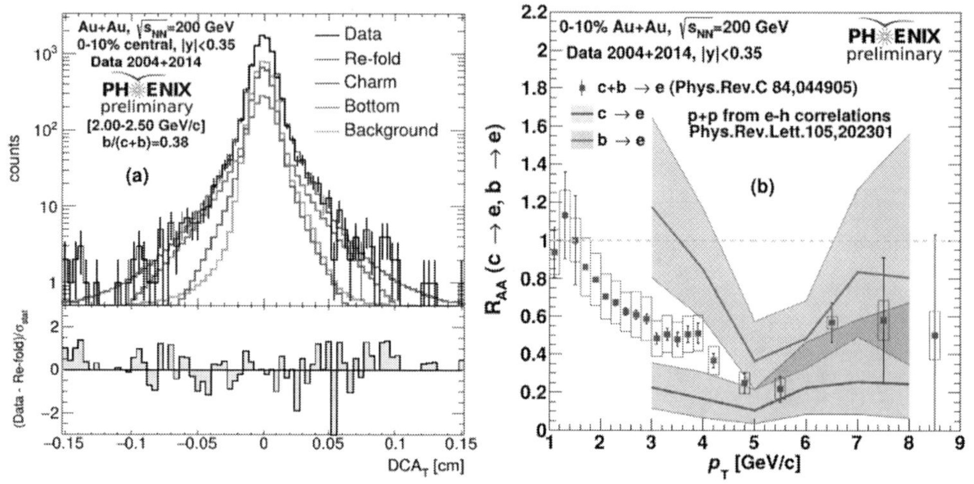

Figure 3. Panel (a) shows the DCA$_T$ distribution for electrons with $2.0 < p_T^e < 2.5$ GeV/c in Au + Au 0-10% central collisions at $\sqrt{s_{NN}}$ = 200 GeV. The bottom and charm components from the unfolding method are also plotted. Also shown are the normalized contributions for the various background components. Panel (b) presents the nuclear modification factors R_{AA} of electrons from charm ($c \rightarrow e$), bottom ($b \rightarrow e$) decays and combined heavy flavor as a function of p_T^e in Au + Au 0-10% central collisions at $\sqrt{s_{NN}}$ = 200 GeV.

order to extract the heavy flavor contributions, all background components as shown in figure 3(a), were fully accounted for and their DCA$_T$ shapes as a function of p_T incorporated. Details of the method of subtracting all background components from DCA$_T$ shape can be found in Ref. [13].

The nuclear modifications, $R_{AA}^{c\rightarrow e}$, and $R_{AA}^{b\rightarrow e}$ gather in 0-10% central Au + Au collisions at $\sqrt{s_{NN}}$ = 200 GeV, for charm and bottom hadron decays respectively are presented in figure 3(c). We find that electrons from bottom hadron decays are less suppressed than those from charm for the region $p_T < 5$ GeV/c. We observe also that the result is consistent with the expectation of the mass ordering of energy loss. It should be point out that mass scale for the quark energy-loss may be relevant only for $p_T < m_q$, as figure 3(b) suggests. he work to finalize the results and provide more centrality classes are in progress.

3.2 Measurement of B-meson → J/ψ in $p + p$ and Cu + Au collisions

The production of b-quarks are mainly from gluon fusion and flavor excitation at RHIC energy, and from split gluons at the LHC energy [17]. This characteristic may provide fundamental informations on how b-quarks lose energy in the QGP. While b-quarks from gluon splitting behave as a gluon when crossing the QGP plasma. However, b-quarks from fusion traverse the medium with a definite quantum number and their mass comes to be important for the amount of energy they lose in the medium [18]. Also plasma induced gluon splitting into b-quark pairs should be examined. At the LHC energy, the preeminence of gluon splitting may be the feasible reasons of why nuclear modifications of b-jets and inclusive jets are in good agreement [19, 20]. To verify how initial state effects influence the b-quark production in heavy ion collisions is to measure B-meson yields integrated over momentum and centrality. One way to do this is to measure dimuon pairs produced from the B-mesons decay.

Datasets used in this analysis, measurements of the B-mesons → J/ψs, including VTX and FVTX trackers in PHENIX experiment are from 2012 and 2015 RHIC runs. The 3D vertex position provided

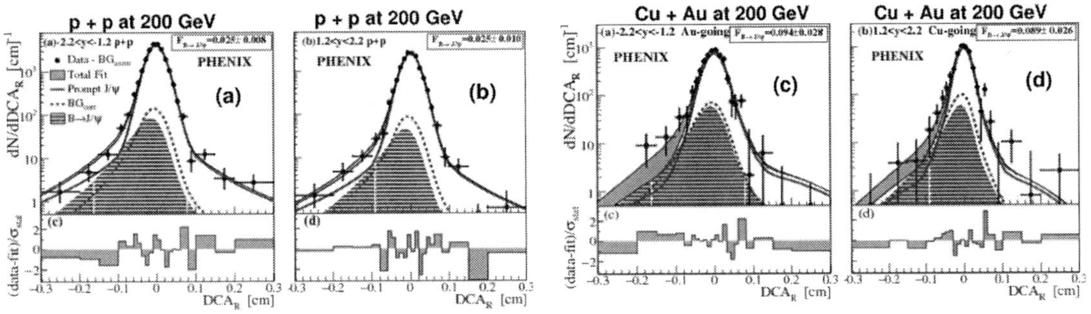

Figure 4. Radial distance of closest approach of single muons from dimuons in the J/ψ mass region in p+p and Cu+Au collisions at $\sqrt{s_{NN}}$ = 200 GeV. Combinatorial background and incorrect FVTX-MuTr associations are subtracted [15].

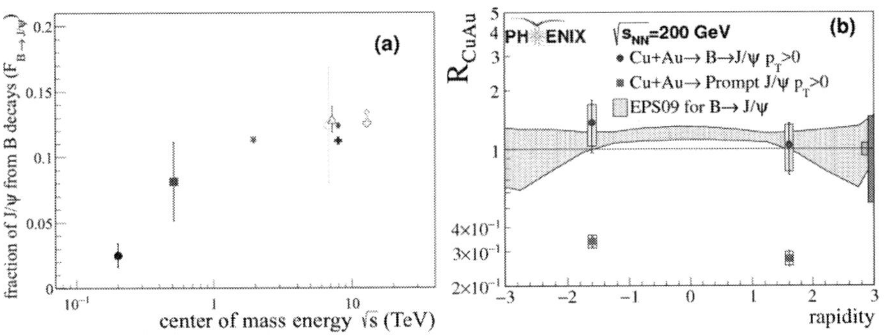

Figure 5. Panel (a) presents collision energy dependence of the fraction of B-meson $\to J/\psi$ [19]. Panel (b) shows nuclear modification factor of B-mesons and prompt J/ψ in Cu + Au collisions [15, 21] compared to an estimate from nuclear parton distribution modification (EPS09) [22].

by all VTX and FVTX tracks has a precision ranging from 200 μm in $p+p$ collisions to 30 μm in central Cu+Au collisions. $J/\psi \to \mu^+\mu^-$ decays are identified by the dimuon invariant mass peak in the muon arms, where each arm is composed of approximately one-meter thick hadron absorber, a muon tracker with three stations (MuTr) and a muon identifier (MuID). A MuTr+MuID identified J/ψ can have one or both muons in the pair matching FVTX, all matchings are considered in the analysis. The muon track displacements are measured by the FVTX. The track displacement is defined by the radial projection of the distance of the closest approach (DCAR), in which direction the FVTX provides the best DCAR precision between 100-200 μm. The DCAR distribution has contributions from the following sources: (a) dimuon combinatorial background; (b) incorrect matchings between the FVTX and the MuTr standalone tracks; (c) correlated dimuons in the J/ψ mass region coming from $c\bar{c} \to \mu^+\mu^-$ and $b\bar{b} \to \mu^+\mu^-$; and (d) prompt and non-prompt J/ψ. The fraction of non-prompt J/ψ in the total J/ψ yield ($F_B \to J/\psi$) is obtained from a likelihood fit of the DCAR distribution as shown in Fig 4. The fraction $F_B \to J/\psi$ is the only free parameter in the fit. For more details about the analysis procedure see Refs. [15, 19, 21]. Figure 5(a) shows measured $F_B \to J/\psi$ in $p+p$ collisions at $\sqrt{s_{NN}}$ = 200 and 510 GeV compared to the world data. The trend exhibits an increase from $\sqrt{s_{NN}}$ = 200 GeV to 510 GeV at RHIC and almost constant at Tevatron and LHC energies. Figure 5(b) presents the B-meson nuclear modification $R_{CuAu}^{B-meson}$. It was obtained using $F_B \rightarrow J/\psi$ results in $p+p$ and Cu + Au collisions, and measured J/ψ nuclear modification factor in Cu + Au collisions ($R_{CuAu}^{J/\psi}$) all at the same energy, $\sqrt{s_{_{NN}}}$ = 200 GeV. The $R_{CuAu}^{B-meson}$ is in good agreement with binary scaling from $p+p$ collisions for B-mesons. Calculation of nuclear modification factor from initial state effects in EPS09 model [22] is in good accordance with experimental result. However, the $R_{CuAu}^{B-meson}$ for prompt J/ψ exhibits a strong suppression, caused by breakup in the medium.

3.3 Conclusions

PHENIX measurements of hadrons conveying strange quarks or heavy quarks in heavy-ion collisions at RHIC have led to a broad and deep understanding of the properties of hot QCD matter. The PHENIX measurements of ϕ meson production in $p+p$ collisions imply an increase as a function of energy from RHIC to LHC, and it follows a power-law. For the first time at RHIC, measurements of charm and bottom separation are obtained in 0-10% central Au + Au at $\sqrt{s_{_{NN}}}$ = 200 GeV via single electrons from semi-leptonic D and B meson decays. We find that electrons from bottom hadron decays are less suppressed than those from charm for the region $p_T < 4$ GeV/c and it is consistent with the expectation of the mass ordering of energy loss. We also presented the first results of B-meson $\rightarrow J/\psi$ in $p+p$ collisions at 200 GeV and 510 GeV, and Cu + Au at 200 GeV. As a function of energy, the trend $F_B \rightarrow J/\psi$ measurements exhibits an increase from $\sqrt{s_{_{NN}}}$ = 200 GeV to 510 GeV at RHIC and almost constant at Tevatron and LHC energies.

References

[1] E. Shuryak, Phys. Rep. **61**, 71 (1980).
[2] L. McLerran, Rev. Mod. Phys. **58**, 1021 (1986).
[3] I. Arsene *et al.* (PHENIX Collaboration), Nucl. Phys. A **757**, 1 (2005).
[4] B. B. Back *et al.* (PHOBOS Collaboration), Nucl. Phys. A **757**, 28 (2005).
[5] J. Adams *et al.* (BRAHMS Collaboration), Nucl. Phys. A **757**, 102 (2005).
[6] K. Adcox *et al.* (STAR Collaboration), Nucl. Phys. A **757**, 184 (2005).
[7] R. Nouicer, European Physical Journal Plus **131**, 70 (2016), and references therein.
[8] A. Adare *et al.* (PHENIX Collaboration), arXiv:1710.01656v1 (2017).
[9] J. Adams *et al.* (STAR Collaboration), Nucl. Phys. A **757**, 102 (2005).
[10] M. Nasim *et al.*, Adv. High Energy Phys. **2015**, 197930 (2015).
[11] R. Nouicer *et al.*, Nucl. Inst. and Meth. in Phys. Res. B **261**, 10671071 (2007).
[12] C. Aidala *et al.*, Nucl. Inst. and Meth. in Phys. Res. A **755**, 44 (2014).
[13] A. Adare *et al.* (PHENIX Collaboration), Phys. Rev. C **93**, 034904 (2016).
[14] C. Aidala *et al.* (PHENIX Collaboration), Phys. Rev. D **95**, 092002 (2017).
[15] A. Adare *et al.* (PHENIX Collaboration), Phys. Rev. C **96**, 064901 (2017)
[16] M. Cacciari *et al.*, Phys. Rev. Lett. **95**, 122001 (2005).
[17] E. Norrbin *et al.*, Phys. Lett. B **442**, 407 (1998).
[18] J. Huang *et al.*, Phys. Lett. B **726**, 251 (2013).
[19] C. L. da Silva *et al.* (PHENIX Collaboration), Nucl. Phys. A **967**, 652 (2017).
[20] S. Chatrchyan *et al.*, Phys. Rev. Lett. **113**, 13 (2014) 132301.
[21] A. Adare *et al.* (PHENIX Collaboration), Phys. Rev. D **95**, 092002 (2017).
[22] K. J. Eskola *et al.*, JHEP **04**, 065 (2009).

Recent highlights from STAR

Wangmei Zha[1,*] for the STAR Collaboration

[1] *Department of modern physics, University of Science and Technology of China, Hefei, 230026, China*

Abstract. The Solenoidal Tracker at RHIC (STAR) experiment takes advantage of its excellent tracking and particle identification capabilities at mid-rapidity to explore the properties of strongly interacting QCD matter created in heavy-ion collisions at RHIC. The STAR collaboration presented 7 parallel and 2 plenary talks at Strangeness in Quark Matter 2017 and covered various topics including heavy flavor measurements, bulk observables, electro-magnetic probes and the upgrade program. This paper highlights some of the selected results.

1 Introduction

Relativistic heavy-ion collisions provide a unique tool to study the most intriguing aspects of the strong interaction in the laboratory. In high energy heavy-ion collisions performed at RHIC and the LHC, the achieved energy densities are extremely high so that a new form of matter, referred to as the quark-gluon plasma (QGP) [1], can be created. This matter has been observed to quickly thermalize and evolve as a perfect fluid with small viscosity over entropy ratio. The STAR heavy-ion program at top RHIC energies is aimed to quantitatively study the properties of the strongly-interacting QGP matter created at vanishingly small baryon density. While at finite baryon density from the Beam Energy Scan (BES) phase-I program at lower energies, the nature of the phase transition is being studied.

At Strangeness in Quark Matter 2017, the STAR collaboration has presented results from two recently installed detectors for heavy-flavor studies, the Heavy Flavor Tracker (HFT) and Muon Telescope Detector (MTD). The HFT, which facilitates topological reconstruction of open heavy flavor hadrons, has enabled the first measurement of D^0 azimuthal anisotropy at RHIC, which helps constrain the charm quark spatial diffusion coefficient in the QGP. The MTD, which enables identification and triggering of muons, opens the door to measure quarkonia via the di-muon decay channel at STAR, providing insights to the understanding of the dissociation and regeneration phenomena of quarkonia in the QGP. With the full BES phase-I datasets, the collectivity of medium has been systematically studied via directed, elliptic and higher order harmonics together with femtoscopy measurements. In non-central relativistic heavy-ion collisions, the created medium possesses strong vorticity, which causes the spin of particles to trend to align with the system's global angular momentum. The first differential measurement of the Λ polarization as a function of Λ azimuthal angle relative to the reaction plan has been reported at this conference. For the electro-magnetic probes, the first dielectron measurement at very low p_T has been shown at this conference.

*e-mail: wangmei@rcf.rhic.bnl.gov

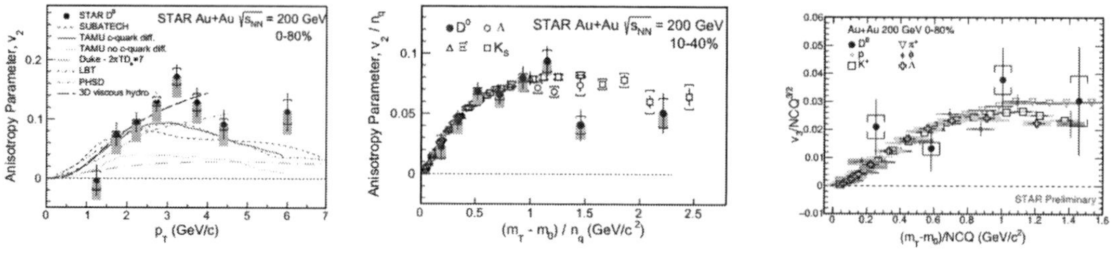

Figure 1. Left: v_2 as a function of p_T for D^0 mesons in 0-80% centrality [3]. Middle (right): v_2/n_q in 10-40% centrality $(v_3/(n_q)^{3/2}$ in 0-80% centrality) as a function of $(m_T - m_0)/n_q$ for D^0 and light flavor hadrons [3].

2 Heavy flavor measurements

In heavy-ion collisions at RHIC, heavy flavor quarks are predominantly produced by initial hard parton scatterings, and subsequently experience the whole evolution of the system. Therefore they can serve as calibrated probes to study the properties of the strongly coupled Quark Gluon Plasma (sQGP). In particular, the anisotropic flow of open heavy flavor hadrons is sensitive to the interactions of heavy flavor quarks with the medium and could also help constrain the magnitude of their spatial diffusion coefficient D_s in sQGP [2]. Shown in the left panel of Fig. 1 are the v_2 of D^0 mesons as a function of p_T in the 0-80% centrality interval [3]. The measured v_2 are compared to theoretical calculations [4–9], which use different approaches to describe the transportation or interaction of charm quarks in the medium. The hydrodynamic model [4] with parameters tuned to reproduce the light flavor hadron v_2 can reasonably describe the D^0 v_2, indicating that charm quarks may have achieved full thermal equilibrium in these collisions. The transport models [6–9] with a temperature dependent diffusion coefficient in the range $2 < 2\pi TD_s < 12$ for $T_c < T < 2T_c$ are also able to reproduce the measured values. The left (middle) panel of Fig. 1 shows the v_2 (v_3) values scaled by the number of constituent quarks (NCQ, n_q) for D^0 mesons and a number of light flavor hadrons, as a fucntion of $(m_T - m_0)/n_q$. The scaled v_2 and v_3 for all the species including D^0 falls on the same trend, which also suggests that charm quarks flow with the medium.

The yield ratios of $D_s^\pm/D^0$ and $\Lambda_c^\pm/D^0$ in $\sqrt{s_{\rm NN}} = 200$ GeV Au+Au collisions at RHIC are shown in Fig. 2. These measurements can help us to understand the hadronization mechanism of charm quarks in the QGP medium [10–12]. Both ratios show significant enhancements compared to those in p+p collisions predicted by PYTHIA, suggesting that charm quarks could hadronize via the coalescence mechanism in the QGP. For $\Lambda_c^\pm/D^0$, model calculations with coalescence of thermalized charm quarks in the QGP medium are compatible with data within uncertainties [10], while for $D_s^\pm/D^0$ the TAMU model calculation for the 10-40% centrality interval with coalescence of charm quarks seems to under-predict the measured values [9].

Quarkonium dissociation by the color-screening effect in the hot and dense medium was initially proposed as signature of the QGP formation [13]. However, the interpretation of J/ψ suppression observed in experiments has remained a challenge due to the regeneration from the coalescence of deconfined heavy quark pairs in the medium, the cold nuclear matter (CNM) effects and feed-down contributions. In order to quantify those effects, we present (i) measurements of nuclear modification factors for inclusive J/ψ production in p+Au collisions at $\sqrt{s_{\rm NN}} = 200$ GeV; (ii) the first measurement of the double ratio of inclusive $\psi(2S)$ and J/ψ production rates at mid-rapidity between p+p and p+Au collisions at $\sqrt{s_{\rm NN}} = 200$ GeV; and (iii)measurements of Υ nuclear modification factors in Au+Au collisions at $\sqrt{s_{\rm NN}} = 200$ GeV.

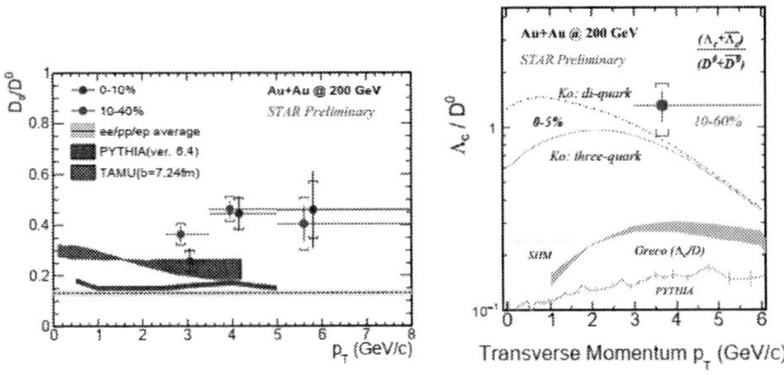

Figure 2. Yield ratios of D_s/D^0 in 0-10% and 10-40% Au+Au collisions (left) and Λ_c/D^0 in 10-60% central Au+Au collisions (right) as a function of p_T.

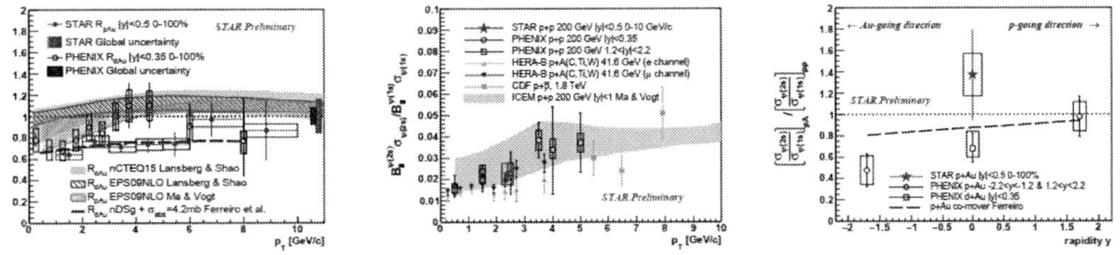

Figure 3. Left: nuclear modification factor R_{pAu} as a function of p_T for inclusive J/ψ in the di-muon decay. Middle: ratio of $\psi(2S)$ to J/ψ production rates as a function of p_T. Right: double ratio of $\psi(2S)$ to J/ψ production rates as a function of rapidity in p+Au and d+Au collisions with respect to that in p+p collisions..

Shown in the left panel of Fig. 3 is the R_{pAu} of inclusive J/ψ in p+Au collisions. The measured R_{pAu} is generally consistent with the R_{dAu} result reported by PHENIX [14] within uncertainties. The theoretical predictions with only the nuclear PDF effect using the nCTEQ15 [15] or EPS09NLO [16] nuclear PDFs sets can touch the upper limit of the data. However, the calculation including an additional nuclear absorption effect [17] is favored by the data. The middle panel of Fig. 3 shows the ratio of inclusive J/ψ and $\psi(2S)$ production cross sections as a function of p_T in p+p collisions at $\sqrt{s} = 200$ GeV. The measurement follows the trend of the world-wide data and could be described by the ICEM calculation [18]. Shown in the right panel of Fig. 3 is the double ratio of $\psi(2S)$ and J/ψ production rates between p+p and p+Au collisions as a function of rapidity. Taking into account the large uncertainties, the measurement is consistent with the published PHENIX results [19] and the co-mover model calculation [20].

The R_{AA} of $\Upsilon(1S)$ as a function of the number of participating nucleons ($N_{\rm part}$) is shown in the left panel of Fig. 4. The measurement is compared to that in Pb + Pb collisions at $\sqrt{s_{\rm NN}} = 2.76$ TeV from CMS [21] and they show similar suppression. Please note that the feed-down contributions from excited states are not excluded from the $\Upsilon(1S)$ R_{AA}. The STAR results are also compared to different theoretical model calculations [22–24], as shown in the middle panel of Fig. 4. Two limiting cases for the in-medium potential between the quark-antiquark pairs that are often used are the internal energy corresponding to a strongly bound scenario (SBS), and the free energy corresponding to a more weakly bound secnario (WBS). Our data seem to favor the SBS scenario toward central collisions. The p_T dependence of $\Upsilon(1S)$ R_{AA} from STAR is shown in the right panel of Fig. 4 and also compared

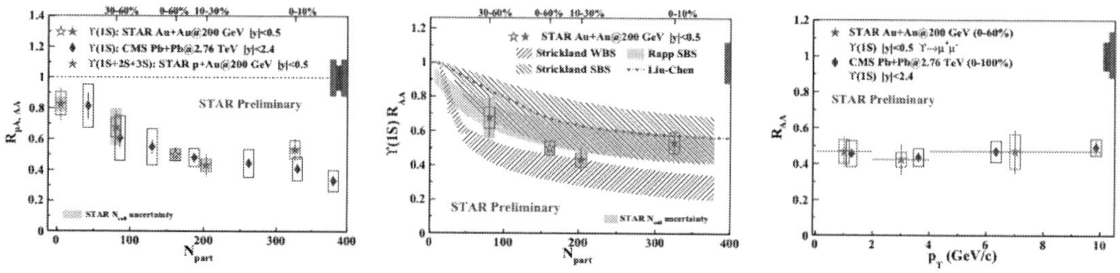

Figure 4. Left: nuclear modification factor, R_{AA}, of $\Upsilon(1S)$ as a function of N_{part} in Au+Au collisions. Middle: the same data as in the left panel compared to different model calculations. Right: R_{AA} of $\Upsilon(1S)$ as a function of p_T in 0-60% central Au+Au collisions at RHIC and compared to that in 0-100% central Pb+Pb collisions at LHC.

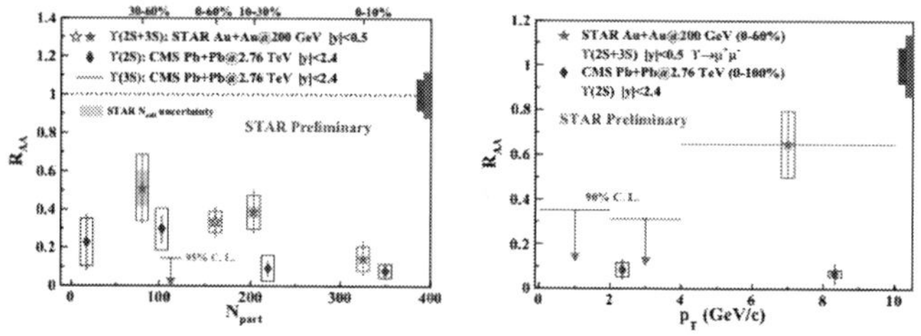

Figure 5. Left: R_{AA} of $\Upsilon(2S+3S)$ as a function of N_{part} in Au+Au collisions. Right: R_{AA} of $\Upsilon(2S+3S)$ as a function of p_T in 0-60% central Au+Au collisions.

to that from CMS. They show similar suppression and possess no significant p_T dependence within uncertainties. The first $\Upsilon(2S+3S)$ R_{AA} measurements from STAR as a function of N_{part} and p_T are shown in the left and right panels of Fig. 5, respectively. In comparison with CMS measurements [21], there seems to be a hint of less suppression at RHIC. Also $\Upsilon(2S+3S)$ shows a significantly stronger suppression in the most central collisions at RHIC than that for $\Upsilon(1S)$, which is consistent with the sequential melting phenomenon.

3 Bulk observables

Experimentally, the dynamics and the collective behaviour of the sQGP have been studied by measuring the azimuthal anisotropy of the produced particles. Strange hadrons, especially multistrange hadrons, due to their small hadronic interaction cross section and freeze-out temperature closer to the phase transition temperature, are believed to be more sensitive to the partonic stage in heavy-ion collisions. Recent developments indicate that higher order flow harmonics can reveal information about initial state fluctuations, which could help us to better extract the QGP properties. Shown in Fig. 6 are $v_n(p_T)$ of K_s^0, ϕ and $\Lambda + \bar{\Lambda}$ at mid-rapidity for 0-10%, 10-40% and 40-80% central U+U collisions at $\sqrt{s_{\text{NN}}} = 193$ GeV and Au+Au collisions at $\sqrt{s_{\text{NN}}} = 200$ GeV [25]. A clear centrality dependence of $v_2(p_T)$ is observed for K_s^0, ϕ and $\Lambda + \bar{\Lambda}$, which is similar to that in Au+Au collisions. Higher order flow harmonics v_3 and v_4 show much weaker centrality dependence, if any.

New measurements of rapidity-even dipolar flow (v_1^{even}) from STAR are presented in this conference. These measurements could serve as constraints to distinguish between different initial-state

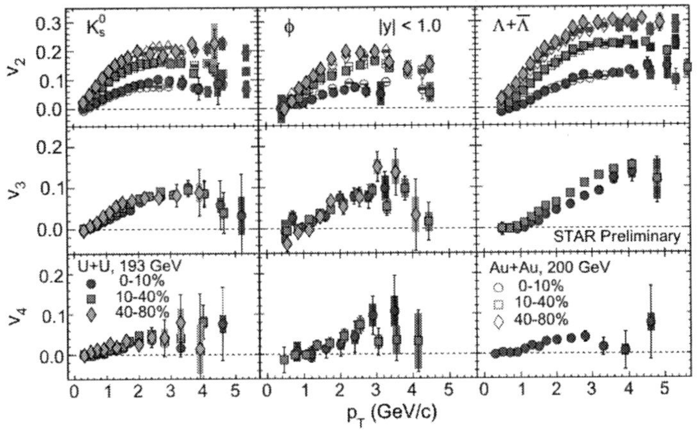

Figure 6. v_2 (top panels), v_3 (middle panels) and v_4 (bottom panels) vs p_T of K_s^0, ϕ and $\Lambda+\bar{\Lambda}$ for 0-10%, 10-40% and 40-80% centrality in U+U collisions at $\sqrt{s_{\rm NN}}$ = 193 GeV and Au+Au collisions at $\sqrt{s_{\rm NN}}$ = 200 GeV [25].

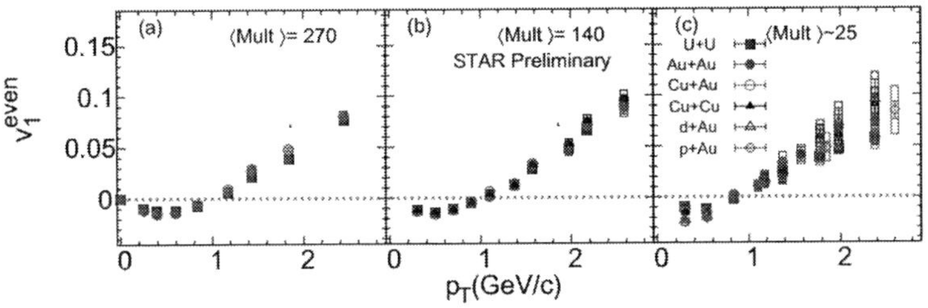

Figure 7. v_1^{even} as a function of p_T for different $\langle N_{Ch}\rangle$ selections for different collision systems at $\sqrt{s_{\rm NN}}$ ~ 200 GeV.

models and aid a more reliable extraction of the viscosity over entropy density ratio η/s. The $v_1^{\rm even}$ for different collision systems are shown in Fig. 7 for different magnitude of $\langle N_{ch}\rangle$. For each selected $\langle N_{ch}\rangle$ class, $v_1^{\rm even}(p_T)$ does not show a significant dependence on the collision system, which suggests that the fluctuation-driven initial-state eccentricity ε_1 is similar for the six collision systems.

Two-particle interferometry measurement, known as femtoscopy or HBT, is a standard method of studying space-time characteristics of the created system from heavy-ion collisions. Recently collected high-statistic datasets from STAR allow measurements with heavier, but less abundant particles. Fig. 8 shows a comparison of kaon and pion [26] source radii as a function of m_T for 0-5% and 30-40% Au+Au collisions. The $R_{\rm out}$ values for kaons are systematically larger than those of pions, while the $R_{\rm side}$ values for kaons and pions are comparable, and the m_T-dependence of $R_{\rm long}$ is different between kaons and pions. These observations suggest different space-time configurations at the kinetic freeze-out for the pions and kaons. However, further understanding of the observed behaviors

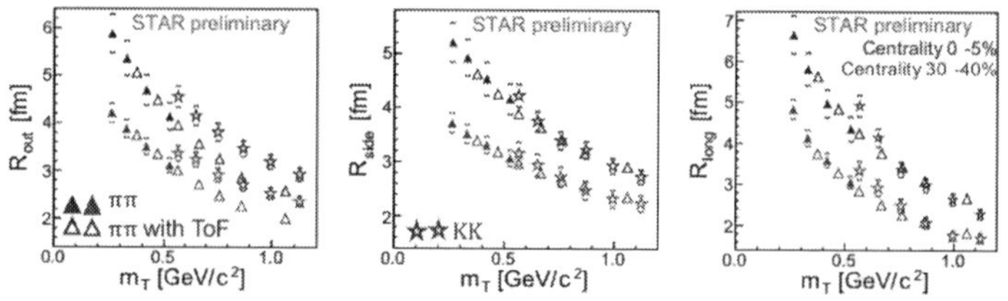

Figure 8. The comparison of the kaon and pion [26] source radii in Au+Au collisions at $\sqrt{s_{\mathrm{NN}}}$ = 200 GeV from STAR.

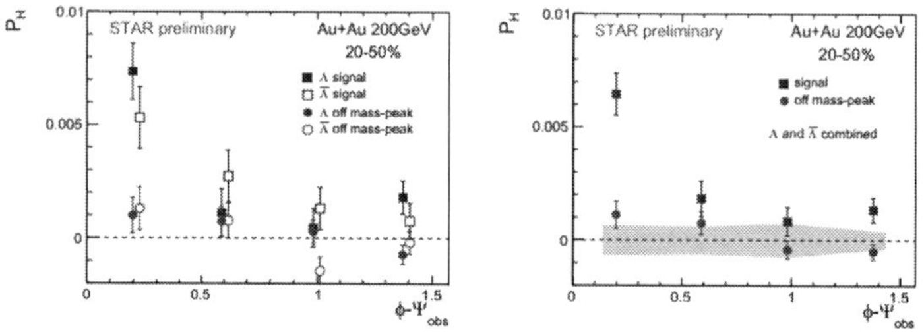

Figure 9. Left: P_H for Λ and $\bar{\Lambda}$ as a function of $\phi - \Psi_{obs}$ in 20-50% central Au+Au collisions at $\sqrt{s_{\mathrm{NN}}}$ = 200 GeV. Right: merged P_H between Λ and $\bar{\Lambda}$ as a function of $\phi - \Psi_{obs}$ in 20-50% central Au+Au collisions at $\sqrt{s_{\mathrm{NN}}}$ = 200 GeV.

require a comparison with the full three-dimensional hydrodynamic calculations with rescattering in the hadronic phase.

A non-zero global Λ polarization has been observed in Au+Au collisions by STAR [27] for the first time, revealing that the matter produced in heavy-ion collisions is by far the most vortical system ever observed. Such vorticity is expected to be maximal at the equator, leads to a larger in-plane than out-of-plane polarization for hyperons. Shown in the left panel of Fig. 9 are the Λ and $\bar{\Lambda}$ polarization (P_H) as a function of $\phi - \Psi_{obs}$ at mid-rapidity in 20-50% central Au+Au collisions at $\sqrt{s_{\mathrm{NN}}}$ = 200 GeV, where Ψ_{obs} is the observed angle of the first-order event plane as an estimate of the reaction plane angle. The right panel of Fig. 9 shows the combined Λ and $\bar{\Lambda}$ P_H as a function of $\phi - \Psi_{obs}$. The significance of ΔP_H, for Λ and $\bar{\Lambda}$ combined, between $\phi - \Psi_{obs}$ bin of $[0, \pi/8]$ and $[3\pi/8, \pi/2]$ is 4.7 σ. The larger in-plane than out-of-plane polarization is consistent with the picture of maximum vorticity in the equator and low viscosity of the system.

4 Electro-magnetic probes

Recently, the ALICE collaboration reported a significant J/ψ excess yield at very low p_T (p_T < 0.3 GeV/c) in peripheral Pb + Pb collisions at $\sqrt{s_{\mathrm{NN}}}$ = 2.76 TeV [28]. This remarkable excess can not be explained by the hadronic production of J/ψ modified by the cold and hot medium effects, however, could be qualitatively explained by coherent photonuclear production mechanism [29]. Shown in the left panel of Fig. 10 are the extracted J/ψ R_{AA} as a function of p_T in Au+Au collisions at $\sqrt{s_{\mathrm{NN}}}$ = 200 GeV and U+U collisions at $\sqrt{s_{\mathrm{NN}}}$ = 193 GeV from STAR. A huge enhancement of J/ψ production has been observed in peripheral (40-80%) Au+Au and U+U collisions for p_T < 0.1 GeV/c,

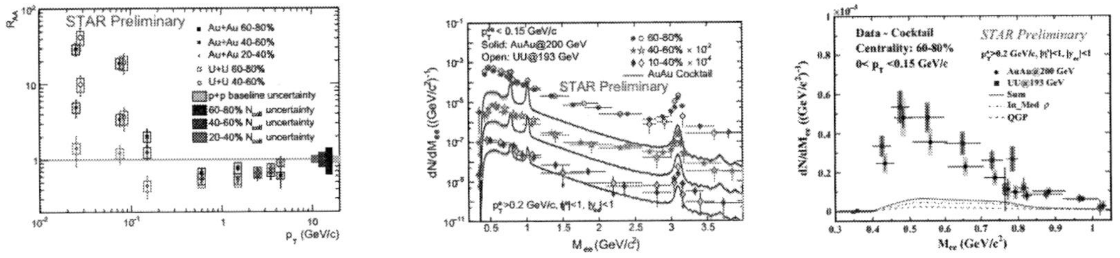

Figure 10. Left: the extracted J/ψ R_{AA} as a function of p_T in Au+Au collisions at $\sqrt{s_{\rm NN}}$ = 200 GeV and U+U collisions at $\sqrt{s_{\rm NN}}$ = 193 GeV. Middle: centrality dependence of dielectron invariant mass spectrum in the STAR acceptance in Au+Au collisions at $\sqrt{s_{\rm NN}}$ = 200 GeV and U+U collisions at $\sqrt{s_{\rm NN}}$ = 193 GeV for dielectron pair p_T < 0.15 GeV/c. The mass spectra are compared to hadronic cocktail simulations shown as solid lines. Right: The low p_T (p_T < 0.15 GeV/c) dielectron excess mass spectra (data-cocktail) within STAR acceptance in 60-80% collisions, compared with a broadened ρ model calculation.

which confirms the ALICE measurement and points to evidence of coherent photonuclear interaction in hadronic collisions. If indeed, the coherent photonuclear interaction in violent hadronic collisions were the underlying mechanism for the observed J/ψ excess, the coherent photon-photon process would be there and contribute to the dielectron pair production. The middle panel of Fig. 10 shows centrality dependence of dielectron invariant mass spectrum in the STAR acceptance in Au+Au collisions at $\sqrt{s_{\rm NN}}$ = 200 GeV and U+U collisions at $\sqrt{s_{\rm NN}}$ = 193 GeV for dielectron pairs with p_T < 0.15 GeV/c. In comparison with the hadronic cocktail calculation, a clear enhancement is observed. The enhancement is larger for more peripheral collisions. To gain more insight into the excess, we subtract the cocktail contributions from the inclusive dielectron data and obtain the excess mass spectra for p_T < 0.15 GeV/c in 60-80% A + A collisions, as shown in the right panel of Fig. 10. Calculation results from an effective many-body model based on in-medium broadening of the ρ-meson spectral function are also shown. The broadened ρ model calculation [30], which has successfully described the dielectron invariant mass spectra integrated over p_T at SPS and RHIC, can not describe the significant enhancement yields observed at very low p_T in peripheral A + A collisions. The enhancement at very low p_T measured in peripheral collisions may originate from the coherent photon-photon interactions.

5 Summary

At this conference, STAR has presented a wide variety of measurements for p+p, p+Au, Au+Au, and U+U collisions, at energies ranging from $\sqrt{s_{\rm NN}}$ = 19.6 to 200 GeV. New results from the HFT include the elliptic and triangular flow of D^0 mesons and D_s/D^0 and Λ_c/D^0 ratios, indicating that charm quarks flow with the medium and hadronize via coalescence mechanism. J/ψ and Υ are measured in the dimuon channel from the MTD to study the dissociation and regeneration of quarkonia in the medium. The dipolar flow, elliptic flow and other higher order flow coefficients are presented to reveal the initial fluctuations and to constrain the transport coefficients of the medium. The source radii of kaon from HBT is presented to probe the space-time evolution of the system. The azimuthal angle dependence of hyperon polarization is studied, which is consistent with the picture of maximum vorticity in the equator and low viscosity of the system. An enhancement of dielectron production at very low p_T in peripheral collisions is observed, which may originate from the coherent photoproduction.

Acknowledgement

The author was funded by the National Natural Science Foundation of China under Grant Nos. 11505180, 11775213 and 11375172, and MOST under Grant No. 2014CB845400.

References

[1] P. Braun-Munzinger, J. Stachel, Nature **448**, 302 (2007)
[2] A. Beraudo et al., Eur. Phys. J. C **75**, 121 (2015)
[3] L. Adamczyk et al. (STAR Collaboration), Phys. Rev. Lett. **118**, 212301 (2017)
[4] L.G. Pang et al., Phys. Rev. D **91**, 074027 (2015)
[5] G.Q. S. S. Cao, S.A. Bass, Phys. Rev. C **92**, 024907 (2015)
[6] M. Nahrgang et al., Phys. Rev. C **91**, 014904 (2015)
[7] S. Cao et al., Phys. Rev. C **94**, 014909 (2016)
[8] T. Song et al., Phys. Rev. C **92**, 014910 (2015)
[9] M. He et al., Phys. Rev. Lett. **110**, 112301 (2013)
[10] Y. Oh et al., Phys. Rev. C **79**, 044905 (2009)
[11] S. Ghosh et al., Phys. Rev. D **90**, 054018 (2014)
[12] V. Greco et al., Phys. Lett. B **595**, 202 (2004)
[13] T. Matsui, H. Satz, Physics Letters B **178**, 416 (1986)
[14] A. Adare et al. (PHENIX Collaboration), Phys. Rev. C **87**, 034904 (2013)
[15] J.P. Lansberg, H.S. Shao, Eur. Phys. J. C **77**, 1 (2017)
[16] H.S. Shao, Comput. Phys. Commun. **198**, 238 (2016)
[17] E.G. Ferreiro et al., Few Body Syst. **53**, 27 (2012)
[18] Y.Q. Ma, R. Vogt, Phys. Rev. D **94**, 114029 (2016)
[19] A. Adare et al. (PHENIX Collaboration), Phys. Rev. Lett. **111**, 202301 (2013)
[20] X. Zhao, R. Rapp, Phys. Rev. C **82**, 064905 (2010)
[21] V. Khachatryan et al. (CMS Collaboration), Phys. Lett. B **770**, 357 (2017)
[22] M. Strickland, D. Bazow, Nucl. Phys. A **879**, 25 (2012)
[23] Y. Liu et al., Phys. Lett. B **697**, 32 (2011)
[24] A. Emerick et al., Eur. Phys. J. A **48**, 72 (2012)
[25] B.I. Abelev et al. (STAR Collaboration), Phys. Rev. C **77**, 054901 (2008)
[26] L. Adamczyk et al. (STAR Collaboration), Phys. Rev. C **92**, 014904 (2015)
[27] L. Adamczyk et al. (STAR Collaboration), Nature **548**, 62 (2017)
[28] J. Adam et al. (ALICE Collaboration), Physical Review Letters **116**, 222301 (2016)
[29] W. Zha et al. (2017), `arXiv: 1705.01460`
[30] R. Rapp, Adv. High Energy Phys. **2013**, 62 (2013)
[31] A.M. Poskanzer, S.A. Voloshin, Phys. Rev. C **58**, 1671 (1998)

ALICE Overview

Alessandro Grelli (the ALICE Collaboration)[1,⋆]

[1] *Utrecht University, Princetonplein 1, 3584CC, The Netherlands*

Abstract. An overview of the recent results obtained by the ALICE Collaboration from the analysis of the pp, p-Pb and Pb-Pb data samples collected during LHC run I and the first half of run II is presented.

1 Introduction

The study of Pb-Pb collisions in the LHC energy regime provides insight on the properties of nuclear matter at extreme conditions of temperature and energy density, where lattice QCD predicts the matter to be in a Quark Gluon Plasma (QGP) state [1]. Properties of the QGP medium, such as shear viscosity to entropy ratio, electromagnetic radiation and energy loss are the focus of past, present and future proposed measurements. In addition, in recent years, the investigation of the properties of small systems (i.e. pp and p-A collisions) has taken up an important role. Since the first experimental evidence of long-range near-side angular correlations in high multiplicity pp [2] and p-Pb [3] interactions a large effort has been made to measure the small system properties as a function of the event multiplicity in order to understand the possible onset of collectivity and its implications.
ALICE [4] is the dedicated heavy ion experiment at the LHC. It is equipped with excellent particle identification (PID) and tracking capabilities down to very low transverse momentum (p_T). The aforementioned properties make ALICE unique in measuring identified particles spectra, (anti-)(hyper-) nuclei production and in reconstructing decays of heavy-flavour hadrons by resolving their secondary decay vertices both in small systems and in heavy-ion collisions.
In the next sections, an overview of recent ALICE results from LHC run I and II data is presented with particular focus on strangeness, hard probes, nuclei production and collectivity. It has to be noted that, at the moment this manuscript was written, the LHC run II is still ongoing. At present ALICE collected an integrated Pb-Pb luminosity of $\sim 250\,\mu b^{-1}$ during the 2015 data taking while an additional $\sim 750\,\mu b^{-1}$ is expected by the end of 2018 for a total run II Pb-Pb integrated luminosity of $L_{\rm int} \sim 1\,nb^{-1}$.
Finally, the ALICE Collaboration is preparing the upgrade of the detector, which will take place during LHC long shutdown 2 in 2019, when, among other upgrades, the inner tracking system and the readout chambers of the Time Projection Chamber will be fully replaced [5].

2 Highlights on strangeness production

Strangeness enhancement (i.e. higher abundance of strangeness per participating nucleon in A-A collisions with respect to pp collisions at the same energy) was one of the first proposed signatures

⋆e-mail: a.grelli@uu.nl

of the formation of the QGP, suggested by Rafelski and Muller in 1982 [6]. The effect was indeed observed for the first time at the SPS [7] and then at RHIC [9, 10]. The investigation of strangeness production at RHIC and LHC, revealed several features, such as the baryon to meson anomaly and the enhancement decreasing with the increase in collision energy, which triggered further theoretical investigations [8]. Quark coalescence has been used to explain the intermediate-p_T baryon to meson enhancement [11], while statistical hadronization models succeeded in describing hadron yields using a grand-canonical approach [12]. In the statistical model, the lower production of strangeness in small systems (pp collisions) is understood as the consequence of the small phase-space volume available, and it is referred to as canonical suppression [13] . The ALICE experiment detects K^0_S, Λ, $\Xi^\pm$ and $\Omega^\pm$

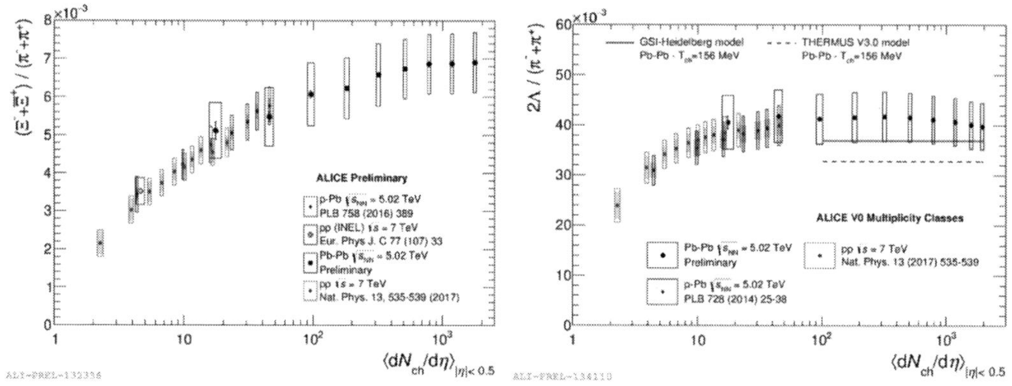

Figure 1. Left: Integrated production yield of $\Xi^- + \Xi^+$ normalized to charged pions is shown versus $\langle dN_{ch}/d\eta \rangle_{|y|<0.5}$. Right: same plot for the Λ compared with Thermal Model predictions [14, 15].

through the reconstruction of the daughter tracks coming from their weak decay in the rapidity region $|y| < 0.5$. The Inner Tracking System and the Time Projection Chamber are used for secondary vertex determination, tracking and PID through specific energy loss measurements. Recent measurements by the ALICE experiment [16] show that strangeness enhancement is also present in high-multiplicity pp and p-Pb events. Moreover, the Λ/K^0_S ratio shows a qualitatively similar behavior in p-Pb and Pb-Pb collisions [17]. Therefore, the study of strangeness production in small and large systems acquires a fundamental importance in the investigation of the origin of the heavy-ion-like effects found in small systems [18].
The preliminary measurement of the production yield of strange and multi-strange hadrons in Pb-Pb at $\sqrt{s_{NN}} = 5.02$ TeV was shown for the first time at this conference. As example, in the left panel of figure 1 the integrated production yield of $\Xi^- + \Xi^+$ normalized to charged pions is shown as a function of the charged-particle multiplicity $\langle dN_{ch}/d\eta \rangle_{|y|<0.5}$ for the three different collision systems. The right panel shows the same ratio for the Λ baryon. The results provide clear evidence that in high multiplicity pp interactions strangeness enhancement is comparable with p-Pb and peripheral Pb-Pb. While the general behavior of the finding can be understood using the grand canonical approach, all the generators and models fail to reproduce the shape and/or the magnitude of the effect [16, 19].

3 Highlights on heavy-flavour production

Heavy-flavour hadrons, which are produced in large-virtuality parton scatterings in the early stages of the collision [20], are sensitive probes of the medium formed in heavy-ion collisions. They are expected to be sensitive to the energy density through the in-medium energy loss of their heavy quark constituents [21, 22]. Furthermore, the in-medium thermal production is expected to be negligible [23]. In addition, if in-medium hadronization contributes to charm hadron formation at low-p_T, then strange charm hadrons (i.e. D_s^+) should be enhanced relative to non-strange charm hadrons [24]. Recently ALICE released the first measurement of two charmed baryon to meson ratios in pp and p-Pb collisions at central rapidity at $\sqrt{s}$ = 7 TeV and $\sqrt{s_{NN}}$ = 5.02 TeV respectively (Λ_c/D^0 and Ξ_c/D^0) [38]. Both results challenge models and point toward something not well understood in the fragmentation of charm quark into baryons. The measured Λ_c/D^0 ratio shown in the left panel of figure 2 is a factor 2-5 higher than all the models considered. In this regard it is interesting to note that only PYTHIA8 with color reconnection can qualitatively reproduce the shape of the measured ratio. The right panel of figure 2 reports the ratio of the average D meson (D^0, D^{*+}, D^+) production in central to peripheral p-Pb collisions (Q_{CP}) obtained at $\sqrt{s_{NN}}$ = 5.02 TeV. The shape of the ratio shows a hint of deviation from unity at intermediate p_T (at ~ 1.5σ level). This deviation could be caused by nuclear modification of the PDFs or by final state effects on D meson production.

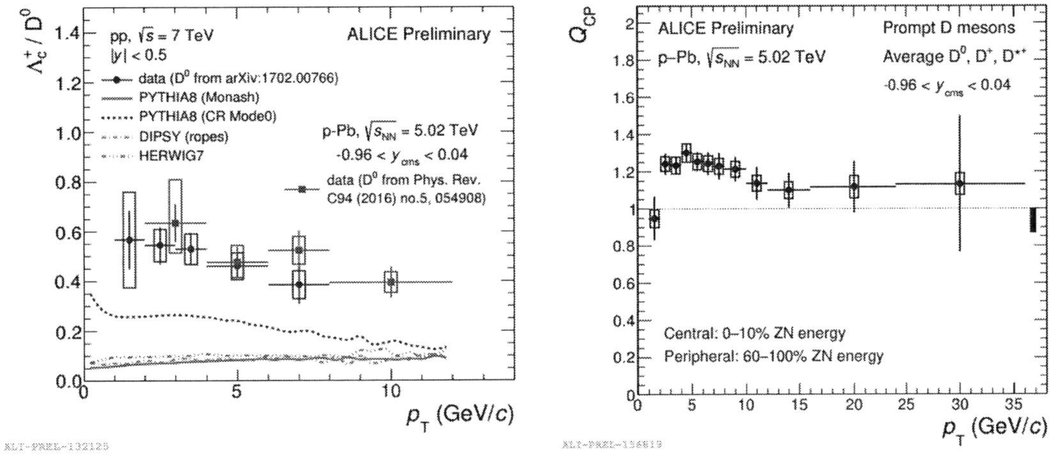

Figure 2. Left panel: Λ_c/D^0 ratios at central rapidity (|y|<0.5) as measured in pp and p-Pb interactions using LHC run I data at $\sqrt{s}$ = 7 TeV and $\sqrt{s_{NN}}$ = 5.02 TeV respectively. Right panel: Average Q_{CP} of non-strange D mesons (D^0, D^+ and D^{*+}) as measured in p-Pb LHC run II data at $\sqrt{s_{NN}}$ = 5.02 TeV.

In Pb-Pb collisions the interactions with the produced medium can be studied via the nuclear modification factor, R_{AA}, defined as the ratio between the yields measured in Pb-Pb and pp collisions after normalizing the Pb-Pb yield to the average number of nucleon-nucleon collisions in the considered centrality class. It is expected that the medium created in the collision affects the momentum spectra of the originally produced hard probes, resulting in a value of R_{AA} different from unity. It has however to be considered that other effects related to the presence of nuclei in the initial state can break the expected binary scaling due to the cold-nuclear matter effects (CNM) (e.g. nuclear modifications of the PDFs, Cronin enhancement). By measuring an R_{pPb} of D mesons compatible with unity in a wide p_T range [26] ALICE suggests that CNM effects are small at the LHC. In the left panel of figure 3 the non-strange D mesons R_{AA} and the D_s^+ R_{AA} are shown for the 10% most central Pb-Pb collisions. The large suppression found, factor 5-6 at p_T ~ 10 GeV/c cannot be explained by CNM effects and therefore is attributed to a final state effect. It has to be noted that the data show

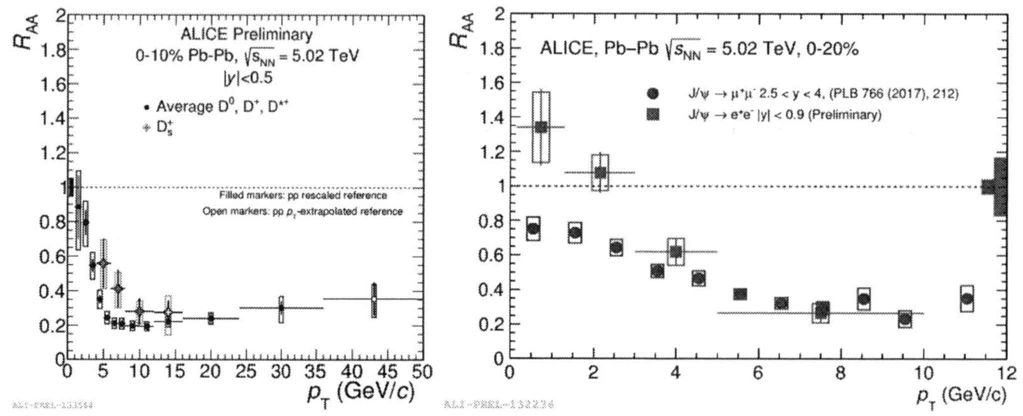

Figure 3. Left panel: R_{AA} of non-strange D mesons (D^0, D^+ and D^{*+}) in black compared with the D_s^+ R_{AA} in orange as obtained from the most central 10% Pb-Pb collisions. Right panel: R_{AA} of J/Ψ at central and forward rapidity in the 20% most central collisions.

hint of relative enhancement of the D_s^+ in the low-p_T region. However, this observation is not yet significant due to the large uncertainties.

Quarkonium states are also expected to be suppressed ($R_{AA} < 1$) in the QGP, due to the color screening of the force which binds the $c\bar{c}$ (or $b\bar{b}$) state. Quarkonium suppression is expected to occur sequentially according to the binding energy of each meson: strongly bound states like J/ψ and $\Upsilon(1S)$ should melt at higher temperatures with respect to more loosely bound states [27, 28]. At LHC energies, it is predicted that the more abundant production of charm quarks would lead to charmonium regeneration from recombination of c and $\bar{c}$ quarks at the hadronization level, resulting in an enhancement in the number of observed J/ψ [29]. The right panel of figure 3 shows the R_{AA} of J/ψ measured at central rapidity (|y|<0.9) via the decay channel $J/\psi \to e^+e^-$ compared with the same observable measured at forward rapidity (2.5<y<4) via the decay channel $J/\psi \to \mu^+\mu^-$. The measurement at central rapidity suggests a different behavior at low-p_T with respect to the forward rapidity measurement. This is consistent with the possibility of a stronger regeneration of central rapidity [30]. However, the large uncertainties of the central-rapidity measurement prevent a clear conclusion on this point.

4 Highlights on collectivity

The collective motions arising from the large pressure gradients generated by compressing and heating the nuclear matter is a typical feature of the medium produced in heavy-ion collisions. The radial flow, generated by the collective expansion of the fireball, is studied by measuring the transverse momentum spectra of identified hadrons [32]. The radial flow velocity at the thermal freeze-out, is estimated via a blast-wave fit. In the Blast Wave model [41], radial flow breaks the m_T-scaling present in the non-flowing case. Heavier particles gain more momentum from the common collective flow field and this leads to a mass ordering of elliptic flow at low-p_T. Indeed a mass ordering in the measurement of elliptic flow of $\pi^\pm$, $K^\pm$, $p + \bar{p}$, ϕ, $\Omega^\pm$ and $\Xi^\pm$ was found by ALICE with data from the first LHC Pb-Pb run at $\sqrt{s_{NN}} = 2.76$ TeV [31]. The study also provided evidence for constituent quark scaling [33–35] to be violated in the intermediate p_T region at the level of ±20%. Profiting from the larger sample collected during the second LHC Pb-Pb run at $\sqrt{s_{NN}} = 5.02$ TeV, ALICE released

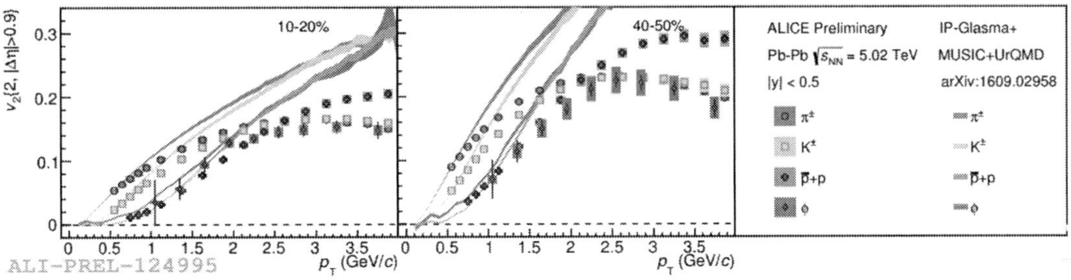

Figure 4. v_2 for pions, protons, kaons, and ϕ mesons for various centrality classes, measured with the scalar product method in Pb-Pb at $\sqrt{s_{NN}}$ = 5.02 TeV. Comparison with hydrodynamics simulations is added.

a new, more precise, measurement of the elliptic flow of $\pi^{\pm}$, $K^{\pm}$, p + $\bar{p}$, and ϕ (see Fig. 4) as well as measurement of the higher order harmonics v_3 and v_4. Hydrodynamical simulations [25] based on IP-Glasma+ MUSIC+UrQMD can reproduce the measured ordering qualitatively rather well in the p_T region below 1-2 GeV/c for all the v_n considered. In figure 5 on the left panel, the elliptic flow of

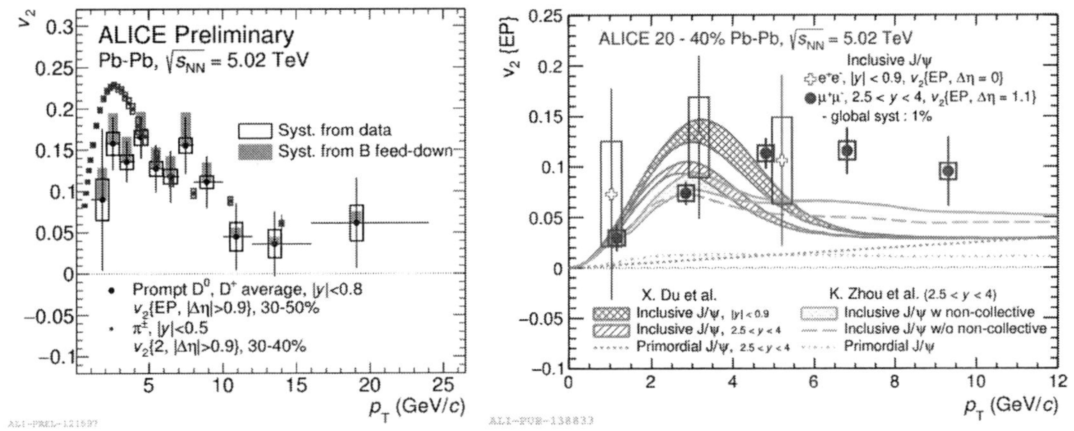

Figure 5. Left panel: Average non-strange D-mesons v_2 measured in the centrality class 30 – 50% compared with pion v_2 measured in the centrality class 30 – 40%. Right panel: J/Ψ v_2 measured at central (|y| < 0.9) and forward (2.5<y<4) rapidity.

D-mesons in the centrality range 30 – 50% , measured with event plane method [36], is compared to charged pion v_2. The measurement [37] shows that D-mesons have a positive v_2 in range $2 < p_T < 10$ GeV/c indicating that charm quark takes part in the collective motion of the system. At higher p_T, v_2 is compatible with 0 within uncertainties.

In the right panel of the same figure the J/ψ elliptic flow is shown as measured by ALICE at forward (2.5 < y < 4) and central (|y| < 0.9) rapidities. While the large error bars of the measurement at central rapidity do not allow for a quantitative comparison with theory calculations [39], for the forward rapidity measurement, at low p_T, the theory calculation [40] is able to reproduce the magnitude of the measured v_2 by adding a strong J/ψ regeneration component while at high-p_T the magnitude of the v_2 is underestimated.

5 Highlights on (Anti-)(Hyper-) nuclei production

The production of (anti-)(hyper-)nuclei [46] was measured in all the available collision systems at LHC at central rapidity ($|y| < 0.5$). The production spectra of deuterons in Pb-Pb collisions at $\sqrt{s_{\rm NN}} = 5.02$ TeV and in pp collisions at $\sqrt{s} = 13$ TeV are shown in the left panel of figure 6. The typical

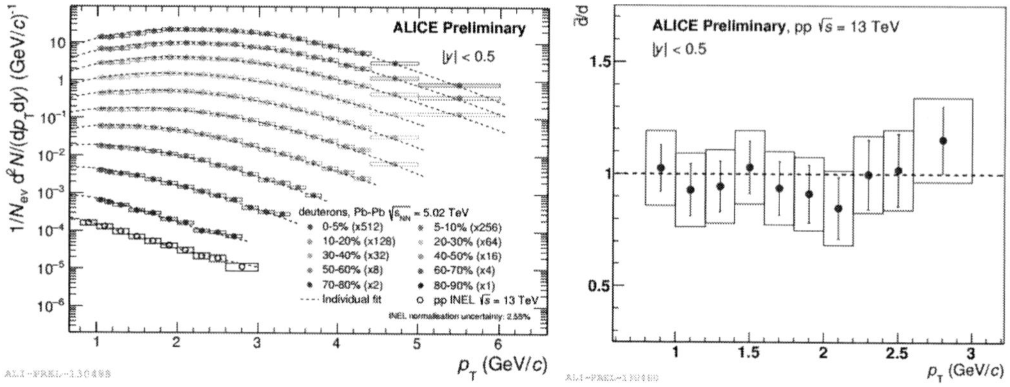

Figure 6. Left panel: deuteron production in pp at $\sqrt{s} = 13$ TeV and in Pb-Pb at $\sqrt{s_{NN}} = 5.02$ TeV. Right panel: $\overline{\rm d}$/d ratio in pp at $\sqrt{s} = 13$ TeV.

hardening of the spectra with increasing centrality, already observed for lighter particles [42] and for deuterons [43] at lower energies, is visible. The observation suggests an increasing radial flow with increasing centrality. Moreover, in the right panel of the same figure, the ratio between the production spectra of deuteron and anti-deuteron is shown to be compatible with unity. This observation is compatible with expectations from the statistical hadronisation [44] and the hadron coalescence models [45].

Finally, ALICE presented one of the most precise measurements of the ${}^{3}_{\Lambda}$H lifetime using the Pb-Pb data sample at $\sqrt{s_{\rm NN}} = 5.02$ TeV collected in 2015. The preliminary measured value of $\tau = 237^{+33}_{-36}$(stat.) $\pm$ 17(syst.) ps is compatible with both the previously computed world average of $\tau = 216^{+18}_{-16}$ ps [47] and with the free Λ lifetime.

6 Conclusions

ALICE is the dedicated heavy ion experiment at the CERN LHC and it is equipped with excellent tracking and particle identification capabilities. The large sample collected during the second Pb-Pb run at LHC allowed ALICE to enter a precision hera for many observables (e.g. strangeness production and collectivity in Pb-Pb). In terms of azimuthal anisotropy, measurements of the v_3 and v_4 of identified particles set additional tight constraints to models. In the heavy-flavour sector ALICE released the first measurement of the charmed baryon to meson ratio in pp and p-Pb interactions at central rapidity at LHC. Such a measurement constitutes, at present, a challenge for models of charm fragmentation. There are interesting hints for collective behavior in high-multiplicity pp and p-Pb collision, one of them being the measurement of the $Q_{\rm CP}$ of non-strange D mesons. However, the low significance and the fact that the effect of a Color Glass Condensate or other cold matter effects may yet provide an alternative explanation for the observed behavior make the measurement not yet conclusive. The excellent PID capabilities enable ALICE to perform high precision nuclei measurements. At this conference ALICE presented precision measurements of deuteron and anti-deuteuteron production in pp and Pb-Pb as well as one of the most precise measurements so far of the hyper-triton lifetime.

References

[1] D. J. Gross and F. Wilczek, Phys. Rev. Lett. **30**, 1343, H.D. Politzer, Phys. Rev. Lett. **30**, 1346 (1973).
[2] V. Khachatryan *et al.*, (CMS), JHEP **1009** (2010) 091.
[3] B. Abelev *et al.*, (ALICE), Phys.Lett. B**719** (2013) 29-41.
[4] K. Aamodt et al. (ALICE), JINST **3**, S08002 (2008).
[5] G. Martinez (ALICE), These proceedings (2017).
[6] J. Rafelski, B. Muller, Phys. Rev. Lett. **48** (1982) 1066-1069.
[7] Andersen, E. *et al.*, (WA97), Physics Letters B **449** (3-4) (1999) 401-406.
[8] C. Blume, C. Markert, Progress in Particle and Nuclear Physics **66** (4) (2011) 834-879.
[9] Abelev B. *et al.*, (STAR), Phys. Rev. C **77**, 044908 (2008).
[10] D. Tlusty, These proceedings (2017).
[11] R. Fries, V. Greco, P. Sorensen, Annual Review of Nuclear and Particle Science **58** (1) (2008) 177-205.
[12] P. Braun-Munzinger, D. Magestro, K. Redlich, J. Stachel, Physics Letters B **518** (12) (2001) 41-46.
[13] Hamieh S., Redlich K. and Tounsi A., Phys. Lett. B **486**, 61-66 (2000).
[14] A. Andronic, P. Braun-Munzinger, and J. Stachel, Physics Letters B **673** (2009) 142-145.
[15] S. Wheaton, J. Cleymans, and M. Hauer, Computer Physics Communications **180** (2009) 84-106.
[16] K. Aamodt *et al.* (ALICE), Nature Physics **13**, 535-539 (2017).
[17] K. Aamodt *et al.* (ALICE), Physics Letters B **728** (2014) 25-38.
[18] L. Bianchi, (ALICE), These proceedings (2017).
[19] M. Sefcik, (ALICE), These proceedings (2017).
[20] F.-M. Liu and S.-X. Liu, Phys. Rev. **C 89** no. 3, (2014) 034906.
[21] R. Baier, Y. L. Dokshitzer, et al, Nucl.Phys. **B 484** (1997) 265282.
[22] M. Gyulassy and M. Plumer, Phys. Lett. **B 243** (1990) 432438.
[23] P. Lvai and R. Vogt, Phys. Rev. C **56**, 2707.
[24] V. Greco, C. M. Ko, and R. Rapp, Phys. Lett. B **595** (2004) 202208. A. Andronic, P. Braun-Munzinger, K. Redlich, and J. Stachel,Phys. Lett. B **571** (2003) 3644.
[25] S. McDonald *et al.* Phys. Rev. C **95**, 064913 (2017).
[26] F. Grosa, (ALICE), These proceedings (2017).
[27] S. Chatrchyan *et al.* (CMS), Phys. Rev. Lett. **109** (2012) 222301.
[28] A. Abelev *et al.* (ALICE), Phys. Lett. B **738** (2014) 361-372.
[29] P. da Costa. (ALICE), Nucl.Phys. A **956** (2016) 705-708.
[30] D. Weiser (ALICE), These proceedings (2017).
[31] B. Abelev *et al.*, (ALICE), JHEP **06** 190 (2015).
[32] R. A. Bertens. (ALICE), These proceedings (2017).
[33] D. Molnar, S.A. Voloshin, Phys. Rev. Lett. **91**, 092301 (2003).

[34] A. Adare *et al.*, (PHENIX), Phys. Rev. Lett. **98**, 162301 (2007).
[35] B. Abelev *et al.*, (STAR), Phys. Rev. C**75**, 054906 (2007).
[36] A. M. Poskanzer, S.A. Voloshin, Phys. Rev. **C58**, 1671 (1998).
[37] S. Acharya *et al.*, (ALICE), ALICE-PUBLIC-2017-003, (2017).
[38] J. Normal, (ALICE), These proceedings (2017).
[39] X. Du and R. Rapp, Nucl. Phys. A **943** (2015) 147-158.
[40] K. Zhou, N. Xu, Z. Xu, and P. Zhuang, Phys. Rev. C **89** no. 5, (2014) 054911.
[41] E. Schnedermann, J. Sollfrank, U.W. Heinz, Phys. Rev. **C48**, 2462 (1993).
[42] J. Adam *et al.*, (ALICE), Phys. Rev. C **93**, 034913 (2016), 1506.07287.
[43] J. Adam *et al.*, (ALICE), Phys. Rev. C **93**, 024917 (2016), 1506.08951.
[44] A. Andronic, P. Braun-Munzinger, J. Stachel, H. Stocker, Phys. Lett. B **697**, 203 (2011).
[45] J. I. Kapusta, Phys. Rev. C **21**, 1301 (1980).
[46] M. Puccio (ALICE), These proceedings (2017).
[47] J. Adam *et al.*, (ALICE), Phys. Lett. B **754**, 360 (2016).

Recent heavy-flavour results from ATLAS

Jing Chen On behalf of the ATLAS Collaboration[1,*]

[1] *University of Science and Technology of China, Department of Modern Physics, 96 Jinzhai Road, Anhui Hefei, China*

Abstract. The ATLAS heavy-ion program utilizes heavy-flavour hadrons to probe the hot, dense matter formed at the LHC. Quarkonium measurements have been performed in pp, p+Pb and Pb+Pb systems to study medium effects. The Pb+Pb results show a strong suppression of charmonium productions in more central events. Proton-lead interactions show little modification of the 1S charmonium state, but seem to indicate a centrality dependence of the 2S state. Upsilons have been studied in p+Pb and are found to show decreasing behaviour in more central collisions. Inclusive muons with p_{T} above 4 GeV have been studied to provide insight on open-flavour production, and are found to be strongly suppressed in Pb+Pb collisions with a substantial and significant elliptic flow signal. Muon-hadron correlations have also been studied in the 2016 $\sqrt{s_{NN}} = 8.16$ TeV p+Pb data. There are clear indications of a near-side ridge, suggesting that similar mechanisms may be relevant for both the small and large systems.

1 Introduction

Measurements of quarkonium production and the azimuthal anisotropy of hadron production in p+Pb and Pb+Pb collisions provide opportunites to understand the cold nuclear matter (CNM) effects and the properties of hot and dense medium, quark gluon plasma (QGP), created in heavy-ion collisions. This paper presents a few measurements of this kind carried out using datasets collected by the ATLAS detector [1] from 2013 to 2016 in pp, p+Pb and Pb+Pb collisions at the LHC. In this paper, the following topics are discussed: 1) the measurement of nuclear modification factor of quarkonium production in p+Pb and Pb+Pb collisions [2, 3]; 2) the measurement of azimuthal anisotropy of charged particles and muons [4]. Among CNM effects, modification of nPDFs, nuclear absorption, induced medium energy loss and gluon saturation can be quantified by the nuclear modification factor, the ratio of quarkonium production cross sections in A+A relative to the cross sections measured in pp collisions at the same centre-of-mass energy. For excited quarkonium states, double ratio (the ratio of the nuclear modification factor in excited quarkonium states to the nuclear modification factor in the ground state) often to be used to quantify the CNM effects. The azimuthal anisotropy in particle $dN/d\phi$ distributions, usually described by a Fourier coefficient with n^{th} order harmonic $\langle \cos[n(\phi - \Psi_n)] \rangle$, v_n, where ϕ is the azimuthal angle of the partical momentum, and Ψ_n is the experimental event plane angle of nth order. These measurements can provide valuable constraints on QGP transport property.

*e-mail: jing.chen@cern.ch

2 Nuclear modification factor

It is well-known that J/ψ and $\psi(2S)$ can be produced from short-lived QCD decays including feed-down from other charmonium states (prompt) and produced in the decays of long lived b-hadrons (non-prompt). In this paper the measurements of prompt and non-prompt J/ψ, $\psi(2S)$ and Υ production are presented. Measurement of J/ψ, $\psi(2S)$ and $\Upsilon(nS)$ production using 28 nb^{-1} of p+Pb data at $\sqrt{s_{NN}}$ =5.02 TeV and 25 pb^{-1} of pp data at centre-of-mass energy of $\sqrt{s}$ =5.02 TeV [2]. Measurement of J/ψ and $\psi(2S)$ productions in 0.42 nb^{-1} of Pb+Pb data at $\sqrt{s_{NN}}$ =5.02 TeV and 25 pb^{-1} of pp data at centre-of-mass energy of $\sqrt{s}$ =5.02 TeV are also presented [3]. The nuclear modification factor:

$$R_{\mathrm{AA}} = \frac{N_{\mathrm{AA}}}{\langle T_{\mathrm{AA}} \rangle \times \sigma^{pp}} \tag{1}$$

$$R_{p\mathrm{Pb}} = \frac{N_{p\mathrm{Pb}}}{\langle T_{p\mathrm{Pb}} \rangle \times \sigma^{pp}} \tag{2}$$

where N_{AA} ($N_{p\mathrm{Pb}}$) is the per event yield of quarkonium states measured in A+A (p+Pb) collisions, $\langle T_{\mathrm{AA}} \rangle$ ($\langle T_{p\mathrm{Pb}} \rangle$) is the mean nuclear thickness function and σ^{pp} is the cross section for the corresponding quarkonium states in pp collisions at the same energy. Figure 1 and Figure 2 show the modification factor for J/ψ. In p+Pb collisions, no obvious p_{T} and y^*(centre-of-mass rapidity) dependence is observed. In Pb+Pb results, for prompt J/ψ, there is a trend of slight increase in R_{AA} with increasing p_{T}. For non-prompt J/ψ, R_{AA} is almost a constant as a function of p_{T} within the uncertainties. Both prompt and non-prompt J/ψ R_{AA} are essentially flat as a function of rapidity. J/ψ production is most strongly suppressed in central collisions.

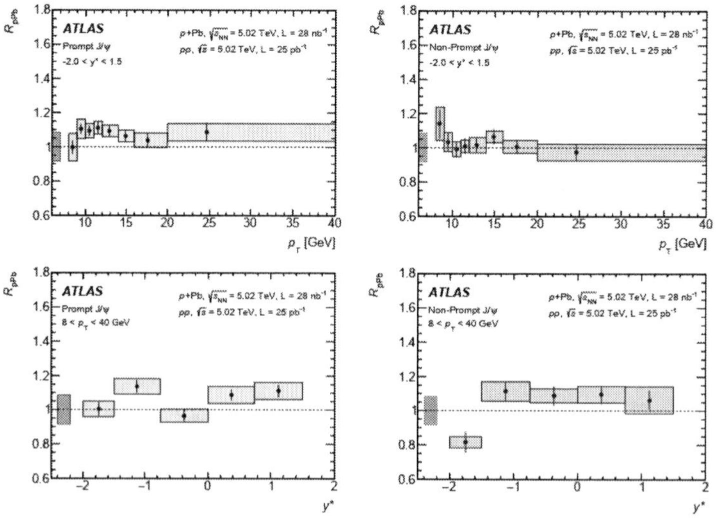

Figure 1. The nuclear modification factor, $R_{p\mathrm{Pb}}$, as a function of p_{T} (top) and y^* (bottom) for the production of prompt J/ψ, non-prompt J/ψ. The vertical error bars correspond to the statistical uncertainties [2].

In Figure 3, the $\Upsilon(nS)$ ($n = 2, 3$) to $\Upsilon(1S)$ double ratio provides information on the difference of the CNM effects on states with different binding energies. Both the integrated double ratio are found to be less than unity by two standard deviations. For the double ratio as a function of centrality, both

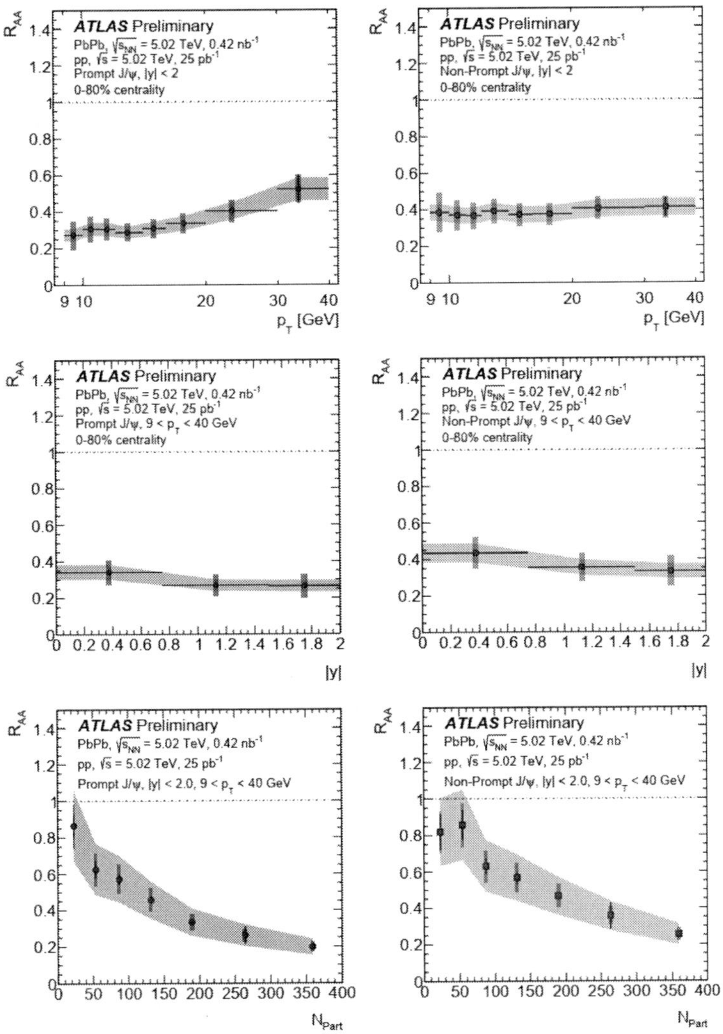

Figure 2. The nuclear modification factor as a function of p_T (top), rapidity y (middle) and centrality (bottom) for the prompt J/ψ (left) and non-prompt J/ψ (right). The statistical uncertainty on each point is indicated by a narrow error bar which is smaller than the plotting symbol for most points. The wide error bar plotted with each point represents the point-to-point systematic uncertainty, while the error band connecting the points represents correlated scale uncertainties [3].

are found to decrease slightly with increasing centrality at the significance level of one-standard-deviation. The suppression of double ratio for prompt $\psi(2S)$ and J/ψ indicates the more tightly bound quarkonium system, the J/ψ , survives the temperature of the hot dense medium with a higher probability than the more loosely bound quarkonium, the $\psi(2S)$. Non-prompt mesons are always formed outside the medium, thus, the double ratio is expected to be unity, as is observed.

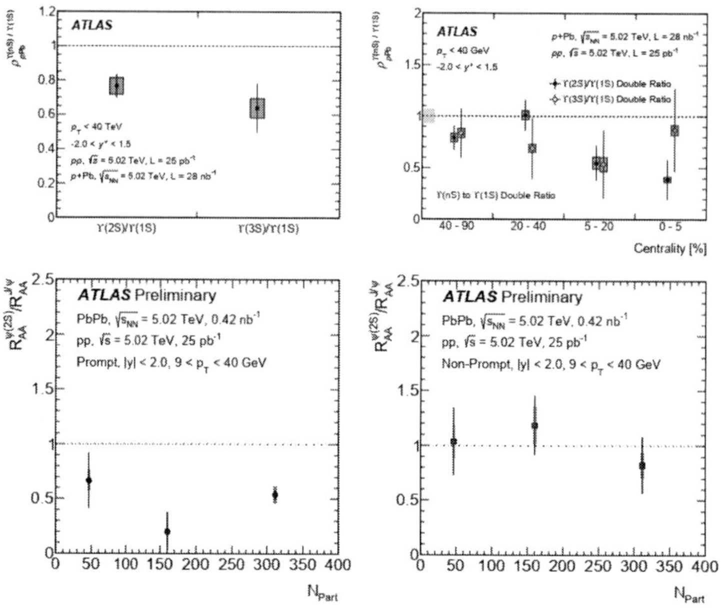

Figure 3. Top: The $\Upsilon(n\mathrm{S})$ $(n = 2, 3)$ to $\Upsilon(1\mathrm{S})$ double ratio, integrated in the whole measured p_T and y* range (left). The $\Upsilon(n\mathrm{S})$ $(n = 2, 3)$ to $\Upsilon(1\mathrm{S})$ double ratio, as a function of event centrality in p+Pb collisions [2]. Bottom: double ratio for $\psi(2\mathrm{S})$ to J/ψ, as a function of centrality expressed as N_{Part}, for prompt meson production (left) and non-prompt meson production (right). This ratio has the advantage that corrections are similar for the numerator and denominator, and thus systematic uncertainties are reduced. The narrow error bar represents the statistical uncertainties while the wider error bar represents the total systematic uncertainty [3].

3 Azimuthal anisotropy

For a given event class, the two particle correlation function is defined as [4]:

$$C(\Delta\eta, \Delta\phi) = \frac{S(\Delta\eta, \Delta\phi)}{B(\Delta\eta, \Delta\phi)} \tag{3}$$

where $\Delta\phi \equiv \phi^a - \phi^b$ is the relative azimuthal angle, $\Delta\eta \equiv \eta^a - \eta^b$ is the relative pseudorapidity. S and B represent pair distributions constructed from the same event and from mixed event. One-dimensional correlation function $C(\Delta\phi)$ is obtained by integrating Eq. (3). Figure 4 shows the two-particle h-μ correlation functions in 8.16 TeV p+Pb collisions. A significant contribution to the two-particle corelations arise from back to back dijets is observed. In order to estimate the contribution of such dijets and other process, a template fit $C^{\mathrm{templ}}(\Delta\phi)$ is used:

$$C^{\mathrm{templ}}(\Delta\phi) = FC^{\mathrm{periph}}(\Delta\phi) + C^{\mathrm{ridge}}(\Delta\phi) \tag{4}$$

$$C^{\mathrm{ridge}}(\Delta\phi) = G(1 + \sum_{n=2}^{\infty} 2v_{n,n}\cos(n\Delta\phi)) \tag{5}$$

where F is the scale factor, $C^{\mathrm{periph}}(\Delta\phi)$ is the correlation measured in low multiplicity or "peripheral" events, $C^{\mathrm{ridge}}(\Delta\phi)$ is the genuine long-range harmonic modulation. In Eq. (5), G represents the magnitude of the combinatoric component of $C^{\text{ridge}}(\Delta\phi)$. The harmonics $v_{n,n}$ quantify the strengh of the genuine long-range correlation. The single-particle anisotropies v_n can be obtained from $v_{n,n}$:

$$v_{n,n}(p_{\mathrm{T}}^{a}, p_{\mathrm{T}}^{b}) = v_n(p_{\mathrm{T}}^{a})v_n(p_{\mathrm{T}}^{b}) \tag{6}$$

where a and b are two particles in the long-range correlations.

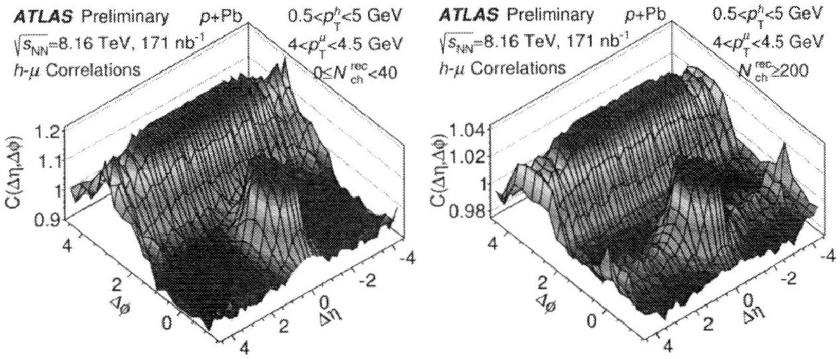

Figure 4. Two-particle h-μ correlation functions in $\sqrt{s_{NN}}$ = 8.16 TeV p+Pb collisions [4].

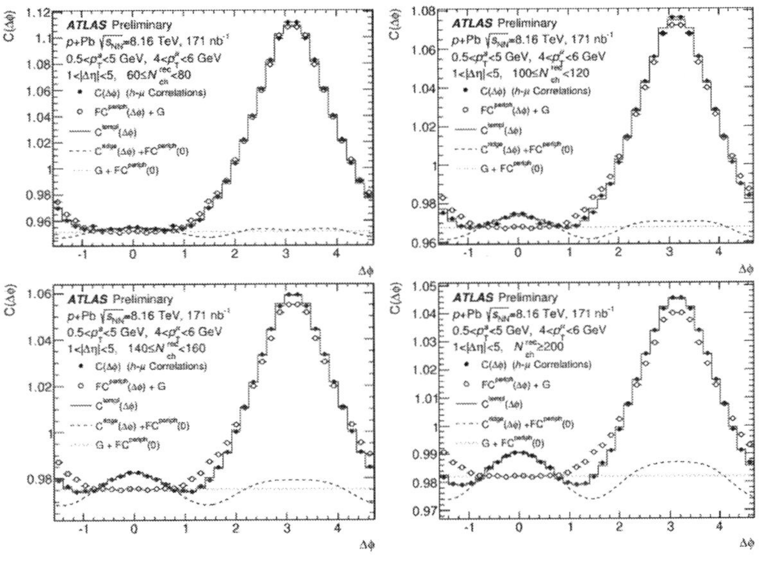

Figure 5. Template fits to h-μ correlations for $|\Delta\eta|$>1. Each panel corresponds to a different multiplicity interval [4].

Figure 5 shows template fits to h-μ correlations for $|\Delta\eta|$>1. Each panel corresponds to a different multiplicity interval. The template fitting includes second-order, third-order and fourth-order harmonics. Measurements of heavy flavour muons in Pb+Pb and pp collisions show an elliptic anisotropy [5].

Such correlations were understood to be the result of azimuthal anisotropies of the single-particle distributions. In p+Pb analyses of long-range correlations, the event activity is quantified by $N_{\mathrm{ch}}^{\mathrm{rec}}$: the total number of reconstructed charged-particle tracks passing the track selections [4]. In Figure 6, the measured v_2 for charged particles increases with $N_{\mathrm{ch}}^{\mathrm{rec}}$ at low $N_{\mathrm{ch}}^{\mathrm{rec}}$ and eventually saturates for $N_{\mathrm{ch}}^{\mathrm{rec}}$ >150. However even at the lowest measured multiplicity, a large v_2 is observed. Over the p_{T} range shown, the measured v_2 for muons are consistent with having no $N_{\mathrm{ch}}^{\mathrm{rec}}$ dependence.

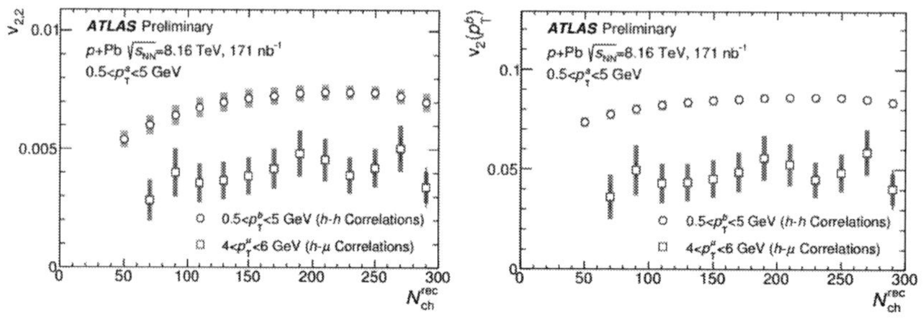

Figure 6. The $v_{2,2}$ to h-h correlations (circles), h-μ correlations (squares) and the corresponding v_2 values for the associated particle [4].

4 Conclusion

ATLAS has measured the production of quarkonia and heavy flavour muons in p+Pb and Pb+Pb systems. For quarkonium production, stronger suppression in more central Pb+Pb collisions is observed. Proton-lead interactions show little modification of the ground charmonium state. Upsilons also have been studied in p+Pb and are found to show decreasing behaviour in more central collisions. Significant azimuthal flow modulation of heavy flavour muon is observed via heavy flavour muon-hadron correlation. These results can provide insight into nuclear parton distributions, allow quantitative studies of heavy quarkoium production, and also provide the opportunity to improve our understanding of the propagation of heavy quarks in the QGP.

References

[1] ATLAS Collaboration, JINST **3**, S08003 (2008)
[2] ATLAS Collaboration (2017), `arXiv:1709.03089`
[3] ATLAS Collaboration, ATLAS-CONF-2016-109 (2016), `https://cds.cern.ch/record/2220771`
[4] ATLAS Collaboration, ATLAS-CONF-2017-006 (2017), `https://cds.cern.ch/record/2244808`
[5] ATLAS Collaboration, ATLAS-CONF-2015-053 (2015), `https://cds.cern.ch/record/2055674`

Recent results from CMS

Émilien Chapon[1,★], on behalf of the CMS Collaboration

[1] *Experimental Physics Department, CERN, CH-1211 Geneva 23, Switzerland*

Abstract. Recent results from the CMS Collaboration are reported, covering the wide variety of probes accessible to the experiment.

The strongly interacting matter created in nuclear collisions at the LHC, in pPb and PbPb collisions but also high multiplicity pp collisions, can be studied with the means of a large variety of probes, from identified hadrons to jets, from light flavours and strangeness to heavy flavours, charm and beauty. The CMS collaboration [1] uses many different observables in these studies, starting with the production cross section. Nuclear modification factors (R_{AA}) can also be built, from the ratio of the cross section in heavy ions collisions (HIC), pPb or PbPb, to that in pp collisions. In this way, one can clearly see if the production of a given probe is suppressed or enhanced in HIC: for instance, in the case of charmonia, colour charge Debye screening in the quark-gluon plasma (QGP) induces a suppression in the production in PbPb collisions, while recombination can also play a role and enhance the yields at low transverse momentum (p_{T}). Much information can also be gained by focusing on collective properties of each event, such as N-particle correlations, or event-by-event fluctuations of such quantities. In particular, the Fourier coefficients of the azimuthal correlations (v_2, v_3, etc.) provide information on the anisotropic flow in the event (resp. elliptic, triangular, etc). A collective behaviour is observed not only in PbPb collisions, but also in small systems: pPb and even pp collisions. In dijet production, the imbalance of the two jets can be measured and compared between different systems, pp and PbPb. Heavy flavour jets, such as b jets, then allow to probe the flavour dependence of parton energy loss.

1 Charmonia

The different charmonium states (J/ψ, $\psi(2S)$) can be reconstructed through their decay to a pair of oppositely charged muons. The invariant mass distribution $m_{\mu\mu}$ of these pairs is then fitted to extract the observed yields. However, further work is needed to disentangle the different sources of charmonium production. Some fraction of it, called nonprompt, originates from the decay of B mesons, which can be resolved using the pseudo-proper decay length $\ell_{\mathrm{J}/\psi}$, reconstructed using the distance between the primary and dimuon vertices. In the case of the J/ψ meson, a two-dimensional fit allows to simultaneously extract the prompt and nonprompt production, while in the case of the $\psi(2S)$ meson, a cut-based method was used to remove the nonprompt contribution. The prompt component of J/ψ meson production receives also a contribution from the feed-down from excited charmonium states, χ_c and $\psi(2S)$ mesons. This cannot be separated from direct J/ψ meson production in the results reported here.

★e-mail: emilien.chapon@cern.ch

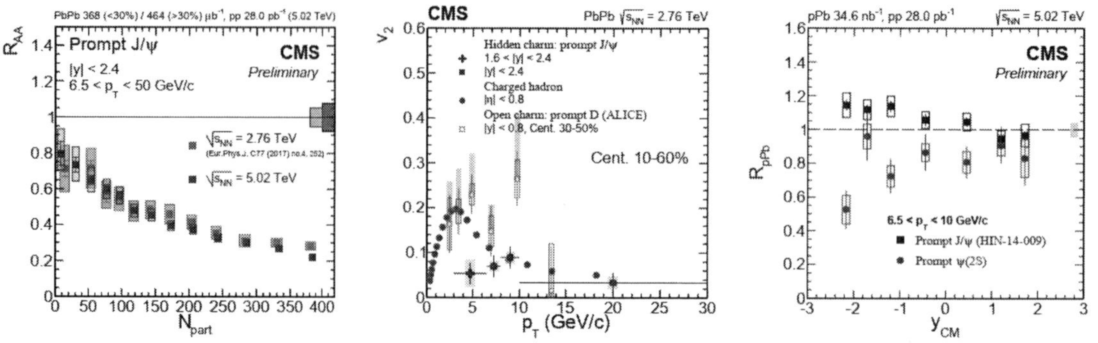

Figure 1. Left: R_{AA} of prompt J/ψ mesons as a function of the number of participants to the collision, at $\sqrt{s_{\mathrm{NN}}} = 2.76$ [2] and 5.02 TeV [3]. Center: v_2 of prompt J/ψ mesons as a function of p_{T}, at $\sqrt{s_{\mathrm{NN}}} = 2.76\,\mathrm{TeV}$, compared to that of other hadrons. Right: R_{pPb} of prompt $\psi(2S)$ [4] and prompt J/ψ [5] mesons at $\sqrt{s_{\mathrm{NN}}} = 5.02\,\mathrm{TeV}$, as a function of rapidity in the centre-of-mass frame.

The nuclear modification factor of prompt J/ψ mesons has been measured in pPb collisions at $\sqrt{s_{\mathrm{NN}}} = 5.02\,\mathrm{TeV}$ [5], as well as in PbPb collisions at $\sqrt{s_{\mathrm{NN}}} = 2.76\,\mathrm{TeV}$ [2] and 5.02 TeV [3], always compared to pp measurements at the same collision energy. In pPb collisions, nuclear modifications of the parton distribution functions are expected to impact the prompt J/ψ meson production. The CMS measurement of the R_{pPb} as a function of the rapidity of the J/ψ meson is consistent with this expectation, but other effects could also contribute, such as energy loss in cold nuclear matter. In PbPb collisions and at high p_{T} ($p_{\mathrm{T}} > 6.5\,\mathrm{GeV}$), the creation of a QGP further suppresses J/ψ meson production, because of parton energy loss, especially at very high p_{T}, and melting induced by Debye colour charge screening in the deconfined medium. This suppression is even more pronounced in central collisions, where the QGP is hotter and longer lived, as can be seen in Fig. 1. The elliptic flow v_2 of prompt J/ψ mesons has also been measured and is found to be non-zero, even at relatively high p_{T} (up to 10 GeV, see Fig. 1)), possibly pointing to a path length dependence of parton energy loss. On the other hand, the v_2 measurement is found to be compatible at high p_{T} with that of charged hadrons, dominated by light hadrons, which could be pointing to a flavour independence of this path length dependence.

Additional information can be gained by comparing the behaviour of the ground and excited charmonium states, the J/ψ and $\psi(2S)$ mesons. A higher suppression of the $\psi(2S)$ meson, compared to the J/ψ meson, is observed both in pPb [4] and PbPb [6] collisions at $\sqrt{s_{\mathrm{NN}}} = 5.02\,\mathrm{TeV}$. This difference seems to be more pronounced at backward rapidity in pPb collisions, as shown in Fig. 1, and it is observed at all p_{T} between 3 and 30 GeV in PbPb collisions. This points to the importance of final state effects, which have a different impact on the ground and excited states. Understanding such different behaviours of the J/ψ and $\psi(2S)$ mesons is challenging to understand theoretically, especially in pPb where final state effects should be of lesser importance.

Bottomonia are measured in the CMS experiment, inclusively from $p_{\mathrm{T}} = 0\,\mathrm{GeV}$. In addition, as opposed to J/ψ and $\psi(2S)$ meson production, $\Upsilon(nS)$ meson production is free from any nonprompt contribution, because they are heavier than B mesons. Feed-down from P-wave and heavier S-wave states is however also present in the $\Upsilon(1S)$ and $\Upsilon(2S)$ meson measurements. The comparison of the R_{AA} of the different $\Upsilon(nS)$ states [7] shows a clear ordering, with the heavier, weaker bound excited states being more suppressed, as expected from sequential suppression because of the lower

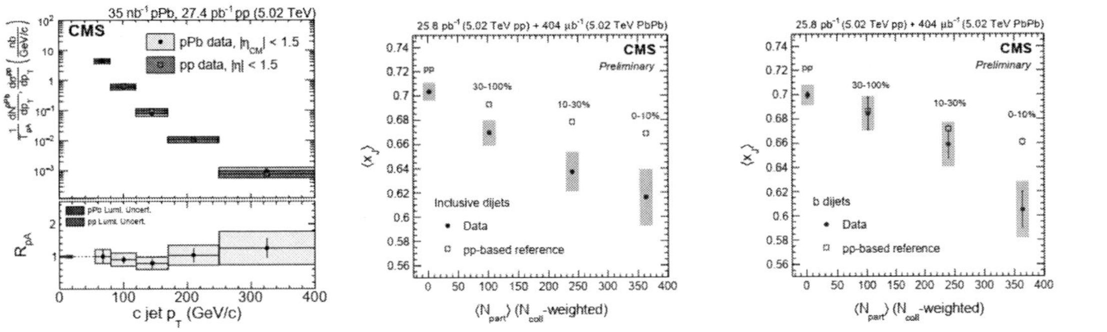

Figure 2. Left: differential cross section for c-jet production in pp and pPb collisions at $\sqrt{s_{\mathrm{NN}}} = 5.02\,\mathrm{TeV}$, as a function of p_{T} [9]. The R_{pPb} is also shown in the bottom panel. Center: dijet momentum imbalance of inclusive dijets in pp and PbPb collisions at $\sqrt{s_{\mathrm{NN}}} = 5.02\,\mathrm{TeV}$, as a function of the number of participants weighted by the number of binary collisions [10]. Right: same as center, for b dijets.

dissociation temperature of the $\Upsilon(2S)$ and $\Upsilon(3S)$ mesons with respect to the $\Upsilon(1S)$ meson. This suppression is also higher in central than in peripheral events. The $\Upsilon(3S)$ meson is not observed in the PbPb dataset at $\sqrt{s_{\mathrm{NN}}} = 5.02\,\mathrm{TeV}$, even in the most peripheral events [8].

2 Heavy flavour jets

Tagged jets provide another probe in the study of heavy flavour production in HIC. Tagging relies on several properties of the jet, such as the secondary vertex mass, which allow to discriminate jets originating from a charm or bottom quark from those initiated by a light parton. The contribution from these different categories of jets is then extracted thanks to templates of this discriminating variable. Heavy flavour jets allow to probe charm and bottom production at high p_{T}, above 55 GeV, and include contributions from all hadron species. The comparison with light jets can provide information on the importance of the quark mass in energy loss mechanisms and other phenomena.

The production of both charm [9] and bottom [11] jets has been measured in both pp and pPb collisions at $\sqrt{s_{\mathrm{NN}}} = 5.02\,\mathrm{TeV}$. In the case of c jets, the first measurement of this kind in pp and pPb collisions, the spectra have been found to be compatible between the two systems (see Fig. 2). The b jet measurement is also the first in pPb collisions, and is compared to expectations from PYTHIA: again no significant differences are found between the data and the pp model.

The measurement of jet pair production, and of the momentum imbalance between the two jets, $\langle x_J \rangle = \langle p_{\mathrm{T},1}/p_{\mathrm{T},2} \rangle$, is another tool in the understanding of heavy quark energy loss in HIC. In the case of light jets, a smaller $\langle x_J \rangle$ is found in PbPb than in pp collisions at $\sqrt{s_{\mathrm{NN}}} = 5.02\,\mathrm{TeV}$ [10], possibly more pronounced in the most central collisions. A similar behaviour is observed when comparing the $\langle x_J \rangle$ of b jets in PbPb and pp collisions (see Fig. 2). Several processes give a significant contribution to b jet production, from gluon splitting to flavour creation, but the former is suppressed by selecting back-to-back b dijets.

3 Open heavy flavour

Numerous measurements of open heavy flavour production have been performed by the CMS Collaboration, from pp to pPb and PbPb collisions, using D mesons (down to a p_{T} of 2 GeV) and B mesons.

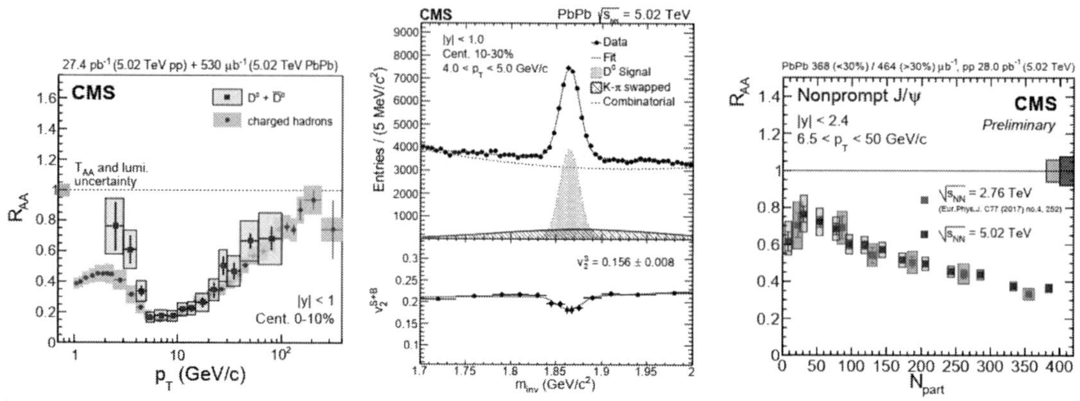

Figure 3. Left: R_{AA} of prompt D^0 mesons and charged hadrons in pp and PbPb collisions at $\sqrt{s_{NN}} = 5.02$ TeV, as a function of p_T [12]. Center: example of simultaneous fit of the invariant mass spectrum and the v_2, for the extraction of the v_2 of prompt D^0 mesons in PbPb collisions at $\sqrt{s_{NN}} = 5.02$ TeV [13]. Right: R_{AA} of nonprompt J/ψ mesons as a function of the number of participants to the collision, at $\sqrt{s_{NN}} = 2.76$ [2] and 5.02 TeV [3].

The decay channels used are $D^0 \rightarrow K\pi$ and $B^{\pm} \rightarrow J/\psi K$. Similarly to J/ψ, they also include a nonprompt component from the decay of B mesons, which can be separated from the prompt component using the impact parameter of the D^0 meson decay vertex with respect to the primary vertex.

When comparing the nuclear modification factor of D^0 mesons [12] to that of charged hadrons, as shown in Fig. 3, a similar suppression is observed, especially at high p_T. Within the larger uncertainties, the R_{AA} of $B^{\pm}$ mesons [14] is also comparable. Such a comparison of the behaviour of the suppression of different flavours of hadrons in PbPb collisions at $\sqrt{s_{NN}} = 5.02$ TeV, compared to pp at the same collision energy, provides information on the flavour dependence of parton energy loss. The v_2 of prompt D^0 mesons [13] has also been measured in PbPb as a function of p_T, in several centrality intervals. At low p_T, up to about 7 GeV, the v_2 of D^0 mesons is smaller than that of charged hadrons, hinting to a mass ordering in the flow of hadrons. At high p_T, on the other hand, the charmed mesons feature a similar v_2 as the light hadrons, consistent with a flavour independence of the path length dependence of energy loss.

Coming back to J/ψ measurements, their nonprompt component provides another way to study B meson (open beauty) production, up to 30 GeV. In pPb collisions [5], no large modification is observed, in any rapidity or p_T bin. On the other hand, a strong suppression is observed in PbPb collisions, both at $\sqrt{s_{NN}} = 2.76$ TeV [2] and 5.02 TeV [3], even stronger in central events, which can be seen in Fig. 3. The elliptic flow of nonprompt J/ψ in PbPb collisions at $\sqrt{s_{NN}} = 2.76$ TeV [2] has also been measured, but still suffers from large uncertainties at the moment, preventing from drawing firm conclusions about the collective behaviour of the bottom quark.

4 Collectivity

Detailed studies of the collective behaviour of particle production in HIC help better understand the properties of the medium and the fluctuations in the initial state. Thus, the event-by-event fluctuations in the v_2 have been quantified with the measurement of the probability distribution of this quantity [15], in several centrality classes in PbPb data at $\sqrt{s_{NN}} = 5.02$ TeV. These distributions are fitted with two analytic functions, and the data is found to be better described by an elliptic power than

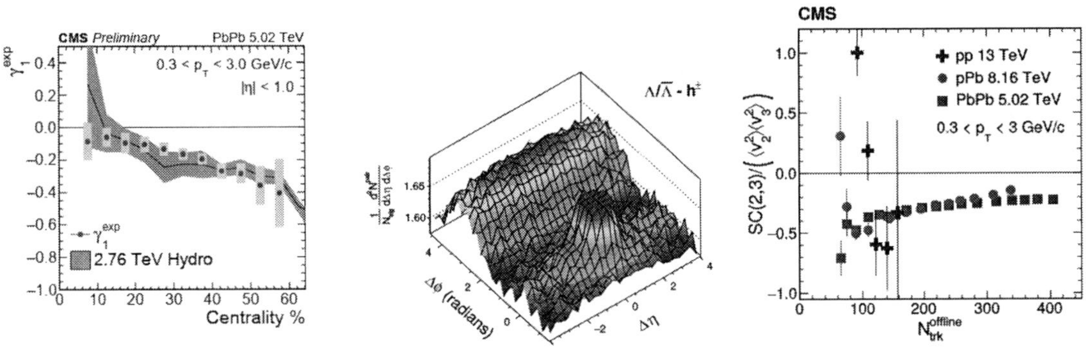

Figure 4. Left: the estimated skewness for the unfolded $p(v_2)$ in PbPb data at $\sqrt{s_{\mathrm{NN}}} = 5.02\,\mathrm{TeV}$, as a function of centrality [15]. Center: correlations between the hyperon and a charged hadron in high multiplicity pp collisions at $\sqrt{s} = 13\,\mathrm{TeV}$, as a function of $\Delta\eta$ and $\Delta\phi$ [16]. Right: normalised symmetric cumulant between the second and third harmonics, as a function of the number of offline tracks, in pp, pPb and PbPb collisions [17].

by a Bessel-Gaussian function. The skewness of the fluctuations, γ_1^{exp}, is also extracted and found to be in good agreement with predictions from an hydrodynamic model, as can be seen in Fig. 4. Such observations suggest non-Gaussian fluctuations of the eccentricity of the collisions.

Evidence for a collective behaviour is also found in particle production in smaller systems, in pp and pPb collisions. One striking feature is the observation of long-range correlations, the so-called ridge effect, even in high multiplicity pp collisions [16]. A correlation (the "ridge") is found at $\Delta\phi = 0$, even at large $\Delta|\eta|$, in pairs of charged hadrons, but also with strange particles, such as short kaons $\mathrm{K}_{\mathrm{S}}^0$ or hyperons Λ (see Fig. 4). The v_2 of charged particles has also been measured as a function of track multiplicity, in pp, pPb and PbPb collisions, using different methods. In all three systems, the v_2 computed from two-particle correlations is found to be larger than that obtained with other correlators, but four-, six- and eight-particle correlations, as well as the Lee-Yang zeroes method, all yield to similar v_2 values, as expected from a collective origin for the correlations.

Another useful observable in the study of collectivity is the event-by-event correlation between Fourier harmonics of different orders, v_n and v_m. The CMS Collaboration has measured these normalised symmetric cumulants in pp, pPb and PbPb collisions, as a function of track multiplicity [17]. Similar observations are made in all three systems. In the case of SC(2, 3), which gauges the correlation between v_2 and v_3, an anti-correlation is found at high track multiplicity, as shown in Fig. 4. On the contrary, $\mathrm{SC}(2,4) > 0$: the v_2 and v_4 values are positively correlated event-by-event. Similar trends are observed in pPb and PbPb collisions, and high multiplicity pp collisions, regarding the trend of these observables as a function of track multiplicity.

Characteristics of strangeness production have been compared between pp, pPb and PbPb collisions. A higher average kinematic energy $\langle KE_{\mathrm{T}}\rangle$ has been found at higher multiplicities, for all three systems, and for $\mathrm{K}_{\mathrm{S}}^0$, Λ and Ξ^- particles alike [18]. In addition, the increase of $\langle KE_{\mathrm{T}}\rangle$ is faster in smaller systems, and for heavier particles. A similar mass ordering is also seen in pp collisions [16], when comparing the v_2 of charged hadrons and of $\mathrm{K}_{\mathrm{S}}^0$ and Λ particles as a function of p_{T} at high multiplicity, after correcting for back-to-back jet correlations estimated from low-multiplicity data, as shown in Fig. 5. Blast-wave fits have also been performed to the p_{T} spectra of strange particles [18], in several multiplicity bins (see Fig. 5). The interpretation of the parameters of these fits, $\mathrm{T}_{\mathrm{kin}}$ and β_{T}, is model dependent. However, similar trends are found in all three systems: β_{T} increases with the

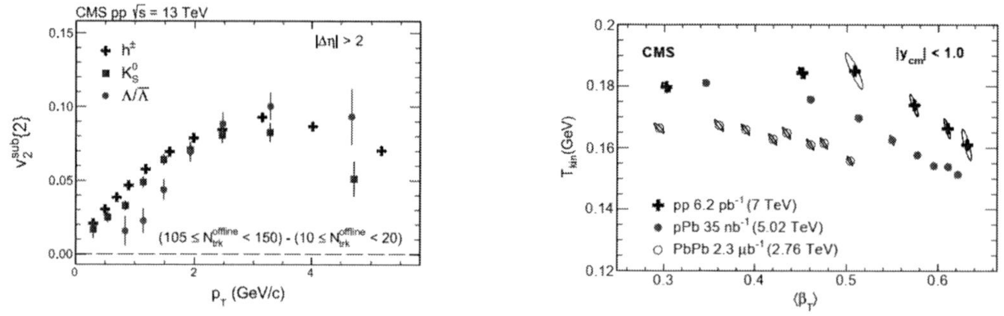

Figure 5. Left: the v_2 results of inclusive charged particles, K_S^0 and $\Lambda/\bar{\Lambda}$ particles as a function of p_T for high multiplicity in pp collisions, after correcting for back-to-back jet correlations estimated from low-multiplicity data [16]. Right: the extracted kinetic freeze-out temperature, T_{kin}, versus the average radial-flow velocity, $\langle\beta_T\rangle$, from a simultaneous blast-wave fit to the K_S^0 and Λ particles p_T spectra for different multiplicity intervals in pp , pPb , and PbPb collisions [18].

multiplicity in the event, while T_{kin} tends to decrease. In addition, a larger radial flow velocity (large β_T) is found in smaller systems.

References

[1] CMS Collaboration, JINST **3**, S08004 (2008)
[2] CMS Collaboration, Eur. Phys. J. C **77**, 252 (2017)
[3] CMS Collaboration, CMS-PAS-HIN-16-025 (2017), https://cds.cern.ch/record/2274286
[4] CMS Collaboration, CMS-PAS-HIN-16-015 (2017), https://cds.cern.ch/record/2244643
[5] CMS Collaboration, Eur. Phys. J. C **77**, 269 (2017)
[6] CMS Collaboration, Phys. Rev. Lett. **118**, 162301 (2017)
[7] CMS Collaboration, CMS-PAS-HIN-16-023 (2017), https://cds.cern.ch/record/2244680
[8] CMS Collaboration (2017), arXiv:1706.05984
[9] CMS Collaboration, Phys. Lett. B **772**, 306 (2017)
[10] CMS Collaboration, CMS-PAS-HIN-16-005 (2016), https://cds.cern.ch/record/2202805
[11] CMS Collaboration, Phys. Lett. B **754**, 59 (2016)
[12] CMS Collaboration (2017), arXiv:1708.04962
[13] CMS Collaboration (2017), arXiv:1708.03497
[14] CMS Collaboration, Phys. Rev. Lett. **119**, 152301 (2017)
[15] CMS Collaboration, CMS-PAS-HIN-16-019 (2017), https://cds.cern.ch/record/2244666
[16] CMS Collaboration, Phys. Lett. B **765**, 193 (2017)
[17] CMS Collaboration (2017), arXiv:1709.09189
[18] CMS Collaboration, Phys. Lett. B **768**, 103 (2017)

Highlights from LHCb

Albert Bursche[1,2,*] on behalf of the LHCb Collaboration

[1] *Università degli studi di Cagliari, Italy*
[2] *INFN Cagliari, Italy*

Abstract. The recent highlights from LHCb in soft QCD and Heavy Ion physics are presented. This includes measurements from collisions of proton and lead $^{208}_{82}Pb^{82+}$ ion beams with other beams as well as noble gas targets. An outlook on future analyses of $^{129}_{54}Xe^{54+}$ collisions is presented.

1 Introduction

The LHCb detector is a unique apparatus for soft QCD and Heavy Ion physics. With the forward acceptance at high rapidity(y) $2 < y < 5$; with excellent tracking; particle identification capabilities and a system for noble gas injection - the SMOG system - for fixed target collisions [1] with the LHC beam LHCb can do measurements no other experiment can do. Additional forward scintillators (HeRSCheL) in the region $5 < \eta < 9$ can be used to select central exclusive events.

2 Fixed Target Physics

For fixed target physics in LHCb the SMOG system is used [1]. This system allows for the injection of noble gases in the accelerator vacuum. LHCb's acceptance then corresponds to a central rapidity range. LHCb has recorded the collisions of protons with Helium; Neon and Argon as well as collisions of lead $^{208}_{82}Pb^{82+}$ ions with Argon.

2.1 Antiproton production in proton helium collisions at $\sqrt{s_{NN}} = 110\ GeV$

Antiproton production in proton helium collisions is currently not well understood and is an important ingredient to the transport models of high energy cosmic rays [2, 3]. The double differential cross section of antiproton production has been measured as a function of antiproton momentum and transverse momentum [4]; compared to EPOS LHC [5] and is presented in figure 1. This result is currently being updated with an improved determination of the luminosity for publication.

2.2 Open Charm and Charmonium Production in Proton Argon Collisions at $\sqrt{s_{NN}} = 110\ GeV$

Production of J/ψ and D^0 mesons is observed in proton Argon collisions at $\sqrt{s_{NN}} = 110$ GeV. Cross Section and yields are measured differentially in transverse momentum and rapidity [8]. There is evidence for D^+, D^{*+}, D_s^+ and Λ_c^+ production.

*e-mail: albert.bursche@cern.ch

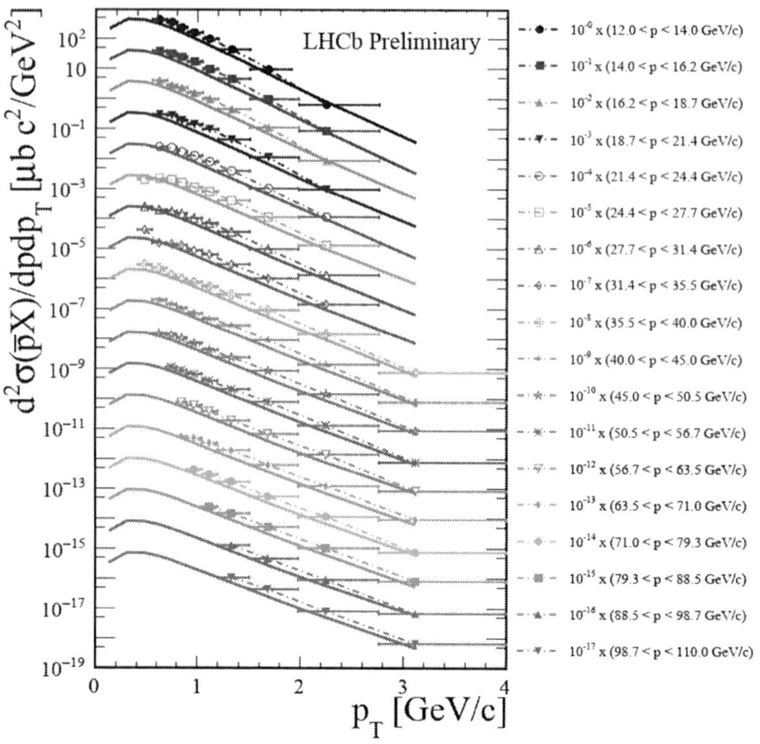

Figure 1. Antiproton production cross section in pHe collisions[4] measured double differentially in momentum and transverse momentum. Compared to EPOS LHC absolute predictions [5]. EPOS LHC predictions are systematically underestimating the data by a factor 1.5. Better agreement is found with EPOS 1.99 and HIJING.

3 Collider Mode

3.1 First Observation of Ξ_{cc}^{++}

The first observation of the double charmed baryon Ξ_{cc}^{++} with the mass of

$$m_{\Xi_{cc}^{++}} = 3621.40 \pm 0.72 \pm 0.27 \pm 0.140 \text{ MeV}/c^2$$

is reported [6]. The first uncertainty is statistical; the second is systematic and the third originates from the limited knowledge of the Λ_c mass. This measurement was performed using a data set corresponding to 2 fb^{-1} at $\sqrt{s} = 8$ TeV and checked using 1fb^{-1} at $\sqrt{s} = 7$ TeV of proton proton collisions. This observation is in tension with a Ξ_{cc}^{+} state reported by the SELEX collaboration [7]. The difference to the mass reported by SELEX,

$$m_{\Xi_{cc}^{+}} = 3519 \pm 2 \text{ MeV}/c^2$$

is inconsistent with the presumption that those two states are separated by isospin only.

3.2 Open Charm Production at $\sqrt{s} = 5$ TeV in Proton Proton Collisions

Measurements of the double differential cross section for D^0, D^+,D_s^+ and D^{+*} production and the production ratios between 13 TeV and 5 TeV have been reported [9]. The data generally agree well with the FONLL, POWHEG+NNPDF3.0L and GMVFNS predictions for the ratios of cross-sections at $\sqrt{s}$ = 13 TeV and $\sqrt{s}$ = 5 TeV.

3.3 Open Charm Production in Collisions of $^{208}_{82}Pb^{82+}$ Ions and Protons at $\sqrt{s_{NN}} = 5$ TeV

Prompt D^0 production was measured in proton $^{208}_{82}Pb^{82+}$ collisions. Double differential cross sections (y, p_T); forward backward ratios and nuclear modification factors have been measured [10]. The nuclear modification factor is shown in figure 2.

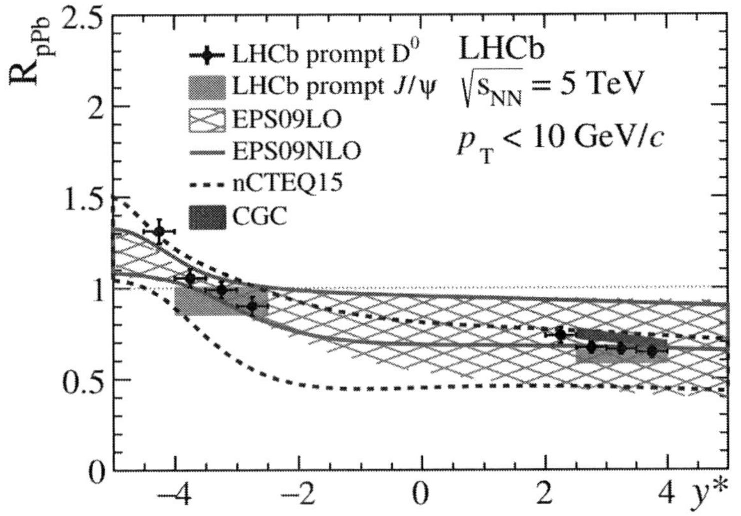

Figure 2. Nuclear modification factor for D^0 production as a function of the rapidity in the centre of mass frame (y^*) at $\sqrt{s_{NN}} = 5$ TeV compared to predictions using nuclear PDFs (EPS09, EPS09NLO, nCTEQ15)[11–13] and a colour glass condensate model (CGC) [14].

3.4 Charmonium Production in Collisions of $^{208}_{82}Pb^{82+}$ Ions and Protons at $\sqrt{s_{NN}} = 8.16$ TeV

Charmonium production in collisions of $^{208}_{82}Pb^{82+}$ ions and protons at $\sqrt{s_{NN}} = 8.16$ TeV has been measured. The prompt component and the J/ψ component originating from beauty quark decays are separated in this measurement. Nuclear modification factor and forward backward ratio have been measured. The nuclear modification factor is shown in figure 3.

There is evidence for D^+, D_s^+, Λ_c^+ production as well as exclusive ρ^0 production in ultraperipheral collisions.

3.5 Collisions of $^{208}_{82}Pb^{82+}$ Ions at $\sqrt{s_{NN}} = 5$ TeV

The analyses of the data collected in $^{208}_{82}Pb^{82+}$ collisions are still ongoing. Evidence for open charm, open strange and charmonium production in ion collisions in both mesonic and barionic final states up to 50% centrality was presented. There is also clear evidence of exclusive charmonium production in ultraperiphereal collisions which is currently under study.

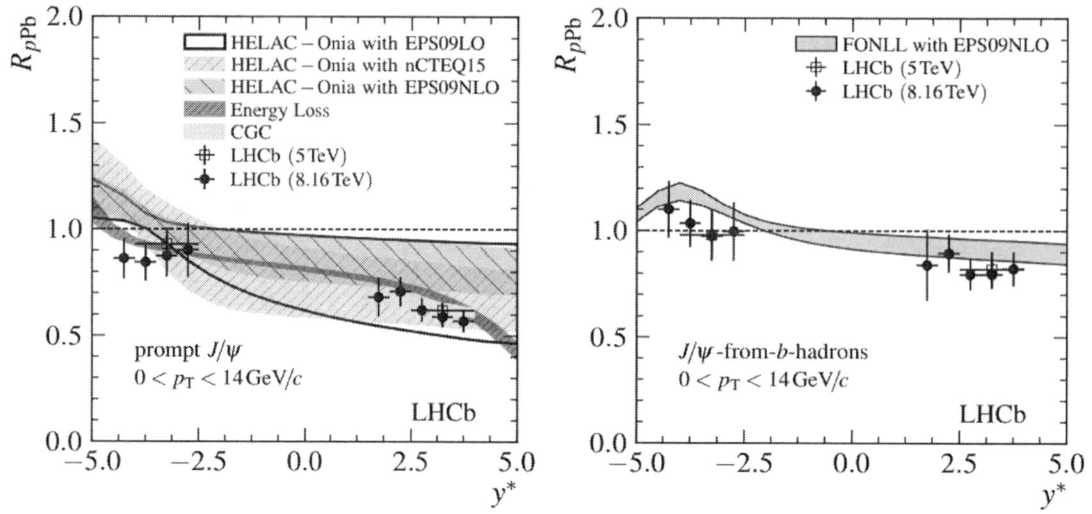

Figure 3. Nuclear modification factor as a function of the rapidity in the centre of mass frame (y^*) for prompt (left) and non-prompt (right) J/ψ production at $\sqrt{s_{NN}}$ = 5 TeV and at $\sqrt{s_{NN}}$ = 8.16 TeV compared to predictions from HELAC-Onia [11, 12] using EPS09LO [15]; nCTEQ15 [16] and EPS09NLO [15] nuclear parton distribution functions and GCC and an energy loss model in the prompt case and to FONLL [17, 18] using EPS09NLO [15] in the detached case.

4 Conclusions

The efforts to analyse ion collisions in LHCb start to bear fruits. Despite being geared for heavy flavour production in proton collisions and missing out on the first lead run of the LHC the LHCb collaboration is gaining momentum in heavy ion physics. Especially in the asymmetric collisions of protons with lead more interesting results are in preparation. The recently finished LHC run with xenon beams confirmed the LHCb capability to observe heavy flavour production in heavy ion collisions. Evidence for charm production shown in figure 4.

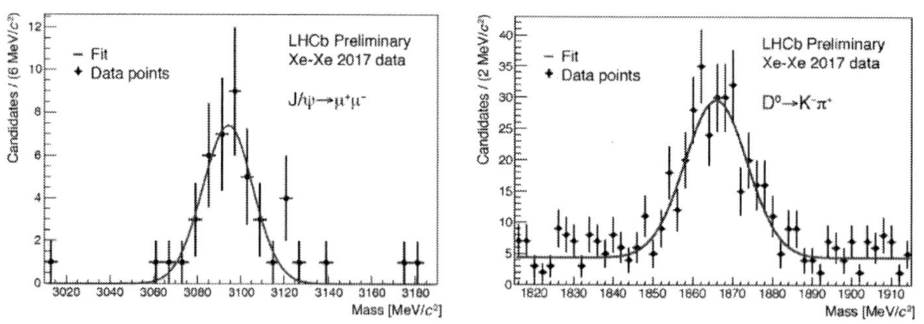

Figure 4. Invariant mass of dimuons and $K^{\pm}\pi^{\mp}$ measured in $\left({}^{129}_{54}Xe^{54+}\right)$ collisions at $\sqrt{s_{NN}}$ = 5.44 TeV.

References

[1] LHCb Collaboration JINST **9** P12005 (2014)
[2] PAMELA Collaboration JETP Letters **96** 621 (2013)
[3] AMS Collaboration Phys. Rev. Lett. **117** 091103 (2016)
[4] LHCb Collaboration preliminary LHCb-CONF-2017-002
[5] T. Pierog *et. al.* Phys. Rev. **C95** 034906 (2015)
[6] LHCb Collaboration Phys. Rev. Lett. **119** 112001 (2017)
[7] SELEX Collaboration Phys. Rev. Lett. **89** 112001 (2002)
[8] LHCb Collaboration preliminary LHCb-CONF-2017-001 (2017)
[9] LHCb Collaboration JHEP **06** 147 (2017)
[10] LHCb Collaboration JHEP **10** 090 (2017)
[11] J.-P. Lansberg and H.-S. Shao Eur. Phys. J. **C77** 1 (2017)
[12] H.-S. Shao Comput. Phys. Commun. **184** 2562 (2013)
[13] H.-S. Shao Comput. Phys. Commun. **189** 238 (2016)
[14] B. Ducloué, T. Lappi, and H. Mäntysaari Phys Rev. **D91** 114005 (2015)
[15] K. J. Eskola, H. Paukkunen, and C. A. Salgado JHEP **04** 065 (2009)
[16] K. Kovarik *et. al.* Phys. Rev. **D93** 085037 (2016)
[17] M. Cacciari, M. Greco, and P. Nason JHEP **05** 007 (1998)
[18] M. Cacciari, M. Greco, and P. Nason JHEP **03** 006 (2001)

Strangeness at high μ_B

Recent data from FOPI and HADES

Yvonne Leifels[1,*]

[1] *GSI Helmholtzzentrum für Schwerionenforschung, Planckstr. 1, 64291 Darmstadt, Germany*

Abstract. Strangeness production in heavy-ion reactions at incident energies at or below the threshold in NN collisions gives access to the characteristics of bulk nuclear matter and the properties of strange particles inside the hot and dense nuclear medium, like potentials and interaction cross sections. At these energies strangeness is produced in multi-step processes potentially via excitation of intermediate heavy resonances. The amount of experimental data on strangeness production at these energies has increased substantially during the last years due to the FOPI and the HADES experiments at SIS18 at GSI. Experimental data on K^+ and K^0 production support the assumption that particles with an $\bar{s}$ quark feel a moderate repulsive potential in the nuclear medium. The situation is not that clear in the case of K^-. Here, spectra and flow of K^- mesons is influenced by the contribution of ϕ mesons which are decaying into K^+K^- pairs with a branching ratio of 48.9 %. Depending on incident energy upto 30 % of all K^- mesons measured in heavy-ion collisions are originating from ϕ-decays. Strangeness production yields - except the yield of Ξ^- - are described by thermal hadronisation models. Experimental data not only measured for heavy-ion collisions but also in proton induced reactions are described with sets of temperature T and baryon chemical potential μ_B which are close to a universal freeze-out curve which is fitting also experimental data obtained at lower baryon chemical potential. Despite the good description of most particle production yields, the question how this is achieved is still not settled and should be the focus of further investigations.

1 Introduction

The phase diagram of QCD usually presented in terms of temperature T and baryon chemical potential μ_B is in the focus of many experimental and theoretical investigations. At small baryonic chemical potentials the hadronic phase is assumed to be separated from the quark-gluon phase by a cross over transition. At a critical point the cross over transition is expected to evolve with rising baryonic chemical potential into a first order phase transition. At even higher baryonic chemical potential exotic states may appear.

Highest chemical potentials μ_B and moderate temperatures T at freeze-out can be accessed experimentally with heavy-ion collisions in the beam energy regime $E_{beam} < 6$ GeV/u (fixed target). In the course of these reactions, densities up to several times normal nuclear matter density ρ_0 are reached for relatively long times. The compressed nucleons are excited to heavier resonances, but the matter is still baryon dominated. Meson production is present but not prevailing. Most of the strange particles

*e-mail: Y.Leifels@gsi.de

are produced at or below their production thresholds in NN collisions. Production of strangeness at sub-threshold energies occurs in multi-step processes. Resonances or intermediate mesons (mainly π's) are utilized as energy reservoirs. Consequently, the production rates are very sensitive to densities reached in the course of those collisions and to in-medium effects, in particular the modification of life times and masses in the hot and dense nuclear matter.

Strangeness production at threshold has been studied at the BEVALAC in Berkeley and the SIS18 accelerator at GSI. Despite the low production yields of strange particles at or below threshold energies, modern detector set-ups capable to sustain high data rates are able to accumulate enough statics even on very rare probes.

After the pioneering experiments in Berkeley with the Streamer Chamber, the first set-up with which it was possible to gather high statistics on strange particles production below threshold was the KaoS spectrometer at SIS18. Equipped with a highly selective trigger and particle identification by magnetic rigidity it was possible to measure spectra of positive and negative Kaons within a small angular acceptance. The KaoS collaboration measured various systems in the energy regime available at SIS18. In particular, the Au+Au and C+C systems between 0.6 GeV/u and the highest energies available at GSI for these nuclei (1.5 Gev/u for Au and 2.0 GeV/u for C) have been investigated [1]. Due to the limited acceptance and the characteristics of the KaoS spectrometer, neutral mesons with open or hidden strangeness (like the K^0 or ϕ) or hyperons were not accessible. This was possible with the FOPI detector system at GSI having full azimuthal acceptance for particles emitted in the region of target to mid-rapidity. Measurement of directed flow of strange particles in a large acceptance and reconstruction on strange resonances decaying into charged products became available [2]. The HADES set-up is capable to record events with the highest rates achieved so far for heavy ion collision experiments in this energy regime. With the upgraded HADES detector high statistics data in a large acceptance is available, which allows to access observables which have been beyond the detection limit [3] of the former experiments.

Transport models are an essential tool to assess the physics of strange particles in the dense medium, and more precise measurements of strange particles help to confine the input parameters of the models. One particular result of the KaoS measurements is rather tight constraints on the nuclear matter equation of state at densities between 2-3 ρ_0. Theoretical models predict a strong dependence of the positive Kaon production on the nuclear equation of state. Since Kaons are mostly produced in multi-step processes at sub-threshold energies, the production is very sensitive to the number of collisions and – consequently – the density reached during the reaction. Theoretical models employing a soft or a hard nuclear equation of state predict quite different production yields for Kaons as function of beam energy. Experimental data from the KaoS collaboration and comparison to various model predictions yield a nuclear matter equation of state [4–6] which is rather soft.

In this contribution, the most relevant results on strange particle production of the GSI experiments are reviewed with special emphasis on the latest FOPI and HADES results

2 In-medium potentials

Modifications of hadron properties in dense baryonic matter are a current subject of intensive research in hadron physics. Various theoretical approaches [5, 8–10] agree qualitatively in predicting modifications of masses and coupling constants for kaons and anti-kaons. Kaons (K^+ and K^0) are expected to be subject to a repulsive potential, whereas antikaons (K^- and $\overline{K}^0$) should be attracted, see Fig. 1, where theoretical predictions for the density dependence of the kaon masses are presented [7]. The $\overline{K}^0$ is mixing with the K^0 to K^0_S and K^0_L. At the energies discussed here, the $\overline{K}^0$ contributes only little the K^0_S yield, and the measured K^0_S are considered to result from K^0 only. Mass and production threshold energy for kaons should increase as a result of the in-medium KN interaction, whereas for antikaons the corresponding values should decrease substantially [7]. The in-medium modifications of kaonic properties have been already studied experimentally in terms of KN potentials and reported by several experiments focused on strangeness production at near-threshold energies [11–15].

2.1 Kaons

The first moment v_1 of the Fourier expansion to the azimuthal distributions $dN/d\Phi \propto 1 + 2v_1 cos(\Phi - \Phi_R) + 2v_2 cos(\Phi - \Phi_R) + ...$, where the azimuthal angle Φ is measured with respect to the reaction plane Φ_R is quantifying the directed side flow in a heavy ion reactions. In Fig. 2 values for v_1 measured by the FOPI collaboration in central Ni+Ni collisions at 1.9 AGeV are plotted as function of rapidity for positive and negative kaons and protons together with predictions of HSD [17] and IQMD [11] models. Near target rapidity K^+ - mesons show a collective in-plane deflection in the direction opposed to that of protons. This finding is in agreement with other data [18]. Models are able to describe this result when incorporating a slightly repulsive kaon nucleon (KN) potential of $U_{pot,KN}(\rho_0, \mathbf{k} = 0) \approx 20$ MeV, which is linearly depending on density (see Fig. 1). Calculations without an in-medium KN potential predict a smaller, close to isotropic directed flow. It should be noted that both models describe the proton flow in heavy-ion collisions in this energy regime and give consistent results for the strangeness production yields, despite the fact that the programs are conceptually different: IQMD is based on quantum molecular dynamics and HSD on a Boltzmann-Uehling-Uhlenbeck approach.

A similar result is obtained when comparing model predictions with K^0 spectra. In contrast to K^+ where the Coulomb interaction contribute to the repulsive potential, K^0 spectra are unbiased by this effect. In addition, they are detected by their decay into charged pions and can be measured down to very small momenta. In Fig. 3 HADES data for K^0 is presented. The momentum spectra of K^0 emitted close to mid-rapidity are shown and confronted with predictions of the IQMD model [11]. Again, the spectra are best described with a repulsive potential. The potential is estimated to be $U_{pot,KN} \approx +40$ MeV. Hence, HADES and FOPI results are showing the same trend, but are preferring slightly different potential heights. This discrepancy may be due to the different observables chosen or due to other input parameters of the models, e.g. production or re-scattering cross sections of Kaons. Those need to be confined by further systematic measurements.

The HADES collaboration recently measured Au+Au collisions at 1.25 AGeV with very high statistics, which allowed to reconstruct phase space distributions of charged and uncharged Kaons. The low momentum part of K^0 spectra is well described by IQMD model predictions employing the same input parameters as for the light system and the same kaon nucleon potential [19]

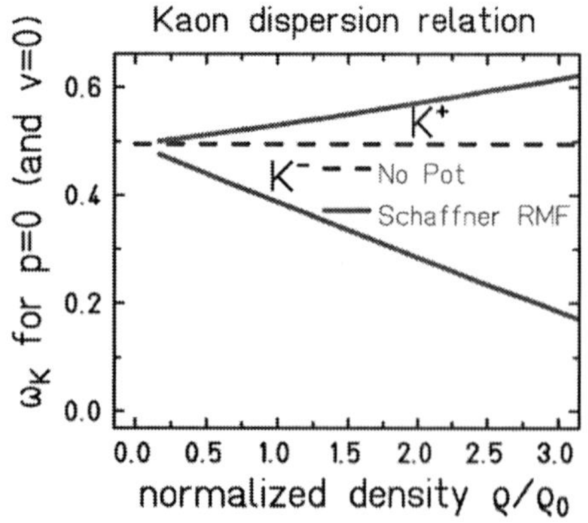

Figure 1. In-medium kaon energy $\omega(\mathbf{k} = 0)$ as function of baryon density fitted to reproduce the calculations of [7]: $\omega(\mathbf{k}, \rho) = \sqrt{(\mathbf{k} - \Sigma_v)^2 + m^2 + m\Sigma_s} \pm \Sigma_v^0$ with a scalar self energy Σ_s and a vector self energy (Σ_v^0, Σ_v), where the sign of the vector term $\pm\Sigma_v^0$ is positive for kaons and negative for antikaons.

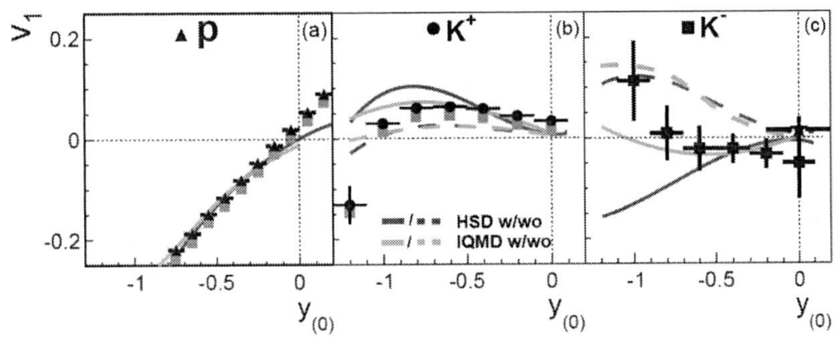

Figure 2. FOPI Results for directed flow v_1 of kaons in comparison to the flow of protons as a function of the center-of-mass rapidity $y_{c.m.}$. Solid lines are predictions using the HSD transport code.The v_1 - values are obtained within the acceptance of the CDC, for details see text. Data taken from [16]

2.2 Antikaons

In contrast to the situation for kaons where the question of the in-medium potential seems to be essentially settled, the antikaon nucleon potential is not so easy to assess. Due to the existence of a resonance, the Λ(1405), close to the K$^-$p threshold a non-perturbative treatment is required in theoretical models. Theoretical predictions for the $\bar{\text{K}}$N potential differ substantially, therefore. Kaonic atoms confirm that the $\bar{\text{K}}$N interaction is attractive at small momenta and low densities but its strength at larger densities is less constrained. On the other hand, antikaons are predicted to be produced via the strangeness exchange reaction $\Lambda + \pi \rightarrow \text{K}^- + N$. Hence, their production is closely following that of the K$^+$. Additionally, K$^-$ are strongly absorbed in the nuclear medium, a process which partially covers the effect of the potential. Microscopic transport models employing an attractive density dependent with $\bar{\text{K}}$N potential $U_{pot,\bar{K}N}(\rho_0, \mathbf{k} = 0) = 50 - 90$ MeV predict: a) the spectra show a steeper slope as compared to kaons, i.e. a smaller apparent temperature, b) the flow signals v_1 and v_2 are modest. Those predictions indeed describe the experimental data reasonably well, antikaon multiplicities as well as antikaon spectra. In the right panel of Fig. 2, directed flow v_1 of K$^-$ measured by the FOPI collaboration is compared to model predictions with and without employing an attractive, density dependence $\bar{\text{K}}$N potential $U_{pot,\bar{K}N}(\rho_0) = 50$. In HSD, a G-Matrix approach is used for obtaining the $\bar{\text{K}}$N potential and particles are propagated off-shell, whereas in IQMD the antikaons are treated as quasi-particles subject to a potential taken from [21] and during the production process the free production cross section is taken at $\sqrt{s}$ modified by the in-medium mass of the K$^-$. Both models

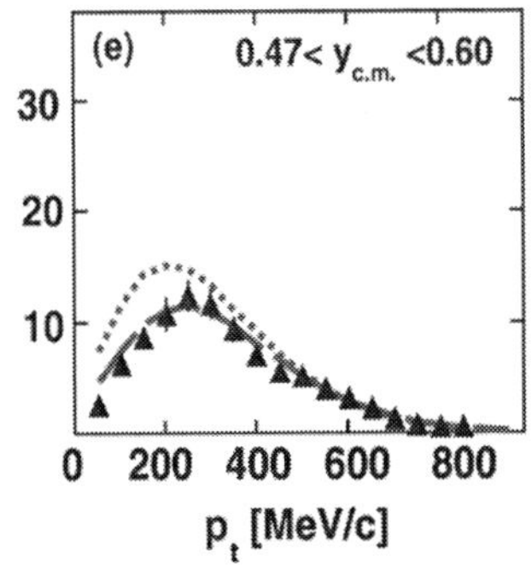

Figure 3. HADES results on K^0 production at mid-rapidity in Ar+KCl collisions at 1.76 GeV/u in comparison to IQMD model calculations. The blue line represents model predictions employing a KN potential with the height of $U_{pot,KN} = +40$ MeV and the green dotted line predictions without an explicit KN potential.

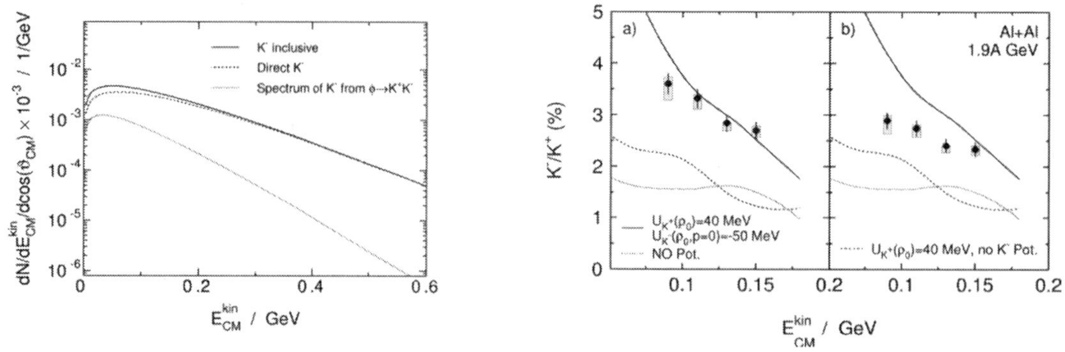

Figure 4. Left panel: Kinetic energy distributions of K^- mesons obtained by the inclusive fit to the experimental data (solid line), K^- from ϕ meson decays (dotted line), and "direct" K^- mesons generated from the thermal source (dashed line). See text for details. Right panel: a) the K^-/K^+ ratio as a function of $E_{k,CM}$ in the Al+Al experiment. Error bars represent statistical uncertainties. Grayed rectangles represent the estimation of systematical errors. Lines represent results of HSD model predictions including different values of KN potentials. b) the K^-/K^+ ratio distribution corrected for K^- mesons from ϕ decays. Grayed rectangles do not include the uncertainties of T_{eff,K_ϕ}. Data from [20].

differ in predicting v_1 in the target rapidity region, but the large statistical error of the experimental data prevents to distinguish between the two.

However, precise in-medium modification studies have to account for feeding of the K^- emission yield by the $\phi(1020)$ meson decays. A comparison of the mean free path of the latter particles, $c\tau_\phi \approx 50$ fm, to the average duration of the hot and dense collision zone (20-30 fm/c) suggests that most ϕ mesons decay outside this zone. It should be also noted, that the ϕ meson mass is only 32 MeV/c^2 larger than that of a K^+K^- pair. Thus, a substantial coupling of the ϕ and K^- dynamics is expected. In addition, K^- originating from the ϕ meson decays are governed by different kinematics than those emitted directly from the collision zone.

For light to intermediate heavy systems colliding at beam energies of 1.7–1.9 GeV/u it is found that about 20% of the K^- mesons originate from ϕ meson decays [22–24]. The contribution of the ϕ is seemingly larger at lower beam energies. It measured to be 30% for 1.25 AGeV Au+Au collisions [19]. Hence, the contribution of ϕ mesons has to be considered in the models. Neither HSD nor IQMD are reproducing the high ϕ yields measured experimentally. Instead, a simple two source model for antikaon production is used to disentangle the two contributors to the K^- phase space distributions.

- the inclusive K^- emission spectrum is isotropic (an assumption which is justified by experimental data) and described by a Boltzmann distribution with the parameter T_{eff,K^-}
- K^- mesons from ϕ meson decays and observed in the experiment are produced outside the collision zone.
- both the "direct" and the ϕ meson sources of K^- are characterized by Boltzmann distributions, with parameters: $T_{\mathrm{eff},K_{\mathrm{direct}}}$ (unknown, to be found), T_{eff,K_ϕ} (from experimental data), and the normalization factors weighting the inclusive K^- emission probability according to the found ϕ/K^- ratio.

The results of this procedure are presented in the left panel of Fig. 4 which shows the total kinetic energy distribution of K^- mesons in Al+Al collisions at 1.9 AGeV, and its two respective components. The "direct" component was obtained by subtracting the contribution from the ϕ meson decays from the inclusive spectrum.

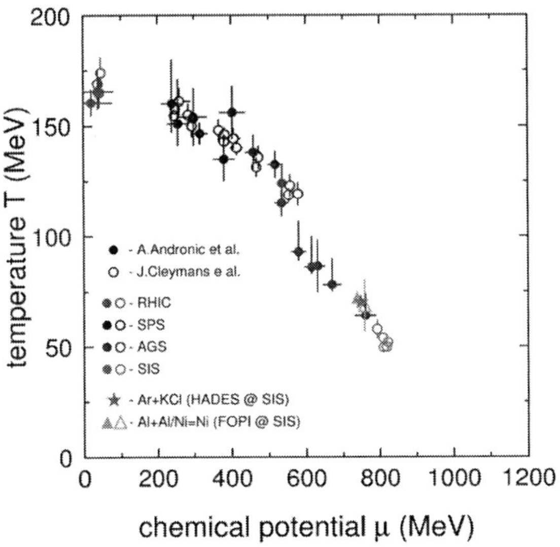

Figure 5. Statistical hadronisation model results for various incident energies and systems. Data are taken from various references. The low energy (SIS18) data are from [20, 24, 27].

The experimentally observed difference between K^+ and K^- inverse slope parameters can be partially explained by this effect. HADES claims that the ϕ decay accounts for the whole T_{eff} difference in Ar+KCl collisions at 1.7 AGeV [22] and Au+Au at 1.2 AGeV [19], in Al+Al and Ni+Ni collisions at higher energies it only accounts for approximately half the temperature difference [20, 24]. As consequence, the determination of the $\bar{K}N$ requires models which also reproduce the ϕ production yields. With the UrQMD model [25] it is possible to reproduce measured ϕ yields, but it is lacking potentials.

3 Chemical and kinetic freeze-out

Statistical hadronisation models are able to describe production yields of strange particles with essentially three parameters: temperature T, baryon chemical potential μ_B and a parameter describing the volume of the source R_V. In the limit of high temperature and/or large systems, the grand-canonical treatment of strangeness is adequate, but at low energies strangeness has to be implemented exactly in the canonical ensemble. This is achieved by introducing a strangeness correlation radius, $R_C < R_V$,

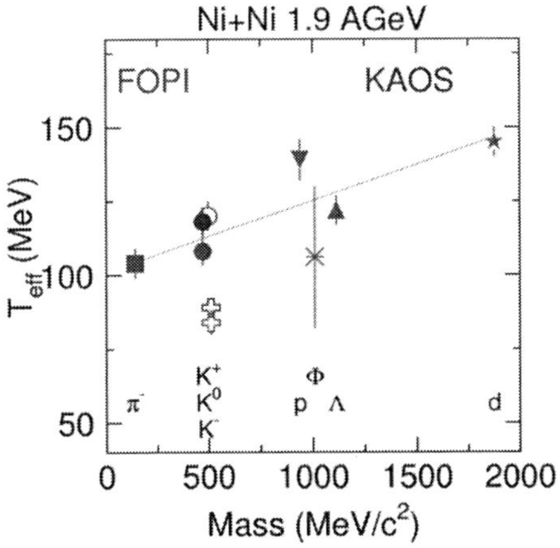

Figure 6. Effective temperatures T_{eff} measured for various particles emitted at mid-rapidity in Ni+Ni collisions at 1.9 AGeV as function of the particle mass. Data from KaoS (blue symbols) and FOPI (red symbols) are shown.

which is determined from experimental data. With the exception of the double strange Ξ^- hyperon [26, 27], the production yields can be reproduced within statistical models with parameters which fall reasonably well on the universal freeze-out curve (see Fig. 5). However, some of the investigated systems (e.g. Al+Al by FOPI) are rather small and the assumption of complete equilibration questionable.

Kinetic freeze-out shows a more complicated pattern which can be seen in Fig. 6, where inverse slope parameters T_{eff} of spectra at mid-rapidity are shown for various particle species emitted in central Ni+Ni collisions at 1.9 GeV/u. Data from the FOPI and KAOS collaborations are compiled [24, 28], for π's only the inverse slope parameter of the high momentum part of the spectra are shown because the low momentum part is influenced by the decay of Δ resonances. The inverse slope parameters for π, protons and deuterons is rising with mass, which is interpreted as signal of a collective radial flow. In lighter systems this effect vanishes because of the diminished pressure built-up, whereas it is more pronounced in heavier systems.

It is not obvious that the strange particles do follow this trend apart from the K^+: kaons with anti-strangeness $K^{+,0}$ and kaons with strangeness K^- have significantly different temperatures. Three explanation have been offered to account for the lower K^- temperatures: the influence of the $\bar{K}N$ potential, the contribution of the ϕ decay, and - inspired by a thermal picture - different freeze-out times. The latter is motivated by the fact that most of the K^- production at sub-threshold energies will proceed via strangeness exchange in hyperon-pion reactions. Hence, K^- will appear after the production of K^+ when the collision system already have cooled down. Within a microscopic picture, both the decay of the ϕ and the attractive $\bar{K}N$ potential will yield lower T_{eff} for the K^- than for the K^+ (see discussion in Section 2.2)

The experimental T_{eff} of the ϕ is of the same magnitude as that of the K^+ but not as high as the inverse slope parameter of the protons. Regarding the large experimental error bars it is necessary to measure the ϕ mesons with higher statistics in the future, as well as to study the interaction processes of the ϕ with nucleons, i.e. re-scattering and absorption cross sections.

The inverse slope parameter of the Λ is smaller than that of the protons. This in accordance with the predictions of microscopic models: the Λ is produced when the compressed collision zone has already expanded and the temperature has slightly decreased, and the subsequent elastic collisions of the Λ with the surrounding nucleons are insufficient to rise the inverse slope parameter T_{eff} to the level of that of the nucleons. The spectral distributions of Λ measured at different energies and system sizes may give access not only to the in-medium ΛN cross sections but also to ΛN potential in dense matter.

4 Conclusion

New experimental data and observables in comparison to state-of-the-art microscopic models give access to the KN in-medium potential. On the other hand the $\bar{K}N$ potential is not easily accessible in heavy-ion reactions, new experimental observables and approaches have to be found. One possibility is to investigate correlated K^+K^- production in elementary and heavy-ion reactions and exclude the ϕ from the analysis. Such a procedure has been suggested by the ANKE collaboration [29]. Additional high statistics data of strangeness production as a function of energy and system size is necessary for continuing such type of analyses and constraining further the input parameters in the microscopic models. Supplementary elementary p+A and π+A data are important to investigate the role of heavy resonances.

There are two model approaches to strangeness production at sub-threshold energies leading to puzzling results: in-medium modifications of the kaon masses are required to describe the experimental data in microscopic models, whereas statistical hadronisation models reproduce production yields with vacuum masses. Generally, all measured particle yields in heavy-ion reactions at SIS energies can be described within thermal models with the exception of the Ξ^- data. On the other hand, it has been shown that the particle ratios measured in heavy-ion collisions can also be reproduced by microscopic models like UrQMD [25]. If microscopic transport models also reproduce the difference between kinetic and chemical temperatures (including the gross properties of the reactions like stopping and flow), one might investigate the mechanisms leading to the apparent thermal equilibration in heavy-ion reactions within those models. A key issue for this might be the analysis of interdependencies in the production of different particle species. For instance, it has been suggested that strangeness exchange may cause the experimentally observed proportionality between the beam energy dependence of the K^-/K^+ and π/A ratios below 10 GeV/u; here A is the system size [11].

The investigation of strange particles produced close to their threshold in NN collisions will be continued at SIS18 with HADES and later at FAIR with the Compressed Baryonic Matter CBM.

References

[1] P. Senger et al., Nucl. Instr. Meth. Phys. Res. A **327**, 393 (1993).
[2] J. Ritman *et al.*, Nucl. Phys. B (Proc. Suppl.) **44**, 708 (1995)
[3] G. Agakishiev *et al.*, Eur. Phys. J. A **41**, 243 (2008).
[4] J. Aichelin and C.M. Ko, Phys. Rev. Lett. **55**, 2661 (1985)
[5] C. Fuchs, Prog. Part. Nucl. Phys. **56**, 1 (2006).
[6] C. Hartnack, H. Oeschler, and J. Aichelin, Phys. Rev. Lett. **90**, 102302 (2003)
[7] J. Schaffner-Bielich, J. Bondorf, I. Mishustin,
[8] G.E. Brown, and M. Rho, Phys. Rev. Lett. **66**, 2720 (1991)
[9] W. Weise, Nucl. Phys. A **610**, 35c (1996)
[10] M.F.M. Lutz, Prog. Part. Nucl. Phys. **53**, 125 (2004).
[11] C. Hartnack, H. Oeschler, Y. Leifels, E. Bratkovskaya, J. Aichelin, Phys. Rep. **510**, 119 (2012).
[12] Y. Shin *et al.* [KaoS Collaboration], Phys. Rev. Lett. **81**, 1576 (1998)
[13] K. Wiśniewski *et al.*, Eur. Phys. J. A **9**, 515 (2000).
[14] F. Uhlig *et al.* [KaoS Collaboration], Phys. Rev. Lett. **95**, 012301 (2005)
[15] M.L. Benabderrahmane *et al.* [FOPI Collaboration], Phys. Rev. Lett. **102**, 182501 (2009).
[16] V. Zinyuk *et al.*, Phys. Rev. C **90**, 025210 (2014).
[17] W. Cassing, L. Tolos, E. L. Bratkovskaya and A. Ramos, Nucl. Phys. A **727**, 59 (2003)
[18] P. Crochet *et al.* [FOPI Collaboration], Phys. Lett. B **486**, 6 (2000)
[19] H. Schuldes *et al.*, Contribution to this conference
[20] P. Gasik *et al.*, Eur. Phys. J. A **52**, 177 (2016)
[21] C. M. Ko and G. Q. Li, J. Phys. G **22** (1996) 1673
[22] G. Agakishiev *et al.*, Phys. Rev. C **80**, 025209 (2009).
[23] M. Lorenz, PoS (BORMIO 2010), 038.
[24] K. Piasecki *et al.*, Phys. Rev. C **91**, 054904 (2015).
[25] J. Steinheimer and M. Bleicher, J. Phys. G **43** 015104 (2015)
[26] M. Merschmeyer *et al.* [FOPI Collaboration], Phys. Rev. C **76** 024906 (2007)
[27] G. Agakishiev et al., Eur. Phys. J. A **52**, 178 (2016).
[28] A. Förster *et al.*, Phys. Rev. Lett **91** 152301 (2003)
[29] E. Paryev *et al.*, J. Phys. G **42**, 075107 (2015)

Strangeness at Intermediate Baryon Density

David Tlusty[1,★]

[1] *Physics & Astronomy Department, Rice University, Houston TX, USA*

Abstract.
Exploration of the QCD phase diagram has been one of the main programs of contemporary nuclear physics. The intermediate baryon density region covers a broad range of the baryon chemical potential, between 100 and 700 MeV, and is expected to include a possible critial point at the end of a phase equilibrium curve between the hadron gas and quark gluon plasma phases. Experimental programs at the SPS and RHIC facilities have provided valuable insights in this range. These proceedings motivate the exploration of the QCD phase diagram through the use of strangeness. A selection of relevant experimental results from RHIC and SPS beam energy scan programs with associated theoretical predictions is presented along with a discussion of possible physical conclusions and future plans.

1 Introduction

Finite temperature lattice Quantum Chromodynamics (QCD) calculations predict [1] a cross-over from a hadronic to a QGP phase at a vanishing baryon chemical potential μ_B and temperature $T = 154 \pm 9$ MeV [2]. Several QCD-based calculations [3, 4] show that at lower T and higher μ_B a first-order phase transition may take place. The point in the QCD phase diagram, where the first order phase transition ends, is the QCD critical point [5]. In the low-T, and low and intermediate μ_B phase, interacting hadronic matter in the ground state can be well described in terms of a gas of non-interacting hadrons and resonances - the Hadron Resonance Gas (HRG) model [6].

The availability of more precise experimental data now allows to better discriminate between models. The lattice QCD prediction power is still very limited at non-zero μ_B, hence, experimental exploration becomes especially vital. These proceedings will focus on discussion of the intermediate μ_B region, which covers a wide μ_B range, roughly spanning between 100 MeV and 700 MeV. The intermediate baryon density μ_B region is accessible through experiments at Relativistic Heavy Ion Collider (RHIC), Super Proton Synchrotron (SPS), and in the future Facility for Antiproton and Ion Research (FAIR) which will be able to access also high μ_B region. Both SPS and RHIC have been running centre-of-mass energy $\sqrt{s_{NN}}$ and system size scans to fully explore the region.

One of the most important experimental tools to probe the QCD phase diagram is the study of strange particle production. Strange quarks are strongly interacting particles from the second quark flavor generation. In a QGP, strangeness can be easily produced as strange-antistrange quark pairs via gluon-gluon or quark-antiquark fusion [7], where the minimum momentum transfer Q for this

★e-mail: tlusty@rice.edu

process is $Q \approx 200$ MeV, while the lowest Q process in a hadron gas, $n + n \longrightarrow n + \Lambda + K$, needs 670 MeV. Strangeness is also abundant since nearly 20% of all energy content of QGP is transferred to the production of strangeness when chemical equilibrium is reached [8].

Studies of strange hadron production in the energy and system size scans help to understand many crucial questions; to name a few: what is the temperature and density of strongly interacting matter formed in heavy ion collisions, what types of phase transitions occur, is there a critical point in the QCD phase diagram, and at which collision energy do the QGP signals turn off.

The next sections will present a selection of experimental results and theoretical calculations relevant to the discussion.

2 Selection of Experimental Results and Theoretical Calculations

2.1 Flavor Hierarchy in the Deconfinement Transition of QCD

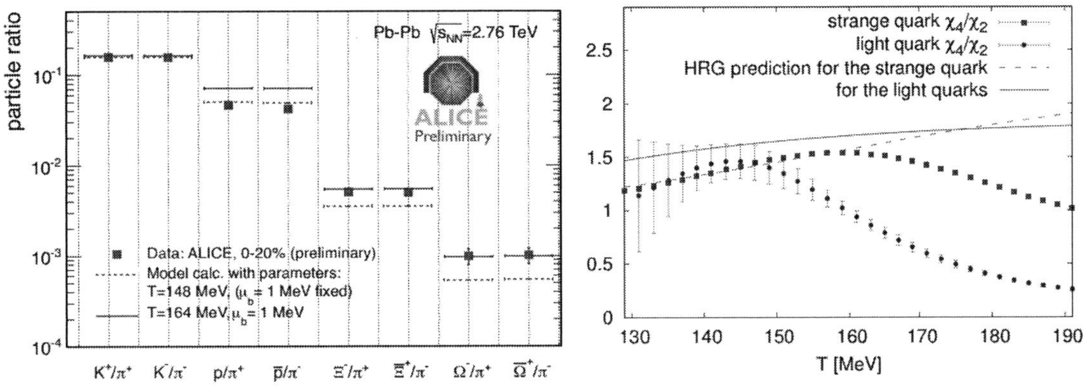

Figure 1. Left panel: Hadron-production ratios from Pb+Pb collisions at $\sqrt{s_{NN}} = 2.76$ TeV compared to the statistical hadronization model predictions with parameters $T = 148, 164$ MeV and $\mu_B = 1$ MeV. Right Panel: The T dependence of the susceptibility ratios χ_2/χ_4 for light and strange quarks in the continuum limit. The lattice data are compared to HRG calculations. Left panel taken from Ref [9] and right panel from Ref [15].

At the Strangeness in Quark Matter conference 2011, the ALICE Collaboration has presented results of particle production ratios [9]: $K^+/\pi^+, K^-/\pi^-, p/\pi^+, \bar{p}/\pi^-, \Xi^-/\pi^+, \overline{\Xi}^+/\pi^-, \Omega^-/\pi^+$, and $\overline{\Omega}^+/\pi^-$, (i.e., ratios of strange-to-light and light-to-light hadrons) together with their thermal statistical model predictions [10, 11] at chemical freeze-out temperatures of $T_{\mathrm{ch}} = 164$ and 148 MeV. As shown in Fig. 1 (left panel), the thermal model prediction with $T_{\mathrm{ch}} = 164$ MeV describes the strange-to-light ratios while the prediction with $T_{\mathrm{ch}} = 148$ MeV describes $p/\pi^+, \bar{p}/\pi^-, K^+/\pi^+$, and K^-/π^-. This is indicative of a separation of chemical freeze-out temperatures between light and strange quark hadrons. Since the ALICE results came from Pb+Pb collisions at $\sqrt{s_{NN}} = 2.7$ TeV, the μ_B was nearly 0 and the results are suited for comparisons with lattice QCD calculations of higher order quark number susceptibilities χ_n [12–14] for light and strange quarks [15]. Figure 1 (right panel) shows the dependency of the susceptibility ratios χ_4/χ_2 for light and strange quarks on temperature T together with corresponding HRG model predictions. The critical temperature T_c is assumed to be the temperature where the lattice QCD calculated χ_4/χ_2 and HRG prediction begin to differ. The choice of susceptibility ratio has two advantages. The susceptibility ratio cancels finite volume effects and is proportional to some experimental observables which will be discussed later. As one can observe from from Fig. 1 (right panel), the $T_c \approx 160$ MeV for strange quarks differ from the $T_c \approx 145$ MeV for the light quarks and both are very close to the corresponding $T_{\rm ch}$ from the thermal statistical model predictions for the ALICE data. This suggests a flavor hierarchy in the deconfinement transition of QCD.

The previous paragraph introduces a flavor hierarchy in the deconfinement transition of QCD at low μ_B. At intermediate μ_B a pattern has been observed similar to ALICE's observations at low μ_B. The STAR Collaboration recently published [16] its results on bulk properties from the RHIC BES. Figure 2 shows the comparison of extracted freeze-out parameters in Au+Au collisions at $\sqrt{s_{NN}} = 39$ GeV for a Grand Canonical Ensemble (GCE) using particle yields as input to the THERMUS model [17]. Results were compared for four different sets, specified in the legend of Fig. 2, of particle yields used as input for fitting. When only π, K, and p yields are used in the fit, the temperature obtained is lower compared to other sets that include strange hadron yields.

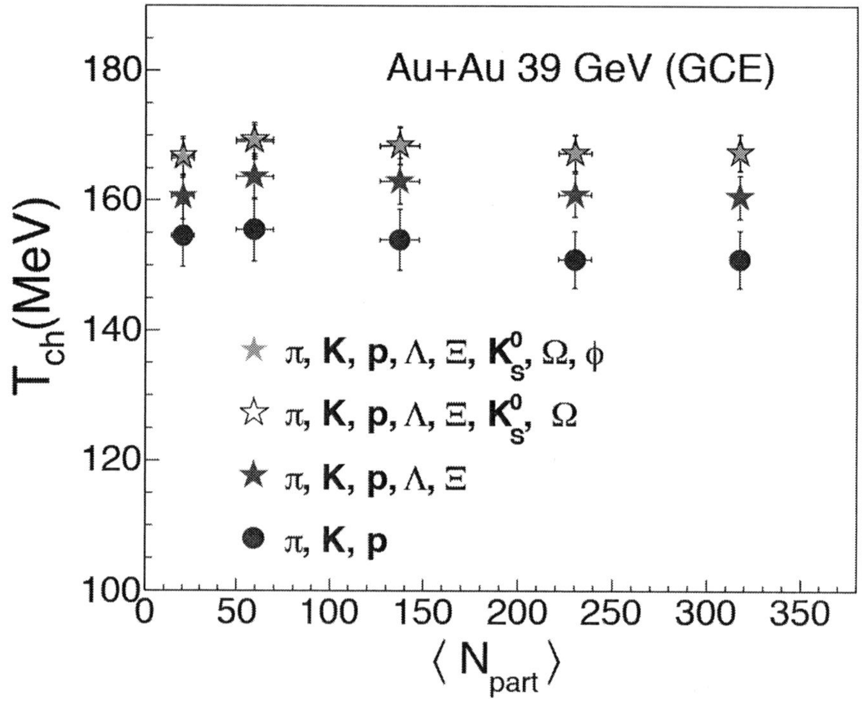

Figure 2. Extracted chemical freeze-out temperature for GCE using particle yields as input for THERMUS model. Results are compared for Au+Au collisions at $\sqrt{s_{NN}} = 39$ GeV for four different sets of particle yields used in fitting. Figure taken from Ref [16].

2.2 Cumulant Ratios of Net-Particle Multiplicity

Fluctuations of conserved quantities are sensitive observables in the study of phase transitions in the QCD matter and critical point [18]. UrQMD model [19] based studies of centrality and energy dependence of various order cumulants and cumulant ratios (up to fourth order) of net-proton, net-charge, and net-kaon multiplicity distributions in Au+Au collisions at $\sqrt{s_{NN}} = 7.7$, 11.5, 19.6, 27,

39, 62.4, and 200 GeV were presented in [18]. These calculations were compared with the data from RHIC Beam Energy Scan (BES) published in [20, 21].

A distribution function can be characterized by the various moments, such as mean M, variance σ^2, skewness S, and kurtosis κ. The moments products $\kappa\sigma^2$ and $S\sigma$ are directly related to the ratios of various order susceptibilities as

$$\kappa\sigma^2 = \frac{\chi_4}{\chi_2} \qquad S\sigma = \frac{\chi_3}{\chi_2}$$

As mentioned in Sect. 2.1, using the ratios allows finite volume effects to nearly cancel out and provides for a direct comparison with lattice QCD predictions.

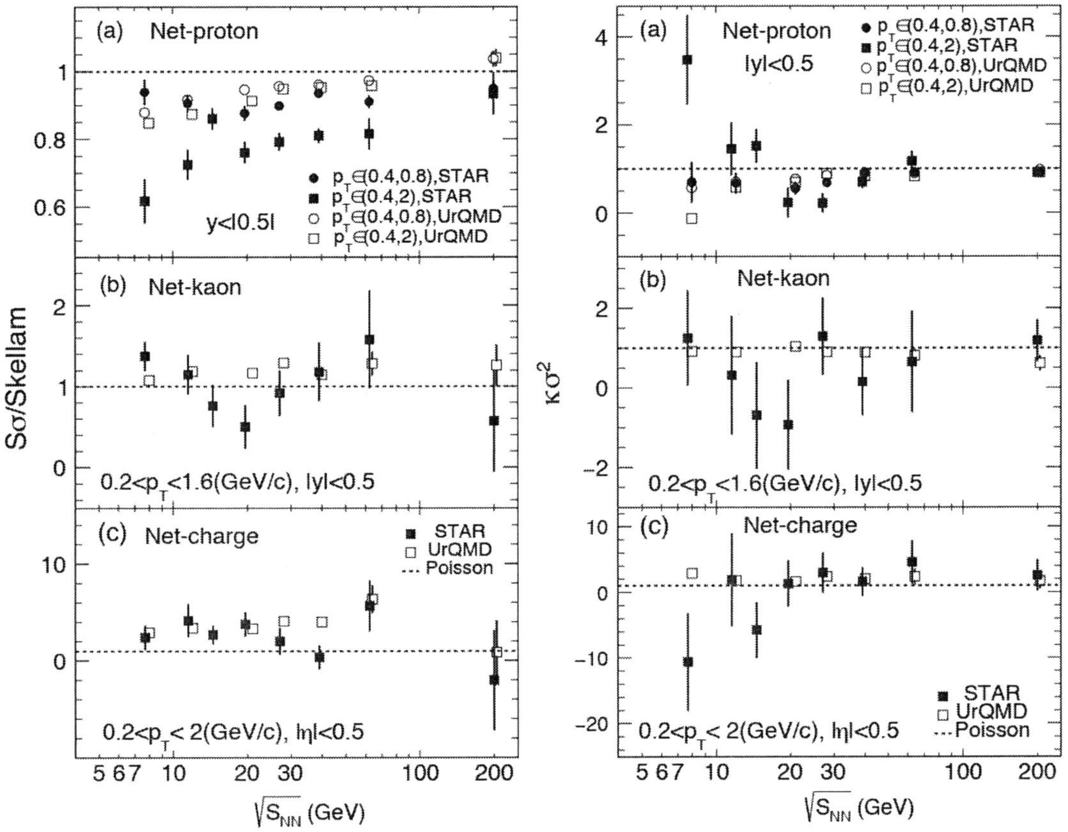

Figure 3. Energy dependence of cumulant ratios $S\sigma$ (left panel), $\kappa\sigma^2$ (right panel) of net-proton, net-charge, and net-kaon multiplicity distributions for the Au+Au collision at $\sqrt{s_{NN}}$ = 7.7 to 200 GeV. The solid markers represent the results from the STAR measurement, the open markers represent results from the UrQMD calculation. The dashed lines denote the Poisson expectations for the STAR data. Figure taken from Ref [18].

Figure 3 shows the energy dependence of cumulant ratios $S\sigma, \kappa\sigma^2$ of net-proton, net-charge, and net-kaon multiplicity distributions of the 5% most central Au+Au collisions in RHIC BES energies from the STAR experiment [20, 21] and UrQMD calculations [18]. The non-monotonic energy dependence of the net-proton $\kappa\sigma^2$ and net-kaon $S\sigma$ contrasts with the monotonic behavior of UrQMD predictions. This discrepancy might suggest critical behavior since the critical physics was not implemented in the UrQMD. It is important to reduce uncertainties of these measurements, which is expected in RHIC BES phase II [22], to get to more conclusive results.

2.3 Study of the Onset of Deconfinement

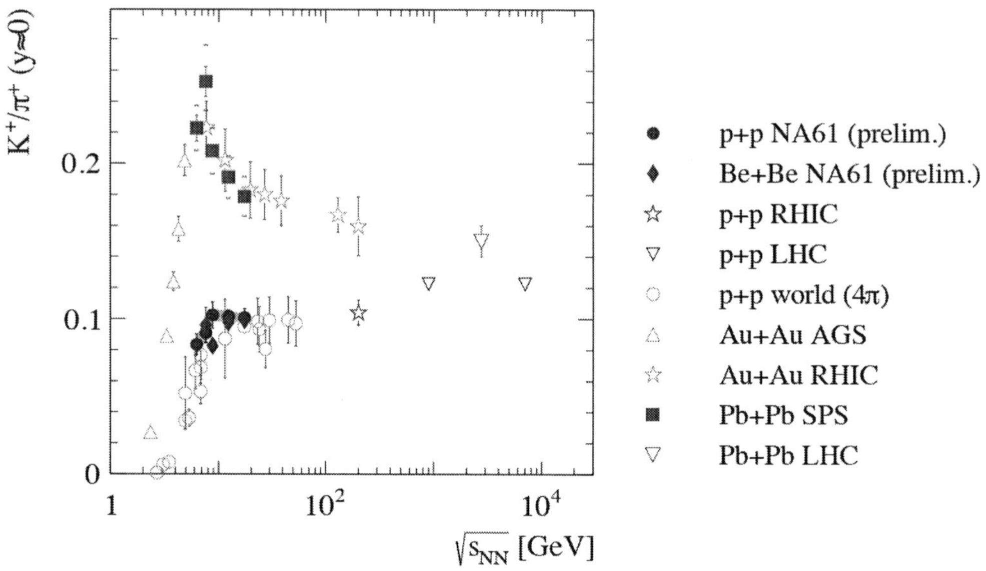

Figure 4. Energy dependence of the positively charged kaon multiplicity divided by corresponding charged pion multiplicity at mid-rapidity. Figure taken from Ref [23]

At the SPS, measurements of hadron production in a two-dimensional scan in beam momentum (13–158 AGeV/c) and system size (p+p, p+Pb, ^{7}Be+^{9}Be, Ar+Sc, Xe+La and Pb+Pb) were conducted. New results from p+p and ^{7}Be+^{9}Be were shown at Quark Matter 2017 conference [23], presenting the energy dependence of the K^+/π^+ multiplicity ratio at mid-rapidity. As shown in Fig. 4, the data from ^{7}Be+^{9}Be do not show a "horn" structure as data from heavier collision system do (average number of participants in ^{7}Be+^{9}Be collisions is around 10). This observation constrains size of system to study location of onset of deconfinement energy. The energy dependence of the inverse slope parameter of the transverse mass distribution of charged kaons, as shown in Fig. 5, shows the step structure in both p+p and ^{7}Be+^{9}Be. The step structure was predicted by the Statistical Model of the Early Stage (SMES) [24] and 3+1 Hydro model [25] as a signature of the onset of deconfinement.

The UrQMD based transport approach [26] provides further insights. It concludes that the "step" is only reproduced if a first order phase transition with a large latent heat is applied or the equation of state is effectively softened due to non-equilibrium effects in the hadronic transport calculation.

2.4 Ratios of The Freeze-out Parameters

The experimental results of ratios of the freeze-out parameters μ_S/μ_B and μ_B/T (μ_S is strange chemical potential) from the NA57 experiment [27] and the STAR experiment [28] can be reproduced by

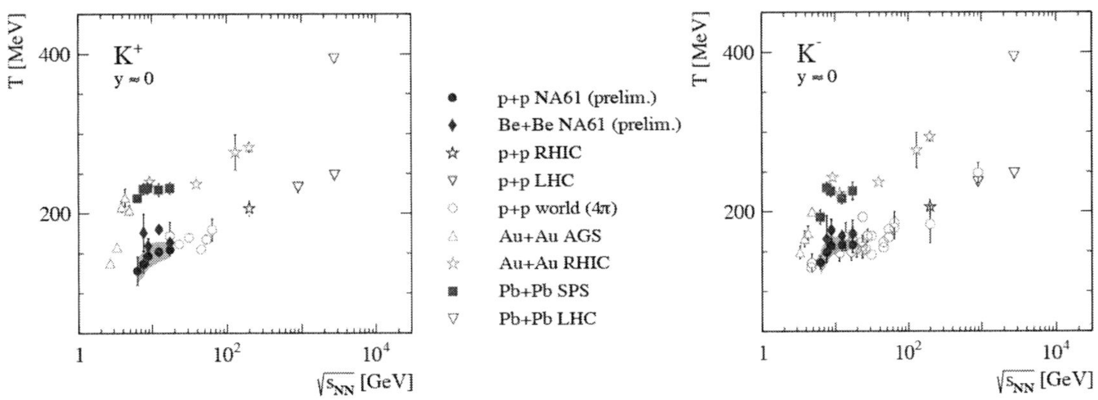

Figure 5. Energy dependence of the inverse slope parameter of the transverse mass distribution at mid-rapidity for charged kaons. The NA61/SHINE results on p+p interactions (full blue circles) and new results on Be+Be (full green diamonds) collisions are compared with world data on p+p and heavy ion (Pb+Pb and Au+Au) reactions. Figure taken from Ref [23]

a HRG model, as discussed in [29]. The HRG model prediction is in better agreement with lattice QCD calculations [29] if additional (strange) hadron resonances are included which are predicted by the quark model [30, 31] but until today unobserved (such a setup of the HRG model is denoted as QM-HRG). The μ_S/μ_B and μ_B/T ratios were obtained by a statistical thermal model [32] fitting data for strange-baryon ratios $\overline{\Lambda}/\Lambda, \Xi^+/\Xi^-$, and Ω^+/Ω^- and are shown in Fig. 6 together with lattice QCD predictions, and HRG predictions. One HRG predictions assumes only experimentally discovered hadrons (PDG-HRG) and another is the QM-HRG. As Fig. 6 presents, the QM-HRG predictions match the experimental data and the lattice QCD at the temperatures consistent with the lattice QCD and the data while the PDG-HRG matches at temperatures significantly higher. Hence, it appears that additional, experimentally yet-to-be observed strange hadrons become thermodynamically relevant in the vicinity of the QCD crossover.

3 Summary and Future Plans

Recent results shown at this conference have underlined once again the relevance of the study of strangeness at intermediate μ_B. There are indications of flavor hierarchy with respect to the chemical freeze-out [33]. Hadrons containing strange quarks appear to hadronize earlier than the hadrons containing only quarks from the first generation. This has been predicted by lattice QCD at $\mu_B = 0$ and experimentally observed by the ALICE Collaboration [9] as well as at the intermediate μ_B by the STAR Collaboration [16].

In order to describe the freeze-out parameter ratios consistently with the experimental results from STAR and NA57, a Hadron Resonance Gas needs to assume additional, yet-to-be observed strange hadrons. The experimental results are also predicted by lattice QCD.

The NA61 Collaboration has shown intriguing results from ^{7}Be+^{9}Be collisions. A comparison with its results on K^+/π^+ ratios shows a remarkably similar structure, which departs from the "horn".

The RHIC BES Phase II is scheduled in years 2019 and 2020. It will provide substantially higher statistics thanks to the accelerator upgrades [22]. Detector upgrades will further reduce systematic uncertainties of measurements. The fixed target program at STAR will allow for the QCD phase

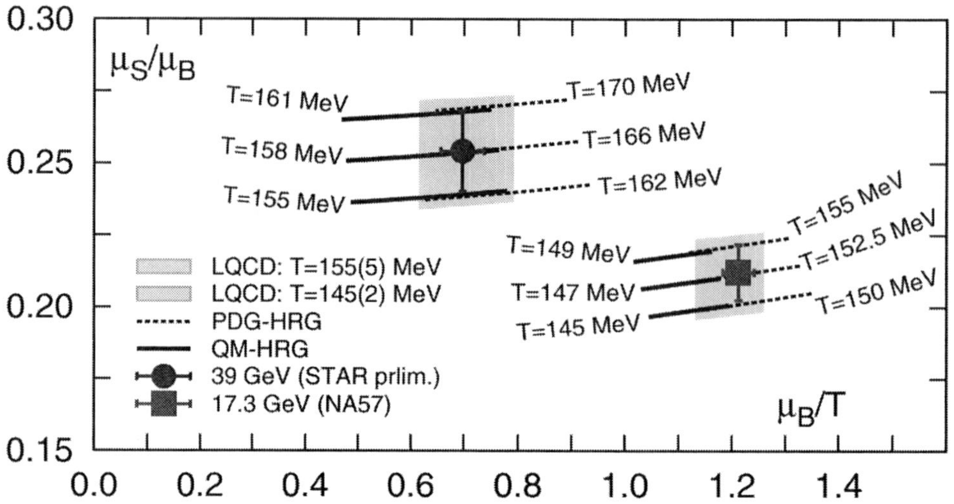

Figure 6. Values of μ_S/μ_B and μ_B/T extracted from fits to multiple strange hadrons yields (blue circle at $\sqrt{s_{NN}}$ = 39 GeV, red square at $\sqrt{s_{NN}}$ = 17.3 GeV) are compared to the lattice QCD calculations (shaded bands) as well as to the QM-HRG (solid lines) and PDG-HRG (dotted lines) models. Figure taken from Ref [29].

space diagram exploration to be extended to $\mu_B \simeq 720$ MeV. At the SPS, NA61 will upgrade its data acquisition system to allow for a 1 kHz readout. This is expected to be ready in 2020 and will allow a high statistics beam momentum scan with Pb+Pb collisions for precise measurement of multi-strange hyperon production. All these upgrades will help further constrain the search for a critical point, and the type of phase transition in the intermediated baryon density region.

Acknowledgements This work is part of and supported by U.S. Department of Energy under contract number DE-FG02-10ER41666.

References

[1] Y. Aoki *et al.*, Nature **443**, 675 (2006)
[2] A. Bazavov *et al.*, Phys. Rev. D **80**, 054503 (2012)
[3] S. Ejiri, Phys. Rev. D **78**, 074507 (2008)
[4] E. S. Bowman and J. I. Kapusta, Phys. Rev. C **79**, 015202 (2009)
[5] S. Gupta *et al.*, Science **332**, 1525 (2011)
[6] P. Braun-Munzinger, K. Redlich, and J. Stachel, *Quark-Gluon Plasma*, edited by R. C. Hwa and X.-N. Wang (World Scientific, Singapore, 2004), Vol. 3, p. 491.
[7] J. Rafelski, B. Müller, Phys. Rev. Lett. **48**, 1066 (1982)
[8] J. Rafelski, Acta Physica Polonica B **43**, 829 (2012)
[9] R. Preghenella (ALICE Collaboration), Acta Physica Polonica B **43**, 555 (2012)
[10] J. Cleymans *et al.*, Phys. Rev. C **74**, 034903 (2006)
[11] A. Andronic *et al.*, Phys. Lett. B **673**, 142 (2009)
[12] S. Gottlieb *et al.*, Phys. Rev. Lett **59**, 2247 (1987)
[13] R. V. Gavai and S. Gupta, Phys. Rev. D **68**, 034506 (2003)

[14] S. Datta *et al.*, Phys. Rev. D **95**, 054512 (2017)
[15] R. Bellwied *et al.*, Phys. Rev. Lett. **111**, 202302 (2013)
[16] L. Adamczyk *et al.* (STAR Collaboration), Phys. Rev. C **96**, 044904 (2017)
[17] S. Wheaton *et al.*, Comput. Phys. Commun. **180**, 84 (2009)
[18] J. Xu *et al.*, Phys. Rev. C 94, 024901 (2016)
[19] M. Bleicher *et al.*, J. Phys. G: Nucl. Part. Phys. **25**, 1859 (1999)
[20] L. Adamczyk *et al.* (STAR Collaboration), Phys. Rev. Lett. **113**, 092301 (2014)
[21] L. Adamczyk *et al.* (STAR Collaboration), Phys. Rev. Lett. **112**, 032302 (2014)
[22] STAR Collaboration, *Studying the Phase Diagram of QCD Matter at RHIC*, Star Note **0598** (2014)
[23] A. Aduszkiewicz (NA61/SHINE Collaboration), Nucl. Phys. A **967**, 35 (2017)
[24] M. Gazdzicki and M. I. Gorenstein, Acta Phys. Polon. B **30**, 2705 (1999)
[25] M. Gazdzicki *et al.*, Braz. Jour. Phys. **34**, 322 (2004)
[26] H. Petersen *et al.*, J. Phys. G **36**, 055104 (2009)
[27] F. Antinori *et al.* (NA57 Collaboration), Phys. Lett. B **595**, 68 (2004)
[28] F. Zhao (STAR Collaboration), *Proc. Sci.*, CPOD2013 (2013) 036
[29] A. Bazavov *et al.*, Phys. Rev. Lett. **113**, 072001 (2014)
[30] S. Capstick and N. Isgur, Phys. Rev. D **34**, 2809 (1986)
[31] D. Ebert *et al.*, Phys. Rev. D **79**, 114029 (2009)
[32] J. Cleymans, arXiv:nucl-th/9704046v1
[33] R. Belwied, *Sequential Strangeness Freeze-out*, SQM2017 proceedings

EPJ Web of Conferences **171**, 02003 (2018) https://doi.org/10.1051/epjconf/201817102003
SQM 2017

Energy dependence of strangeness production and event-by-event fluctuations

Anar Rustamov[1,2,3,4,5,⋆]

[1] *GSI Helmholtzzentrum für Schwerionenforschung, Darmstadt, Germany*
[2] *Physikalisches Institut, Universität Heidelberg, Heidelberg, Germany*
[3] *Baku State University, Baku, Azerbaijan*
[4] *National Nuclear Research Center, Baku, Azerbaijan*
[5] *Joint Institute for Nuclear Research, Dubna, Russia*

Abstract.
We review the energy dependence of strangeness production in nucleus-nucleus collisions and contrast it with the experimental observations in pp and p-A collisions at LHC energies as a function of the charged particle multiplicities. For the high multiplicity final states the results from pp and p-Pb reactions systematically approach the values obtained from Pb-Pb collisions. In statistical models this implies an approach to the thermodynamic limit, where differences of mean multiplicities between various formalisms, such as Canonical and Grand Canonical Ensembles, vanish. Furthermore, we report on event-by-event net-proton fluctuations as measured by STAR at RHIC/BNL and by ALICE at LHC/CERN and discuss various non-dynamical contributions to these measurements, which should be properly subtracted before comparison to theoretical calculations on dynamical net-baryon fluctuations.

1 Strangeness enhancement

The production of strange particles has always been at the focus of high energy nucleus-nucleus collisions. The enhancement of strange quark production, relative to light u and d quarks, in heavy-ion collisions, normalized to the corresponding signals from elementary reactions has been among the first signals for probing the Quark Gluon Plasma (QGP) formation. Indeed, in QGP, the production of a strange-antistrange quark pair can proceed by the fusion of two gluons or massless light quarks; $q + \bar{q} \leftrightarrows s + \bar{s}$, $g + g \leftrightarrows s + \bar{s}$, with q denoting u and d quarks. For the latter only the mass excess of $Q_{QGP} = 2m_s \approx 200$ MeV is needed. Moreover, in QGP the equilibration of strangeness is more efficient due to a large gluon density. In hadronic gas (HG), on the other hand, strangeness production proceeds in free space via e.g. associated production channels $N + N \rightarrow N + \Lambda + K$, with a considerably larger Q value of $Q_{HG} = m_\Lambda + m_K - m_N \approx 670$ MeV. Based on these considerations it was advocated that strangeness production should be significantly enhanced in QGP relative to that of a free hadron gas [1–3]. Furthermore, the enhancement was predicted to depend on the strangeness content of the (anti-)baryons and to appear in a typical hierarchy; $E_\Lambda < E_\Xi < E_\Omega$. The enhancement,

⋆e-mail: a.rustamov@cern.ch

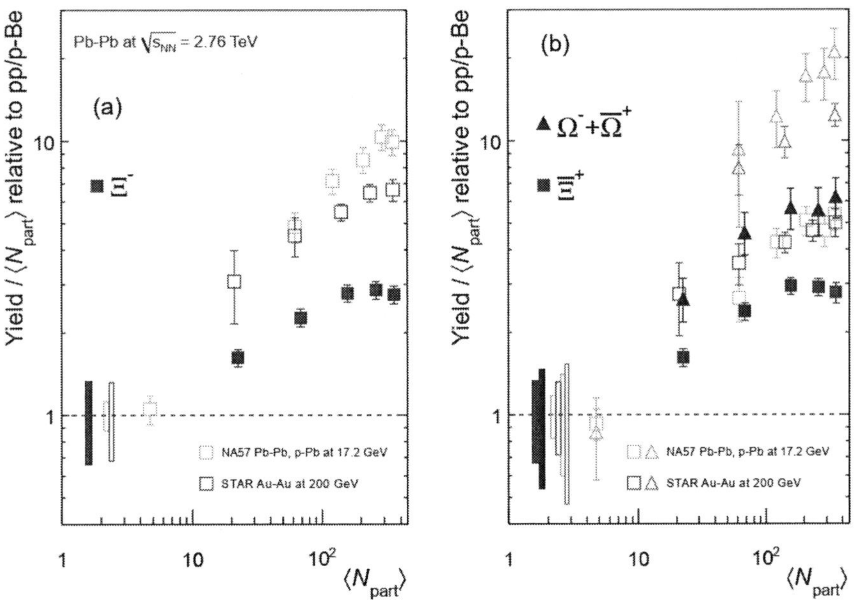

Figure 1. Strangeness enhancements at LHC(ALICE, full symbols), RHIC and SPS (open symbols) in the rapidity range $|y| < 0.5$ as a function of the mean number of participant nucleons $\langle N_{part} \rangle$. Boxes on the dashed line at unity indicate statistical and systematic uncertainties on the pp or p-Be reference. Error bars on the data points represent the corresponding uncertainties for all the heavy-ion measurements and those for p-Pb at the SPS. For more details cf. [4] and references therein.

more in particular its predicted hierarchy, was observed experimentally as presented in the right panel of Figure 1, where it is quantified as the yield of strange particles measured in A-A collisions, per mean number of participants, $\langle N \rangle_{part}$, relative to pp or p-A reactions. In the case of pp reference the enhancement is therefore defined as:

$$E = \frac{Yield_{(A-A)}}{Yield_{(pp)}} \frac{2}{\langle N \rangle_{part}} \tag{1}$$

Intuitively the above prescription suggests an increase of the enhancement with increasing energy of colliding nuclei, which is at odds with the experimental measurements presented in Figure 1.

1.1 Strangeness suppression as apparent enhancement

As discussed in the previous section the production of strange hadrons appears to be quite different in pp reactions compared to heavy-ion collisions. This is in particular the case at low energies. The remarkable success of statistical approach in describing the hadron yields produced in nucleus-nucleus collisions clearly indicates that for a large statistical systems the conservation laws can be treated within a Grand Canonical Ensemble (GCE) formulation of statistical mechanics [5]. In this approach all hadron multiplicities are proportional to the system volume. For a small systems, however, with small (anti-)strange particle multiplicities, Canonical Ensemble (CE) have to be exploited, in which the conservation of strangeness is implemented in each micro state, i.e. on an event-by-event basis [6].

As a consequence the mean multiplicities of strange particles in CE decrease faster with decreasing volume compared to corresponding multiplicities in GCE, which means that the denominator in Eq. 1 becomes smaller due to local conservation of strangeness in smaller systems. Hence, the apparent enhancement observed experimentally is an artefact of canonical suppression in small systems, which are used as a baseline. Quantitatively it was demonstrated that, for a small values of x, with a good accuracy the canonical multiplicities are suppressed by a factor of $F = I_s(x)/I_0(x)$ with respect to the corresponding Grand Canonical values [7, 8]. Here $I_s(x)$ denotes the modified Bessel functions with order s and argument x is proportional to the volume of the system. This implies a number of important predictions: (i) as the strangeness content of a hadron enters into suppression factor F as an order of the Bessel function, for a fixed value of x, the larger the strangeness content the smaller the suppression factor and hence the larger the enhancement. (ii) the suppression factor decreases with decreasing energy which leads to apparent increasing trend of the enhancement with decreasing energy and (iii) for large particle multiplicities the thermodynamic limit should be reached where the suppression factor approaches unity and the differences between CE and GCE multiplicities diminish. These predictions are consistent with the experimental observations plotted in Figure 1. To shed more light on this, we present in Figure 2 the yields of K_s^0, Λ, Ξ and Ω normalized to the pion $(\pi^+ + \pi^-)$ yield as a function of $dN_{ch}/d\eta$ for various systems measured by the ALICE collaboration [9]. A significant enhancement of strange to non-strange hadron production is observed with increasing particle multiplicity in pp collisions. Moreover, the behaviour in pp and p-Pb results, albeit the latter measured at slightly lower center-of-mass energy, are consistent in terms of both the values of the ratios and their evolutions with multiplicity. At high multiplicities, the yield ratios in pp and p-Pb collisions reach values observed in Pb-Pb collisions, thus approaching the GCE limit. This observation is in line with the canonical

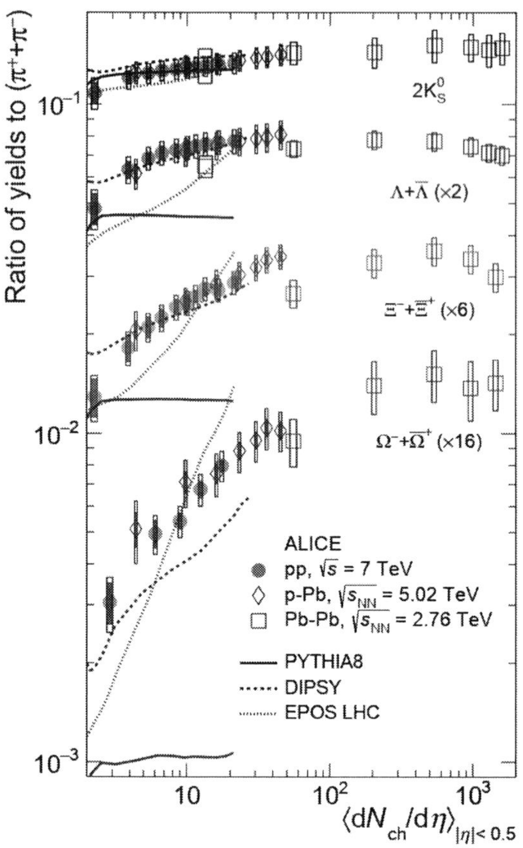

Figure 2. pT-integrated yield ratios to pions as a function of $\langle dN_{ch}/d\eta\rangle$ measured in $|y| < 0.5$. The error bars show the statistical uncertainty, whereas the empty and dark-shaded boxes show the total systematic uncertainty and the contribution uncorrelated across multiplicity bins, respectively. The values are compared to calculations from MC models and to results obtained in p-Pb and Pb-Pb collisions at the LHC. The indicated uncertainties all represent standard deviations. For more details cf. [9] and references therein.

strangeness suppression picture and shows that the origin of the strangeness enhancement is indeed driven by the final state multiplicities rather by the collision system and/or energy. We note that the concept of canonical strangeness suppression does not apply for a strangeness neutral ϕ meson. Nonetheless, experimental results for a ϕ meson also exhibit an enhancement as a function of event multiplicity (cf. Ref. [10]), i.e., it behaves like a particle with the finite net-strangeness.

2 Event-by-Event net-proton fluctuations

Perhaps the most challenging problem in our understanding of strongly interacting nuclear matter is its phase structure and possible existence of a critical point, at which matter undergoes a second-order phase transition. At the BNL/RHIC and CERN/SPS substantial experimental efforts are devoted to searches for a QCD critical point in ultra-relativistic heavy-ion collisions. This also serves as a motivations for the research programs at the future facilities, such as FAIR in Darmstadt and NICA in Dubna. Phase transitions can be addressed by investigating the response of the system to external perturbations via fluctuations of conserved charges. Moreover, one of the crucial issues is to establish the relation between freeze-out and the QCD chiral transition at vanishing net-baryon densities. This makes fluctuation studies at LHC of particular importance, especially in view of the reliable theoretical calculations in lattice QCD for vanishing values of baryon chemical potential [11–14]. Althoug LQCD predicts a Skellam behaviour for the second cumulants of net-baryon distributions at a pseudo-critical temperature of about 155 MeV, this is not the case for higher cumulants. Even on the level of second cumulants differences between HRG and LQCD results are significant for the ratio of off-diagonal susceptibilities [13, 14]. At this point we remind that, fluctuations of conserved charges are predicted in the Grand Canonical Ensemble formulation of thermodynamics [15]. In order to compare theoretical calculations within GCE, such as the Hadron Resonance Gas (HRG) model [16] and Lattice QCD (LQCD) [11], to experimental results, the requirements of GCE have to be achieved in experiments. This is typically done by analysing the experimental data in a finite acceptance by imposing cuts on rapidity and/or transverse momentum of detected particles. However, if the selected acceptance window is too small, the possible dynamical correlations we are after will also be strongly reduced [17] and consequently, net-baryons will be distributed according to the difference of two independent Poisson distributions [18]. We remind that for the Poisson distribution, all its cumulants are equal to its mean. The probability distribution of the difference X_1 - X_2 of two random variables, each generated from statistically independent Poisson distributions, is called the Skellam distribution. According to the additivity of cumulants, the cumulants of the Skellam distribution will then be $\kappa_n(Skellam) = \langle X_1 \rangle + (-1)^n \langle X_2 \rangle$, where $\langle X_1 \rangle$ and $\langle X_2 \rangle$ are mean values of X_1 and X_2 respectively. On the other hand, the increase of the acceptance will enlarge significance of correlations due to baryon number conservation. In order to be more sensitive to dynamical fluctuations, the better approach is to study the fluctuation of conserved charges in a larger acceptance and subtract the correlation part caused by the global conservation laws. This is actually the appropriate way to address the fluctuation physics when both, trivial and dynamical fluctuations are close to Poisson probability distributions. To proceed further we provide the necessary definitions. The first four cumulants of net-baryon $\Delta n_B = n_B - n_{\bar{B}}$ distribution are defined as:

$$\begin{aligned}
\kappa_1(\Delta n_B) &= \langle \Delta n_B \rangle \\
\kappa_2(\Delta n_B) &= \left\langle (\Delta n_B - \langle \Delta n_B \rangle)^2 \right\rangle \\
\kappa_3(\Delta n_B) &= \left\langle (\Delta n_B - \langle \Delta n_B \rangle)^3 \right\rangle \\
\kappa_4(\Delta n_B) &= \left\langle (\Delta n_B - \langle \Delta n_B \rangle)^4 \right\rangle - 3\kappa_2^2(\Delta n_B)
\end{aligned} \tag{2}$$

The second cumulant can be represented as a sum of corresponding cumulants for single baryons plus the correlation term for joint probability distributions of baryons and antibaryons

$$\kappa_2(\Delta n_B) = \kappa_2(n_B) + \kappa_2(n_{\bar{B}}) - 2\left(\langle n_B n_{\bar{B}}\rangle - \langle n_B\rangle\langle n_{\bar{B}}\rangle\right). \quad (3)$$

Eq. 3 shows that, in the case of missing correlations between baryons and antibaryons, $\langle n_B n_{\bar{B}}\rangle = \langle n_B\rangle\langle n_{\bar{B}}\rangle$, the second cumulant of net-baryons is exactly equal to the sum of the corresponding second cumulants for baryons and antibaryons. A correlation can emerge, in addition to that originating from critical fluctuations, from conservation laws. If the experimental acceptance is large enough the correlation term in Eq. 3 acquires a finite value. Hence we expect that the second cumulant of net-baryons becomes smaller than the sum of baryons and antibaryons.

We note that in this work net-protons are used as a proxy for net-baryons, which is justified at higher center-of-mass energies ($\sqrt{s_{NN}} > 10$ GeV) [19].

2.1 Results and discussions

Recently the ALICE collaboration reported on the second cumulants of net-proton distributions, measured in Pb-Pb collisions at $\sqrt{s_{NN}} = 2.76$ TeV [20]. The analysis of experimental data was performed with the Identity Method [21–25]. Futhermore, based on the recently proposed model, it was demonstrated that non-dynamical fluctuations, such as those stemming from unavoidable fluctuations of participant nucleons are immaterial for the second cumulants of net-protons measured at LHC energies [18]. The obtained results for the second cumulants of net-protons, normalised to the Skellam baseline, are presented in the left panel of Figure 3. For pseudorapidity ranges of $|\eta| < 0.4$, which corresponds to $\Delta\eta < 0.8$, the experimentally measured net-proton distributions follow a Skellam distribution. This agreement is due to the small acceptance as discussed above. Beyond $\Delta|\eta| > 0.8$, however, some deviations from the Skellam baseline are observed. It is argued that these deviations, as a functions of acceptance, are due to the global baryon conservation, which should be corrected for before comparison to theoretical calculations within GCE. For the second cumulants the contribution from the global baryon number conservation depends only on the acceptance factor $\alpha = \langle n_p\rangle / \langle N_B^{4\pi}\rangle$ with $\langle n_p\rangle$ and $\langle N_B^{4\pi}\rangle$ referring to the mean number of protons inside the acceptance and the mean number of baryons in the full phase space respectively [18]:

$$\frac{\kappa_2\,(p - \bar{p})}{\kappa_2\,(Skellam)} = 1 - \alpha. \quad (4)$$

For each rapidity range the acceptance factor was then estimated using the total number of baryons as measured by the ALICE experiment in the pseudorapidity range of $|\eta| < 0.5$ [26]. Next, from HIJING and AMPT simulations, total number was obtained of baryons in the full phase space . The number of protons, used in the definition of α (cf. Eq. 4), was taken from the experimental analysis for each rapidity range. Finally, using these values of α the red band in the left panel of Figure 3 was calculated with Eq. 4. The finite width of the band reflects the difference between the two event generators. We hence conclude that the reported deviation from Skellam is caused by global baryon number conservation and that other possible dynamical fluctuations are not visible in the second cumulants of net-protons measured at LHC energies.

In the right panel of Figure 3 the ratio of fourth to second cumulants, $k\sigma^2 = \kappa_4(p - \bar{p})/\kappa_2(p - \bar{p})$ of net-proton, proton and anti-proton for the 5% most central Au+Au collisions are presented, as measured by the STAR Collaboration. The data shows strong enhancement at lowest center-of-mass energy of 7.7 GeV, while results at 19.6, 27 and 39 GeV are below the HRG baseline. For higher energies, however, the data, within uncertainties, is consistent with unity. Moreover, the transport

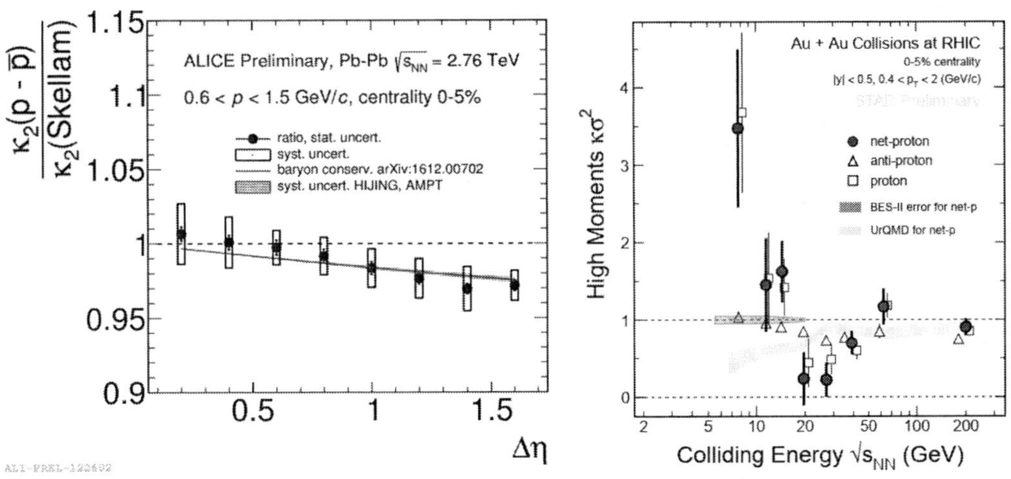

Figure 3. Left panel: Pseudorapidity dependence of the normalised second cumulants of net-protons as measured by the ALICE experiment at LHC. The red solid line shows the effect of the baryon number conservation. Right panel: Ratio of fourth to second cumulants as measured by the STAR Collaboration for Au+Au collisions at different center-of-mass energies.

model UrQMD exhibits smooth decrease with decreasing energy. Few comments are in order at this point; (i) in order to get rid of participant fluctuations [18, 27], the experimental data of STAR was corrected with the so called Centrality Bin Width Correction (CBWC) procedure [28, 29]. The essential idea behind the CBWC is to get rid of the participant fluctuations by subdividing a given centrality bin into smaller ones and then merging them together incoherently [30]. However, although CBWC procedure reduces the overall level of fluctuations significantly it cannot fully eliminate the participant fluctuations; (ii) the same rapidity interval of $|y| < 0.5$ is used for all energies. This leads to significant decrease of the fraction of accepted protons with increasing energy and increases the artefacts of global baryon number conservation at low energies, which drives the results at $\sqrt{s_{NN}}$ = 19.6, 27 and 39 GeV below unity. This explanation is also consistent with the energy dependence of the UrQMD results. Obviously baryon number conservation cannot explain the increase observed at lower energies. Recently it was argued that independent combination of baryon stopping and participant fluctuations also cannot explain this increase, though assumption of the collective stopping of multiple baryons catches the essential part of the observed structure [31].

3 Conclusions

In summary we reviewed the energy dependence of so-called strangeness enhancement in A-A collisions relative to small interacting systems. We argued that originally proposed idea of the strangeness enhancements contradicts with the energy dependence of enhancement observed experimentally. On the other hand, calculations within statistical models in Canonical Ensemble indicates that the observed enhancement is in fact the artefact of strangeness suppression in small systems, which are used as a baseline to quantify the observed apparent enhancement. Although this explains nearly all measured signals, including the reported energy dependence, the procedure fails to reproduce the centrality dependence of strangeness neutral ϕ meson production. In this case the canonical suppression is irrelevant, alhough the data indicates a suppression similar to other particles with non-zero net strangeness. Next, we reviewed first measurements of net-proton fluctuations from the ALICE experiment at LHC and discussed preliminary results form the STAR experiment. The measured second cumulants of net-protons by the ALICE Collaboration, which are used as a proxy for net-baryons, are, after accounting for baryon number conservation, in agreement with the corresponding second cumulants of the Skellam distribution. We note that LQCD predicts a Skellam behaviour for the second cumulants of net-baryon distributions at a pseudo-critical temperature of about 155 MeV, which is very close to the freeze-out temperature from the HRG model applied to the ALICE data. Critical behaviour is predicted by LQCD starting from second off-diagonal cumulants, the analysis of which is ongoing in ALICE. The non-monotonic energy excitation of fourth to second ratio of cumulants of net-protons, reported by STAR, can be partially explained by the global baryon number conservation, while the increase of fluctuations at lower energies still to be understood. It is however expected that non-dynamical fluctuations, such as those stemming from participant fluctuations are still present in the final results, because the applied CBWC procedure cannot eliminate them entirely. There is no general consensus whether or not the observed non-monotonic energy dependence is indeed connected to the critical phenomena associated with the second order phase transition at the critical point.

Acknowledgments

The author acknowledges stimulating discussions with Peter Braun-Munzinger. This work is part of and supported by the DFG Collaborative Research Centre "SFB 1225 (ISOQUANT)". The research was also supported in part by the ExtreMe Matter Institute EMMI at the GSI Helmholtzzentrum fuer Schwerionenphysik, Darmstadt, Germany

References

[1] J. Rafelski and B. Müller, Phys. Rev. Lett. 48 (1082) 1066.
[2] P. Koch, J. Rafelski, and W. Greiner, Phys. Lett. B 123 (1983) 151.
[3] P. Koch, B. Müller, and J. Rafelski, Phys. Rep. 142 (1986) 167.
[4] ALICE Collaboration, Phys.Lett. B728 (2014) 216-227, Erratum: Phys. Lett. B734 (2014) 409-410.
[5] J. Stachel, A. Andronic, P. Braun-Munzinger and K. Redlich, J. Phys. Conf. Ser. 509 (2014) 012019
[6] M.I. Gorenstein, W. Greiner, A. Rustamov, Phys.Lett. B731 (2014) 302-306.
[7] K. Redlich, A. Tounsi, Eur.Phys.J. C24 (2002) 589-594.
[8] S. Hamieh, K. Redlich, A. Tounsi, Phys.Lett. B486 (2000) 61-66
[9] ALICE Collaboration, Nature Phys. 13 (2017) 535-539.
[10] V. Vislavicius, A. Kalweit, arXiv:1610.03001.
[11] A. Bazavov et al., Phys. Rev. D85 (2012) 054503.
[12] B. Friman, F. Karsch, K. Redlich, and V. Skokov, Eur. Phys. J. C71 (2011) 1964.
[13] F. Karsch, Nucl.Phys. A967 (2017) 461-464.
[14] O. Kaczmarek, Nucl.Phys. A967 (2017) 137-144.
[15] L. D. Landau and E. M. Lifshitz, Statistical Physics, Pergamon Press, 1980.
[16] A. Andronic, P. Braun-Munzinger, K. Redlich, and J. Stachel, J.Phys.Conf.Ser. 779 (2017).
[17] V. Koch, in "Relativistic Heavy Ion Physics", R. Stock (ed.) (Springer, Heidelberg, 2010), (Landolt-Boernstein New Series I, v. 23) p. 626, arXiv:0810.2520v1.
[18] P. Braun-Munzinger, A. Rustamov, J. Stachel, Nucl. Phys. A960 (2017), 114.

[19] M. Kitazawa, and M. Asakawa, Phys. Rev. C86 (2012) 024904 and erratum, ibidum 069902, arXiv:1205.3292 [nucl-th].
[20] A. Rustamov for the ALICE Collaboration, Nucl.Phys. A967 (2017) 453-456.
[21] M. Gazdzicki et al., Phys. Rev. C83 (2011) 054907 .
[22] M. I. Gorenstein, Phys. Rev. C84 (2011) 024902.
[23] A. Rustamov, M. I. Gorenstein, Phys. Rev. C86 (2012) 044906.
[24] T. Anticic et al. (NA49 Collab.), Phys.Rev. C89 (2014) 054902.
[25] M. Arslandok for the ALICE Collaboration, Nucl.Phys. A956 (2016) 870-873.
[26] P. Braun-Munzinger, A. Kalweit, K. Redlich, and J. Stachel, Phys.Lett. B747 (2015) 292.
[27] V. Skokov, B. Friman, and K. Redlich, Phys.Rev. C88 (2013) 034911.
[28] X. Luo, PoS (CPOD2014) 019, arXiv:1503.02558.
[29] X. Luo, N. Xu, Nucl.Sci.Tech. 28 (2017) no.8, 112.
[30] X. Luo, J. Xu, B. Mohanty, N. Xu, J. Phys. G 40 (2013) 105104.
[31] A. Bzdak, V. Skokov, Eur.Phys.J. C77 (2017) no.5, 288.

EPJ Web of Conferences **171**, 02004 (2018) https://doi.org/10.1051/epjconf/201817102004
SQM 2017

Signatures of chiral symmetry restoration and its survival throughout the hadronic phase interactions

E.L. Bratkovskaya[1,2,*], P. Moreau[1], A. Palmese[3], W. Cassing[3], E. Seifert[3], and T. Steinert[3]

[1] *Institute for Theoretical Physics, University of Frankfurt, Frankfurt, Germany*
[2] *GSI Helmholtzzentrum für Schwerionenforschung GmbH, Darmstadt, Germany*
[3] *Institute for Theoretical Physics, University of Giessen, Giessen, Germany*

Abstract. The effect of the chiral symmetry restoration (CSR) on observables from heavy-ion collisions is studied in the energy range $\sqrt{s_{NN}}$=3–20 GeV within the Parton-Hadron-String Dynamics (PHSD) transport approach. The PHSD includes the deconfinement phase transition as well as essential aspects of CSR in the dense and hot hadronic medium, which are incorporated in the Schwinger mechanism for the hadronic particle production. We adopt different parametrizations of the nuclear equation of state from the non-linear $\sigma - \omega$ model, which enter in the computation of the quark scalar density for the CSR mechanism, in order to estimate the uncertainty in our calculations. For the pion-nucleon Σ-term we adopt $\Sigma_\pi \approx 45$ MeV which corresponds to a 'world average'. Our systematic studies show that chiral symmetry restoration plays a crucial role in the description of heavy-ion collisions at $\sqrt{s_{NN}}$=3–20 GeV, realizing an increase of the hadronic particle production in the strangeness sector with respect to the non-strange one. We identify particle abundances and rapidity spectra to be suitable probes in order to extract information about CSR, while transverse mass spectra are less sensitive ones. Our results provide a microscopic explanation for the "horn" structure in the excitation function of the K^+/π^+ ratio: the CSR in the hadronic phase produces the steep increase of this particle ratio up to $\sqrt{s_{NN}} \approx 7$ GeV, while the drop at higher energies is associated to the appearance of a deconfined partonic medium.

1 Introduction

The strange particle production has always been suggested as one of the most sensitive observables in heavy-ion collisions (HIC) that could indicate the creation of a Quark-Gluon-Plasma (QGP) during the early stages of the reaction. The earliest suggested signature is the strangeness enhancement in A+A collisions with respect to elementary p+p collisions [1, 2]. Later on, Gazdzicki and Gorenstein [3] proposed that a sharp rise and drop in the excitation function of the K^+/π^+ ratio (so called "horn") should show up due to the appearance of a QGP phase at a center-of-mass energy $\sqrt{s_{NN}} \sim 7$ GeV. Several statistical models [4–6] have succeeded in reproducing the trend of the experimental observation of the K^+/π^+ ratio and other strange to non-strange particle ratios, but they can provide only a statistical description of the heavy-ion collision process. On the other hand there was no conclusive interpretation of the "horn" from dynamical approaches for HIC, like microscopic transport models

*e-mail: E.Bratkovskaya@gsi.de

[7, 8]. Only recently, the Parton-Hadron-String Dynamics (PHSD), a transport approach describing HIC on the basis of partonic, hadronic and string degrees-of-freedom, obtained a striking improvement on this issue when including chiral symmetry restoration (CSR) in the string decay for hadronic particle production [9–11].

Apart from deconfinement the chiral symmetry restoration addresses another aspect of the QCD phase diagram in the (T, μ_B)-plane as an additional transition between a phase with broken and a phase with restored chiral symmetry. As in case of the QCD deconfinement phase transition, the boundaries of the CSR phase transition line are not well known. Lattice QCD (lQCD) calculations show that at vanishing baryon chemical potential (μ_B=0) the CSR takes place at roughly the same critical temperature and energy density as the deconfinement phase transition which is a crossover. At finite baryon chemical potential lQCD calculations cannot be performed due to the sign problem and one must rely on effective models (or extrapolations) in order to study the QCD phase transitions [12–14]. Different models support the idea that at finite chemical potential a partially restored phase is achieved before the deconfinement occurs [15]. In order to distinguish the two phases of such a transition, effective models use the scalar quark condensate $\langle \bar{q}q \rangle$ as an order parameter. As the baryon density and temperature increase, the scalar quark condensate $\langle \bar{q}q \rangle$ is expected to decrease from a non-vanishing value in the vacuum to $\langle \bar{q}q \rangle \approx 0$ which corresponds to CSR. Since $\langle \bar{q}q \rangle$ is not a measurable quantity, it is crucial to determine experimental observables which are sensitive to this quantity. The dilepton spectroscopy has been in the focus in this respect since in a chirally restored phase the spectral functions of the the ρ- and the a_1-meson should become identical. However, no clear evidence has been achieved so far [16]. On the other hand, the strangeness production at Alternating-Gradient Synchrotron (AGS) and lower SPS energies was suggested to be a signature of CSR [9].

In this contribution we will report on the main results from Refs. [9–11] where a systematic study has been performed on effects of the CSR on final particle distributions within the PHSD approach.

2 Reminder of the PHSD transport approach

The Parton-Hadron-String Dynamics (PHSD) is a microscopic covariant dynamical approach for strongly interacting systems in and out-of equilibrium [17–20] incorporating also the self-generated retarded electromagnetic fields [21]. It is a transport approach which goes beyond the quasi-particle approximation, since it is based on Kadanoff-Baym equations for the Green's functions in phase-space representation in first-order gradient expansion [22, 23]. Including both hadronic and partonic phase as well as a transition between the effective degrees-of-freedom, PHSD is capable to describe the full time evolution of a relativistic heavy-ion collision. The theoretical description of the partonic degrees-of-freedom (quarks and gluons) is realized in line with the Dynamical-Quasi-Particle Model (DQPM) [23] which reproduces lQCD results in thermodynamical equilibrium and provides the properties of the partons, i.e. masses and widths in their spectral functions. In equilibrium the PHSD reproduces the partonic transport coefficients such as shear and bulk viscosities or the electric conductivity from lQCD calculations as well. For a review on PHSD we refer the reader to Ref. [24]. We further note that the pure hadronic phase in PHSD is equivalent to the Hadron-Strings Dynamics (HSD) approach [25]. Accordingly, the comparison between PHSD and HSD calculations allows us to disentangle the role of the QGP phase in heavy-ion collisions.

The PHSD approach has been tested for different colliding systems (p+p, p+A, A+A) in a wide range of bombarding energy, from AGS to LHC energies, and has been able to describe a large number of experimental observables, such as charged particle spectra, collective flow coefficients v_n as well as electromagnetic probes such as photons and dileptons [24]. More recently, it has been also shown to provide a microscopic description of the maximum in the K^+/π^+ ratio in central nucleus-nucleus collisions [9–11].

2.1 String fragmentation in PHSD

The string formation and decay represents the dominant particle production mechanism in nucleus-nucleus collisions for bombarding energies from 2 AGeV to 160 AGeV. In PHSD, the primary hard scatterings between nucleons are described by string formation and decay in the FRITIOF Lund model [26]. A string is an excited color-singlet state, which is composed of two string ends corresponding to the leading constituent quarks of the colliding hadrons and a color flux tube in between. As the string ends recede, virtual $q\bar{q}$ or $qq\bar{q}\bar{q}$ pairs are produced in the uniform color field, causing the breaking of the string. Finally, the string decays into mesons or baryon-antibaryon pairs with formation time $\tau_f \sim 0.8$ fm/c (in the rest-frame of the string).
In the string decay, the flavor of the produced quarks is determined via the Schwinger formula [27], which defines the production probability of massive $s\bar{s}$ pairs with respect to light-flavor production of $(u\bar{u}, d\bar{d})$ pairs:

$$\frac{P(s\bar{s})}{P(u\bar{u})} = \frac{P(s\bar{s})}{P(d\bar{d})} = \gamma_s = \exp\Big(-\pi\frac{m_s^2 - m_{u,d}^2}{2\kappa}\Big), \tag{1}$$

with $\kappa \approx 0.176\,\text{GeV}^2$ representing the string tension and $m_{u,d,s}$ denoting the constituent quark masses for strange and light quarks. For the constituent quark masses $m_u \approx 0.35$ GeV and $m_s \approx 0.5$ GeV in the vacuum, the production of strange quarks is suppressed by a factor of $\gamma_s \approx 1/3$ with respect to the light quarks, which is the default setting in the FRITIOF routines.

2.2 Modeling of the chiral symmetry restoration

In Refs. [9, 10] the PHSD has been extended to include CSR in the string decay in a hadronic environment of finite baryon and meson density. Here we recall the main aspect of this extension which is based on the Hellman-Feynman theorem for the scalar quark condensate [28]. Accordingly, a linear decrease of the scalar quark condensate $\langle\bar{q}q\rangle$ – which is nonvanishing in the vacuum due to a spontaneous breaking of chiral symmetry – is expected with baryon density ρ_B towards a chiral symmetric phase characterized by $\langle\bar{q}q\rangle \approx 0$ [29, 30]. This decrease of the scalar quark condensate is expected also to lead to a change of the hadron properties with density and temperature, i.e. in a chirally restored phase the vector and axial vector currents should become equal [31]. Since the scalar quark condensate $\langle\bar{q}q\rangle$ is not a direct observable, its manifestations should also be found indirectly in different hadronic abundances and spectra or particle ratios like K^+/π^+, $(\Lambda+\Sigma^0)/\pi^-$ etc. as advocated in Ref. [9].

In leading order the scalar quark condensate $\langle\bar{q}q\rangle$ can be evaluated in a dynamical calculation as follows [32]:

$$\frac{\langle\bar{q}q\rangle}{\langle\bar{q}q\rangle_V} = 1 - \frac{\Sigma_\pi}{f_\pi^2 m_\pi^2}\rho_S - \sum_h \frac{\sigma_h \rho_S^h}{f_\pi^2 m_\pi^2}, \tag{2}$$

where σ_h stands for the σ-commutator of the relevant mesons h, $\langle\bar{q}q\rangle_V$ represents the vacuum condensate, $\Sigma_\pi \approx 45$ MeV is the pion-nucleon Σ-term and f_π and m_π are the pion decay constant and pion mass, respectively.

In Eq. (2), the quantities ρ_S and ρ_S^h denote the nucleon scalar density and the scalar density for a meson of type h, respectively. The scalar density of mesons h is evaluated in the independent-particle approximation as:

$$\rho_S^h(x) = \frac{(2s+1)(2\tau+1)}{(2\pi)^3}\int \mathrm{d}^3p \frac{m_h}{\sqrt{\mathbf{p}^2 + m_h^2}} f_h(x, \mathbf{p}), \tag{3}$$

where $f_h(x, \mathbf{p})$ denotes the meson phase-space distribution ($x = (\mathbf{r}, t)$) and s, τ refer to the discrete spin and isospin quantum numbers, respectively. Moreover, the vacuum scalar condensate $\langle \bar{q}q \rangle_V = \langle \bar{u}u \rangle_V + \langle \bar{d}d \rangle_V \approx 2\langle \bar{u}u \rangle_V$ can be computed according to the Gell-Mann-Oakes-Renner (GOR) relation [33, 34],

$$f_\pi^2 m_\pi^2 = -\frac{1}{2}(m_u^0 + m_d^0)\langle \bar{q}q \rangle_V \,, \tag{4}$$

and gives $\langle \bar{q}q \rangle_V \approx -3.2\,\mathrm{fm}^{-3}$ for the bare quark masses $m_u^0 = m_d^0 \approx 7\,\mathrm{MeV}$. Finally, in Eq. (2) the nucleon scalar density ρ_S has to be determined in a suitable model with interacting degrees-of-freedom in order to match our knowledge on the nuclear EoS at low temperature and finite density. A proper (and widely used) approach is the non-linear $\sigma - \omega$ model for nuclear matter where ρ_S is defined as:

$$\rho_S(x) = \frac{d}{(2\pi)^3} \int \mathrm{d}^3p \frac{m_N^*}{\sqrt{\mathbf{p}_N^{*2} + m_N^{*2}}} f_N(x, \mathbf{p}) \,, \tag{5}$$

where m_N^* and p_N^* denote the effective mass and four-momentum, respectively, and $f_N(x, \mathbf{p})$ the phase-space occupation of a nucleon while the degeneracy factor is d=4. In fact, in the non-linear $\sigma - \omega$ model the nucleon mass is modified due to the scalar interaction with the medium:

$$m_N^*(x) = m_N^v - g_s\sigma(x) \,, \tag{6}$$

where m_N^v denotes the nucleon mass in vacuum and $\sigma(x)$ is the scalar field which mediates the interaction between the nucleons and the medium with the coupling g_s. In order to calculate ρ_S, we need to determine the value of the scalar field $\sigma(x)$ at each space-time point x. This is done via the non-linear gap equation [35, 36]:

$$m_\sigma^2\sigma(x) + B\sigma^2(x) + C\sigma^3(x) = g_s\rho_S(x) = g_s d \int \frac{\mathrm{d}^3p}{(2\pi)^3} \frac{m_N^*(x)}{\sqrt{\mathbf{p}^2 + m_N^{*2}}} f_N(x, \mathbf{p}) \,, \tag{7}$$

since for matter at rest we have $\mathbf{p}^* = \mathbf{p}$. In Eq. (7) the self-interaction of the σ-field is included up to the forth order. The parameters g_s, m_σ, B, C are fixed in order to reproduce the values of the nuclear matter quantities at saturation, i.e. the saturation density, the binding energy per nucleon, the compression modulus, and the effective nucleon mass. Actually, there are different sets for these quantities (NL1,NL2,NL3) that lead to slightly different saturation properties. For details we refer the reader to Ref. [10].

The main idea in Refs. [9, 10] is to consider effective masses for the dressed quarks in the Schwinger formula (1) for the string decay in a hot and dense medium. The effective quark masses can be expressed in terms of a scalar coupling to the quark condensate $\langle \bar{q}q \rangle$ in first order as follows:

$$m_s^* = m_s^0 + (m_s^v - m_s^0)\frac{\langle \bar{q}q \rangle}{\langle \bar{q}q \rangle_V} \,, \qquad m_q^* = m_q^0 + (m_q^v - m_q^0)\frac{\langle \bar{q}q \rangle}{\langle \bar{q}q \rangle_V} \,, \tag{8}$$

with $m_s^0 \approx 100\,\mathrm{MeV}$ and $m_q^0 \approx 7\,\mathrm{MeV}$ for the bare quark masses. In Eq. (8) the effective masses decrease from the vacuum values with decreasing scalar condensate $\langle \bar{q}q \rangle$ to the constituent masses. This adaptation of the Schwinger formula in case of a hot and dense medium implies a modification of the flavor production factors in Eq. (1). In an actual nucleus-nucleus collision, PHSD incorporates a dynamical calculation of all these features for each cell in space-time:

- the scalar density ρ_S is determined by solving the gap equation (7) for the σ-field;
- the scalar condensate $\langle \bar{q}q \rangle$ is then computed via Eq. (2);

- the effective masses m_q^*, m_s^* are calculated according to Eqs. (8) and plugged in the Schwinger formula (1) in order to compute the flavor production ratios for the string decay.

We stress that, once the nucleon scalar density ρ_S and $\Sigma_\pi(\approx 45$ MeV) are fixed, there is no additional 'parameter' in the PHSD3.3 compared to the previous version PHSD3.2 that has been employed for a couple years for the analysis of relativistic heavy-ion reactions [24]. We note that the sets NL1 and NL3 have the same compression modulus K but differ in the effective mass m^*/m at saturation density whereas NL1 and NL2 have the same effective mass but differ in the compression modulus K. By comparing the results from NL1, NL2 and NL3 we are able to explore separately the effects from the effective mass and compression modulus. For a detailed discussion of the flavor ratio γ_s versus energy density for the different parameter sets NL1, NL2 and NL3 we refer the reader to Ref. [10].

3 Application to nucleus-nucleus collisions

In this section we present results for the most abundant particles from HIC at AGS and SPS energies for different nuclear equations of state in order to estimate the uncertainties of our approach. We briefly recall the main results on the hadron rapidity spectra discussed in detail in Ref. [10]. In general, the restoration of chiral symmetry gives an enhancement of the strange particle yields both for mesons and baryons at AGS and lower SPS energies. On the other hand, it produces a slight decrease in the number of pions at midrapidity due to the suppression of pions in the string decays in favor of strange hadrons. The proton rapidity spectra do not show any sensible variation, in fact the CSR as implemented in PHSD modifies essentially the chemistry of the newly produced particles in the string decay and has a minor impact the dynamics of the nucleons, which in the string picture are associated to the string ends of the primary interactions in the system. The inclusion of the CSR is essential in order to correctly reproduce the strange particle rapidity spectra especially for $(\Lambda + \Sigma^0)$ hyperons and K^+ mesons. We note that our calculations for the hadron rapidity spectra are in good agreement with experimental observation [10].

The results from PHSD with CSR using three different parametrizations for the nuclear equation of state, i.e. NL3, NL2 and NL1 show that the general features of the strangeness enhancement hold for all parametrizations; in particular the NL1 set provides larger values for the strange particle rapidity spectra at midrapidity since the scalar nucleon density is larger in this case. The difference between the parametrizations represents the uncertainty of our results related to CSR as implemented in PHSD (see below). At the top SPS energy E_{Lab} = 158 AGeV the CSR does not play a significant role, since the dynamics is dominated by the QGP phase. Thus, there is no appreciable difference between the results with and without CSR for the different nuclear EoS. Our results for K^+ are slightly lower with respect to the experimental data [10], however, the $(\Lambda + \Sigma^0)$, π^- and K^- as well as the protons are correctly reproduced.

We, furthermore, recall that in earlier HSD calculations (without a partonic phase) the slopes of the transverse mass distributions of kaons and antikaons have been severely underestimated [8]. This improves drastically when a partonic phase is incorporated as in PHSD. However, the effect from CSR on the transverse mass spectra is only very moderate since the CSR acts directly on the chemistry and not so much on the dynamics of the Schwinger mechanism [10].

We now summarize our findings with respact to the excitation function of the particle ratios K^+/π^+, K^-/π^- and $(\Lambda + \Sigma^0)/\pi$ at midrapidity from 5% central Au+Au collisions. In Fig. 1 (l.h.s.) we show the calculations for the following three scenarios: the default PHSD without CSR (blue dotted line), PHSD including CSR with NL3 and NL1 as parameter sets for the nuclear EoS from the non-linear σ–ω model (red solid and green dashed lines, respectively). The shaded area displays the uncertainties of our calculations from the two scenarios for the nuclear EoS since the results from the parameter set NL2 are always in between those from NL1 and NL3 [10].

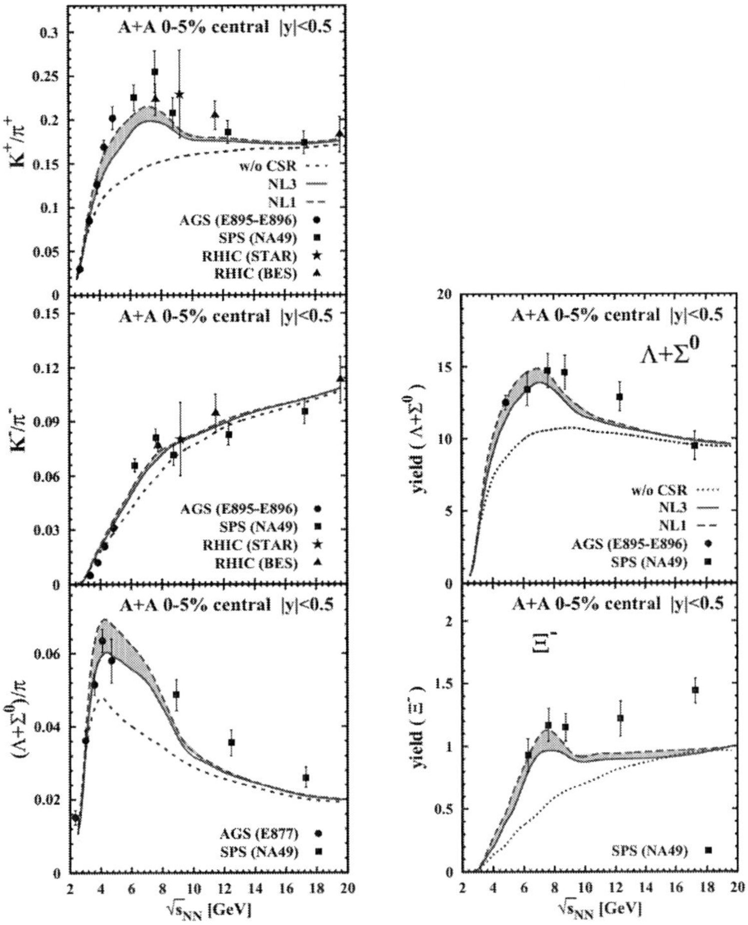

Figure 1. (l.h.s.) The ratios K^+/π^+, K^-/π^- and $(\Lambda + \Sigma^0)/\pi$ at midrapidity from 5% central Au+Au collisions as a function of the invariant energy $\sqrt{s_{NN}}$ up to the top SPS energy in comparison to the experimental data from [38–40]. The grey shaded area represents the results from PHSD including CSR taking into account the uncertainty from the parameters of the $\sigma - \omega$-model for the EoS. (r.h.s.) The yields of $(\Lambda + \Sigma^0)$ and Ξ^- at midrapidity from 5% central Au+Au collisions as a function of the invariant energy $\sqrt{s_{NN}}$ up to the top SPS energy in comparison to the experimental data from Refs. [37, 40].

As described in Refs. [9, 10], the inclusion of CSR in PHSD is responsible for the strong strangeness enhancement at AGS and low SPS energies. The experimental observations of the ratios K^+/π^+ and $(\Lambda + \Sigma^0)/\pi$ show the well-known "horn" structure, which is reproduced by the PHSD calculations with CSR. In fact, CSR gives rise to a steep increase of these ratios at energies lower than $\sqrt{s_{NN}} \approx 7$ GeV, while the drop at larger energies is associated to the appearance of a deconfined partonic medium. The NL1 parameter set produces a sharper peak both in the K^+/π^+ and in the $(\Lambda+\Sigma^0)/\pi$ excitation functions with a $\approx$ 10% maximum increase with respect to the NL3 result. We point out that even adopting different parametrizations for the $\sigma - \omega$ model, we recover the same "horn" feature. This supports the reliability of the CSR mechanism as implemented in the PHSD model.

At AGS energies, the energy dependencies of the ratios K^+/π^+ and $(\Lambda + \Sigma^0)/\pi$ are closely connected, since K^+ and Λ (or Σ^0) are mostly produced in pairs due to strangeness conservation. On the other hand, the excitation function of the K^-/π^- ratio does not show any peak, but smoothly increases as a function of $\sqrt{s_{NN}}$. In fact, especially at AGS energies, the antikaon production differs substantially from the production of K^+ and Λ, which occurs predominantly via string formation. The antikaons are produced dominantly via secondary meson-baryon interactions by flavor exchange and their production is suppressed with respect to the Λ hyperons that carry most of the strange quarks. This is the reason why the inclusion of chiral symmetry restoration provides a substantial enhancement of the K^+/π^+ and $(\Lambda + \Sigma^0)/\pi$ excitation functions and a smaller change on the K^-/π^- ratio. We also notice that there is no sizeable difference between the NL1 and NL3 results for the K^-/π^- ratio. At top SPS energies the strangeness is produced predominantly by the hadronization of partonic degrees-of-freedom, thus our results for all the ratios do not show an appreciable sensitivity to the nuclear EoS and the calculations with and without CSR tend to merge at $\sqrt{s_{NN}} \approx 20\,\text{GeV}$.

Finally, in Fig. 1 (r.h.s.) we present the yields of $(\Lambda + \Sigma^0)$ and Ξ^- at midrapidity from 5% central Au+Au collisions as a function of the invariant energy $\sqrt{s_{NN}}$ in comparison to the available data from Refs. [37, 40]. We recover a "horn" structure, similar to that shown in Fig. 1 (l.h.s.) for the energy dependence of the strange to non-strange particle ratios. A sensitivity on the nuclear model parametrizations persists at low energy, while in the top SPS energy regime the results corresponding to the different scenarios merge. The comparison with the available data at $\sqrt{s_{NN}} < 8\,\text{GeV}$ supports the validity of the CSR picture, while at larger energies we under-estimate the experimental observations for Ξ^- baryons. We mention that this discrepancy is not due to the CSR mechanism, since it does not play an essential role in the high-energy regime.

4 Summary

We point out that our PHSD calculations provide a microscopic interpretation of the "horn" structure in the excitation function of the K^+/π^+ ratio in central Au+Au (or Pb+Pb) collisions. The steep rise of this ratio at AGS energies is associated to chiral symmetry restoration (CSR) in the hadronic phase, while the drop at higher SPS energies is due to the appearance of the QGP phase in an increasing volume of the interaction region. We have found an analogous energy dependence for the $(\Lambda + \Sigma^0)/\pi$ ratio, while the excitation function of the K^-/π^- ratio does not show any explicit peak. In general, the PHSD results obtained with the inclusion of CSR are in a good agreement with the available data for all observables analyzed, while calculations without CSR fail substantially.

The microscopic PHSD studies support the idea that CSR occurs in hadronic systems with high temperatures and densities before the deconfinement phase transition takes over. We suggest that the strange particle spectra and yields are suitable signatures to study the properties of CSR in HICs in future also as a function of system size and centrality as advocated in Ref. [10].

Acknowledgments

The authors acknowledge inspiring discussions with J. Cleymans, M. Gazdzicki, M. Gorenstein and O. Linnyk and thank the Helmholtz International Center for FAIR (HIC for FAIR), the Helmholtz Graduate School for Hadron and Ion Research (HGS-HIRe), the Helmholtz Research School for Quark Matter Studies in Heavy-Ion Collisions (H-QM), and the Bundesministerium für Bildung und Forschung (BMBF) for support. The computational resources have been provided by the LOEWE-CSC.

References

[1] J. Rafelski and B. Müller, Phys. Rev. Lett. **48**, 106 (1982).
[2] R. Stock, J. Phys. **G 28**, 1517 (2002).
[3] M. Gazdzicki and M.I. Gorenstein, Acta Phys. Polon. **B 30**, 2705 (1999).
[4] J. Cleymans *et al.*, Phys. Lett. **B 615**, 50 (2005).
[5] A. Andronic, P. Braun-Munzinger, and J. Stachel, Phys. Lett. **B 673**, 142 (2009).
[6] K. A. Bugaev *et al.*, Europhys. Lett. **104**, 22002 (2013).
[7] J. Geiss, W. Cassing, and C. Greiner, Nucl. Phys. **A 644**, 107 (1998).
[8] E. L. Bratkovskaya *et al.*, Phys. Rev. **C 69**, 054907 (2004); Phys. Rev. Lett. **92**, 032302 (2004).
[9] W. Cassing, A. Palmese, P. Moreau, and E.L. Bratkovskaya, Phys. Rev. **C 93**, 014902 (2016).
[10] A. Palmese *et al.*, Phys. Rev. **C 94**, 044912 (2016).
[11] E.L. Bratkovskaya, W. Cassing, P. Moreau, and A. Palmese, Eur. Phys. J. **A 52**, 219 (2016).
[12] P. Senger *et al.*, Lect. Notes Phys. **814**, 681 (2011).
[13] C. S. Fischer, L. Fister, J. Luecker, J. M. Pawlowski, Phys. Lett. **B 732**, 273 (2014).
[14] T. K. Herbst, J. M. Pawlowski, and B. J. Schaefer, Phys. Rev. **D 88**, 014007 (2013).
[15] L. McLerran and R. D. Pisarski, Nucl. Phys. **A 796**, 83 (2007).
[16] R. Rapp, Adv. High Energy Phys. **2013**, 148253 (2013).
[17] W. Cassing and E. L. Bratkovskaya, Phys. Rev. **C 78**, 034919 (2008).
[18] W. Cassing and E. L. Bratkovskaya, Nucl. Phys. **A 831**, 215 (2009).
[19] E. L. Bratkovskaya *et al.*, Nucl. Phys. **A 856**, 162 (2011).
[20] V. P. Konchakovski *et al.*, Phys. Rev. C **85**, 044922 (2012); Phys. Rev. **C 85**, 011902 (2012).
[21] V. D. Toneev *et al.*, Phys. Rev. C **95**, 034911 (2017).
[22] L. P. Kadanoff and G. Baym, *Quantum Statistical Mechanics*, Benjamin, New York, 1962.
[23] W. Cassing, Eur. Phys. J. ST **168**, 3 (2009); Nucl. Phys. **A 795**, 70 (2007).
[24] O. Linnyk, E. Bratkovskaya, and W. Cassing, Prog. Part. Nucl. Phys **87**, 50 (2016).
[25] W. Cassing and E. L. Bratkovskaya, Phys. Rep. **308**, 65 (1999).
[26] B. Andersson, G. Gustafson, and H. Pi, Z. Phys. **C 57**, 485 (1993).
[27] J. Schwinger, Phys. Rev. **83**, 664 (1951).
[28] T. D. Cohen, R. J. Furnstahl, and D. K. Griegel, Phys. Rev. **C 45**, 1881 (1992).
[29] U. Vogl and W. Weise, Prog. Part. Nucl. Phys. **27**, 195 (1991).
[30] M. C. Birse, J. Phys. **G 20**, 1537 (1994).
[31] V. Koch, Int. J. Mod. Phys. **E 6**, 203 (1997).
[32] B. Friman, W. Nörenberg, and V. D. Toneev, Eur. Phys. J. **A 3**, 165 (1998).
[33] M. Bando, T. Kugo, and K. Yamawaki, Phys. Rep. **164**, 217 (1988).
[34] T. D. Cohen, R. J. Furnstahl, D. K. Griegel, and X. Jin, Prog. Part. Nucl. Phys. **35**, 221 (1995).
[35] J. Boguta and A. R. Bodmer, Nucl. Phys. **A 292**, 413 (1977).
[36] A. Lang *et al.*, Z. Phys. **A 340**, 287 (1991).
[37] S. Albergo *et al.*, Phys. Rev. Lett. **88**, 062301 (2002).
[38] J. L. Klay *et al.*, Phys. Rev. **C 68**, 054905 (2003).
[39] B. I. Abelev *et al.*, Phys. Rev. **C 81**, 024911 (2010); M. M. Aggarwal *et al.*, Phys. Rev. **C 83**, 024901 (2011).
[40] C. Alt *et al.*, Phys. Rev. **C 78**, 034918 (2008).

Strangeness Suppression and Color Deconfinement*

Helmut Satz[1,**]

[1] *Fakultät für Physik, Universität Bielefeld, Germany*

Abstract. The relative multiplicities for hadron production in different high energy collisions are in general well described by an ideal gas of all hadronic resonances, except that under certain conditions, strange particle rates are systematically reduced. We show that the suppression factor γ_s, accounting for reduced strange particle rates in pp, pA and AA collisions at different collision energies, becomes a universal function when expressed in terms of the initial entropy density s_0 or the initial temperature T of the produced thermal medium. It is found that γ_s increases from about 0.5 to 1.0 in a narrow temperature range around the quark-hadron transition temperature $T_c \simeq 160$ MeV. Strangeness suppression thus disappears with the onset of color deconfinement; subsequently, full equilibrium resonance gas behavior is attained.

The relation of color deconfinement and strange hadron production in high energy collisions has attracted much attention over more than thirty years. It had been observed that in elementary (pp, e^+e^-) collisions as well as in low energy nucleus-nucleus interactions, the production of strange particles was suppressed relative to the rates expected for an equilibrium resonance gas. In this context, it was argued early on that quark gluon plasma formation would enhance strange particle production in nucleus-nucleus collisions [1, 2], bringing the produced system closer to equilibrium resonance gas rates. Over the years, different mechanisms for such an effect were discussed [3–6, 8]. The aim of the present paper is to provide clear experimental evidence showing that in AA, pA and pp collisions, strangeness production as function of the initial temperature of the produced thermal medium follows a universal form, with suppression ending in a narrow temperature band around the color deconfinement temperature.

High energy proton-proton, proton-nucleus and nucleus-nucleus collisions lead to abundant multihadron production. The relative rates of the secondaries thus produced are well accounted for by an ideal gas of all hadrons and hadronic resonances at fixed temperature T and baryochemical potential μ, with one well-known caveat. Strangeness production is reduced with respect to the rates thus predicted. This suppression can, however, be taken into account by one further parameter, $0 < \gamma_s \leq 1$, if the predicted rate for a hadron species containing $\nu = 1, 2, 3$ strange quarks is suppressed by the factor γ_s^ν [7].

The basic quantity for the resonance gas description is the grand-canonical partition function for an ideal gas at temperature T in a spatial volume V

$$\ln Z(T) = V \sum_i \frac{d_i \gamma_s^{\nu_i}}{(2\pi)^3} \phi(m_i, T), \qquad (1)$$

*based on joint work [arXiv:1709.02706] with P. Castorina and S. Plumari, U. di Catania, Italy

**e-mail: satz@physik.uni-bielefeld.de

with d_i specifying the degeneracy (spin, isospin) of species i, and m_i its mass; the sum runs over all species. For simplicity, we assume for the moment that the baryochemical potential vanishes, $\mu = 0$. In eq. (1)

$$\phi(m_i, T) = \int d^3p \ \exp\{\sqrt{p^2 + m_i^2}/T\} \sim \exp-(m_i/T) \tag{2}$$

is the Boltzmann factor for species i, so that the ratio of the production rates N_i and N_j for hadrons of species i and j is given by

$$\frac{N_i}{N_j} = \frac{d_i \gamma_s^{\nu_i} \phi(m_i, T)}{d_j \gamma_s^{\nu_j} \phi(m_j, T)}, \tag{3}$$

where $\nu_i = 0, 1, 2, 3$ specifies the number of strange quarks in species i. We note that in the grand-canonical formulation the volume cancels out in the form for the relative abundances.

Both temperature T and strangeness suppression factor γ_s have been measured at various collision energies and for different collision configurations. The resulting temperature of the emerging resonance gas is found to have a universal value $T_c \simeq 160 \pm 10$ MeV for all (high) collision energies and all collision configurations, see Fig. 1.

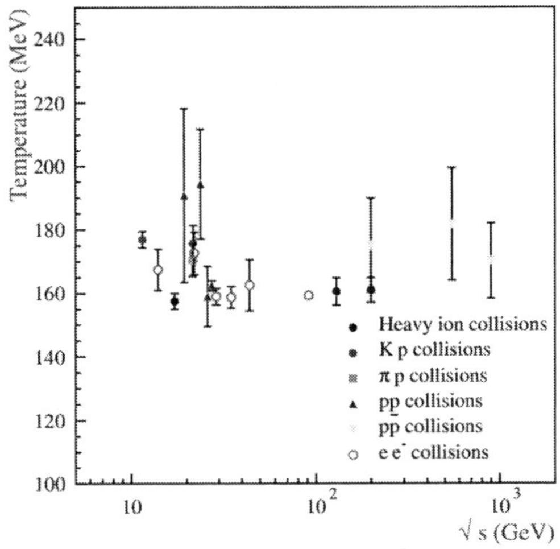

Figure 1. Hadronization temperatures for different collision energies and different collision configurations [9].

In contrast, the strangeness suppression factor γ_s depends over a wide range on the collision energy and configuration; in particular, it differs considerably in pp and AA interactions, see Fig. 2. We note here that for pp data, strangeness conservation generally is generally taken into account exactly, rather than in a grand canonical scheme with a strangeness chemical potential.

The curves in Fig. 2 are interpolation fits, leading to

$$\gamma_s^p(s) = 1 - c_p \exp(-d_p s^{1/4}), \tag{4}$$

for pp and

$$\gamma_s^A(s) = 1 - c_A \exp(-d_A \sqrt{A\sqrt{s}}) \tag{5}$$

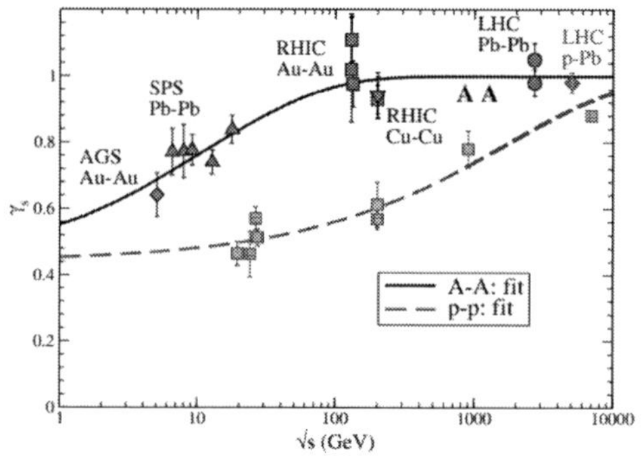

Figure 2. Strangeness suppression in pp and AA collisions at cms energy $\sqrt{s}$ [10].

for AA collisions, with$c_p = 0.5595$; $d_p = 0.0242$; $c_A = 0.606$, $d_A = 0.0209$. It is evident from these fits that for sufficiently high collision energies, one expects γ_s to converge to unity for pp as well as for AA collisions, so that strangeness suppression then also disappears in proton-proton interactions. This restoration of an equilibrium grand-canonical strangeness distribution has also been suggested on theoretical grounds [8].

The results shown in Fig. 2 suggest an obvious question: is there a unified description for the strangeness suppression observed in AA and as well as in pp collisions? Our aim here is to that if we consider the collision energy $\sqrt{s}$ by more natural thermal variables, then γ_s in fact becomes one universal function.

The canonical view of a high energy collision is shown in Fig. 3. At proper time τ_0, a thermal medium is formed, which at proper time τ_h hadronizes, i.e., freezes out into an ideal resonance gas.

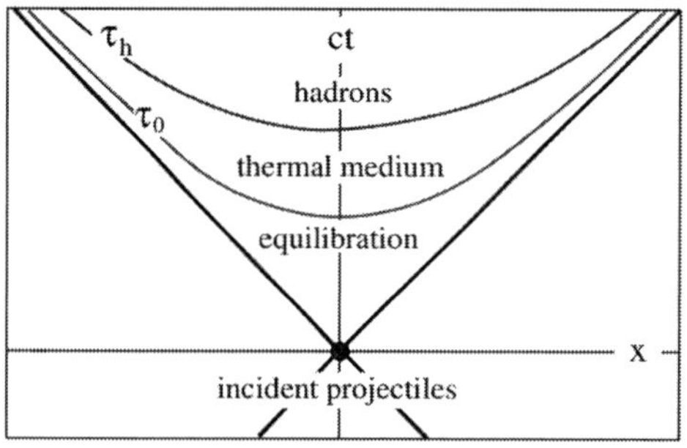

Figure 3. High energy collision in terms of longitudinal space x and time t in the center of mass.

The initial entropy density s_0 at the thermalisation time τ_0 is for one-dimensional hydrodynamic expansion given by the Bjorken relation

$$s_0 \tau_0 \simeq \frac{1.5A^x}{\pi R_x^2} \left(\frac{dN}{dy}\right)^x_{y=0}, \text{ with } x \sim pp, pA, AA, \tag{6}$$

where $(dN/dy)^x(s)$ denotes the average multiplicity per unit central rapidity in the corresponding reaction x at collision energy $\sqrt{s}$, while R_x gives the radius of the associated transverse area. This multiplicity has been measured at different collision energies for pp as well as for AA collisions, with the result shown in Fig. 4

Using $(dN_{ch}/d\eta)/0.5\langle N_{part}\rangle$ for $(dN/dy)_{y=0}$, the fit curves shown in this figure are given by

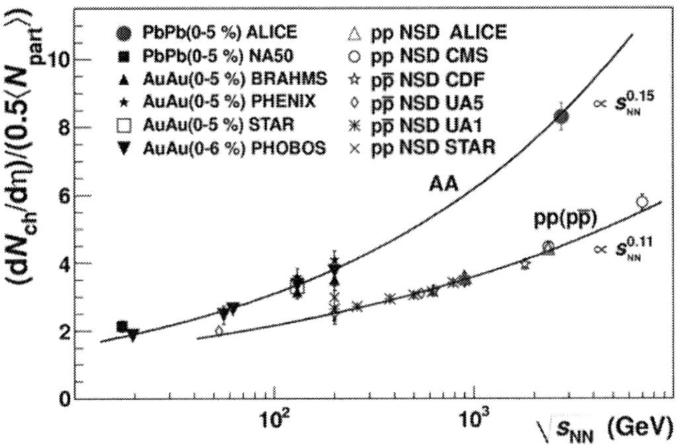

Figure 4. Average multiplicities in pp and AA collisions as function of cms energy $\sqrt{s}$ [11].

$$\left(\frac{dN}{dy}\right)^{AA}_{y=0} = a_A(\sqrt{s})^{0.3} + b_A \tag{7}$$

and

$$\left(\frac{dN}{dy}\right)^{pp}_{y=0} = a_p(\sqrt{s})^{0.22} + b_p \tag{8}$$

with $a_A = 0.7613$, $b_A = 0.0534$; $a_p = 0.797$; $b_p = 0.04123$. We can use these results to obtain the initial entropy density $s_0(s)$ in eq. (6) as function of the collision energy. Next we then use eq's. (4), (5) and (6) to express the strangeness suppression factor γ_s as function of the initial entropy density of the medium produced in the corresponding collision. For the relevant parameters in eq. (6) we take $R_{pp} = 0.8$ fm for pp, $R_{pPb} = R_{pp}(0.5\bar{N}_{part})^{1/3}$ with $\bar{N}_{part} = 8$ for pPb [12], and $R_{AA} = 1.25A^{1/3}$ fm for AA. For the equilibration time, we follow the common choice $\tau_0 = 1$ fm. The result is shown in Fig. 5.

We see that we now indeed obtain a universal form of strangeness suppression, in which the pp and AA interpolation curves coincide and with which pp, pA and AA data agree. The relevant variable for the amount of strangeness suppression is thus the initial entropy density. As already indicated above, the pronounced strangeness suppression observed in pp collisions at lower entropy densities disppears when collision energy and hence s_0 increase. Such an effect had been predicted some time ago as a consequence of the increasing life-time of the thermal medium with increasing collision energy [6, 8].

Since finite temperature lattice QCD calculations [13] provide us with the dependence of s_0 on the initial temperature T, we can use the results of Fig. 5 to express γ_s as a function of T. The resulting universal strangeness suppresson as function of the initial temperature is shown in Fig. 6. Here we should keep in mind that the lattice results hold for small values of the baryochemical potential μ, so that the AA points at lowest temperatures (AGS) are to be considered estimates only.

From Fig. 6 it is evident that if the initial temperature of the thermal medium produced in the collision lies below the deconfinement temperature, i.e., if the medium is in a hadronic state, there is strangeness suppression, with $\gamma_s \simeq 0.5$. For media of initial temperatures above the transition region, such a suppression is no longer present; it disappears in a very narrow temperature range around

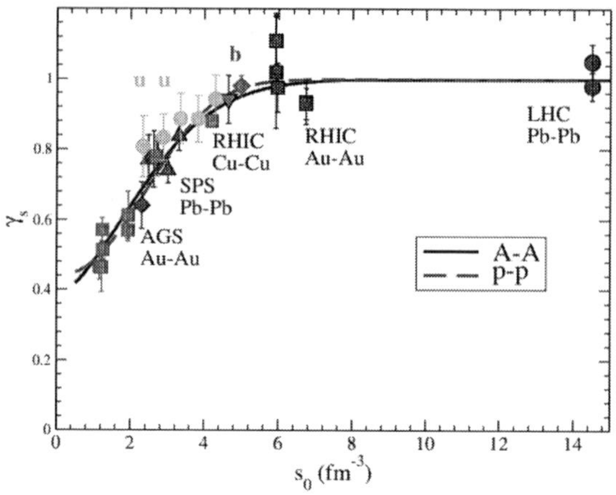

Figure 5. Strangeness suppression $\gamma_s(s_0)$ as function of the initial entropy density s_0.

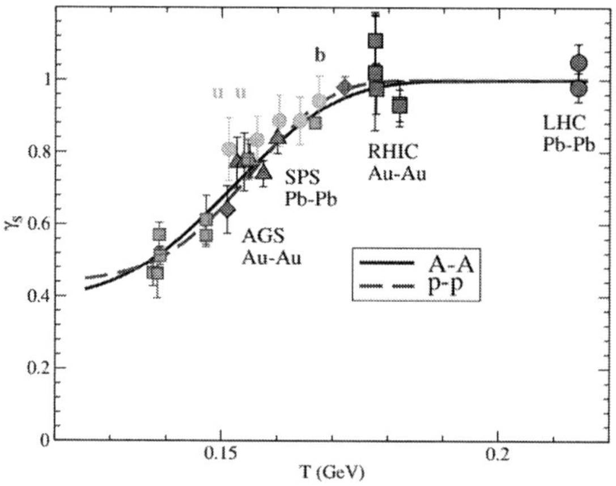

Figure 6. Strangeness suppression $\gamma_s(T)$ as function of the initial temperature T.

$T_c \simeq 160$ MeV, for pp as well as for nuclear collisions. In other words, strangeness suppression disappears with deconfinement.The same relation to the confinement/deconfinement point is of course also present in Fig. 5: here the transition occurs in a narrow band around the corresponding entropy density at the transition point, $s_0 \simeq 3$ fm^{-3}.

To obtain the universal form of strangeness suppression, we had used the average multiplicities. We could, however, use the above formalism also to derive predictions for $\gamma_s(s)$ at a fixed cms energy as function of the associated multiplicity. In particular, we could in this way obtain the multiplicity dependence of strangeness suppression in pp collisions at the LHC. This, however, requires as input the transverse area, which also depends on the corresponding final multiplicity, since high multiplicity events arise from large transverse area fluctuations. This has been estimated in specific approaches, such as the color glass condensate [14], but as such remains model-dependent. We therefore postpone such an analysis to later work.

In conclusion, we have shown that in terms of thermal variables, strangeness suppression is a universal phenomenon in all high energy hadronic interactions, pp, pA and AA. It decreases with increasing initial entropy density of the produced thermal medium, or equivalently, with increasing initial temperature, and it vanishes with the onset of deconfinement. Above the deconfinement transition regime, full equilibration of strange as well as non-strange secondaries is reached for all high energy multihadron production.

References

[1] J. Rafelski and B. Müller, Phys. Rev. Lett. **48**, 1066 and erratum ...(1982).
[2] J. Rafelski, Nucl. Phys. A **418**, 215 (1984).
[3] J. S. Hamieh, K. Redlich and A. Tounsi, Phys. Lett. B **486**, 61 (2000).
[4] For further references to early work, see
P. Braun-Munzinger, K. Redlich and J. Stachel, in *Quark-Gluon Plasma 3*, R. C. Hwa and X.-N Wang (Eds.), World Scientific, Singapore 2003.
[5] P. Castorina and H. Satz, Int. J. Mod. Phys. E **23**, 1450019 (2014).
[6] P. Castorina, S. Plumari and H. Satz, J. Mod. Phys. E **25**, 1650058 (2016).
[7] J. Letessier, J. Rafelski and A. Tounsi, Phys. Rev. C **64**,406 (1994).
[8] P. Castorina and H. Satz, Eur. Phys. J. CA **52**, 200 (2016).
[9] F. Becattini et al., Eur. Phys. J. C**56**, 493 (2008).
F. Becattini and R. Fries, Landolt-Börnstein **23**, 208 (2010).
F. Becattini, J. Manninen and M. Gazdzicki, Phys. Rev. C **73**, 044905 (2006).
F. Becattini and J. Manninen, Phys. Rev. C **78**, 054901 (2008)
[10] The γ_s values were obtained in the following data analyses and in ref. [9], where also the original experimental references are given:
F. Becattini et al. Eur. Phys. J. C**66**, 377 (2010).
M. Floris et al., Nucl. Phys. A **931**, 103 (2014).
F. Becattini, hep-ph/9701275.
F. Becattini et al., Phys. Rev. Lett. **111**, 082302 (2013).
The fits eq's (4) and (5) were made by S.Plumari.
[11] K. Aamodt et al. (ALICE Coll.), Phys. Rev. Lett. **105**, 252301 (2010).
[12] B. Abelev et al. (ALICE Coll.), Phys. Rev. Lett. **110**, 052301 (2013).
[13] A. Bazazov et al. (HotQCD), Phys. Rev. D **90**, 094503 (2014).
[14] L. McLerran, M. Praszalowicz and B. Schenke, Nucl. Phys. A **916**, 210 (2013);
L. McLerran and P. Tribedy, Nucl. Phys. A **945**, 216 (2016).

Sequential Strangeness Freeze-out

Rene Bellwied[1,*]

[1] *Department of Physics, University of Houston, Houston, TX 77204, USA*

Abstract. I will describe the latest results from lattice QCD pertaining to a potential flavour hierarchy in the hadronic freeze-out from the QCD crossover region. I will compare these results to a variety of improved hadronic resonance gas calculations and to experimental data of fluctuations of net-charge, net-proton and net-kaon multiplicity distributions, which serve as a proxy for the susceptibilities of conserved quantum numbers on the lattice. I will conclude that there is intriguing evidence for a flavour dependent freeze-out, and I will suggest expansions to the experimental program at RHIC and the LHC that could potentially demonstrate the impact of a flavour separation during hadronization.

1 Introduction

The detailed determination of a pseudo-critical temperature based on continuum extrapolations of the temperature dependence of the chiral susceptibility on the lattice, in comparison to calculations of the chemical freeze-out temperature using particle yields at RHIC and the LHC, seems to indicate that hadronization and freeze-out coincide near the phase boundary in the QCD phase diagram. The question arises whether this transition from quark to hadron degrees of freedom occurs at the same temperature for all particle species and/or quark flavours. The application of statistical hadronization models is successful in describing hadronic particle yields over many orders of magnitude. From the abundant pion yields to the rare alpha particle, the thermally equilibrated system can be broadly defined with two common freeze-out parameters, namely the chemical freeze-out temperature and the baryo-chemical potential. These calculations were applied over a wide range of collision energies from the SPS to the LHC, and system sizes from pPb to PbPb. However, recent high resolution measurements of the particle yields in ALICE at the LHC and STAR at RHIC, as well as the net-particle fluctuations in STAR, seem to indicate that there might be evidence for a sub-structure in the common freeze-out picture. The early LHC measurements attributed the tension in a common fit to a 'proton anomaly', because, at the time, only the proton yields seemed to deviate from the anticipated particle yields. As of late, though, i.e. with the Run-2 data from ALICE, also the multi-strange Ξ baryons show a significant deviation from the common temperature fit. These results can be related to attempts to deduce the freeze-out temperatures of particular conserved quantum numbers independently from fluctuations of net-particle distributions and lattice QCD.

*e-mail: bellwied@uh.edu

2 Susceptibilities from lattice QCD

Susceptibilities are defined as the derivatives of the pressure with respect to the chemical potential. Continuum extrapolated susceptibility calculations of single flavour quantum numbers showed that there is a difference between flavours in the crossover region [1], see Fig.1 (left). Fig. 1 (right) shows the flavour specific susceptibility ratio χ_4/χ_2 [2], which was suggested as a specific observable to deduce chemical freeze-out temperatures directly [3], from a comparison of experimental data to first principle calculations. The lattice data themselves show a peak at different temperatures and their agreement with Hadron Resonance Gas (HRG) model calculations begins to deviate at these temperatures as well. This is, without a direct comparison to experimental data, not yet proof of a flavour hierarchy in the crossover region, but it is suggestive of different freeze-out temperatures for light and strange flavour particles.

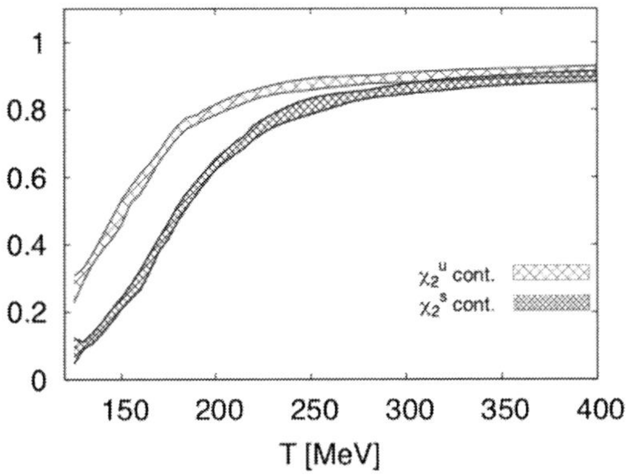

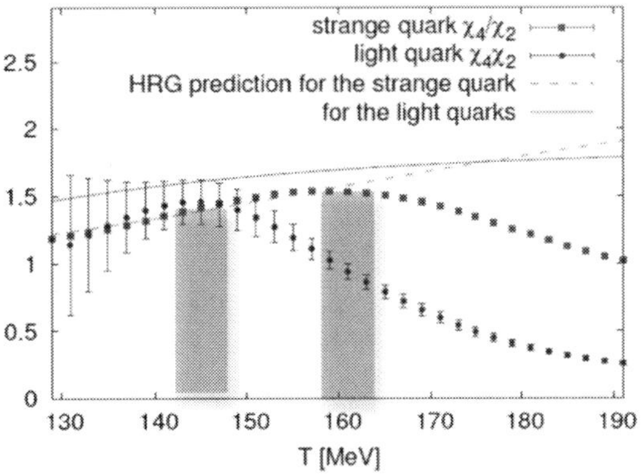

Figure 1. (left): Continuum extrapolated lattice QCD results for χ_2^u and χ_2^s [1], (right): Continuum extrapolated lattice QCD results for χ_4/χ_2 for light and strange quarks in comparison to HRG model calculations [2].

3 Comparison to the Hadron Resonance Gas Model

The relevance of the hadronic spectrum included in the HRG calculations was highlighted early on by Bazavov et al. [4] in order to explain a potential deviation from the lattice curves at different temperatures for different flavours. It was shown that the inclusion of certain states significantly improves the agreement between HRG and lattice QCD for certain susceptibilities in the QCD crossover region. The initial study simply expanded the list of available states by expectations from the non-relativistic Quark Model [5]. More refined Quark Model variations, which include the possibility of quark-quark interactions in the hadron, significantly reduce the number of expected states, while at the same time the number of experimentally verified states in the listing of the particle data group, e.g. PDG-2016 [6], continues to increase. A detailed study of the HRG approach including the latest findings, was presented at this conference and in a recent publication [7, 8] and indicated that a compromise between too many and too few excited states in the high mass hadronic spectrum can be found, based on all states listed in PDG-2016. With this input the HRG calculation can describe most features of a multitude of susceptibility ratios calculated with lattice QCD up to the pseudo-critical temperature(s). As an example, Fig. 2 shows the two most relevant susceptibility ratios that can be measured to determine flavour dependencies, namely the $(\chi_S/\chi_B)_{LO}$ and the χ_4/χ_2 for strange quarks [8]. It is evident that the so-called PDG-2016+ list gives the best compromise between old PDG and QM listings.

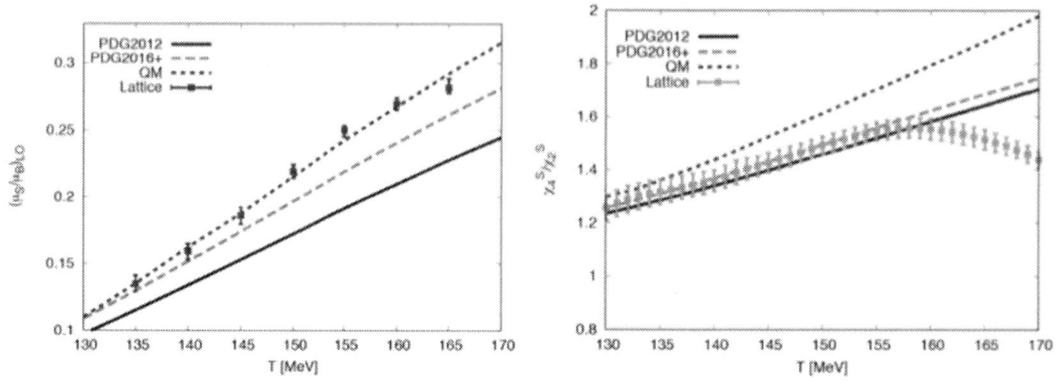

Figure 2. Continuum extrapolated lattice QCD results for $(\chi_S/\chi_B)_{LO}$ (left) and for χ_4^S/χ_2^S (right) in comparison to HRG model calculations with varying number of resonant states based on PDG-2012, PDG-2016+ (i.e. incl. one star states), and non-relativistic Quark Model predictions [8].

A more provocative extension of the standard HRG approach was also presented at this conference and in a recent publication, expanding on the question whether a non-interacting resonance gas is indeed the most realistic proxy for the hadronic interaction strength near the phase transition [9]. As an alternative the authors proposed to either use a parametrization of the low energy van-der-Waals interactions or an excluded volume to describe the attractive and repulsive hadron-hadron interactions. These changes to the HRG approach seemingly extend the agreement between HRG and lattice QCD to higher temperatures, but a.) the agreement now reaches beyond the pseudo-critical temperature and b.) the rather unconstrained parameter base for the interactions and the excluded volume allows for rather large variations in the fit. Nevertheless the approach is intriguing and, in the future, could also take into account potential flavour dependent differences in the parameters.

4 Fits based on experimental studies of fluctuations and yields

Recent results published by the STAR collaboration [10] regarding the evolution of the chemical freeze-out temperature for the various energies of the RHIC beam energy scan showed that although a common freeze-out temperature can be found through fits to the measured yields of all particles, this temperature will be about 15-20 MeV lower, if only light flavour particles (and kaons) are included in the fit (see Fig. 35 in [10]). The kaon yield is included, but was shown to be rather insensitive to the freeze-out temperature [11]. Independently the work by Chatterjee et al. [12] showed that the fits to ALICE yields significantly improve if two separate chemical freeze-out temperatures are assumed for light and strange particles. This requirement for two temperatures seemingly disappears when the yields in small systems (pp and pPb collisions) are fitted [13]. Regarding the use of net-particle fluctuations rather than the yields, our studies that tried to determine net-electric charge and net-baryon number fluctuations, by using the net-charged particle and net-proton distributions as a proxy, found a chemical freeze-out temperature for these particles, based on a HRG fit, that was consistently 15-20 MeV below the expected value from a common fit to all measured particle yields, see Fig. 3 (left) [14]. The resulting temperatures are in agreement with lattice QCD calculations based on the related susceptibility ratios [15, 16]. One should note that all HRG fits are applied to the χ_2/χ_1 ratios for a particular particle species. We found that the error bars for the higher moments are still too large and that the higher moments might potentially have contributions from critical fluctuations at the RHIC beam energies, which will negatively impact the chemical freeze-out fits [14]. In our conclusions, we pointed out that both, the net-charged particles and the net-protons, are dominated by light quark particles (pions and protons, respectively), and that a fit to a strangeness proxy, i.e. the fluctuations of net-kaons, might shed light on the question whether the chemical freeze-out could indeed be flavour dependent. In Fig. 3 (right) I am presenting preliminary results of a HRG fit to the recently published net-kaon data from STAR [17]. The HRG model is the same used in Fig. 3 (left), in fact the baryo-chemical potential was adopted from the fit to net-protons and net-charges and only the temperature was left as a free parameters. As one can see the net-kaons tend towards higher freeze-out temperatures than the other net-particle distributions. This might be surprising because as mentioned before, the kaon yields are rather insensitive to the temperature in a statistical hadronization fit, but our group also showed that the higher order moments of the net-kaon distributions are more sensitive to the freeze-out parameters than the yields themselves [18]. In Fig. 3 (right) we also show that the final result is rather insensitive to the number of hadronic states included in the HRG calculation. Lattice calculations assuming a Boltzmann approach to the partial pressure in order to isolate the contribution of the kaons to the strangeness susceptibilities confirm this result [19].

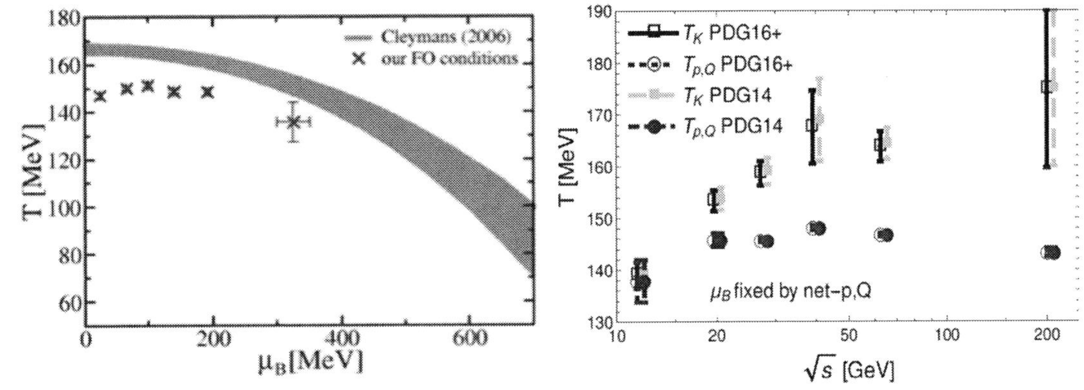

Figure 3. (left): Results from a combined HRG fit to the χ_2/χ_1 measurements for net-charges and net-protons from STAR (blue points) [14]. The extracted freeze-out parameters are compared to the curve by Cleymans et al. [20], which tried to parametrize all freeze-out results from statistical hadronization models to yields from SIS, SPS, RHIC and LHC. (right): Preliminary results from fitting the χ_2/χ_1 measurements for net-kaons from STAR with the same model used in Fig. 3 (left). The μ_B was fixed by the net-p,Q results. The figure also shows a comparison between results using different PDG lists in the HRG calculation.

One should note that the usage of net-kaons as a proxy for net-strangeness is less justified than the proxies for net-charge and net-baryon number [21]. Certainly the inclusion of fluctuation data for strange baryons is very important in order to gauge the relative contributions of flavour and baryon number to the final result. Studies of these fluctuations are underway in STAR and ALICE.

5 Conclusions and outlook

There is intriguing evidence that the flavour composition of the produced hadrons might play a role in their freeze-out parameters and thus in their hadronization dynamics. Particularly it seems that the quark mass plays a significant role in calculating the transition of flavour specific susceptibilities on the lattice as long as the mass is not negligible compared to the temperature of the equilibrated system. Experimentally this could mean that heavier quark particles prefer to freeze-out at a higher temperature. I have presented evidence to that effect based on yield and fluctuation measurements from STAR and ALICE. Studies of the baryon mass evolution in the crossover region, based on PNJL [22] and lattice calculations [23], also point at finite quark mass, and thus flavour, dependencies. Certainly more evidence is needed, in particular from strange baryon fluctuation measurements, but in theory this flavour dependence should also be measurable in the charm sector as long as the charm quarks thermalize with the system. A direct impact of a higher freeze-out temperature would be an enhancement of strange particles relative to the yield obtained at a common lower freeze-out temperature. Indirect evidence can be found in the ALICE data based on the latest Ξ yields in central PbPb collisions and on the comparison of PbPb yields to pp yields for strange particles. The latter point is often attributed to canonical suppression in the small system [24], but a comparison between preliminary RHIC-BES data [25] and ALICE data [26] leads us to believe that the energy dependence of the canonical suppression makes this effect almost negligible at LHC energies, see Fig. 4. Based on the data it seems that the strange anti-baryon yields follow the same trend from $\sqrt{(s)} = 62$ GeV on up. In addition, HRG fits, assuming a thermal system is produced in pp collisions, show that the temperature in the small systems is about 15 MeV lower than the strange particle freeze-out temperature in heavy ion collisions [12], which could explain at least part of the strangeness enhancement. Further studies are needed, though.

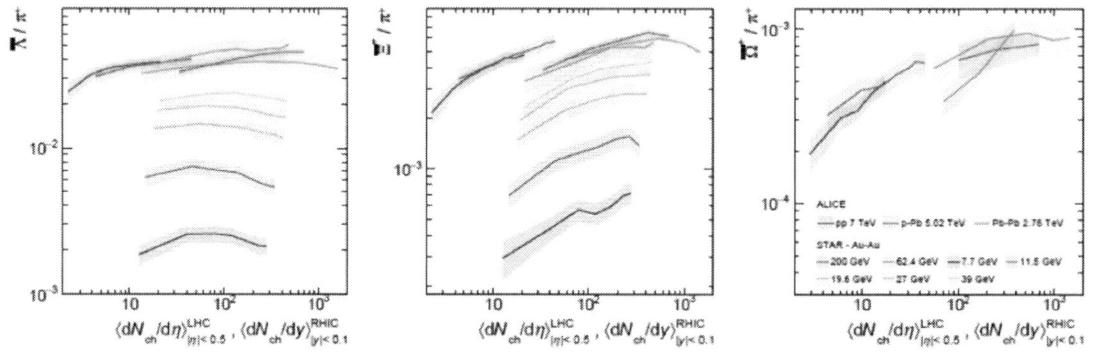

Figure 4. Anti-baryon over p- production as a function of collision energy and charged particle density (for LHC: $\langle dN_{ch}/d\eta\rangle$ in $|\eta| < 0.5$, for RHIC: $\langle dN_{ch}/dy\rangle$ in $|y| < 0.1$) based on STAR [25] and ALICE [26] data.

Regarding a more speculative aspect of particle production, this enhancement in strange quarks could lead to strangeness clustering, which might manifest itself in strange multi-quark states. The discovery of charmed tetra-and penta-quark states by LHCb certainly has triggered renewed interest in exotica searches in the strangeness sector. In terms of more dynamic quantities, a higher freeze-out temperature could potentially lead to a shortened partonic phase for strange baryons. This could reduce dynamic quantities, such as R_{AA} and v_2, as long as a significant contribution to the flow or the suppression is generated close to pseudo-critical temperature.

Acknowledgements

I would like to thank my collaborators for various contributions to this work: Paolo Alba, Livio Bianchi, Szabolcs Borsanyi, Zoltan Fodor, Jana Guenther, Anders, Knospe, Valentina Mantovani-Sarti, Jackie Noronha-Hostler, Paolo Parotto, Attila Pasztor, and Claudia Ratti. I also thank Kryzstof Redlich and Frithjof Karsch for some inspiring discussions. This work was supported by the U.S. Department of Energy under DE-FG02-07ER41521.

References

[1] C. Ratti, R. Bellwied, M. Cristoforetti and M. Barbaro, Phys. Rev. D **85**, 014004 (2012)
[2] R. Bellwied, S. Borsanyi, Z. Fodor, S. D. Katz and C. Ratti, Phys. Rev. Lett. **111**, 202302 (2013)
[3] F. Karsch, Central Eur. J. Phys. **10**, 1234 (2012)
[4] A. Bazavov et al, Phys. Rev. Lett. **113**, 072001 (2014)
[5] S. Capstick and N. Isgur, Phys. Rev. D **34**, 2809 (1986)
[6] C. Patrignani et al. (Particle Data Group), Chin. Phys. C **40**, 10001 (2016)
[7] C. Ratti, contribution to this conference (2017)
[8] P. Alba et al., Phys. Rev. D **96**, 034517 (2017)
[9] V. Vovchenko et al., Phys. Rev. Lett. **118**, 182301 (2017)
[10] L. Adamczyk et al., Phys. Rev. C **96**, 044904 (2017)
[11] D. Magestro, J. Phys. G **28**, 1745 (2002)
[12] S. Chatterjee, A.K. Dash and B. Mohanty, J. Phys. G **44**, 105106 (2017)
[13] S. Chatterjee, contribution to this conference (2017)
[14] P. Alba et al., Phys. Lett. B **738**, 305 (2014)
[15] S. Borsanyi et al., Phys. Rev. Lett. **113**, 052301 (2014)
[16] F. Karsch, Nucl. Phys. A **956**, 352 (2016)
[17] L. Adamczyk et al. (STAR Collaboration), arXiv:1709.00773
[18] P. Alba et al., Phys. Rev. C **92**, 064910 (2015)
[19] J. Noronha-Hostler et al., arXiv :1607.02527
[20] J. Cleymans, H. Oschler, K. Redlich, S. Wheaton, Phys. Rev. C **73**, 034905 (2006)
[21] C. Zhou, J. Xu, X. Luo, F. Liu, Phys. Rev. C **96**, 014909 (2017)
[22] J.M. Torres-Rincon, contribution to this conference (2017)
[23] G. Aarts, contribution to this conference (2017) and arXiv:1710.08294
[24] K. Redlich and A. Tounsi, Eur. Phys. J. C **24**, 589 (2002)
[25] X. Zhu for the STAR Collaboration, PoS (CPOD 2014), 008 (2015)
[26] J. Adam et al. (ALICE Collaboration), Nature Phys. **13**, 535 (2017)

EPJ Web of Conferences **171**, 03001 (2018) https://doi.org/10.1051/epjconf/201817103001
SQM 2017

Open Questions in the Understanding of Strangeness Production in HIC – Experiment Perspective

Christoph Blume[1,*]

[1] *Institut für Kernphysik, Goethe-Universität Frankfurt am Main, Max-von-Laue-Str. 1, 60438 Frankfurt am Main, Germany*

Abstract. Open questions concerning strangeness production in heavy-ion collisions are discussed with a focus on the experimental aspects. The open points are presented in the context of recent measurements.

1 Introduction

The following is an attempt to compile a list of the most important open topics concerning strangeness production in heavy-ion physics. This list is seen from an experimentalist point-of-view (the theory perspective is discussed in [1]) and is also naturally incomplete and biased. In order to not get lost in the many facets of strangeness physics, the questions will be limited to those belonging to four main subjects: energy dependence of strangeness enhancement in nucleus-nucleus collisions, understanding of small systems, strangeness production at low energies and hyperon-interaction & hypernuclei.

2 Energy dependence of strangeness enhancement

It has been established already quite a while ago that the production of strange particles is significantly enhanced in heavy-ion reactions relative to elementary proton-proton collisions (for a review see [2]). Usually, the enhancement factor E is defined as:

$$E = \frac{2}{N_{\text{part}}} \left(\left. \frac{dN(\text{AA})}{dy} \right|_{y=0} \right) \Bigg/ \left(\left. \frac{dN(\text{pp})}{dy} \right|_{y=0} \right) \tag{1}$$

It is a remarkable fact that E decreases significantly when going from SPS energies to the very high energies available at the LHC. E.g. for the Ω^- E is found to be around 20 at $\sqrt{s_{\text{NN}}}$ = 17.3 GeV [3], $\sim$ 12 at $\sqrt{s_{\text{NN}}}$ = 200 GeV [4] and only $\sim$ 6 at $\sqrt{s_{\text{NN}}}$ = 2.76 TeV [5], as illustrated in Fig. 1. While the enhancement factor has been measured for essentially all multi-strange (anti-)particles at these energies, data are still quite scarce for these rare particles at energies below $\sqrt{s_{\text{NN}}}$ = 17.3 GeV. An interesting exception is the measurement of Ξ^- production in Ar+KCl collisions at 1.76A GeV beam energy by the HADES collaboration (left panel of Fig. 2) [6]. HADES also has studied the production of ϕ mesons at these sub-threshold energies (right panel of Fig. 2) [7]. The data are in so far remarkable as also here an enhancement of the rare strange particles is observed. The $\Xi^-/(\Lambda + \Sigma^0)$-ratio is found to be much larger than the statistical model expectation and the ϕ/K^--ratio rises dramatically towards very low energies.

*e-mail: blume@ikf.uni-frankfurt.de

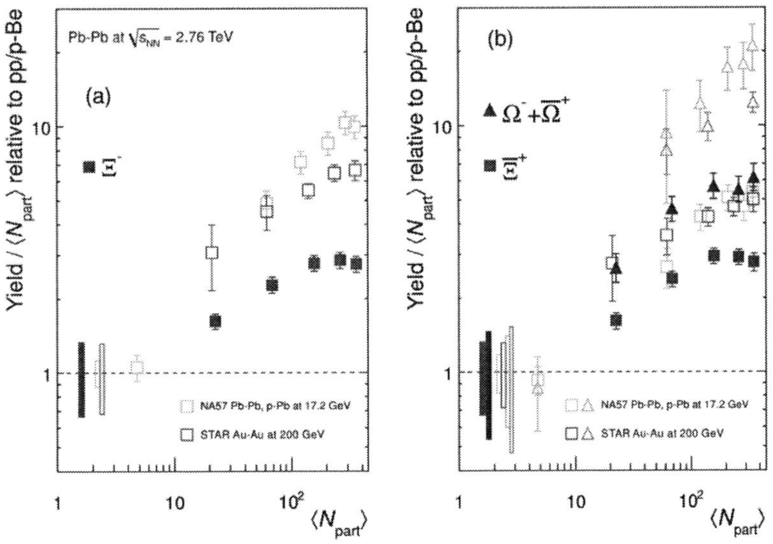

Figure 1. The enhancement factors for multi-strange (anti-)particles as a function of the number of participants $\langle N_{\text{part}} \rangle$ as measured at the SPS [3], RHIC [4] and LHC [5].

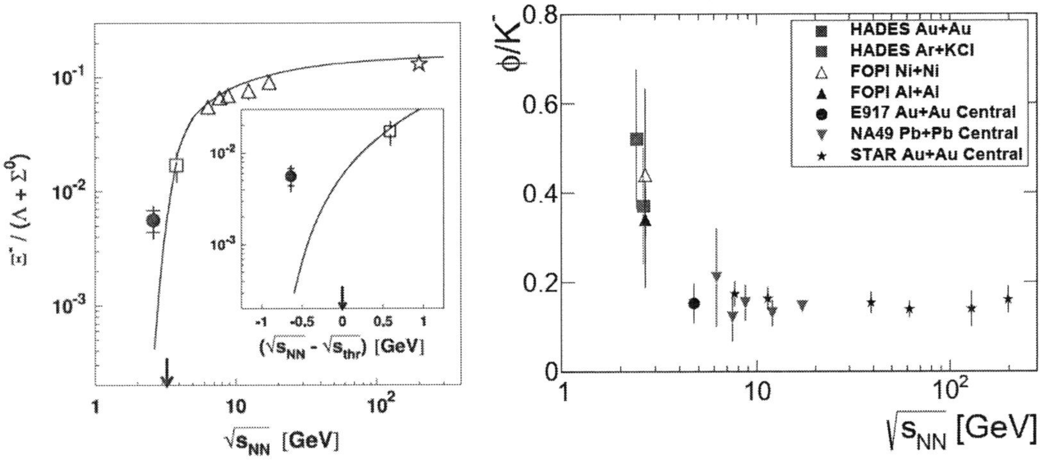

Figure 2. The $\Xi^-/(\Lambda + \Sigma^0)$-ratio (left panel, [6]) and the ϕ/K^--ratio (right panel, [7]) as a function of the centre-of-mass energy as measured in central heavy-ion collisions.

Taking these observations together, one finds that the strangeness enhancement, in particular for rare multi-strange (anti-)particles, exhibits a quite complicated energy dependence. As sketched in Fig. 3 there is evidence for a strong increase at sub-threshold energies and it has been established that there is a strong enhancement at intermediate (i.e. SPS energies) energies, which slowly decreases again towards high energies (i.e. RHIC and LHC). It remains an open question if there is any evidence for an onset of strangeness enhancement between low and intermediate energies. This should happen in a region where sub-threshold phenomena (dashed line) do not play a role any more and the partonic

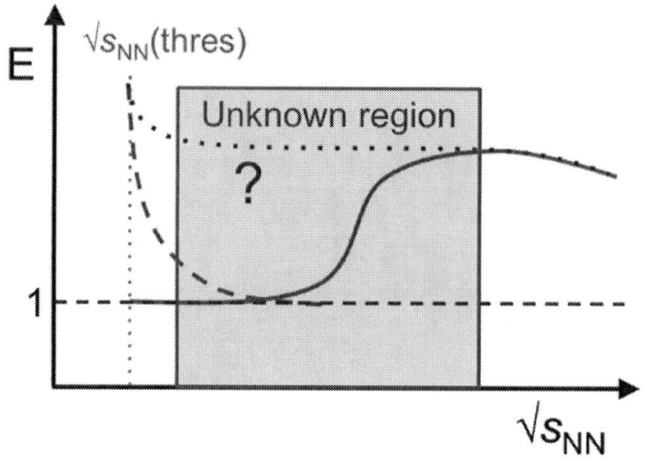

Figure 3. Sketch of the energy dependence of the enhancement factor of rare multi-strange (anti-)particles.

degrees of freedom (solid line) are becoming more and more effective. It might also very well be that there is no discernible onset visible, due to additional hadronic medium effects (dotted line) (e.g. multi-step hadronic reactions, resonances, multi-meson fusion processes, etc.). However, it is obvious that high quality data of rare particles in the energy range below $\sqrt{s_{NN}} \sim 10$ GeV would be highly relevant (see also [8]) in order to answer the following questions:

- What is the energy dependence of strangeness enhancement over the whole energy region, in particular for multi-strange (anti-)particles?
- Could there be an onset somewhere?
- To what extend can hadronic effects cause a strangeness enhancement at intermediate energies (SPS and below)?
- Do we understand the dramatic effects at sub-threshold energies?
- Or, in other words, can finally a direct connection between strangeness enhancement and QGP formation be established?

3 Understanding of small systems

Small systems play a key role for the understanding of strangeness production in heavy-ion collisions. The decrease of the strangeness enhancement factor between SPS and LHC energies, as discussed in the previous section, is actually not caused by a decrease of the yields in heavy-ion collisions. In fact, they are already close to the statistical model expectation at top SPS-energies and increase only slightly towards higher energies. However, the yields of multi-strange particles increases much stronger in pp collisions than in AA (see left panel of Fig. 4). For instance, one finds that the Ξ/π ratios are almost the same at RHIC and LHC in central AA collisions, while this ratios increase significantly in pp collisions. Thus, the decrease of the strangeness enhancement is due to a release of the strangeness suppression in pp with increasing energies.

This phenomena can also be investigated by comparing pp collisions of different reaction violence at high energies. This is done by selecting pp event classes of different charged particle multiplicity

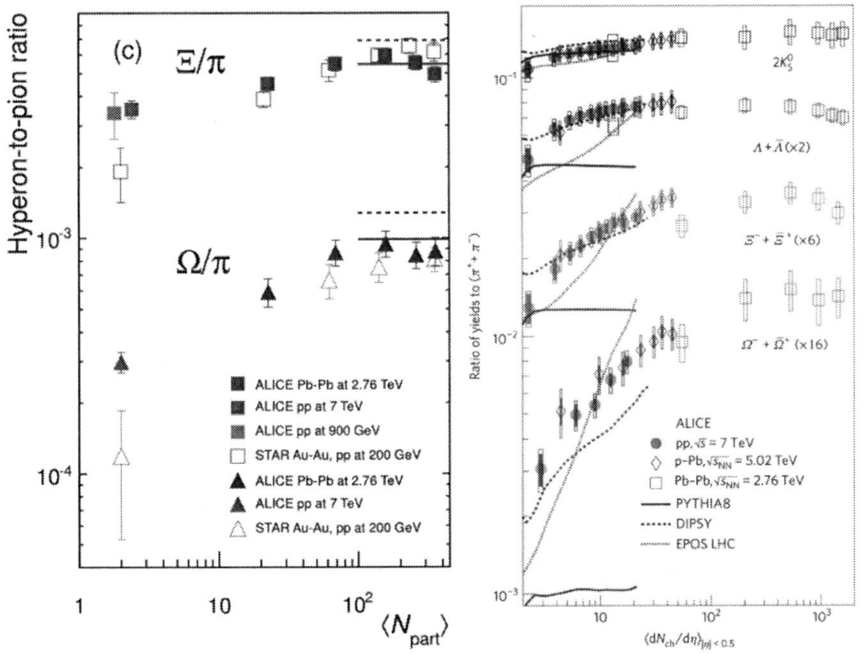

Figure 4. Left: the Ξ^-/π and Ω/π ratios in pp and AA collisions as a function of $\langle N_{\text{part}} \rangle$ [5]. The lines correspond to statistical model predictions (solid line [9], dashed line [10]). Right: the yield ratios of strange particles and pions measured in pp, pA and AA collisions at the LHC as a function of $\mathrm{d}N_{\text{ch}}/\mathrm{d}\eta$ [11].

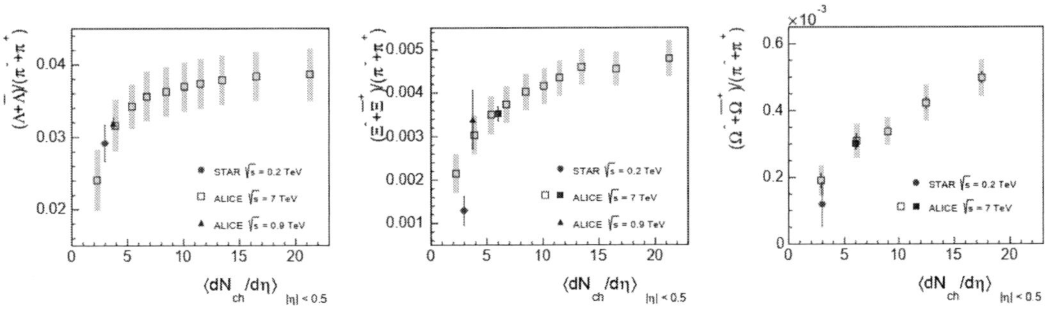

Figure 5. The yield ratios of strange particles to pions measured in multiplicity selected pp collisions at $\sqrt{s}$ = 7 TeV [11] and in minimum bias pp collisions measured at various energies by ALICE [12, 13] and STAR [14, 15] as a function of $\mathrm{d}N_{\text{ch}}/\mathrm{d}\eta$.

$\mathrm{d}N_{\text{ch}}/\mathrm{d}\eta$. Such a study has recently be performed by the ALICE collaboration [11] (see right panel of Fig. 5). Here it was found that the yield ratios of strange particles to pions observed in very high multiplicity pp collisions are on the same level as the ones measured in peripheral heavy-ion reactions. Also, the multiplicity dependence of these ratios turns out to be very similar in pp and pA collisions. This is not at all trivial since the physics in a high multiplicity pp collision is quite different from the one in pA at the same multiplicity. While in the first case a very rare and violent pp interaction has to be involved, in the latter case the same multiplicity can be achieved by the much more likely superposition of several soft pp collisions.

In the statistical model approach strangeness enhancement is described by the transition from a canonical to a grand-canonical ensemble, which depends on the volume V_0 of the system. In this picture a strangeness hierarchy is expected [16], i.e. the volume dependence gets stronger with increasing strangeness content. Since such a hierarchy was observed in the multiplicity dependence [11], it is natural to interpret the pp data in this way by assuming a relation $V_0 \propto \langle dN_{ch}/d\eta \rangle$. In fact, as reasonable description of all particle ratios, with the notable exception of the ϕ/π-ratio (other peculiarities related to the ϕ-meson are discussed in Sect. 4), can be achieved within this model [17].

If this interpretation holds, the particle ratios should saturate for pp collisions at very high multiplicities when the grand-canonical limit is reached. As shown in Fig. 5, this might indeed already be the case for the Λ/π and Ξ/π ratios, while for the Ω/π it is rather still a continuous increase. Also, it would be interesting to establish whether $dN_{ch}/d\eta$ does provide an universal scaling variable for all energies. Fig. 5 shows a comparison of the multiplicity selected pp collisions at $\sqrt{s_{NN}}$ = 2.76 TeV to minimum bias pp reactions at lower energies. The minimum bias data points roughly follow the trend, however, it would be worthwhile to examine the multiplicity dependence also at low energies with good accuracy, in order to test whether really an universal scaling holds. In summary, the following questions would require further investigations:

- Does $dN_{ch}/d\eta$ provide an universal scaling for system size dependencies (pp $\rightarrow$ pA $\rightarrow$ AA)?
- Is the relation of $dN_{ch}/d\eta$ to the reaction volume the only relevant factor (look at other observables)?
- Does the multiplicity dependence match the transition from a canonical to a grand-canonical ensemble at all energies?
- Is a saturation of particle ratios observed in very high multiplicity pp collisions?

4 Low energies and the ϕ-meson

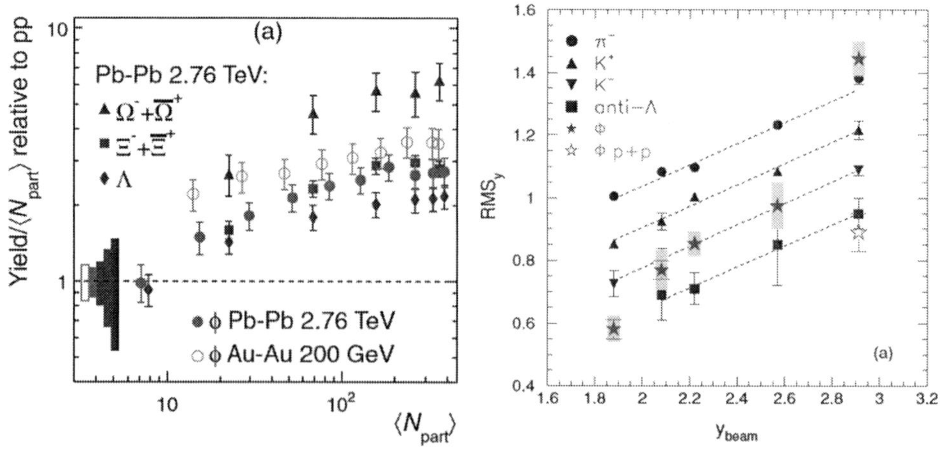

Figure 6. Left: the yield ratios of strange particles to pions measured in pp and AA collisions at different centre-of-mass energies as a function of $\langle N_{part} \rangle$ [18]. Right: the widths of the rapidity distributions of the ϕ-meson measured at the SPS as a function of the beam rapidity y_{beam} [19].

Also at the lower end of the energy spectrum many unresolved questions are remaining. These are particularly important, since they are concerned with the behaviour of strange particles in a hadronic medium, which need to be understood before any conclusions about the partonic phases of heavy-ion collisions can be made.

A topic discussed since quite a while is the propagation of kaons in the medium. In low energy AA collisions a significant difference between the inverse slope parameters of the p_t spectra of K^- and K^+ has been observed ($T(K^+) > T(K^-)$). Since the cross section of reactions with nucleons are different ($\sigma(K^-) > \sigma(K^+)$) due to strangeness exchange reactions of the K^-, the rescattering with the hadronic medium should cause different freeze-out conditions for the two kaon species and thus can explain the different slope parameters [20]. Also, the kaon-nucleon potential, which is expected to be attractive for K^- and repulsive for K^+, can modify the inverse slope parameters. On the other side, the unexpectedly high ϕ/K^--ratio at low energies (see right panel of Fig. 2) can also provide a natural explanation for this observation. Since the feed-down from ϕ-decays at these energies is a substantial contribution to the kaon spectra, it will also modify their shape. This affects more strongly the rarer K^- than K^+ and thus will result in the different spectral shapes [21].

With its $s\bar{s}$ valence quark structure the ϕ-meson is effectively a strangeness neutral particle ($S = 0$). Nevertheless, it behaves in many ways as if it would have a non-zero strangeness. The enhancement factor for the ϕ is found to be between the ones for the Λ and Ξ (see left panel of Fig. 6), thus rather behaving as a particle with an effective strangeness in the range $1 - 2$ [18, 22]. Another so far unexplained observation related to the ϕ is the energy dependence of the widths of its rapidity distributions (right panel of Fig. 6). In heavy-ion collisions, it broadens much stronger than for π, K^- and K^+, which is difficult to reconcile with kaon-coalescence being the main production mechanism for ϕ-mesons [19]. Some important points to be clarified concerning heavy-ion collisions at low energies and the ϕ-mesons are therefore:

- Do we understand the production and propagation of strangeness at low energies?
- Is there any evidence for a sequential freeze-out due to different cross sections?
- Does the medium also at low energy behave macroscopically and can fully be described by the statistical model?
- Why does a non-strange particle behave so strange?

5 Hyperon interaction and hypernuclei

The investigation of hyperon-interactions is a crucial ingredient for the theoretical description of neutron stars. It may in particular be relevant for the understanding of high mass neutron stars with $M > 2\ M_{\odot}$. One way to obtain informations is via two-particle correlations. The STAR collaboration recently managed to extract a $\Lambda\Lambda$-correlation function in heavy-ion collisions [23] (see left panel of Fig. 8). As the strong interaction between the Λ-pairs causes a deviation of the correlation function from the quantum-statistical expectation of $C_{\Lambda\Lambda}(Q = 0) = 0.5$, one can infer information on the scattering lengths and effective interaction ranges by comparing it to corresponding models. Using the one by Lednický and Lyuboshitz [24], a weak repulsive interaction was inferred by the STAR collaboration [23]. However, an alternative analysis [25] rather favours a weak attractive interaction.

Another important source of information on this subject are hypernuclei and more and more data is becoming available. For instance, with the measurement of anti-hypertritons the STAR collaboration achieved the first observation of an anti-hypernucleus [26]. Generally, it is found that the (anti-)hypertriton yields agree very well with statistical model expectations [27]. Since their binding energy is very small, it is surprising that their yields are fixed in a chemical freeze-out environment

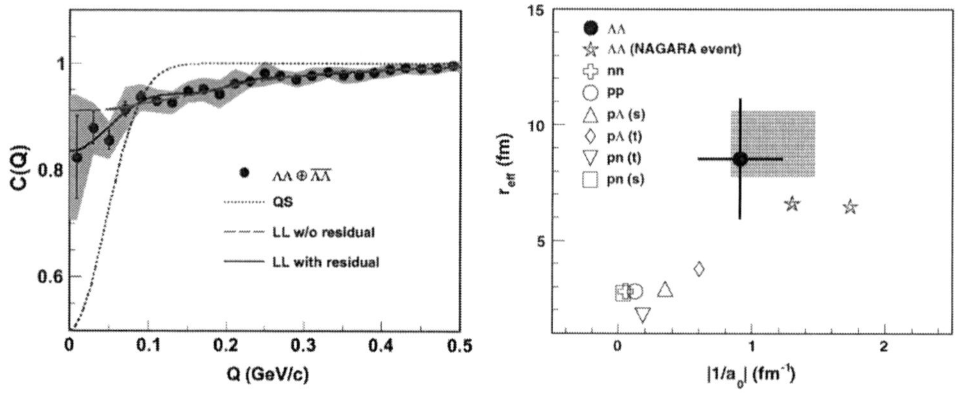

Figure 7. Left: the ΛΛ-correlation function measured in Au+Au collisions at RHIC [23]. Right: the S-wave scattering length a_0 and effective interaction range r_{eff}, as extracted from this correlation function with the model by Lednický and Lyuboshitz [24].

with a temperature higher by about two orders of magnitude. Current measurements of the Λ-lifetime from the decay of ${}^{3}_{\Lambda}$H seem to indicate that it is slightly lower than the one of a free Λ [28], which would be an indication for a modification of hyperon properties inside a nuclear medium.

The study of hyperon-interactions and their properties in nuclei has regained quite some momentum recently with the measurements at RHIC and LHC. High statistics data, in particular on double-hypernuclei, which will become available in the near future with facilities such as FAIR, will allow to address the questions listed below with much more precise information:

- What do we really know about hyperon-hyperon interactions?
- What is the possible contribution to the understanding of large-mass neutron stars?
- Why are the yields of very weakly bound objects (e.g. ${}^{3}_{\Lambda}$H) so well described by the statistical model ("snowball in hell")?
- Are the properties of hyperons modified inside nuclei?

6 Acknowledgements

The author would like to acknowledge many helpful discussions with B. Dönigus, D. Cebra, C.-M. Ko, M. Lorenz, H. Ströbele and many others.

References

[1] C.M. Ko, *Open Questions in the Understanding of Strangeness Production in HIC – Theory Perspective*, these proceedings
[2] C. Blume, C. Markert, Prog. Part. Nucl. Phys. **66**, 834 (2011), 1105.2798
[3] F. Antinori et al. (NA57), J. Phys. **G32**, 427 (2006), nucl-ex/0601021
[4] B.I. Abelev et al. (STAR), Phys. Rev. **C77**, 044908 (2008), 0705.2511
[5] B.B. Abelev et al. (ALICE), Phys. Lett. **B728**, 216 (2014), [Erratum: Phys. Lett.B734,409(2014)], 1307.5543

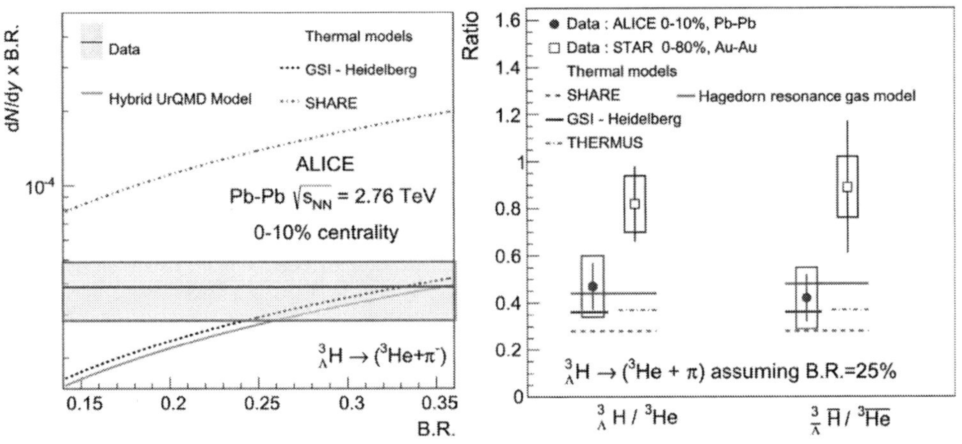

Figure 8. Left: the $^3_\Lambda$H yield as measured at the LHC in comparison to several model predictions [27]. Right: the $^3_\Lambda$H/^{3}He and $^3_{\bar\Lambda}\overline{\text{H}}/^3\overline{\text{He}}$ ratios measured at RHIC and LHC in comparison to statistical model predictions [27].

[6] G. Agakishiev et al. (HADES), Phys. Rev. Lett. **103**, 132301 (2009), 0907.3582
[7] J. Adamczewski-Musch et al. (HADES) (2017), 1703.08418
[8] D. Cebra, *Open Questions in the Understanding of Strangeness Production in HIC – Future Directions*, these proceedings
[9] A. Andronic, P. Braun-Munzinger, J. Stachel, Phys. Lett. **B673**, 142 (2009), [Erratum: Phys. Lett.B678,516(2009)], 0812.1186
[10] J. Cleymans, I. Kraus, H. Oeschler, K. Redlich, S. Wheaton, Phys. Rev. **C74**, 034903 (2006), hep-ph/0604237
[11] J. Adam et al. (ALICE), Nature Phys. **13**, 535 (2017), 1606.07424
[12] B. Abelev et al. (ALICE), Phys. Lett. **B712**, 309 (2012), 1204.0282
[13] K. Aamodt et al. (ALICE), Eur. Phys. J. **C71**, 1594 (2011), 1012.3257
[14] B.I. Abelev et al. (STAR), Phys. Rev. **C75**, 064901 (2007), nucl-ex/0607033
[15] B.I. Abelev et al. (STAR), Phys. Rev. **C79**, 034909 (2009), 0808.2041
[16] S. Hamieh, K. Redlich, A. Tounsi, Phys. Lett. **B486**, 61 (2000), hep-ph/0006024
[17] V. Vislavicius, A. Kalweit (2016), 1610.03001
[18] B.B. Abelev et al. (ALICE), Phys. Rev. **C91**, 024609 (2015), 1404.0495
[19] C. Alt et al. (NA49), Phys. Rev. **C78**, 044907 (2008), 0806.1937
[20] C. Hartnack, H. Oeschler, Y. Leifels, E.L. Bratkovskaya, J. Aichelin, Phys. Rept. **510**, 119 (2012), 1106.2083
[21] H. Schuldes, *Recent results from HADES*, these proceedings
[22] B.I. Abelev et al. (STAR), Phys. Lett. **B673**, 183 (2009), 0810.4979
[23] L. Adamczyk et al. (STAR), Phys. Rev. Lett. **114**, 022301 (2015), 1408.4360
[24] R. Lednicky, V.L. Lyuboshits, Sov. J. Nucl. Phys. **35**, 770 (1982), [Yad. Fiz.35,1316(1981)]
[25] K. Morita, T. Furumoto, A. Ohnishi, Phys. Rev. **C91**, 024916 (2015), 1408.6682
[26] B.I. Abelev et al. (STAR), Science **328**, 58 (2010), 1003.2030
[27] J. Adam et al. (ALICE), Phys. Lett. **B754**, 360 (2016), 1506.08453
[28] L. Adamczyk et al. (STAR) (2017), 1710.00436

Theoretical perspective on strangeness production

Che Ming Ko[1,*]

[1] *Cyclotron Institute and Department of Physics, Texas A&M University, College Station, Texas 77843-3366, USA*

Abstract. A brief review of some highlights and puzzles on strangeness production in heavy ion collisions is given. These include strangeness production and the nuclear equation of state; deeply subthreshold strangeness production; mean-field potentials on strange hadrons; phi meson in dense matter; anomalous strange hadron to pion ratios; density fluctuations on particle production; Λ hyperon polarization and the vorticity field, and exotic hadrons.

1 Introduction

Strangeness production in heavy ion collisions has been a topic of continuous interest since the first experiment in 1982 at Bevalac to study kaon production in various collision systems at 2.1 AGeV [1]. The motivation at that time was to use kaon as a probe of the dense matter produced in high energy heavy ion collisions because it has a much smaller scattering cross section with nucleon than the pion. Based on a simple row-on-row cascade model, the kaon yield was calculated in Ref. [2] using as input the differential cross sections for kaon production in elementary baryon-baryon collsions estimated in a simple model, and the predicted result was later found to be comparable to that measured in Ref. [1]. This was in contrast with the result from a thermal model study [3], which assumes that kaons are produced from a thermally and chemically equilibrated fireball. In this study, the predicted kaon yield was about a factor of 40 larger than that from Ref. [2], although the thermal model could describe the measured pion yield. The failure of the thermal model used in Ref. [3] was later known to be due to the overlook of exact strangeness conservation in its grand canonical treatment of particle distributions [4]. Including strangeness conservation through the canonical treatment, which is naturally included in the cascade model study of Ref. [2] through the pair production of kaon and lambda, would give a significantly smaller number of strange hadrons if their mean number is small, called the canonical suppression. It was later shown in Ref. [5] via the kinetic theory for kaon production in heavy ion collisions at the Bevalac energy that taking into account strangeness conservation would lead to the same equilibrium number as that from the canonical formulation of the conservation law in the thermal model.

With increasing collision energy than that available at Bevalac, a quark-gluon plasma (QGP) would be produced during the early stage of heavy ion collisions. Since the mass

*e-mail: ko@comp.tamu.edu

of strange quark is below the temperature of the QGP, strange quarks are expected to be abundantly produced in QGP and reach chemical equilibrium with light quarks as shown in Ref. [6]. In heavy ion collisions, these strange quarks will be converted to strange hadrons, leading to an enhancement of their production than the case without the formation of a QGP. Enhanced production of strange hadrons, such as the phi meson [7], in heavy ion collisions has thus been considered as a signal for the formation of the QGP.

Because of the unique feature of strangeness production in heavy ion collisions, strange hadron production has since been studied at SIS, AGS, SPS, RHIC and LHC. The present talk will review some highlights from these studies and discuss a few puzzles that remain to be explained. These include strangeness production and the nuclear equation of state, deeply subthreshold strangeness production and baryon resonances, mean-field potentials on strange hadrons, phi meson in dense matter, anomalous strange hadron to pion ratios, density fluctuations on particle production, Λ hyperon polarization and the vorticity field, and exotic hadrons.

2 Strangeness production and the nuclear equation of state

Determining the nuclear equation of state, which is important for understanding the properties of neutron stars, has been one of the main motivations for studying heavy ion collisions, in which nuclear density of many times of that in normal nuclei can be reached. Besides the collective flow of charged particles [8], the production of kaon in heavy ion collisions at energy below its production threshold in a nucleon-nucleon collision, which is 1.56 GeV, was shown in Ref. [9] to be sensitive to the stiffness of nuclear equation of state at high density. For central Nb + Nb collisions at 700 AMeV studied in Ref. [9], the kaon production probability from a soft nuclear equation of state with the compressibility of 200 MeV was a factor of 2.5 larger than that from a stiff equation of state with the compressibility of 380 MeV. In both cases, the kaon is mostly produced at the compressional stage of the collision when the density is about three times that of normal nuclear matter. Comparing results from the transport model study with the experimental data from SIS for Au + Au collisions has led to the conclusion that the nuclear equation of state is soft, similar to the conclusion from the study of the collective flow of charged particles [8].

Subthreshold strange hadron production is also sensitive to the density dependence of the nuclear symmetry energy, which is the energy needed per nucleon to convert a symmetric nuclear matter to a pure neutron matter and is essentially unknown at densities above the normal nuclear matter density [10]. Studies based on isospin-dependent transport models have shown that besides the π^-/π^- ratio [11], the K^+/K^0 [12, 13] and the Σ^-/Σ^+ [14] ratio in heavy ion collisions with neutron-rich nuclei depend on the stiffness of the nuclear symmetry energy at high density, with the soft one giving larger values than the stiff one. Measuring these ratios of strange hadrons thus provides a potential probe to the nuclear symmetry energy at high density.

3 Deeply subthreshold strangeness production and baryon resonances

Strange hadron production in heavy ion collisions at subthreshold energies further provides the possibility to find the signals for the missing baryon resonances predicted by the quark model and the decay channels of known resonances that are not yet observed in experiments. A recent study of Ξ^- production using the UrQMD model [15] for Ar + KCl collisions at

https://doi.org/10.1051/epjconf/201817103002

1.76 AGeV at SIS, which is substantially below the threshold energy of 3.74 GeV in the nucleon-nucleon collision, has shown that to describe the experimental data from the HADES Collaboration [16] requires contributions not only from the strange-exchange reactions $\bar{K}Y \to \pi\Xi$ [17] and $YY \to N\Xi$ [18] but also from decays of $N^*(1990)$, $N^*(2080)$, $N^*(2190)$, $N^*(2220)$, and $N^*(2250)$ to ΞKK with a branching ratio of 10%, which is obtained from fitting the experimental data on Ξ^- production in nucleon-nucleon collisions. This is very different from Ξ^- production in Au + Au collisions at 6 AGeV at AGS, which is much above the threshold energy. In this case, the experimental data can be satisfactorily described by the transport model that includes only the reaction $\bar{K}Y \to \pi\Xi$ [19]. Including contributions from the decay of above baryon resonances to $N\phi$ with a branching ratio of 0.2%, which is also determined from fitting ϕ meson production in nucleon-nucleon collisions, can also improve the description of the UrQMD for ϕ meson production in Ar + KCl collisions at 1.76 AGeV at SIS [15].

4 Phi meson in nuclear medium

The ϕ meson is detected in experiments through its decay either to $K\bar{K}$ or e^+e^- ($\mu^+\mu^-$). While the former allows the determination of the number of phi mesons at the kinetic freeze-out in a heavy ion collision, the latter can further provide information on the phi meson number during the expansion of the hot dense hadronic matter [20, 21]. Both earlier [22, 23] and recent [24] calculations have indicated that the phi meson absorption cross sections by the nucleon, pion and kaon are appreciable, which is consistent with the experimental results that phi meson production from nuclei is suppressed due to absorption in the nuclear medium [25–27]. Since experiments from both SPS [28] and RHIC [29] have shown that the phi meson yields determined from $K\bar{K}$ and e^+e^- ($\mu^+\mu^-$) channel are similar, the phi meson number thus does not vary much during the hadronic evolution. The appreciable absorption and regeneration of phi mesons in the hadronic stage then indicates that they flow with the matter. The observation that the elliptic flow of phi mesons at RHIC follows approximately the number-of-constituent quark scaling (NCQ) by the STAR Collaboration [30] is thus consistent with the picture that the elliptic flow in heavy ion collisions at RHIC is largely produced during the partonic phase [31]. The observed violation of NCQ for the phi meson flow in heavy ion collisions at LHC by the ALICE Collaboration [32] is then the result of their scattering during the longer hadronic phase than at RHIC [33].

Although phi meson production in heavy ion collisions has been generally understood, recent experimental data from Pb + Pb collisions at SPS show an anomalous energy dependence of the width of its rapidity distribution compared to those of other hadrons [34]. The values of the root-mean-squared rapidity width of π^-, K^+, K^-, and Λ at a given collision energy decrease with increasing hadron mass, and all increase linearly with the collision energy. Both the larger width for lighter hadrons and the increase of width with collision energy are expected. The larger K^+ than K^- rapidity width may be due to their respective repulsive and attractive potentials in baryon-rich matter produced at these collision energies. The anomalous behavior of the phi meson rapidity width, which is similar to that of Λ at low collision energy but that of π^- at high collision energy, is at present not understood.

5 Mean-field potentials on strange hadrons

Effects of mean-field potentials on strange hadrons have been well studied in heavy ion collisions at SIS energies [35, 36]. In particular, the kaon and antikaon potentials at normal nuclear matter density are about 20 MeV and −120 MeV, respectively, based on the chiral effective Lagrangian that fits the empirical data on kaon- and antikaon-nucleus scattering [35]. It was shown in Ref. [37] that the repulsive kaon potential would lead to a smaller direct flow of kaons relative to that of nucleons. This effect was indeed seen in experiments at both SIS [38] and AGS [39]. The kaon direct flow almost vanishes in collisions at SIS energy and even becomes negative in collisions at AGS due to the much higher baryon density reached at higher collision energy. The repulsive kaon potential may provide an explanation for the decreasing net kaon directed flow with deceasing collision energy observed in the preliminary data from the STAR Collaboration in the beam energy scan (BES) program at RHIC. The kaon and anitkaon mean-field potentials have also been shown to contribute to the splitting between the kaon and antikaon elliptic flows in Au + Au collisions [40, 41], which is found by the STAR Collaboration to be appreciable at $\sqrt{s_{NN}} = 7.7$ GeV and decreases with increasing collision energy [42].

6 Anomalous strange hadron to pion ratios

One anomalous structure in the collision energy dependence of the strange hadron to pion ratios, such as the K^+/π^+ and $(\Lambda + \Sigma^0)/\pi^+$, measured from heavy ion experiments at AGS, SPS and RHIC is the peak at $\sqrt{s_{NN}} \approx 8$ GeV, called the horn [43]. In the statistical model description of heavy ion collisions, this has been attributed to the onset of QGP formation [43]. A recent study based on the PHSD model [44] has shown, however, that it is due to the restoration of the chiral symmetry, which leads to an enhanced production of strange quark pairs than light quark pairs from the string fragmentation during the initial stage of the collisions. It is assumed in this study that the chiral symmetry restoration proceeds before the deconfinement transition in baryon-rich matter. Such a scenario has been suggested in Ref. [45] from arguments based on the large N_C limit of QCD. In the absence of lattice QCD calculations at finite baryon chemical potential, it is not known if nature behaves this way.

7 Density fluctuations on strange hadron production

The search for the critical endpoint (CEP), where the first-order phase transition from the QGP to the hadronic matter ends, of the QCD phase diagram in heavy ion collisions is currently of great interest. Studies based on both the hydrodynamic approach [46] and the microscopic transport model [47] have shown that the spinodal instability during the first-order phase transition between the QGP and hadronic matter can induce large baryon density fluctuations. This large density fluctuation can be probed by studying ratios of particle yields. Similar to the derivation in Ref. [48] based on the coalescence model for light nuclei production, which shows that the yield ratio $\frac{N_{^3\mathrm{H}}N_p}{N_d^2}$ in heavy-ion collisions is sensitive to the neutron relative density fluctuation $\Delta n = \langle(\delta n)^2\rangle/\langle n\rangle^2$ at kinetic freeze-out, the yield ratio $\frac{\Xi^- K^+}{\Lambda\phi}$ is sensitive to the s quark relative density fluctuation $\Delta s = \langle(\delta s)^2\rangle/\langle s\rangle^2$ at the QGP to hadronic matter phase transition if they are produced through quark coalescence. It is known from the success of the statistical model in describing the yield ratios of hadrons that chemical freeze-out in heavy ion collisions occurs at the phase transition temperature, which has recently been shown to be related to the constancy of the entropy per particle during evolution from the chemical to the kinetic freeze-out [49]. Therefore, the yield ratios of hadrons are not expected to be modified during the hadronic evolution, and studying their dependence on collisions energy could provide a unique probe to the fluctuations of quark densities during the first-order phase transition from the QGP to the hadronic matter, which would help locate the CEP in the QCD phase diagram.

8 Λ hyperon polarization and the vorticity field

Recently, the STAR Collaboration has observed that the Λ hyperon produced in non-central relativistic heavy ion collisions is partially polarized along the direction perpendicular to the reaction plane [50]. This confirms an earlier suggestion in Ref. [51] that due to their spin-orbit interactions, quarks and antiquarks in the produced QGP can be polarized by the large orbital angular momentum in non-central heavy ion collisions, and they then lead to the production of polarized Λ hyperons after hadronization. In more recent studies based on the hydrodynamic [52–55] and transport [56] model, the measured spin polarization of Λ hyperons can be understood by assuming that Λ hyperons are in thermal equilibrium in the final rotating fireball. Also, based on the chiral kinetic approach using initial conditions from a multiphase transport model [57], the spin polarizations of quarks and antiquarks in non-central heavy ion collisions have been studied in Ref. [58]. Because of the non-vanishing vorticity field due to the large orbital angular momentum in these collisions, quarks and antiquarks are found to acquire appreciable spin polarizations in the direction perpendicular to the reaction plane. Converting quarks and antiquarks to hadrons via the coalescence model, the resulting spin polarizations of Λ and $\bar{\Lambda}$ hyperons, which are the same in this study, are comparable to those measured in experiments by the STAR Collaboration. To explain the seemingly larger $\bar{\Lambda}$ than Λ polarization seen in the experimental data requires, however, additional mechanisms such as the magnetic field. Also, the effect of hadronic evolution on Λ and $\bar{\Lambda}$ polarizations is not included in this study, which may also contribute to the difference in the Λ and $\bar{\Lambda}$ polarizations because of their different dynamics in baryon-rich matter.

9 Exotic strange hadrons

Because of the abundant number of strange quarks present in the produced QGP, relativistic heavy ion collisions provides the opportunity to study exotic strange hadrons [59–61]. For the most known H dibaryon ($uuddss$) [62], which has the same quark structure as two Λ hyperons, analyses of measured $\Lambda\Lambda$ correlation functions in heavy ion collisions at RHIC seem to indicate that the attractive interaction between two Λ is not strong enough to lead to a bound state [63, 64]. This conclusion is consistent with results from the lattice QCD calculations. Although lattice calculations show that the H dibaryon is bound in the massive pion cases [65], it evolves to a resonance near the ΞN threshold when the baryon-baryon potentials obtained from the lattice QCD calculations with a smaller pion mass are used in calculations based on the coupled-channel Schrödinger equation [66]. On the other hand, various theoretical models have indicated that $\Xi^-\Xi^0$ might be bound [67]. Even the diomega $\Omega\Omega$ has been shown in the chiral quark model to be bound [68], which seems to be supported by the strong attractive $\Omega\Omega$ potential from the latest lattice QCD calculation [69]. Experimental measurements of $\Xi^-\Xi^0$ and $\Omega\Omega$ correlations, as well as others such as the $p\Omega$ [70, 71] and $p\Xi^-$ correlations [72], will be of great interest to confirm these predictions

10 Summary

Since the first experiment at Bevalac on kaon production from heavy ion collisions and the theoretical study using a simple cascade model, a lot of progress has been made in the study of strangeness production in heavy ion collisions from energies below the kaon production threshold in nucleon-nucleon collisions to ultrarelativistic energies at the LHC. These studies have led to a better understanding of the nuclear equation of state at high density and the in-medium properties of strange hadrons. They also provide the possible evidence for some missing baryons resonances and their decay channels, the restoration of chiral symmetry in dense nuclear matter, and the large vorticity field in the produced QGP. The study of strange particle production in heavy ion collisions further provides the possibility to probe the QCD phase diagram through the effect of quark density fluctuations on the hadron yield ratios. However, there remain many puzzles such as the abnormal behavior in the collision energy dependence of the phi meson rapidity width, the relation between the chiral and deconfinement transition of baryon-rich QGP, and the origin of different Λ and $\bar{\Lambda}$ polarizations. Many more progresses and puzzles, which are not discussed here, have made the study of strangeness production in heavy ion collisions one of the most fascinating topics in high energy nuclear physics. More surprises are expected in future experiments.

ACKNOWLEDGEMENTS

The author would like to thank Christoph Blume, Daniel Cebra, André Mischke and Herbert Stroebele for helpful communications and suggestions. This work was supported in part by the US Department of Energy under Contract No. DE-SC0015266 and the Welch Foundation under Grant No. A-1358.

References

[1] S. Schnetzer, M.C. Lemaire, R. Lombard, E. Moller, S. Nagamiya, G. Shapiro, H. Steiner, I. Tanihata, Phys. Rev. Lett. **49**, 989 (1982)
[2] J. Randrup, C.M. Ko, Nucl. Phys. **A343**, 519 (1980), [Addendum: Nucl. Phys.A411,537(1983)]
[3] F. Asai, H. Sato, M. Sano, Phys. Lett. **98B**, 19 (1981)
[4] R. Hagedorn, K. Redlich, Z. Phys. **C27**, 541 (1985)
[5] C.M. Ko, V. Koch, Z.w. Lin, K. Redlich, M.A. Stephanov, X.N. Wang, Phys. Rev. Lett. **86**, 5438 (2001)
[6] J. Rafelski, B. Muller, Phys. Rev. Lett. **48**, 1066 (1982), [Erratum: Phys. Rev. Lett.56,2334(1986)]
[7] A. Shor, Phys. Rev. Lett. **54**, 1122 (1985)
[8] P. Danielewicz, R. Lacey, W.G. Lynch, Science **298**, 1592 (2002)
[9] J. Aichelin, C.M. Ko, Phys. Rev. Lett. **55**, 2661 (1985)
[10] B.A. Li, L.W. Chen, C.M. Ko, Phys. Rept. **464**, 113 (2008)
[11] B.A. Li, Phys. Rev. Lett. **88**, 192701 (2002)
[12] Q.f. Li, Z.x. Li, S. Soff, R.K. Gupta, M. Bleicher, H. Stoecker, J. Phys. **G31**, 1359 (2005)
[13] X. Lopez et al. (FOPI), Phys. Rev. **C75**, 011901 (2007)
[14] Q.f. Li, Z.x. Li, E.g. Zhao, R.K. Gupta, Phys. Rev. **C71**, 054907 (2005)
[15] J. Steinheimer, M. Bleicher, J. Phys. **G43**, 015104 (2016)
[16] G. Agakishiev et al., Phys. Rev. Lett. **114**, 212301 (2015)
[17] C.H. Li, C.M. Ko, Nucl. Phys. **A712**, 110 (2002)
[18] F. Li, L.W. Chen, C.M. Ko, S.H. Lee, Phys. Rev. **C85**, 064902 (2012)

[19] S. Pal, C.M. Ko, J.M. Alexander, P. Chung, R.A. Lacey, Phys. Lett. **B595**, 158 (2004)
[20] W.S. Chung, C.M. Ko, G.Q. Li, Nucl. Phys. **A641**, 357 (1998)
[21] S. Pal, C.M. Ko, Z.w. Lin, Nucl. Phys. **A707**, 525 (2002)
[22] W.S. Chung, G.Q. Li, C.M. Ko, Nucl. Phys. **A625**, 347 (1997)
[23] L. Alvarez-Ruso, V. Koch, Phys. Rev. **C65**, 054901 (2002)
[24] D. Cabrera, A.N. Hiller Blin, M.J. Vicente Vacas, P. Fernandez De Cordoba, Phys. Rev. **C96**, 034618 (2017)
[25] T. Ishikawa et al., Phys. Lett. **B608**, 215 (2005)
[26] M.H. Wood et al. (CLAS), Phys. Rev. Lett. **105**, 112301 (2010)
[27] A. Polyanskiy et al., Phys. Lett. **B695**, 74 (2011)
[28] D. Adamova et al. (CERES), Phys. Rev. Lett. **96**, 152301 (2006)
[29] M. Wada (STAR), Nucl. Phys. **A904-905**, 1019c (2013)
[30] B.I. Abelev et al. (STAR), Phys. Rev. Lett. **99**, 112301 (2007)
[31] J.H. Chen, Y.G. Ma, G.L. Ma, X.Z. Cai, Z.J. He, H.Z. Huang, J.L. Long, W.Q. Shen, C. Zhong, J.X. Zuo, Phys. Rev. **C74**, 064902 (2006)
[32] B.B. Abelev et al. (ALICE), JHEP **06**, 190 (2015)
[33] S. Choudhury, D. Sarkar, S. Chattopadhyay, Phys. Rev. **C95**, 024904 (2017)
[34] C. Alt et al. (NA49), Phys. Rev. **C78**, 044907 (2008)
[35] C.M. Ko, G.Q. Li, J. Phys. **G22**, 1673 (1996)
[36] C.M. Ko, V. Koch, G.Q. Li, Ann. Rev. Nucl. Part. Sci. **47**, 505 (1997)
[37] G.Q. Li, C.M. Ko, B.A. Li, Phys. Rev. Lett. **74**, 235 (1995)
[38] J.L. Ritman et al. (FOPI), Z. Phys. **A352**, 355 (1995)
[39] S. Pal, C.M. Ko, Z.w. Lin, B. Zhang, Phys. Rev. **C62**, 061903 (2000)
[40] J. Xu, L.W. Chen, C.M. Ko, Z.W. Lin, Phys. Rev. **C85**, 041901 (2012)
[41] J. Xu, T. Song, C.M. Ko, F. Li, Phys. Rev. Lett. **112**, 012301 (2014)
[42] L. Adamczyk et al. (STAR), Phys. Rev. Lett. **110**, 142301 (2013)
[43] M. Gazdzicki, M. Gorenstein, P. Seyboth, Acta Phys. Polon. **B42**, 307 (2011)
[44] W. Cassing, A. Palmese, P. Moreau, E.L. Bratkovskaya, Phys. Rev. **C93**, 014902 (2016)
[45] L. McLerran, R.D. Pisarski, Nucl. Phys. **A796**, 83 (2007)
[46] J. Steinheimer, J. Randrup, Phys. Rev. Lett. **109**, 212301 (2012)
[47] F. Li, C.M. Ko, Phys. Rev. **C95**, 055203 (2017)
[48] K.J. Sun, L.W. Chen, C.M. Ko, Z. Xu, Phys. Lett. **B774**, 103 (2017)
[49] J. Xu, C.M. Ko, Phys. Lett. **B772**, 290 (2017)
[50] L. Adamczyk et al. (STAR), Nature **548**, 62 (2017)
[51] Z.T. Liang, X.N. Wang, Phys. Rev. Lett. **94**, 102301 (2005), [Erratum: Phys. Rev. Lett.96,039901(2006)]
[52] I. Karpenko, F. Becattini, Eur. Phys. J. C **77**, 213 (2017)
[53] Y. Xie, D. Wang, L.P. Csernai, Phys. Rev. C **95**, 031901 (2017)
[54] L.G. Pang, H. Petersen, Q. Wang, X.N. Wang, Phys. Rev. Lett. **117**, 192301 (2016)
[55] Yu.B. Ivanov, A.A. Soldatov, Phys. Rev. **C95**, 054915 (2017)
[56] H. Li, L.G. Pang, Q. Wang, X.L. Xia, Phys. Rev. **C96**, 054908 (2017)
[57] Z.W. Lin, C.M. Ko, B.A. Li, B. Zhang, S. Pal, Phys. Rev. C **72**, 064901 (2005)
[58] Y. Sun, C.M. Ko, Phys. Rev. **C96**, 024906 (2017)
[59] S. Cho et al. (ExHIC), Phys. Rev. Lett. **106**, 212001 (2011)

[60] S. Cho et al. (ExHIC), Phys. Rev. **C84**, 064910 (2011)
[61] S. Cho et al. (ExHIC), Prog. Part. Nucl. Phys. **95**, 279 (2017)
[62] R.L. Jaffe, Phys. Rev. Lett. **38**, 195 (1977), [Erratum: Phys. Rev. Lett.38,617(1977)]
[63] K. Morita, T. Furumoto, A. Ohnishi, Phys. Rev. **C91**, 024916 (2015)
[64] L. Adamczyk et al. (STAR), Phys. Rev. Lett. **114**, 022301 (2015)
[65] S.R. Beane et al. (NPLQCD), Phys. Rev. Lett. **106**, 162001 (2011)
[66] T. Inoue, S. Aoki, T. Doi, T. Hatsuda, Y. Ikeda, N. Ishii, K. Murano, H. Nemura, K. Sasaki (HAL QCD), Nucl. Phys. **A881**, 28 (2012)
[67] G.A. Miller, Chin. J. Phys. **51**, 466 (2013)
[68] Z.Y. Zhang, Y.W. Yu, C.R. Ching, T.H. Ho, Z.D. Lu, Phys. Rev. **C61**, 065204 (2000)
[69] S. Gongyo et al. (2017), `arXiv:1709.00654 [hep-lat]`
[70] F. Etminan, H. Nemura, S. Aoki, T. Doi, T. Hatsuda, Y. Ikeda, T. Inoue, N. Ishii, K. Murano, K. Sasaki (HAL QCD), Nucl. Phys. **A928**, 89 (2014)
[71] K. Morita, A. Ohnishi, F. Etminan, T. Hatsuda, Phys. Rev. **C94**, 031901 (2016)
[72] T. Hatsuda, K. Morita, A. Ohnishi, K. Sasaki, Nucl. Phys. **A967**, 856 (2017)

Heavy flavour production at RHIC and LHC

Gian Michele Innocenti[1,*]

[1]Massachusetts Institute of Technology

Abstract. In this proceedings, I present selected experimental results on heavy-flavour production at RHIC and at the LHC, which were presented at the Strangeness in Quark Matter 2017 conference. I will present a brief introduction to the heavy-flavour physics in heavy ion collisions and I will focus on recents measurements of in-medium energy loss and and collective properties of heavy-flavour particles, which provided important information on the mechanisms of heavy flavour interaction with the hot and dense medium created in ultra-relativistic heavy-ion collisions.

1 Introduction

Heavy quarks are effective probes to study the properties of the deconfined medium created in heavy ion collisions[1]. These quarks, as a consequence of their large masses, are mostly produced in primary hard QCD scatterings with a production timescale that is shorter than the formation time of the QGP. During their propagation through the medium, heavy quarks lose energy via radiative and collisional interactions with the medium constituents. QCD predicts that quarks lose less energy than gluons as a consequence of their smaller colour factor. In addition, the so-called "dead-cone effect" is expected to reduce small-angle gluon radiation of heavy quarks when compared to both gluons and light quarks. Precise measurements of the nuclear modification factor R_{AA} of particles containing both light and heavy quarks can thus provide important tests for the predicted flavour-dependence of in-medium energy loss. In addition, the measurement of the R_{AA} of strange D mesons can also help estimating the relevance of the coalescence mechanism in which charmed hadrons are formed by the combination of charm quarks with light quarks from the medium. If coalescence plays a relevant role in the charm hadronisation, the yield of D_s^+ particles are expected to be significantly enhanced with respect to the one of D^0 mesons at low-intermediate transverse momenta (p_T), as a consequence of the increased abundance of strange quarks in the deconfined medium. The production ratio of Λ_c/D^0 is also expected to be enhanced with respect to that of charmed mesons under the assumption that di-quarks bounded states in the quark-gluon plasma can be formed. Finally, the study of the collective properties of D and B mesons through detailed measurements of their azimuthal coefficienct v_n at low-intermediate p_T can also help quantifying the extent to which charm and beauty quarks flow with the medium, which is a good measure of their interaction strength and of their level of thermalisation.

The production of open heavy-flavours is measured at RHIC and at the LHC using

*e-mail: ginnocen@cern.ch

[1]More details and a complete bibliography for this section can be found in [1]

several experimental techniques [1]. Charm and beauty hadrons can me measured via the exclusive reconstruction of their hadronic or semi-leptonic decay channels (e.g. $D^0 \rightarrow K^-\pi^+$ or $B^+ \rightarrow J/\Psi\, K^+ \rightarrow \mu^+\mu^- K^+$) or via the inclusive reconstruction of semi-leptonic decays (e.g. $B \rightarrow J/\Psi\ + X$). The production of heavy-flavour particles can also be studied up to very high p_T by measuring jets of hadrons produced in the fragmentation of a charm or beauty quark. Charm and beauty decays are characterised by relativetely small cross sections and by decay vertices which are displaced few hundreds μm from the main interaction vertices. Therefore the study of these observables requires very large statistics of minimum-bias and triggered events and precise tracking and vertexing detectors that can operate under conditions of very large detector occupancy.

2 Proton-proton measurements

The study of heavy-flavour production in proton-proton collisions is considered a fundamental reference for nucleus-nucleus measurements but at the same time it provides an important benchmark for perturbative QCD calculations based on a factorisation approach [1]. The possibility of measuring the charm and beauty production both at RHIC and the LHC is crucial to constrain the relative relevance of different production mechanisms (e.g. flavour creation vs gluon fusion processes), which are expected to play a different role at different energies. The p_T- and y-differential cross sections of charm and beauty quarks were measured by several collaborations both at RHIC and at the LHC [2–13] and were found to be consistent with perturbative QCD calculations (e.g. FONLL[16]) that include a consistent description of next-to-leading order processes. As an example, in the left panel of Fig. 1, the D^0 production cross sections as a function of p_T measured by the ALICE experiment at LHC at 7 TeV and by the STAR experiment at RHIC at 200 and 500 GeV are presented. In the central panel of the same figure, the B^+ production measured by the ATLAS experiment at 7 TeV is shown. Although theoretical calculations are able to describe successfully single-differential observables, more differential measurements like $D\bar{D}$ or $B\bar{B}$ azimuthal correlations [14, 15] showed that the current understanding of next-to-leading order processes at least at the LHC is still not satisfactory. In particular, the contribution of the gluon-splitting production mechanism ($g \rightarrow c\bar{c}$ or $g \rightarrow b\bar{b}$), which is dominant at small angles between the particle-anti-particle pairs is sizeably understimated by theoretical calculations (right panel of Fig. 1). A better description of the gluon splitting processes would be extremely important since charm and beauty quarks produced via these mechanisms are mostly sensitive to the energy loss of gluons instead of that of the heavy parton.

3 Proton-nucleus measurements

In presence of a nuclear environment, the yield of production of light and heavy particles can be modified in absence of any deconfined medium. In particular, modifications of the parton distribution functions (PDFs) in the nucleus with respect to nucleon PDFs (e.g. due to shadowing) are expected to be responsable for sizeable modifications of the heavy-flavour cross section at low transverse momenta [1]. In order to quantify the effects of these phenomena and validate the hypothesis that the suppression observed in nucleus-nucleus collisions is a consequence of the presence of a a hot and dense medium, one can study the heavy-flavour production in proton-nucleus collisions, where the formation of the Quark-Gluon Plasma is not expected. In the last years, the role of "reference" of proton-nucleus measurements have been deeply questioned. However, at the present moment, no measurement has highlighted the presence of sizeable jet quenching phenomena in proton-nucleus or proton-proton collisions, suggesting that measurements of production yields in smaller collision systems can still

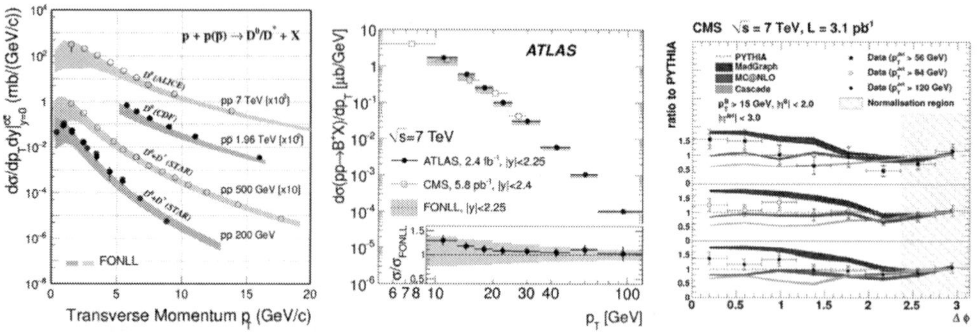

Figure 1. (Left) D^0 production cross sections measured by the ALICE experiment at LHC in pp collisions at 7 TeV and by the STAR experiment at RHIC at 200 and 500 GeV. (Middle) B^+ production measured by the ATLAS experiment in pp collisions at 7 TeV. (Right) $B\bar{B}$ azimuthal correlation measured by the CMS experiment in pp collisions at 7 TeV.

be used to isolate the effect of cold nuclear matter effects. The nuclear modification factors $R_{p\mathrm{A}}$ of heavy-flavoured particles have been measured both at RHIC and at the LHC [17–23]. Thanks to the very high statistics 2016 pPb run, the $R_{p\mathrm{A}}$ of D^0 mesons have been measured at the LHC with very high precision (Fig. 2). The measurements performed by LHCb at forward and backword rapidity have highlighted for the first time significant modifications of the production yields of D mesons for $p_T < 10$ GeV in pPb collisions, which are compatible with the predictions of models that include the effect of modifications of nuclear PDFs. The measurement performed by ALICE at central rapidity and an a simple rapidity extrapolation of the LHCb measurement show that at central rapidity the effect of cold nuclar matter effects at low p_T is smaller or equal to about 15%. At higher p_T, as expected, no significant deviations from binary scaling ($R_{p\mathrm{A}}$=1) was observed. In the beauty sector at the LHC, measurements of heavy-flavour electrons performed by ALICE and exclusive B-meson measurements from CMS have measured $R_{p\mathrm{A}}$ consistent with unity from 1 to 60 GeV/c at mid rapidity with uncertainties of about 30-40%. A recent CMS measurement of the non-prompt J/Ψ $R_{p\mathrm{A}}$ has measured at central rapidity $R_{p\mathrm{A}}$=1 with uncertainty smaller than 15%. A new impressive LHCb measurement of non-prompt J/Ψ $R_{p\mathrm{A}}$ performed down to 0 GeV/c has highlighted for the first time a significant deviation from unity for the $R_{p\mathrm{A}}$ at forward rapidity as expected in presence of nuclear shadowing also in the beauty sector. Also at the RHIC energies, the $R_{p\mathrm{A}}$ of beauty has been recently measured via the reconstruction of non-prompt J/Ψ decays. The PHENIX collaboration measured a $R_{p\mathrm{A}}$ consistent with unity with uncertainties of the order of 40–50%.

4 Nucleus-nucleus measurements

Indications of a strong charm suppression at the LHC are observed by the ALICE and the CMS collaborations using the R_{AA} of promptly produced D mesons at mid-rapidity at a nucleon-nucleon centre-of-mass energy $\sqrt{s_{\mathrm{NN}}}$ = 2.76 and $\sqrt{s_{\mathrm{NN}}}$ = 5.02 TeV [24–26]. The results of the two collaborations are fully consistent if compared in the same p_T and centrality region (left and middle panel of Fig. 3). The R_{AA} of D^0 mesons at the LHC shows a maximum suppression of a factor 5–6 with respect to the pp reference for the 10% most central collisions in the p_T range of 6–10GeV/c. For D^0 mesons in the high-p_T range of

[1] Biased opionion of the author.

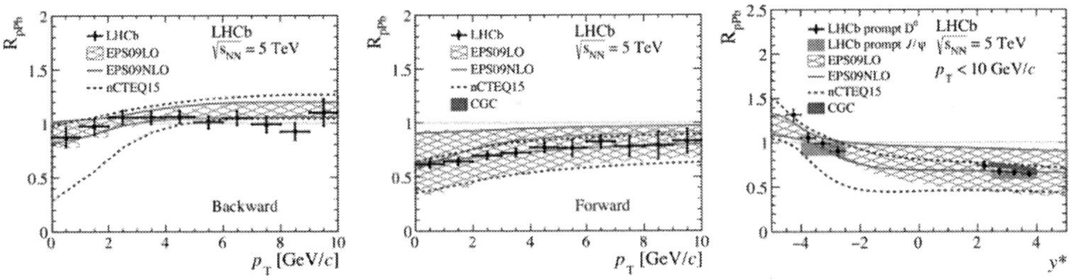

Figure 2. D^0 R_{pA} in the backward (left) and forward (right) rapidity as a function of p_T and as a function of rapidity for p_T <10 GeV/c measured by the LHCb collaboration in pp collisions at 7 TeV.

60–100GeV/c, a significantly smaller suppression is observed. The R_{AA} as a function of p_T is quite well described by several theoretical calculations that include collisional or radiative energy loss or a mixture of the two mechanisms, confirming that the R_{AA} alone is not powerful enough to discriminate between models even if measured in a very wide kinematic range. An increasing suppression is observed as a function of centrality as expected in case of the creation of a denser and hotter medium in central PbPb collisions. This trend is also well described by theoretical calculations. Quite surprisingly, at the LHC the R_{AA} of D^0 mesons was found to be almost completely consistent with the charged particle R_{AA} from 1 to 100 GeV/c [32]. The same R_{AA}, however, in presence of very different p_T-spectra and fragmentation functions does not imply the same energy loss. On the contrary, the evidence of a very similar suppression for charged particles and D mesons seems to be a very fortunate (too fortunate?) cancellation of the effects of different energy loss, p_T-spectra and fragmentation functions. Although described by several theoretical calculations, this observation does not present a clear and unique physics interpretation[1]. At RHIC energies, the D^0 R_{AA} was also measured by the STAR Collaboration. The D-meson R_{AA} measured at 200 GeV presents values larger than 1.5 even if with large uncertaintes at p_T <2 GeV/c, being significantly higher than the charged particle R_{AA} in the same transverse momentum region (right panel of Fig. 3). Such a difference between the ratio of the light and D-meson R_{AA} at the two energies might be attributed to differences in radial flow or nuclear PDF modifications at the two energies.

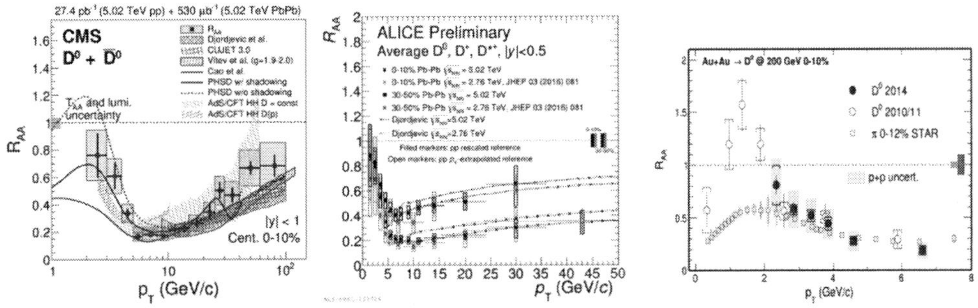

Figure 3. (Left) D^0 nuclear modification factor for central PbPb collisions at 5.02 TeV measured by CMS. (Center) D^0 nuclear modification factor for central and semi-peripheral collisions measured by ALICE. (Right) D^0 nuclear modification factor for central AuAu collisions at 200 GeV measured by STAR compared to the one of charged particles in a similar centrality range.

The suppression of beauty particles was also measured by different collaborations both at RHIC and at the LHC [27–31]. At the LHC, the $R_{\rm AA}$ of beauty particles was measured down to 2 GeV by the ALICE experiment with heavy-flavour electron measurements and from 3 GeV up to 50 GeV using exclusively reconstructed B mesons and non-prompt B meson decays (B $\rightarrow$ J/Ψ +X). As shown in the left and middle panel Fig. 4, the current measurements indicate a hint of different suppression of beauty and charmed particles at low-intermediate $p_{\rm T}$ that vanishes when going to higher transverse momenta. Several reasonable concerns were raised with respect to the CMS comparison plot. The first one is related to the fact that in this plot the $p_{\rm T}$ of the non-prompt J/Ψ is not corrected for the difference between the $p_{\rm T}$ of the beauty hadrons and the $p_{\rm T}$ of the non-prompt J/Ψ. This is a valid point that should be addressed in future version of this result expecially for what concerns the low $p_{\rm T}$ region. At higher $p_{\rm T}$, the fact that the $R_{\rm AA}$ does not show a strong $p_{\rm T}$-dependence should make this missing correction less relevant. The second concern is related to the possible discrepancy between the non-prompt J/Ψ and the exclusive B^+ meson result. In this case, I would like to stress the fact that the uncertainties on the current B^+ meson $R_{\rm AA}$ are quite sizeable and the two measurements are still compatible within the uncertainties. In addition, it has to be noticed that in principle the non-prompt J/Ψ result might be affected by effects of enhancement of strange beauty mesons and baryons that could be not completely negligible in this $p_{\rm T}$ range. The last concerns is related to the fact that the results reported here mix measurements performed in different rapidity regions. To addess this point, more differential analysis of the $R_{\rm AA}$ of the different species as a function of rapidity should be performed[2]. However, the recent measurement of non-prompt J/Ψ of CMS as function of the rapidity does not indicate a strong y-dependence of the suppression. This evidence seems to indicate that the different rapidity coverage of the measurements should not affect the physics conclusions we draw with the current version of this plot. At RHIC the suppression of beauty hadrons was also measured by the STAR and the PHENIX collaborations using non-prompt J/Ψ and heavy-flavour electrons. Also at this energies, as shown in the right panel Fig. 4 in the PHENIX heavy-flavour electron results, a hint of different suppression is observed for $p_{\rm T}$ smaller than about 6 GeV/c.

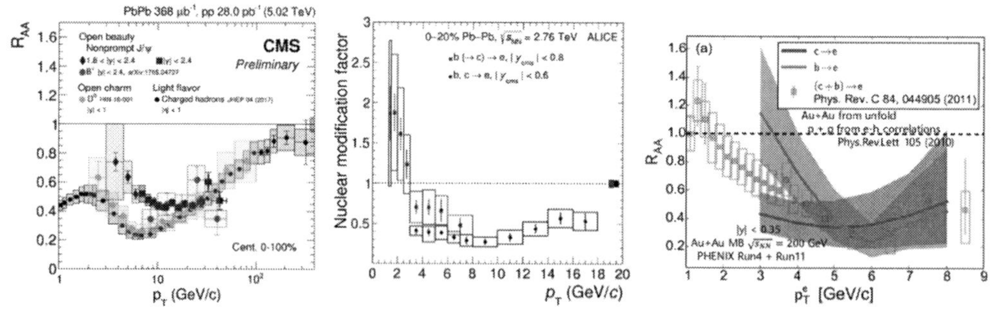

Figure 4. (Left) Nuclear modification factor of charged particles, D^0 mesons, non-prompt J/Ψ and B^+ mesons performed by CMS at in PbPb collisions at 5.02 TeV. (Center) Nuclear modification factor of heavy flavoured electrons performed by ALICE in central PbPb collisions at 5.02 TeV. (Right) Nuclear modification factor of heavy flavoured electrons performed by PHENIX in minimum-bias PbPb collisions at 200 GeV.

[2]The reason why this was not done yet is related to the fact that the a B-meson measurement in a smaller rapidity region is currently not possible due to the limited statistics and the non-prompt J/Ψ measurement at central rapidity cannot reach the very low-momentum region as a consequence of the kinematic cuts applied to the muons and to the characteristics of the CMS muon detectors.

A first indication of a relevant contribution of recombination in the open charm sector was recently observed both at RHIC and at the LHC [33, 34]. A hint of a larger R_{AA} of $\mathrm{D_s^+}$ mesons with respect to non-strange D-meson one was observed by the ALICE and the STAR collaboration in central and semi-central collisions in the low-intermediate p_{T} region. This is currently interpretated as an indication that coalescence is playing a relevant role in the hadronisazion of charm quarks in heavy-ion collisions. The observation of this phenomenon seems to confirm the presence of free color charges in the medium. In absence of color connectivity indeed, recombination processes cannot take place. A precise estimation of the effect of recombination is also very important as an input for theoretical calculations that describe the suppression of heavy-flavoured particles. A wrong estimation of recombination effects can indeed bias our understanding of the magnitude of energy loss processes. The STAR Collaboration also recently reported a very interesting first measurement of the Λ_c R_{AA} in semi-peripheral gold-gold collisions. Also in this case, an ehnancement with respect to the pp baseline was observed. This result can be intepreted theoretically if we assume that charm recombines in a medium where di-quark bound states are formed.

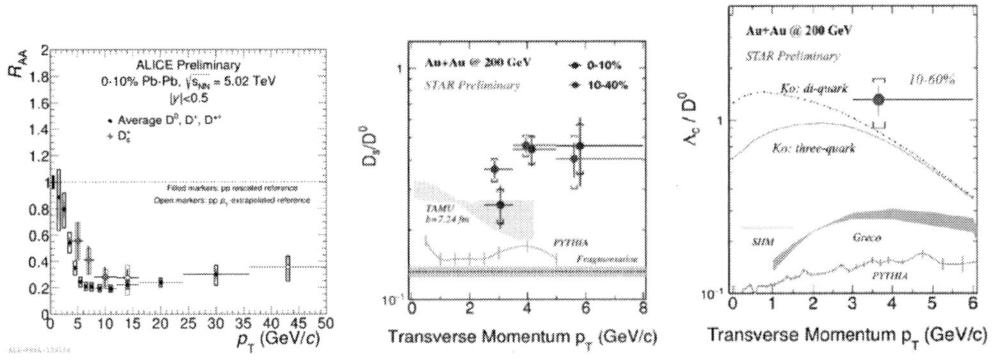

Figure 5. (Left) Nuclear modification factor of $\mathrm{D_s^+}$ mesons performed by ALICE in PbPb collisions at 5.02 TeV compared to the result for non-strange D mesons. (Center) $\mathrm{D_s^+/D^0}$ meson ratio performed by STAR in central and semi-central AuAu collisions at 200 GeV. (Right) $\Lambda_c/\mathrm{D^0}$ meson ratio performed by STAR in AuAu collisions at 200 GeV in the centrality range 0–80%.

Strong indications of charm flow were also observed by several collaborations both at RHIC and at the LHC [35–37]. Large and positive values of the elliptic flow coefficients v_2 were measured by ALICE and CMS at low-intermediate p_{T} in semi-pheripheral PbPb collisions and by STAR in AuAu collisions (left and middle panel of Fig. 6). Due to the limited statistics available, the STAR result was performed in an integrated centrality interval 0–80%. Therefore a direct comparison between the magnitude of the elliptic flow at the two energies cannot be currently done. At the LHC, where more differential measurement were performed, the magnitude of the D-meson v_2 coefficient was found to decrease when going from peripheral to more central events as expected in case the elliptic flow is generated as a consequence of the initial anysotropy of the fireball. A non-zero v_2 is also observed at larger p_{T} (>10 GeV/c). In this kinematic region, the v_2 measurement is no longer sensitive to collective phenomena but it is sensitive to the in-medium path dependence of the energy loss and can be therefore use to contraint jet quenching theoretical calculations. At the LHC, a significant indication of smaller v_2 of D mesons with respect to charged-particle v_2 was observed and a first hint of non-zero v_3 of D mesons was also measured by CMS in non-central collisions. These high-precision results provided strong constraints on theoretical calculations. An important theoretical effort is ongoing to use these measurements to quantify the level of charm thermalisation in the medium, to understand if charm quarks can be described by a hydro-dynamic description and to extract the difficusion parameters of charm quarks in the medium. The shape of the v_2 D-meson measurement, in addition, can provide important information on the relative relevance of collisional and radiative energy loss processes, being therefore complementare to the the R_{AA} measurement. At the LHC, a first attempt was also made by the CMS collaboration to measure the v_2 coefficient of beauty by the measurement of non-prompt J/Ψ in semi-peripheral PbPb events. The current result (right panel of Fig. 6), even if still not conclusive, represents a very promising attempt to estimate the collective behavior of beauty charms in heavy-ion collisions.

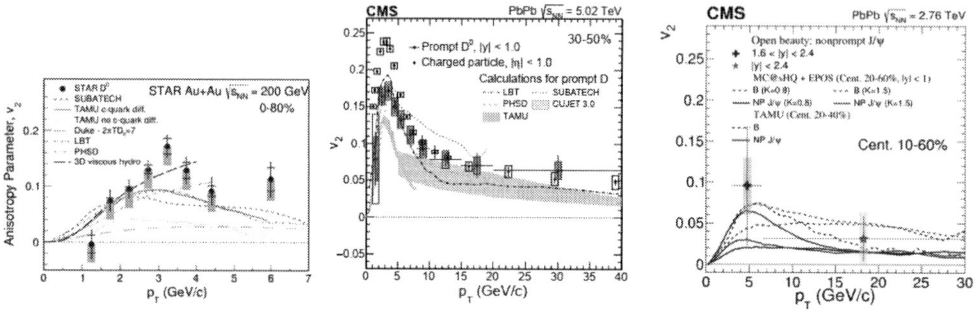

Figure 6. (Left) v_2 coefficient of D^0 mesons performed by STAR in AuAu collisions at 200 GeV in the centrality range 0–80%. (Center) v_2 coefficient of D^0 mesons performed by CMS in semi-peripheral PbPb collisions at 5.02 TeV. (Right) v_2 coefficient of non-prompt J/Ψ performed by CMS in semi-peripheral PbPb collisions at 5.02 TeV.

5 Conclusions

In the last few years, impressive results were obtained in the sector of heavy-flavour observables in heavy-ion collisions, moving our field to an exciting era of high-precision measurements. Although a first quantitative understanding of heavy-flavour phenomena was achieved, a long way is still ahead in order to be able to use heavy-flavour observables to improve our understanding of the the properties of the Quark-Gluon Plasma. In the coming future, the new high-luminosity runs at RHIC and LHC and the upgraded detectors will allow us to improve the precision of the traditional observables but also to perform new more differential measurements (like D-hadron, D-jet, D-$\bar{D}$ correlations, D-meson event-shape engineering measurements) that can contraint better the mechanisms of heavy-flavour production and the mechanisms of in-medium energy loss. A better and quantitative understanding of the microscopic mechanisms of heavy-quark interaction with the medium (energy loss mechanisms and hadronisation mechanisms in particular) is a pre-requisite in order to use theoretical calculations to extract fundamental properties like the transport coefficients or the the $\hat{q}$ coefficient. At the same time, measurements of heavy-flavour productions in smaller systems should be exploited to better understand the properties and the nature of the collective phenomena that were recently observed in both pp and proton-nucleus collisions.

References

[1] A. Andronic et al., Eur. Phys. J. C 76 (2016) 107, doi:10.1140/epjc/s10052-015-3819-5, arXiv:1506.03981.

[2] STAR Collaboration, Phys.Rev.D86(2012)072013.
[3] CDF Collaboration,etal.,Phys.Rev.Lett.91(2003)241804.
[4] ALICE Collaboration, et al., JHEP 01 (2012) 128.
[5] ALICE Collaboration, et al., arXiv:1702.00766.
[6] CMS Collaboration, Phys.Rev.Lett.106(2011)112001.
[7] ATLAS Collaboration, JHEP 10 (2013) 042.
[8] CMS Collaboration, et al., arXiv:1609.00873.
[9] C. Aidala, et al., arXiv:1701.01342.
[10] LHCb Collaboration, JHEP 03 (2016) 159.
[11] LHCb Collaboration, JHEP 08 (2013) 117.
[12] PHENIX Collaboration, arXiv:1405.3301
[13] LHCb Collaboration, JHEP 1603 (2016) 159, Erratum: JHEP 1705 (2017) 074.
[14] CMS Collaboration, JHEP03(2011)136.
[15] LHCb Collaboration, JHEP06(2012)141.
[16] M. Cacciari,P.Nason,R.Vogt,Phys.Rev.Lett.95(2005)122001.
[17] ALICE Collaboration, Phys.Rev.Lett.113(23)(2014)232301.
[18] ALICE Collaboration,ALICE-PUB-2017-008.
[19] LHCb Collaboration, arXiv:1707.02750v1.
[20] ALICE Collaboration, EPJC 77 (2017) 269.
[21] ALICE Collaboration, arXiv:1609.03898.
[22] LHCb Collaboration, these proceedings.
[23] PHENIX Collaboration, arXiv.1702.01085.
[24] CMS Collaboration, arXiv:1708.04962v1.
[25] ALICE Collaboration, ALICE-PUBLIC-2017-003.
[26] STAR Collaboration, PRL118, 212301.
[27] CMS Collaboration, arXiv 1705.04727,
[28] CMS Collaboration, arXiv: 1610.00613 EPJC 77 (2017) 252.
[29] ALICE Collaboration, arXiv:1609.03898.
[30] STAR Collaboration, these proceedings.
[31] PHENIX Collaboration, these proceedings.
[32] CMS Collaboration, arXiv: 1611.01664 JHEP 04 (2017) 039.
[33] ALICE Collaboration, ALICE-PUBLIC-2017-003.
[34] STAR Collaboration, these proceedings.
[35] CMS Collaboration, arXiv:1708.03497v1.
[36] ALICE Collaboration, arXiv:1707.01005v1.
[37] STAR Collaboration, PRL118, 212301.

EPJ Web of Conferences **171**, 04002 (2018) https://doi.org/10.1051/epjconf/201817104002
SQM 2017

Hidden heavy flavour production in heavy-ion collisions

Javier Castillo Castellanos[1,⋆]

[1] *IRFU, CEA, Université Paris-Saclay, F-91191Gif-sur-Yvette, France*

Abstract. An overview of recent experimental results on quarkonium production in heavy-ion collisions at RHIC and LHC energies is presented. Their implications in the theoretical understanding of the production of quarkonia is discussed.

1 Introduction

Heavy quarks, charm (c) and beauty (b), are important probes of the Quark-Gluon Plasma (QGP). Because of their large mass, heavy-quark pairs are produced by the primary hard collisions of the incoming partons. A small fraction of those initial pairs will bind together to form a quarkonium. Heavy quarks and quarkonia will interact with the QGP throughout its evolution and probe its properties.

Originally, it was proposed that the production of J/ψ mesons will be suppressed in the QGP by a color-screening mechanism [1]. Then, this was extended to all quarkonium states leading to the sequential suppression scenario: a hierarchy of suppression depending on the binding energy of the quarkonia [2]. Other dynamical mechanisms such as gluon dissociation [3] and quasi-free scattering [4, 5] were also proposed. The sequential suppression could inform about the QGP temperature.

More recently, new mechanisms of quarkonium production by the QGP have been proposed. In the statistical hadronization model [6] all heavy quarks and antiquarks, produced in the initial hard-scatterings, thermalize in the QGP. Then, all quarkonia are produced by statistical hadronization at the phase boundary of the QGP. Here full suppression of the initially produced quarkonia is assumed. In [5] the approach is extended to consider partial charmonium suppression. Additional quarkonia could also be regenerated throughout the QGP evolution by kinetic recombination of heavy-quarks [7]. In this type of models, called transport models, continuous dissociation and regeneration of quarkonia throughout the life time of the QGP is implemented by a kinetic rate equation. The QGP evolution is described by a hydrodynamical-like expansion. For more recent implementations and calculations of the transport model see [8–10] and [4, 11–13]. These additional production mechanism will therefore compete with the initial quarkonium production and their suppression by the hot medium. The higher initial temperatures of the QGP and the larger heavy-quark production cross sections reached at LHC energies should favor the regeneration mechanism.

Cold nuclear matter (CNM) effects also affect the production of quarkonia in heavy-ion collisions. These can be separated into initial and final-state effects. Initial-state effects occur prior to the formation of the heavy-quark pair, including the modification of the kinematical distribution of the partons in the nuclei (nPDF) compared to that in free nucleons [14–16] and parton energy loss [17, 18]. Final-state effects are those that affect the heavy-quark pair during the finite time it needs to form a

⋆e-mail: Javier.Castillo@cea.fr

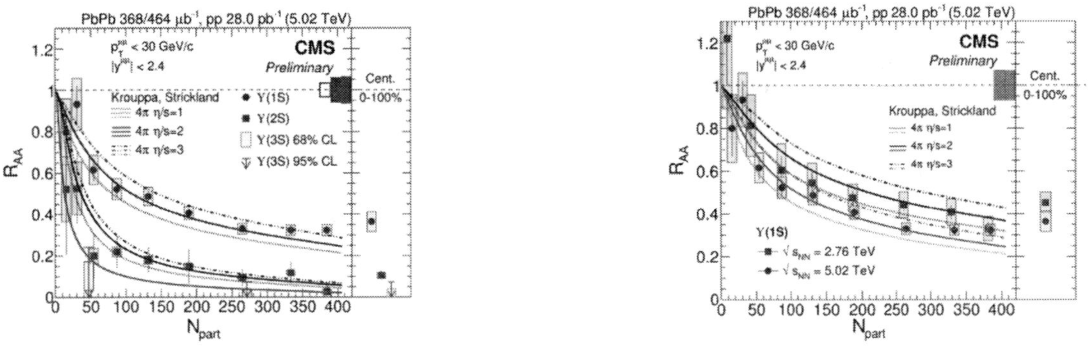

Figure 1. (Left) R_{AA} of Υ(nS) versus $N_{\rm part}$ in Pb–Pb collisions at $\sqrt{s_{\rm NN}} = 5.02$. (Right) R_{AA} of Υ(1S) versus $N_{\rm part}$ in Pb–Pb collisions at $\sqrt{s_{\rm NN}} = 2.76$ and 5.02 TeV. The measurements by CMS [22, 23] are compared to the calculations of a hydrodynamical model [24, 25].

quarkonium state or after the state has been formed [19], such as the $Q\bar{Q}$ pair break up by the nuclear matter (expected to be small at high energy) [19, 20], and the final-state resonance interaction with surrounding comovers [21]. Finally, parton energy loss could be coherent and thus not factorized into initial and final state effects [18]. Cold nuclear matter effects can be studied in proton–nucleus (p–A) or deuteron–nucleus (d–A) collisions, where the QGP is not expected to be formed.

We will review recent measurements of quarkonium production and their implications in terms of suppression or regeneration mechanisms in the QGP, or CNM effects.

2 Sequential suppression

In the sequential suppression scenario, the suppression of a given quarkonium state should increase with increasing temperature or energy density of the QGP. Furthermore, the suppression should increase with decreasing binding energy of the quarkonium state, e.g. the exited Υ(3S) will be more suppressed than the Υ(2S) than the ground Υ(1S).

In Fig. 1 (left), the recent measurement by the CMS Collaboration of the nuclear modification factor R_{AA} of Υ(nS) in Pb–Pb at $\sqrt{s_{\rm NN}} = 5.02$ TeV is shown versus the collisions centrality, quantified by the number of participating nucleons $N_{\rm part}$. First, one clearly observes that the R_{AA} of Υ(1S) and Υ(2S) decrease from peripheral to central collisions, showing that the suppression of Υ is larger in central collisions where the QGP energy density is maximum. Second, for the most central collisions, the Υ(3S) appears to be more suppressed (or fully suppressed) than the Υ(2S), which in turn is more suppressed than the Υ(1S), confirming the expectations. The comparison of the Υ(1S) R_{AA} in Pb–Pb collisions at $\sqrt{s_{\rm NN}} = 2.76$ TeV [22] and 5.02 TeV [23] is shown in Fig. 1 (right). A hint of a stronger Υ(1S) suppression at the highest energy can be seen. The behaviors of the Υ(nS) versus the Υ state, the collision centrality and energy are naively expected by the scenarios in which the suppression is solely driven by the QGP energy density. These Υ(nS) measurements are well reproduced by a hydrodynamical model [24, 25]. In this model, the suppression of the bottomonium states by the QGP, described by a 3+1d hydrodynamical model, is based on a complex potential.

Figure 2 (right) shows the STAR measurement of the ratio of the yields of Υ(2S+3S) divided by those of Υ(1S) versus $N_{\rm part}$ in Au–Au collisions at $\sqrt{s_{\rm NN}} = 0.2$ TeV [26]. The horizontal band shows the average of the same ratio for pp collisions. As it was the case at LHC energies, at RHIC energies the excited states of the Υ appear to be more suppressed than ground Υ(1S) state.

The $\Upsilon(1S)$ R_{AA} measured by ALICE at forward rapidity in Pb–Pb collisions at $\sqrt{s_{NN}} = 5.02$ TeV [27] is shown in Fig. 2 as a function of rapidity together with the measurement by CMS. The two experiments complement each other and cover a wide rapidity range ($0 < y < 4$). The experimental measurements are compared to the calculations of the hydrodynamical model [25]. The model is able to reproduce the $\Upsilon(1S)$ R_{AA} over the full rapidity range, although at forward y the trends of the model and the ALICE R_{AA} seem opposite. Such different trend was more evident at the lower LHC energy (see e.g. [28]). In this context, a more precise measurement at forward y will be important. Indeed, in the hydrodynamical model [25] only suppression of bottomonia by the QGP is considered, neglecting CNM effects and any possible regeneration component. Transport models [12, 29], which include, in addition to a suppression by the QGP, CNM effects and a small regeneration component obtain an inverted rapidity dependence for the $\Upsilon(1S)$ R_{AA}. Finally, to answer the question if direct $\Upsilon(1S)$ are suppressed, a precise understanding of the CNM effects is needed.

3 Regeneration

Due to the large initial charm-quark density, the charmonium family is best suited to investigate any possible quarkonium regeneration mechanism. Figure 3 shows the recent J/ψ R_{AA} at forward (left) and mid- (right) y versus N_{part} measured by ALICE in Pb–Pb collisions at $\sqrt{s_{NN}} = 5.02$ TeV [31, 32], compared to the earlier measurements at $\sqrt{s_{NN}} = 2.76$ TeV [30, 33, 34] and in Au–Au collisions at $\sqrt{s_{NN}} = 0.2$ TeV by PHENIX [35, 36]. At RHIC energies, the J/ψ R_{AA} exhibits a decreasing trend from peripheral to central collisions, whereas at LHC energies a saturation of the R_{AA} is seen for $N_{part} \gtrsim 100$. At mid-y even a hint of an increased R_{AA} for the most central collisions is seen. For the most central collisions, the J/ψ R_{AA} is larger at LHC than at RHIC energies, i.e. the J/ψ suppression is smaller where the QGP energy density is higher. Such observations clearly establish the case for an important contribution of the regeneration mechanism at the LHC. Models implementing full [37] or partial [10, 11, 13, 21, 38, 39] J/ψ regeneration are able to reproduce these results.

The regeneration of J/ψ is expected to contribute mostly at low p_T. Figure 4 (left) shows the J/ψ R_{AA} at forward y versus p_T in Pb–Pb collisions at $\sqrt{s_{NN}} = 5.02$ TeV for three centrality classes [40]. The bottom panel shows the ratio of the R_{AA} at $\sqrt{s_{NN}} = 5.02$ and 2.76 TeV. For most central collisions, the R_{AA} increases with decreasing p_T confirming the contribution of J/ψ regeneration at low p_T. The calculations of a transport model [11, 13, 38] including J/ψ regeneration can reproduce the

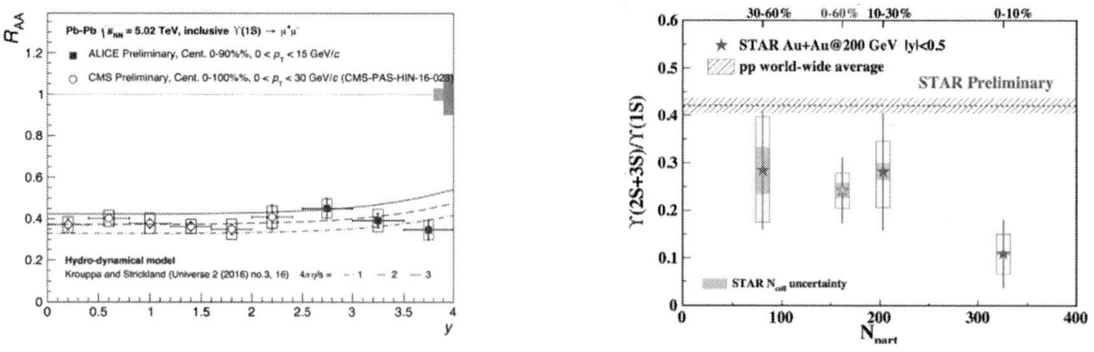

Figure 2. (Left) R_{AA} of $\Upsilon(1S)$ versus rapidity in Pb–Pb collisions at $\sqrt{s_{NN}} = 5.02$ TeV [23, 27] compared to the calculations of a hydrodynamical model [25]. (Right) Ratio of yields of Υ(2S+3S) over $\Upsilon(1S)$ in Au–Au collisions at $\sqrt{s_{NN}} = 0.2$ TeV [26] compared to the same ratio in pp collisions shown as a band.

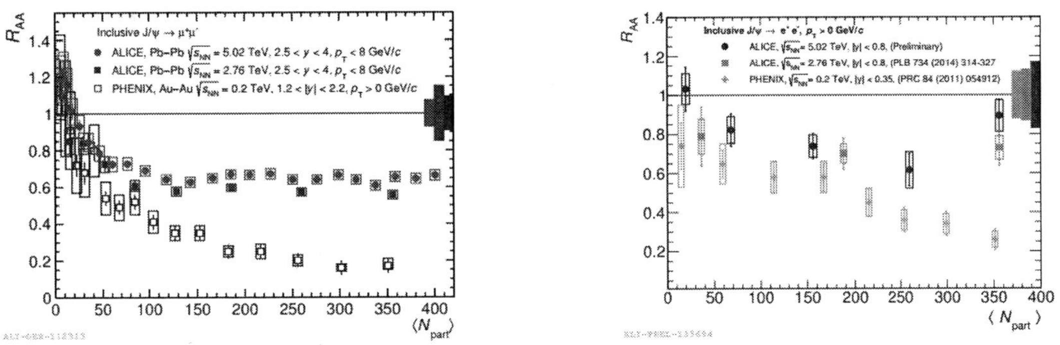

Figure 3. J/ψ $R_{\rm AA}$ at forward (left) and mid- (right) y versus $N_{\rm part}$ in Pb–Pb collisions at $\sqrt{s_{\rm NN}}$ = 5.02 TeV and $\sqrt{s_{\rm NN}}$ = 2.76 TeV [30–34] and in Au–Au collisions at $\sqrt{s_{\rm NN}}$ = 0.2 TeV [35, 36].

observed trends. However, given the much improved uncertainties of the experimental measurements, quantitative tensions are observed. The new J/ψ $R_{\rm AA}$ vs. $p_{\rm T}$ at mid-y [32] is compared with the one at forward y [31] in Fig. 4 (right). Although the uncertainties of the mid-y $R_{\rm AA}$ are still large, one can appreciate at low $p_{\rm T}$ a larger $R_{\rm AA}$ at mid- than at forward y, which would indicate a larger contribution of J/ψ regeneration where the QGP energy density and the charm-quark density are larger.

Thermalized charm quarks will be part of the collective motion of the QGP and acquire flow, in particular elliptic flow. Regenerated J/ψ produced by the recombination of thermalized charm quarks will therefore inherit their elliptic flow, resulting in a large v_2 . The new measurements by the ALICE Collaboration of the J/ψ v_2 at forward y in Pb–Pb collisions at $\sqrt{s_{\rm NN}}$ = 5.02 TeV [44] establish, for the first time, an unambiguous non-zero v_2 for J/ψ. Figure 5 shows the J/ψ v_2 versus $p_{\rm T}$ in three classes of centrality, 5-20%, 20-40%, and 40-60%. The J/ψ v_2 increases with $p_{\rm T}$ up to $p_{\rm T}$ ~ 6 GeV/c and decreases or saturates thereafter. The v_2 also appears to be (marginally) maximum in the 20–40% centrality class, in contrast to light hadrons [45]. The v_2 of D-mesons at mid-y and in a somewhat different centrality class (30–50%) [41] is also shown. Clearly, both D and J/ψ mesons show a positive v_2 in the explored $p_{\rm T}$ region, with the J/ψ v_2 being lower than the D-meson at low and intermediate $p_{\rm T}$. Such observation strongly supports the scenario in which both D and J/ψ mesons inherit the v_2

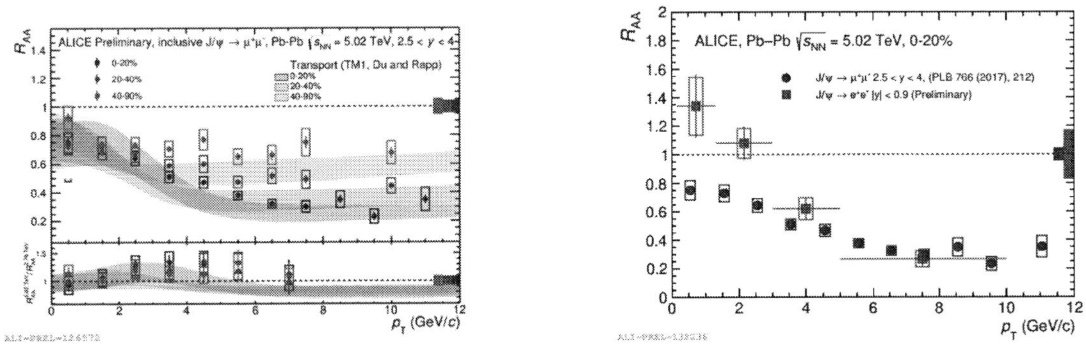

Figure 4. $R_{\rm AA}$ of J/ψ versus $p_{\rm T}$ in Pb–Pb collisions at $\sqrt{s_{\rm NN}}$ = 5.02 TeV: (left) at forward y in three centrality classes [40] and (right) at forward and mid-y in central collisions [31, 32].

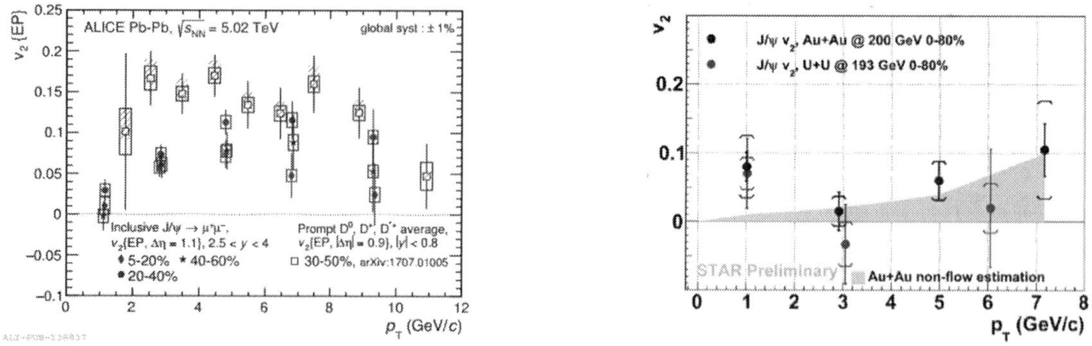

Figure 5. (Left) Inclusive J/ψ $v_2(p_T)$ at forward y in Pb–Pb collisions at $\sqrt{s_{NN}} = 5.02$ TeV for three centrality classes. The D-meson $v_2(p_T)$ at mid-y in the centrality class 30–50% is shown for comparison [41]. (Right) Inclusive J/ψ $v_2(p_T)$ in Au–Au and U–U collisions at $\sqrt{s_{NN}} = 0.2$ and 0.193 TeV, respectively [42, 43].

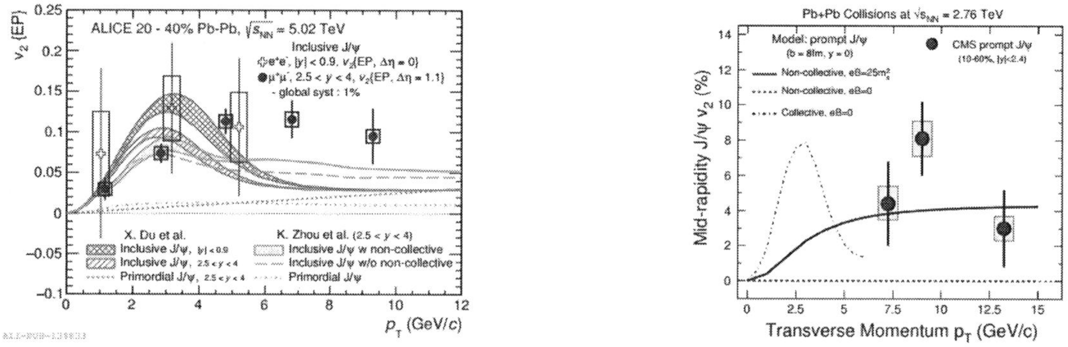

Figure 6. (Left) Inclusive J/ψ $v_2(p_T)$ in semi-central (20–40%) Pb–Pb collisions at $\sqrt{s_{NN}} = 5.02$ TeV [44] compared to transport model calculations by [38] and [10]. (Right) Prompt J/ψ $v_2(p_T)$ in 10–60% non-central Pb–Pb collisions at $\sqrt{s_{NN}} = 2.76$ TeV [46] compared to transport model calculations [10, 47].

from thermalized charm quarks. We note that at RHIC energies, where the regeneration contribution was found to be small, the measured J/ψ v_2 is compatible with zero in Au–Au and U–U collisions at $\sqrt{s_{NN}} = 0.2$ and 0.193 TeV, respectively, as measured by STAR [42, 43] (see Fig. 5 (right)).

Figure 6 (left) shows the J/ψ $v_2(p_T)$ at forward and mid-y in semi-central Pb–Pb collisions at $\sqrt{s_{NN}} = 5.02$ TeV compared to the calculations of two transport models. The model by Du *et al.* [38] reproduces the magnitude of the J/ψ v_2 at forward y and low p_T by a dominant contribution of J/ψ elliptic flow inherited from thermalized charm quarks. However, the overall shape of the $v_2(p_T)$ is missed and the v_2 at high p_T is underestimated. This disagreement suggests a missing mechanism in the model. Similar conclusions can be drawn from the comparison to the model by Zhou *et al.* [10]. The authors add a v_2 arising from a possible strong magnetic field in the early stage of heavy-ion collisions [47], which improves the comparison with the measurement, especially at high p_T. Such non-collective component [47] was able to reproduce the prompt J/ψ v_2 at high p_T measured by CMS in Pb–Pb collisions at $\sqrt{s_{NN}} = 2.76$ TeV [46] (see Fig. 6 (right)).

We note that some tension is also seen between the transport model calculations and the J/ψ R_{AA} measurements in the intermediate region $4 < p_T < 8$ GeV/c. Clearly, the improved precision of both

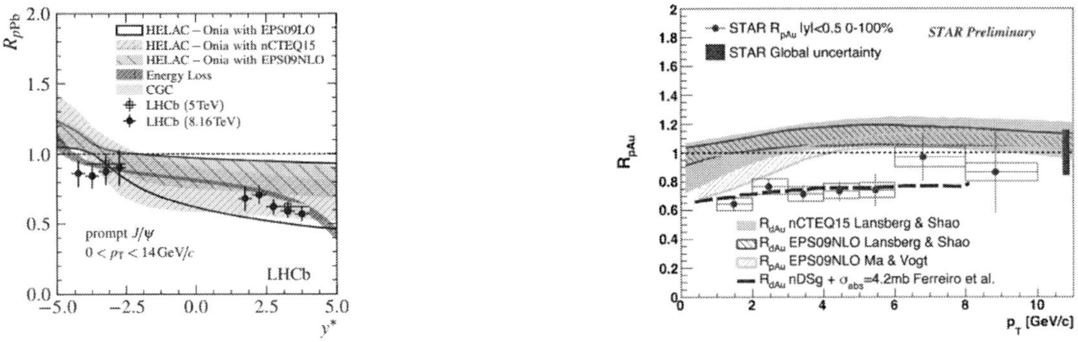

Figure 7. (Left) $R_{\rm pPb}$ of prompt J/ψ versus rapidity in p–Pb collisions at $\sqrt{s_{\rm NN}}$ = 8.16 TeV [52]. (Right) J/ψ $R_{\rm pAu}$ at $\sqrt{s_{\rm NN}}$ = 0.2 TeV [43] compared to various model calculations [48, 50, 51].

$R_{\rm AA}$ and v_2 measurements put strong constraints in the production models and will help to further improve our understanding of the J/ψ production mechanism in heavy-ion collisions.

4 Cold Nuclear Matter effects

In addition to the suppression and regeneration in the QGP, CNM effects also affect the production of quarkonia in heavy-ion collisions. Also here the recent improvement of the measurement uncertainties are seriously challenging the theoretical calculations.

The prompt J/ψ $R_{\rm pPb}$ measured by LHCb in p–Pb collisions at $\sqrt{s_{\rm NN}}$ = 8.16 TeV is shown in Fig. 7 (left) versus rapidity. The results in the p-going ($y > 0$) and Pb-going ($y < 0$) directions are compared to model calculations including shadowing [48], parton energy loss [18], or CGC [49]. While the models are able to qualitatively describe the measured $R_{\rm pPb}$, the quantitative agreement is not satisfactory. Figure 7 (right) shows the new J/ψ $R_{\rm pAu}$ measured by STAR at $\sqrt{s_{\rm NN}}$ = 0.2 TeV compared to model calculations [48, 50, 51]. At RHIC energies, the inclusion of a nuclear absorption cross section $\sigma_{\rm abs}$ = 4.2 mb^{-1} [51] is necessary to describe the $R_{\rm pAu}$.

In the bottomonium sector, we highlight the new measurement by ATLAS of the Υ(1S) $R_{\rm pPb}$ at mid-y in p–Pb collisions at $\sqrt{s_{\rm NN}}$ = 5.02 TeV, which nicely complement the earlier measurements by ALICE [53] and LHCb [54], shown in Fig. 8 (left). A small suppression of the Υ(1S) is seen at mid-y, which could be explained by shadowing (plus comover interactions, as shown in the figure) or parton energy loss (see e.g. [53]). The improvement of the measurements in the backward region will be important to distinguish between nPDF parameterizations with large or small gluon anti-shadowing.

Finally, among the striking results from the p–A programs at RHIC and LHC, we mention the observation of a stronger suppression of the excited states compared to the ground state. Measurements of the production of J/ψ and ψ(2S) at forward and backward y in p–Pb collisions at $\sqrt{s_{\rm NN}}$ = 5.02 TeV have shown that the ψ(2S) is more suppressed than the J/ψ, that this difference is stronger in the Pb-going than in the p-going direction, and that this difference increases with the centrality of the collision [55]. The new measurement by CMS at mid-y shown in Fig. 8 (right) confirms such behavior. The $R_{\rm pPb}$ of the ψ(2S) is smaller than that of the J/ψ, and the difference increases from positive to negative rapidity. The CMS and ATLAS collaborations have also observed a stronger suppression of the Υ(2S) compared to the Υ(1S) at mid-y in p–Pb collisions at $\sqrt{s_{\rm NN}}$ = 5.02 TeV [56, 57]. The observed difference is currently reproduced only by theoretical models including, in addition to initial-state effects, some final-state interaction with the comoving hadronic medium [21, 38].

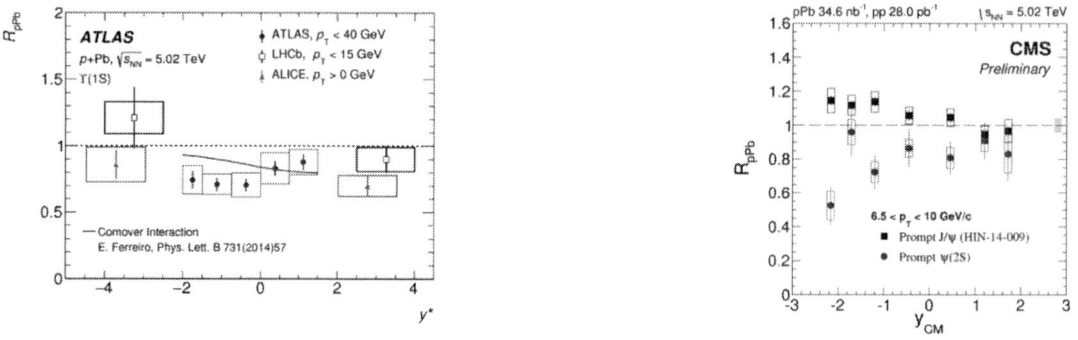

Figure 8. (Left) R_{pPb} of Υ(1S) versus rapidity in p–Pb collisions at $\sqrt{s_{NN}}$ = 5.02 TeV measured by ALICE [53], ATLAS [57] and LHCb [54]. (Right) R_{pPb} of J/ψ and ψ(2S) at mid-y measured by CMS [58, 59].

5 Summary

In summary, the field of quarkonium production in heavy-ion collisions is a very dynamic and prolific one. In the last years the following picture has emerged. The production of quarkonia is suppressed in the QGP, with the loosely bound states being more suppressed than the tightly bound ones. The sequential suppression is evident in the bottomonium sector at both RHIC and LHC energies and for high p_T J/ψ. At high energies, new quarkonia are produced by statistical hadronization or regeneration from thermalized charm quarks in the QGP. Such regeneration mechanism is mostly seen for low p_T J/ψ at the LHC. While this picture is supported by numerous measurements, the high quality of the new results are challenging the available models and calling for a coherent description of all of them.

References

[1] T. Matsui, H. Satz, Phys. Lett. **B178**, 416 (1986)
[2] F. Karsch, M.T. Mehr, H. Satz, Z. Phys. **C37**, 617 (1988)
[3] Y. Park, K.I. Kim, T. Song, S.H. Lee, C.Y. Wong, Phys. Rev. **C76**, 044907 (2007), 0704.3770
[4] X. Zhao, R. Rapp, Phys. Lett. **B664**, 253 (2008), 0712.2407
[5] L. Grandchamp, R. Rapp, Phys. Lett. **B523**, 60 (2001), hep-ph/0103124
[6] P. Braun-Munzinger, J. Stachel, Phys. Lett. **B490**, 196 (2000), nucl-th/0007059
[7] R.L. Thews, M. Schroedter, J. Rafelski, Phys. Rev. **C63**, 054905 (2001), hep-ph/0007323
[8] L. Yan, P. Zhuang, N. Xu, Phys. Rev. Lett. **97**, 232301 (2006), nucl-th/0608010
[9] Y. Liu, Z. Qu, N. Xu, P. Zhuang, Phys. Lett. **B678**, 72 (2009), 0901.2757
[10] K. Zhou, N. Xu, Z. Xu, P. Zhuang, Phys. Rev. **C89**, 054911 (2014), 1401.5845
[11] X. Zhao, R. Rapp, Nucl. Phys. **A859**, 114 (2011), 1102.2194
[12] A. Emerick, X. Zhao, R. Rapp, Eur. Phys. J. **A48**, 72 (2012), 1111.6537
[13] X. Du, R. Rapp, J. Phys. Conf. Ser. **779**, 012042 (2017), 1609.04868
[14] K.J. Eskola, H. Paukkunen, C.A. Salgado, JHEP **04**, 065 (2009), 0902.4154
[15] D. de Florian, R. Sassot, P. Zurita, M. Stratmann, Phys. Rev. **D85**, 074028 (2012), 1112.6324
[16] M. Hirai, S. Kumano, T.H. Nagai, Phys. Rev. **C76**, 065207 (2007), 0709.3038
[17] S. Gavin, J. Milana, Phys. Rev. Lett. **68**, 1834 (1992)
[18] F. Arleo, S. Peigne, JHEP **03**, 122 (2013), 1212.0434

[19] R. Vogt, Phys. Rev. **C81**, 044903 (2010), 1003.3497
[20] E.G. Ferreiro, F. Fleuret, J.P. Lansberg, N. Matagne, A. Rakotozafindrabe, Eur. Phys. J. **C73**, 2427 (2013), 1110.5047
[21] E.G. Ferreiro, Phys. Lett. **B749**, 98 (2015), 1411.0549
[22] V. Khachatryan et al. (CMS), Phys. Lett. **B770**, 357 (2017), 1611.01510
[23] CMS (CMS), CMS-PAS-HIN-16-023 (2017)
[24] B. Krouppa, R. Ryblewski, M. Strickland, Phys. Rev. **C92**, 061901 (2015), 1507.03951
[25] B. Krouppa, M. Strickland, Universe **2**, 16 (2016), 1605.03561
[26] Z. Ye (STAR), Nucl. Phys. **A967**, 600 (2017)
[27] A. Francisco (ALICE), These proceedings (2017)
[28] B.B. Abelev et al. (ALICE), Phys. Lett. **B738**, 361 (2014), 1405.4493
[29] K. Zhou, N. Xu, P. Zhuang, Nucl. Phys. **A931**, 654 (2014), 1408.3900
[30] B. Abelev et al. (ALICE), Phys. Rev. Lett. **109**, 072301 (2012), 1202.1383
[31] J. Adam et al. (ALICE), Phys. Lett. **B766**, 212 (2017), 1606.08197
[32] R.T. Jiménez Bustamante (ALICE), Nucl. Phys. **A967**, 576 (2017)
[33] J. Adam et al. (ALICE), JHEP **05**, 179 (2016), 1506.08804
[34] J. Adam et al. (ALICE), JHEP **07**, 051 (2015), 1504.07151
[35] A. Adare et al. (PHENIX), Phys. Rev. **C84**, 054912 (2011), 1103.6269
[36] A. Adare et al. (PHENIX), Phys. Rev. Lett. **98**, 232301 (2007), nucl-ex/0611020
[37] A. Andronic, P. Braun-Munzinger, K. Redlich, J. Stachel, Nucl. Phys. **A904-905**, 535c (2013), 1210.7724
[38] X. Du, R. Rapp, Nucl. Phys. **A943**, 147 (2015), 1504.00670
[39] E.G. Ferreiro, Phys. Lett. **B731**, 57 (2014), 1210.3209
[40] M. Tarhini (ALICE), Nucl. Phys. **A967**, 588 (2017)
[41] S. Acharya et al. (ALICE) (2017), 1707.01005
[42] L. Adamczyk et al. (STAR), Phys. Rev. Lett. **111**, 052301 (2013), 1212.3304
[43] X. Huang (STAR), These proceedings (2017)
[44] S. Acharya et al. (ALICE) (2017), 1709.05260
[45] B.B. Abelev et al. (ALICE), JHEP **06**, 190 (2015), 1405.4632
[46] V. Khachatryan et al. (CMS), Eur. Phys. J. **C77**, 252 (2017), 1610.00613
[47] X. Guo, S. Shi, N. Xu, Z. Xu, P. Zhuang, Phys. Lett. **B751**, 215 (2015), 1502.04407
[48] H.S. Shao, Comput. Phys. Commun. **198**, 238 (2016), 1507.03435
[49] B. Ducloué, T. Lappi, H. Mäntysaari, Phys. Rev. **D94**, 074031 (2016), 1605.05680
[50] J.P. Lansberg, H.S. Shao, Eur. Phys. J. **C77**, 1 (2017), 1610.05382
[51] E.G. Ferreiro, F. Fleuret, J.P. Lansberg, N. Matagne, A. Rakotozafindrabe, Few Body Syst. **53**, 27 (2012), 1201.5574
[52] R. Aaij et al. (LHCb), Phys. Lett. **B774**, 159 (2017), 1706.07122
[53] B.B. Abelev et al. (ALICE), Phys. Lett. **B740**, 105 (2015), 1410.2234
[54] R. Aaij et al. (LHCb), JHEP **07**, 094 (2014), 1405.5152
[55] J. Adam et al. (ALICE), JHEP **06**, 050 (2016), 1603.02816
[56] S. Chatrchyan et al. (CMS), JHEP **04**, 103 (2014), 1312.6300
[57] M. Aaboud et al. (ATLAS) (2017), 1709.03089
[58] A.M. Sirunyan et al. (CMS), Eur. Phys. J. **C77**, 269 (2017), 1702.01462
[59] CMS (CMS), CMS-PAS-HIN-16-015 (2017)

EPJ Web of Conferences **171**, 04003 (2018) https://doi.org/10.1051/epjconf/201817104003
SQM 2017

Heavy-quark dynamics in a hydrodynamically evolving medium

Marlene Nahrgang[1,⋆], *Jörg* Aichelin[1], *Pol Bernard* Gossiaux[1], and *Klaus* Werner[1]

[1] *SUBATECH, UMR 6457, IMT Atlantique, Université de Nantes, IN2P3/CNRS, 4 rue Alfred Kastler, 44307 Nantes cedex 3, France*

Abstract. In this talk we will discuss the recent advances in describing heavy-quark dynamics in the quark-gluon plasma (QGP), which evolves hydrodynamically. Special emphasis is put on the collective flow of the heavy-quarks with the medium constituents, for which we present our latest results obtained within the MC@sHQ+EPOS2 model at $\sqrt{s} = 5$ TeV.

1 Heavy quarks as probes of the QGP medium

Heavy quarks, especially charm and bottom, have since long been considered valuable probes for properties of the QGP. While the "standard" model of describing the space-time evolution of the bulk medium produced in heavy-ion collisions typically relies on a plasma phase that can be described by hydrodynamics, the heavy-quarks are produced in initial hard scatterings and therefore not equilibrated with the QGP at τ_0, the initial time of hydrodynamics. For the soft and light sector the typical observables are related to collective phenomena on the hydrodynamical hypersurface, for which the memory of microscopic interactions is lost. For heavy quarks, however, some of this memory is kept and we can thus study their dynamics in order to learn about the underlying QCD force.

Since it became possible to produce a significant number of charm and then also bottom quarks at RHIC and the LHC, the theoretical description has produced a variety of models. Although some of them are limited to average energy loss calculations in a medium of average temperature, most approaches follow these general steps in their description:

- **Initial production:** For the initial production one relies on the theoretical results for momentum spectra in proton-proton collisions, like the FONLL [1] formalism. This gives a successful comparison to experimental data for inclusive spectra. For more exclusive spectra, like initial correlations between the produced heavy quark-antiquark pair, one applies event generators, either Pythia [2] or those that couple NLO pQCD matrix elements with parton showers, like POWHEG or MC@NLO [3, 4]. For nuclear collisions, it seems appropriate to include additional cold nuclear matter effects, like shadowing of low-momentum production, via sets of nuclear parton distribution functions [5]. One has to keep in mind that these fits come with relatively large uncertainties.
- **In-medium interaction between heavy-quarks and medium:** After the heavy-quark formation time or the equilibration time of the QGP medium, τ_0, the heavy-quarks start interacting with the medium. Depending on the dynamical evolution equation, see below, the interaction is either described effectively via Fokker-Planck transport coefficients or from scattering cross sections with

⋆e-mail: marlene.nahrgang@subatech.in2p3.fr

the medium constituents. For both approaches, the local temperature and the velocities are obtained from a hydrodynamical evolution. This hydrodynamical evolution might also be replaced by a full partonic microscopic transport model. In order to obtain reliable results it is important that the space-time evolution of the QGP is well tested against the ample experimental data available for the bulk observables.

- **Hadronization:** Since comparison to experimental data occurs on the level of hadronic particles, it is necessary to perform a hadronization around the transition temperature of the confinement-deconfinement phase transition. There are basically two different mechanism applied: coalescence of a heavy quark with a light quark of the medium, which is most likely to happen at small momenta, or fragmentation of a heavy quark, predominantly happening at larger momenta. While high-momentum fragmentation is rather well constraint from comparisons to proton-proton data, the modeling of coalescence in particular contains many unkown non-perturbative effects, which can at best be modeled in an effective way. In the resonance recombination model a hadronization process stemming from the same underlying interaction as the heavy-quark medium interaction is realized [6, 7].
- **and eventually final hadronic interactions:** For a long time, hadronic final interactions seemed not so relevant for final D meson spectra as the hadronic cross sections were expected to be small. There is, however, growing awareness that around the pseudo-critical temperature interaction can also be strong on the hadronic side [8, 9]. For entering the era of precision measurement, it will become important to have a good handle on the hadronic final interactions as well.

2 Hydrodynamics and heavy quarks

The hydrodynamical description of the space-time evolution of the QGP has been very successful in reproducing various experimental data from p_T-spectra to flow harmonics - albeit with different combinations of transport coefficients and initial conditions. The relativistic viscous hydrodynamical model describes the evolution of energy-, momentum and charge density according to the conservation equations:

$$\partial_\mu T^{\mu\nu} = 0 \qquad \partial_\mu N^\mu = 0 \tag{1}$$

with the energy-momentum tensor $T^{\mu\nu} = (e + p)u^\mu u^\nu - pg^{\mu\nu} - \Pi\Delta^{\mu\nu} + \pi^{\mu\nu}$ and $N^\mu = nu^\mu + j^\mu$, which include the ideal part and viscous corrections. The first approach to couple particles to a hydrodynamical evolution is motivated by the non-relativistic Brownian motion and uses Fokker-Planck dynamics. It has the advantage that a knowledge about the nature of the (quasi-)particles of the QGP, or even the assumption of quasiparticles, is not needed and the interaction is encoded in three transport coefficients, the drag coefficient A and the longitudinal and transverse momentum diffusion coefficients $B_\|$ and $B_\perp$. These coefficients depend on the heavy-quark momentum and the medium temperature. The evolution equation for the heavy-quark distribution f_Q is then

$$\frac{\partial}{\partial t} f_Q(t,\vec{p}) = \frac{\partial}{\partial p^i}\left(A^i(\vec{p}) f_Q(t,\vec{p}) + \frac{\partial}{\partial p^j}\left[B^{ij}(\vec{p}) f_Q(t,\vec{p})\right]\right). \tag{2}$$

For a numerical solution of a finite number of heavy-quarks in the medium, the Fokker-Planck equation can be recast to the Langevin equation for individual particles

$$\frac{\mathrm{d}}{\mathrm{d}t}\vec{p} = -\eta_D(p)\vec{p} + \vec{\xi} \quad \text{with} \quad \langle \xi^i(t)\xi^j(t')\rangle = \kappa\delta^{ij}\delta(t-t'), \tag{3}$$

where the transport coefficients in the two equations are related to each other differently depending on whether a pre-point, mid-point or post-point prescription for the time discretization in the stochastic integral of the Langevin process is applied [12].

In order to satisfy detailed balance and thus to describe a system, which in the long-time limit reaches thermal equilibrium [10, 11], the transport coefficients need to fulfill the fluctuation-dissipation theorem, which gives $\eta_D = \frac{\kappa}{2m_Q T}$ and for the spatial diffusion coefficient $D_s = \frac{T}{m_Q \eta_D}$.

The Fokker-Planck or Langevin equation is a second moment approximation of the Boltzmann equation for assumed small momentum transfer

$$\frac{\mathrm{d}}{\mathrm{d}t} f_Q(t, \vec{x}, \vec{p}) = C[f_Q] , \tag{4}$$

with the collision integral

$$C[f_Q] = \int \mathrm{d}\vec{k} [w(\vec{p} + \vec{k}, \vec{k}) f_Q(\vec{p} + \vec{k}) - w(\vec{p}, \vec{k}) f_Q(\vec{p})] \tag{5}$$

with a gain term (first term) and a loss term (second term). The transition probabilities are obtained from the underlying interaction model.

In order to obtain similar results for the Fokker-Planck and the Boltzmann equation it is important to consistently calculate one transport coefficient from the same underlying interaction model and fix the other two from the Einstein relations. It is an interesting question if there is a general consistency between the two approaches and at which point they produce different results for heavy-quark dynamics [13].

3 The heavy-quark diffusion coefficient

The spatial diffusion coefficient is an important quantity to compare different model approaches. It can be obtained with varying precision and accuracy from lattice QCD calculations, various interaction models and from a model-to-data analysis.

3.1 D_s from lattice QCD

Typical temperatures reached in heavy-ion collisions are close to Λ_{QCD} and thus within the nonperturbative regime of QCD. A natural choice would therefore be to turn to lattice QCD calculations. Lattice QCD calculations at finite T are performed in Euclidean space, which makes it very difficult to extract dynamical quantities. The relevant transport coefficients need be inferred from calculable quantities on the lattice, such as the correlation function of conserved currents. This means that integral equations of the following type with kernel K need to be inverted

$$G(\tau; T) = \int_0^\infty \frac{\mathrm{d}\omega}{2\pi} \rho_E(\omega; T) K(\tau, \omega; T) \tag{6}$$

Here, a large precision of the lattice QCD result and a well-motivated ansatz for the spectral function is needed. One then obtains the transport coefficient from the slope of the spectral function ρ_E at $\omega = 0$. For the momentum diffusion

$$\frac{\kappa}{T^3} = \lim_{\omega \to 0} \frac{2T \rho_E(\omega; T)}{\omega} \tag{7}$$

and the spatial diffusion is then $D_s = \frac{2T^2}{\kappa}$.

The range of D_s observed in lattice QCD calculations [14–16] is $D_s \sim (2 - 7)(2\pi T)$. The current uncertainties are still large and certain approximations, concerning quenched QCD, heavy quark vs. charm mass, and/or the missing continuum extrapolation, still limit the reliability of these results.

3.2 D_s from interaction models

There are a variety of different interaction models being used in the study of heavy-quark dynamics in the QGP. A reasonable understanding includes pQCD (inspired models) at large momenta and some effective modeling of non-perturbative effects at lower momenta and temperatures. We will not go into the details of all of these models, but give only a short explanation. The reader may turn to the references for more information.

- **pQCD (inspired models):** MC@sHQ+EPOS2: collisional energy loss with running coupling α_s and one gluon-exchange approximation; radiative energy loss for intermediate momenta, where incoherent processes dominate, and an effective LPM reduction of the emission spectra, Boltzmann evolution coupled to a 3 + 1d ideal fluid dynamical calculation from EPOS2 initial conditions [17–21]; BAMPS, similar energy loss models as MC@sHQ, but the QGP evolution and heavy quarks are both treated in a full Boltzmann approach [22, 23]; Djordjevic et al.: collisional and radiative energy loss in a finite size medium with dynamical scattering centers, running coupling and magnetic masses, but no space-time dynamics of the medium [24–27].
- **nonperturbative approaches:** AdS/CFT [28–30]; PHSD: off-shell transport of light quasiparticles with masses and width obtained from fits to the lattice QCD equation of state, full Boltzmann propagation of heavy-quarks [31, 32]; POWLANG: Langevin dynamics with transport coefficients inspired from lattice QCD [33]; TAMU: thermodynamic T-matrix approach using the lattice QCD internal/free energy of a static $Q\bar{Q}$ pair as input, comprehensive sQGP approach for the equation of state, light quark & gluon spectral functions, quarkonium correlators and heavy quark diffusion, resonance correlations in the T-matrix naturally lead to recombination near T_c from the same underlying interactions [6, 7, 12, 34–36].

3.3 D_s from model-to-data analysis

In most of the above models an explicit or implicit dependence on the parameters of the model is included, in both the description of the soft/light and the heavy sector. Bayesian analysis allows to tune several of these parameters simultaneously by a systematic comparison to the available data. This has been performed with the Duke model of heavy-quark Langevin propagation + UrQMD hadronic rescattering of the D mesons including a parametrized spatial diffusion coefficient in the QGP phase [37]

$$D_s(T,p) = \frac{1}{1+(\gamma^2 p)^2}(D_s 2\pi T)^{\text{lin}}(T;\alpha,\beta) + \frac{(\gamma^2 p)^2}{1+(\gamma^2 p)^2}(D_s 2\pi T)^{\text{pQCD}}(T,p)\,, \tag{8}$$

with $(D_s 2\pi T)^{\text{lin}}(T;\alpha,\beta) = \alpha(1+\beta(T/T_c-1))$. This parametrization contains a linear part at small temperature and momenta and smoothly couples to the pQCD result at higher temperatures and momenta. The background 2d fluid dynamical evolution is tuned via Bayesian analysis to reproduce the available experimental data for the bulk observables, which in addition to some technical parameters gives the temperature dependence of the shear and bulk viscosities [38]. For the heavy-quark analysis various observables from different beam energies and experimental collaborations are used. The final result for the heavy-quark diffusion coefficient as a function of temperature and of momentum is obtained from a multi-step statistical analysis, which not only provides the best fit for α, β and γ, but also provides the confidence level and correlations between these parameters. The obtained diffusion coefficient lies in the range of lattice QCD results and is at the lower end of the range covered by models. Please refer to [37] for details.

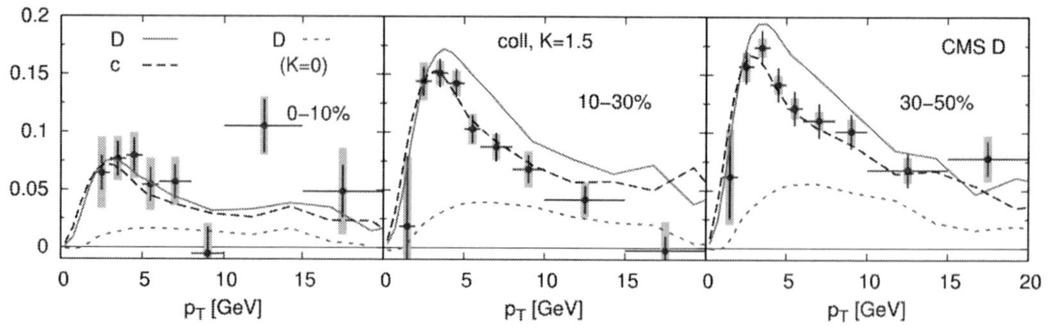

Figure 1. The centrality dependence of the elliptic flow of D mesons and charm quarks as a function of transverse momentum for a purely collisional energy loss scenario, scaled by $K = 1.5$. The solid (orange) curve is for final D mesons, the long dashed (black) curve is for charm quarks at the hadronization temperature of $T_c = 155$ MeV and the short dashed (green) curve is a scenario without any in-medium interactions (corresponding to $K = 0$), where only coalescence contributes to the D meson flow. Experimental data is for D_0 mesons from the CMS collaboration [39].

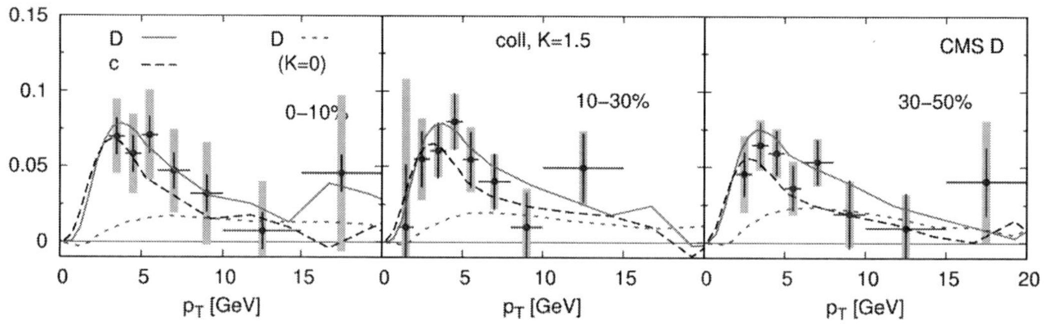

Figure 2. Same as Fig. 1 but for the triangular flow.

4 Collective flow of heavy-quarks

In the remainder of this work we will discuss the possible origins of collective flow of the heavy quarks. For this we will show latest results obtained within the MC@sHQ+EPOS2 model for the highest LHC energy at $\sqrt{s} = 5$ TeV in comparison to the (preliminary) experimental data from the CMS collaboration [39]. Previously, we had published the first predictions for triangular flow of heavy quarks at $\sqrt{s} = 200$ GeV and $\sqrt{s} = 2.76$ TeV in [20]. We compare two different models of energy loss, the purely collisional and the collisional plus radiative energy loss. Both include an overall parameter K, which is fitted to reproduce central $\sqrt{s} = 2.76$ TeV data for the nuclear modification factor R_{AA} in the intermediate p_T range. For this we find $K = 1.5$ for the purely collisional case and $K = 0.8$ for the collisional + radiative scenario.

In Figs 1 - 4 we compare the respective flow harmonics of D mesons to the ones of charm quarks. During hadronization via coalescence, the charm quarks pick up flow from the light quark they recombine with. Therefore the D meson v_n is larger and somewhat shifted in p_T. Another way to look at this contribution is to switch off all in-medium interaction of the charm quarks in the QGP, which

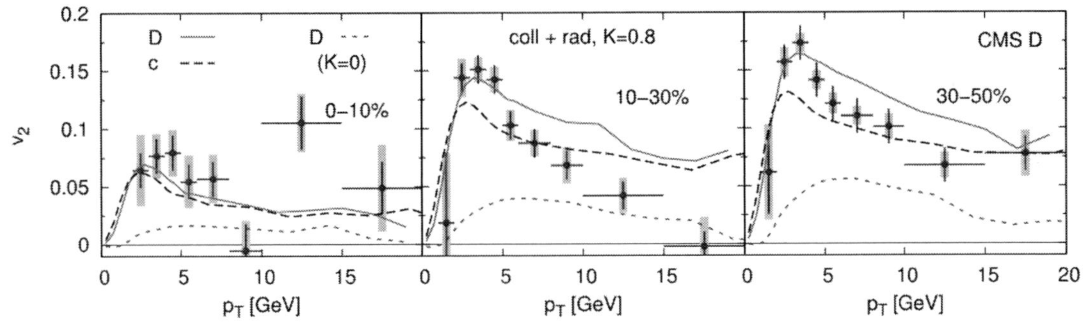

Figure 3. Same as Fig. 1 but for the collisional + radiative energy loss mechanism with $K = 0.8$.

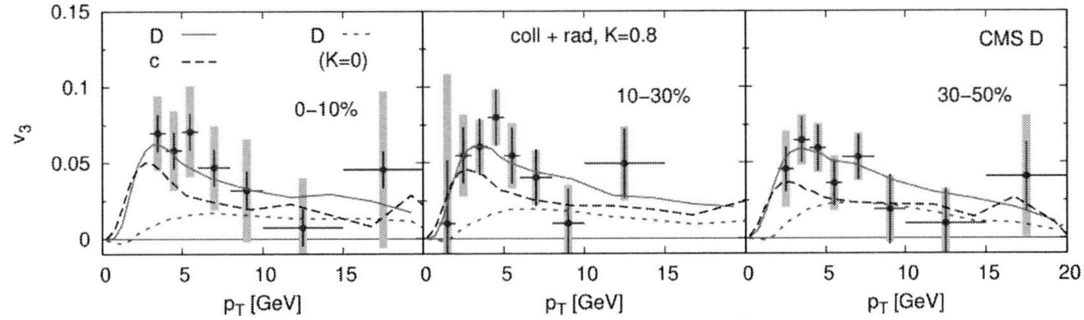

Figure 4. Same as Fig. 3 but for the triangular flow.

means their flow vanishes. The corresponding D meson flow is then only stemming from coalescence, since hadronic interactions are not included in our model.

In the case of elliptic flow we see a clear centrality dependence of both the in-medium and the coalescence contributions, which is clear from the changing geometry. The origin of the triangular flow is, however, due to fluctuations in each centrality class and already the charged particle v_3 shows a very small centrality dependence. Since the medium size and temperature decrease with increasing centrality, it is expected that the charm flow picks up less and less flow from the medium in more peripheral collisions. We can observe in Fig. 2 and 4 that the maximum of the charm quark flow is indeed decreasing for larger centralities. The final D meson flow, is however the same, because the centrality dependence of the light parton flow.

In comparison to the experimental data we conclude that the tuned purely collisional energy loss mechanism seems to be favored. This is a consistent conclusion, which is also obtained from high-momentum R_{AA}. It can be explained by the missing ingredients in the radiative contribution to the energy loss at higher p_T in our model, which has its strengths in the intermediate p_T range. At high momentum finite path length effects lead to additional coherent suppression of energy loss. Moreover, the current radiative vertex has fixed coupling of $\alpha_s = 0.3$. Work in both of these direction is in progress.

Finally, we investigate the sensitivity of the v_2 and the v_3 on the underlying diffusion coefficient, which scales directly with K. For different values of K we calculate the integrated v_2 and v_3 for the

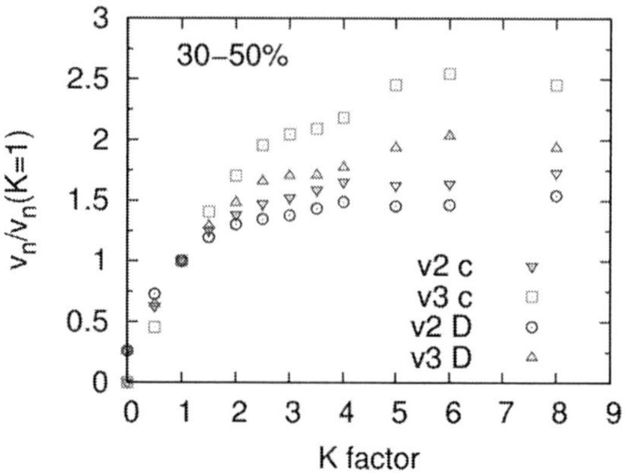

Figure 5. The dependence of the integrated flow harmonics v_2 and v_3 of charm quarks and D mesons on the diffusion coefficient, tuned by a scaling factor K, which multiplies the scattering cross sections. Here, we focus on a purely collisional energy loss mechanism. The curves are scaled to their respective values at $K = 1$ to visualize the relative spread.

collisional energy loss scenario in the 30 – 50% centrality class. The integrated flow harmonics are dominated by the abundant low-momentum charm quarks, where collisional energy loss is expected to the main mechanism. As expected Fig. 5 shows that with increasing K the charm quarks pick up more and more flow from the medium, while the coalescence contribution is independent of K and therefore the D meson flow does not show the same strong increase. Since the charm quarks can, however, not pick up more flow than is carried by the underlying medium expansion, one observes a saturation toward larger values of K. We normalize the curves by their respective values at the nominal $K = 1$ in order to compare the relative sensitivity on K and see that the higher-order triangular flow is more sensitive than the elliptic flow. We conclude that measuring the v_3 with high precision provides very helpful additional information in constraining the heavy-quark diffusion coefficient.

5 Summary

We have reviewed the general approaches to describe the heavy-quark dynamics by a coupling of either the Fokker-Planck or the Boltzmann equation with a (mostly) hydrodynamical evolution of the underlying QGP medium. The input transport coefficients for these calculations are obtained either from lattice QCD or from an appropriate interaction model or from a model-to-data Bayesian analysis.

Finally, we presented result for the elliptic and triangular flow harmonics of charm quarks and D mesons, obtained in our model MC@sHQ+EPOS2. We showed the different contributions stemming from the in-medium interaction of charm quarks with the QGP (quasi)particles and during hadronization via coalescence. The relatively flat centrality dependence of the D meson v_3 as seen in the experimental data, can be explained by a slightly decreasing charm flow and a slightly increasing component from coalescence. The relative sensitivity of v_3 on the transport coefficient is larger than that of the standard v_2. Measurements of v_3 with good precision are therefore valuable additional sources of information about the QCD interaction between heavy flavor and the QGP.

References

[1] M. Cacciari, M. Greco and P. Nason, JHEP **9805** (1998) 007; M. Cacciari, S. Frixione and P. Nason, JHEP **0103** (2001) 006; M. Cacciari, S. Frixione, N. Houdeau, M. L. Mangano, P. Nason and G. Ridolfi, JHEP **1210** (2012) 137.

[2] T. Sjostrand, S. Mrenna and P. Z. Skands, JHEP **0605** (2006) 026

[3] S. Frixione and B. R. Webber, JHEP **0206** (2002) 029.

[4] S. Frixione, P. Nason and B. R. Webber, JHEP **0308** (2003) 007.
[5] K. J. Eskola, H. Paukkunen and C. A. Salgado, JHEP **0904** (2009) 065.
[6] M. He, R. J. Fries and R. Rapp, Phys. Rev. C **82** (2010) 034907.
[7] M. He, R. J. Fries and R. Rapp, Phys. Rev. C **86** (2012) 014903.
[8] L. Tolos and J. M. Torres-Rincon, Phys. Rev. D **88** (2013) 074019.
[9] V. Ozvenchuk, J. M. Torres-Rincon, P. B. Gossiaux, L. Tolos and J. Aichelin, Phys. Rev. C **90** (2014) 054909.
[10] D. B. Walton and J. Rafelski, Phys. Rev. Lett. **84** (2000) 31.
[11] G. D. Moore and D. Teaney, Phys. Rev. C **71** (2005) 064904.
[12] M. He, H. van Hees, P. B. Gossiaux, R. J. Fries and R. Rapp, Phys. Rev. E **88** (2013) 032138.
[13] S. K. Das, F. Scardina, S. Plumari and V. Greco, Phys. Lett. B **747** (2015) 260.
[14] H. T. Ding *et al.*, Phys. Rev. **D86**, 014509 (2012).
[15] D. Banerjee, S. Datta, R. Gavai, and P. Majumdar, Phys. Rev. **D85**, 014510 (2012).
[16] A. Francis, O. Kaczmarek, M. Laine, T. Neuhaus and H. Ohno, Phys. Rev. D **92**, no. 11, 116003 (2015).
[17] P. B. Gossiaux and J. Aichelin, Phys. Rev. C **78** (2008) 014904.
[18] M. Nahrgang, J. Aichelin, P. B. Gossiaux and K. Werner, Phys. Rev. C **90** (2014) no.2, 024907.
[19] M. Nahrgang, J. Aichelin, P. B. Gossiaux and K. Werner, Phys. Rev. C **89** (2014) no.1, 014905.
[20] M. Nahrgang, J. Aichelin, S. Bass, P. B. Gossiaux and K. Werner, Phys. Rev. C **91** (2015) no.1, 014904.
[21] M. Nahrgang, J. Aichelin, P. B. Gossiaux and K. Werner, Phys. Rev. C **93** (2016) no.4, 044909.
[22] O. Fochler, J. Uphoff, Z. Xu and C. Greiner, Phys. Rev. D **88** (2013) no.1, 014018.
[23] J. Uphoff, O. Fochler, Z. Xu and C. Greiner, Phys. Lett. B **717** (2012) 430.
[24] M. Djordjevic and M. Gyulassy, Phys. Rev. C **68** (2003) 034914.
[25] M. Djordjevic, Phys. Rev. C **74** (2006) 064907.
[26] M. Djordjevic and U. W. Heinz, Phys. Rev. Lett. **101** (2008) 022302.
[27] M. Djordjevic and M. Djordjevic, Phys. Lett. B **709** (2012) 229.
[28] S. S. Gubser, Nucl. Phys. B **790** (2008) 175
[29] W. A. Horowitz, Phys. Rev. D **91** (2015) no.8, 085019
[30] R. W. Moerman and W. A. Horowitz, arXiv:1605.09285 [hep-th].
[31] T. Song, H. Berrehrah, D. Cabrera, J. M. Torres-Rincon, L. Tolos, W. Cassing and E. Bratkovskaya, Phys. Rev. C **92** (2015) no.1, 014910.
[32] T. Song, H. Berrehrah, D. Cabrera, W. Cassing and E. Bratkovskaya, Phys. Rev. C **93** (2016) no.3, 034906.
[33] A. Beraudo, A. De Pace, M. Monteno, M. Nardi and F. Prino, Eur. Phys. J. C **75** (2015) no.3, 121.
[34] H. van Hees, M. Mannarelli, V. Greco and R. Rapp, Phys. Rev. Lett. **100**, 192301 (2008).
[35] F. Riek and R. Rapp, Phys. Rev. C **82** (2010) 035201
[36] S. Y. F. Liu and R. Rapp, arXiv:1612.09138 [nucl-th].
[37] Y. Xu, M. Nahrgang, S. Cao, J. E. Bernhard and S. A. Bass, [arXiv:1710.00807 [nucl-th]].
[38] J. E. Bernhard, J. S. Moreland, S. A. Bass, J. Liu and U. Heinz, Phys. Rev. C **94** (2016) no.2, 024907.
[39] CMS Collaboration [CMS Collaboration], CMS-PAS-HIN-16-007.

Update on Heavy Flavor Production in Cold Matter

R. Vogt[1,2,a]

[1]*Nuclear and Chemical Sciences Division, Lawrence Livermore National Laboratory, Livermore, CA 94550, USA*
[2]*Physics Department, University of California at Davis, Davis, CA 95616, USA*

Abstract. I will discuss areas of heavy flavor theory where new progress has been made.

1 Introduction

I first focus on open heavy flavor and quarkonium production in $p + p$ collisions before going on to discuss some effects due to cold nuclear matter. The length of a conference proceeding is too short for a complete review of these topics in $p + p$ and $p + A$ interactions. For a somewhat more detailed discussion of open heavy flavor and quarkonium production in $p + p$ collisions, as well as additional references, see Ref. [1].

2 Open heavy flavor production

There are currently two approaches to heavy flavor production at colliders: collinear factorization and low x k_T-factorization approach. I briefly describe each one as applied to single inclusive heavy flavor observables. I then turn to exclusive observables and correlated heavy quark pair production.

There are two main methods of calculating the spectrum of single inclusive open heavy flavor production in perturbative QCD assuming collinear factorization. The underlying idea is similar but the technical approach differs. Both note that large logarithms of p_T/m arise at all orders of the perturbative expansion, spoiling the convergence. The first terms in the expansion are the leading (LL) and next-to-leading logarithmic (NLL) terms, $\alpha_s^2[\alpha_s \log(p_T/m)]^k$ and $\alpha_s^3[\alpha_s \log(p_T/m)]^k$ respectively.

It is worth noting that the single inclusive heavy flavor p_T distribution is finite at leading order (LO) as $p_T \rightarrow 0$ because of the finite quark mass scale. A next-to-leading order (NLO) calculation that assumes production of massive quarks but neglects LL terms, a "massive" formalism, can result in large uncertainties at high p_T [2–4]. (This massive formalism is sometimes referred to as a fixed-flavor-number (FFN) scheme.) If, instead, the heavy quark is treated as "massless" and the LL and NLL corrections absorbed into the fragmentation functions, the approach breaks down as p_T approaches m even though it improves the result at high p_T. The massless formalism is also sometimes referred to as the zero-mass, variable-flavor-number (ZM-VFN) scheme [5, 6]. There are schemes which interpolate smoothly between the FO/FFN scheme at low p_T and the massless/ZM-VFN scheme. The fixed-order next-to-leading logarithm (FONLL) approach is one such scheme. In

[a]e-mail: rlvogt@lbl.gov

FONLL, the fixed order and fragmentation function approaches are merged so that the mass effects are included in an exact calculation of the leading (α_s^2) and next-to-leading (α_s^3) order cross section while also including the LL and NLL terms [7]. The NLO fixed order (FO) result is combined with a calculation of the resumed (RS) cross section in the massless limit. The FO and RS approaches need to be calculated in the same renormalization scheme. The FONLL result is then, schematically,

$$\text{FONLL} = \text{FO} + (\text{RS} - \text{FOM0})G(m, p_T) . \tag{1}$$

The interpolating function $G(m, p_T) \sim p_T^2/(p_T^2 + (cm)^2)$ is arbitrary but must approach unity for $m/p_T \to 0$. Fragmentation functions for D and D^* mesons have been calculated in an approach consistent with an FONLL calculation [8]. The second interpolation scheme is the generalized-mass variable-flavor-number (GM-VFN) scheme [9]. The large logarithms in charm production for $p_T \gg m$ are absorbed in the charm parton distribution function and can thus be evolved by the evolution equations for the parton distributions. The logarithmic terms can be incorporated into the hard cross section to achieve better accuracy for $p_T \geq m$. By adjusting the mass-dependent subtraction terms, no interpolating function is required [9].

In the k_T-factorization approach, off-shell leading order matrix elements for $g^*g^* \to c\bar{c}$ are used together with unintegrated gluon densities that depend on the transverse momentum of the gluon, k_T, as well as the usual dependence on x and μ_F. The motivation for choosing the k_T-factorized approach is that, at sufficiently low x, collinear factorization should no longer hold.

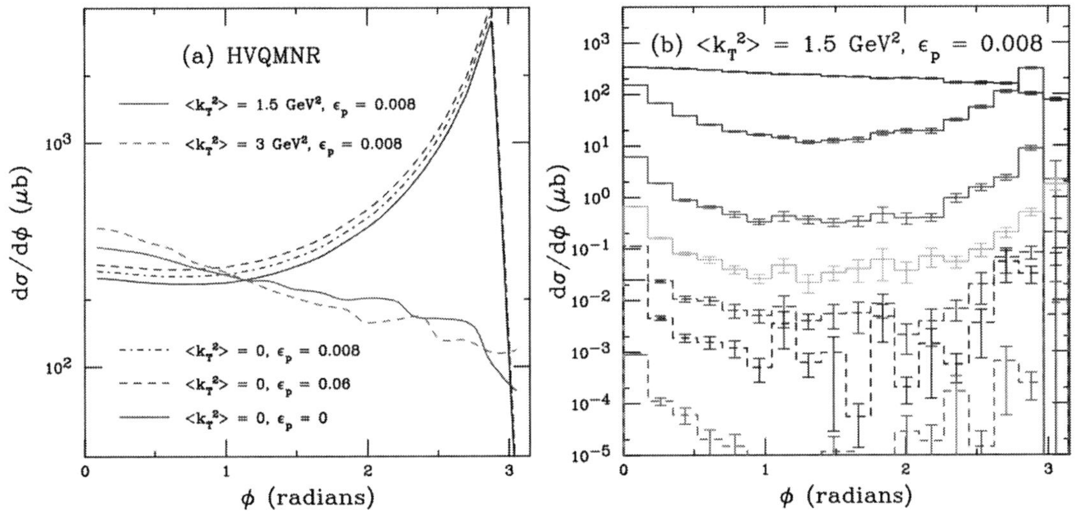

Figure 1. Calculations of the NLO azimuthal distribution between two charm quarks using the HVQMNR code at $\sqrt{s} = 7$ TeV and forward rapidity. (a) The blue solid, dashed and dot-dashed curves, all without k_T broadening, show the effect of modifying the charm fragmentation function. The red solid curve uses a k_T broadening consistent with J/ψ production at 7 TeV [14] while the dashed magenta curve shows the effect of doubling the k_T broadening. (b) Using $\langle k_T^2 \rangle = 1.5$ GeV^2 and $\epsilon_p = 0.008$ from the solid red curve in (a), the p_T cuts on the charm quarks are varied. From top to bottom the results are: $p_T < 10$ GeV; $p_T > 10$ GeV; $p_T > 20$ GeV; $p_T > 30$ GeV; $p_T > 40$ GeV; $p_T > 50$ GeV; $p_T > 75$ GeV. The curves for all but the lowest p_T cut are scaled up by 10^3.

The LHC data has been compared to calculations in both approaches and those assuming collinear factorization compare well with the LHC data. Recent ALICE data [10], at $0 < p_T < 2$ GeV supports collinear factorization. Their D^0 results at 7 TeV for $|y| < 0.5$, were compared to FONLL, GV-VFNS and LO k_T-factorization calculations. Only the k_T-factorized calculation is inconsistent with the shape of the data in the p_T range where the calculation should best apply. The forward rapidity data of LHCb at 7 TeV [11] and 13 TeV [12], also agree well with the collinear factorization assumption. The recent 13 TeV data from LHCb is within the uncertainty bands of FONLL and POWHEG (discussed below) down to $p_T \to 0$, albeit with large uncertainty bands and near the upper limit of the band. The GM-VFNS calculation is given only for $p_T > 3$ GeV but agrees well with the data with small uncertainties [12]. Perhaps a NLO k_T-factorized result could lead to improved agreement with the data but so far collinear factorization appears to still work well for low x charm production.

A further challenge to theory may arise from measurements of correlated production, specifically of the azimuthal opening angle of heavy flavors, either by direct reconstruction of both D mesons or of a $\overline{D}$ meson and the decay product of its partner, either a light hadron or a lepton [13]. Naively, at LO $Q\overline{Q}$ pairs are produced back-to-back with a peak at $\Delta\phi = \pi$. Higher order production will, however, result in a more isotropic distribution in $\Delta\phi$ due to light parton emission in the final state.

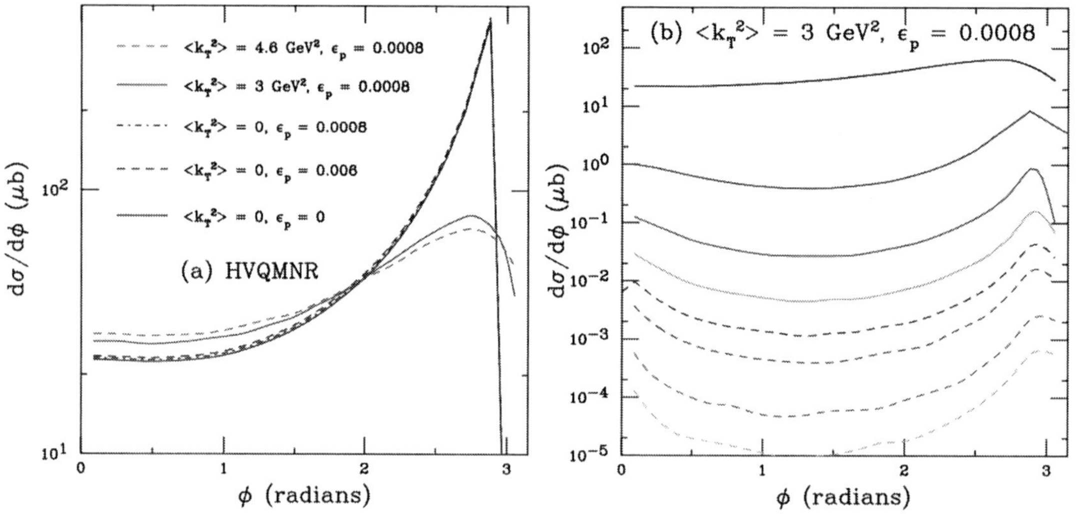

Figure 2. Calculations of the NLO azimuthal distribution between two bottom quarks using the `HVQMNR` code at $\sqrt{s} = 7$ TeV and forward midrapidity. (a) The blue solid, dashed and dot-dashed curves, all without k_T broadening, show the effect of modifying the charm fragmentation function. The red solid curve uses the k_T broadening for Υ production at 7 TeV while the dashed magenta curve shows the effect of doubling the k_T broadening. (b) Using $\langle k_T^2 \rangle = 3$ GeV^2 and $\epsilon_p = 0.0008$ from the solid red curve in (a), the p_T cuts on the bottom quarks are varied. From top to bottom the results are: $p_T < 10$ GeV; $p_T > 10$ GeV; $p_T > 20$ GeV; $p_T > 30$ GeV; $p_T > 40$ GeV; $p_T > 50$ GeV; $p_T > 75$ GeV and $p_T > 100$ GeV. No scale factor is included on the results.

The FONLL and GM-VFNS approaches are for single inclusive production only. There are NLO heavy flavor codes that, in addition to inclusive heavy flavor production, also calculate exclusive $Q\overline{Q}$ pair production. `HVQMNR` [15] uses negative weight events used to cancel divergences numerically. Smearing the parton momentum through the introduction of intrinsic transverse momenta, k_T, reduces the importance of the negative weight events. `HVQMNR` does not include any resummation. `POWHEG – hvq` [16] is a positive weight generator that includes leading-log resummation. The entire event is available since `PYTHIA` [17] and `HERWIG` [18] are employed to produce the complete event. In addition to these NLO codes, heavy flavor correlations can also be simulated employing LO event generators such as PYTHIA. The k_T-factorization approach can also be employed to calculate correlated $c\overline{c}$ production since the unintegrated parton densities have a transverse component, giving finite p_T and $\Delta\phi$ distributions even at LO.

There are LHC data, particularly from ALICE [19] and LHCb [20], on charm pair correlations to test these approaches.

LHCb measured $c\overline{c}$, cc, and cJ/ψ correlations in $p + p$ collisions at 7 TeV for $2 < y < 4$. About 10% of the $c\overline{c}$ rate was found in the cc and cJ/ψ channels [20]. Only the $c\overline{c}$ pairs can be produced in a single hard scattering. The other events are more consistent with production through double parton scattering. The azimuthal and rapidity distributions for the cc events are consistent with isotropic emission [20], as might be expected from two hard scatterings.

The ALICE collaboration recently presented an analysis of azimuthal correlations between reconstructed D mesons and a light hadron trigger particle [19]. The light hadrons were primary particles, emitted from the collision points. These particles include those from heavy flavor decays, such as the unreconstructed partner D meson. The data were binned according to the transverse momentum of both the D meson and the light hadron. The minimum light hadron p_T was soft, $p_T > 0.3$ GeV. These data were further subdivided into two p_T ranges, $0.3 < p_T < 1$ GeV and $p_T > 1$ GeV. The D meson p_T was considerably higher: $3 < p_T < 5$ GeV, $5 < p_T < 8$ GeV, and $8 < p_T < 16$ GeV. To improve statistics, the "D meson" is an average over the D^0, D^+ and $D*^+$. The ALICE measurements cover the central region, $|y| < 0.5$ for the D and $|\Delta\eta| < 1$ for the light hadron. The general behavior is, however, the same as the LHCb $D^0\overline{D}^0$ pairs, a peak at $\Delta\phi = 0$ and a smaller enhancement at $\Delta\phi = \pi$. The peak at $\Delta\phi = 0$ increases with increasing trigger particle p_T and also with increasing D meson p_T. The data were compared to simulations with various PYTHIA tunes and also POWHEG+PYTHIA. All simulations reproduced the trends of the data [19].

Figure 1(a) shows how the ϕ distribution between the c and $\overline{c}$ changes with the fragmentation parameter ϵ_p and the strength of the k_T broadening chosen. Note that the NLO result is not a delta function at $\phi = \pi$, as at LO, due to the emission of light partons in the final state, even when $\epsilon_p = 0$. Changing ϵ_p does not have a strong effect on the shape of $d\sigma/d\phi$. However, introducing k_T broadening changes $d\sigma/d\phi$ from peaked at $\phi \sim \pi$ to more isotropic with a maximum at $\phi = 0$ instead. No p_T cut is made on the momentum of the charm quarks. The effect of a p_T-dependent cut on the charm quarks is explored in Fig. 1(b) using $\langle k_T^2 \rangle = 1.5$ GeV^2, consistent with the broadening needed to describe the J/ψ p_T dependence in Ref. [14], and $\epsilon_p = 0.008$, softer than the traditional Peterson function. With these choices, the HVQMNR p_T distribution is in agreement with the FONLL D^0 p_T distribution for the same choice of charm mass and scale parameters. When only low p_T charm quarks are considered, $p_T < 10$ GeV, the azimuthal distribution is rather flat. Increasing the p_T cut on the charm quarks results in distributions that are peaked at $\phi = 0$ and π: either the charm quarks are aligned opposite a high momentum light parton ($\phi = 0$) or the charm quarks are effectively back-to-back with a soft light parton emitted ($\phi \sim \pi$).

Figure 2 shows a similar calculation for bottom quarks. The effect of changing the fragmentation function is reduced for the heavier bottom quarks, as is the effect of k_T broadening. The harder p_T dependence for b quarks allows the azimuthal distribution to be calculated at higher p_T, up to $p_T > 100$ GeV with higher statistics. The same trends are observed for bottom and charm quarks.

It appears that, so far, the open charm results presented by the LHC collaborations, both single inclusive production and charm pair correlations, are in agreement with calculations based on collinear factorization with single hard scatterings. The exception, the cc and cJ/ψ events at LHCb, are consistent with double parton scattering.

3 Quarkonium Production

I now turn to quarkonium production. Perturbative QCD is employed to calculate the production characteristics. A number of models have been used since the discovery of the J/ψ just over 40 years ago. One of the first was the color evaporation model (CEM) [21–23] which does not specify the color or spin state of the produced $Q\overline{Q}$ pairs. The next was the color singlet model (CSM) [24] which only considered quarkonium states produced as color neutral objects. Despite the addition of a number of higher order corrections, a good description of the data has not been achieved. The current most successful approach, non-relativistic QCD (NRQCD) [25, 26], is based on an expansion of the cross section in both the strong coupling constant and $Q\overline{Q}$ velocity with a separation of the hard and soft scales so that contributions to the cross section are divided into different color states with different weight factors, long distance matrix elements (LDMEs), assumed to be universal, which are adjusted to data. Most calculations have been done in the collinear factorization approach. However, some work has also been done in k_T-factorization both in the context of the CSM and NRQCD. In the remainder of this section, I will discuss only recent results in NRQCD and the CEM.

According to the NRQCD factorization theorem for J/ψ production, the production cross section can be written as

$$\sigma_{J/\psi} = \Sigma_n \sigma_{c\overline{c}[n]} \langle O^{J/\psi}[n] \rangle \qquad (2)$$

where the sum over n includes all Fock states, including color octet states [25]. The cross section $\sigma_{c\overline{c}[n]}$ is the production rate of a $c\overline{c}$ pair in the color and spin state n, calculated in perturbative QCD. Finally, $\langle O^{J/\psi}[n] \rangle$ represents the LDMEs which describe the conversion of the $c\overline{c}[n]$ state into a final state J/ψ, assuming that the hadronization does not change the spin or momentum. The full cross section includes convolution with the parton densities appropriate to the process in question.

In comparisons of the NRQCD calculations to data, the color octet LDMEs are determined above some p_T cut, along with uncertainty on their values. The uncertainty band is determined by fixing the LDMEs to their central fit values and varying the renormalization and factorization scales by a factor of two relative to the central values of $\mu_R = \mu_F = m_T = (p_T^2 + 4m_c^2)^{1/2}$ with $m_c = 1.5$ GeV. No variation of the charm quark mass or the LDMEs are included in the uncertainty band [26].

One of the most important questions for NRQCD to resolve is whether or not the color octet LDMEs are indeed universal. Their universality has been called into increasing question recently. Different groups arrive at different values of the LDMEs depending on the chosen p_T cut, the data sets included in the analysis, and whether or not the polarization data are included in the fits, see Ref. [26] and references therein. In addition, the LDMEs fit to collider data do not agree with the integrated cross sections, $d\sigma/dy|_{y=0}$ [27]. Recent measurements of the η_c at forward rapidity in LHCb [28] strongly disagree with the p_T distributions calculated assuming heavy quark spin symmetry [29].

If one goes beyond current next-to-leading order (NLO) calculations, it is not clear that the p_T dependence would substantially change. The p_T distributions already include the powers giving the hardest p_T distributions. Adopting more fit parameters, including the mass and scales, will not have a strong effect on the p_T distributions. Changing the factorization scale has the largest effect on the shape but modifying the renormalization scale or the quark mass has a larger effect on the magnitude than the shape of the distributions.

It has been suggested that NRQCD factorization does not hold for polarization. However, if factorization holds for the p_T distributions, then, under the assumption that the spin and momentum are unaffected by hadronization, the initial polarization should survive as well.

The difficulties that have plagued the NRQCD description of J/ψ production are reduced for Υ production. The larger mass, higher scale and lower velocity could make the Υ a better candidate for an accurate description by NRQCD. In addition, since there are more Υ states, there are more color octet LDMEs available for fitting, allowing a description of the p_T-dependent yields and the polarization simultaneously.

Some new developments to the color evaporation model make it worth a second look. In the CEM, heavy flavor and quarkonium production is treated on an equal footing. In the CEM, the quarkonium production cross section is some fraction of all $Q\overline{Q}$ pairs below the $H\overline{H}$ threshold where H is the lowest mass heavy-flavor hadron. Thus the CEM cross section is simply the $Q\overline{Q}$ production cross section with a cut on the pair mass. The color and spin have been integrated over so that the color of the state is said to have been 'evaporated' away without changing the kinematics of the pair.

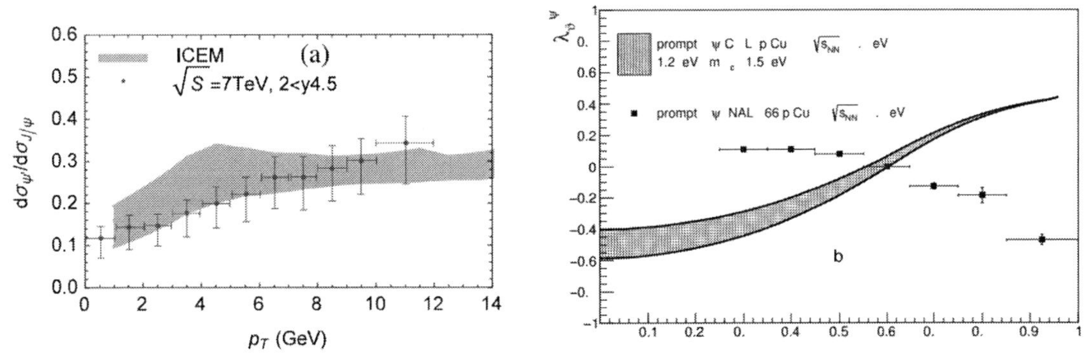

Figure 3. (a) Results for ratio of the ψ' to J/ψ production cross sections at forward rapidity for 7 TeV $p+p$ collisions [30]. (b) The x_F dependence of the polarization parameter λ_θ for prompt production of J/ψ in p+Cu collisions at $\sqrt{s_{NN}} = 38.8$ GeV is compared to the E866/NuSea data. The horizontal uncertainties are the experimental bin widths. From [31].

At leading order, the production cross section of quarkonium state C in a $p+p$ collision is

$$\sigma_C^{\rm CEM}(s_{_{NN}}) = F_C \sum_{i,j} \int_{M^2}^{4m_H^2} d\hat{s} \int dx_1\, dx_2\, f_i^p(x_1,\mu_F^2)\, f_j^p(x_2,\mu_F^2)\, \mathcal{J}(\hat{s})\hat{\sigma}_{ij}(\hat{s},\mu_F^2,\mu_R^2)\ , \tag{3}$$

where $ij = q\overline{q}$ or gg and $\hat{\sigma}_{ij}(\hat{s})$ is the $ij \to Q\overline{Q}$ subprocess cross section. Here $\mathcal{J}(\hat{s})$ is a kinematics-dependent Jacobian. In the traditional CEM, the lower limit on the integration over $\hat{s}$ is $M^2 = 4m_c^2$. The fraction F_C must be universal so that, once it is fixed by data, the quarkonium production ratios should be constant as a function of $\sqrt{s}$, y and p_T. The actual value of F_C depends on the heavy quark mass, m, the scale parameters, the parton densities and the order of the calculation. The same values of the charm quark mass and scale parameters as found in Ref. [14] are employed to obtain the normalization F_C for the J/ψ,

Two recent updates of the CEM are worth noting. The first is a calculation of the ψ'/ψ ratio as a function of p_T in an "improved" CEM calculation (ICEM) [30]. In this calculation, the p_T of the quarkonium state is modified by the ratio of the pair invariant mass to the quarkonium mass, M_C. In addition, the lower limit on the integral is now $M^2 = M_C^2$. Thus a rise in the ratio with p_T consistent with measurement is found [30], see Fig. 3(a). The second is the first calculation of polarization in the CEM. If one assumes that the spin of the quarks are either aligned or anti-aligned with the momentum, $J_z = -1, 0$ or 1, and this alignment, like the momentum of the quarks, survives hadronization, the polarization can be obtained. The first calculation separated the longitudinal and transverse components [32] while the second separated the S and P states to calculate the polarization parameter, λ_ψ

in the ICEM [31]. The results so far are at LO. Therefore, only comparison to fixed-order results is possible so far, as shown for p+Cu collisions at $\sqrt{s}$ = 38.8 GeV. Work is underway to calculate the p_T dependence in the k_T-factorization approach before attempting a full NLO calculation.

4 Heavy Flavors in Cold Nuclear Matter

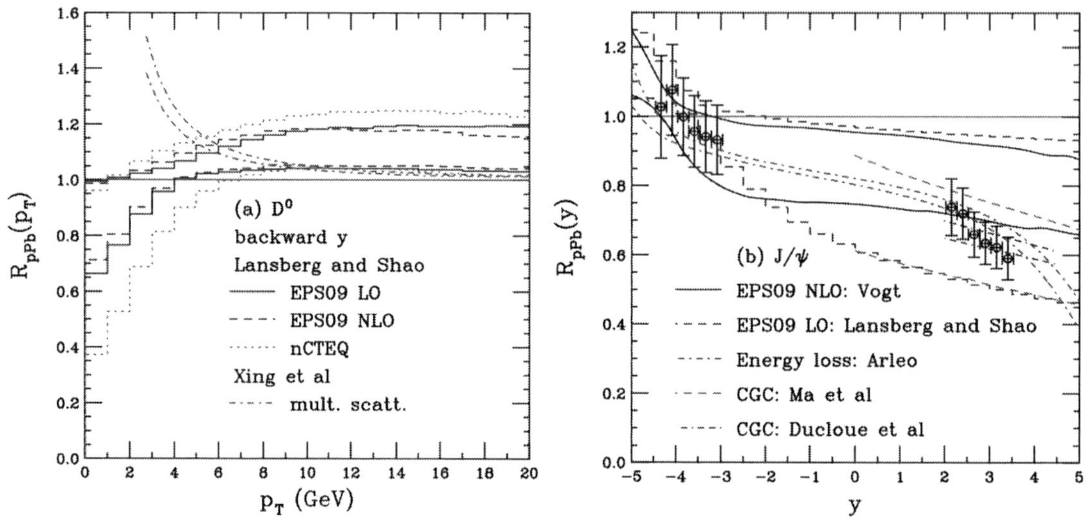

Figure 4. (a) The D^0-meson $R_{pPb}(p_T)$ at $-4 < y < -2.96$. The red band is by Xing *et al.*. The data-driven shadowing results of Lansberg and Shao at $-4.46 < y < -2.96$ are shown as histograms for EPS09 NLO (solid blue), EPS09 LO (dashed blue) and nCTEQ15 (dotted magenta). (b) The J/ψ $R_{pPb}(y)$ for EPS09 NLO by Vogt (black), the data-driven result from Lansberg and Shao for EPS09 LO (blue dashed histogram), the energy loss calculation by Arleo (dot-dashed red), and the CGC calculations by Doucloué *et al.* (blue dot-dashed) and Ma *et al.* (dashed magenta). The ALICE data [34] are shown in black. For details, see Albacete *et al.* [35].

There are several important cold nuclear matter (CNM) effects that need to be taken into account when determining the strength of deconfinement effects on a particular final state. These include modifications of the parton distribution functions in nuclei, either by shadowing in collinear factorization or through the color glass condensate (CGC) approach; energy loss and/or k_T broadening in the Cronin effect; and isospin effects, negligible for heavy flavor production. In addition, for quarkonium effects such as nuclear absorption and comover breakup may play a role. See Ref. [35] for a summary of these effects as well as predictions for the 8 TeV LHC p+Pb run and Refs. [33] for a summary for 5 TeV predictions as well as comparison to data.

Figure 4 shows results for (a) D^0 and (b) J/ψ modifications in p+Pb relative to $p+p$ collisions at 8 TeV. The multiple scattering model of D^0 modifications gives a significantly different ratio $R_{pPb}(p_T)$ at low p_T than the results with shadowing alone: one produces an enhancement while the other gives effective suppression. The J/ψ results as a function of rapidity are compared to preliminary ALICE data [34]. All the models give reasonable agreement with the data so far. Note that the CGC predictions are only shown at forward rapidity.

Acknowledgments

I thank the organizers for the invitation to give this talk. This work was performed under the auspices of the U.S. Department of Energy by Lawrence Livermore National Laboratory under Contract DE-AC52-07NA27344 and supported by the U.S. Department of Energy, Office of Science, Office of Nuclear Physics (Nuclear Theory) under contract number DE-SC-0004014.

References

[1] R. Vogt, EPJ Web Conf. **137** (2017) 01022.
[2] P. Nason *et al.*, Nucl. Phys. B **303** (1988) 607; B **327** (1988) 49; B **335** (1990) 260(E).
[3] W. Beenakker, H. Kuijf, W. L. van Neerven and J. Smith, Phys. Rev. D **40**, (1989) 54.
[4] W. Beenakker *et al.*, Nucl. Phys. B **351** (1991) 507.
[5] M. Cacciari and M. Greco, Nucl. Phys. B **421** (1994) 530.
[6] J. Binnewies, B. A. Kniehl and G. Kramer, Phys. Rev. D **58** (1998) 014014.
[7] M. Cacciari, M. Greco and P. Nason, JHEP **05** (1998) 007.
[8] M. Cacciari and P. Nason, JHEP **0309** (2003) 006.
[9] B. A. Kniehl, G. Kramer, I. Schienbein and H. Spiesberger, Phys. Rev. D **71** (2005) 014018.
[10] J. Adam *et al.* [ALICE Collaboration], Phys. Rev. C **94** (2016) 054908.
[11] R. Aaij *et al.* (LHCb Collaboration), Nucl. Phys. B **871** (2013) 1.
[12] R. Aaij *et al.* [LHCb Collaboration], JHEP **1603** (2016) 159, Erratum: [JHEP **1609** (2016) 013], Erratum: [JHEP **1705** (2017) 074]
[13] A. Mischke, Phys. Lett. B **671** (2009) 361.
[14] R. E. Nelson, R. Vogt and A. D. Frawley, Phys. Rev. C **87** (2013) 014908.
[15] M. L. Mangano, P. Nason, and G. Ridolfi, Nucl. Phys. B **373** (1992) 295.
[16] S. Frixione, P. Nason, and G. Ridolfi, JHEP **0709** (2007) 126; arXiv:0707.3081 [hep-ph].
[17] T. Sjostrand *et al.*, Comput. Phys. Commun. **135** (2001) 238; arXiv:hep-ph/0308153.
[18] G. Corcella *et al.*, JHEP **0101** (2001) 010.
[19] J. Adam *et al.* [ALICE Collaboration], Eur. Phys. J. C **77** (2017) 245.
[20] R. Aaij *et al.* [LHCb Collaboration], JHEP **1206** (2012) 141, [JHEP **1403** (2014) 108].
[21] R. Gavai *et al.*, Int. J. Mod. Phys. A **10** (1995) 3043.
[22] G. A. Schuler and R. Vogt, Phys. Lett. B **387** (1996) 181.
[23] J. F. Amundson, O. J. P. Eboli, E. M. Gregores and F. Halzen, Phys. Lett. B **390** (1997) 323.
[24] R. Baier and R. Rückl, Z. Phys. C **19** (1983) 251.
[25] G. T. Bodwin, E. Braaten and G. P. Lapage, Phys. Rev. D **51** (1995) 1125.
[26] N. Brambilla *et al.*, Eur. Phys. J. C **74** (2014) 2981.
[27] Y. Feng, J. P. Lansberg and J. X. Wang, Eur. Phys. J. C **75** (2015) 313.
[28] R. Aaij *et al.* [LHCb Collaboration], Eur. Phys. J. C **75** (2015) 311.
[29] M. Butenschön, Z.-G. He and B. A. Kniehl, Phys. Rev. Lett. **114** (2005) 092004.
[30] Y. Q. Ma and R. Vogt, Phys. Rev. D **94** (2016) 114029.
[31] V. Cheung and R. Vogt, Phys. Rev. D **96** (2017) 054014.
[32] V. Cheung and R. Vogt, Phys. Rev. D **95** (2017) 074021.
[33] J. L. Albacete *et al.*, Int. J. Mod. Phys. E **25** (2016) 1630005.
[34] ALICE Collaboration, CERN-ALICE-PUBLIC-2017-001.
[35] J. L. Albacete *et al.*, arXiv:1707.09973 [hep-ph].

Overlap between Lattice QCD and HRG with in-medium effects and parity doubling*

Kenji Morita[1,2], *Chihiro* Sasaki[1], *Pok Man* Lo[1,3], and *Krzysztof* Redlich[1,2,**]

[1] *Institute of Theoretical Physics, University of Wrocław, PL-50204, Wrocław, Poland*
[2] *iTHES Research Group, RIKEN, Saitama 351-0198, Japan*
[3] *ExtreMe Matter Institute EMMI, GSI, D-64291 Darmstadt, Germany*

Abstract. We investigate the fluctuations and correlations involving baryon number in hot hadronic matter with modified masses of negative-parity baryons, in the context of the hadron resonance gas. Temperature-dependent masses are adopted from the recent lattice QCD results and from a chiral effective model which implements the parity doubling structure with respect to the chiral symmetry. Confronting the baryon number susceptibility, baryon-charge correlation, and baryon-strangeness correlation and their ratios with the lattice QCD data, we show that the strong downward mass shift in hyperons can accidentally reproduce some correlation ratios, however it also tends to overshoot the individual fluctuations and correlations. This indicates, that in order to correctly account for the influence of the chiral symmetry restoration on the fluctuation observables, a consistent framework of in-medium effects beyond hadron mass shifts is required.

1 Introduction

Fluctuations and correlations of conserved charges provide diagnostic tools for the nature of strongly interacting matter described by Quantum Chromodynamics (QCD). The first-principles calculations by lattice QCD (LQCD) have provided not only equation of state at the physical quark masses but also the fluctuations and correlations of the net-baryon, net-electric charge, and net-strangeness [1–6]. The fluctuations and correlations can also be measured in heavy-ion collisions to identify the state of created matter. In particular, non-Gaussian (higher-order) fluctuations have been expected to probe critical properties of the system, such as the QCD critical point in the beam-energy scan program at RHIC and a remnant of the $O(4)$ criticality at small baryon density [7, 8].

For a physical interpretation of the thermodynamic properties, the hadron resonance gas (HRG) model [9, 10] has been used as a reference. The equation of state and fluctuations have been well described by the model below the chiral crossover temperature $T = 154 \pm 9$ MeV [11, 12], and the experimental data of fluctuations have been analyzed in terms of a gas of hadronic states [13–16]. At vanishing baryon density, the critical behavior due to the chiral phase transition appears at the sixth order of net-baryon fluctuations [8]. This is because the singular contribution to the free energy is suppressed in the lower-order fluctuations even in the vicinity of the expected second order transition.

*Presented at *Strangeness in Quark Matter*, Utrecht, the Netherlands, 10-15 July 2017
**Speaker, e-mail: redlich@ift.uni.wroc.pl

Therefore, one expects that the lower order fluctuations and correlations may be well described by the hadronic degrees of freedom, and they would provide a reliable baseline for exploring the critical behavior [7].

According to the recent LQCD calculations, however, some correlations between conserved charges cannot be explained by a conventional HRG model. Particularly interesting quantities are those involving net-baryon number, baryon-charge (BQ) and baryon-strangeness (BS) correlations. Since mesons do not contribute to these quantities, one may gain access to a role of baryonic degrees of freedom and interactions which are not visible in the equation of state and other meson-dominated quantities due to their heavy masses [17].

In the HRG model the interaction of hadrons is replaced by resonances, and in the first approximation their widths can be neglected. Then, the partition function of the interacting hadronic system can be written as a mixture of free gases of all stable and resonant hadrons [18]. The validity of the vanishing-width approximation for the fluctuations has been recently examined based on the S-matrix formalism for K–π [19] and π–N–Δ [20] systems. It has been found that an explicit treatment of the width can have a substantial effect on the fluctuations.

On the other hand, at finite temperature and density one expects changes in the spectral property of hadrons. In particular, a search for those modifications because of the partial restoration of the QCD chiral symmetry has been one of the central subjects in heavy-ion collisions [21]. Since the vacuum masses are used in the HRG model, the observed agreement of the equation of state in the HRG and LQCD suggests no need for a strong mass reduction in the dominant degrees of freedom such as light mesons. Although chiral symmetry predicts substantial medium modifications of low-lying mesons [22, 23], their reliable estimates in LQCD have been limited to the screening masses [24].

Recently, masses of the non-strange and strange baryons with positive and negative parity have been extracted from the temporal correlation functions by the FASTSUM collaboration [25, 26]. The negative-parity states clearly show downward mass shifts, whereas the positive-parity baryons stay insensitive to temperature. The obtained spectra follow an expectation from the parity doublet picture of the chiral symmetry, indicating that the masses of a negative- and positive-parity partners tend to degenerate when approaching the chiral crossover. The above medium modification was used to possibly explain a missing contribution in the correlations between the conserved charges [26].

In the following we investigate the fluctuations and correlations of the net-baryon with net-charge and net-strangeness on the basis of the HRG model implementing in-medium mass modifications. We employ temperature dependent masses of the negative-parity octet and decouplet baryons from the lattice QCD and from a chiral effective model with parity doubling [27]. We show that the strong downward mass shift in hyperons can accidentally reproduce some correlation ratios, however it also tends to overshoot the individual fluctuation and correlation observables. This indicates that to quantify fluctuation and correlation of conserved charges in the presence of chiral symmetry restoration it is not sufficient to implement in-medium hadron masses in the statistical sum of the hadron resonance gas.

2 Baryon masses in parity doublet picture

2.1 Lattice QCD

The FASTSUM collaboration presented the temperature dependent masses of N, Δ, and Ω states, extracted from imaginary time correlators of the corresponding interpolating operators in $N_f = 2 + 1$ lattice simulations [25]. Their calculations were performed for heavier light-quark mass than the physical one; $m_\pi = 384$ MeV, while the strange quark is set to the physical one. Thus, the mass of baryons except Ω are heavier than the physical ones. This yields the nucleon with positive parity

Table 1. Assignment of negative parity states in in-medium HRG

P^- State	N	Λ	Σ	Ξ	Δ	Σ^*	Ξ^*	Ω
Mass LGT [25, 26]	1779	1899	1823	1917	2138	2131	2164	2193
$M_-^i(T_c)$ [MeV]	1254	1172	1329	1295	1405	1398	1426	1383
b_i	0.338	0.369	0.257	0.275	0.312	0.257	0.246	0.213
Assignment A	1535	1405	1750	1690	1700	1670	1820	2250
Assignment B	1535	1670	1750	1950	1700	1940	1820	2250
Mass [MeV]	1535	1790	1880	2090	1710	1930	2150	2380
Assignment C	1535	1800	1880	2120	1700	1940	2250	2380

N_+ has $m_{N_+} = 1158$ MeV (939 MeV in PDG), while the positive parity Ω_+ has $m_{\Omega_+} = 1661$ MeV (1672 MeV in PDG). The results of other octet and decuplet states have been shown in [26]. Here we use the masses at $T/T_c = 0.24$, 0.76, 0.84, and 0.95. Note that owing to the heavier pion mass, the crossover temperature $T_c = 185$ MeV, which was determined from the renormalized Polyakov loop, is also higher than the physical one, $T_c = 154$ MeV.

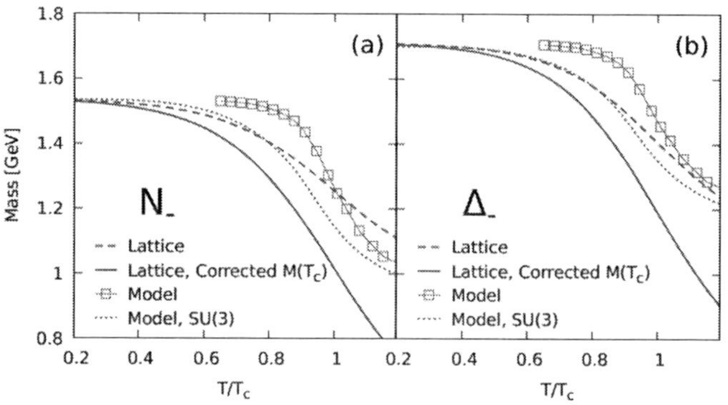

Figure 1. Temperature dependence of the mass of N_- (a) and Δ_- (b) from lattice-motivated parameterization Eq. (1) (lines) and from a chiral effective model (squares).

In Table 1, we list the masses of the octet and decuplet negative-parity states at $T = 0.24T_c$. They are well parameterized, for the given the state i, by [25, 26]

$$M_-^i(T) = M_-^i(T=0)\omega(T, b_i) + M_-^i(T_c)(1 - \omega(T, b_i)) \quad (1)$$

$$\omega(T, b_i) = \tanh[(1 - T/T_c)/b_i]/\tanh(1/b_i) \quad (2)$$

where $M_-^i(T = 0)$ is fixed to be the value obtained at $T = 0.24T_c$ and $M_-^i(T_c)$ and b_i are the fitting parameters fixed for each channel. The data are given as a function of T/T_c; the value of T_c does not affect the fit but later it is set to the physical value 154 MeV. The parameter b_i corresponds to the width of the chiral crossover. The results of the fits are shown in Table 1.

To correct the unphysical effect from the heavy up and down quarks, $M_-^i(T = 0)$ is set to its PDG mass in the following. The value of $M_-(T_c)$ needs to be corrected as well, and we shall re-scale it by multiplying the factor $M_+^{\rm PDG}(T = 0)/M_+^{\rm lattice}(T = 0)$ with $M_-(T_c)$.

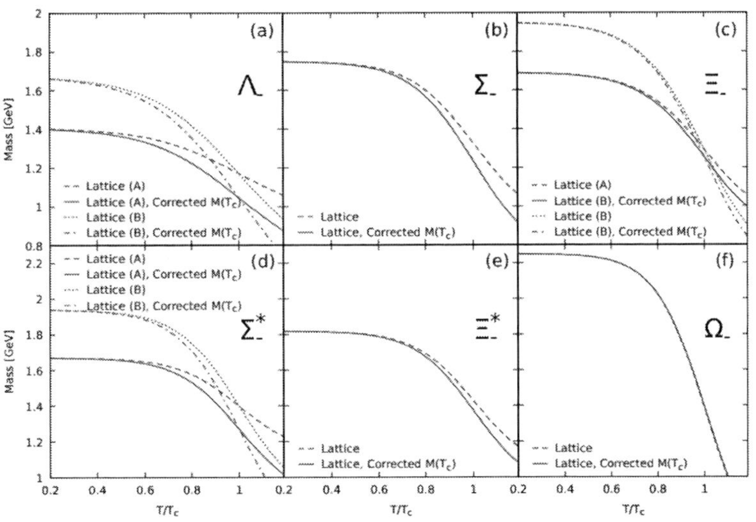

Figure 2. Temperature dependence of octet (upper row) and decuplet (lower row) hyperons from lattice-motivated parameterization Eq. (1).

It is not clear how to assign the negative-parity states to the observed ones because of unknown quantum numbers of some of the candidates. We follow the suggestion in [26] and further adopt two different assignments, A and B. The set A takes the lighter state for Λ_-, Ξ_-, and Σ^*_- while the set B does the heavier ones.

The obtained temperature-dependent masses are displayed in Figs. 1 and 2. The masses of negative-parity states start to drop around $T/T_c \simeq 0.6$, then approach those of their positive-parity partners. In comparing the non-strange baryons with the hyperons, one finds the correction of $M(T_c)$ significantly affects the baryons with more light quarks, as expected.

2.2 Chiral model

The parity-doubled baryons can be modeled in an effective chiral approach [28], where a chiral-invariant mass is naturally introduced. In [27], the mass relations to the light quark σ_q and strange quark σ_s condensates are given for the octet and decuplet states as [1]

$$\begin{aligned}
m_{N_\pm} &= (a_N \mp b_N)\,3\sigma_q + m_{N0}\,, & m_{\Delta_\pm} &= (a_\Delta \mp b_\Delta)\,3\sigma_q + m_{\Delta 0} \\
m_{\Sigma_\pm} &= (a_N \mp b_N)\left(2\sigma_q + \sqrt{2}\sigma_s\right) + m_{N0} + m_1\,, & m_{\Sigma^*_\pm} &= (a_\Delta \mp b_\Delta)\left(2\sigma_q + \sqrt{2}\sigma_s\right) + m_{\Delta 0} + m_s \\
m_{\Lambda_\pm} &= (a_N \mp b_N)\left(2\sigma_q + \sqrt{2}\sigma_s\right) + m_{N0} + m_3\,, & m_{\Xi^*_\pm} &= (a_\Delta \mp b_\Delta)\left(\sigma_q + 2\sqrt{2}\sigma_s\right) + m_{\Delta 0} + 2m_s \\
m_{\Xi_\pm} &= (a_N \mp b_N)\left(\sigma_q + 2\sqrt{2}\sigma_s\right) + m_{N0} + m_2\,, & m_{\Omega_\pm} &= (a_\Delta \mp b_\Delta)\,3\sqrt{2}\sigma_s + m_{\Delta 0} + 3m_s\,.
\end{aligned} \tag{3}$$

The parameters $m_{1,2,3}$ are related via the Gell-Mann–Okubo relation as

$$m_1 = 2m_2 - 3m_3\,. \tag{4}$$

The parameters in the above expressions are determined at zero temperature as in Table 2.

[1] In [27] the same m_0 was assumed common to the octet and decuplet baryons. Here we lift this constraint and introduce two independent masses. Their splitting can be deduced from the spin-spin interaction.

a_N	b_N	m_1 [GeV]	m_2 [GeV]	a_Δ	b_Δ	m_s [GeV]
1.22	1.08	0.248	0.367	1.16	0.862	0.139

Table 2. Set of parameters in the baryon-mass relations. We set $m_{N0} = 0.9$ GeV and $m_{\Delta 0} = 1.15$ GeV.

The masses at $T = 0$ are also summarized in Table 1. Thermal modifications of the baryon masses are driven by the quark condensates. Following [27], we shall use the in-medium condensates measured in LQCD by the HotQCD collaboration [12]. One readily finds that some of the hyperon states have larger masses than those expected from the lattice baryon-spectrum in [25, 26]. Thus, we assign these hyperons to the PDG states as in the bottom line of Table 1 and refer to set C. For a comparison with LQCD, we also apply the set C to the lattice-motivated parametrization (1).

The non-strange baryons exhibit the thermal behavior as in Fig. 1. To some extent, their trends are similar to the results by the FASTSUM collaboration, whereas the parametrization (2) leads to a sizable difference in the mass dropping near T_c. This is particularly evident when the re-scaling of $M_-(T_c)$ is made.

Figure 3 displays the mass of the hyperons in the assignment C. The most distinct difference from the non-strange baryons lies in the mass of the hyperons near T_c; the model calculation shows a rather moderate downward-shift, while the lattice QCD results lead to a strong mass drop. Comparing Λ (Σ), Ξ, and Ω states between the model and the lattice-inspired scaling, one clearly finds that the difference becomes more significant as the baryon contains more strangeness. This may be attributed to the difference in the masses between the light and strange quarks ; The FASTSUM setup with a large pion mass $m_\pi = 384$ MeV and the physical kaon mass is rather close to the flavor SU$_f$(3) limit, while the model calculation done with the physical pion mass is dominated by SU$_f$(2). Therefore, it will be intriguing to see whether the strong mass reduction of Ω and Ξ would still persist in the lattice simulations with a lighter pion.

3 Fluctuations and correlations from in-medium HRG

We employ the hadron resonance gas model to explore the fluctuations and correlations in the hot hadronic matter. In this work we entirely rely on the zero-width approximation; all the resonances are treated as point-like particles and their spectral functions are given by the delta function. Consequently, the thermodynamic pressure at temperature T and chemical potentials μ_j ($j = B, Q, S$) is given by the summation of ideal gas pressure of each particle species as

$$p^{\mathrm{HRG}}(T, \mu_B, \mu_Q, \mu_S) = \sum_{i=\mathrm{hadrons}} p^{\mathrm{ideal}}(T, \mu_i; m_i) \tag{5}$$

where $\mu_i = B_i\mu_B + Q_i\mu_Q + S_i\mu_S$ is the chemical potential of particle i with baryon number B_i, electric charge Q_i, and strangeness S_i and the ideal gas pressure with degeneracy d reads

$$p^{\mathrm{ideal}}(T, \mu; m) = \mp\frac{dT}{2\pi^2}\int_0^\infty dp p^2 \ln[1 \mp e^{-(\sqrt{p^2+m^2}-\mu)/T}]. \tag{6}$$

The signs are negative for mesons and positive for baryons. The fluctuations and the correlations at vanishing chemical potentials can be expressed as the generalized susceptibilities

$$\chi_{ijk}^{BQS} \equiv \frac{\partial^{i+j+k} p^{\mathrm{HRG}}/T^4}{\partial^i(\mu_B/T)\partial^j(\mu_Q/T)\partial^k(\mu_S/T)} \tag{7}$$

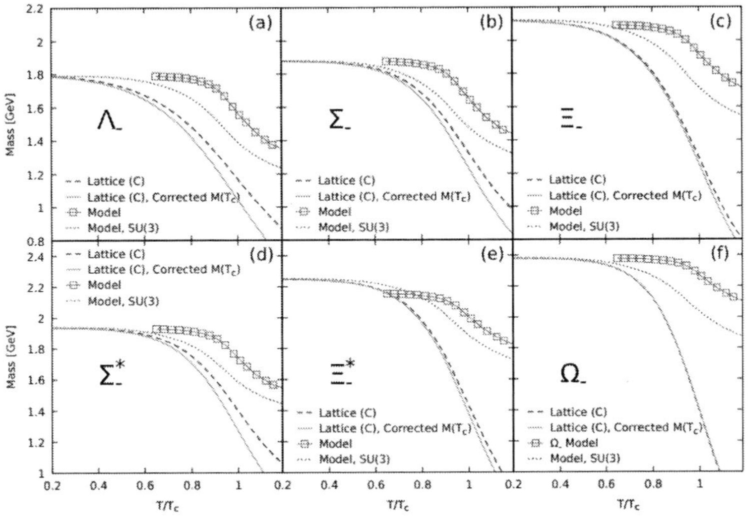

Figure 3. Comparison of hyperon masses between the lattice-motivated parametrization (lines) and the chiral effective model (squares) in the assignment C. As in Fig. 2, the octet and decuplet hyperons are shown in upper and lower panels, respectively.

Although heavy particles ($m \gg T$) are thermally suppressed, the existence of missing states are expected likely in the strange sectors [29, 30]. In fact, the assignment C in Table 1 contains several 1-star states whose spins are undetermined in the latest PDG [31]. For those state we shall assume the same spin with its partner. For consistency we also include other unconfirmed state; we have in total 28 nucleons, 22 Δ baryons, 18 Λ baryons, 21 Σ baryons, 8 Ξ baryons, 3 Ω baryons, and their isospin multiplets. We also include a deuteron, triton, ^{3}He and ^{4}He as they are considered to be thermal ingredients in QCD thermodynamics and heavy ion collisions [10].

In this work we focus on χ_2^B, χ_{11}^{BQ}, and χ_{11}^{BS} at $\mu_B = \mu_Q = \mu_S = 0$. Since $\chi_{ijk}^{BQS} \propto (B/T)^i(Q/T)^j(S/T)^k$, mesons do not contribute to these susceptibilities. Thus they are good measures of in-medium effects in the baryon sector on the QCD thermodynamics. In general, particle masses in a medium can depend not only on temperature but also on chemical potentials. We neglect such an intrinsic chemical-potential dependence in the baryon masses in the present calculations. This is well justified at small chemical potential since the quark condensates are not much affected.

We display χ_2^B, χ_{11}^{BQ}, and $-\chi_{11}^{BS}$ in Fig. 4. In the baryon number susceptibility χ_2^B, the HRG model agrees with the LQCD results below T_c. [2] With the in-medium mass shift, a decrease in the negative-parity baryon masses lead to an enhancement of the susceptibility. One sees a moderate enhancement in all the three scenarios, Lattice (A) (B) and (C), and it becomes stronger when $M(T_c)$ is corrected, since the correction further reduces the masses just below T_c. On the other hand, the different assignments of the parity-partners lead to a minor change. The results with the masses from the chiral effective model follows the same trend, but the amount of the enhancement is much smaller than the tanh-parameterization, owing to the weaker downward mass shifts in the hyperon sectors (see Fig. 3.) In any case, all the results with mass reduction overshoot the lattice data of χ_2^B.

[2]The present HRG leads to a slightly larger χ_2^B than the conventional one because of including the 1-star states and the multi-baryons.

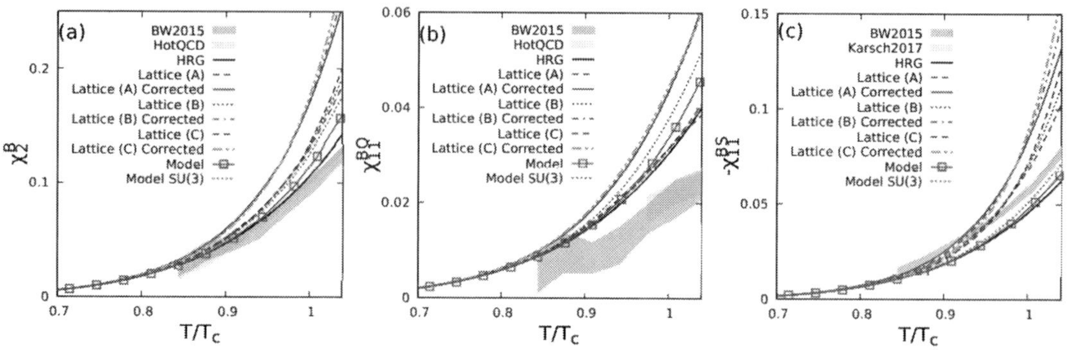

Figure 4. Fluctuations and correlations from the HRG model with and without the mass shifts. From left to right, (a) Baryon number susceptibility, (b) Baryon-charge correlation, and (c) Baryon-strangeness correlation are shown together with the LQCD results from HotQCD [6, 32] and Budapest-Wuppertal Collaboration [33]

The χ_{11}^{BQ} shows a different tendency in Fig. 4(b); the results of Lattice (A)-(C) without the correction of $M(T_c)$ do not exhibit any enhancement from the HRG, even below the "Model" result. In the χ_{11}^{BQ}, $I = 1$ states do not contribute because neutral particles do not contribute and the positively-charged states cancel with the negatively-charged states. Charged N states and doubly charged Δ^{++} states contribute positively, while negatively-charged Ξ_-^-, Ξ_-^{*-} and Ω_-^- states contribute oppositely. Without a strong mass-shift of these multi-strange states, the χ_{11}^{BQ} becomes larger due to the smallness of the hyperon contribution suppressed by the Boltzmann factor, as in the "Model" case. With the strong mass shift of the hyperons, they tend to cancel a part of the positive contribution, and suppress the χ_{11}^{BQ}. The strong enhancement with the corrected $M(T_c)$ clearly signals that the effect of the correction is smaller for Ξ states and larger for N and Δ states, as seen in Figs. 1-3.

The behavior of $-\chi_{11}^{BS}$ in Fig. 4(c) can be understood in a similar way to the χ_2^B; the "Model" result does not differ much from the HRG, because of a weaker mass reduction. Owing to the different trend in the hyperon sectors, the curves deviate more sensitively depending on the fate of the masses.

The observed differences can be more pronounced by taking the ratio between the susceptibilities. In Fig. 5(a), we display the χ_{11}^{BQ}/χ_2^B together with the LQCD data. The results from Lattice (A) and (B) reproduce the one shown in [26]; they follows the trend seen in the HotQCD data. As explained above, however, the individual susceptibilities cannot be reproduced. The coincidence comes from the reduction of χ_{11}^{BQ} owing to the strong mass reductions in the charge asymmetric states. The "Model" case shows the opposite trend that the ratio is above the HRG, due to the much milder mass-reduction in the hyperon sectors. Therefore, the data cannot be explained solely by the mass reduction.

Also shown in Fig. 5(b) is the $-\chi_{11}^{BQ}/\chi_{11}^{BS}$. Again, while the sets (A) and (B) follow the trend of the HotQCD data, the "Model" shows the opposite. This supports the observation that naive insertion of the in-medium masses of parity doublers into the HRG results in fluctuation observables which are inconsistent with LGT data. Nevertheless, our results do not rule out the manifestation of parity doublers. Several studies [19, 20] have revealed that the conventional treatment of the particles in the HRG is not sufficient and a proper inclusion of the width improves the thermodynamics. Although those analysis have been limited to the S-matrix approaches where interactions are incorporated via the two-body scattering phase shift in the vacuum, a consistent treatment of the resonance widths at finite temperature and densities should clarify the consequences of the partial restoration of the chiral symmetry in the baryonic sector of correlations and fluctuations.

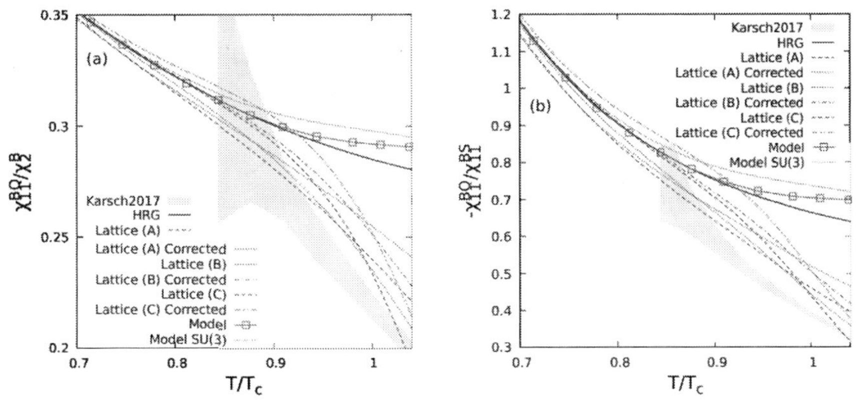

Figure 5. Ratio of the baryon-charge correlation to the baryon number fluctuation (a) and the baryon-strangeness correlation (b). The legends are the same as Fig. 4.

4 Concluding remarks

We have studied the fluctuations and correlations of conserved charges by making use of the hadron resonance gas model with in-medium masses of the low-lying parity-doubled baryons. Motivated by the recent LQCD result on the baryon masses at finite temperature, we have examined the consequences of several scenarios with in-medium masses for the negative-parity octet and decouplet baryons, adopting a lattice-motivated parametrization and that from a chiral effective theory. We have computed the baryon number fluctuation χ_2^B, baryon-charge correlation χ_{11}^{BQ}, and baryon-strangeness correlation χ_{11}^{BS}. The results have been understood on the basis of the presence or absence of the strong mass reduction in the hyperon sectors. We have pointed out that the reproduction of the LQCD results of the ratios χ_{11}^{BQ}/χ_2^B and $-\chi_{11}^{BQ}/\chi_{11}^{BS}$ is just accidental when the in-medium masses are naively introduced in the conventional HRG approach. Since the strong mass shifts of the hyperons observed in LQCD can be regarded as a consequence of the approximate $SU_f(3)$ nature due to a heavy pion mass, it is a critical issue to be examined whether the hyperon mass-shift persists for the physical pion mass.

Although our results clarify the interplay of the different particle species in the behavior of the baryon-charge and baryon-strangeness correlations, we emphasize that the treatment of the resonances in the conventional HRG model is insufficient to quantify correlation and fluctuations of conserved charges. This indicates, that in order to correctly account for the influence of the chiral symmetry restoration on the fluctuation observables, a consistent framework of in-medium effects beyond hadron mass shifts is required.

Acknowledgments

We acknowledge stimulating discussions with Gábor Almási and Bengt Friman. This work has been partly supported by the Polish Science Center NCN under Maestro grant 2013/10/A/ST2/00106 and RIKEN iTHES project.

References

[1] F. Karsch, Nucl. Phys. A **967**, 461 (2017).

[2] A. Bazavov *et al.* [HotQCD Collaboration], Phys. Rev. D **96**, no. 7, 074510 (2017).
[3] H. T. Ding, F. Karsch and S. Mukherjee, Int. J. Mod. Phys. E **24**, no. 10, 1530007 (2015).
[4] S. Borsányi, G. Endrödi, Z. Fodor, A. Jakovác, S. D. Katz, S. Krieg, C. Ratti, and K. K. Szabó, JHEP **1011**, 077 (2010).
[5] A. Bazavov, T. Bhattacharya, C. Detar, H. T. Ding, S. Gottlieb, R. Gupta, P. Hegde, U. M. Heller, F. Karsch, E. Laermann, et al., Phys. Rev. D **90**, 094503.
[6] A. Bazavov et al. (HotQCD Collaboration), Phys. Rev. D **86**, 034509 (2012).
[7] S. Ejiri, F. Karsch, and K. Redlich, Phys. Lett. B **633**, 275 (2006).
[8] B. Friman, F. Karsch, K. Redlich, and V. Skokov, Eur. Phys. J. C **71**, 1694 (2011).
[9] P. Braun-Munzinger, K. Redlich, and J. Stachel, in *Quark-Gluon Plasma 3*, edited by R. C. Hwa and X. N. Wang (World Scientific, 2004), p. 491.
[10] A. Andronic, P. Braun-Munzinger, K. Redlich and J. Stachel, arXiv:1710.09425 [nucl-th].
[11] F. Karsch and K. Redlich, Phys. Lett. B **695**, 136 (2011).
[12] A. Bazavov, T. Bhattacharya, M. Cheng, C. DeTar, H. T. Ding, S. Gottlieb, R. Gupta, P. Hedge, U. M. Heller, F. Karsch, et al., Phys. Rev. D **85**, 054503 (2012).
[13] P. Garg, D. K. Mishra, P. K. Netrakanti, B. Mohanty, A. K. Mohanty, B. K. Singh, and N. Xu, Phys. Lett. B **726**, 691 (2013).
[14] P. Alba, W. Alberico, R. Bellwied, M. Bluhm, V. Mantovani Sarti, M. Nahrgang and C. Ratti, Phys. Lett. B **738**, 305 (2014).
[15] P. Braun-Munzinger, A. Kalweit, K. Redlich, and J. Stachel, Phys. Lett. B **747**, 292 (2015).
[16] F. Karsch, K. Morita, and K. Redlich, Phys. Rev. C **93**, 034907 (2015).
[17] P. M. Lo, B. Friman, K. Redlich and C. Sasaki, arXiv:1710.02711 [hep-ph].
[18] R. Dashen, S. K. Ma, and H. J. Bernstein, Phys. Rev. **187**, 345 (1969).
[19] B. Friman, P. M. Lo, M. Marczenko, K. Redlich, and C. Sasaki, Phys. Rev. D **92**, 074003 (2015).
[20] P. M. Lo, B. Friman, M. Marczenko, K. Redlich, and C. Sasaki, Phys. Rev. C **96**, 015207 (2017).
[21] R. S. Hayano and T. Hatsuda, Rev. Mod. Phys. **82**, 2949 (2010).
[22] T. Hatsuda, Y. Koike, and S. H. Lee, Nucl. Phys. **B394**, 221 (1993).
[23] P. M. Hohler and R. Rapp, Phys. Lett. B **731**, 103 (2014).
[24] Y. Maezawa, F. Karsch, S. Mukherjee, and P. Petreczky, PoS **LATTICE2015**, 199 (2016).
[25] G. Aarts, C. Allton, D. De Boni, S. Hands, B. Jäger, C. Praki, and J.-I. Skullerud, JHEP **06**, 034 (2017).
[26] G. Aarts, *Baryons in the plasma: in-medium effects and parity doubling* (2017), talk given at SQM2017.
[27] C. Sasaki, *Parity doubling of baryons in a chiral approach with three flavors* (2017), arXiv: 1707.05081.
[28] C. E. Detar and T. Kunihiro, Phys. Rev. D **39**, 2805 (1989).
[29] A. Bazavov, H.-T. Ding, P. Hegde, O. Kaczmarek, F. Karsch, E. Laermann, Y. Maezawa, S. Mukherjee, H. Ohno, P. Petreczky, et al., Phys. Rev. Lett. **113**, 072001 (2014).
[30] P. M. Lo, M. Marczenko, K. Redlich, and C. Sasaki, Phys. Rev. C **92**, 055206 (2015).
[31] C. Patrignani et al. (Particle Data Group), Chin. Phys. **C40**, 100001 (2016).
[32] F. Karsch, *Conserved charge fluctuations at vanishing and non-vanishing chemical potential* (2017), talk given at Quark Matter 2017.
[33] R. Bellwied, S. Borsanyi, Z. Fodor, S. D. Katz, A. Pasztor, C. Ratti, and K. K. Szabo, Phys. Rev. D **92**, 114505 (2015).

Lattice QCD results on chemical freeze-out

Claudia Ratti[1,*]

[1] *Department of Physics, University of Houston, Houston, TX 77204, USA*

Abstract. The latest lattice QCD results on fluctuations of conserved charges at finite temperature and density are reviewed, and connected to the experimental moments of their net-multiplicity distributions, in order to describe the chemical freeze-out from first principles.

1 Introduction

The chemical freeze-out is the moment, in the evolution of a heavy-ion collision, at which all inelastic collisions between particles cease. The chemical composition is fixed at this point, which means that both measured particle multiplicities and fluctuations carry information about this particular moment in the evolution of the system. In particular we are interested in determining the freeze-out temperature and chemical potential, and establishing how far the freeze-out line is from the transition line of QCD.

The first thing to look at, to characterize the chemical freeze-out, is therefore particle yields. The Hadron Resonance Gas (HRG) model has been very successful in fitting particle yields and ratios over nine orders of magnitude [1–5]. However, the spectrum used as an input in the HRG model is still being regularly updated, and many missing states are expected to be relevant to the thermodynamics around the transition temperature [6–9]. This introduces an uncertainty in the freeze-out parameters, since the decays modes of these additional states are not known. For this reason, a model-independent approach would be desirable.

All of this motivates the idea of finding a way to extract the freeze-out parameters from first principles, namely by comparing experimental results to lattice QCD directly. Lattice simulations are the most established tool to address QCD in its non-perturbative regime. Lattice QCD is a numerical method, and as such affected by statistical and systematic uncertainties. Due to a steady and continuous improvement in computer resources, numerical algorithms and our physical understanding which manifests itself in physical techniques (e.g. the Wilson-flow scale setting introduced in Ref. [10]), the lattice results which are being produced today reach an unprecedented level of accuracy. This allows a quantitative comparison to experimental observables for the first time in heavy ion physics.

2 Thermal fits to particle yields and ratios

The first thing to look at, to characterize the chemical freeze-out, is therefore particle yields and ratios. The particle yields are given by this simple formula, plus a contribution coming from resonance

*e-mail: cratti@uh.edu

decays, in the HRG model:

$$N_i = -\frac{\partial \ln Z_i}{\partial \mu} = \frac{g_i V}{2\pi^2} \int_0^\infty \frac{p^2 dp}{\exp[(E_i - \mu_i)]/T \pm 1} \tag{1}$$

where $E_i = \sqrt{p^2 + m_i^2}$, $\mu_i = \mu_B B_i + \mu_S S_i + \mu_Q Q_i$, and g_i is the degeneracy of particle i. This kind of fits has been performed for a long time and has been very successful in describing the multiplicity of particles produced in heavy-ion collisions. Changing the collision energy, it is possible to draw the freeze-out line in the (T, μ_B)-plane. An example of such a fit for the LHC energies, together with the chemical freeze-out line, is shown in Fig. 1. Recently, the possibility of a flavor hierarchy in

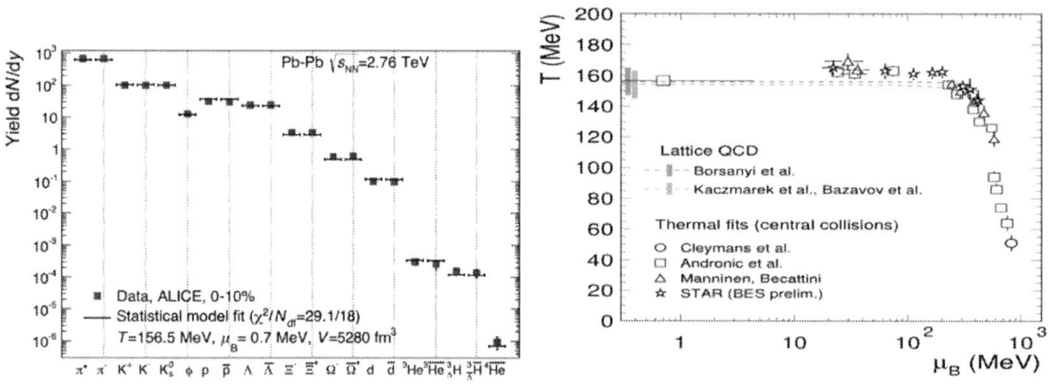

Figure 1. Left: Best fit of hadron multiplicities in central (0-10%) Pb-Pb collisions at the LHC ALICE experiment [11]. Right: The phase diagram of strongly interacting matter with the points representing the thermal fits of hadron yields at various collision energies [1–5]. The crossover T values from lattice QCD [12, 13] are shown as bands at small μ_B values. The dashed lines represent lattice QCD calculations of the curvature parameter [14, 15].

the freeze-out temperatures has been discussed [16, 17], with the freeze-out temperature for strange hadrons being larger than the one for light hadrons. Among the explanations of this phenomenon, I will focus here on the one proposed in Ref. [8]. A disagreement was found between the ratio of strange to baryon chemical potential obtained on the lattice and the one predicted by the HRG model based on the resonance spectrum listed in the particle data book. The agreement can be reached when additional states are added to the spectrum, which have not yet been detected but are predicted by the quark model. Since the strange chemical potential enters the thermal fits, this will affect the freeze-out parameters, potentially reducing the strangeness freeze-out temperature and bringing it closer to the light quark one. In Ref. [9] we performed a systematic study of the contribution of different hadronic species to the QCD pressure, separated according to their flavor and baryon number content. They are shown in Fig. 2. This has allowed us to determine which hadronic families are still missing resonances, and indeed most groups of hadron need more states than the ones listed as "known" in the PDG. This will impact the thermal fits, since the decay modes of these states are not known.

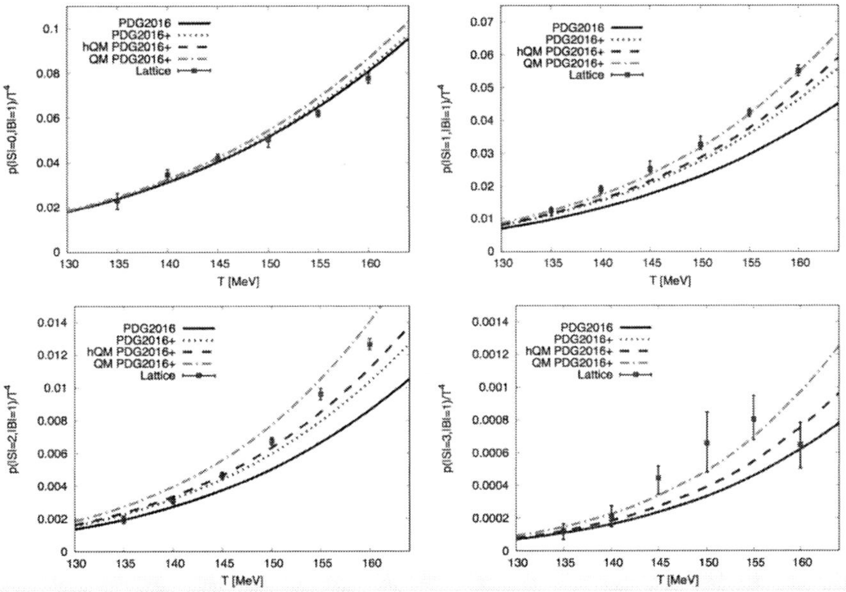

Figure 2. Comparison between the lattice results for the partial pressures and the HRG model predictions. Upper panels: non-strange baryons (left), |S| = 1 baryons (right). Lower panels: |S| = 2 baryons (left), |S| = 3 baryons (right). The points are the lattice results, while the curves are PDG2016 (solid black), PDG2016+ (including one star states, red dotted), PDG2016+ and additional states from the hQM (blue, dashed) [18, 19], PDG2016+ and additional states from the QM (cyan, dash-dotted) [20, 21].

3 Fluctuations of conserved charges

Fluctuations of conserved charges are defined as

$$\chi_{lmn}^{BSQ} = \frac{\partial^{l+m+n}(p/T^4)}{(\partial\mu_B/T)^l(\partial\mu_S/T)^m(\partial\mu_Q/T)^n}. \quad (2)$$

They can be calculated on the lattice as combinations of quark flavor fluctuations, through the following relationship between chemical potentials:

$$\begin{aligned}\mu_u &= \frac{1}{3}\mu_B + \frac{2}{3}\mu_Q \\ \mu_d &= \frac{1}{3}\mu_B - \frac{1}{3}\mu_Q \\ \mu_s &= \frac{1}{3}\mu_B - \frac{1}{3}\mu_Q - \mu_S. \end{aligned} \quad (3)$$

The relevance of fluctuations for the physics of heavy ion collisions has been increasing in recent years. The first continuum-extrapolated second order fluctuations of conserved charges, evaluated at physical values for the quark masses, were presented in Ref. [22] (almost-physical quark mass results are shown in Ref. [23], heavier quark mass results are shown in Ref. [24]) and later extended to selected fourth-order fluctuations and correlations [25, 26].

The most interesting feature of fluctuations is that they can be related to the moments of the distribution of the corresponding conserved charges (mean M, variance σ^2, skewness S and kurtosis

κ) through the following formulas

$$M = \chi_1 \qquad \sigma^2 = \chi_2$$
$$S = \chi_3/\chi_2^{3/2} \qquad \kappa = \chi_4/\chi_2^2. \tag{4}$$

Usually ratios are defined, so that the volume factor cancels out in the theoretical definition of fluctuations, and they become functions only of T and μ_B:

$$M/\sigma^2 = \chi_1/\chi_2 \qquad S\sigma = \chi_3/\chi_2$$
$$S\sigma^3/M = \chi_3/\chi_1 \qquad \kappa\sigma^2 = \chi_4/\chi_2. \tag{5}$$

If the detector could cover the entire solid angle, none of the conserved charges would fluctuate. But if we look at a sub-system, limited by the acceptance of the detector, the conserved charges can fluctuate in and out of this sub-system, which then can be treated as a Grand Canonical Ensemble. Therefore, the detector should have a rapidity window small enough that charge conservation does not suppress the signal, but large enough that the system can be treated as a thermodynamic ensemble.

The rapidity dependence is only one of the effects that we need to study in order to have a meaningful comparison between lattice QCD and experiments. Other possible experimental sources of non-thermal fluctuations are corrected for in the STAR data analysis: the centrality-bin-width correction method minimizes effects due to volume variation because of finite centrality bin width (such effects have been studied e.g. in [27–29]); the moments are corrected for the finite reconstruction efficiency based on binomial probability distribution [30]. Interactions in the hadronic phase and non-equilibrium effects might affect fluctuations [32–35]; the consistency between the freeze-out parameters obtained from the analysis of fluctuations of different quantum numbers, such as electric charge and baryon number, represent a fundamental test of the equilibrium scenario.

One should also notice that, experimentally, only the net-proton multiplicity distribution is measured, as opposed to the lattice net-baryon number fluctuations. However, taking into account the effects of resonance feed-down and isospin randomization [36, 37], one can show that the net-proton and net-baryon number fluctuations are numerically very similar, at least in the case of low-order fluctuations [38].

The first comparisons between lattice and experiment were performed by the WB collaboration in 2013 [39, 40] and used χ_3/χ_1 for baryon number to extract an upper limit for the freeze-out temperature and χ_2/χ_1 from the same conserved charge to get the baryonic chemical potential for the top 4 collision energies at RHIC [41, 42]. From this comparison, a maximum freeze-out temperature from fluctuations of 151 ± 4 MeV was found. The chemical potentials from χ_2/χ_1 for baryon number were also obtained. An independent analysis based on a comparison between lattice QCD results and experimental data for the electric charge χ_2/χ_1 found perfect consistency between the freeze-out chemical potentials. These results also agree with an independent study of the same fluctuations in the framework of the Hadron Resonance Gas model [43]. A freeze-out temperature of 147 ± 2 MeV at the highest RHIC collision energy was obtained in Ref. [44] from a study of the ratio χ_1/χ_2 for electric charge divided by the baryon number one. This analysis also provided an upper limit on the curvature for the freeze-out line, which turns out to be very small.

More recently, preliminary results have been shown for a combined fit of χ_1/χ_2 for electric charge and baryon number, which allows to extract both freeze-out T and μ_B at a given collision energy [45]. The results are shown in the left panel of figure 3. The colored lines are the trajectories in the (T, μ_B) plane which satisfy the experimental value for χ_1/χ_2 for the two conserved charges. The points where they cross yield the freeze-out T and μ_B from fluctuations. Notice that the black points in the plot are obtained from the independent analysis of the same data done in the HRG model in Ref. [43], which perfectly agrees with the lattice results. On the same plot, the current understanding of the

QCD phase diagram from lattice simulations is shown [15], together with the isentropic trajectories that were reconstructed starting from the freeze-out points [46].

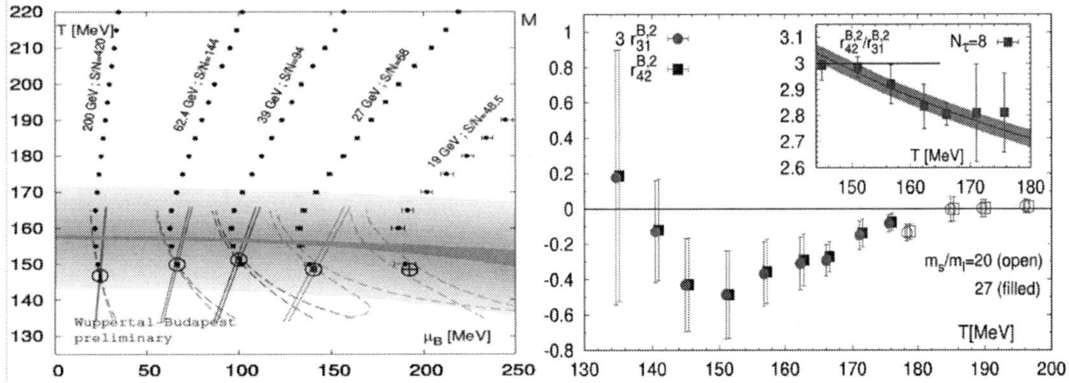

Figure 3. Left: Preliminary results of the WB collaboration [45]. The colored full and dashed lines are the contours at constant mean/variance ratios of the net electric charge from lattice simulations. The contours that correspond to STAR data intersect in the freeze-out points of Ref. [43]. The red band is the QCD phase diagram shown in Ref. [15]. Also shown are the isentropic contours that match the chemical freeze-out data [46]. Right: Taylor expansion coefficients for χ_3/χ_1 multiplied by 3 (red), and χ_4/χ_2 (black) as functions of the temperature from Ref. [49].

Preliminary results for net-proton fluctuations up to second order at the LHC have been presented recently [47]. The authors of Ref. [48] have performed an analysis, assuming that the lower moments follow the Skellam distribution. In that case, it is possible to express the second-order moments in terms of particle yields, for which experimental data exist for a long time. The analysis was focused on the ratios χ_2^B/χ_2^S, χ_2^B/χ_{11}^{QS} and χ_{11}^{BS}/χ_2^S. The obtained freeze-out temperature was slightly larger than then one obtained from the fluctuation analysis at RHIC.

In general, the higher order fluctuations are more affected by some of the non-thermal sources of fluctuations discussed above. Nevertheless, they can be calculated on the lattice at finite density by means of a Taylor expansion and compared to the experimental results, to test whether QCD can explain some of the trends seen in the data. For example, the authors of Ref. [49] have expanded the baryonic χ_3/χ_1 and χ_4/χ_2 in Taylor series in powers of μ_B/T. The Taylor expansion coefficients for the two are plotted in the right panel of Fig. 3: the red points are the coefficient for χ_3/χ_1 multiplied by 3, while the black points are the coefficient for χ_4/χ_2. It is evident that χ_4/χ_2 falls three times faster than χ_3/χ_1 with increasing chemical potential. The experimental data for these quantities show the same trend.

The next question is whether the third conserved charge, namely strangeness, shows the same consistency which has been found between electric charge and baryon number with respect to the chemical freeze-out parameters. Full strangeness fluctuations are difficult to measure, due to the multi-strange particles. The STAR collaboration recently finalized the results for the kaon fluctuations [50]. Kaon fluctuations are very sensitive to the freeze-out parameters [51] and could represent a first step to extract the strangeness freeze-out temperature. Recently, a method to isolate the kaon fluctuations on the lattice, based on the HRG model, was proposed [52]. This will allow to determine the kaon freeze-out temperature by comparing the experimental data to lattice QCD simulations, once they become available.

4 Conclusions and outlook

The precision of the present lattice results allows a comparison to experimental results for the first time in our field. From the fluctuations of conserved charges it is possible to gain information about the evolution of matter created in heavy ion collision, in particular on the chemical freeze-out. Many effects need to be understood in order to have a meaningful comparison: the theoretical and experimental efforts are progressing towards an understanding of these effects and their magnitudes. Various comparisons between theory and experiment have been performed in order to extract the freeze-out parameters of heavy ion collisions from first principles, or to predict the behavior of higher order fluctuations. While a consistent picture between baryon number and electric charge seems to emerge, an unambiguous determination of the freeze-out temperature for strangeness will help to better clarify the remaining issues.

Acknowledgements

The author acknowledges useful discussions with her colleagues and collaborators Paolo Alba, Rene Bellwied, Szabolcs Borsanyi, Zoltan Fodor, Jana Günther, Sandor Katz, Volker Koch, Valentina Mantovani-Sarti, Jorge Noronha, Jacquelyn Noronha-Hostler, Paolo Parotto, Attila Pasztor, Israel Portillo, Kalman Szabo. This material is based upon work supported by the National Science Foundation under grants no. PHY-1654219 and OAC-1531814 and by the U.S. Department of Energy, Office of Science, Office of Nuclear Physics, within the framework of the Beam Energy Scan Theory (BEST) Topical Collaboration. An award of computer time was provided by the INCITE program. This research used resources of the Argonne Leadership Computing Facility, which is a DOE Office of Science User Facility supported under Contract DE-AC02-06CH11357. The author gratefully acknowledges the use of the Maxwell Cluster and the advanced support from the Center of Advanced Computing and Data Systems at the University of Houston.

References

[1] J. Cleymans, H. Oeschler and K. Redlich, Phys. Rev. C **59**, 1663 (1999) doi:10.1103/PhysRevC.59.1663 [nucl-th/9809027].

[2] A. Andronic, P. Braun-Munzinger and J. Stachel, Phys. Lett. B **673**, 142 (2009) Erratum: [Phys. Lett. B **678**, 516 (2009)] doi:10.1016/j.physletb.2009.02.014, 10.1016/j.physletb.2009.06.021 [arXiv:0812.1186 [nucl-th]].

[3] J. Manninen and F. Becattini, Phys. Rev. C **78**, 054901 (2008) doi:10.1103/PhysRevC.78.054901 [arXiv:0806.4100 [nucl-th]].

[4] B. I. Abelev *et al.* [STAR Collaboration], Phys. Rev. C **81**, 024911 (2010) doi:10.1103/PhysRevC.81.024911 [arXiv:0909.4131 [nucl-ex]].

[5] M. M. Aggarwal *et al.* [STAR Collaboration], Phys. Rev. C **83**, 034910 (2011) doi:10.1103/PhysRevC.83.034910 [arXiv:1008.3133 [nucl-ex]].

[6] J. Noronha-Hostler, J. Noronha, and C. Greiner, Phys. Rev. Lett. 103, 172302 (2009), arXiv:0811.1571 [nucl-th].

[7] A. Majumder and B. Muller, Phys.Rev.Lett. 105, 252002 401 (2010), arXiv:1008.1747 [hep-ph].

[8] A. Bazavov *et al.*, Phys. Rev. Lett. 113 (2014) 072001.

[9] P. Alba et al., Phys. Rev. D96 (2017) 034517.

[10] S. Borsanyi *et al.*, JHEP **1209** (2012) 010 doi:10.1007/JHEP09(2012)010 [arXiv:1203.4469 [hep-lat]].

[11] A. Andronic, P. Braun-Munzinger, K. Redlich and J. Stachel, J. Phys. Conf. Ser. **779**, no. 1, 012012 (2017) doi:10.1088/1742-6596/779/1/012012 [arXiv:1611.01347 [nucl-th]].
[12] S. Borsanyi *et al.* [Wuppertal-Budapest Collaboration], JHEP **1009**, 073 (2010) doi:10.1007/JHEP09(2010)073 [arXiv:1005.3508 [hep-lat]].
[13] A. Bazavov *et al.* [HotQCD Collaboration], Phys. Rev. D **90**, 094503 (2014) doi:10.1103/PhysRevD.90.094503 [arXiv:1407.6387 [hep-lat]].
[14] O. Kaczmarek *et al.*, Phys. Rev. D **83**, 014504 (2011) doi:10.1103/PhysRevD.83.014504 [arXiv:1011.3130 [hep-lat]].
[15] R. Bellwied, S. Borsanyi, Z. Fodor, J. Guenther, S. D. Katz, C. Ratti and K. K. Szabo, Phys. Lett. B **751**, 559 (2015) doi:10.1016/j.physletb.2015.11.011 [arXiv:1507.07510 [hep-lat]].
[16] R. Bellwied *et al.*, Phys.Rev.Lett. 111 (2013) 202302.
[17] M. Floris, Nucl Phys. A931 (2014) 103.
[18] M. Ferraris, M. M. Giannini, M. Pizzo, E. Santopinto, and L. Tiator, Phys. Lett. B364, 231 (1995).
[19] R. Bijker, J. Ferretti, G. Galata, H. Garcia-Tecocoatzi, and E. Santopinto, Phys. Rev. D94, 074040 (2016), arXiv:1506.07469 [hep-ph].
[20] S. Capstick and N. Isgur, Phys.Rev. D34, 2809 (1986).
[21] D. Ebert, R. Faustov, and V. Galkin, Phys.Rev. D79, 114029 (2009), arXiv:0903.5183 [hep-ph].
[22] S. Borsanyi, Z. Fodor, S. D. Katz, S. Krieg, C. Ratti and K. Szabo, JHEP **1201**, 138 (2012) doi:10.1007/JHEP01(2012)138 [arXiv:1112.4416 [hep-lat]].
[23] A. Bazavov *et al.* [HotQCD Collaboration], Phys. Rev. D **86**, 034509 (2012) doi:10.1103/PhysRevD.86.034509 [arXiv:1203.0784 [hep-lat]].
[24] R. V. Gavai and S. Gupta, Phys. Rev. D **73**, 014004 (2006) doi:10.1103/PhysRevD.73.014004 [hep-lat/0510044].
[25] R. Bellwied, S. Borsanyi, Z. Fodor, S. D. Katz, A. Pasztor, C. Ratti and K. K. Szabo, Phys. Rev. D **92**, no. 11, 114505 (2015) doi:10.1103/PhysRevD.92.114505 [arXiv:1507.04627 [hep-lat]].
[26] H.-T. Ding, S. Mukherjee, H. Ohno, P. Petreczky and H.-P. Schadler, Phys. Rev. D **92**, no. 7, 074043 (2015) doi:10.1103/PhysRevD.92.074043 [arXiv:1507.06637 [hep-lat]].
[27] V. Skokov, B. Friman and K. Redlich, Phys. Rev. C **88** (2013) 034911
[28] P. Braun-Munzinger, A. Rustamov and J. Stachel, Nucl. Phys. A **960**, 114 (2017) doi:10.1016/j.nuclphysa.2017.01.011 [arXiv:1612.00702 [nucl-th]].
[29] V. Begun and M. Mackowiak-Pawlowska, arXiv:1705.01110 [nucl-th].
[30] A. Bzdak and V. Koch, Phys. Rev. C **86** (2012) 044904
[31] S. Jeon and V. Koch, Phys. Rev. Lett. **85** (2000) 2076
[32] F. Becattini, J. Steinheimer, R. Stock and M. Bleicher, Phys. Lett. B **764**, 241 (2017) doi:10.1016/j.physletb.2016.11.033 [arXiv:1605.09694 [nucl-th]].
[33] J. Steinheimer, J. Aichelin and M. Bleicher, Phys. Rev. Lett. **110**, no. 4, 042501 (2013) doi:10.1103/PhysRevLett.110.042501 [arXiv:1203.5302 [nucl-th]].
[34] F. Becattini, M. Bleicher, T. Kollegger, T. Schuster, J. Steinheimer and R. Stock, Phys. Rev. Lett. **111**, 082302 (2013) doi:10.1103/PhysRevLett.111.082302 [arXiv:1212.2431 [nucl-th]].
[35] J. Steinheimer, V. Vovchenko, J. Aichelin, M. Bleicher and H. Stoecker, arXiv:1608.03737 [nucl-th].
[36] M. Kitazawa and M. Asakawa, Phys. Rev. C **85** (2012) 021901
[37] M. Kitazawa and M. Asakawa, Phys. Rev. C **86**, 024904 (2012) Erratum: [Phys. Rev. C **86**, 069902 (2012)] doi:10.1103/PhysRevC.86.024904, 10.1103/PhysRevC.86.069902

[arXiv:1205.3292 [nucl-th]].

[38] M. Nahrgang, M. Bluhm, P. Alba, R. Bellwied and C. Ratti, Eur. Phys. J. C **75** (2015) 12, 573

[39] S. Borsanyi, Z. Fodor, S. D. Katz, S. Krieg, C. Ratti and K. K. Szabo, Phys. Rev. Lett. **111** (2013) 062005

[40] S. Borsanyi, Z. Fodor, S. D. Katz, S. Krieg, C. Ratti and K. K. Szabo, Phys. Rev. Lett. **113** (2014) 052301

[41] L. Adamczyk *et al.* [STAR Collaboration], Phys. Rev. Lett. **112**, 032302 (2014) doi:10.1103/PhysRevLett.112.032302 [arXiv:1309.5681 [nucl-ex]].

[42] L. Adamczyk *et al.* [STAR Collaboration], Phys. Rev. Lett. **113**, 092301 (2014) doi:10.1103/PhysRevLett.113.092301 [arXiv:1402.1558 [nucl-ex]].

[43] P. Alba, W. Alberico, R. Bellwied, M. Bluhm, V. Mantovani Sarti, M. Nahrgang and C. Ratti, Phys. Lett. B **738**, 305 (2014) doi:10.1016/j.physletb.2014.09.052 [arXiv:1403.4903 [hep-ph]].

[44] A. Bazavov *et al.*, Phys. Rev. D **93**, no. 1, 014512 (2016) doi:10.1103/PhysRevD.93.014512 [arXiv:1509.05786 [hep-lat]].

[45] C. Ratti, Nucl. Phys. A **956**, 51 (2016) doi:10.1016/j.nuclphysa.2016.02.022 [arXiv:1601.02367 [hep-lat]].

[46] J. Gunther, R. Bellwied, S. Borsanyi, Z. Fodor, S. D. Katz, A. Pasztor and C. Ratti, EPJ Web Conf. **137**, 07008 (2017) doi:10.1051/epjconf/201713707008 [arXiv:1607.02493 [hep-lat]].

[47] A. Rustamov [ALICE Collaboration], Nucl. Phys. A **967**, 453 (2017) doi:10.1016/j.nuclphysa.2017.05.111 [arXiv:1704.05329 [nucl-ex]].

[48] P. Braun-Munzinger, A. Kalweit, K. Redlich and J. Stachel, Phys. Lett. B **747**, 292 (2015) doi:10.1016/j.physletb.2015.05.077 [arXiv:1412.8614 [hep-ph]].

[49] A. Bazavov *et al.* [HotQCD Collaboration], arXiv:1708.04897 [hep-lat].

[50] L. Adamczyk *et al.* [STAR Collaboration], arXiv:1709.00773 [nucl-ex].

[51] P. Alba, R. Bellwied, M. Bluhm, V. Mantovani Sarti, M. Nahrgang and C. Ratti, Phys. Rev. C **92**, no. 6, 064910 (2015) doi:10.1103/PhysRevC.92.064910 [arXiv:1504.03262 [hep-ph]].

[52] J. Noronha-Hostler, R. Bellwied, J. Gunther, P. Parotto, A. Pasztor, I. P. Vazquez and C. Ratti, arXiv:1607.02527 [hep-ph].

Final state hadronic rescattering with UrQMD

J. Steinheimer[1,⋆], *V.* Vovchenko[1,2], *J.* Aichelin[1,3], *M.* Bleicher[1,2,5], and *H.* Stöcker[1,2,4]

[1] *Frankfurt Institute for Advanced Studies, Ruth-Moufang-Str. 1, 60438 Frankfurt am Main, Germany*
[2] *Institut für Theoretische Physik, Goethe Universität Frankfurt, Max-von-Laue-Strasse 1, D-60438 Frankfurt am Main, Germany*
[3] *SUBATECH, UMR 6457, Université de Nantes, Ecole des Mines de Nantes, IN2P3/CNRS. 4 rue Alfred Kastler, 44307 Nantes cedex 3, France*
[4] *GSI Helmholtzzentrum für Schwerionenforschung GmbH, D-64291 Darmstadt, Germany*
[5] *John von Neumann Institut für Computing, FZ Jülich, 52425 Jülich, Germany*

Abstract. In this talk we discuss the effects of the hadronic rescattering on final state observables in high energy nuclear collisions. We do so by employing the UrQMD transport model for a realistic description of the hadronic decoupling process. The rescattering of hadrons modifies every hadronic bulk observable. For example apparent multiplicity of resonances is suppressed as compared to a chemical equilibrium freeze-out model. Stable and unstable particles change their momentum distribution by more than 30% through rescattering. The hadronic rescattering also leads to a substantial decorrelation of the conserved charge distributions. These findings show that it is all but trivial to conclude from the final state observables on the properties of the system at an earlier time where it may have been in or close to local equilibrium.

1 Introduction

The aim of ultrarelativistic heavy ion experiments at the Relativistic Heavy Ion Collider (RHIC) and the Large Hadron Collider (LHC) is is the study of the properties of the socalled quark gluon plasma (QGP) which is created for a short time in these reactions [1–5]. This includes the determination of the yet unknown transport properties of QCD matter as well as the study of its phase structure at large net baryon densities. It involves the study of hadron multiplicities but also fluctuations of conserved quantities in a given rapidity interval which have been compared with lQCD results

In the theoretical description of these reactions we are confronted with rapidly expanding systems which may or may not hadronize at (or near) the crossover line (chemical freeze out) and continue to expand until the interactions cease. This latter phase is called hadronic phase and is characterized by hadronic interactions which govern the time period when the system drops out of equilibrium. To draw meaningfull conclusions one has to assess how the hadronic phase has modified the observables.

To do so sophisticated model simulations are usually employed. The current state-of-the-art of such models are so called hybrid models where a (viscous) fluid dynamical simulation [6–9] is followed by an expanding gas of hadrons. The evolution of such a hadron gas can be described with the relativistic Boltzmann equation:

$$p^{\mu} \cdot \partial_{\mu} f_i(x^{\nu}, p^{\nu}) = C_i \quad . \tag{1}$$

Here the change of the single particle distributions $f_i(x^{\nu}, p^{\nu})$ (left hand side) is caused by the collision term C_i (right hand side). Using a transport model (based on the Boltzman equation) as an afterburner which follows the fluid dynamical evolution has become the standard method of treating the interactions during the hadronic phase of nuclear collisions [6, 8–15]. Recently several studies have pointed out the importance of the hadronic rescattering, during the freeze-out phase of a nuclear reaction, on the description of observables like particle numbers and spectra, resonance production, momentum anisotropies and particle number fluctuations [16–19].

This talk will give a summary of how the hadronic interactions affect the final state observables. The results presented can be found in [20, 21] where a more detailed description and discussion is presented.

2 UrQMD

To study the dynamics of the hadronic phase we will use the UrQMD hybrid model [9]. This model combines a 3+1D ideal fluid dynamical simulation with a state-of-the-art transport description of the non-equilibrium hadronic decoupling phase. In the present investigation we use the

⋆e-mail: steinheimer@fias.uni-frankfurt.de

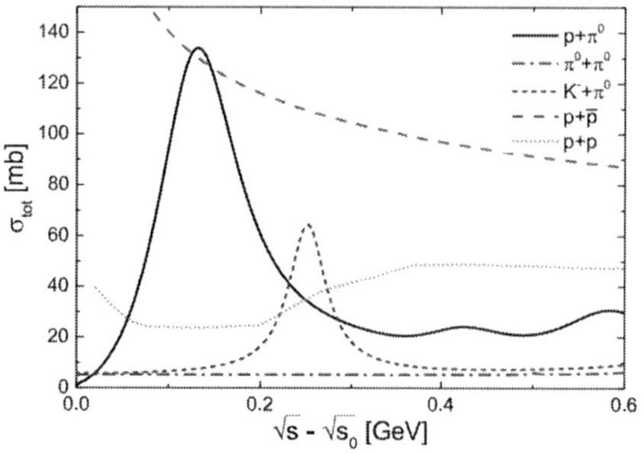

Figure 1. Different total (elastic + inelastic) hadronic cross sections as implemented in UrQMD. We show the cross sections as function of the relative energy, i.e. we subtract the rest masses $\sqrt{s_0}$ of both hadrons from the total invariant mass.

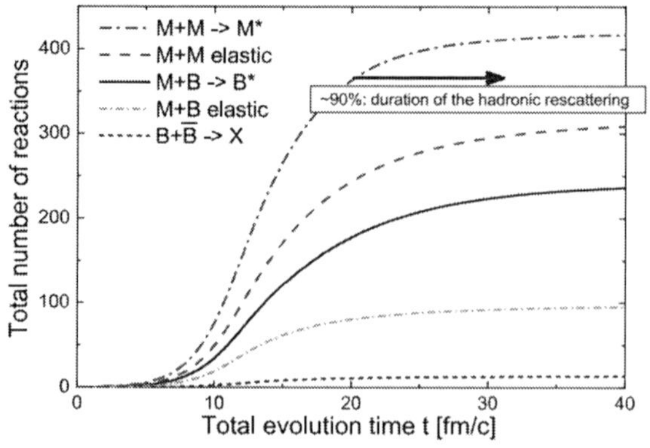

Figure 2. Total number of different hadronic interactions occurring until the time t, during the hadronic phase of central collisions of Au+Au at $\sqrt{s_{NN}} = 200$ GeV.

Cooper-Frye equation,

$$E\frac{dN}{d^3p} = \int_\sigma f(x,p)p^\mu d\sigma_\mu \qquad (2)$$

to transform the fluid dynamical fields to discrete hadrons on an iso-energy density hypersurface [22] $e_{CF} \approx 350$ MeV/fm^3. Here $f(x, p)$ corresponds to the grand canonical Bose- or Fermi- distribution functions, depending essentially on the local temperature $T(x)$ and chemical potentials $\mu(x)$. To obtain a finite number of particles the integral is randomly sampled, taking into account the global cosnervation of baryon number strangeness and electric charge.

The non-equilibrium transport part of the UrQMD model is based on $2 \rightarrow n$ hadronic scattering, according to measured reaction cross sections [23], which serve as the main input to the model. The model includes nearly 60 different baryonic species with their anti-particles as well as about 40 mesonic states [24, 25]. The possible interactions between these hadrons include elastic scattering, resonance excitations, string excitations as well as strangeness exchange reactions. To present an example of the cross sections implemented in the UrQMD model we show the total cross sections for several processes in Fig. 1. As one can see there are several processes with very large cross sections, mainly involving protons. In particular the proton+anti-proton annihilation cross section is large. The probabilities of certain reactions do not only depend on the magnitude of the cross section but also on the phase space densities of the involved hadrons, making the microscopic description of the hadronic phase so important.

3 Results

In the following the dynamics of the hadronic interactions in the final phase of collisions of Au+Au at beam energies of $\sqrt{s_{NN}} = 200$ GeV are investigated, focusing on the mid rapidity region $-0.5 < y < 0.5$ of most central ($b < 3.4$ fm) collisions where experimental data are available.

The total number of hadronic scatterings, up to a time t, defined as the time since the first initial collisions have taken place, is shown in figure 2. Here different possible hadronic interactions, meson+meson resonance excitation and elastic scattering, meson+baryon resonance excitation and elastic scattering as well as baryon+antibaryon annihilation are compared. The final rescattering is dominated by meson+meson interactions and only very few annihilation processes take place. The number of scatterings changes most drastically in the time between hadronization, around 10 fm/c, and a time of roughly 20 fm/c, after which the number of hadronic scatterings does essentially not change anymore. This means that the hadronic rescattering mainly takes place in the 10 fm/c after hadronization.

3.1 Resonance dynamics

If one wants to experimentally verify the existence and duration of the hadronic phase, the study of the production and absorption rates of hadronic resonances in nuclear collisions is a useful tool [26]. Resonances which decay early during the hadronic phase may not be identified experimentally because the decay products rescatter. Therefore the hadronic intercations are reflected in the number of identified resonances. The states discussed in this work, are summarized in table 1.

It is obvious that the listed hadronic resonances have widely varying properties and in particular different lifetimes which should be reflected in the dynamics of these resonances during the hadronic phase. In figure 3, the net resonance gain rate as a function of time for the different hadronic resonances is shown. The net gain rate is defined as

$$\frac{1}{N_{CF}}\left(\frac{dN_{gain}}{dt} - \frac{dN_{loss}}{dt}\right) \qquad (3)$$

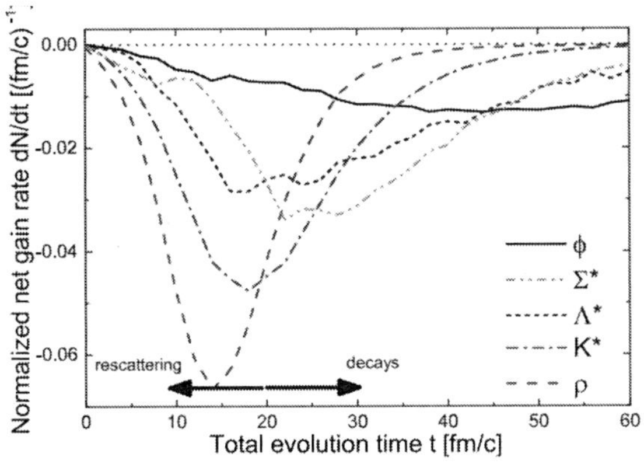

Figure 3. Net gain rates of different hadronic resonances as function of time and normalized to the initial yield at hadronization (Cooper-Frye). The results are for central collisions of Au+Au at $\sqrt{s_{NN}} = 200$ GeV.

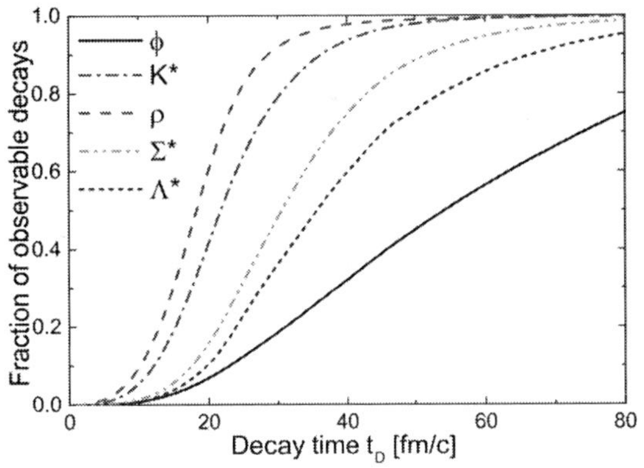

Figure 4. Fraction of observable resonances which have decayed until a time t. The results are for central collisions of Au+Au at $\sqrt{s_{NN}} = 200$ GeV.

where N_{CF} is the resonance yield at hadronization (Cooper-Frye), $\frac{dN_{\text{gain}}}{dt}$ is the number of reactions that create a resonance and $\frac{dN_{\text{loss}}}{dt}$ is the number of reactions that destroy a resonance, e.g. decay or inelastic scattering, per unit of time. The results presented in figure 3 clearly reflect the different lifetimes of the resonances, as the ρ meson shows its largest loss rate during early times, while the ϕ decays only slowly and therefore shows a slow loss. However there is an interesting comparison between the Λ^* and Σ^*. Even though there is a lifetime difference of a factor 2 between these states, their net gain rate appears to be very similar. This hints to the fact that the Σ^* has a quite substantial regeneration rate at early times of the hadronic phase.

The lifetime hierarchy can also be observed in figure 4 where we show the fraction of observable/reconstructable resonances which have decayed until a time t_D, as function of time. As expected, the ϕ meson decays at a very late time and the short lived resonances decay at an earlier time.The difference of the Σ^* and Λ^* is smaller than expected due to the significant Σ^* regeneration at early times. An important observation is that even though short lives resonances like the ρ or K^* are the first to decay, most of the observable resonances still decay at times $t > 15$ fm/c, i.e. a rather dilute hadronic system and hence may not carry much information on the dense system where they where born.

Not only do these resonances have different lifetimes, but they also decay into different daughter particles which have to be identified in order to reconstruct the original resonance. As these daughter particles also undergo rescattering they may be lost to reconstruction, thus also the information on the mother resonance would be lost. The probability of the daughter particle rescattering does depend on the time of the resonance decay and on the rescattering cross section of the daughter particles which is different for example for pions and kaons.

Not all resonances decay at the same time and therefore the detection probability for the different resonances will depend on their lifetime. In figure 5 the time integrated detection probability is shown as a function of the transverse momentum p_T of the resonance. There is a clear hierarchy of the detection probability of the resonances where the ϕ with its long lifetime and small hadronic cross section has the highest and the ρ has the lowest probability of being detectable.

Will the final state interactions lead to observable differences in resonance multiplicities and properties? To answer this question we study the ratio of the p_T integrated multiplicity of detectable resonances and the corresponding multiplicity of their ground state hadron in figures 6 and 7. Here the results are compared with experimental data from the STAR experiment [27–29]. A very good agreement of our results with the experimental data in the case of the full model is observed. When the final hadronic rescattering is neglected there is a deviation from the experimental results, especially for the K^* and Λ^* ratios, indicating that they are most sensitive to the rescattering phase.

Table 1. Resonances and their decay channels

Resonance	decay	branching ratio	lifetime [fm/c]
$\rho(770)^0$	$\pi^+ + \pi^-$	1	1.335
$K^*(892)^0$	$\pi^- + K^+$	0.67	4.16
$\phi(1020)$	$K^+ + K^-$	0.489	46.26
$\Sigma(1385)^+$	$\pi^+ + \Lambda$	0.870	5.48
$\Sigma(1385)^-$	$\pi^- + \Lambda$	0.870	5.01
$\Lambda(1520)$	$K^- + p$	0.225	12.54
$\Xi(1530)^0$	$\pi^+ + \Xi^-$	0.67	22

3.2 Stable hadrons

In the following we will discuss whether we can also expect an impact of the rescattering on stable hadron yields

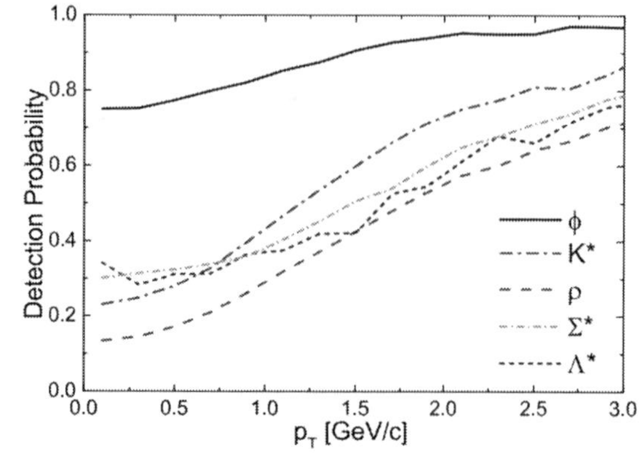

Figure 5. Detection probability of different hadronic resonances as function of their transverse momentum. A resonance is detectable if none of their decay products rescatter in the hadronic phase. The results are for central collisions of Au+Au at $\sqrt{s_{NN}}$ = 200 GeV.

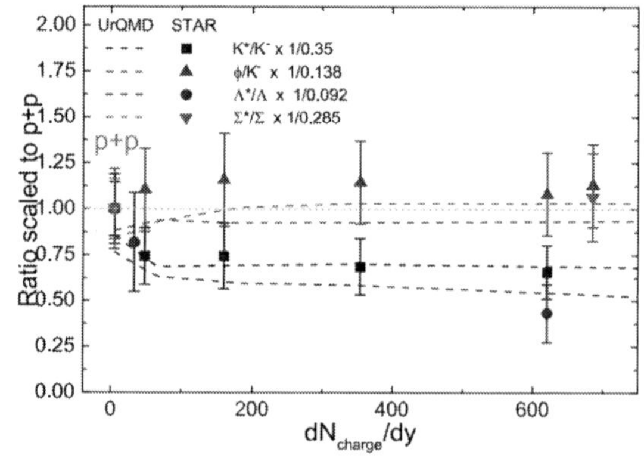

Figure 7. Ratios of detectable resonances to their ground state after the hadronic phase as function of centrality (lines) compared to experimental data (symbols). The results are for collisions of Au+Au at $\sqrt{s_{NN}}$ = 200 GeV. The quoted temperature T_{FO} corresponds to the $\epsilon \approx 350$ MeV/fm^3 criterion.

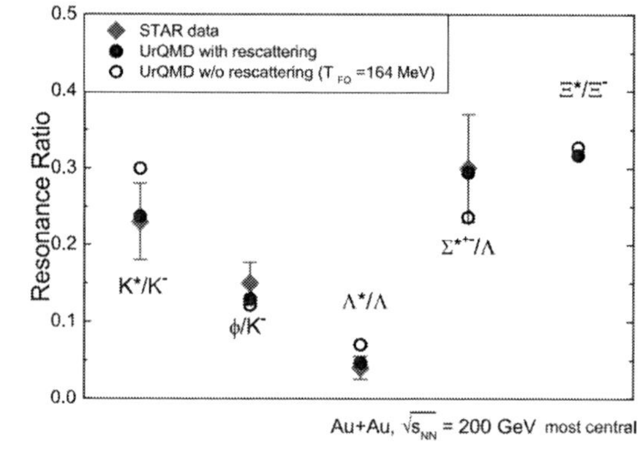

Figure 6. Ratios of detectable resonances to their ground state after the hadronic phase (full circles) and before hadronic rescattering (open circles), compared to data from the STAR experiment (magenta diamonds).

and properties like spectra and collective flow. Here one can distinguish two types of hadronic scatterings. One is the (pseudo-)elastic scattering which changes only the momentum distribution of the present hadrons but not their abundances. There is also be inelastic scatterings which change the flavor and hadronic abundances during the hadronic rescattering. The most relevant processes her is the annihilation of baryons and anti-baryons. But also strangeness exchange reactions and multi-body decays of resonances increasing the number of pions. Figure 8 we show the time dependence of stable hadron yields (including contributions from resonance decays) as a function of time, for most central collisions of Au+Au at a beam energy of $\sqrt{s_{NN}}$ = 200 GeV. The multiplicities are scaled to the final (observable) value to quantify the effect of the hadronic rescattering and the clear time dependence confirms that the hadronic rescattering changes the chemical composition of the stable hadrons to times as large as 20 fm/c. For example the protons show the largest decrease in yield due to annihilation, even though the total baryon number is always exactly conserved. The proton annihilation is stronger than that of Λ hyperons due to the larger annihilation cross section. In fact a clear hierarchy due to the decrease of the cross section with increasing strangeness is visible.

The effect of hadronic rescattering on the final observed momentum spectra is shown in figure 9. Here the change of the spectra is shown as the ratio of the final p_T spectrum over the spectrum at chemical freeze out (Cooper-Frye). We compare three distinct hadrons, pions, protons and Ω. Each species shows a very different characteristic change of the spectrum. The proton spectrum is mainly enhanced at large momenta and is depleted at low momenta, which reflects the generation of transverse collective velocity flow which has been observed in these reactions. This change is mainly caused by elastic and pseudo-elastic scatterings, while baryon antibaryon annihilation decreases the spectrum mostly independent of p_T. The pions also obtain an additional radial flow, however also the low p_T part of the pion spectrum is enhanced. This is due to contributions from resonance decays which populate mainly the low transverse momenta. The Ω baryon receives the smallest change in the spectrum due to its small hadronic rescattering cross section. It is little affected from the transverse flow. That the average value of the ratio is below one is likely a consequence of the strangeness exchange reaction and longitudinal expansion.

3.3 Annihilation

An important and much discussed contribution to the final state hadronic rescattering is the baryon+anti-baryon annihilation reaction. In this reaction the baryons, e.g. proton and anti-proton, annihilate to form multiple mesons.

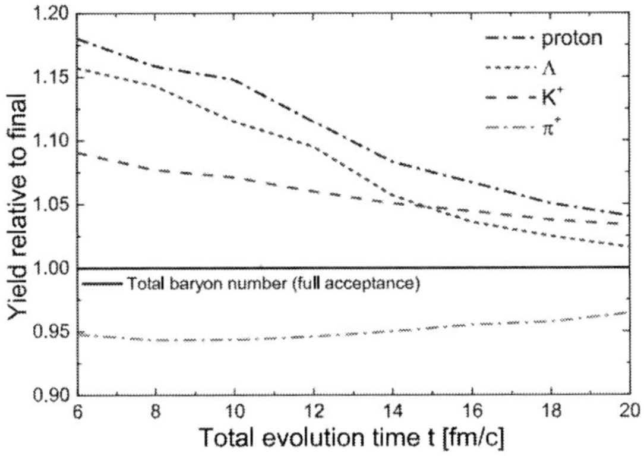

Figure 8. Time dependence of the stable hadron yields (including their resonance feed down), normalized to the final yields, during the hadronic phase. The results are for central collisions of Au+Au at $\sqrt{s_{NN}} = 200$ GeV.

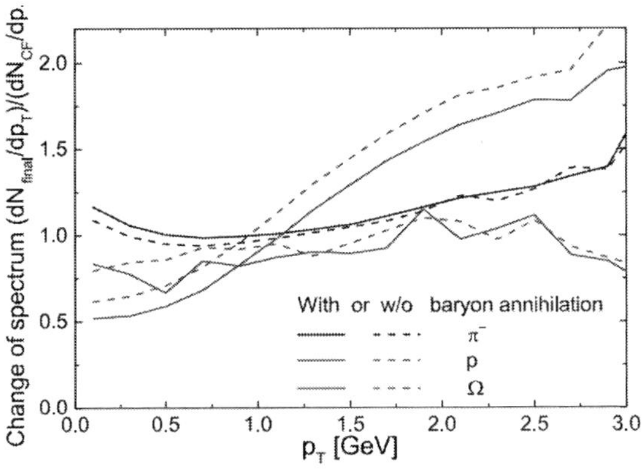

Figure 9. Change of the spectrum of stable hadrons due to the finals state rescattering. We compare two scenarios, one including the baryon+antibaryon annihilation and one excluding the annihilation reaction. The results are for central collisions of Au+Au at $\sqrt{s_{NN}} = 200$ GeV.

For the case of nucleons and anti-nucleon on average about 5 pions are created. Since most transport models (as UrQMD) do not include multi particle scatterings, the inverse reaction, 5 pions creating a baryon + anti-baryon pair, is not taken into account [30–33]. This violates detailed balance but in an expanding system which hadronizes at an energy density of $350 MeV/fm^3$ this reaction, which decreases with ρ^5, is rare [34]. This has also been shown in studies where the detailed balance for this reaction has been explicitly included and the resulting baryon and anti-baryon loss in the hadronic phase is comparable to our scenario [35]. Nevertheless it is worthwhile to further pursue the implementation of a hadronic rescattering phase that includes such multi-particle scattering in order to minimize the systematic errors on the model side.

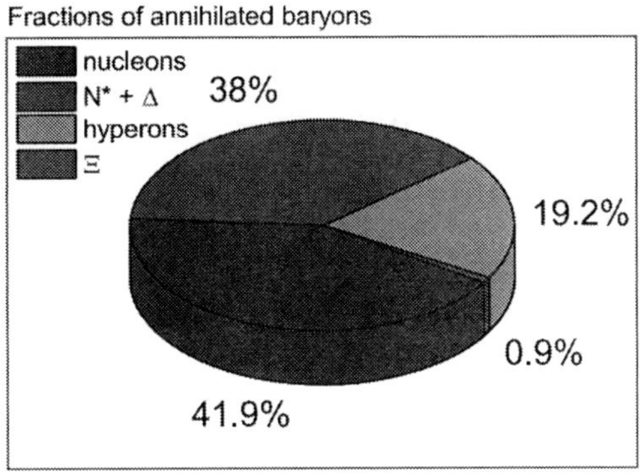

Figure 10. Relative fraction of different baryonic species lost due to baryon+antibaryon annihilation in the hadronic phase. Only 42% of all annihilated baryons are actual nucleons. The rest is composed of excited nucleonic states as well as hyperons and their resonance states. As a result less than 25 percent of all annihilations actually takes place between two nucleons in their ground state. The results are for central collisions of Au+Au at $\sqrt{s_{NN}} = 200$ GeV.

For simplicity one usually argues about the annihilation of ground state nucleons, and possibly hyperons, but generally one neglects that a significant portion of baryons consists of excited resonance states with significant larger masses. For example at LHC and RHIC at chemical freeze out almost half of all baryon exist in form a baryonic resonance. In our approach the absolute value and the $\sqrt{s}$ dependence of the annihilation cross section for the resonances is assumed to be identical to that of the ground state with the same quantum numbers. Figure 10 depicts the percentage of the different annihilation channels which contribute to the annihilation. One can clearly see that less than half of all annihilated baryons are actually in their ground state. This also means that in less than 25% of all annihilation reactions two ground state baryons annihilate. The majority of these reactions will have a larger invariant mass and therefore produce on average more than 5 pions and eventually even heavier mesons like kaons. Thus the back reaction of these annihilation reactions will be even more suppressed by the phase space density and also the effective implementation of all possible processes is much more involved.

3.4 Effects on fluctuations of conserved charges

As a last point we will discuss the effects on the distribution of conserved charges. We set up a time-dependent correlation function to explore the diffusion of conserved charges during the hadronic phase. Following the standard definition of the correlation coefficient this function reads as

$$r_{\text{IF}}(t) = \frac{\sum_n (I_n(t) - \bar{I}(t))(F_n - \bar{F})}{\sqrt{\sum_n (I_n(t) - \bar{I}(t))^2 \sum_n (F_n - \bar{F})^2}}, \qquad (4)$$

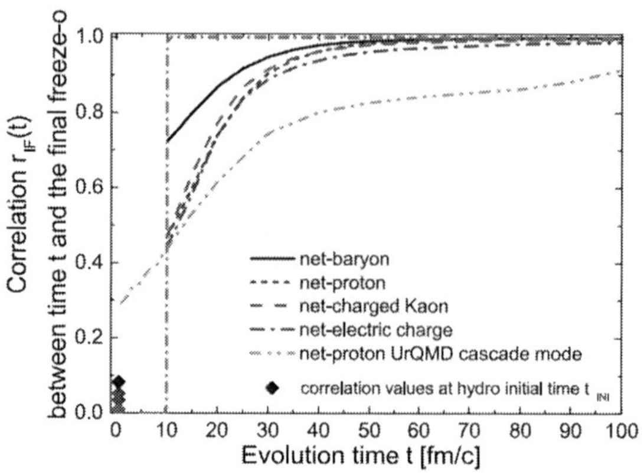

Figure 11. Correlation of the net-charge, net-kaon, net-proton and net-baryon number in the rapidity interval $-0.5 < y < 0.5$ with an earlier time t in that same rapidity interval.

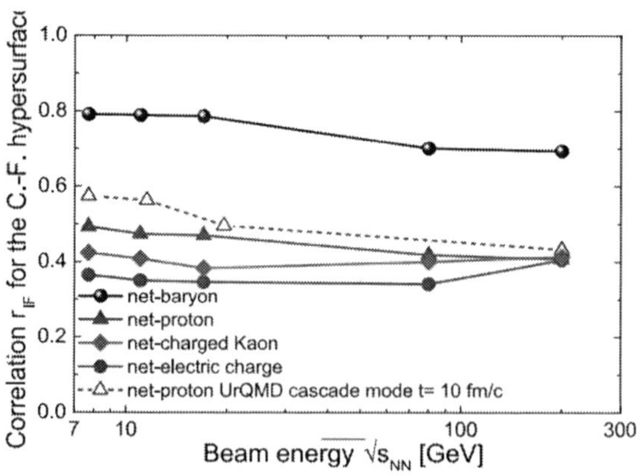

Figure 12. Correlation of the finally observed net-charge, net-kaon, net-proton and net-baryon number in the rapidity interval $-0.5 < y < 0.5$, at the Cooper-Frye hypersurface, with the final net-charge density as a function of beam energy.

where $I_n(t)$ denotes the number of a given charge (or a net-particle number) in a given rapidity and momentum window at the time t in the event number n. F_n is the final number of that charge after all interactions have ceased, in the same event. Finally, the $\overline{I}(t)$ and $\overline{F}$ are the corresponding averages. The sum runs over all events in the sample.

It is evident that if the net-charge in a given bin at a given time t is perfectly correlated with the net-charge at the end of the evolution, then the value of this correlation function $r_{IF}(t)$ will be equal to 1 whereas if the net-charge number at given time t is completely uncorrelated with the final value, then the $r_{IF}(t)$ will be equal to zero.

In the standard scenario [36], which is employed in most comparisons of the (normalized) cumulants with the HRG or lattice QCD results one essentially assumes that this correlation is equal to zero before the chemical freeze-out of hadrons and then instantly jumps to one at the chemical freeze-out, a value which does not change until the cumulants are observed by experiment. This scenario is shown in figure 11 by a gray dash dotted line for an exemplary freeze-out time of 10 fm/c. In such a scenario the final value of the cumulants correspond to that at chemical freeze-out, without any information about the QGP phase before freeze-out or about the hadronic rescattering which follows hadronization.

To estimate the effect of the full local equilibration on the decorrelation of the different charges we will also contrast the hybrid model simulations with results obtained from the standard UrQMD model in its cascade version. In the cascade version of the model, the system remains essentially out of local equilibrium for a long time and, due to the finite cross sections, will retain more information and therefore show a stronger correlation with the early times.

Figure 11 shows the dependence of the correlation $r_{IF}(t)$, for different net-charge and net-particle numbers, as a function of time. Unlike in the standard scenario assuming instant freeze-out, the correlation function strongly depends on the duration of the rescattering phase for all considered quantities. At very late times ($t > 40$ fm/c) the correlation function approaches 1, which is expected as the system has essentially frozen out and interactions as well as resonance decays have ceased by that time. However, the correlation changes rapidly in the time interval between particle production (on the Cooper-Frye hypersurface) at around 10 fm/c and decoupling of the system at about 25 fm/c. At the time of interest for most studies, namely the particle production time, the correlation is approximately 0.5 for most net-charges. Only the net-baryon number correlation is slightly stronger. However, the quantity that is measured experimentally, namely the net-proton number, shows a similar correlation as the net-charge, as significant iso-spin changing reactions take place during the rescattering phase. One should keep in mind that the fact that the correlation still changes at rather late times does not mean that the lifetime of the hadronic phase is on the order of 40 fm/c. At a time of 20fm/c about 90% of all 2-body reactions have already occurred. After that we observe mainly resonance decays.

There is a crucial point in time at which it is worthwhile to study the correlation $r_{IF}(t)$ in more detail. It is the time at which hadrons are produced according to chemical equilibrium distributions. At this point, at the transition between the hadronic and quark phase, the hadrons are sampled from the transition (hadronization) hypersurface according to the Cooper-Frye equation. In fact since the iso-energy hypesurface does not correspond to a fixed time we would not compare the net charge at two different fixed times, but on the hypersurface volume and in the final state. The second one is an even earlier time at which the fluid dynamical evolution starts. Therefore we show in figure 12 the correlation $r_{IF}(t = t_{CF})$ which is defined as the correlation between the net-particle number going through the Cooper-Frye hypersurface and the final state particle number. Again this correlation does not compare two fixed times but two 4-volumes in which the relevant quantities are conserved globally. The correlation

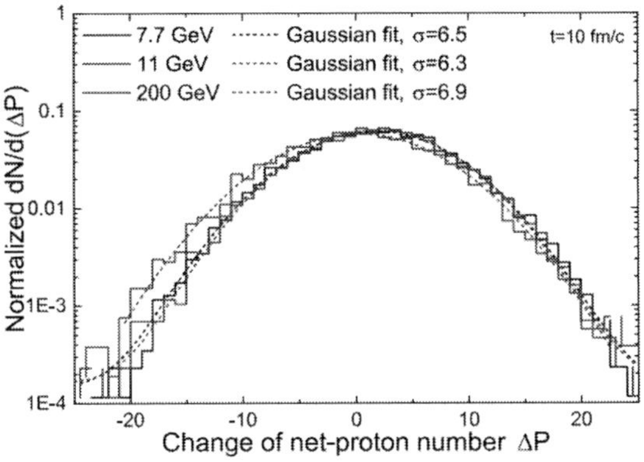

Figure 13. Distribution of the change of the net proton number in a given event, for different beam energies.

between the Cooper-Frye surface and the final state apparently shows only a very weak beam energy dependence down to $\sqrt{s_{NN}}$ = 7 GeV. This means that the importance of the hadronic rescattering does not change much within the beam energy range of the beam energy scan program at RHIC.

As a last step we will demonstrate the effect of the de-correlation of the different net-charge and net-particle numbers on their final number distributions. To do so we show the distribution of the change of a particular net-particle number (i.e. the net-proton number) for different beam energies. The change of the net-proton number ΔP is defined as the final net-proton number in a single event minus the net-proton number at a given time t in that single event. Figure 13 shows that this distribution function for the change of the net-proton number has a broad distribution. We were able to fit this distribution with a Gaussian of width $\sigma = 6.5$, which is only slightly smaller than the width of the final net-proton number distribution which is 7.5. The width of the Gaussian smearing is approximately constant for all beam energies, which is consistent with the observation in figure 12 that the de-correlation is almost independent of beam energy.

3.5 Relevance for observations

At this point we want to put our results in context with previous publications [37, 38] on interpreting data from heavy ion collisions. It is important to keep in mind that, even though we have shown that the actual net-charge number in an acceptance window will change in every event due to the rescattering, this does not necessarily mean that also the measured cumulant of that net-charge will also change. Let us take for example a system which does not entail any correlations between particles. Then the cumulants of any order are merely the mean number of that charge (see e.g. [39]). So if the mean particle number is fixed at some point during the evolution the cumulants for a non correlated system will be fixed by definition at that same point, called the chemical freeze out point (not to be mistaken with the latest point of chemical equilibrium).

If the system has strong correlations present at the point where the mean particle numbers are fixed, due to a phase transition or critical endpoint, and these correlations are washed out by the hadronic rescattering, again the cumulants will simply reflect the mean values. Note that there is a small contribution to the cumulants from conservation laws. These anti-correlation effects are small for high beam energies and they essentially cancel when taking ratios of same order cumulants, thus they are not relevant for the discussion presented here.

In consequence, if indeed the determination of the 'freeze-out' point by use of net-charge cumulants is consistent with the 'freeze-out' point of the mean particle values, one can only conclude that either no correlations where present near the 'freeze-out' or they have been washed out by the rescattering. If the measurement shows practically independent particle production then no new information can be gained from the measurement of higher order cumulants as compared to the mean particle multiplicities.

4 Summary

We have presented a summary of the dynamics in the hadronic re-scattering phase of the UrQMD transport model. The hadronic re-scattering has a substantial (~ 30%) impact on most hadronic observables. In particular the calculated resonance yields which reproduce well the experimental data verify the existence of the hadronic phase experimentally. Furthermore the spectra and yields of stable hadrons are changed significantly, e.g. the proton number decreases by about 20% due to the annihilation of nucleons and excited baryonic states with their anti-particles. We also observe significant (~ 20%) modifications for the elliptic flow v_2 during the hadronic phase. We found that the hadronic rescattering phase leads to a substantial decorrelation of the conserved charge distributions. This effect is present for all investigated conserved charges and is mostly energy independent. This means that the final distribution function, the only one which can be observed, is to a large extent uncorrelated to that of the newly born hadrons which may carry information on the quark phase of QCD. These observations show that the experimental results cannot assert, or be directly compared to, the properties of the system at a conjectured hadronization or chemical freeze-out point.

5 Acknowledgments

V.V. appreciates the support from HGS-HIRe for FAIR. H.St. appreciates the support from J.M. Eisenberg Laureatus chair. The computational resources were provided by the LOEWE Frankfurt Center for Scientific Computing (LOEWE-CSC), and by the Kronos computing cluster at GSI.

References

[1] M. Gyulassy and L. McLerran, Nucl. Phys. A **750**, 30 (2005)

[2] J. Adams *et al.* [STAR Collaboration], Nucl. Phys. A **757**, 102 (2005)
[3] B. B. Back *et al.*, Nucl. Phys. A **757**, 28 (2005)
[4] I. Arsene *et al.* [BRAHMS Collaboration], Nucl. Phys. A **757**, 1 (2005)
[5] K. Adcox *et al.* [PHENIX Collaboration], Nucl. Phys. A **757**, 184 (2005)
[6] C. Gale, S. Jeon and B. Schenke, arXiv:1301.5893 [nucl-th].
[7] K. Werner, I. Karpenko, T. Pierog, M. Bleicher and K. Mikhailov, Phys. Rev. C **82**, 044904 (2010)
[8] C. Shen, Z. Qiu, H. Song, J. Bernhard, S. Bass and U. Heinz, Comput. Phys. Commun. **199**, 61 (2016)
[9] H. Petersen, J. Steinheimer, G. Burau, M. Bleicher and H. Stocker, Phys. Rev. C **78**, 044901 (2008)
[10] S. A. Bass and A. Dumitru, Phys. Rev. C **61**, 064909 (2000)
[11] D. Teaney, J. Lauret and E. V. Shuryak, Phys. Rev. Lett. **86**, 4783 (2001)
[12] T. Hirano, U. W. Heinz, D. Kharzeev, R. Lacey and Y. Nara, Phys. Lett. B **636**, 299 (2006)
[13] C. Nonaka, AIP Conf. Proc. **1235**, 165 (2010).
[14] T. Hirano, P. Huovinen and Y. Nara, Phys. Rev. C **83**, 021902 (2011)
[15] S. Ryu, S. Jeon, C. Gale, B. Schenke and C. Young, arXiv:1210.4588 [hep-ph].
[16] A. G. Knospe, C. Markert, K. Werner, J. Steinheimer and M. Bleicher, Phys. Rev. C **93**, no. 1, 014911 (2016)
[17] J. Steinheimer, J. Aichelin and M. Bleicher, Phys. Rev. Lett. **110**, 042501 (2013)
[18] J. Steinheimer, J. Aichelin and M. Bleicher, EPJ Web Conf. **36**, 00002 (2012).
[19] A. Ilner, D. Cabrera, C. Markert and E. Bratkovskaya, Phys. Rev. C **95**, no. 1, 014903 (2017)
[20] J. Steinheimer, J. Aichelin, M. Bleicher and H. Stöcker, Phys. Rev. C **95**, no. 6, 064902 (2017)
[21] J. Steinheimer, V. Vovchenko, J. Aichelin, M. Bleicher and H. Stöcker, arXiv:1608.03737 [nucl-th].
[22] P. Huovinen and H. Petersen, arXiv:1206.3371 [nucl-th].
[23] C. Patrignani *et al.* [Particle Data Group], Chin. Phys. C **40**, no. 10, 100001 (2016).
[24] S. A. Bass *et al.*, Prog. Part. Nucl. Phys. **41**, 255 (1998) [Prog. Part. Nucl. Phys. **41**, 225 (1998)]
[25] M. Bleicher *et al.*, J. Phys. **G25**, 1859 (1999)
[26] M. Bleicher and J. Aichelin, Phys. Lett. B **530**, 81 (2002)
[27] J. Adams *et al.* [STAR Collaboration], Phys. Rev. C **71**, 064902 (2005)
[28] B. I. Abelev *et al.* [STAR Collaboration], Phys. Rev. C **78**, 044906 (2008)
[29] B. I. Abelev *et al.* [STAR Collaboration], Phys. Rev. Lett. **97**, 132301 (2006)
[30] R. Rapp and E. V. Shuryak, Phys. Rev. Lett. **86**, 2980 (2001)
[31] C. Greiner and S. Leupold, J. Phys. **G27**, L95 (2001)
[32] W. Cassing, Nucl. Phys. A **700**, 618 (2002)
[33] L. M. Satarov, I. N. Mishustin and W. Greiner, Phys. Rev. C **88**, 024908 (2013)
[34] P. Braun-Munzinger, J. Stachel and C. Wetterich, Phys. Lett. B **596**, 61 (2004)
[35] Y. Pan and S. Pratt, arXiv:1210.1577 [nucl-th].
[36] F. Karsch and K. Redlich, Phys. Lett. B **695**, 136 (2011)
[37] P. Braun-Munzinger, A. Kalweit, K. Redlich and J. Stachel, Phys. Lett. B **747**, 292 (2015)
[38] P. Alba, R. Bellwied, M. Bluhm, V. Mantovani Sarti, M. Nahrgang and C. Ratti, Phys. Rev. C **92**, no. 6, 064910 (2015)
[39] A. Bzdak, V. Koch and N. Strodthoff, Phys. Rev. C **95**, no. 5, 054906 (2017)

A warm and friendly memorial session for Helmut Oeschler

Jean Cleymans[1,⋆], *Boris* Hippolyte[2], *Alexander* Kalweit[3], *Christian* Müntz[4], and *Joachim* Stroth[4,5]

[1] *Department of Physics, University of Cape Town, ZA-7701 Rondebosch, South Africa*
[2] *Institut Pluridisciplinaire Hubert Curien, Institut d'Études Avancées de l'Université de Strasbourg, France*
[3] *European Organization for Nuclear Research (CERN), CH-1211 Geneva 23, Switzerland*
[4] *Institut für Kernphysik (IKF), Goethe-Universität, Max-von-Laue-Str. 1, D-60438 Frankfurt, Germany*
[5] *GSI Helmholtzzentrum für Schwerionenforschung, Planckstr. 1, D-64921 Darmstadt, Germany*

Abstract. A full session was organized in memory of Helmut Oeschler during the 2017 edition of the Strangeness in Quark Matter Conference. It was heart-warming to discuss with the audience his main achievements and share anecdotes about this exceptionally praised and appreciated colleague, who was also a great friend for many at the conference. A brief summary of the session is provided with these proceedings.

1 Introduction

Trying to summarise the studies performed by Helmut Oeschler during his long and fruitful career, a couple of months only after he passed away, was a difficult and emotional task. During a dedicated memorial session at the Strangeness in Quark Matter edition of 2017, four of his closest colleagues and friends presented selected studies, moments and anecdotes, illustrating his successful career and how cheerful and enthusiastic Helmut was for many of us. We hope that, with these proceedings, we manage to convey how positive his presence was among us and how profound and influential his ideas were.

⋆e-mail: Jean.Cleymans@uct.ac.za

https://doi.org/10.1051/epjconf/201817106001

2 Studies in Saclay with SATURNE

Already in the early eighties Helmut worked on studying reactions of ions at relativistic energies. The goal of Helmut's research was to measure the energy and momentum transfer in reactions of light ions on heavy ones at energies up to 1 AGeV. Truly large detector system and respective collaborations did not exist yet. The method applied by him and his colleagues was as clever as it was simple in terms of the instrumentation needed (see Fig. 1). When a fissile nucleus is hit by a relativistic projectile energy and momentum is transferred through knock-out of target nucleons or through stopping of projectile nucleons. In case the damage to the fissile nucleus is moderate, the imparted energy is equilibrated and a compound nucleus is formed. As a consequence, the target nucleus will fission and the two fragments will separate back-to-back with a relative velocity essentially determined by the saddle point configuration. Yet, in the laboratory frame the two fragments are boosted due the momentum transfer to the target nucleus. By measuring the laboratory emission angle and the velocities of the two fragments, the momentum transfer can be reconstructed. Moreover, due to the strong change of the fragment mass distribution for increasing excitation energy (or temperature) of the compound system, or angular distortions due to post-scission particle evaporation from the fission fragments, also the excitation energy could be accessed in the experiment. A series of runs were conducted

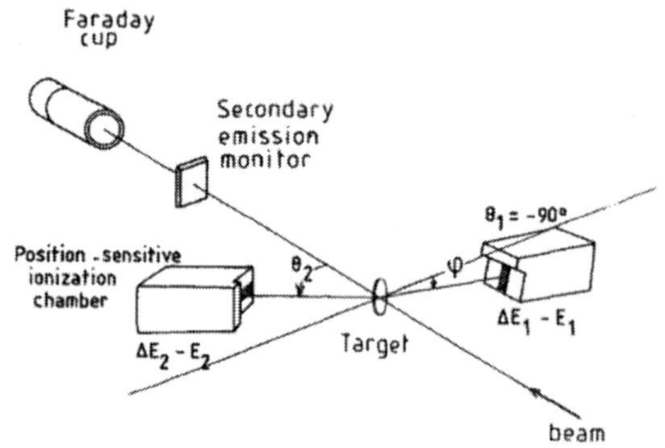

Figure 1. Experimental set-up used to measure linear momentum transfer at SATURNE. The beam hits a target foils tilted by 45° degrees w.r.t. the beam direction to minimize energy loss in the target. Fission fragments are detected in the ionization chambers, one movable in the reaction plane, the other one out-of-plane. Luminosity was monitored by a Faraday cup. Taken from [1].

at the SATURNE proton synchrotron operated at Saclay, France [1]. Beams of protons, deuterium and helium where used to bombard a Thorium (^{232}Th) target. The fission fragments where detected by two position sensitive ionization chambers providing total energy and energy-loss information. The position resolution was provided in one dimension only. Therefore, one detector was rotated with respect to the other to allow measuring out-of-plane angles as well. Moreover, the ionization chambers where mounted on moving supports to change the global angular range.

One of the derived observables, the correlation angle $\Theta_{corr} = \Theta_2 - \Theta_1$, is displayed in Fig. 2. This correlation angle amounts to 180° if zero momentum is transferred. The three panels show results for three different projectiles (p, d, α) and various beam energies. At first sight, no clear trend is observable. It is important to note, however, that (a) centrality of the collision is not accessible and (b) the method is sensitive only to reactions which result in binary fission. Indeed, the distributions for a proton beam does not change very much except for a growth in width and an emerging asymmetry towards smaller correlation angles. The small vertical arrow at each spectrum indicates the correlation angle if full stopping is achieved. Here a clear trend is visible: full stopping gets less likely as the beam energy increases. The change in reaction mechanism from fusion-fission, the dominant process below the Fermi energy regime, to nucleon knock-out is best visible for argon projectiles. At the lowest energy of $17AGeV$, a clear maximum at full transfer is visible with the suppressed shoulder

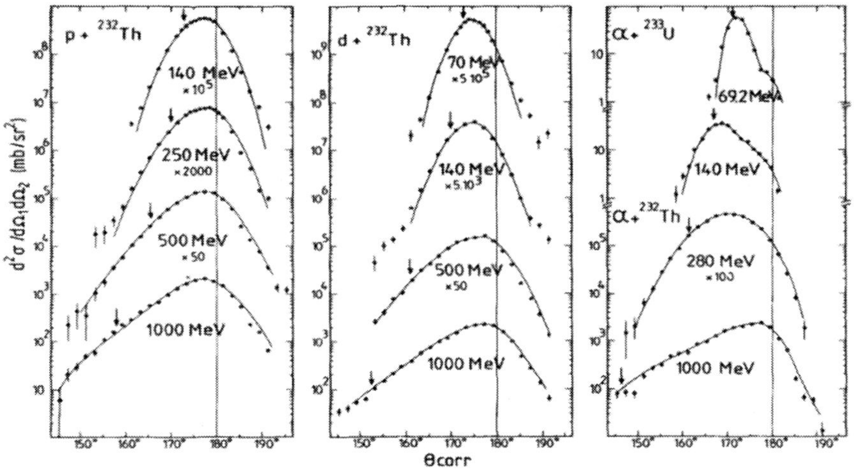

Figure 2. Distribution of measured correlation angles for three different collision systems and different kinetic energies of the projectiles. The lines are drawn to guide the eyes. The vertical lines indicates $\Theta = 180°$, the case if zero momentum is transferred. Small vertical arrows each mark the correlation angle expected at full stopping.

at larger correlation angles resulting from peripheral transfer reactions. While the energy rises, the maximum of the distributions move to smaller correlation angles as expected for this reaction class but then separates from full momentum transfer and moves back to small transfer.

Helmut was always very much interested in developing his own understanding of the underlying physical principles. In the case of the linear momentum transfer studies, he launched a project in his group at TU Darmstadt to simulate such reaction with a microscopic transport code [2]. At that time available was the Intra Nuclear Cascade Code (INC) and the code of Cugnon. Respective studies revealed that linear momentum transfer primarily results from stopping of projectile nucleons while excitation energy was by large driven by knockout of deeply bound nucleons. Excitation energy finally also defined the limitation of the method (binary fission) but at the same time the discovery of a particular decay of the struck nucleus later called multi fragmentation. This decay channel wheeled substantial cross section to reactions with final states with showing more than two decay products. Consequently, Helmut and his collaborators had to use more and more complex set-ups. On the other hand, Helmut proposed experiments at other laboratories to have heavier beams (GANIL) [3] or higher energies (JINR) [4], setting up such experiments for a small group turned out to be at the edge of what one would consider feasible. Yet he loved these experiments and he continued measurements until the early nineties when already SIS18 and the KaoS experiment (see below) needed most of his attention. Actually, the last measurements of this type he conducted at the Synchrophasotron with Helium beam in 1991, exactly the time, when Russia experienced a historic change with the "Agreement on the Establishment of the Commonwealth of Independent States" signed in Viskuli, Belarus.

Among the many photographs which Helmut took while visiting the laboratories, in particular rousing are those showing him walking around the White House in Moscow in the Fall of 1991. Helmut liked taking photographs a lot. When remembering the great times with Helmut, two attributes come to mind which should not be missed, among many: Helmut took most attention to involve young students in his scientific activities. He enjoyed a lot taking them to the experiments and teaching them not only all the important aspects about doing experiments and the physics addressed. He also gave his students the opportunity to share with him all the coming from the cultural aspects of working together with colleagues from other countries and in laboratories there. If one says that science can help bringing nations together, Helmut is a prime example of a colleague who was living this with all his grandeur. Two students of the Technical University of Darmstadt (then TH Darmstadt) graduated with the topic discussed here with doctoral degree. Namely, Gerold Klotz-Engmann covering the earlier experiments and Volker Lips who was analysing data they had taken in Dubna. The authors would like to thank both of them for providing photographs, memories and anecdotes. Like the one about the first student, who was attracted by Helmut to conduct experiments at Saclay. Unfortunately, it never came to it since this student did not really obey the rules for entering the Saclay campus while coming back from the evening excursion in Paris. Officials decided that he had entered Saclay for the last time.

3 Studies at GSI within the KaoS Collaboration

Helmut Oeschler joined the KaoS collaboration in the mid 80s. Back in these days, he was still strongly involved in experiments on multi-fragmentation and he was managing Prof. Dr. E. Kankeleit's group at GSI Darmstadt, working on electron and positron production in heavy ion collisions at the Coulomb barrier. Helmut has started his contribution to KaoS by taking responsibility for the particle tracking, in close collaboration with the GSI detector laboratory, where large-area multi-wire proportional counters were built. They were equipped with dedicated readout electronics, developed by the electronics workshop at the nuclear physics department of Technische Hochschule Darmstadt, Helmut's home institute. At that time, the physics agenda of relativistic heavy ion collisions in the 1-AGeV projectile energy regime was mainly driven by questions related to the equation of state of hot and dense nuclear matter. In the early 90s KaoS started taking data. Helmut's curiosity was excited by the non-monotonic shape of charged pion spectra first reported by the LBL streamer chamber. Hence, he devoted himself and his students to a systematic study of pions produced in heavy ion collisions. This episode was characterized by measuring apparent temperatures of pion spectra, by the emission dynamics of pions during the course of the reaction and by investigating the relevance of exciting and de-exciting of broad baryonic resonances. In this context, Helmut made first contact with the statistical model with the emphasis of compiling a more global picture of particle production in heavy ion collisions. Consequently, at the end of the 90's Helmut's interest shifted towards the KaoS core program, the production of charged kaons and together with his group he made significant contributions on the incompressibility of nuclear matter, as well as emission times and patterns of charged kaons. During that time span with KaoS, Helmut has supervised eight Diploma and five PhD theses, needless to say, accompanied by numerous conference contributions and publications in renowned journals. Both, the urge to continuously document scientific progress and at the same time mentoring his students with his distinct openhearted and cheerful nature is Helmut's signature feature, which his students and colleagues will always remember.

4 Studies in Cape Town

In 1998 Helmut Oeschler spent several weeks at the University of Cape Town. This was the start of an incredibly fruitful scientific collaboration with physicists from this University, resulting in over 30 publications with an average of 82 citations per paper and about 2000 citations. The introduction to one of his papers, written over ten years ago, is still relevant today:
"One of the most remarkable results to emerge from heavy-ion collisions over the past two decades is the striking regularity shown by particle yields at all energies. This has led to several very successful

proposals describing particle yields over a very wide range of beam energies, reaching from 1 A GeV up to 200 A GeV, using only one or two parameters. A systematic comparison of these proposals is presented here. The conditions of fixed energy per particle, baryon+anti-baryon density, normalized entropy density as well as percolation model are investigated. The results are compared with the most recent chemical freeze-out parameters obtained in the thermal-statistical analysis of particle yields. The sensitivity and dependence of the results on parameters is analyzed and discussed. It is shown that in the energy range above the top AGS energy, within present accuracies, all chemical freeze-out criteria give a fairly good description of the particle yields." [5]

His first visit to South Africa was followed by many subsequent ones, several times with his family.

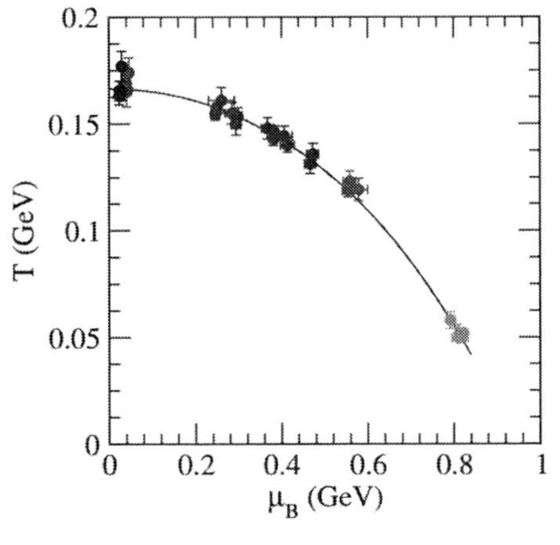

Figure 3. Values of μ_B and T for different energies. The solid line is a parameterization obtained from fits to data. Taken from [5].

Helmut went way beyond the writing of papers and was involved at a personal level in helping the South African community of physicists to be part of the ALICE Collaboration at CERN as expressed by Zinhle Buthelezi: *Still can't believe about Helmut! He was a good friend and ally of South Africa. He helped us tremendously to navigate CERN and the ALICE collaborations when we joined. He will be sorely missed!*

In South Africa we remember Helmut Oeschler as a person full of enthusiasm, full of energy, with new ideas, trying to understand physics at an intuitive level, never happy with complicated explanations. So full of life that it is hard comprehend what happened so suddenly.

5 Studies at CERN within the ALICE Collaboration

In 2000, Helmut joined the ALICE Collaboration as team leader of the Technical University Darmstadt. From the beginning, he was one of the leading scientists in the study of light flavour hadrons at LHC energies. He prepared an important part of the ALICE Collaboration's physics programme associated with the production of these hadrons and anticipate their thermal analysis [6]. He was also one of the key contributors for topics related to particle identification using the central detectors and lead the efforts for the preparation of the publication on identified hadrons (illustrated by Fig. 4) for the very first proton-proton collisions recorded at the LHC at the center of mass energy of 0.9 TeV [7].

Among many roles he had in the collaboration, he served as chair of the ALICE Editorial Board for several years. As of today, the ALICE Collaboration has more than 1800 members from more than 90 countries and produced more than 180 publications for many of which Helmut was instrumental with shaping and organising the scientific content. Helmut truly enjoyed the very international and intercultural environment, being a mentor not only for his own students and post-docs, but for many young

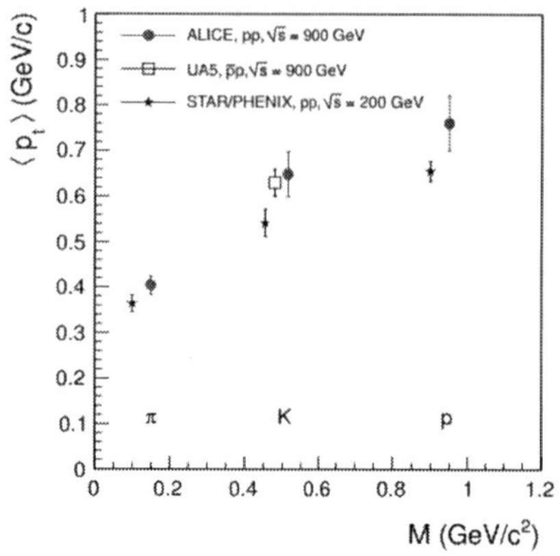

Figure 4. Mean p_t as a function of the mass of the emitted hadron in proton-proton collisions at $\sqrt{s} = 900$ GeV with ALICE at the LHC (red solid circles) and compared to results at 200 GeV obtained at RHIC. Taken from [7].

scientists from all over the world. His scientific interests covered a broad range of topics, but his main focus was on the production of strangeness from low to high energies as well as from small to large colliding systems. Helmut strongly believed in the applicability of the thermal-statistical approach to high multiplicity proton-proton and proton-Pb collisions and thus it was extraordinary for him to witness the observation of strangeness enhancement in these collision systems by ALICE. In addition, he was fascinated by the complex study of anti-nuclei and hadronic resonances. In particular, Helmut advanced our understanding of the production of the ϕ-meson and the description of light anti- and hyper-nuclei formation via coalescence and with thermal-statistical models.
For many years, Helmut served as a member of the International Advisory Committee of Strangeness in Quark Matter Conference. This was his favourite conference and he always played a pivotal role in the preparation and presentation of the ALICE results. We will certainly follow his path in addressing some of the most important unresolved problems in this field.

6 Acknowledgements

The authors are extremely grateful to the SQM 2017 Organising Committee who decided to dedicate time for such a memorial session and to the members of the International Advisory Committee for their unanimous support. Many thanks to Oeschler's family for the heart-warming presence at the Conference and the pleasant moments, dinners and memories that were shared. We would also like to apologise to the numerous colleagues and friends who contacted us while preparing the memorial session, providing many wonderful stories about Helmut that we did not have the time to present.

References

[1] F. Saint-Laurent *et al.*, Nucl. Phys. A **422**, 307 (1984). doi:10.1016/0375-9474(84)90519-0
[2] G. Klotz-Engmann *et al.*, Nucl. Phys. A **499**, 392 (1989). doi:10.1016/0375-9474(89)90054-7
[3] M. Conjeaud *et al.*, Phys. Lett. **159B**, 244 (1985). doi:10.1016/0370-2693(85)90243-6
[4] V. Lips *et al.*, Phys. Rev. Lett. **72**, 1604 (1994). doi:10.1103/PhysRevLett.72.1604
[5] J. Cleymans *et al.*, Phys. Rev. C **73**, 034905 (2006). doi:10.1103/PhysRevC.73.034905
[6] B. Alessandro *et al.*, J. Phys. G **32** 1295 (2006). doi:10.1088/0954-3899/32/10/001
[7] K. Aamodt *et al.* (ALICE Collaboration), Eur. Phys. J. C**71** 1655 (2011) doi:10.1140/epjc/s10052-011-1655-9

Polarization in relativistic heavy ion collisions: a theoretical perspective

Francesco Becattini[1]*

[1] *Dipartimento di Fisica e Astronomia, University of Florence, Via G. Sansone 1, I-50019, Sesto Fiorentino (Firenze), Italy*

Abstract. We review the theoretical framework for the calculation of particle polarization in relativistic heavy ion collisions within the hydrodynamical model. The covariant decomposition of the mean spin vector is presented and open theoretical issues addressed.

1 Introduction

Global polarization of Λ hyperons in semi-peripheral relativistic heavy-ion collisions has been recently observed by the STAR experiment over a center-of-mass energy range between 7.7 and 200 GeV [1]. This finding confirms quantitative predictions based on local thermodynamic equilibrium of spin degrees of freedom [2] which provides a relation between polarization and relativistic vorticity, and it agrees quantitatively with the hydrodynamic model calculations to a very good degree of accuracy. In the hydrodynamic framework, the distinctive feature of polarization is its proportionality to the gradients of the combined temperature and velocity fields, so that its measurement is a stringent test of the hydrodynamic picture which is distinct and complementary to momentum spectra. Moreover, hydrodynamics and local thermodynamic equilibrium predict that polarization must be the same for particles and antiparticles, that is a C-even effect, in agreement with the observations. Conversely, if electromagnetic or other C-odd fields were responsible for this phenomenon, polarization would be *opposite* for particles and antiparticles.

The experimental efforts, thus far, focussed on the search of the average global polarization of Λ hyperons along the direction of the angular momentum of the plasma. This measurement requires the identification of the reaction plane in peripheral collision as well as its orientation, that is the direction of the total angular momentum vector **J**. The average global polarization along **J** is also found to decrease rapidly as a function of center-of-mass energy [1], from few percent at $\sqrt{s_{\rm NN}} = O(10)$ GeV to few permille at $\sqrt{s_{\rm NN}} = O(100)$ GeV. In the TeV energy range, at the LHC, the global polarization along **J** is not seen [3] as it is most likely beyond experimental sensitivity.

2 The theoretical approaches to global polarization: a summary

Particles produced in relativistic heavy ion collisions are expected to be polarized in peripheral collisions because of angular momentum conservation. At finite impact parameter, the Quark Gluon

*e-mail: becattini@fi.infn.it

Plasma (QGP) has a finite angular momentum perpendicular to the reaction plane and some fraction thereof may be converted into spin of final state hadrons. Therefore, measured particles may show a finite mean *global* polarization along the angular momentum direction.

Early estimates of this effect [4] were based on the general idea that polarized quarks in the QGP stage of the production process would eventually give rise to polarized hadrons, making it possible to predict qualitative features of the final hadrons' polarization. It was then proposed [5, 6] that polarization can be calculated assuming that the spin degrees of freedom are at local thermodynamical equilibrium at the hadronization stage in much the same way as the momentum degrees of freedom. In other words, polarization can be predicted by extending the familiar Cooper-Frye formula to particles with spin. A specific derivation was presented in refs. [2, 5] where it was pointed out that the hydrodynamical quantity steering the polarization is the *thermal vorticity*, that is (minus) the antisymmetric part of the gradient of the four-temperature field $\beta = (1/T)u$ where T is the proper temperature and u the hydrodynamic four-velocity:

$$\varpi_{\mu\nu} = -\frac{1}{2}\left(\partial_\mu\beta_\nu - \partial_\nu\beta_\mu\right) \tag{1}$$

Particularly, the first-order expansion of the polarization in terms of thermal vorticity was obtained in ref. [2] for hadrons with spin $1/2$ (lately recovered with a different method in ref. [7]), yet its extension to higher spins could be derived from the corresponding global equilibrium expression [8].

This theoretical work made it possible to make definite quantitative predictions of global Λ polarization in nuclear collisions from hydrodynamic calculations, with a resulting mean value ranging from some permille to some percent [9–11], with an apparently strong dependence on the initial conditions, particularly on the initial longitudinal velocity field. Calculations of vorticity in relativistic heavy ion collisions - which could be then turned into a polarization map - were also recently presented in ref. [12]. To complete the theoretical overview on the subject, it should be pointed out that different approaches, as well as additional mechanisms, to the Λ polarization in relativistic nuclear collisions were proposed in refs. [13–16].

Notation

In this paper we use the natural units, with $\hbar = c = K = 1$.
The Minkowskian metric tensor is diag$(1, -1, -1, -1)$; for the Levi-Civita symbol we use the convention $\epsilon^{0123} = 1$.
Operators in Hilbert space will be denoted by a large upper hat, e.g. $\widehat{T}$ while unit vectors with a small upper hat, e.g. $\hat{v}$.

3 Hydrodynamics and the local equiibrium density operator

The proper relativistic extension of the spin concept, for massive particles, requires the introduction of a spin four-vector operator. This is defined as follows:

$$\widehat{S}^\mu = -\frac{1}{2m}\epsilon^{\mu\nu\rho\lambda}\widehat{J}_{\nu\rho}\widehat{p}_\lambda \tag{2}$$

where $\widehat{J}$ and $\widehat{p}$ are the angular momentum operator and four-momentum operator of a single particle. As it can be easily shown, the spin four-vector operator commutes with the four-momentum operator (hence it is a compatible observable) and it is space-like on free particle states as it is orthogonal to the four-momentum:

$$\widehat{S}^\mu\widehat{p}_\mu = 0 \tag{3}$$

and has thus only three independent components.

The mean spin S^μ and polarization P^μ four-vectors can now be defined as the mean values with the suitable density operators, that is:

$$S^\mu = \langle \widehat{S}^\mu \rangle \equiv \mathrm{tr}(\widehat{\rho}\,\widehat{S}^\mu) \tag{4}$$

and

$$P^\mu = \langle \widehat{S}^\mu \rangle / S \tag{5}$$

The crucial role in the calculation of (4) for the fluid produced in relativistic heavy ion collisions is played by the density operator. For a system at local thermodynamic equilibrium, this reads [17]:

$$\widehat{\rho}_{\mathrm{LE}} = (1/Z)\exp\left[-\int_\Sigma \mathrm{d}\Sigma_\mu \left(\widehat{T}^{\mu\nu}\beta_\nu - \zeta\widehat{j}^\mu\right)\right] \tag{6}$$

where β is the four-temperature vector, $\widehat{T}$ the stress-energy tensor, $\widehat{j}$ a conserved current - like the baryon number - and $\zeta = \mu/T$.

If we wish to calculate the mean value of a local operator $\widehat{O}(x)$ (such as, for instance the stress-energy tensor $\widehat{T}$, or the current $\widehat{j}$)

$$O(x) = \mathrm{tr}(\widehat{\rho}\widehat{O}(x)) \tag{7}$$

and if the fields β,ζ vary significantly over a distance which is much larger than the typical microscopic length (indeed the *hydrodynamic limit*), then they can be Taylor expanded in the density operator starting from the point x where the mean value $O(x)$ is to be calculated. The leading terms in the exponent of (6) then become [17]:

$$\widehat{\rho}_{\mathrm{LE}} \simeq \frac{1}{Z_{\mathrm{LE}}}\exp\left[-\beta_\nu(x)\widehat{P}^\nu + \xi(x)\widehat{Q} - \frac{1}{4}(\partial_\nu\beta_\lambda(x) - \partial_\lambda\beta_\nu(x))\widehat{J}_x^{\lambda\nu} + \frac{1}{2}(\partial_\nu\beta_\lambda(x) + \partial_\lambda\beta_\nu(x))\,\widehat{L}_x^{\lambda\nu} + \nabla_\lambda\xi(x)\,\widehat{d}_x^\lambda\right]. \tag{8}$$

where the last two terms with the shear tensor and the gradient of ζ are dissipative and vanish at equilibrium. The ∇_λ operator stands for:

$$\nabla_\lambda = \partial_\lambda - u_\lambda u \cdot \partial$$

as usual in relativistic hydrodynamics. The term which is responsible for polarization is the one involving the generators of the Lorentz group $\widehat{J}_x$.

The polarization of particles in a fluid was then obtained by means of the spin tensor $\widehat{S}$ which is the local operator associated to spin density along with an ansatz about the form of the covariant Wigner function [2]. The mean spin vector of $1/2$ particles with four-momentum p, produced around point x at the leading order in the thermal vorticity is then found to be:

$$S^\mu(x,p) = -\frac{1}{8m}(1 - n_F)\epsilon^{\mu\rho\sigma\tau}p_\tau\varpi_{\rho\sigma} \tag{9}$$

where $n_F = (1 + \exp[\beta(x)\cdot p - \nu(x)Q/T(x)] + 1)^{-1}$ is the Fermi-Dirac distribution and $\varpi(x)$ is given by eq. 1. The eq. (9) has been recovered with a different approach in ref. [7]. This formula is suitable for the situation of relativistic heavy ion collisions, where one deals with a local thermodynamic equilibrium hypersurface Σ where hydrodynamic stage ceases and particle description sets in.

4 Covariant decomposition of the spin vector in a relativistic fluid

To gain insight into the physics of polarization in a relativistic fluid, it is very useful to decompose the gradients of the four-temperature vector in the eq. (9). We start off with the seperation of the gradients of the comoving temperature and four-velocity field:

$$\partial_\mu \beta_\nu = \partial_\mu \left(\frac{1}{T}\right) + \frac{1}{T}\partial_\mu u_\nu$$

Then, we can introduce the acceleration and the vorticity vector ω^μ with the usual definitions:

$$\begin{aligned} A^\mu &= u \cdot \partial u^\mu \\ \omega^\mu &= \frac{1}{2}\epsilon^{\mu\nu\rho\sigma}\partial_\nu u_\rho u_\sigma \end{aligned}$$

The antisymmetric part of the tensor $\partial_\mu u_\nu$ can then be expressed as a function of A and ω:

$$\frac{1}{2}\left(\partial_\nu u_\mu - \partial_\mu u_\nu\right) = \frac{1}{2}\left(A_\mu u_\nu - A_\nu u_\mu\right) + \epsilon_{\mu\nu\rho\sigma}\omega^\rho u^\sigma$$

therafter plugged into the (9) to give:

$$\begin{aligned} S^\mu(x,p) &= \frac{1}{8m}(1-n_F)\epsilon^{\mu\nu\rho\sigma}p_\sigma \nabla_\nu(1/T)u_\rho && (10) \\ &+ \frac{1}{8m}(1-n_F)\,2\,\frac{\omega^\mu u\cdot p - u^\mu \omega \cdot p}{T} && (11) \\ &- \frac{1}{8m}(1-n_F)\frac{1}{T}\epsilon^{\mu\nu\rho\sigma}p_\sigma A_\nu u_\rho && (12) \end{aligned}$$

Hence, polarization stems from three contributions: a term proportional to the gradient of temperature, a term proportional to the vorticity ω, and a term proportional to the acceleration. Further insight into the nature of these terms can be gained by choosing the particle rest frame, where $p = (m, \mathbf{0})$ and restoring the natural units. The eq. (10) then certifies that the spin in the rest frame is proportional to the following combination:

$$\mathbf{S}^*(x,p) \propto \frac{\hbar}{KT^2}\gamma\mathbf{v}\times\nabla T + \frac{\hbar}{KT}\gamma(\omega - (\omega\cdot\mathbf{v})\mathbf{v}/c^2) + \frac{\hbar}{KT}\gamma\mathbf{A}\times\mathbf{v}/c^2 \qquad (13)$$

where $\gamma = 1/\sqrt{1 - v^2/c^2}$ and all three-vectors, including vorticity, acceleration and velocity, are observed in the particle rest frame.

The three independent contributions are now well discernible in eq. (13). The second term scales like $\hbar\omega/KT$ and is the one already known from non-relativistic physics, proportional to the vorticity vector seen by the particle in its motion amid the fluid, with an additional term vanishing in the non-relativistic limit. The third term is a purely relativistic one and scales like $\hbar A/KTc^2$; it is usually overwhelmingly suppressed, except in heavy ion collisions where the acceleration of the plasma is huge ($A \sim 10^{30}g$ at the outset of hydrodynamical stage). The first term, instead, is a new non-relativistic term [2] and applies to situations where the velocity field is not parallel to the temperature gradient. For ideal uncharged (thus relativistic) fluids, this term is related to the acceleration term because the equations of motion reduce to:

$$\nabla_\mu T = T A_\mu / c^2$$

Therefore, being the QGP a quasi-ideal fluid and almost uncharged at very high energy, the first and third term are tightly related. It can be shown that they contribute non-trivially to the final predicted polarization [18].

5 Open theoretical issues

Specifying the mean spin vector at some point x requires the spin tensor $\widehat{\mathcal{S}}$, what was used in ref. [2] to obtain the function in eq. (9). However, in relativistic heavy ion collisions, we do not measure spin at some spacetime point, but only as a function of particle momentum. In formulae, what we can do is to measure the mean spin after an integration of (9) over the freeze-out or particlization hypersurface Σ.

$$S^\mu(p) = \frac{\int d\Sigma_\lambda p^\lambda f(x,p) S^\mu(x,p)}{\int d\Sigma_\lambda p^\lambda f(x,p)} \tag{14}$$

It is presently an unsettled issue whether a formula like (14) can be obtained without using the spin tensor, that is in a quantum field theory where the stress-energy tensor is the symmetrized Belinfante tensor. The issue is a relevant one for if (14) was dependent on a spin tensor, $\widehat{\mathcal{S}}$ would acquire a full physical meaning and relativistic hydrodynamics - to say the least - should be extended [19] to fluids with spin.

Another major theoretical problem is to find an exact solution for the polarization at global thermodynamic equilibrium with $\varpi = \text{const} \neq 0$, that is when the (6) becomes:

$$\rho = \frac{1}{Z} \exp\left[-b_\mu \widehat{P}^\mu + \frac{1}{2}\varpi_{\mu\nu}\widehat{J}^{\mu\nu} + \zeta\widehat{Q}\right] \tag{15}$$

Indeed, the formulae (9),(14) are first-order expansions in thermal vorticity and based on an educated *ansatz* on the Wigner function of the Dirac field with the density operator (15). Even though the formula (9) appears to be the only reasonable first-order expression, finding the exact solution would certainly be a benchmark in the theory of relativistic fluids with polarization and a crucial tool to settle open issues.

Acknowledgments

I am greatly indebted with I. Karpenko for his invaluable help in preparing this talk. The author would like to thank the organizers for their excellent work in arranging a very enjoyable conference and for their patience in waiting for my contribution to the proceedings.

References

[1] L. Adamczyk *et al.* [STAR Collaboration], Nature **548**, 62 (2017)
[2] F. Becattini, V. Chandra, L. Del Zanna and E. Grossi, Annals Phys. **338**, 32 (2013).
[3] M. Konyushikhin for the ALICE Coll., *Global polarization of Λ hyperons in Pb-Pb collisions at 2.76 TeV*, poster presented at the conference Quark Matter 2017, Chicago, USA.
[4] Z. T. Liang and X. N. Wang, Phys. Rev. Lett. **94**, 102301 (2005); erratum ibidem **96**, 039901(E) (2006).
[5] F. Becattini and F. Piccinini, Annals Phys. **323**, 2452 (2008).
[6] F. Becattini, F. Piccinini and J. Rizzo, Phys. Rev. C **77**, 024906 (2008).
[7] R. h. Fang, L. g. Pang, Q. Wang and X. n. Wang, Phys. Rev. C **94**, no. 2, 024904 (2016).
[8] F. Becattini, I. Karpenko, M. Lisa, I. Upsal and S. Voloshin, Phys. Rev. C **95**, no. 5, 054902 (2017).
[9] F. Becattini, L. Csernai and D. J. Wang, Phys. Rev. C **88**, no. 3, 034905 (2013); Erratum: [Phys. Rev. C **93**, no. 6, 069901 (2016)]

[10] F. Becattini *et al.*, Eur. Phys. J. C **75**, no. 9, 406 (2015).
[11] L. G. Pang, H. Petersen, Q. Wang and X. N. Wang, Phys. Rev. Lett. **117**, no. 19, 192301 (2016).
[12] Y. Jiang, Z. W. Lin and J. Liao, Phys. Rev. C **94**, no. 4, 044910 (2016) Erratum: [Phys. Rev. C **95**, no. 4, 049904 (2017)].
[13] A. Ayala, E. Cuautle, G. Herrera and L. M. Montano, Phys. Rev. C **65**, 024902 (2002).
[14] C. d. C. Barros, Jr. and Y. Hama, Phys. Lett. B **699**, 74 (2011).
[15] M. I. Baznat, K. K. Gudima, A. S. Sorin and O. V. Teryaev, Phys. Rev. C **93** 031902 (2016).
[16] A. Aristova, D. Frenklakh, A. Gorsky and D. Kharzeev, arXiv:1606.05882 [hep-ph].
[17] F. Becattini, L. Bucciantini, E. Grossi and L. Tinti, Eur. Phys. J. C **75**, no. 5, 191 (2015).
[18] I. Karpenko, talk given at this conference; I. Karpenko and F. Becattini, in preparation.
[19] W. Florkowski, B. Friman, A. Jaiswal and E. Speranza, arXiv:1708.04035 [hep-ph].

Vorticity and particle polarization in heavy ion collisions (experimental perspective)

Sergei A. Voloshin[1,⋆]

[1] *Wayne State University, 666 W. Hancock, Detroit 48201, Michigan, U.S.A.*

Abstract. The recent measurements of the global polarization and vector meson spin alignment along the system orbital momentum in heavy ion collisions are briefly reviewed. A possible connection between the global polarization and the chiral anomalous effects is discussed along with possible experimental checks. Future directions, in particular those aimed on the detailed mapping of the vorticity fields, are outlined. The Blast Wave model is used for an estimate of the anisotropic flow effect on the vorticity component along the beam direction. We also point to a possibility of a circular pattern in the vorticity field in asymmetric, e.g. Cu+Au, central collisions.

1 Introduction

The idea of the global polarization in heavy ion collisions, the phenomenon characterized by the polarization of the secondary particles along the global system orbital momentum, is almost 15 years old. It went "on-shell" in 2004 [1, 2] with the initial predictions for the hyperon polarization as high as "in the order of tens of a percent" [1]. The first measurements [3] by the STAR Collaboration of the lambda hyperon polarization in Au+Au collisions at 200 GeV put an upper limit on hyperon polarization of about $|P_\Lambda| \leq 0.02$. Subsequently, the theoretical predictions have been improved [4], especially with a better understanding of the statistical mechanics of vortical fluid with non-zero spin particles [5], and development of the hydrodynamical calculations assuming local angular momentum equilibrium. Rough estimate of the polarization can be obtained with the help of a simple nonrelativistic expression for a particle distribution in a fluid with nonzero vorticity [6] (for a strict relativistic consideration see [7]):

$$p \propto \exp\left[-E/T - \omega(\mathbf{s}+\mathbf{l})/T - \mu B/T\right], \tag{1}$$

where $\omega = \frac{1}{2}\nabla\times\mathbf{v}$ is the nonrelativistic vorticity, and $\mathbf{v}$ is the fluid velocity. Figure 1 shows a cartoon of a non-central nuclear collision with arrows indicating the velocity field of the matter at the plane $z = 0$. Just "guessing" that the difference in velocities in the "upper" and "lower" parts of the system is about a few tenths of the speed of light and that the transverse size of the system is about 10 fm, one would conclude that the vorticity might be at the level of a few percent of fm^{-1}. Then the nonrelativistic formula (1) yields for the spin 1/2 particle polarization, $P \approx \omega/(2T)$, in the range of a few percent (assuming $T \sim 100$ MeV).

⋆e-mail: sergei.voloshin@wayne.edu

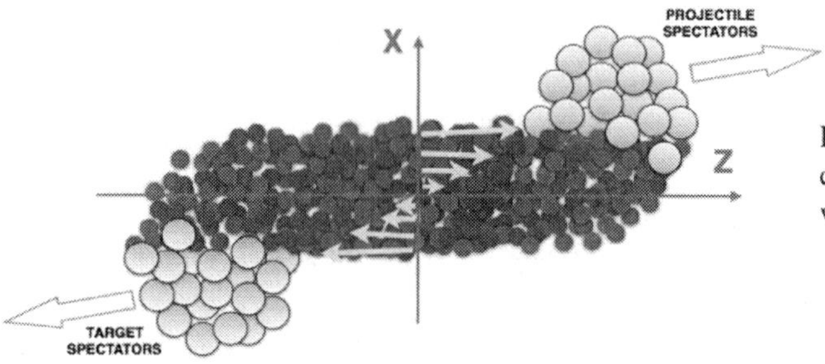

Figure 1. Cartoon of a non-central nuclear collision. The arrows indicate the collective velocity of the matter at $z = 0$ plane.

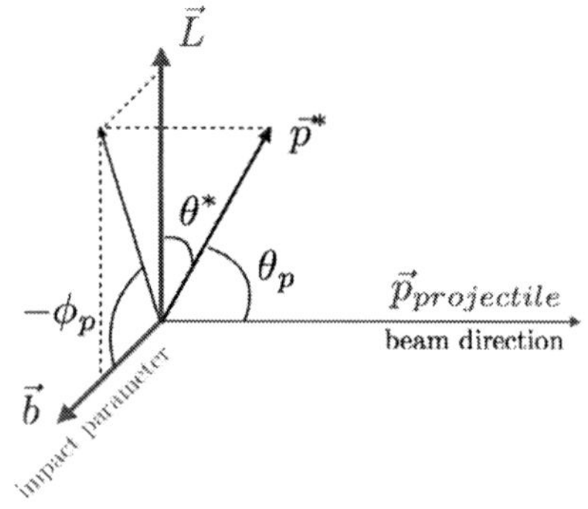

Figure 2. Diagram explaining the notations for different angles discussed in the text. $\mathbf{p}^*$ is the proton momentum in the hyperon rest frame. Vertical direction is the direction of the global orbital momentum – the global polarization direction. ϕ_p is the proton emission azimuth in the system with $x - z$ plane aligned with the reaction plane.

The simplest way to measure the global polarization is via analysis of the angular distributions of the products of weakly decaying hyperons. Weak interaction violates parity, and, e.g. in the lambda hyperon decay the protons are emitted preferentially in the direction of the lambda's spin:

$$\frac{dN}{d\cos\theta^*} \propto 1 + \alpha_H P_H \cos\theta^*, \tag{2}$$

where θ^* is the polar angle of the proton emission relative to the polarization direction in the hyperon rest frame, $-1 \leq P_H \leq 1$ is the hyperon polarization, and the parameter $\alpha_\Lambda = -\alpha_{\bar{\Lambda}} \approx 0.624$. To measure the polarization of strongly decaying particles is obviously significantly more difficult. It is not at all possible for spin 1/2 particles; for the vector mesons one can hope to measure the deviation from 1/3 of the probability for the spin projection to be zero [8]. The angular distribution (averaged over the azimuthal distribution around the polarization direction) of the decay products in this case reads:

$$\frac{dN}{d\cos\theta^*} \propto \rho_{0,0}|Y_{1,0}|^2 + \rho_{1,1}|Y_{1,-1}|^2 + \rho_{-1,-1}|Y_{1,1}|^2 \propto \rho_{0,0}\cos^2\theta^* + \frac{1}{2}(\rho_{1,1} + \rho_{-1,-1})\sin^2\theta^* \tag{3}$$

$$\propto (1 - \rho_{0,0}) + (3\rho_{0,0} - 1)\cos^2\theta^* \tag{4}$$

where $\rho_{0,0}$, $\rho_{1,1}$, and $\rho_{-1,-1}$ are the probabilities for the particle to have spin projection on the direction of polarization to be zero, $+1$, and -1, respectively. The deviation from the non-polarized state value $\rho_{0,0} = 1/3$ is in this case a second order effect [8]. For example, an estimate based on Eq. 1 yields $\rho_{00} = 1/[3 + (\omega/T)^2]$.

To measure the hyperon global polarization or vector meson spin alignment experimentally, one can either analyze directly the distributions in θ^*, or, in case of the global polarization direction defined by the one of the flow event planes, analyze the azimuthal distribution of the decay products (in the resonance rest frame) relative to that flow plane. The azimuthal distribution analysis can be often simpler in terms of applying necessary corrections for the reaction plane resolution and detector acceptance. In the case of the azimuthal analysis, the polarization can be calculated as:

$$P_H = -\frac{8}{\pi \alpha_H} \langle \sin(\phi_p^* - \Psi_{\rm RP}) \rangle, \tag{5}$$

$$\rho_{00} = \frac{1}{3} - \frac{8}{3} \langle \cos[2(\phi_p^* - \Psi_{\rm RP})] \rangle. \tag{6}$$

For the global polarization, the analysis has to be performed using one of the first harmonic event planes, with the reaction plane (and correspondingly, the angular momentum) direction to be determined by the deflection direction of the projectile spectators (which on average deflect outward of the collision [9]). For the spin alignment measurements it is possible to use the second order event plane (which typically has much better resolution).

2 Results

The progress in vector spin alignments measurements was presented at this conference in talks by the STAR and ALICE Collaborations [10, 11]. The uncertainties in these measurements are still relatively large, and the results are rather inconclusive; below I concentrate on the discussion of the global polarization results.

Figure 3 shows a compilation of published[3, 12] and presented at this conference [10, 11] results on the average global polarization of lambda and anti-lambda hyperons at mid-rapidity in mid-central collisions as a function of collision energy. The blue solid and dashed lines are the results of hydrodynamic calculation [7, 13], with and without accounting for the hyperon feed-down contribution, respectively. The procedure for the feed-down correction is outlined in [6] with Eq. 1 used for an estimate of the polarization of the higher spin resonances. Note that the effect of the feed-down correction is rather modest – at the level of $\sim 15\%$.

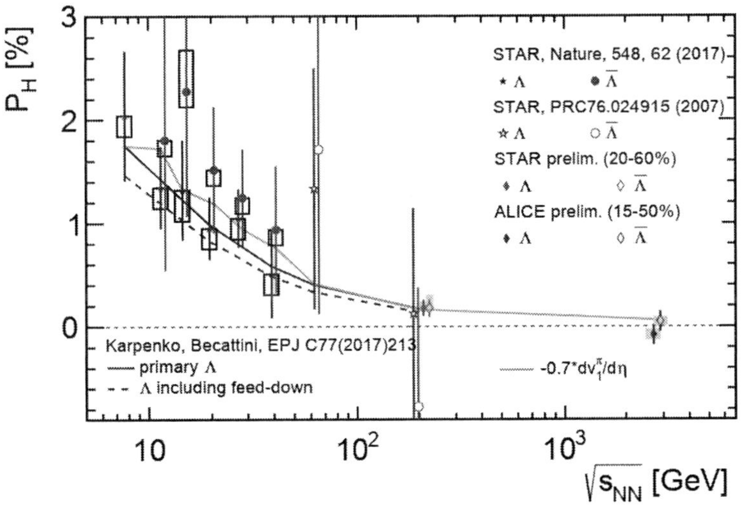

Figure 3. Average global polarization of lambda hyperons as a function of collision energy. Boxes indicate the systematic uncertainties.

At RHIC beam energy scan collision energies the global polarization of lambdas and antilambdas seems to be slightly different. Calculations [14] show that the effect of nonzero baryon chemical potential could result in a split of the polarization values for particle and antiparticle, but numerically (assuming the equilibrium in the hadronic gas) this effect expected to be small. More realistic explanation for the difference would be the effect of the magnetic field that is strongly aligned with the direction of the orbital momentum. Then the global polarization analysis might become a very sensitive tool for the measurement of the magnetic field in heavy ion collisions. The present uncertainties in polarization measurements allow to put only an upper limit on the magnitude of the magnetic field, $eB \lesssim 0.01 m_\pi^2$, but even that is of great interest, as at present, the uncertainty in the magnetic field calculations spans several orders of magnitude [15].

There exist no hydrodynamic calculations for the global polarization at the LHC energies. For that we can develop an empirical prediction based on the observation that the slope of the directed flow at mid-rapidity is likely strongly correlated with the vorticity. Indeed, the calculation [16] support such a hypothesis. Then one can use the available data on v_1 slope [17, 18] (we used that of charged pions) for an estimate of the global polarization at higher energies. This is shown by a green line in Fig. 3. The ALICE global polarization measurements [11, 19], which within the error-bars at present are consistent with zero, do not contradict such an estimate. It will be required to improve the uncertainties about factor of 3–5 to be able to measure the polarization at the level of an estimate from the directed flow slopes.

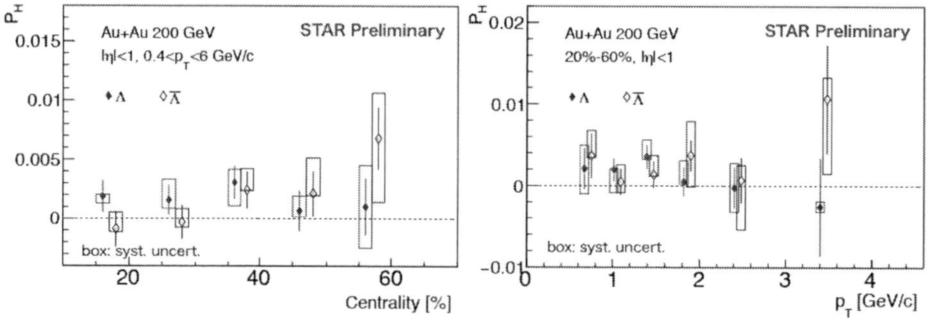

Figure 4. Centrality and p_T dependence of the global polarization in Au+Au collisions at 200 GeV [10, 20].

Global polarization dependence on centrality, the hyperon transverse momentum, and the orientation of the hyperon emission relative to the reaction plane have been also of great interest as different models predict significantly different dependencies. Within the available statistic, both, the centrality and p_T dependencies, see the preliminary STAR results [10, 20] presented in Fig. 4, are rather modest. The azimuthal angle dependence is discussed in more detail in [10]. The preliminary results indicate stronger polarization of the hyperons emitted in-plane than those emitted out-of-plane, opposite to the expectations from the hydrodynamical calculations [21].

2.1 Global polarization and chiral anomalous effects

The exact mechanism relating the vorticity of the medium to the global polarization is not yet firmly established. It was argued in [22] that the global polarization appears as a results of the axial anomalous current:

$$\mathbf{J}_5 = \left(\frac{\mu^2 + \mu_5^2}{4\pi^2} + \frac{T^2}{12} \right) \omega \tag{7}$$

For a more detailed discussion of this possibility, see [23] and references therein.

Axial current can be also "generated" by the magnetic field in the system with nonzero vector charge chemical potential (for a review of chiral anomalous effects in heavy ion collisions see [24]):

$$\mathbf{J}_5 = \frac{1}{2\pi^2}\mu_V(Qe)\mathbf{B}, \tag{8}$$

which in particular relates the direction of the axial current to the sign of the chemical potential. This provides an interesting possibility for the experimental check of the existence of such a relationship [25] by measuring the global polarization as a function of the event charge asymmetry or the net kaons, hoping that those are proportional to either electrical charge or strangeness chemical potentials. In some sense such a measurement would constitute a measurement of the first "half" of the chiral magnetic wave mechanism (in which the first step is the separation of the axial charge and the second step includes the electric currents rearranging the electric charge and finally leading to the quadrupole charge configuration, see review [24]). The preliminary results from the STAR Collaboration [20] presented in Fig. 5 are not yet conclusive due to the large statistical uncertainties.

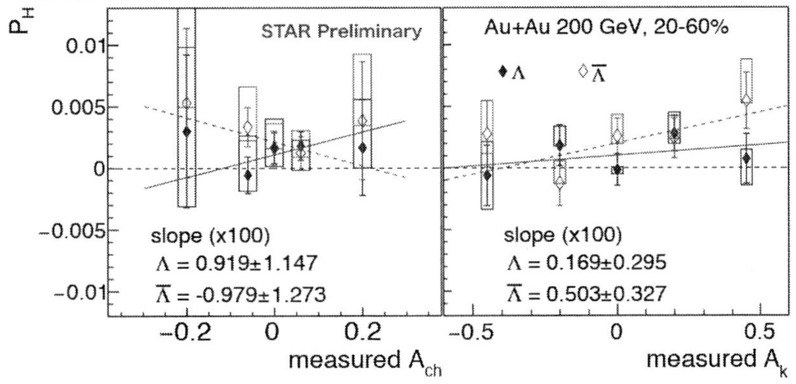

Figure 5. Global polarization as a function of the event charge asymmetry $A_{ch} = (N_+ - N_-)/(N_+ + N_-)$, and the event net kaon ratio $A_K = (N_{K^+} - N_{K^-})/(N_{K_+} + N_{K^-})$ [20].

2.2 Local vorticity and polarization along the beam direction

While the system orbital momentum provides an obvious choice for the average direction of the vorticity (and particle polarization), locally, the vorticity have in general nonzero all three components. Those might be generated by a particular process, e.g. jet propagation [26] (not yet studied experimentally) or simply by the expansion of the initially anisotropic medium, see, for example, the results in [16] (including those presented only in the arXiv version of this publication), as well as calculations in [14, 22]. In particular, the vorticity z-component, along the beam direction, might reach large values [7, 13, 16].

Below I discuss a simple Blast Wave (BW) model estimates for the vorticity z component. In a simplest version of BW model that includes anisotropic flow, the particle production source at freeze-out is parameterized with 4 parameters: temperature T, maximum radial flow velocity (rapidity),

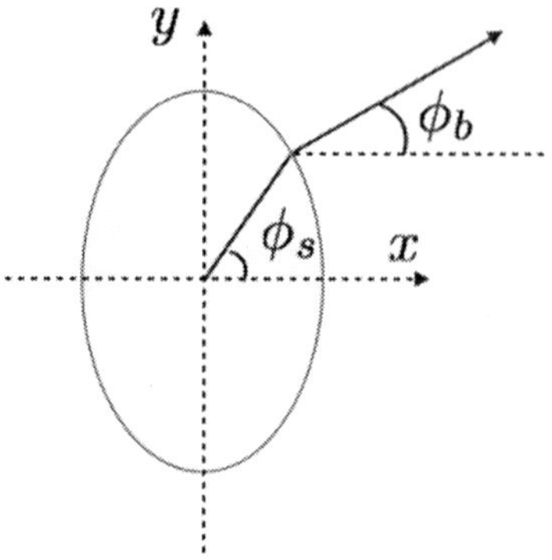

Figure 6. The collective velocity of the source element at angle ϕ_s at the surface is along the boost angle ϕ_b, perpendicular to the surface described by Eq. 9. The boost velocity is given by Eq. 10.

notated as ρ_0, amplitude of azimuthal modulation in expansion velocity, noted below as b, and the spatial anisotropy parameter a. The source (see Fig. 6) is then described by the following equations:

$$r_{max} = R[1 - a\cos(2\phi_s)], \tag{9}$$

$$\rho_t = \rho_{t,max}[r/r_{max}(\phi_s)][1 + b\cos(2\phi_s)] \approx \rho_{t,max}(r/R)[1 + (a+b)\cos(2\phi_s)]. \tag{10}$$

It is assumed that the collective velocity of the source element located at azimuthal angle ϕ_s is boosted with velocity ρ_t perpendicular to the surface of the ellipse similar to that of Eq. 9. Assuming that $a \ll 1,\ b \ll 1$, the difference $\phi_s - \phi_b \approx 2a\sin(2\phi_s)$ and the vorticity:

$$\omega_z = 1/2(\nabla \times \mathbf{v})_z \approx (\rho_{t,nmax}/R)\sin(n\phi_s)[b_n - a_n]. \tag{11}$$

The estimates above should be valid for anisotropic flow of any harmonics - which is the reason we have changed in Eq. 11 the harmonic order from 2 to n. It is obviously quite a rough approximation (which in principle can be improved) as it leads to a discontinuity at the origin. It provides the following estimate for the hyperon polarization:

$$P_z \approx \omega_z/(2T) \approx 0.1\sin(n\phi_s)[b_n - a_n], \tag{12}$$

where we assumed that $\rho_{t,nmax} \sim 1$, $R \approx 10$ fm, and $T \approx 100$ MeV. In practice, the coefficients b_n and a_n are both of the order of a few percent, often close to each other. That results in the values for z-polarization not greater than a few per-mill, almost an order of magnitude lower than obtained in hydrodynamics calculations [7, 13].

The measurements of the z component of polarization could be relatively simple as they do not require the knowledge of the first harmonic event plane. The acceptance effects should be also readily accounted for requiring that the z component of the polarization averaged over all azimuthal angles to be zero.

We note that vorticity fields due the anisotropic flow are formed closer to the freeze-out, unlike the ones due to the "shear" in the initial velocity fields (as shown in Fig 1). Having in mind the finite relaxation time for establishing the equilibrium the relation between these two vorticities and the final polarizations can be different.

Finally, we mention another very interesting possibility for vorticity studies in asymmetric nuclear collisions such as Cu+Au. For relatively central collisions, when during the collision a smaller nucleus is fully "absorbed" by the larger one (e.g. such collisions can be selected by requiring no signal in the zero degree calorimeter in the lighter nucleus beam direction), one can easily imagine a configuration with toroidal velocity field, and as a consequence, a vorticity field in the form of a circle. The direction of the polarization in such a case would be given by $\hat{\mathbf{p}}_{\mathrm{T}} \times \hat{\mathbf{z}}$, where $\hat{\mathbf{p}}_{\mathrm{T}}$ and $\hat{\mathbf{z}}$ are the unit vectors along the particle transverse momentum and the (lighter nucleus) beam direction.

3 Summary

The hyperon polarization measurements provide a unique possibility to study the velocity fields in heavy ion collisions. The difference in particle//antiparticle polarization might provide a tool for the magnetic field measurements. Studing the global polarization as a function of event net electrical charge or strangeness could establish the relation between the global polarization and the chiral anomalous effect. The future 28 GeV Au+Au RHIC run, as well as the future LHC runs, will bring a high statistics data sets suitable for many differential studies outlined in this presentation.

Very fruitful discussions with F. Becattini, Yu. Karpenko, M. Konyushikhin, M. Lisa, T. Niida, I. Upsal, and X.-N. Wang are greatly appreciated. This material is based upon work supported in part by the U.S. Department of Energy Office of Science, Office of Nuclear Physics under Award Number DE-FG02-92ER-40713.

References

[1] Z. T. Liang and X. N. Wang, "Globally polarized quark-gluon plasma in non-central A+A collisions," Phys. Rev. Lett. **94**, 102301 (2005) Erratum: [Phys. Rev. Lett. **96**, 039901 (2006)] doi:10.1103/PhysRevLett.94.102301 [nucl-th/0410079].

[2] S. A. Voloshin, "Polarized secondary particles in unpolarized high energy hadron-hadron collisions?," nucl-th/0410089.

[3] B. I. Abelev *et al.* [STAR Collaboration], "Global polarization measurement in Au+Au collisions," Phys. Rev. C **76**, 024915 (2007) doi:10.1103/PhysRevC.76.024915

[4] J. H. Gao, S. W. Chen, W. t. Deng, Z. T. Liang, Q. Wang and X. N. Wang, "Global quark polarization in non-central A+A collisions," Phys. Rev. C **77**, 044902 (2008) doi:10.1103/PhysRevC.77.044902 [arXiv:0710.2943 [nucl-th]].

[5] F. Becattini and F. Piccinini, "The Ideal relativistic spinning gas: Polarization and spectra," Annals Phys. **323**, 2452 (2008) doi:10.1016/j.aop.2008.01.001

[6] F. Becattini, I. Karpenko, M. Lisa, I. Upsal and S. Voloshin, "Global hyperon polarization at local thermodynamic equilibrium with vorticity, magnetic field and feed-down," Phys. Rev. C **95**, no. 5, 054902 (2017) doi:10.1103/PhysRevC.95.054902

[7] F. Becattini, these proceedings

[8] Z. T. Liang and X. N. Wang, "Spin alignment of vector mesons in non-central A+A collisions," Phys. Lett. B **629**, 20 (2005) doi:10.1016/j.physletb.2005.09.060

[9] S. A. Voloshin and T. Niida, "Ultrarelativistic nuclear collisions: Direction of spectator flow," Phys. Rev. C **94**, no. 2, 021901 (2016) doi:10.1103/PhysRevC.94.021901

[10] Biao Tu, for the STAR Collaboration, these proceedings

[11] B. Mohanty, for the ALICE Collaboration, these proceedings

[12] L. Adamczyk *et al.* [STAR Collaboration], "Global Λ hyperon polarization in nuclear collisions: evidence for the most vortical fluid," Nature **548**, 62 (2017) doi:10.1038/nature23004

[13] Yu. Karpenko, these proceedings

[14] R. h. Fang, L. g. Pang, Q. Wang and X. n. Wang, "Polarization of massive fermions in a vortical fluid," Phys. Rev. C **94**, no. 2, 024904 (2016) doi:10.1103/PhysRevC.94.024904

[15] L. McLerran and V. Skokov, "Comments About the Electromagnetic Field in Heavy-Ion Collisions," Nucl. Phys. A **929**, 184 (2014) doi:10.1016/j.nuclphysa.2014.05.008

[16] F. Becattini *et al.*, "A study of vorticity formation in high energy nuclear collisions," Eur. Phys. J. C **75**, no. 9, 406 (2015) doi:10.1140/epjc/s10052-015-3624-1

[17] S. Singha, P. Shanmuganathan and D. Keane, "The first moment of azimuthal anisotropy in nuclear collisions from AGS to LHC energies," Adv. High Energy Phys. **2016**, 2836989 (2016) doi:10.1155/2016/2836989

[18] B. Abelev *et al.* [ALICE Collaboration], "Directed Flow of Charged Particles at Midrapidity Relative to the Spectator Plane in Pb-Pb Collisions at $\sqrt{s_{NN}}$=2.76 TeV," Phys. Rev. Lett. **111**, no. 23, 232302 (2013) doi:10.1103/PhysRevLett.111.232302

[19] M. Konyushikhin for the ALICE Collaboration, poster at Quark Matter 2017 conference

[20] T. Niida for the STAR Collaboration, talk at QCD Chirality workshop, UCLA, 2017.

[21] F. Becattini, L. Csernai and D. J. Wang, "Λ polarization in peripheral heavy ion collisions," Phys. Rev. C **88**, no. 3, 034905 (2013) Erratum: [Phys. Rev. C **93**, no. 6, 069901 (2016)] doi:10.1103/PhysRevC.93.069901, 10.1103/PhysRevC.88.034905

[22] M. I. Baznat, K. K. Gudima, A. S. Sorin and O. V. Teryaev, "Femto-vortex sheets and hyperon polarization in heavy-ion collisions," Phys. Rev. C **93**, no. 3, 031902 (2016) doi:10.1103/PhysRevC.93.031902

[23] O. Teryaev, these proceedings

[24] D. E. Kharzeev, J. Liao, S. A. Voloshin and G. Wang, "Chiral magnetic and vortical effects in high-energy nuclear collisions—A status report," Prog. Part. Nucl. Phys. **88**, 1 (2016) doi:10.1016/j.ppnp.2016.01.001

[25] S. Schlihting and S. Voloshin, in preparation

[26] B. Betz, M. Gyulassy and G. Torrieri, "Polarization probes of vorticity in heavy ion collisions," Phys. Rev. C **76**, 044901 (2007) doi:10.1103/PhysRevC.76.044901

Vorticity and Λ polarization in baryon rich matter

Mircea Baznat[1], *Konstantin* Gudima[1], *George* Prokhorov[2], *Alexander* Sorin[2,3], *Oleg* Teryaev[2,3,⋆], and *Valentin* Zakharov[4,5]

[1] *Institute of Applied Physics, Academy of Sciences of Moldova, MD-2028 Kishinev, Moldova*
[2] *Joint Institute for Nuclear Research, 141980 Dubna, Russia*
[3] *National Research Nuclear University (MEPhI), 115409 Moscow, Russia*
[4] *Alikhanov Institute for Theoretical and Experimental Physics (ITEP) B. Cheremushkinskaya ul. 25, RU-117218 Moscow, Russia*
[5] *School of Biomedicine, Far Eastern Federal University, Vladivostok, 690950, Russia*

Abstract. The polarization of Λ hyperons due to axial chiral vortical effect is discussed. The effect is proportional to (strange) chemical potential and is pronounced at lower energies in baryon-rich matter. The polarization of $\bar{\Lambda}$ has the same sihn and larger magnitude. The emergence of vortical structures is observed in kinetic QGSM models. The hydrodynamical helicity separation receives the contribution of longitudinal velocity and vorticity implying the quadrupole structure of the latter. The transition from the quark vortical effects to baryons in confined phase may be achieved by exploring the axial charge. At the hadronic level the polarization corresponds to the cores of quantized vortices in pionic superfluid. The chiral vortical effects may be also studied in the frmework of Wigner function establishing the relation to the thermodynamical approach to polarization.

1 Introduction

The experimental evidences for polarization of hyperons in heavy-ion collisions found by STAR collaboration [1] attracted recently much attention [2–8].

The studies of polarization are often performed [9] in the framework of approach exploring local equilibrium thermodynamics [10] and hydrodynamical calculations of vorticity [11–13].

There is another approach to polarization first proposed in [14] and independently in [15]. The so-called axial vortical effect (see e.g. [16] and references therein) being the macroscopic manifestation of axial anomaly [17]leads to induced axial current of strange quarks which may be converted to polarization of Λ-hyperons [14, 15].

The effect is proportional to vorticity and helicity of the strong interacting medium which was calculated [18] in the kinetic Quark-Gluon-String Model(QGSM)[19–21]. In this way, in fact, some particular case of fundamental relation between kinetic and hydrodynamic description was considered.

This calculation were later made more detailed[22], including the spatial and temporal dependence of strange chemical potential which is also the ingredient of anomalous approach to polarization.

⋆e-mail: teryaev@jinr.ru

2 Helicity separation and quadrupole structure

The effect of helicity separation was found in [18]. It was also found [18] that it receives contributions from both transverse and longitudinal velocity and vorticity. The latter, as the longitudinal velocity changes sign in each half-space, should lead to the quadrupole structure of longitudinal vorticity, in accordance to [2, 3].

Indeed, the longitudinal vorticity averaged over z in the symmetric cylinder $-z, z$ should change sign when $x \longrightarrow -x$ in order to have non-zero helicity in the lower and upper half-planes. Also, it should change the sign when $y \longrightarrow -y$ in order to guarantee that the siigns of helicity in these half-planes are different.

We now supplement this conclusion following already from observations in [18] by explicit calculation (Fig.1).

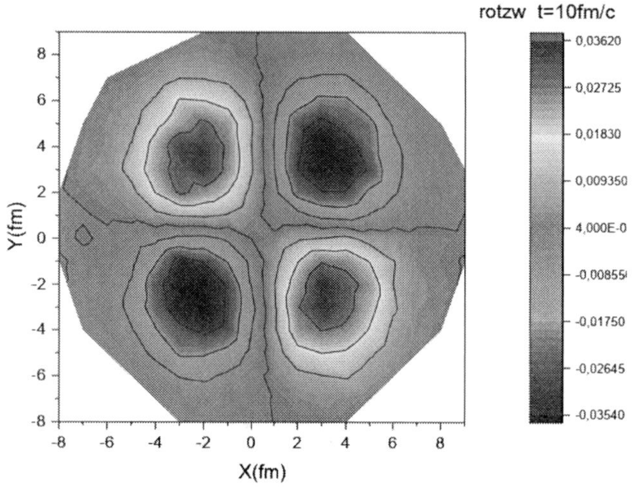

Figure 1. Quadrupole structure of longitudinal vorticity.

3 Anomalous mechanism of hyperon polarization

Anomalous mechanism of polarization makes this effect qualitatively similar to the local violation [23] of discrete symmetries in strongly interacting QCD matter. The most well known here is the Chiral Magnetic Effect (CME) which uses the (C)P-violating (electro)magnetic field emerging in heavy ion collisions in order to probe the (C)P-odd effects in QCD matter.

The polarization, in turn, is similar to Chiral Vortical Effect (CVE)[24] due to coupling to P-odd medium vorticity leading to the induced electromagnetic and all conserved-charge currents [14], in particular the baryonic one.

Recently the pioneering preliminary results on global polarization of Λ and $\bar{\Lambda}$ hyperons in $Au-Au$ collisions in Beam Energy Scan at RHIC were released [1] and the qualitative tendency of polarization decrease with energy in agreement with the prediction [14] was revealed. The recent theoretical analysis [25] suggested that decrease of polarization with energy is related to the suppression of Axial Magnetic effect contribution in strongly correlated QCD matter found in lattice simulations.

Indeed, the chiral vorticity coefficient describing the axial vortical effect

$$c_V = \frac{\mu_s^2 + \mu_A^2}{2\pi^2} + \frac{T^2}{6}, \quad (1)$$

contains temperature dependent term related to gravitational anomaly [27], and naively it can be quite substantial and increase with energy. However, lattice simulations [28] lead to the zero result in the confined phase and to the suppression by one order of magnitude at high temperatures. Neglecting axial chemical potential, the coefficient c_V takes the form

$$c_V = \frac{\mu_s^2}{2\pi^2} + k\frac{T^2}{6}, \quad (2)$$

As soon as for free fermion gas the $T^2/6$ term is recovered [29] for large lattice volume at fixed temperature, the above-mentioned suppression should be attributed to the correlation effects. It was suggested[25], that the accurate measurements of polarization energy dependence may serve a sensitive probe of strongly correlated QCD matter. Here we discuss the numerical simulations to implement this suggestion.

4 Numerical simulations of axial anomaly contributions to (anti)hyperon polarization

We performed the numerical simulations in QGSM model [19–21]. We decomposed the space-time to the cells, allowing to define velocity and vorticity in the kinetic model, as described in detail in [18]. To define the strange chemical potential (assuming that Λ polarization is carried by strange quark) we used the matching procedure [22] of distribution functions to its (local) equilibrium values. We also determine in this way the values of temperature [26]. In general, let us stress that we realized in our particular case the relation between kinetics, hydrodynamics and thermodynamics.

We first neglect the gravitational anomaly contribution and start by considering the energy dependence of polarization (described in detail in [25]) for three values of impact parameter. The results are presented at Fig.2. The curves correspond to $b = 8.0fm, 6.4fm, 4.8fm$.

We continue by the inclusion of contribution related to gravitational anomaly[26]. The results are presented at Fig.3. We consider as a starting point the original value of anomaly coefficient[27] $T^2/6$ which is reproduced for large lattice volume at fixed temperature[29]. We present the curves following from the coefficients suppressed by factor k (2) resulting from the lattice calculations [28]. We compare values of $k = 1$ with $k = 0, 1/15, 1/10$. As one can see, the lattice-supported value $1/15$ is most close to the behavior of preliminary data which may be considered as a signal of strongly correlated matter formation. The closeness of $k = 0$ curve to the experimental points may be related to the contribution of confinement phase, where lattice calculations [28] lead to zero temperature-dependent effect. At the same time, already $k = 1/10$ leads to the curve growing with energy, while $k = 1$ leads to extremely strong growth.

Such an interplay of the temperature and chemical potential effects opens the possibility of detailed experimental and numerical studies. One may, in particular, explore thiis interplay at various impact parameters and investigate the possibility of compensation of the growth of temperature and decrease of chemical potential with energy.

The $\bar{\Lambda}$ polarization is emerging due to the polarization of $\bar{s}$–quarks, which has the same sign, as the axial current and charge are C-even operators. The magnitude of the $\bar{\Lambda}$ is larger as the same axial charge is distributed between the polarizations of the smaller number of particles [25]. It is mandatory to take into account the contribution of K^* mesons. In the case of Λ the $K^{*-}, \bar{K}^0$ mesons contain two

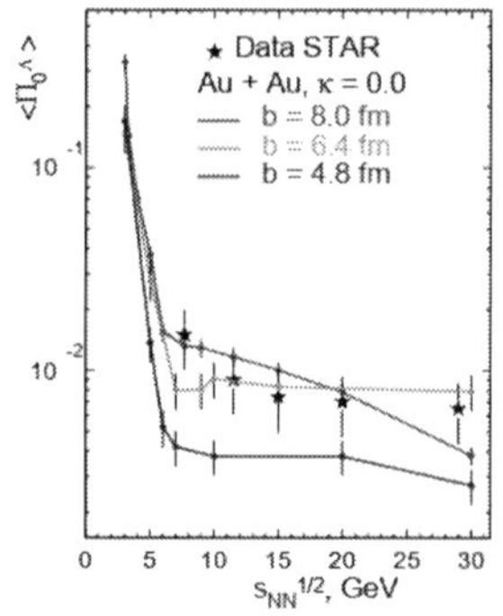

Figure 2. Energy dependence of polarization for different values of impact parameter

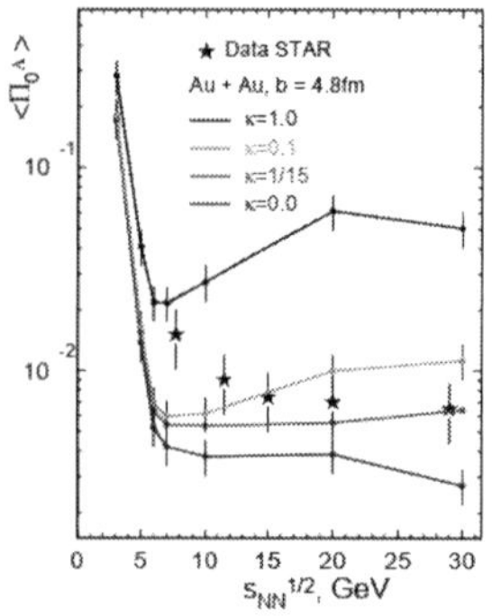

Figure 3. Energy dependence of polarization for different values of gravitational anomaly contribution

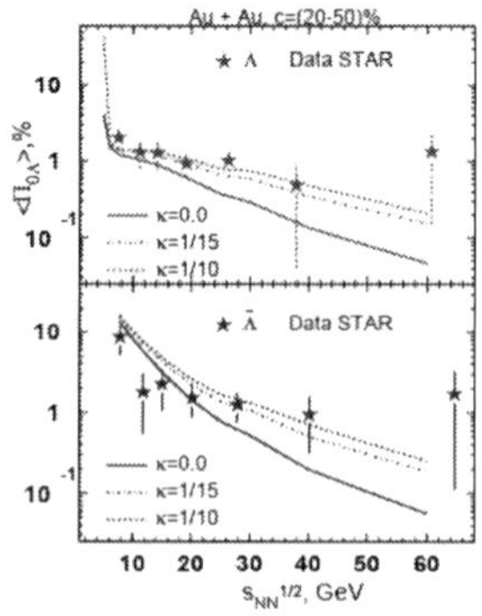

Figure 4. Comparison of Λ and Λ̄ polarizations for different values of gravitational anomaly contribution.

sea(anti)quarks and does not change the polarization significantly. At the same time, for $\bar{\Lambda}$ the relevant K^{*+}, K^0 mesons are more numerous and make the excess of $\bar{\Lambda}$ polarization less pronounced.

Note that this excess is larger for smaller energies, where suppression of $\bar{\Lambda}$ is larger. This differs from the (C-odd) effect of magnetic field, which is increasing with energy, although more detailed studies taking into account the finite time of magnetic field existence are required.

The quantitative analysis of these effects, taking into account the gravitational anomaly contribution, is presented at Figure 4. The result is in reasonable agreement with STAR data, although further analysis is required, in particular, the detailed study of the interplay of growing temperature and decreasing chemical potential, mentioned above.

5 Quantized vortices in pionic superfluid

The description of polarization at the ahdronic level may be achieved by consideration of quantized vortices in mesonic superfluid [30]. The baryonic polarization emerges as a radiative correction at their cores leading to the current

$$(j^5_B)_\alpha \;=\; \frac{N_c}{36\pi^2 f_\pi^2}\epsilon_{\alpha\beta\gamma\delta}(\partial^\beta \pi^0)(\partial^\gamma \partial^\delta \pi^0)\,. \tag{3}$$

One may also explain that in such a way: as the quantized circulation should lead to the growing velocity for small distances, the pointlike spin contribution at the vortex core is required in order not to exceed the velocity of light. In the liquid helium, to the contrary, the velocity of sound is not exceeded because the quantized circulation is not manifested at the distances smaller than interatomic ones.

6 Chiral vortical effect from Wigner function

The approach based on chiral vortical effects and making therefore the polarization a kind of anomalous transport seems to be rather different from the thermodynamical approach [2]. However, the recent study [31] presented also as a poster at this conference shows that the covariant Wigner function should lead to the expression analogous to induced anomalouys current. This may be related to the fact that the in-medium current correlators should include triangle diagrams when the expansion over the external parameters is performed. This however does not imply the anomaloud non-conservation of axial current. The appearance of extra contributions to axial current [31] may be due to the ambiguity in Wigner function definition [32]. One can even guess that their absence may be used as a constraint for the proper definition of Wigner function.

7 Conclusions

We found that the anomalous mechanism may naturally explain:

- Decrease of polarization with energy.
- The polarization of antihyperons of the same sign and larger magnitude as hyperons with the difference dereasing with energy.

The polarization at the hadronic level may be attributed to the cores of quantized vortices in pionic superfluid.

The anomalus contribution may be obtained in the framework of Wigner function approach.

The further experimental theoretical and experimental studies should be performed. In particular, this is related to the polarization os other (Σ, Ξ) hyperons, which should have the same (up to the mass rescaling) polarization in thermodynamical approach, but may differ significantly due to the different chemical potentials in the anomalous approach.

Acknowledgments

G.P., A.S. and O.T. are indebted to Organizers and, especially, to A. Mischke for warm hospitality at Utrecht. They are also grateful to F. Becattini and S. Voloshin for useful discussions. The work was supported by RFBR grant 17-02-01108.

References

[1] L. Adamczyk *et al.* [STAR Collaboration], Nature **548**, 62 (2017) doi:10.1038/nature23004 [arXiv:1701.06657 [nucl-ex]].

[2] F. Becattini, These Proceedings.

[3] S. Voloshin, These Proceedings.

[4] F. Becattini, I. Karpenko, M. Lisa, I. Upsal and S. Voloshin, Phys. Rev. C **95**, no. 5, 054902 (2017) doi:10.1103/PhysRevC.95.054902 [arXiv:1610.02506 [nucl-th]].

[5] I. Karpenko and F. Becattini, Eur. Phys. J. C **77**, no. 4, 213 (2017) doi:10.1140/epjc/s10052-017-4765-1 [arXiv:1610.04717 [nucl-th]].

[6] I. Karpenko and F. Becattini, J. Phys. Conf. Ser. **779**, no. 1, 012068 (2017) doi:10.1088/1742-6596/779/1/012068 [arXiv:1611.08136 [nucl-th]].

[7] Y. Xie, R. C. Glastad and L. P. Csernai, Phys. Rev. C **92** (2015) no.6, 064901 doi:10.1103/PhysRevC.92.064901 [arXiv:1505.07221 [nucl-th]].

[8] R. h. Fang, J. y. Pang, Q. Wang and X. n. Wang, arXiv:1611.04670 [nucl-th].

[9] F. Becattini, L. Csernai and D. J. Wang, "Λ polarization in peripheral heavy ion collisions," Phys. Rev. C **88**, no. 3, 034905 (2013) [arXiv:1304.4427 [nucl-th]].

[10] F. Becattini, L. Bucciantini, E. Grossi and L. Tinti, Eur. Phys. J. C **75**, no. 5, 191 (2015) doi:10.1140/epjc/s10052-015-3384-y [arXiv:1403.6265 [hep-th]].

[11] B. Betz, M. Gyulassy and G. Torrieri, Phys. Rev. C **76** (2007) 044901 doi:10.1103/PhysRevC.76.044901 [arXiv:0708.0035 [nucl-th]].

[12] L. P. Csernai, V. K. Magas and D. J. Wang, Phys. Rev. C **87** (2013) no.3, 034906 doi:10.1103/PhysRevC.87.034906 [arXiv:1302.5310 [nucl-th]].

[13] L. P. Csernai, D. J. Wang and T. Csorgo, Phys. Rev. C **90** (2014) no.2, 024901 doi:10.1103/PhysRevC.90.024901 [arXiv:1406.1017 [hep-ph]].

[14] O. Rogachevsky, A. Sorin and O. Teryaev, "Chiral vortaic effect and neutron asymmetries in heavy-ion collisions," Phys. Rev. C **82**, 054910 (2010) [arXiv:1006.1331 [hep-ph]].

[15] J. -H. Gao, Z. -T. Liang, S. Pu, Q. Wang and X. -N. Wang, "Chiral Anomaly and Local Polarization Effect from Quantum Kinetic Approach," Phys. Rev. Lett. **109**, 232301 (2012) [arXiv:1203.0725 [hep-ph]].

[16] T. Kalaydzhyan, "Temperature dependence of the chiral vortical effects," Phys. Rev. D **89**, no. 10, 105012 (2014) [arXiv:1403.1256 [hep-th]].

[17] D.T. Son and P. Surowka, Phys. Rev. Lett. **103**, 191601 (2009)

[18] M. Baznat, K. Gudima, A. Sorin and O. Teryaev, "Helicity separation in Heavy-Ion Collisions," Phys. Rev. C **88**, 061901 (2013) [arXiv:1301.7003 [nucl-th]].

[19] V.D. Toneev, K.K. Gudima, Nucl. Phys. A **400**, 173c (1983).

[20] V.D. Toneev, N.S. Amelin, K.K. Gudima, S.Yu. Sivoklokov, Nucl. Phys. A **519**, 463c (1990).

[21] N.S. Amelin, E.F. Staubo, L.S. Csernai et al., Phys.Rev. C **44**, 1541 (1991).

[22] M. I. Baznat, K. K. Gudima, A. S. Sorin and O. V. Teryaev, Phys. Rev. C **93**, no. 3, 031902 (2016) doi:10.1103/PhysRevC.93.031902 [arXiv:1507.04652 [nucl-th]].

[23] K. Fukushima, D.E. Kharzeev and H.J. Warringa, "The Chiral Magnetic Effect," Phys. Rev. D **78**, 074033 (2008) [arXiv:0808.3382 [hep-ph]].

[24] D. Kharzeev and A. Zhitnitsky, "Charge separation induced by P-odd bubbles in QCD matter," Nucl. Phys. A **797**, 67 (2007) [arXiv:0706.1026 [hep-ph]].

[25] A. Sorin and O. Teryaev, Phys. Rev. C **95**, no. 1, 011902 (2017) doi:10.1103/PhysRevC.95.011902 [arXiv:1606.08398 [nucl-th]].

[26] M. Baznat, K. Gudima, A. Sorin and O. Teryaev, arXiv:1701.00923 [nucl-th].

[27] K. Landsteiner, E. Megias, L. Melgar and F. Pena-Benitez, "Holographic Gravitational Anomaly and Chiral Vortical Effect," JHEP **1109** (2011) 121

[28] V. Braguta, M.N. Chernodub, K. Landsteiner, M.I. Polikarpov, M.V. Ulybyshev Phys.Rev. D88 (2013) 071501 DOI: 10.1103/PhysRevD.88.071501 e-Print: arXiv:1303.6266 [hep-lat]; V. Braguta, M.N. Chernodub, V.A. Goy, K. Landsteiner, A.V. Molochkov, M.I. Polikarpov, Phys.Rev. D89 (2014) no.7, 074510 DOI: 10.1103/PhysRevD.89.074510 e-Print: arXiv:1401.8095 [hep-lat]; V. Braguta, M. N. Chernodub, V. A. Goy, K. Landsteiner, A. V. Molochkov and M. Ulybyshev, "Study of axial magnetic effect," AIP Conf. Proc. **1701**, 030002 (2016); doi:10.1063/1.4938608; V. Goy, "Investigation of SU(2) gluodynamics in the framework of the latice approach", PhD Thesis (in Russian), Vladivostok, 2015.

[29] P. V. Buividovich, "Axial Magnetic Effect and Chiral Vortical Effect with free lattice chiral fermions," J. Phys. Conf. Ser. **607**, no. 1, 012018 (2015) doi:10.1088/1742-6596/607/1/012018 [arXiv:1309.4966 [hep-lat]].

[30] O. V. Teryaev and V. I. Zakharov, Phys. Rev. D **96**, no. 9, 096023 (2017); arXiv:1705.01650 [hep-th].

[31] G. Prokhorov and O. Teryaev, arXiv:1707.02491 [hep-th].

[32] F. Becattini, Private communication.

Strongly interacting matter under rotation

Yin Jiang[1,⋆], *Zi-Wei* Lin[2,3,⋆⋆], *Xu-Guang* Huang[4,5,⋆⋆⋆], and *Jinfeng* Liao[6,⋆⋆⋆⋆]

[1]*School of Physics and Nuclear Energy Engineering, Beihang University, Beijing 100191, China.*
[2]*Department of Physics, East Carolina University, Greenville, NC 27858, USA.*
[3]*Key Laboratory of Quarks and Lepton Physics (MOE) and Institute of Particle Physics, Central China Normal University, Wuhan 430079, China.*
[4]*Physics Department and Center for Particle Physics and Field Theory, Fudan University, Shanghai 200433, China.*
[5]*Key Laboratory of Nuclear Physics and Ion-beam Application (MOE), Fudan University, Shanghai 200433, China.*
[6]*Physics Department and Center for Exploration of Energy and Matter, Indiana University, 2401 N Milo B. Sampson Lane, Bloomington, IN 47408, USA.*

Abstract. The vorticity-driven effects are systematically studied in various aspects. With AMPT the distributions of vorticity has been investigated in heavy ion collisions with different collision parameters. Taking the rotational polarization effect into account a generic condensate suppression mechanism is discussed and quantitatively studied with NJL model. And in chiral restored phase the chiral vortical effects would generate a new collective mode, i.e. the chiral vortical wave. Using the rotating quark-gluon plasma in heavy ion collisions as a concrete example, we show the formation of induced flavor quadrupole in QGP and estimate the elliptic flow splitting effect for Λ baryons.

1 Introduction

Pseudovector field-driven effects have attracted more and more interests recently [1–4]. As a parity-even and time-reversal-odd field, numbers of novel and non-trivial phenomena would be driven by it,such as chiral magnetic effect [1, 2, 5, 6]. As another pseudovector field which is familiar to us, vorticity plays an important role in our daily life as well as in the high energy physics [7, 8]. In relativistic heavy ion collision experiments, the typical collision events are off-central and the created QCD matter will carry a nonzero angular momentum on the order of $10^4\hbar \sim 10^5\hbar$ with local angular velocity in the range $0.01 \sim 0.1$GeV [9, 10]. There has also been impressive progress to study the rotating QCD matter using lattice gauge theory simulations [11]. Both first principle calculations and model studies suggest that vorticity would drive the phase transition [12] and anomalous transport [13–15]. In the following we will first simulate the collisions to extract the vorticity with AMPT model in the particle scenario [10]. With the results we would see the vorticity is so remarkable that we could

⋆e-mail: jiang_y@buaa.edu.cn
⋆⋆e-mail: linz@ecu.edu
⋆⋆⋆e-mail: huangxuguang@fudan.edu.cn
⋆⋆⋆⋆e-mail: liaoji@indiana.edu

not simply neglect its impact on the phase transition and chiral transport. Hence we discuss the rotational polarization effects for quark pairings, which would change the chiral restoration and color superconductivity critical behaviors [12]. In chiral restored phase the vorticity would also induce two anomalous chiral transport relations for the left and right-handed quarks respectively [13, 14]. Combining these two effects there is a new collective mode called chiral vortical wave(CVW) [15]. The CVW would reshuffle the final state hadrons' distributions so that provide a new contribution to the elliptic flow splitting of hadrons.

2 Vorticity in heavy ion collisions

As a transport model AMPT would output trajectories of all the partons in the phase space during collisions. For a large enough system after proper coarse-graining (e.g. like the fluid being made of many fluid cells), it can be considered as a continuous medium characterized by a series of locally defined quantities like momentum density, energy density and particle number density $\vec{p}(\vec{r})$, $\epsilon(\vec{r})$ and $n(\vec{r})$ respectively. Here we define the 3-velocity field as $\vec{v}(\vec{r}) = \vec{p}(\vec{r})/\epsilon(\vec{r})$. And the non-relativistic vorticity $\vec{\omega} = \nabla \times \vec{v}$ is adopted in this work. The detailed patterns of vorticity shown in [10] could be roughly understood by a radial flow profile $\vec{v}$ that can be well parameterized in the form $\vec{v}(\rho, \phi, \eta) = \hat{e}_\rho v_0(\rho, \eta)[1 + 2c_2(\rho, \eta)\cos 2\phi]$, where $\hat{e}_\rho$ is the unit vector along the transverse radial direction, ρ and ϕ are transverse radial and azimuthal coordinates. Noting that this flow profile has no global rotation. It suggests that, to extract the component of local vorticity that is truly associated with the global rotation, one needs to perform an average over the fireball. Upon such averaging, the background flow contributions to local vorticity would cancel out, and what remains can be attributed to the rotational motion. The averaged vorticity over the whole fireball depends on the weight we choose. In the

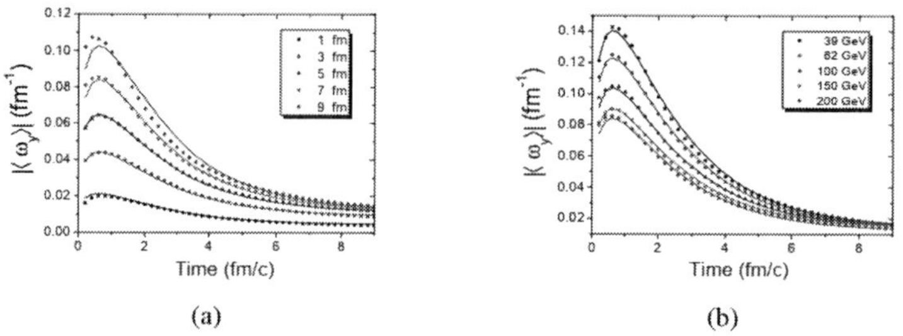

Figure 1. Averaged vorticity $\langle \omega_y \rangle$ from the AMPT model as a function of time at various impact parameter b for fixed beam energy $\sqrt{s_{NN}} = 200$ GeV. (a). Averaged vorticity $\langle \omega_y \rangle$ from the AMPT model as a function of time at varied beam energy $\sqrt{s_{NN}}$ for fixed impact parameter $b = 7$ fm. (b).

following we choose $\mathcal{W}(\vec{r}) = \rho^2 \epsilon(\vec{r})$. Different weights dependence has also been investigated in [10]. The centrality dependence of $\langle \omega_y \rangle$ at given beam energy $\sqrt{s_{NN}} = 200$ GeV: see Fig. 1(a). The $\langle \omega_y \rangle$ briefly increases with time which is most likely due to parton scatterings during the early stage (when the transverse radial expansion is not developed yet) that in certain way decrease the fluid moment of inertia. The averaged vorticity reaches peak value at an almost universal time around 1 fm/c and then follows a steady decrease with time. The decrease is due to the system's expansion which increases total moment of inertia at the price of reduced vorticity due to the constraint of constant angular momentum. The results also clearly demonstrate that the averaged vorticity increases from central to peripheral collisions: this trend is different from the angular momentum. Such difference again can be understood as follows: while the vorticity increases with b, the fluid moment of inertia (pertinent to rotation) in the fireball decreases with b, thus the angular momentum shows a non-monotonic behavior due to the two competing trends.

We next show the beam energy dependence of $\langle\omega_y\rangle$ at given impact parameter $b = 7$ fm: see Fig. 1(b). Similar time evolution patterns are observed at all energies. We notice that the averaged vorticity increases with decreasing beam energy, in quite the opposite trend to the angular momentum. This may be understood as follows: with increasing beam energy, the fluid moment of inertia (pertinent to rotation) increases more rapidly than the decrease of vorticity, thus the total angular momentum is still increasing. We have numerically checked that this is indeed the case. And the trend has also been observed in STAR's results [9]

3 Rotation-induced phase transition

From the simulation we have already seen that the local vorticity can be several tens or one hundred of MeV. Considering the polarization effect the quark energy shift should be $E \pm \vec{\omega}\cdot\vec{J}$ which behaves as a chemical potential. Therefore it would induce the non-negligible modification to the fermi surface of the quarks, which further affects the pairing probability. Theoretically it is more intuitive to study a fermionic system under a global rotation $\vec{\omega}$ to show the generic rotation-induce effect.

In this proceeding we only consider the $J = 0$ pairing cases. For a pair of spin-$\frac{1}{2}$ fermions, there are different ways of forming a spin-0 pairing state: either, the pair could have both nonzero orbital angular momentum L and nonzero total spin S, with L and S being opposite thus resulting in total $J = 0$; or the pair could have zero orbital angular momentum, and have opposite individual spin configurations for the two fermions. When such a system is under rotation, there will be a generic rotational suppression effect on the scalar pairing states. Intuitively this can be understood as follows. The global rotation, implying a nonzero macroscopic angular momentum of the whole system, will induce a *rotational polarization effect* which tends to "force" all microscopic angular momentum to be aligned with the global angular momentum. So for a pair of fermions, their relative orbital angular momentum L as well as their individual spins would prefer to be parallel to the global angular momentum rather than to arrange themselves into a scalar state with zero angular momentum. This therefore leads to a generic suppression effect on the scalar pairing states. It also implies that pairing states with nonzero angular momentum could become more favorable. In the following, we quantitatively demonstrate this effect in the Nambu–Jona-Lasinio model for the chiral condensate (of quark-anti-quark pairing states with $L = S = 1$ but $J = 0$). And the diquark condensate (of quark-quark pairing state with $L = S = 0$) case could be referred to [12].

In the chiral condensate case, the mean-field approximation gives the grand potential as

$$\begin{aligned}\Omega &= \int d^3\vec{r}\Big\{\frac{(M-m)^2}{4G} - \frac{N_f N_c}{16\pi^2}\sum_n \int dk_t^2 \int dk_z \times \quad [J_n(k_t r)^2 + J_n(k_t r)^2] \\ &\times T\Big[\ln\left(1+e^{(\epsilon_n-\mu)/T}\right) + \ln\left(1+e^{-(\epsilon_n-\mu)/T}\right) + \ln\left(1+e^{(\epsilon_n+\mu)/T}\right) + \ln\left(1+e^{-(\epsilon_n+\mu)/T}\right)\Big]\Big\} \quad (1)\end{aligned}$$

In the above the mean-field quasiparticle dispersion ϵ_n is given by $\epsilon_n = \sqrt{k_z^2 + k_t^2 + M^2} - (n + \frac{1}{2})\omega$. Note that because of the rotation around the z-axis, the $O(3)$ symmetry has been broken into $O(2)$ here. The mean-field chiral condensate (or equivalently the mass gap M) at given values of temperature T, chemical potential μ and rotation ω, can then be determined from the usual gap equation through minimization the potential: $\frac{\delta\Omega}{\delta M(r)} = 0$ and $\frac{\delta^2\Omega}{\delta M(r)^2} > 0$. We will numerically solve the gap equation for the case of $N_f = 2$ and $N_c = 3$ and present the results below. For the parameters G, G_d and a cutoff scale Λ of this model, we choose the standard values (see e.g. [16]).

The chiral restorations are observed in Fig 2(a) as the angular velocity increases at different temperatures. It is tempting to envision a new phase diagram of the chiral phase transition on the $T - \omega$ parameter space: see Fig. 2(b) (as computed from the present model). It features a chiral-symmetry-broken phase at low temperature and slow rotation while a chiral-symmetry-restored phase at high temperature and/or rapid rotation. A smooth crossover transition region at high T and low T and a first-order transition line at low T and high ω are connected by a new *critical end point*. Given the present model parameters, this critical point is located at $T_{CEP} = 0.020$GeV and $\omega_{CEP} = 0.644$GeV. As already discussed previously, the "rotational suppression" of the scalar condensate is a quite generic effect. It is conceivable that similar phase transition behaviors under rotation would also occur in other dynamical models for studying chiral condensate. In [12] we also studied the diquark

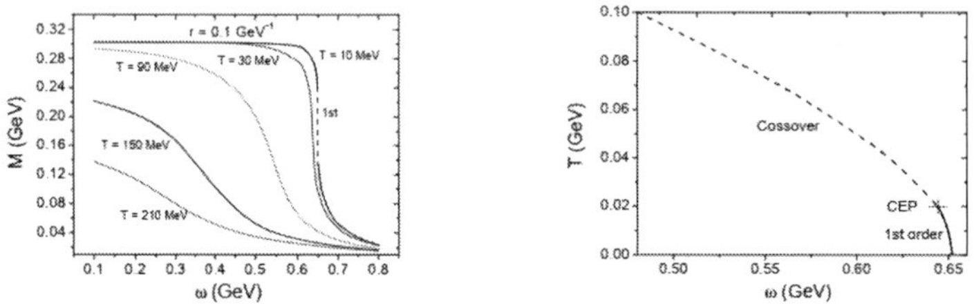

(a) (b)

Figure 2. The mean-field mass gap M (at radius $r = 0.1\text{GeV}^{-1}$) as a function of ω for various fixed value of T (a). The phase diagram on T-ω plane (b).

condensate case. As our analysis in above the rotational suppression would emerge again in that system. It should be mentioned that the rotational movement would always try to push matters far away from the center, which means it would be an inhomogeneous system. The radial distance-induced phase transition, which can not be smooth enough, is being studied self-consistently.

4 Chiral vortical wave

The CVW is the direct result of Chiral vortical effects

$$\vec{J}_V = \frac{1}{\pi^2}\mu\mu_5\vec{\omega}, \quad \vec{J}_A = \left[\frac{1}{6}T^2 + \frac{1}{2\pi^2}(\mu^2 + \mu_5^2)\right]\vec{\omega}\ . \tag{2}$$

Using the chiral basis and considering the current conservation relations for the left and right-handed quark currents $\vec{J}_{L/R} = \frac{1}{2}(\vec{J}_V \mp \vec{J}_A)$, we would obtain the wave equations

$$\partial_t n_{L/R} = \pm\frac{1}{4\pi^2}\omega\partial_x(\mu_{L/R}^2) = \pm\frac{\omega\mu_{L/R}}{2\pi^2}\partial_x\mu_{L/R}, \tag{3}$$

where the background vorticity is $\vec{\omega} = \omega\hat{x}$ and CVW speed is $V_\Omega = \frac{\mu_0\,\omega}{2\pi^2\chi_{\mu_0}}$ in the linearized approximation. The χ_{μ_0} is the susceptibility and μ_0 the background chemical potential. Obviously the CVW will transport charges along or against the vorticity direction. The detailed process depends on the initial condition [15]. If the initial condition is a purely axial charge density fluctuation $F_i^A(x)$ at $t = 0$, a *charge dipole moment* would be generated eventually with the size $(\delta n)_t^V \approx [-\partial_x F_i^A(x)]V_\Omega\, t$. On the other hand a purely vector charge density fluctuation $F_i^V(x)$ at $t = 0$ would result a vector charge quadrupole $(\delta n)_t^V \approx F_i^V(x) + [\partial_x^2 F_i^V(x)/2](V_\Omega\, t)^2$.

To find possible experimental manifestation, we consider the rotating quark-gluon plasma created in off-central heavy ion collisions. In such a QGP, CVW occurs for each light flavors, e.g. u, d quarks and possibly s quarks as well. The global rotation points in the out-of-plane direction. Given initial fluctuations in their respective vector densities, each flavor develops its CVW along ω. This transports flavor charges toward the two "tips" of the QGP fireball, thus leading to a quadrupole charge distribution on the transverse plane, which is shown in Fig. 3(a). In [15] we made the first estimate of such effect. The measurement may be done by the elliptic flow splitting of identified

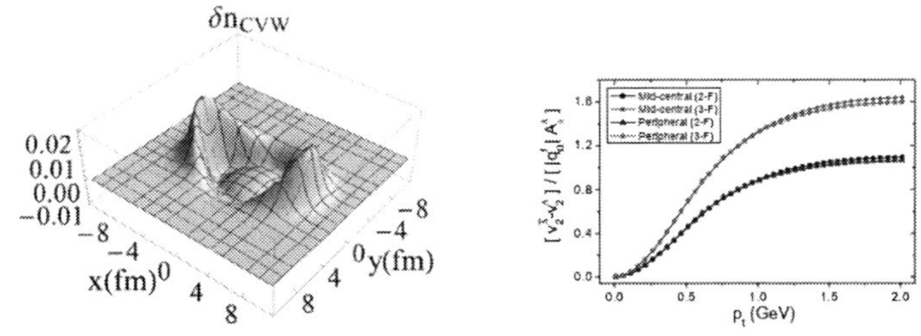

(a) (b)

Figure 3. CVW-induced flavor charge density profile (a). Normalized $\bar{\Lambda}$ and Λ elliptic flow splitting, $[v_2^{\bar{\Lambda}} - v_2^{\Lambda}]/[|q_\Omega^f|A_\pm^\Lambda]$, for symmetric 2-flavor(2-F) and 3-flavor(3-F) cases. The mid-central and the peripheral correspond to for $15-30\%$ and $60-92\%$ centrality class (see [17]) (b).

particles/anti-particles with nonzero baryonic charges. Flavor quadrupole implies more quarks on the out-of-plane tips of fireball and less on the in-plane equator , and therefore more baryons will be formed on the tips than on the equator. The stronger in-plane radial flow will thus translate the quadrupole into baryon/anti-baryon v_2 splitting. This mechanism is in analogy to the electric charge quadrupole induced by CMW [6]. To quantify such baryon flow spitting, one needs to combine quarks of varied flavors into hadrons. Suppose at the freeze-out, the flavor-wise chemical potential for quarks contains the CVW-induced quadrupole contribution $\delta\mu_f \propto 2q_\Omega^f \cos(2\phi_s)$ (with $f = u, d, s$). The corresponding chemical potential for a given type of hadron can be determined from its constituent quark content, e.g. for Λ baryon $\delta\mu_\Lambda \propto 2(q_\Omega^u + q_\Omega^d + q_\Omega^s)\cos(2\phi_s)$. We particularly propose to use Λ baryon which is electric charge neutral thus unaffected by possible CMW effect. We then use the STAR blast-wave model [17] to compute the resulting differential flow splitting. As it is unclear how much the s quark mass may reduce their chiral effects, we consider two extreme cases: a symmetric two-flavor (2-F) case $q_\Omega^u = q_\Omega^d = q_\Omega^f$ with $q_\Omega^s = 0$, or a symmetric three-flavor (3-F) case $q_\Omega^u = q_\Omega^d = q_\Omega^s = q_\Omega^f$. From Cooper-Frye scheme it is easy to see $\Delta v_2 = v_2^{\bar{\Lambda}} - v_2^{\Lambda} \propto |q_\Omega^f|A_\pm^\Lambda$ with $A_\pm^\Lambda = (N^\Lambda - N^{\bar{\Lambda}})/(N^\Lambda + N^{\bar{\Lambda}})$ the Λ-asymmetry that is directly related to background density μ_0 (in analogy to a similar relation in CMW case [6]). The results for normalized flow splitting $\Delta v_2/[|q_\Omega^f|A_\pm^\Lambda]$ are shown in Fig.3(b). Such effect in heavy ion collisions could be quantitatively investigated in the framework of anomalous fluid dynamics similarly to the recent development for the study of chiral magnetic effect [18].

5 Summary

In summary we systematically studied the vorticity-induced effects in various aspects. With AMPT it has been shown that the QGP not only carries large angular momentum and local vorticity, but also remarkable global rotation speed. Motivated by the large vorticity and taking the rotational polarization effect into account the spin zero condensate would be suppressed. As a generic suppression mechanism it may have non-negligible impacts in both QCD phase diagram and numbers of low energy systems. Finally in the chiral restored phase the vorticity would excite the chiral vortical wave. Using the rotating quark-gluon plasma in heavy ion collisions as a concrete example, we show the formation of induced flavor quadrupole in QGP and estimate the elliptic flow splitting effect for Λ baryons. Such proposal could be tested with future experimental data.

Acknowledgments. This material is based upon work supported by the U.S. Department of Energy, Office of Science, Office of Nuclear Physics, within the framework of the Beam Energy Scan Theory (BEST) Topical Collaboration. The work is also supported in part by the NSF Grant No. PHY-1352368 (JL), by the DFG Collaborative Research Center "SFB 1225 (ISOQUANT)" (YJ), by the NSFC of China under Grants No. 11628508(ZWL), by Young 1000 Talents Program of China, NSFC with Grant No.11535012 and No.11675041, and the Scientific Research Foundation of State Education Ministry for Returned Scholars(XGH).

References

[1] D. Kharzeev and A. Zhitnitsky, Nucl. Phys. A **797**, 67 (2007).

[2] D. E. Kharzeev, L. D. McLerran and H. J. Warringa, Nucl. Phys. A **803**, 227 (2008).

[3] D. E. Kharzeev, J. Liao, S. A. Voloshin and G. Wang, Prog. Part. Nucl. Phys. **88**, 1 (2016) doi:10.1016/j.ppnp.2016.01.001 [arXiv:1511.04050 [hep-ph]].

[4] J. Liao, Pramana **84**, no. 5, 901 (2015) doi:10.1007/s12043-015-0984-x [arXiv:1401.2500 [hep-ph]].

[5] D. E. Kharzeev and H. U. Yee, Phys. Rev. D **83**, 085007 (2011) doi:10.1103/PhysRevD.83.085007 [arXiv:1012.6026 [hep-th]].

[6] Y. Burnier, D. E. Kharzeev, J. Liao and H. U. Yee, Phys. Rev. Lett. **107**, 052303 (2011); arXiv:1208.2537 [hep-ph].

[7] L. P. Csernai, V. K. Magas, H. Stocker and D. D. Strottman, Phys. Rev. C **84**, 024914 (2011) doi:10.1103/PhysRevC.84.024914 [arXiv:1101.3451 [nucl-th]].

[8] L. P. Csernai, D. D. Strottman and C. Anderlik, Phys. Rev. C **85**, 054901 (2012) doi:10.1103/PhysRevC.85.054901 [arXiv:1112.4287 [nucl-th]].

[9] L. Adamczyk *et al.* [STAR Collaboration], Nature **548**, 62 (2017) doi:10.1038/nature23004 [arXiv:1701.06657 [nucl-ex]].

[10] Y. Jiang, Z. W. Lin and J. Liao, Phys. Rev. C **94**, no. 4, 044910 (2016) Erratum: [Phys. Rev. C **95**, no. 4, 049904 (2017)] doi:10.1103/PhysRevC.94.044910, 10.1103/PhysRevC.95.049904 [arXiv:1602.06580 [hep-ph]].

[11] A. Yamamoto and Y. Hirono, Phys. Rev. Lett. **111**, 081601 (2013) doi:10.1103/PhysRevLett.111.081601 [arXiv:1303.6292 [hep-lat]].

[12] Y. Jiang and J. Liao, Phys. Rev. Lett. **117**, no. 19, 192302 (2016) doi:10.1103/PhysRevLett.117.192302 [arXiv:1606.03808 [hep-ph]].

[13] D. T. Son and P. Surowka, Phys. Rev. Lett. **103**, 191601 (2009).

[14] D. E. Kharzeev and D. T. Son, Phys. Rev. Lett. **106**, 062301 (2011).

[15] Y. Jiang, X. G. Huang and J. Liao, Phys. Rev. D **92**, no. 7, 071501 (2015).

[16] S. P. Klevansky, Rev. Mod. Phys. **64**, 649 (1992). doi:10.1103/RevModPhys.64.649

[17] F. Retiere and M. A. Lisa, Phys. Rev. C **70**, 044907 (2004).

[18] Y. Jiang, S. Shi, Y. Yin and J. Liao, arXiv:1611.04586 [nucl-th].

Strange matter in compact stars

Thomas Klähn[1,2,⋆] and *David B.* Blaschke[2,3,4,⋆⋆]

[1] *Department of Physics and Astronomy, California State University Long Beach, California 90840, U.S.A*
[2] *Institute of Theoretical Physics, University of Wrocław, pl. M. Borna 9, 50-204 Wroclaw, Poland*
[3] *Bogoliubov Laboratory of Theoretical Physics JINR Dubna, Joliot-Curie str. 6, 141980 Dubna, Russia*
[4] *National Research Nuclear University (MEPhI), Kashirskoe Shosse 31, 115409 Moscow, Russia*

Abstract. We discuss possible scenarios for the existence of strange matter in compact stars. The appearance of hyperons leads to a hyperon puzzle in ab-initio approaches based on effective baryon-baryon potentials but is not a severe problem in relativistic mean field models. In general, the puzzle can be resolved in a natural way if hadronic matter gets stiffened at supersaturation densities, an effect based on the quark Pauli quenching between hadrons. We explain the conflict between the necessity to implement dynamical chiral symmetry breaking into a model description and the conditions for the appearance of absolutely stable strange quark matter that require both, approximately masslessness of quarks and a mechanism of confinement. The role of strangeness in compact stars (hadronic or quark matter realizations) remains unsettled. It is not excluded that strangeness plays no role in compact stars at all. To answer the question whether the case of absolutely stable strange quark matter can be excluded on theoretical grounds requires an understanding of dense matter that we have not yet reached.

1 Introduction

Typically, neutron stars are thought of as exactly this: a gravitationally bound object with a mass of up to two times the mass of our sun that is made of extremely dense nuclear matter, mostly consisting of neutrons and some minor fraction of protons and electrons. The radius of a neutron star is small, only about 10 to 15 km and consequently the baryon density in a neutron star is huge; it exceeds the nuclear saturation density $n_S \approx 0.16$ fm^{-3} (where nucleons would start to 'touch'). By how much, is not precisely known and certainly depends on the specific object. However, most calculations agree that in the center of the most massive neutron stars, densities of four and more times the saturation density can be reached. This leads to fascinating consequences, among them the possibility that nucleons at these extreme densities might deconfine into a quark–gluon plasma. Whether this is the case or not has not yet been clarified. If confirmed, neutron stars would be the only objects in our Universe where such a transition can take place naturally. A further interesting consequence of the large densities in a neutron star is the appearance of particles more massive than neutrons and protons. In particular, the densities could be large enough to result in particle energies at the Fermi-surface which exceed the rest mass of hyperons. Compact stars could contain strange particles, either in the form of strange hadrons

⋆e-mail: thomas.klaehn@csulb.edu
⋆⋆e-mail: david.blaschke@ift.uni.wroc.pl

or deconfined strange quarks. The most exotic scenario is the existence of absolutely stable strange quark matter [1–3]. In the following we briefly sketch these ideas and review current developments as well as constraints that arise from pulsar measurements.

2 Neutron Star Masses and Radii

The more massive a neutron star is, the higher is the central density of this object. If one therefore assumes interesting things to happen at large densities one should look for the most massive observed neutron stars. Currently, these are two pulsars with a mass of approximately two solar masses which have been reported with sufficient precision. The probably cleanest result has been obtained from the Shapiro delay and orbital data of PSR J1614-2230 [4], a neutron star in a binary system with a white dwarf which results in a neutron star mass of 1.928 ± 0.017 $M_\odot$ [5]. An even larger, precise neutron star mass of 2.01 ± 0.04 $M_\odot$ has been obtained for PSR J0348 + 0432 [6], another neutron star in a binary system with a white dwarf, based on the analysis of orbital data and modeling the structure of the white dwarf. More data indicating massive (even heavier) neutron stars are available, however not with comparable precision. It should be noted that the radius and mass of these heavy neutron stars is almost independent of the symmetry energy and hence the fraction of particles with different isospin as neutrons and protons.

Considering the distance together with the small radii of neutron stars somewhere between 8 and 16 km, it seems evident that precise radius measurements are challenging. This is unfortunate, as the radius of a typical neutron with $1.3 - 1.5$ $M_\odot$ depends significantly on the symmetry energy. Consequently, precise measurements would be helpful to constrain the density behavior of this quantity. Part of the problem is that radius measurements depend on light curve analyses in various forms and hence are typically highly model dependent. The ongoing NICER experiment [7] and future missions as ATHENA+ or LOFT have been designed to gain more precise and conclusive results. Together with the recent first observation of gravitational waves in the binary neutron star merger GW170817 and the electromagnetic signals from the related kilonova event [8] there is hope that our knowledge concerning neutron star masses and radii improves significantly within the next few years.

3 Strange Matter

3.1 Hyperons

At densities of two to three times saturation density the hyperon threshold is expected to be crossed. It has been shown within Brueckner-Bethe-Goldstone calculations [9] that under conservative assumptions for the forces involving hyperons the appearance of these additional degrees of freedom softens the equation of state and hence lowers the maximum mass of a neutron star so that even the well-constrained binary radio pulsar masses of typically $\sim 1.4\ M_\odot$ cannot be reached. Without additional repulsive hyperon interactions this is exactly what all neutron star calculations based on nuclear hyperon model equations of state predict. This is called the "hyperon puzzle" [10]. In general, the equation of state beyond saturation density is not well constrained [11]. For hyperons and their interaction with nucleons and themselves the situation is even worse. However, from the observed existence of hyperons and massive neutron stars it seems evident that nucleon-hyperon and likely hyperon-hyperon repulsion [14] plays an important role (unless hyperons themselves play no decisive role for neutron star structure). Meanwhile, different approaches have been successfully applied which allow for stable neutron stars with up to two solar masses. Repulsion stiffens the hyperon matter equation of state sufficiently to account for massive neutron stars. At the same time the stiffening results in an onset of hyperon degrees of freedom at higher densities. Thus, the fraction of hyperons is reduced. In another scenario this idea is taken to the limit where the hyperon onset density is larger than the densities one would find in a neutron star. Generally, in relativistic mean-field models for hypernuclear matter [15] there is no severe hyperon puzzle, in particular due to the sufficiently repulsive effect of the ϕ−meson mean field [16]. Thos holds also for hypermatter stars with (color supercondicting) quark matter cores [17, 18]. A third solution to the hyperon problem is a transition to sufficiently stiff quark matter [19] which can happen before or after the transition to hyperon matter.

3.2 Quark Matter

Before discussing strange quark matter degrees of freedom we briefly describe the challenges in describing deconfined quark matter. Quarks, gluons and their interactions are synonymous for QCD, the gauge field theory of the strong interaction. At high densities and zero temperature (as in a neutron star) lattice QCD, the ab initio approach to QCD still fails. Although one can hope that perturbative results will constrain effective models at large densities the perturbative domain does not overlap with the densities one expects in a neutron star [20]. Moreover, the QCD phase transition from confined hadrons to deconfined quarks is characterized by features which are not accessible by perturbative approaches to QCD. A standard example for this statement is the description of chiral symmetry breaking which in the light quark sector is a distinctively non-perturbative feature. The most common approach to avoid these difficulties is to apply effective quark matter models which are not derived from QCD but account for certain characteristic features, most notably again the dynamical breaking of chiral symmetry. Prominent examples for effective models are the thermodynamic bag model as the high density limit of the MIT bag model, and effective relativistic mean field models, typically of the Nambu–Jona-Lasinio type [21]. A next approach which is well developed for the study of hadrons is the non-perturbative Dyson-Schwinger formalism which starts from the QCD action and derives gap equations to determine QCD's n-point Green-functions in a medium. As these typically couple to higher order Green-functions, truncation schemes are introduced which, if chosen wisely, preserve key features of QCD [22].

Quark matter at finite densities and zero or small temperature can exhibit an extremely rich phase structure due to different pairing mechanisms which arise from the coupling of color, flavor and spin degrees of freedom and result in a variety of different possible condensates. The importance of condensates is illustrated by the color-flavor locked phase which can appear in three flavor (up, down, strange) matter and is shown to be the asymptotic ground state of quark matter at low temperature [23]. As described for hyperon matter, quark matter would be too soft to account for massive neutron stars if repulsive interactions would not be taken into account. However, vector repulsion arises as naturally as the breaking of chiral symmetry in effective relativistic models and in the Dyson-Schwinger approach as dressing of the fermion propagator's vector and scalar part, respectively. The thermodynamic bag model does not account for vector interactions. Historically, one can argue that this is the heritage of the MIT bag model which has been developed to describe hadron properties in vacuum where terms that scale with the density are not relevant. However, taking perturbative corrections to the thermodynamic bag model into account, it can be made sufficiently stiff to account for massive neutron stars. It should be noted though, that the expansion parameter α_S cannot be considered small in this case.

Chiral symmetry breaking, the generation of an effective quark mass due to quark gluon interaction is a crucial effect as will be shown in more detail later. For now, it is worth to highlight that the restoration of chiral symmetry is flavor dependent. This can result in a sequential appearance of quark flavors at different densities [24, 25]. In particular if the three quark flavors are uncoupled, strange quarks, due to higher dressed masses, will appear at higher densities than light quarks. Even scenarios with a sequential appearance of first up, then down, then strange quarks have been discussed. With the appearance of condensates such a general statement is not feasible. Different studies suggest different scenarios. The transition densities for all flavors can coincide or differ widely. Examples of the first case are the thermodynamic bag model (because it neglects chiral symmetry breaking and thus reduces the onset density of all quark flavors) and a Dyson-Schwinger study [26] where flavor coupling results in a simultaneous appearance of all flavors. NJL type models in general predict a sequential transition. However, in all these scenarios the light quark threshold is a lower bound for the strange quark threshold.

3.3 Strange Quark Matter in Neutron Star Cores

We first discuss the situation where normal nuclear or hyperon matter deconfines at a critical density and forms a quark-gluon plasma. This problem is still far from being solved consistently in a way where hadrons would be taken into account as actually confined quarks which then deconfine dynamically. The standard way to circumvent a detailed description of deconfinement is to choose a two phase approach where nuclear and quark matter are modeled independently and the transition is constructed thermodynamically consistently, viz. in terms of a Maxwell (or similar) construction. In case of the Maxwell construction one simply determines at which baryon-chemical potential the pressure of the nuclear and quark phase are equal and thus defines the transition point. By construction, this implies a first order transition which 'switches' from a given nuclear equation of state to a softer quark matter equation of state. Therefore, this procedure requires a nuclear EoS which is stiff enough to support at least a two solar mass neutron star, and a quark matter EoS which is softer but stiff enough to do the same. How exactly this happens can vary. The quark matter EoS can mimic the nuclear EoS, be generally softer or at some density can turn even stiffer than the underlying nuclear EoS so that a second crossing of the pressure curves occurs ("reconfinement problem" [10]). The latter scenario is justified if one assumes that at densities far enough beyond the transition a comparison of both phases is meaningless as the nuclear EoS does not describe any physical reality anymore. In any of these scenarios, repulsion is a crucial feature to account for the existence of massive neutron stars and, as stated earlier, a natural property of relativistic models. For the onset density of hyperon matter we discussed how it is pushed to increasingly high densities with increasing stiffness or repulsion. The same would be true for quark matter if condensates are not taken into account. However, condensates couple colors and flavors and can lower the transition density. Therefore, quark models can account for transition densities far below the hyperon threshold.

For the appearance of strange matter in neutron stars, a couple of scenarios emerge which we divide into two groups. First, nuclear matter can deconfine directly into (three flavor) strange matter as one would find it for the thermodynamical bag model, as described earlier. Second, a sequential transition from nuclear to two flavor followed by a transition to three flavor matter takes place. Similar to this scenario, there are two more cases which would not result in a strange matter core but are not less realistic. The sequential transition results in neutron star configurations which are stable at all densities below the strange quark threshold but unstable beyond due to the softening of the EoS with this new degree of freedom. This situation has been described for an NJL model with diquark couplings [27]. However, choosing a finite constant as offset to the quark pressure can alter this result [17] and render neutron stars with strange matter core stable (the thermodynamic properties of matter are described in terms of pressure derivatives, hence a constant offset keeps them intact). Another way to stabilize strange core configurations is a strongly density dependent stiffening of the strange matter EoS following the transition [28, 29]. This can result in situations where stable two flavor and three flavor quark core neutron star configurations are separated by a population gap at intermediate central neutron star densities. In extreme scenarios, this can generate separated mass twin configurations, viz. two neutron star families with similar masses but very different radii [30, 31]. If members of both families could be observed, this would be a strong indicator for a first order QCD-like phase transition where (the smaller) neutron stars can carry a core made of strange matter [32].

Another indicator for a first order phase transition in dense matter would be the observation of a delayed second neutrino signal after a supernova. This has been suggested based on simulations which applied a bag model [33]. Although such a measurement would be very exciting it is not clear how it would address the question whether the transition involved strange matter, or quark matter at all.

3.4 Absolutely Stable Strange Matter?

The hypothesis that strange matter could be absolutely stable, bases on the observation that the appearance of strange quarks lowers the energy per baryon. As two flavor quark matter is evidently less stable than Fe (otherwise Fe would decay into it's quark components and so would we) this leaves a window where two-flavor matter is less and strange matter more stable than iron. This hypothesis has been supported by certain parametrizations of the thermodynamic bag model. Choosing the proper bag constant one can indeed find exactly the proposed situation. If this scenario is reality it would have a number of interesting consequences. Therefore, the search for stable strange matter inspired a multitude of experiments and has born many new ideas. Strange matter could form strange nuggets of extreme density with rather small atomic numbers and hence extremely low cross sections. It could form objects very similar to a neutron star, almost entirely made of strange matter, see [12, 13] for reviews of this and other scenarios involving strangeness in compact stars. A seed of strange matter in a neutron star could destabilize the surrounding matter and thus trigger a conversion of the neutron star interior into strange matter. Recently, it has been proposed that muonic bundles that have been observed at ALICE (CERN) are produced by strangelets [34]. Currently, other ways to detect strangelets are actively investigated [35]. The idea of absolutely stable strange matter is certainly appealing.

Theoretically, as mentioned before, the hypothesis has been based on the thermodynamic bag model. It should be noticed, that this model lacks a key feature of QCD, namely chiral symmetry breaking, viz. quarks in the thermodynamic bag model are assumed to have bare masses and therefore extremely small threshold densities (the critical chemical potential scales with the effective mass). NJL-type models, which do generate dressed quark masses, do not confirm that strange matter is absolutely stable. The reason for this is easily found: In the (low) density domain where the bag model predicts absolutely stable strange matter, NJL type models find chiral symmetry to be broken, hence significantly larger quark masses and consequently a higher energy per particle, too high to render strange matter absolutely stable, see Fig. 1. An appealing and sometimes confusing feature of the thermodynamic bag model is that it originates from the MIT bag model which has been developed to describe hadron properties. This evidently seems possible even though the model assumes bare quark masses and makes it easier to believe that massless quarks could form stable strangelets. This confusion can be cleared if one realizes that the MIT bag model does not only assume bare quark masses but a bag constant which is introduced as synonym for 'all we don't know about confinement and further interactions'. The fathers of the MIT bag model stated this explicitly. In [36] we illustrated how to translate a model with chiral symmetry breaking into a bag type model and where this model would break, which is when chiral symmetry is broken. For an earlier investigation of this kind, see [21] for the NJL model case and [37] for a nonlocal chiral quark model. Effects of color superconductivity in this context have been studied and parametrized in [38].

The bag constant mostly originates from chiral symmetry breaking. It is subtracted because the difference between the vacuum pressure of massive quarks and effectively massless quarks is negative. The absolute value is reduced by confinement. Thinking of the absolute value of the bag constant as the energy of a hadron, this makes sense as confinement reduces the energy of the system of chirally

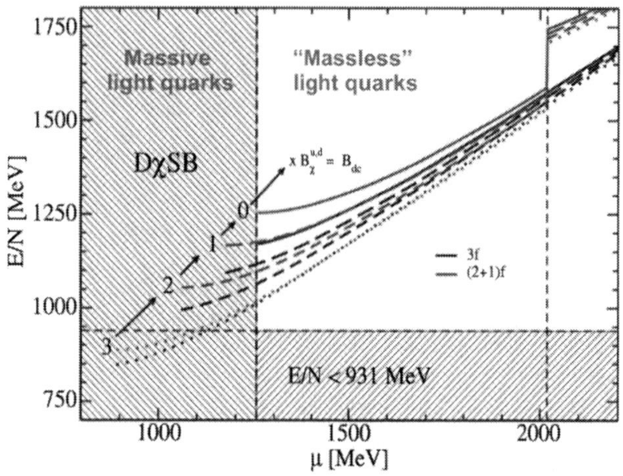

Figure 1. Energy per baryon vs. baryon chemical potential for the vBag quark matter model [36]. Absolutely stable strange quark matter could be obtained if dynamical chiral symmetry breaking (DχSB) is neglected. However: The vertical band marks the approximate region where chiral symmetry is broken. For the shown curves DχSB is ignored for all (black) or all but the s-quark (red). At densities below the s-quark threshold this allows to effectively compare two(red)- and three-flavor(black) quark matter for different effective bag constants. In none of the cases strange matter is more stable than iron (E/N<931 MeV) in a density domain where chiral symmetry is restored.

broken quarks by the binding energy. The MIT bag model is not consistent in the sense that it ignores any relation between quark mass and bag constant. In vacuum this is a reasonable approximation for two reasons: First, the model describes hadrons and does not attempt to predict any dynamical property of individual constituent quarks. Second, it can be fitted to observables and thus repairs the inherent shortcomings. The thermodynamic bag model addresses both effects, chiral symmetry breaking and confinement, in terms of one parameter - the bag constant. Statements regarding quark matter based on the thermodynamic bag model, in particular at low densities where the bag constant affects the total pressure significantly, should be considered with considerable caution as it is all but clear, that the model actually describes deconfined matter.

It should be noticed, that effective quark models with density dependent quark masses have been suggested which indeed would predict absolutely stable strange matter [39]. A distinct feature of these models is a steep *concave* decrease of the strange quark mass at comparably low density opposing to the typically *convex* behavior. This reduces the effective quark mass drastically already at low densities which makes it very similar to to the thermodynamic bag model, evidently with similar results regarding the stability of strange matter. It would be interesting to see, how a microscopic approach which generates this kind of density behavior would perform describing hadron properties.

4 Conclusions

We discussed the existence of strange matter in compact stars in hadronic and quark matter phases. The appearance of hyperons leads to a hyperon puzzle in approaches based on effective baryon-baryon potentials but is not a severe problem in relativistic mean field models. The puzzle is resolved for a stiffening of hadronic matter at supersaturation densities, an effect based on the quark Pauli quenching between hadrons. We further outlined the conflict between the necessity to implement dynamical chiral symmetry breaking for a realistic quark matter model and the condition of undressed, approximate massless quarks for the appearance of absolutely stable strange quark matter. The existence of absolutely stable strange quark matter cannot be excluded on theoretical grounds only. However, we outlined the problems of the reasoning that lead to this hypothesis. In general, the role of strangeness in compact stars in hadronic or quark matter realizations remains unsettled. The possibilities range from the existence of pure strange stars to scenarios where strange particles play no role in compact stars at all.

Acknowledgement

This work was supported by the Polish National Science Center (NCN) under grant No. UMO-2014/13/B/ST9/02621 (T.K.) and by the Russian Science Foundation under grant No. 17-12-01427 (D.B.).

References

[1] A. R. Bodmer, Phys. Rev. D **4** (1971) 1601.
[2] E. Witten, Phys. Rev. D **30** (1984) 272.
[3] E. Farhi and R. L. Jaffe, Phys. Rev. D **30** (1984) 2379.
[4] P. Demorest, T. Pennucci, S. Ransom, M. Roberts and J. Hessels, Nature **467** (2010) 1081.
[5] E. Fonseca *et al.*, Astrophys. J. **832** (2016) 167.
[6] J. Antoniadis *et al.*, Science **340** (2013) 6131.
[7] https://heasarc.gsfc.nasa.gov/docs/nicer
[8] B. P. Abbott *et al.* [LIGO Scientific and Virgo Collaborations], Phys. Rev. Lett. **119** (2017) no.16, 161101.
[9] M. Baldo, G. F. Burgio and H. J. Schulze, Phys. Rev. C **61** (2000) 055801.
[10] J. L. Zdunik and P. Haensel, Astron. Astrophys. **551** (2013) A61.
[11] M. Oertel, M. Hempel, T. Klähn and S. Typel, Rev. Mod. Phys. **89**, no. 1, 015007 (2017)
[12] I. Bombaci, Lect. Notes Phys. **578** (2001) 253.
[13] F. Weber, Prog. Part. Nucl. Phys. **54** (2005) 193
[14] T. A. Rijken and H. J. Schulze, Eur. Phys. J. A **52** (2016) no.2, 21.
[15] F. Hofmann, C. M. Keil and H. Lenske, Phys. Rev. C **64** (2001) 025804.
[16] S. Weissenborn, D. Chatterjee and J. Schaffner-Bielich, Phys. Rev. C **85** (2012) no.6, 065802 Erratum: [Phys. Rev. C **90** (2014) no.1, 019904].
[17] L. Bonanno and A. Sedrakian, Astron. Astrophys. **539** (2012) A16.
[18] R. Lastowiecki, D. Blaschke, H. Grigorian and S. Typel, Acta Phys. Polon. Supp. **5** (2012) 535.
[19] M. Baldo, G. F. Burgio and H.-J. Schulze, astro-ph/0312446.
[20] A. Kurkela, E. S. Fraga, J. Schaffner-Bielich and A. Vuorinen, Astrophys. J. **789** (2014) 127.
[21] M. Buballa, Phys. Rept. **407** (2005) 205.
[22] I. C. Cloet and C. D. Roberts, Prog. Part. Nucl. Phys. **77** (2014) 1.
[23] T. Schäfer and F. Wilczek, Phys. Rev. D **60** (1999) 114033.
[24] D. Blaschke, F. Sandin, T. Klahn and J. Berdermann, Phys. Rev. C **80** (2009) 065807.
[25] M. G. Alford and A. Sedrakian, Phys. Rev. Lett. **119** (2017) no.16, 161104.
[26] D. Nickel, J. Wambach and R. Alkofer, Phys. Rev. D **73** (2006) 114028.
[27] T. Klähn, D. Blaschke, F. Sandin, C. Fuchs, A. Faessler, H. Grigorian, G. Röpke and J. Trümper, Phys. Lett. B **654** (2007) 170.
[28] S. Benic, D. Blaschke, D. E. Alvarez-Castillo, T. Fischer and S. Typel, Astron. Astrophys. **577** (2015) A40.
[29] M. A. R. Kaltenborn, N. U. F. Bastian and D. B. Blaschke, Phys. Rev. D **96** (2017) no.5, 056024.
[30] D. Blaschke, D. E. Alvarez-Castillo and S. Benic, PoS CPOD **2013** (2013) 063 [arXiv:1310.3803 [nucl-th]].
[31] D. E. Alvarez-Castillo and D. B. Blaschke, Phys. Rev. C **96** (2017) no.4, 045809.
[32] N. U. F. Bastian, D. B. Blaschke, M. Cierniak, T. Fischer, M. A. R. Kaltenborn, M. Marczenko and S. Typel, arXiv:1710.09189 [astro-ph.HE].

[33] I. Sagert, T. Fischer, M. Hempel, G. Pagliara, J. Schaffner-Bielich, A. Mezzacappa, F.-K. Thielemann and M. Liebendörfer, Phys. Rev. Lett. **102** (2009) 081101
[34] P. Kankiewicz, M. Rybczynski, Z. Wlodarczyk and G. Wilk, arXiv:1710.08193 [hep-ph].
[35] J. P. VanDevender *et al.*, Scientific Reports 7, 8758 (2017).
[36] T. Klähn and T. Fischer, Astrophys. J. **810** (2015) no.2, 134.
[37] C. Gocke, D. Blaschke, A. Khalatyan and H. Grigorian, [hep-ph/0104183].
[38] H. Grigorian, D. Blaschke and D. N. Aguilera, Phys. Rev. C **69** (2004) 065802.
[39] N. A. Dondi, A. Drago and G. Pagliara, EPJ Web Conf. **137** (2017) 09004.

Dense and hot matter in compact stars and heavy-ion collisions

Stefan Schramm[1,⋆], *Veronica* Dexheimer[2], *Ayon* Mukherjee[1], and *Jan* Steinheimer[1]

[1] *FIAS, Ruth-Moufang-Str. 1, D-60438 Frankfurt am Main, Germany*
[2] *Department of Physics, Kent State University, Kent, OH 44242, USA*

Abstract. We discuss the effect of exotic particles in neutron star matter and the corresponding impact on gross properties of neutron stars within effective models for the strong interaction. Particularly, for the quark-hadron parity-doublet model, we show results for compact star properties and discuss the phase structure of the model and its possible relevance for heavy-ion collision phenomenology.

1 Introduction

Studying the phase structure of hot and dense strongly interacting matter is central to many experimental and theoretical efforts in modern nuclear physics. As a complement to a range of ultrarelativistic heavy-ion collision experiments targeting in particular hot matter, the environment of neutron stars samples high-density matter at very low temperature. In general, one cannot observe the dense matter directly, but the observations of stellar properties like mass, radius and cooling behaviour have to be interpreted using theoretical models in order to constrain the equation of state of stellar matter. In order to calculate strong interaction matter relevant for compact stars as well as heavy-ion physics we have developed and studied hadronic and hadron-quark models within the so-called chiral mean field (CMF) approach.

2 Modeling compact stars

A realistic hadronic model of dense matter as it exists in neutron stars has to contain, in addition to nucleons, hyperons as degrees of freedom. One example is the hadronic chiral mean field model [1], where baryons and mesons from the lowest SU(3) multiplets are coupled linearly:

$$\mathcal{L}_{\rm B} = -\sum_i \bar{\psi}_i[\gamma_0(g_{i\omega}\omega + g_{i\phi}\phi + g_{i\rho}\rho) + g_{i\sigma}\sigma + g_{i\delta}\delta + g_{i\zeta}\zeta + M_{0_i}]\psi_i \,. \qquad (1)$$

The baryon fields ψ_i interact with the mean fields of the vector mesons (the isoscalar ω, the isovector ρ, and the isoscalar field with hidden strangeness ϕ), as well as with the corresponding scalar fields σ, δ, and ζ. M_{0_i} is a small explicit mass term. The coupling strengths g_{BM} follow from a fitting of vacuum properties like baryon masses, nuclear matter properties and hypernuclear binding energies, as discussed in [1, 2].

⋆e-mail: schramm@fias.uni-frankfurt.de

However, independent of the interaction model, including hyperons (or other states) as additional degrees of freedom softens the equation of state and in turn reduces the maximum star mass. Therefore, there has to be a mechanism that suppresses (or removes) the amount of hyperons at densities relevant for compact star physics. One possibility is the vector repulsion of hyperons generated by the exchange of the ϕ meson [3–5]. This finding reinforces the point that when including hyperons as parts of the baryon octet, one should also take into account the corresponding SU(3) multiplets in the meson sector. As illustration of the effect of the ϕ exchange, Fig. 1 shows the maximum stellar mass that can be reached by solving the Tolman-Oppenheimer-Volkov (TOV) equations [6, 7] for a static star using the equation of state from the CMF model including hyperons. The curve illustrates the effect on the stellar mass when the value of the baryon-ϕ meson coupling is decreased down to 0. One can clearly observe the substantial reduction of the stellar mass if one neglects the effects of the ϕ meson.

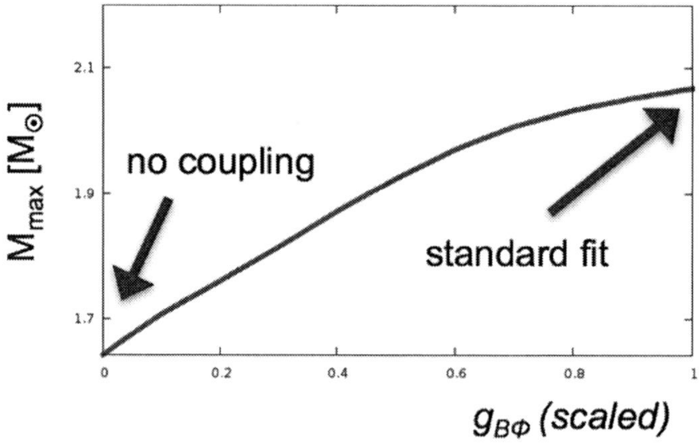

Figure 1. Maximum stellar mass as function of the normalized baryon-ϕ meson coupling strength calculated within the CMF model.

Another possibility to avoid the softening of the equation of state is the change of degrees of freedom through a transition to quark matter. However, in studies using a hadronic model combined with a simple MIT-bag quark model [8] or in combined quark-hadron approaches based on the CMF model [9], a big latent energy is found in the phase transition to quarks. This leads to a softening of the equation of state and, for very strong phase transitions, stars with a quark core are not stable static solutions of the TOV equations. This is a general behaviour for a strong phase transition that can significantly reduce maximum star masses [10]. This effect on the masses might not necessarily be too drastic, however, depending on the transition point. Fig. 2 shows the stellar mass in the purely hadronic CMF model as function of the stellar central density. Here, the maximum mass of about 2.1 solar masses is reached at a density $\rho_{max} \approx 6\rho_0$. The function is rather flat around this value. Consequently, in case exotic components like quarks or other particles appear at some lower density, creating a large latent energy and an instability of the compact star, the maximum mass would still not be reduced significantly, even if this density is, say, a third lower than ρ_{max} due to the flatness of the curve, this is a rather model-independent feature.

Alternatively, a combined quark-hadron approach based on the CMF model in the so-called parity-doublet formulation has been developed (QχP model) [11, 12]. The main feature of such a model is that the octet of baryons us combined with their opposite parity partners [13–16], while quark degrees

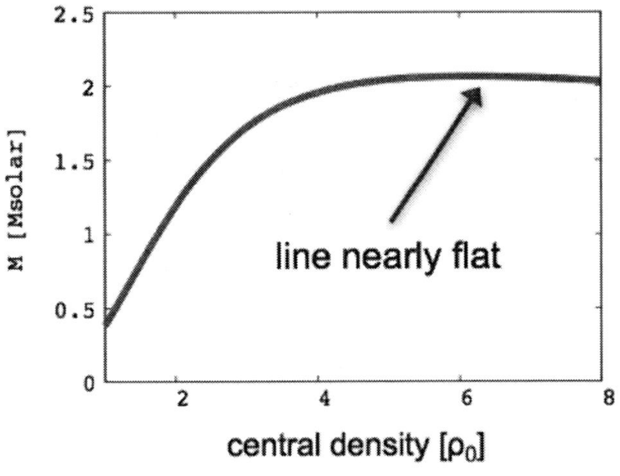

Figure 2. Stellar mass of the hadronic star, calculated within the CMF model, as function of the central density in units of nuclear saturation density.

of freedom that couple linearly to the mean fields as in Eq. 1 are introduced. In the case of the nucleon, this state might be the N(1535) resonance. For the hyperons the situation is not completely clear, here, for simplicity, we assume equal vacuum mass splitting of the various parity partners [13]. Note that related model studies for isospin-symmetric matter have been performed in a 2-flavor quark-meson-nucleon approach [17, 18].

In the parity-doublet approach the baryonic effective masses read

$$m^*_{i\pm} = \sqrt{\left[(g^{(1)}_{\sigma i}\sigma + g^{(1)}_{\zeta i}\zeta)^2 + (m_0 + n_s m_s)^2\right]} \pm g^{(2)}_{\sigma i}\sigma \pm g^{(2)}_{\zeta i}\zeta \,. \tag{2}$$

There are two possible coupling terms of the scalar fields and baryons $g^{(1)}, g^{(2)}$, where the second one generates the splitting of the parity partners. Here, the term $n_s m_s$ is an explicit flavor-breaking term, where n_s is the number of strange quarks in the baryon, and m_0 is a chirally invariant mass term. Note, that the mass of the doublets (in the equation denoted by the ± sign) is degenerate for vanishing scalar fields, similar to the degeneracy of the pion and sigma meson in the chirally restored phase of the linear sigma model. Using the equation of state resulting from this model, one obtains compact star masses and radii as shown in Fig. 3. The maximum mass is about two solar masses in agreement with accurately measured heavy pulsars [19, 20]. As a quite unique feature of this approach, the calculated stars contain an extensive mixed phase of baryons and quarks [12], where the amount of quark matter for specific stars is indicated in the figure. Here, in contrast to the previous discussion of hybrid stars (neutron stars with a quark core), the mixed phase of quarks with baryons that have a strong repulsive interaction is stiff enough to support stable heavy compact stars. Note the rather small values for the stellar radius in agreement with various studies of low-mass X-ray binary systems [21, 22].

3 Hot and dense matter

In calculating the phase structure of dense and hot matter within the QχP model, one obtains the phase diagram shown in Fig. 4. The bands show the transition region of the scalar condensate and

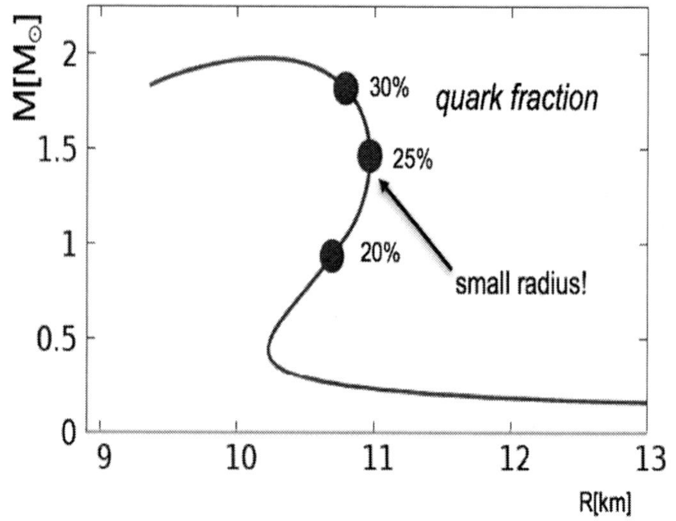

Figure 3. Mass-radius diagram of hybrid stars in the QχP model. The circles indicate different amounts of quark matter in the star.

the Polyakov loop field Φ that effectively describes the effect of the confinement as discussed in Refs. [9, 11, 23]. The full lines indicate first-order phase transitions. There are two first-order transitions: the one at lower chemical potential is the liquid-gas transition, whereas the second one corresponds to the deconfinement transition with a low critical end point temperature of about 20 MeV. While the first-order phase transitions vanish fast with rising temperature, these transitions, nevertheless, can have significant impact on heavy-ion physics phenomenology. This can be seen by calculating the baryon number susceptibilities of order n, χ_n^B, defined as

$$\frac{\chi_{\mathrm{n}}^{\mathrm{B}}}{T^{4-n}} = \frac{\partial^{\mathrm{n}}(P(T,\mu_{\mathrm{B}})/T^4)}{\partial(\mu_{\mathrm{B}}/T)^{\mathrm{n}}} \,. \tag{3}$$

with P denoting the pressure of the system. The susceptibilities are closely connected to particle number fluctuations that can be measured in heavy-ion experiments. Fig. 5 shows the ratio of the fourth- to second-order susceptibility. The black dashed lines indicate the continuation of the first-order transition as crossover transitions. In addition, freeze-out curves [24] for different temperatures are indicated in the plot. As a very interesting feature of this figure, one can observe a strong effect, when the deconfinement and liquid-gas transition merge at high temperatures and low baryochemical potential. Thus, experimental signals are expected to be strongly affected by the interference of those two transitions (see Ref. [25]). Due to the treatment of the matter as homogeneous in this mean-field approach, nucleonic clustering effects have not been considered here. Therefore, an extension of this work along the lines considered in Refs. [26, 27], including clusters as effective degrees of freedom, should be pursued in future work.

In future studies, this approach will be applied to dynamical collision studies by implementing the equation of state in simulation codes for ultrarelativistic heavy-ion collisions. Work along this line is already in progress.

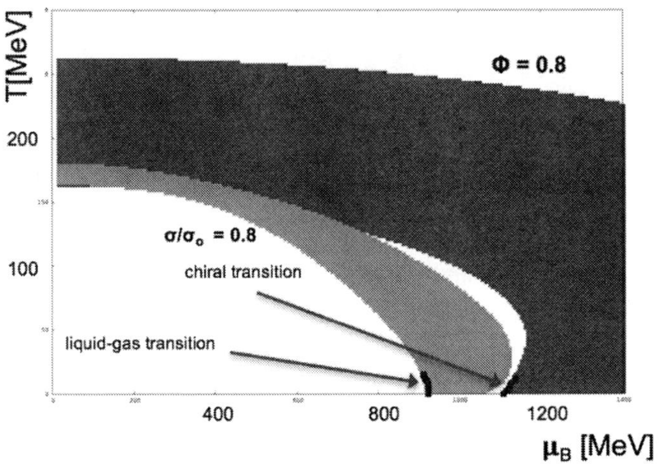

Figure 4. Phase structure of the QχP model in the $\mu_B - T$ plane for isospin symmetric matter with zero net strangeness. The filled areas indicate the range of the scalar condensate with 20% to 80% of the vacuum value, and the Polyakov loop field with values between 0.2 and 0.8 .

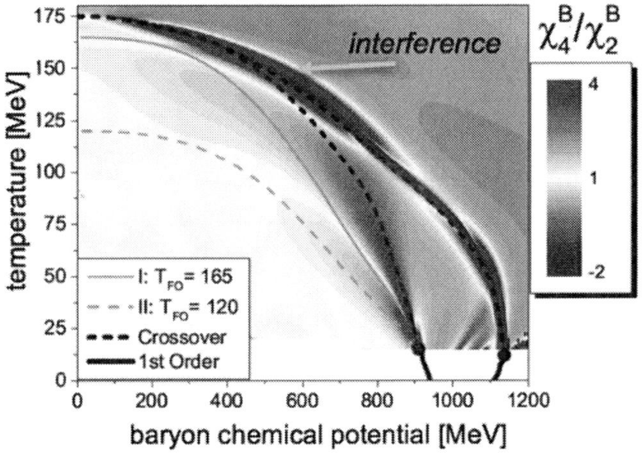

Figure 5. Ratio of baryon number susceptibilities χ_4^B/χ_2^B for heavy-ion matter as function of chemical potential and temperature. Freeze-out curves for different parameters are indicated.

Acknowledgements

The authors acknowledge support from NewCompStar COST Action MP1304, HIC for FAIR, and from BMBF. Computational resources were provided by the Frankfurt Center for Scientific Computing (CSC). The authors thank the referee for pointing out the potential relevance of nuclear clusters in the interference of the phase transitions.

References

[1] P. Papazoglou, D. Zschiesche, S. Schramm, J. Schaffner-Bielich, H. Stoecker, W. Greiner, Phys. Rev. **C59**, 411 (1999), nucl-th/9806087
[2] V. Dexheimer, R. Negreiros, S. Schramm, Phys. Rev. **C91**, 055808 (2015), 1411.4623
[3] V. Dexheimer, S. Schramm, Astrophys. J. **683**, 943 (2008), 0802.1999
[4] I. Bednarek, R. Manka, J. Phys. **G31**, 1009 (2005), hep-ph/0506059
[5] J. Schaffner-Bielich, A. Gal, Phys. Rev. **C62**, 034311 (2000), nucl-th/0005060
[6] J.R. Oppenheimer, G.M. Volkoff, Phys. Rev. **55**, 374 (1939)
[7] R.C. Tolman, Phys. Rev. **55**, 364 (1939)
[8] R. Negreiros, V.A. Dexheimer, S. Schramm, Phys. Rev. **C85**, 035805 (2012), 1011.2233
[9] V.A. Dexheimer, S. Schramm, Phys. Rev. **C81**, 045201 (2010), 0901.1748
[10] M.G. Alford, S. Han, Eur. Phys. J. **A52**, 62 (2016), 1508.01261
[11] A. Mukherjee, J. Steinheimer, S. Schramm, Phys. Rev. **C96**, 025205 (2017), 1611.10144
[12] A. Mukherjee, S. Schramm, J. Steinheimer, V. Dexheimer (2017), 1706.09191
[13] J. Steinheimer, S. Schramm, H. Stocker, Phys. Rev. **C84**, 045208 (2011), 1108.2596
[14] Y. Nemoto, D. Jido, M. Oka, A. Hosaka, Phys. Rev. **D57**, 4124 (1998), hep-ph/9710445
[15] C. DeTar, T. Kunihiro, Phys. Rev. D **39**, 2805 (1989)
[16] T. Hatsuda, M. Prakash, Phys. Lett. **B224**, 11 (1989)
[17] M. Marczenko, C. Sasaki, ArXiv e-prints (2017), 1711.05521
[18] S. Benić, I. Mishustin, C. Sasaki, Phys. Rev. D **91**, 125034 (2015)
[19] P. Demorest, T. Pennucci, S. Ransom, M. Roberts, J. Hessels, Nature **467**, 1081 (2010), 1010.5788
[20] J. Antoniadis et al., Science **340**, 6131 (2013), 1304.6875
[21] S. Guillot, R.E. Rutledge, E.F. Brown, Astrophys. J. **732**, 88 (2011), 1007.2415
[22] S. Guillot, M. Servillat, N.A. Webb, R.E. Rutledge, Astrophys. J. **772**, 7 (2013), 1302.0023
[23] J. Steinheimer, S. Schramm, H. Stocker, J. Phys. **G38**, 035001 (2011), 1009.5239
[24] A. Andronic, P. Braun-Munzinger, J. Stachel, Phys. Lett. **B673**, 142 (2009), [Erratum: Phys. Lett.B678,516(2009)], 0812.1186
[25] A. Andronic, P. Braun-Munzinger, K. Redlich, J. Stachel, ArXiv e-prints (2017), 1710.09425
[26] S. Typel, H.H. Wolter, G. Röpke, D. Blaschke, The European Physical Journal A **50**, 17 (2014)
[27] N.U. Bastian, P. Batyuk, D. Blaschke, P. Danielewicz, Y.B. Ivanov, I. Karpenko, G. Röpke, O. Rogachevsky, H.H. Wolter, The European Physical Journal A **52**, 244 (2016)

Resonance Production in Heavy-Ion Collisions

Anders G. Knospe[1,*]

[1] *The University of Houston, Houston, TX, USA*

Abstract. Hadronic resonances are unique probes that allow the properties of heavy-ion collisions to be studied. Topics that can be studied include modification of spectral shapes, in-medium energy loss of parsons, vector-meson spin alignment, hydrodynamic flow, recombination, strangeness production, and the properties of the hadronic phase. Measurements of resonances in p+p, p+A, and d+A collisions serve as baselines for heavy-ion studies and also permit searches for possible collective effects in these smaller systems. These proceedings present a selection of results related to these topics from experiments at RHIC, LHC, and other facilities, as well as comparisons to theoretical models.

1 Introduction

Measurements of hadronic resonances allow the properties of ion-ion collisions to be studied at various stages in their evolution for both small and large collision systems. In these proceedings I will discuss a selection of such measurements. A particular focus will be the ratios of resonance yields to those of stable particles, which can be used to study the properties of the hadronic phase and for which a large set of measurements of many different resonances now exists. I will also discuss the ways in which resonances can contribute to studies of strangeness production and the various mechanisms that determine the shapes of particle p_{T} spectra. These proceedings will also touch on the nuclear modification factors, spin alignment, and elliptic flow of resonances and the possibility of modifications to their spectral shapes. In the discussion that follows, masses will usually be omitted from particle symbols: $\rho(770)^0$, $K^*(892)$, $\phi(1020)$, $\Delta(1232)^{++}$, $\Sigma(1385)^{\pm}$, and $\Xi(1530)^0$ will be denoted as ρ^0, K^*, ϕ, Δ^{++}, $\Sigma^{*\pm}$, and Ξ^{*0}, respectively. Except where explicitly stated, the results include both particles and antiparticles. Table 1 summarizes the properties [1] of the resonances discussed. Note the variety of masses, lifetimes, and quark contents.

2 Hadronic Phase

While the yields of long-lived particles are expected to be fixed at chemical freeze out, the yields of resonances may continue to be modified by scattering processes during the hadronic phase until kinetic freeze out [2–4]. Resonance yields may be regenerated through the pseudo-elastic scattering of hadrons (*e.g.*, $\pi\pi \rightarrow \rho \rightarrow \pi\pi$). When a resonance decays during the hadronic phase, its decay

*e-mail: anders.knospe@cern.ch

Table 1. Properties of the resonances discussed [1]. The results discussed in these proceedings were obtained by reconstructing the decays of the resonances in the listed channels.

Particle	Quark Content	Mass (MeV/c^2)	Width (MeV/c^2)	Lifetime (fm/c)	Decay Mode	Branching Ratio (%)
ρ^0	$(u\bar{u}+d\bar{d})/\sqrt{2}$	770	150	1.3	$\pi^-\pi^+$	100
K^{*+}	$u\bar{s}$	892	50.3	3.92	$\pi^+K^0_S$	33.3
K^{*0}	$d\bar{s}$	896	47.3	4.17	π^-K^+	66.6
ϕ	$s\bar{s}$	1019	4.247	46.46	K^-K^+	48.9
Δ^{++}	uuu	1232	117	1.69	π^+p	99.4
Σ^{*+}	uus	1383	36.0	5.48	$\pi^+\Lambda$	87
Σ^{*-}	dds	1387	39.4	5.01	$\pi^-\Lambda$	87
$\Lambda(1520)$	uds	1520	15.6	12.6	K^-p	22.5
Ξ^{*0}	uss	1532	9.1	22	$\pi^+\Xi^-$	66.7

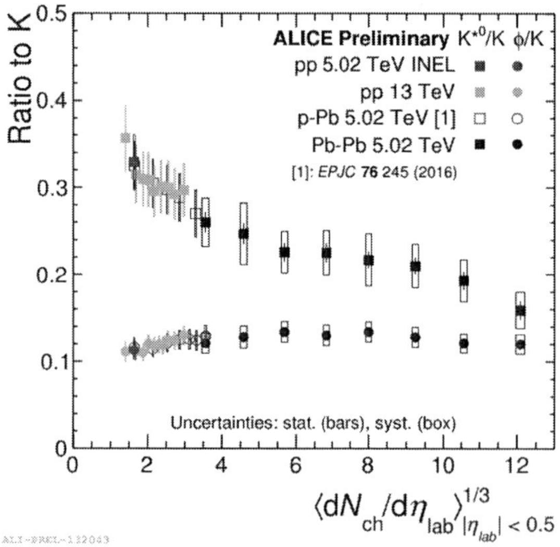

Figure 1. Ratios K^{*0}/K and ϕ/K as a function of the mean charged particle multiplicity at mid-rapidity in p+p, p+Pb, and Pb+Pb collisions at $\sqrt{s_{NN}}$ = 5.02 TeV and in p+p collisions at $\sqrt{s}$ = 13 TeV [7–10]. The yields in the numerator and denominator are the average (not the sum) of particles and antiparticles.

products may under elastic collisions (which can hinder reconstruction of the resonance) or pseudo-elastic scattering through a different resonance state (which removes information about the original resonance); in these cases, the measurable yield of the original resonance is reduced. The final (measurable) resonance yields at kinetic freeze-out will depend on the interplay between regeneration and re-scattering, which is influenced by the temperature and lifetime of the hadronic phase, the resonance lifetime, and the scattering cross-sections of its decay products. Measurements of the ratios of resonance yields to those of long-lived particles can be used, along with models of the hadronic phase, to estimate the temperature and/or lifetime of the hadronic phase [3, 5, 6]. Experiments at RHIC and the LHC have measured many of the resonances listed in Table 1 in various collision systems. The K^{*0} and ϕ mesons have been extensively studied, but measurements of the complete set have only become available recently.

Figure 1 shows the K^{*0}/K and ϕ/K ratios measured by the ALICE Collaboration in p+p collisions at $\sqrt{s}$ = 5.02 and 13 TeV and p+Pb and Pb+Pb collisions at $\sqrt{s_{NN}}$ = 5.02 TeV as functions of the mean charged-particle multiplicity at mid-rapidity [7–10]. In Pb+Pb collisions, the K^{*0}/K ratio decreases with increasing collision centrality, while the ϕ/K ratio does not exhibit any strong evolution with centrality. Similar trends are also observed in Pb+Pb collisions at $\sqrt{s_{NN}}$ = 2.76 TeV and in Au+Au and Cu+Cu collisions at RHIC energies (plot and references in [11]). This behavior may result from the re-scattering of the decay products of the K^{*0} in the hadronic phase, which reduces the measurable yields of that resonance. In contrast, the ϕ lives about 10 times longer and should decay primarily after kinetic freeze-out, meaning that its decay daughters should not be affected by re-scattering. This behavior is qualitatively reproduced by the EPOS model [12], which incorporates UrQMD to describe the hadronic phase; when EPOS is run with UrQMD turned off, the K^{*0} suppression is not as well described. The PHSD model [13] can also describe the suppression qualitatively. For Pb+Pb collisions at $\sqrt{s_{NN}}$ = 2.76 TeV, the central ϕ/K ratio is described by thermal-model calculations [14], while the K^{*0}/K ratio is below the thermal-model values. The measured K^{*0}/K ratio can be used along with models to obtain estimates of the lower limit of the lifetime of the hadronic phase [3, 5, 6]. The same trends are also visible in p+p and Pb+Pb collisions and the ratios at a given multiplicity appear to be independent of the collision system. This raises the possibility that re-scattering in the hadronic phase may also be responsible for the suppression of the K^{*0} yield in high-multiplicity p+p and p+Pb collisions.

Similar studies can be performed for other resonances. The ρ^0/π ratio decreases with centrality in Pb+Pb collisions at $\sqrt{s_{NN}}$ = 2.76 TeV [15] and is below the thermal model value in central Pb+Pb collisions. In contrast, there is no visible suppression of the ρ^0/π ratio in collisions at $\sqrt{s_{NN}}$ = 200 GeV [16], though at this energy the STAR Collaboration only measured p+p, d+Au, and peripheral Au+Au collisions. (The STAR values are also inconsistent with those measured by ALICE at higher energy.) The STAR Collaboration also measured the Δ^{++}/p ratio in p+p and d+Au collisions at $\sqrt{s_{NN}}$ = 200 GeV and observed no visible suppression trend [16]. Measurements of the $\Sigma^{*\pm}/\Lambda$ ratio in various collision systems at RHIC [16] and LHC [17] energies do not indicate any significant dependence on energy or system size; the values in central A+A collisions are consistent with thermal-model values. The ALICE Collaboration has measured the $\Lambda(1520)/\Lambda$ ratio in p+p, p+Pb, and Pb+Pb collisions [8, 9, 18]. The values of this ratio are independent of multiplicity in Pb+Pb collisions, while a decreasing trend with centrality is observed in Pb+Pb collisions. The measured values in central Pb+Pb collisions are below the thermal-model values. The ALICE measurements, including the suppression trend, are consistent with earlier measurements from STAR [16]. The situation is complicated for the Ξ^{*0}/Ξ^- ratio. No significant centrality dependence in this ratio is observed in Pb+Pb collisions at $\sqrt{s_{NN}}$ = 2.76 TeV [8, 9, 18], but the mid-central and central Pb+Pb measurements are lower than the thermal-model values and the values measured in p+p and p+Pb collisions [17]. No significant multiplicity dependence is observed for p+p and p+Pb collisions. Taken together, these results suggest a possible weak suppression of the Ξ^{*0}/Ξ^- ratio in (mid-)central Pb+Pb collisions. When the hadronic phase is described using UrQMD, EPOS [12] is able to qualitatively describe the centrality-dependent trends observed for by ALICE for Pb+Pb collisions at $\sqrt{s_{NN}}$ = 2.76 TeV and p+Pb collisions at $\sqrt{s_{NN}}$ = 5.02 TeV (see Figure 2 for the EPOS values).

In summary, the ρ^0/π, K^{*0}/K, and $\Lambda(1520)/\Lambda$ ratios are suppressed in central A+A collisions, while the ϕ/K and $\Sigma^{*\pm}/\Lambda$ ratios are not. The Ξ^{*0}/Ξ^- ratio may exhibit a weak suppression in A+A collisions, while the Δ^{++}/p ratio in not suppressed in d+Au collisions. It should be noted that the suppression of these resonances is not just dependent on their lifetimes. The $\Sigma^{*\pm}$ is not suppressed even though its lifetime is similar to that of the K^{*0}; EPOS predicts no suppression for the the Δ^{++} in Pb+Pb collisions, even though its lifetime is similar to that of the ρ^0. It is therefore important to remember that the scattering cross-sections of resonance decay products and competition between regeneration and re-scattering can also be important in determining the final resonance yields.

 https://doi.org/10.1051/epjconf/201817109001

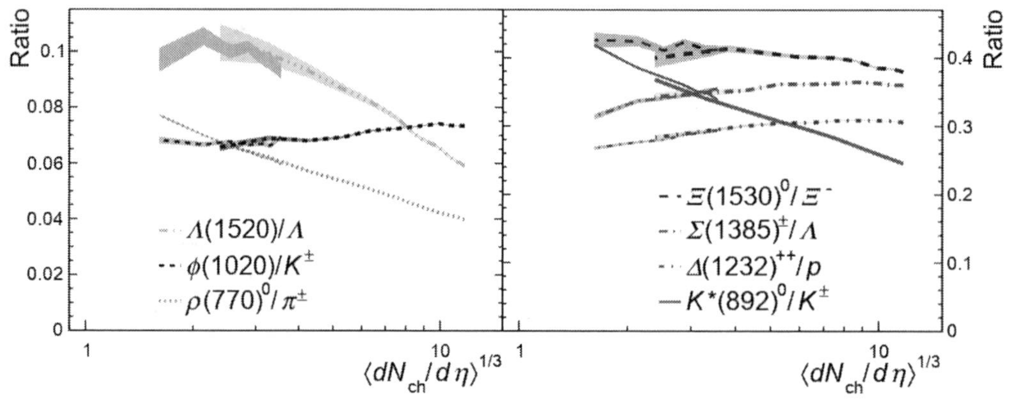

Figure 2. Ratios of resonances to long-lived particles given by the EPOS model (version 3.107) for Pb+Pb [12] and p+Pb collisions. Shaded bands represent statistical uncertainties. The values shown here qualitatively describe the multiplicity evolution of the ratios measured by the STAR and ALICE collaborations, with the possible exception of the Ξ^{*0}/Ξ^- ratio. The yields in the numerator and denominator are always the sum (not the average) of particles and antiparticles.

3 Strangeness Production

The production of strangeness, including its possible enhancement in central A+A collisions and/or canonical suppression in (low-multiplicity) p+p collisions, is usually studied using common long-lived hadrons; see [19] for a discussion. Resonances may also contribute to such studies due to the fact that have different masses than their long-lived counterparts and that some, like the ϕ, have unique quark content.

The Λ/π, Ξ/π, and Ω/π ratios have been observed to increase as functions of multiplicity in p+p, p+Pb, and A+A collisions [20, 21]. The magnitude of the enhancement of these three ratios increases with the strangeness content of the numerator particle, while the p/π ratio is essentially constant in p+Pb collisions. The fact that the $\Sigma^{*\pm}/\Lambda$ and Ξ^{*0}/Ξ^- ratios are constant with multiplicity in p+Pb collisions [17] indicates that the $\Sigma^{*\pm}$ and Ξ^{*0} resonances (which are more massive than the Ξ) are enhanced by the same amount as their ground-state counterparts, suggesting that it is indeed the strangeness, and not the mass, that controls the enhancement.

The ϕ meson has a unique quark content: it is composed only of strange valence quarks, but has no net strangeness. It is therefore expected that while particles with open strangeness should be affected by the canonical suppression of strangeness production in low-multiplicity collisions, the ϕ should be unaffected [22]. In p+p and p+Pb collisions, the ϕ/K ratio has been observed to be fairly constant, with a possible weak increase with multiplicity [8, 9]. Similarly, the Ξ/ϕ ratio exhibits a possible weak increase with multiplicity. In the context of these measurements, the ϕ therefore seems to behave like a particle with 1 to 2 units of strangeness. Why the evolution of ϕ yields seems to follow the evolution of canonically suppressed particles is an open question which requires more input from the theoretical community.

The ϕ/K ratio has been measured in A+A collisions by multiple experiments from $\sim$ 2 GeV to 13 Tev [11, 23]. The ratio is essentially constant over most of that range, but an increase is seen for $\sqrt{s_{NN}}$ < 4 GeV. This increase can be explained by statistical models with a strangeness correlation radius of $R_C \approx 2.2$ fm: strangeness in required to be conserved within this radius, which would suppress kaon yields but not affect the ϕ yields. It can also be noted that the ϕ/K^- ratio approaches

0.5 at these low energies; this indicates a sizable feed-down contribution to the K^- yield and could explain the different slopes observed in the K^- and K^+ p_T spectra.

4 Shapes of p_T Spectra

Resonances, the ϕ in particular, can also be useful for studies of the mechanisms that determine the shapes of hadron p_T spectra. When compared to measurements in p+p collisoions, baryon-to-meson ratios, such as p/π and Λ/K^0_S, exhibit a centrality-depended enhancement for $1.5 \lesssim p_T \lesssim 5$ GeV/c in Pb+Pb collisions [24, 25]. Explanations for this enhancement include hydrodynamic behavior and hadronization through recombination. The hydrodynamic interpretation [26–28] would attribute the enhancement to the differing masses of the numerator and denominator particles, while the recombination interpretation [29, 30] would attribute it to the differing number of valence quarks (which would lead to differently shaped p_T spectra for baryons and mesons). In the p/ϕ ratio, the two particles have very similar masses, permitting the mass and baryon-number effects to be separated. The p/ϕ ratio is observed to be constant in central Pb+Pb collisions for $p_T < 4$ GeV/c [11], which is consistent with the simple hydrodynamic picture (although some recombination models can also describe a near-constant behavior [31]).

The mean transverse momentum $\langle p_T \rangle$ can also be used to study how that shapes of hadron p_T spectra depend on particle properties, including quark content and mass. The similarly shaped proton and ϕ p_T spectra in central A+A collisions result in similar $\langle p_T \rangle$ values for these particles [8, 9, 11] (see Figure 3a), consistent with a hydrodynamic interpretation. In central A+A collisions, hadron $\langle p_T \rangle$ values are ordered with particle mass. However, this is not the case for smaller collision systems, in which the K^{*0} and ϕ take on larger $\langle p_T \rangle$ values than the long-lived baryons with similar masses [7]. The ϕ-meson $\langle p_T \rangle$ values even approach those for the Ξ (which has a 30% greater mass) in p+p and low-multiplicity p+Pb collisions. It is not entirely clear whether this violation of mass ordering is due to differences between baryons and mesons or due to the fact that the ϕ and K^{*0} are resonances. For all light-flavor hadron species (including resonances), the $\langle p_T \rangle$ values in p+Pb collisions follow different trends with multiplicity than the $\langle p_T \rangle$ values observed in Pb+Pb collisions. The p+Pb values increase faster and in the highest multiplicity collisions can reach (or exceed) the $\langle p_T \rangle$ values seen in central Pb+Pb collisions.

5 Other Topics

The nuclear modifications factor of the ρ^0, K^{*0}, and ϕ mesons have been measured in Pb+Pb collisions at $\sqrt{s_{NN}} = 2.76$ TeV [15, 33]. For $p_T \gtrsim 8$ GeV/c, all light-flavor hadron species, including these three resonances, are suppressed by a factor of 4 to 5. Indeed, D mesons are also suppressed by the same factor, indicating that the suppression is independent of hadron properties, including mass, baryon number, and quark content (at least for hadrons containing only u, d, s, and c quarks). At intermediate p_T ($2 \lesssim p_T \lesssim 8$ GeV/c), there appears to be a separation between the R_{AA} values of baryons and mesons, with the mesons (even heavy ones like the D) suppressed more than baryons. There also appears to be a mass ordering among the mesons, with the amount of suppression decreasing with increasing mass; the three resonances mentioned are consistent with this trend.

Non-central A+A collisions will have a large angular momentum and a large magnetic field. One consequence of this could be an observable spin alignment for vector mesons in heavy-ion collisions. The degree of spin alignment is quantified by the matrix element ρ_{00}, which has an expected value of $\frac{1}{3}$ in the absence of any net spin alignment. The STAR Collaboration measured [34] ρ_{00} for ϕ mesons in p+p and Au+Au collisions at $\sqrt{s_{NN}} = 200$ GeV and observed no significant differences between the

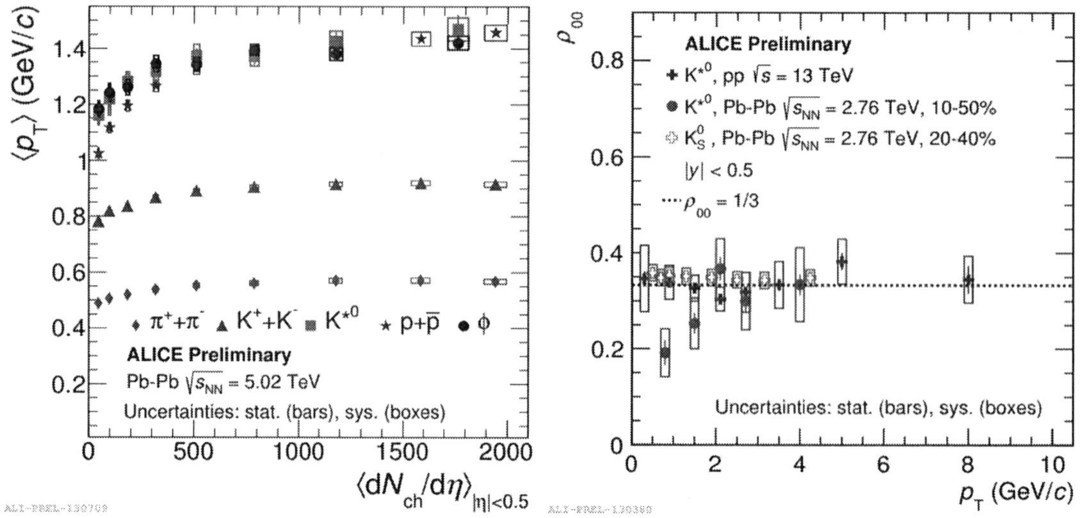

Figure 3. Left: $\langle p_T \rangle$ values for $\pi^{\pm}$, $K^{\pm}$, (anti)protons, K^{*0}, and ϕ in Pb+Pb collisions at $\sqrt{s_{NN}}$ = 5.02 TeV for different centrality classes [8, 9]. **Right:** Spin alignment of the K^{*0} and K^0_S measured by ALICE. [32]

values for the two collision systems. STAR also observed no significant deviation from $\frac{1}{3}$ for the ρ_{00} values for K^{*0} in Au+Au collisions. The ALICE Collaboration has recently measured [32] ρ_{00} for K^{*0} in $p+p$ collisions at $\sqrt{s}$ = 13 TeV and Pb+Pb collisions at $\sqrt{s_{NN}}$ = 2.76 TeV (see Figure 3b). While ALICE observes no significant deviation from $\frac{1}{3}$ for $p+p$ collisions, there are hints of a deviation below $\frac{1}{3}$ (at the 2.5-σ level) at low p_T in non-central (10-50%) Pb+Pb collisions. For comparison, ALICE also measured ρ_{00} for K^0_S (a pseudoscalar meson) in non-central Pb+Pb collisions at the same energy and observed no deviation from $\frac{1}{3}$.

Experiments have also searched for evidence of modifications to the spectral shapes of resonances. Measured excesses of dileptons at low invariant mass in $A+A$ collisions may be due to melting of the ρ^0 meson. The STAR Collaboration has measured [16] a p_T-dependent negative mass shift for ρ^0, K^*, Δ^{++}, and $\Sigma^{*\pm}$ in $p+p$ and d+Au collisions (with a hint of dependence on the collision multiplicity), which may be due to interference, re-scattering, and/or Bose-Einstein correlations involving resonance decay products and other hadrons in the system. The ALICE Collaboration may also observe a hint of a negative mass shift for the K^{*0} meson in Pb+Pb collisions [11] (although the size of the uncertainties on the measurement prevent strong statements from being made). A recent paper on the PHSD model [13] explains this shift in terms of medium modification of the K^{*0} spectral function and re-scattering and absorption in the hadronic phase.

Resonances can also be used in studies of elliptic flow. STAR has observed a violation of v_2 mass ordering for protons and ϕ mesons, with $v_2(\phi)$ much greater than $v_2(p)$ at low p_T ($p_T \lesssim 0.7$ GeV/c) in Au+Au collisions at $\sqrt{s_{NN}}$ = 200 GeV [35]. This may be due to hadronic re-scattering affecting the proton for low p_T. PHENIX [36] and ALICE [37] have measured v_2 of the ϕ meson as functions of KE_T and p_T, respectively. ALICE observes that for $p_T < 2$ GeV/c, the v_2 values of the ϕ are the same as those of the proton, consistent with the mass ordering expected from hydrodynamic behavior. It is observed that for $KE_T \gtrsim 0.8$ GeV/c and $p_T \gtrsim 2$ GeV/c, the v_2 values of the ϕ are consistent with those of other mesons, rather than with the proton. Therefore, there seems to be a transition in the behavior of v_2 values: with mass ordering observed at low p_T and baryon-meson splitting seen at intermediate p_T.

6 Conclusions

Resonances are useful in a wide variety of studies in heavy-ion physics. As a function of collision centrality or multiplicity, the yields of ρ^0, K^{*0}, and $\Lambda(1520)$ are observed to be suppressed; the yields of ϕ, Δ^{++}, and $\Sigma^{*\pm}$ are not suppressed; and the Ξ^{*0} may be weakly suppressed. These trends may be explained as consequences of re-scattering of resonance decay products in the hadronic phase and may be used in future studies to learn more about the properties of the hadronic phase. The observed suppression trends (except possibly that of the Ξ^{*0}) are consistent with values obtained from the EPOS model with UrQMD. The $\Sigma^{*\pm}/\Lambda$ and Ξ^{*0}/Ξ^- ratios are independent of multiplicity in p+Pb collisions, suggesting that the observed increases in strange baryon yields is due to the baryons' strangeness content and not their masses. In small systems, the yields of the ϕ meson appear to increase with multiplicity similar to particles with open strangeness, even though the ϕ should not be subject to canonical suppression. The proton and the ϕ, which have similar masses, have similarly shaped p_T spectra in central A+A collisions (possibly due to hydrodynamic behavior), but mass ordering among hadrons is violated for smaller systems. Measurements of the nuclear modification factors, v_2 parameters, spin alignment, and modified spectral functions of resonances may help us further understand the physics of heavy-ion collisions.

7 Acknowledgements

I thank the ALICE Collaboration for the use of some of its figures, C. Markert and K. Werner for their help with the EPOS model, and R. Bellwied for making this presentation possible.

References

[1] C. Patrignani *et al.* (Particle Data Group), Chin. Phys. C **40**, 100001 (2017)
[2] M. Bleicher, H. Stöcker, J. Phys. G **30**, S111 (2004)
[3] C. Markert *et al.*, in *Proc. of PASI 2002*, edited by H.T. Elze *et al.* (2002), pp. 533–52, arXiv:hep-ph/0206260
[4] S. Vogel, M. Bleicher, in *Proc. of the 43rd International Winter Meeting on Nuclear Physics, Bormio, Italy, 2005*, edited by I. Iori, A. Moroni (INSPIRE, 2005), arXiv:nucl-th/0505027
[5] G. Torrieri, J. Rafelski, Phys. Lett. B **509**, 239 (2001)
[6] J. Rafelski, J. Letessier, G. Torrieri, Phys. Rev. C **64**, 054907 (2001)
[7] J. Adam *et al.* (ALICE Collaboration), Eur. Phys. J. C **76**, 245 (2016)
[8] N. Agrawal (for the ALICE Collaboration), presentation at the 17th International Conference on Strangeness in Quark Matter, Utrecht, The Netherlands (2017)
[9] N. Agrawal (for the ALICE Collaboration), these proceedings (2017)
[10] K. Garg (for the ALICE Collaboration), presentation at the 3rd Resonance Workshop in Bergamo, Bergamo, Italy (2017)
[11] Abelev B *et al.* (ALICE Collaboration), Phys. Rev. C **91**, 024609 (2015)
[12] A.G. Knospe *et al.*, Phys. Rev. C **93**, 014911 (2016)
[13] A. Ilner *et al.* (2017), arXiv:1707.00060

[14] J. Stachel, A. Andronic, P. Braun-Munzinger, K. Redlich, J. Phys.: Conf. Series **509**, 012019 (2014)
[15] A.G. Knospe (for the ALICE Collaboration), J. Phys. Conf. Series **779**, 012072 (2017)
[16] B.I. Abelev *et al.* (STAR Collaboration), Phys. Rev. C **78**, 044906 (2008)
[17] D. Adamová *et al.* (ALICE Collaboration), Eur. Phys. J. C **77**, 389 (2017)
[18] J. Song (for the ALICE Collaboration), presentation at the 26th International Conference on Ultra-Relativistic Nucleus-Nucleus Collisions (Quark Matter 2017), Chicago, USA (2017)
[19] J.Adam *et al.* (ALICE Collaboration), Nature Physics **13**, 535 (2017)
[20] J.Adam *et al.* (ALICE Collaboration), Phys. Lett. B **758**, 389 (2016)
[21] B. Abelev *et al.* (ALICE Collaboration), Phys. Lett. B **728**, 216 (2013)
[22] V. Vislavicius, A. Kalweit (2016), `arXiv:1610.03001`
[23] J. Adamczewski-Musch *et al.* (HADES Collaboration) (2017), `arXiv:1703.08418`
[24] B. Abelev *et al.* (ALICE Collaboration), Phys. Rev. C **88**, 044910 (2013)
[25] B. Abelev *et al.* (ALICE Collaboration), Phys. Rev. Lett. **111**, 222301 (2013)
[26] Z. Qui, C. Shen, U. Heinz, Phys. Lett. B **707**, 151 (2012)
[27] C. Shen, U. Heinz, P. Huovinen, H. Song, Phys. Rev. C **84**, 044903 (2011)
[28] P. Bożek, I. Wyskiel-Piekarska, Phys. Rev. C **85**, 064915 (2012)
[29] R. Fries, B. Muller, C. Nonaka, S. Bass, Phys. Rev. Lett. **90**, 202303 (2003)
[30] R. Fries, V. Greco, P. Sorensen, Annu. Rev. Nucl. Part. Sci **58**, 177 (2008)
[31] V. Minissale *et al.*, Phys. Rev. C **92**, 054904 (2015)
[32] B. Mohanty (for the ALICE Collaboration), these proceedings (2017)
[33] J. Adam *et al.* (ALICE Collaboration), Phys. Rev. C **95**, 064606 (2017)
[34] B.I. Abelev *et al.* (STAR Collaboration), Phys. Rev. C **77**, 061902 (2008)
[35] L. Adamczyk *et al.* (STAR Collaboration), Phys. Rev. Lett. **116**, 062301 (2016)
[36] S. Afanasiev *et al.* (PHENIX Collaboration), Phys. Rev. Lett. **99**, 052301 (2007)
[37] R.A. Bertens (for the ALICE Collaboration) (2017), `arXiv:1704.03028`

Resonance production in high energy collisions from small to big systems

K. Werner[(a)], A. G. Knospe[(b)], C. Markert[(c)], B. Guiot[(a,d)], Iu. Karpenko[(a,e,f)], T. Pierog[(g)], G. Sophys[(a)], M. Stefaniak[(a,h)], M. Bleicher[(i,j)], J. Steinheimer[(i)]

[(a)] SUBATECH, University of Nantes – IN2P3/CNRS – IMT Atlantique, Nantes, France
[(b)] University of Houston, Houston, USA
[(c)] University of Texas, Austin, USA
[(d)] Universidad Tecnica Federico Santa Maria y Centro Cientifico-Tecnologico de Valparaiso, Valparaiso, Chile
[(e)] Bogolyubov Institute for Theoretical Physics, Kiev 143, 03680, Ukraine
[(f)] INFN - Sezione di Firenze, Via G. Sansone 1, I-50019 Sesto Fiorentino (Firenze), Italy
[(g)] Karlsruhe Inst. of Technology, KIT, Campus North, Inst. f. Kernphysik, Germany
[(h)] Warsaw University of Technology, Warsaw, Poland
[(i)] FIAS, Johann Wolfgang Goethe Universitaet, Frankfurt am Main, Germany
[(j)] GSI Helmholtzzentrum, Planckstr. 1, 64291 Darmstadt

Abstract. The aim of this paper is to understand resonance production (and more generally particle production) for different collision systems, namely proton-proton (pp), proton-nucleus (pA), and nucleus-nucleus (AA) scattering at the LHC. We will investigate in particular particle yields and ratios versus multiplicity, using the same multiplicity definition for the three different systems, in order to analyse in a compact way the evolution of particle production with the system size and the origin of a very different system size dependence of the different particles.

The main goal of heavy ion physics at very high energies is the proof of existence of the creation of a quark-gluon plasma, and the study of the properties of this exotic state, by analysing the final state of many thousands of produced hadrons. Unfortunately, many of these final state particles are not directly coming from the decay of the plasma, but they are produced or at least affected by the hadronic cascade, the last stage of the collision before particles freeze out and freely move to the detectors. Here resonance studies come into play : There is a large number of resonances at our disposal, having very different lifetimes, from 1 fm/c to several tens of fm/c, which means that these particles decay with varying probabilities in the hadronic stage, and provide therefore valuable information about the latter one.

We extended this analysis, to also consider small collision systems (pp and pA), in addition to AA. We also consider not only resonances, but also stable particles, providing useful additional information (the lifetime is not the only relevant parameter).

Our tool to analyse particle production is the EPOS model. EPOS3 [1] is a universal model in the sense that for pp, pA, and AA collisions, the same procedure applies, based on several elements:

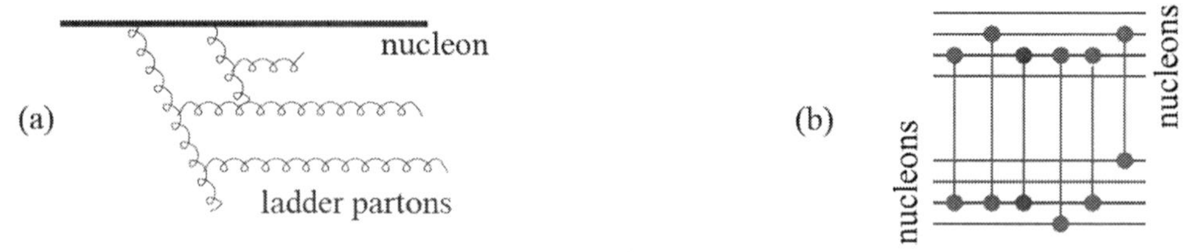

Figure 1. (Color online) (a) Non-linear effect: ladder fusion. (b) Pomerons connected to a given (red) Pomeron

Initial state. A Gribov-Regge multiple scattering approach is employed ("Parton-Based Gribov-Regge Theory" PBGRT [2]), where the elementary object (by definition called Pomeron) is a DGLAP parton ladder. Starting from a multi-Pomeron structure for the elastic scattering S-matrix, one then uses cutting rules to obtain (partial) cross sections of inelastic processes, by employing Markov chain techniques.

Non-linear effects. High parton density effects like gluon fusion during the parton evolution (see fig. 1(a)) are taken into account by using a dynamical saturation scale for each Pomeron, of the form $Q_s = Q_s(N_{\text{Pom}}, \hat{s})$, depending on the number of Pomerons N_{Pom} connected the Pomeron in question (see fig. 1(b)), and its energy $\hat{s}$. The functional form of Q_s is obtained from fitting to the energy dependence of basic cross sections and the multiplicity dependence of particle production in pp at very high transverse momenta.

Core-corona approach. The parton ladders corresponding to Pomerons are treated as classical relativistic (kinky) strings. So in general, we have a large number of (partly overlapping) strings. Based on the momenta and the density of string segments, one separates at some early proper time τ_0 the core (going to be treated as fluid) from the corona (escaping hadrons, including jet hadrons). The core-corona procedure has been first described in [3], a more recent discussion is found in [1]. The corresponding energy-momentum tensor of the core part is transformed into an equilibrium one, needed to start the hydrodynamical evolution. This is based on the hypothesis that equilibration happens rapidly and affects essentially the space components of the energy-momentum tensor.

Viscous hydrodynamic expansion. Starting from the initial proper time τ_0, the core part of the system evolves according to the equations of relativistic viscous hydrodynamics [1, 4], where we use presently $\eta/s = 0.08$. A cross-over equation-of-state is used, compatible with lattice QCD [5, 6]. The "core-matter" hadronizes on some hyper-surface defined by a constant temperature T_H, where a so-called Cooper-Frye procedure is employed, using equilibrium hadron distributions, see [6].

Final state hadronic cascade. After hadronization, there occur still hadron-hadron rescatterings, realized via UrQMD [7].

The above procedure is employed for each event (event-by-event procedure).

We will discuss particle production as a function of the average multiplicity per eta interval $\langle dn/d\eta(0)\rangle$ at central pseudorapidity ($\eta = 0$), whereas forward pseudorapidity intervals are used to define the different multiplicity classes. This is the same procedure as used by the ALICE experiment. In the following $\langle dn/d\eta(0)\rangle$ is simply referred to as "multiplicity".

Crucial for the following discussion is the fact that the core-corona separation behaves quite differently for small and big systems (although we employ the same procedure), see fig. 2. Whereas

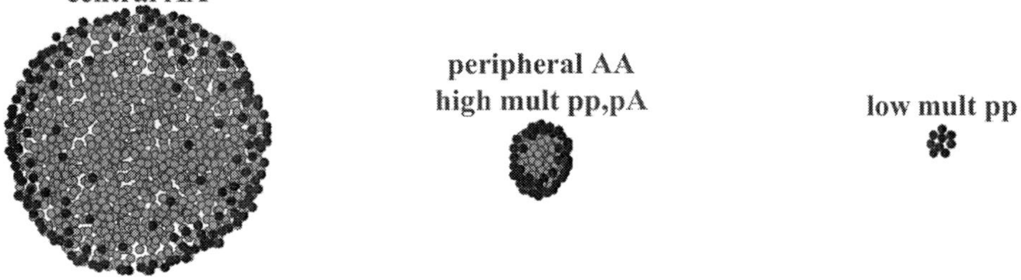

Figure 2. (Color online) Core (red) and corona (blue) in a plane perpendicular to the beam axis, for big (left), small (right) and intermediate (middle) systems.

for very large multiplicity (central AA) most of the matter is core, with small corona contributions at the surface (or some high pt segments further inside), we get at very low multiplicity (in pp) only corona : the system is not dense enough to form a core. High multiplicity pp and pA, or peripheral AA collisions are in between, both core and corona being important.

Another way of seeing the multiplicity dependence of the core-corona separation is shown in fig. 3(a), where we plot the pion yield versus multiplicity, for only core particles ("core", green dashed-dotted), the contribution from core + corona ("co-co", particles from core and corona, yellow dashed), both contributions from EPOS simulation without hadronic cascade. The ratio "core" over "co-co" would be the relative core fraction, which increases continously from zero (at small multiplicity) to unity (at large multiplicity). We also plot the "full" contribution, referring to all particles, for a simulation with hadronic cascade (A summary of color, line, and symbol codes is found in table 1). In all cases we plot curves for the three systems: pp, pA, and AA. There are substantial overlap regions, where different systems contribute, but we observe unique curves, no system size dependence. We also observe that the hadronic cascade has no effect on the pion production ("co-co" and "full" gives identical results).

In fig. 3(b), we plot a quantity relevant for resonance decays in the hadronic stage: The lifetime of this phase, defined as the difference of the average formation time of particles for simulations with and

core	green dashed-dotted	particle from the core only	no hadronic cascade
corona	blue dotted	particles from corona only	no hadronic cascade
co-co	yellow dashed	particles from core and corona	no hadronic cascade
full	red full	all particles	with hadronic cascade
	blue triangles	particles from pure string decay	no cascade, no hydro
	thin lines	pp simulation	
	intermediate lines	pA (pPb) simulation	
	thick lines	AA (PbPb) simulation	
	open circles	pp data	
	open squares	pA (pPb) data	
	open stars	AA (PbPb) data	

Table 1. Color, line, and symbol codes used for the plots in this paper, to accommodate the different contributions and the different systems.

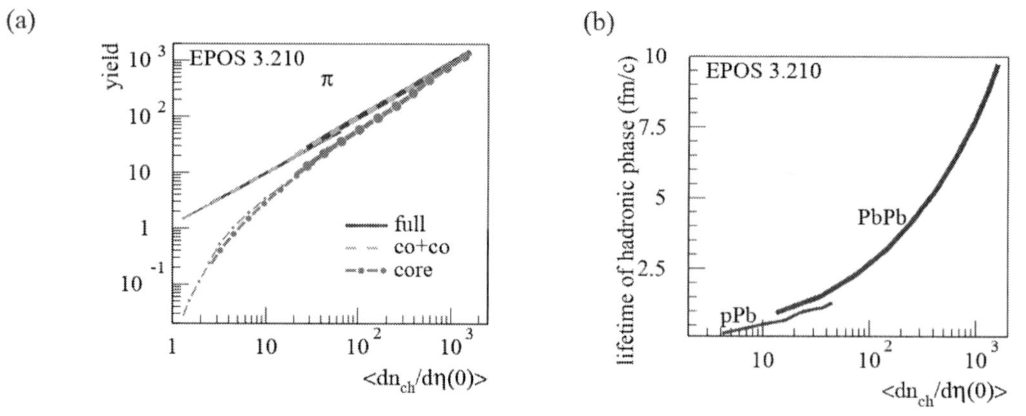

Figure 3. (Color online) (a) Pion yield versus multiplicity. We compare different contributions (core, core plus corona (co-co) and the the full contribution (the latter on including the final state hadronic cascade) for different systems: pp (thin lines), AA (thick lines), and pA (intermediate lines). (A summary of color, line, and symbol codes is found in table 1) (b) Lifetimes of the hadronic phase versus multiplicity.

without hadronic cascade. The bigger the lifetime, the bigger the probablility of a resonance decay inside this phase. The lifetime increases continuously from less than a fm/c to 10fm/c for big systems.

In the following we will investigate the multiplicity dependence (for pp, pA, AA) of particle ratios to pions, for the different contributions from EPOS simulations, compared to data from ALICE [8–17]. We always refer to the color, line, and symbol codes of table 1.

In fig. 4(a), we plot the **kaon** to pion ratio versus multiplicity, for different contributions from the EPOS simulations, for different systems (pp, pA, AA), see table 1 for color, line, and symbol codes. Despite a large overlap for the different systems, we observe universal curves, for all contributions. The core contribution is completely flat, as expected since the particle ratios only depend on the properties of the fluid at freeze out, taken to be the same in all systems. Also the corona contribution (from string segments which escape the fluid) is essentially flat, but with some deviation from pure string fragmentation (blue triangles). But at large multiplicity the corona contribution is small, so we have essentially two flat curves, from core and corona, where here in case of kaons the core contribution is roughly a factor of 2 above the corona. From fig. 3(a), we know that the relative core weight increases continuously from zero to unity, which explains the increase of the core+corona (co-co) contribution which interpolates between the corona level at small multiplicity to the core level at high multiplicity. Considering finally the "full" contribution (including the final state hadronic cascade), we see no difference compared to the case without cascade (co-co), both curves reproducing the experimental data. In fig. 4(b), we consider ϕ **mesons**, and we observe almost the same behavior as seen for the kaons, a continuous increase of roughly a factor of 2 from corona to core level, and little effect of the hadronic cascade. Due to the **long lifetime** of 46.2 fm/c only few ϕ mesons decay in the hadronic phase.

The situation is different for **Ω baryons**, see fig. 4(c), since here core and corona curves are separated by about a factor of 10, leading to a strong increase of the core+corona (co-co) contribution of a factor of 10 with multiplicity. The full curve is slightly reduced at high multiplicity compared

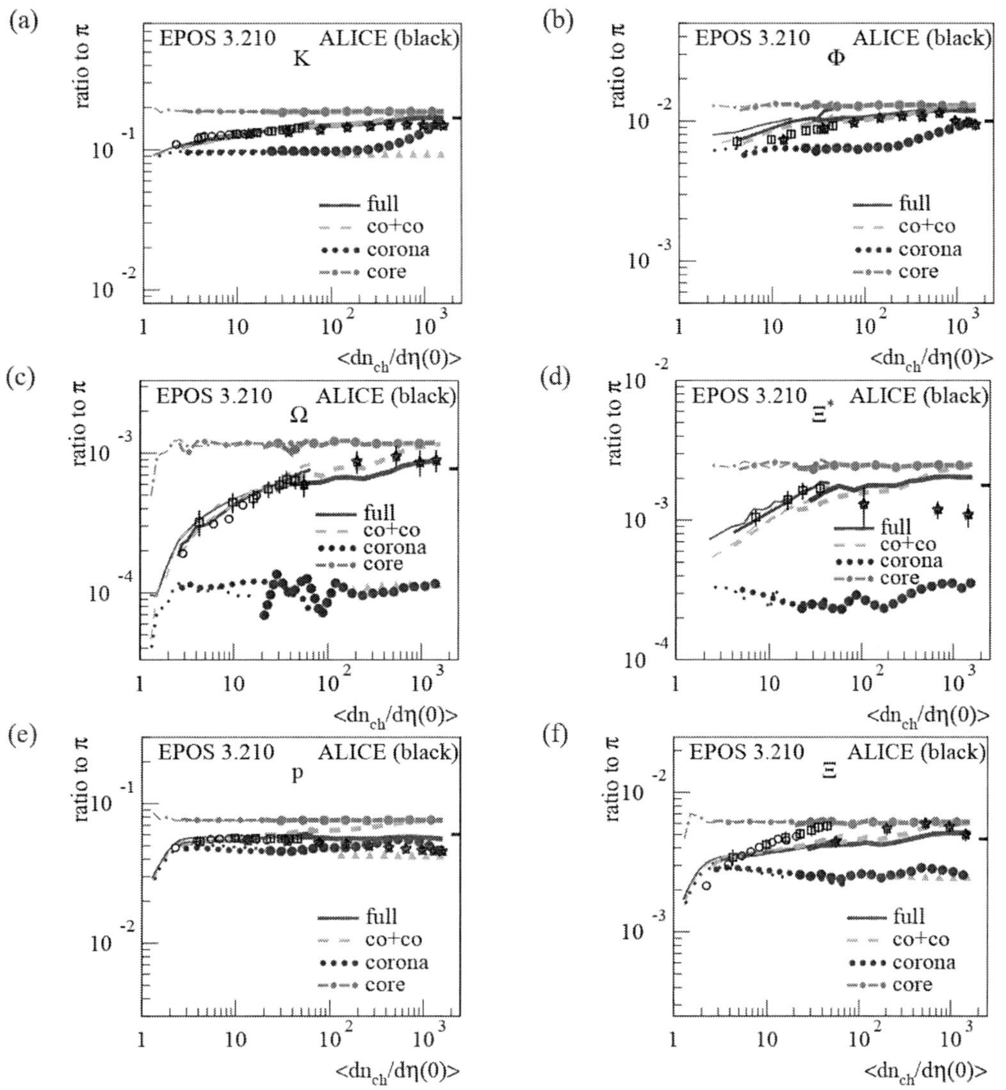

Figure 4. (Color online) Particle to pion ratio for different particle species versus multiplicity, for different contributions from the EPOS simulations, for different systems (pp, pA, AA). We also plot ALICE data from [8–20]. See table 1 for color, line, and symbol codes.

to co-co, due to hadronic final state interactions (baryon-antibaryon annihilation). We observe very similar results for the Ξ^* **resonance** (lifetime 21.7 fm/c), but some discrepancy compared to the data.

Concerning **protons**, we see in fig. 4(e) also first an increase of the co-co curve, form the corona level to the core level, and again (as for the Omega resonance) a reduction at high multiplicity when including final state hadronic rescattering, coming from **baryon-antibaryon annihilation**. We also plot Ξ **baryons** in fig. 4(f), showing a significant increase with multiplicity.

In fig. 5(a), we consider the **kaon resonance** K^*_0. Here core, corona, and as a consequence also core+corona (co-co) are identical (by accident). Interesting: The full results (including the final state hadronic cascade) is considerably lower than the results without cascade (co-co). This is due to

https://doi.org/10.1051/epjconf/201817109002

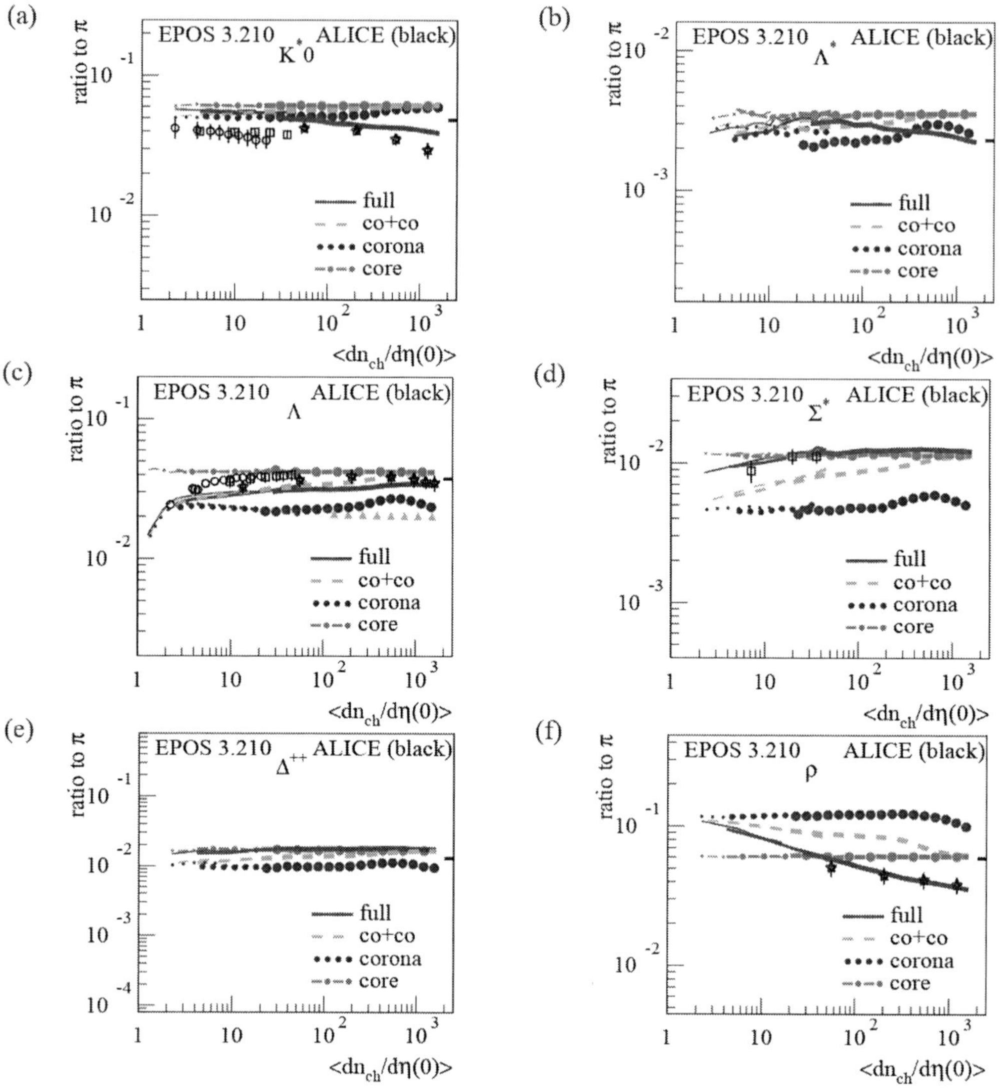

Figure 5. (Color online) Same as fig 4, for other particle species.

the **short lifetime** (4.2 fm/c), which makes the **particles decay within the hadronic stage**. Some of the decay products rescatter (and change momenta) and thus prevent the kaon resonances to be reconstructed via their decay products. So contrary to the cases discussed earlier, we observe here a monotonically decreasing curve, as observed in the data for large multiplicities. The data show some discontinuity at low multiplicity, contrary to what we observe in the simulation. Newer ALICE results show as well a continuous behavior. Also the **Lambda resonance Λ^*** shows in 5(b) a significant decrease of the "full" contribution at high multiplicity, partly due to decays in the hadronic stage (lifetime 12.6 fm/c) and baryon-antibaryon annihilation. But one has to be careful not to draw wrong conclusions. The lifetime is relatively large, so the effect of the decay cannot be very big compared to Lambdas. On the other hand, it seems that the corona and the core level are closer to each other compared to the Lambda (see fig. 5(c)), so the core+corona curve increases less, and therefore the reduction due to annihilation (present in both Λ and Λ^*) provides a bigger decrease in the case of Λ^*.

The **Σ^* resonance,** shown in fig. 5(d), shows also the "usual" monotonic increase of the core+corona contribution, but here we get a strong increase at low multiplicity but little change at high multiplicity, when considering the "full" contribution including the hadronic cascade. We have actually two competing processes: The lifetime is relatively short (5 fm/c), so there should be some **reduction due to decay** at high multiplicity, which is not visible because it is compensated by an increase due to **resonance creation** in the hadronic stage. Whereas the reduction due to decay is only possible for sufficiently big systems (at high multiplicity), resonance creating is possible at any multiplicity, and is therefore visible at small multiplicities. The **Δ^{++} resonance** (lifetime 1.7 fm/c), considered in fig. 5(e), shows a similar behavior, on a smaller scale.

An exceptional case is the **ρ resonance**, shown in fig. 5(f), since here the core level is below the corona one (this is the only case, it happens because ρ production is very frequent in the string breaking procedure), and therefore the **core+corona contribution decreases** with multiplicity. Considering the full contribution (including the hadronic cascade), there is an additional reduction, such that altogether there is a very strong decrease of the ρ over pion yield with multiplicity.

To summarize, we have shown that the study of the multiplicity dependence of particle ratios for different collision systems (pp, pA, AA) provides a wealth of information concerning the production mecanism of particles and about the properties of the medium having produced or affected these particles.

Acknowledgements

This research was carried out within the scope of the GDRE (European Research Group) "Heavy ions at ultrarelativistic energies", and was partially supported by COST Action CA15213 (THOR). This work was supported by U.S. Department of Energy Office of Science under contract number DE-SC0013391. The authors acknowledge the Texas Advanced Computing Center (TACC) at the University of Texas at Austin for providing computing resources that have contributed to the research results reported within this paper. URL: http://www.tacc.utexas.edu. B. G. gratefully acknowledges generous support from Chilean FONDECYT grants 3160493 and Proyecto Basal FB0821.

References

[1] K. Werner, B. Guiot, I. Karpenko, and T. Pierog, Phys.Rev. C89 (2014) 064903, arXiv:1312.1233.
[2] H. J. Drescher, M. Hladik, S. Ostapchenko, T. Pierog and K. Werner, Phys. Rept. 350, 93, 2001
[3] K. Werner, Phys. Rev. Lett. 98, 152301 (2007)
[4] Iu. Karpenko, P. Huovinen, M. Bleicher, arXiv:1312.4160
[5] S. Borsanyi et al., JHEP 1011 (2010) 077, arXiv:1007.2580
[6] K. Werner, Iu. Karpenko, T. Pierog, M. Bleicher, K. Mikhailov, arXiv:1010.0400, Phys. Rev. C 83, 044915 (2011)
[7] M. Bleicher et al., J. Phys. G25 (1999) 1859; H. Petersen, J. Steinheimer, G. Burau, M. Bleicher and H. Stocker, Phys. Rev. C78 (2008) 044901
[8] ALICE Collaboration, Phys. Rev. Lett. 106 032301 (2011), Phys. Rev. C 88 044910 (2013)
[9] ALICE Collaboration, Phys. Rev. Lett. 111 222301 (2013)
[10] ALICE Collaboration, Phys. Lett. B 758 389-401 (2016)
[11] ALICE Collaboration, Phys. Lett. B 728 25-38 (2014)

[12] ALICE Collaboration, Eur. Phys. J. C 76 245 (2016)
[13] ALICE Collaboration, Phys. Lett. B 712 309 (2012)
[14] ALICE Collaboration, Phys. Lett. B 728 216–227 (2014)
[15] ALICE Collaboration, Phys. Rev. C 91 024609 (2015)
[16] ALICE Collaboration, Nature Physics 13 (2017) 535-539
[17] digitized from plenary talk of Rafael Derradi de Souza (for the ALICE Collaboration), SQM2016
[18] derived from talk of Jihye Song (for the ALICE Collaboration) at Quark Matter 2017
[19] digitized from A. G. Knospe (for the ALICE Collaboration) at SQM 2016
[20] ALICE, Eur. Phys. J. C 77 (2017) 389

Medium modification of strange hadronic resonances at SIS, RHIC and LHC energies

Laura Tolos[1,2,⋆]

[1] *Institute of Space Sciences (CSIC-IEEC), Campus Universitat Autònoma de Barcelona, Carrer de Can Magrans, s/n, 08193 Cerdanyola del Vallès, Spain*
[2] *Frankfurt Institute for Advanced Studies, Goethe University Frankfurt, Ruth-Moufang-Str. 1, 60438 Frankfurt am Main, Germany*

Abstract. The properties of strange pseudoscalar and vectors mesons as well as strange baryon resonances in dense matter are reviewed. Some open questions on the properties of strange hadrons in medium are addressed, such as the experimental signatures of in-medium effects coming from the hadronic phase on the final observables in heavy-ion collisions for the experimental conditions at SIS, RHIC and LHC energies.

1 Introduction

Strangeness production and propagation is one of the major research domain in heavy-ion collisions (HICs) for energies ranging from SIS/GSI, RHIC/BNL and LHC/CERN up to the forthcoming CBM/FAIR, BMN/NICA, BESII/RHIC and J-PARC-HI. In particular, understanding the dynamics of strange mesons, such as pseudoscalar (K,$\bar{K}$) and vector (K^*,$\bar{K}^*$) mesons, as well as strange baryon resonances in HICs is still a challenge [1].

The production of K and $\bar{K}$ close to the threshold energy has been thoroughly investigated in low-energy HICs by the KaoS [2], FOPI [3] and HADES Collaborations [4]. The analysis of experimental data together with microscopic transport approaches have permitted drawing several conclusions regarding the production mechanisms and the freeze-out conditions of strange mesons. Still, a simultaneous description of all observables involving K and $\bar{K}$ production is missing.

Recently HADES Collaboration has also reported a deep sub-threshold K^{*0} production in Ar+KCl collisions [5]. The STAR Collaboration has addressed the K^{*0} meson production in Au+Au, Cu+Cu, d+Au and p+p at high-energy HICs of $\sqrt{s_{NN}}$ =62.4 and 200 GeV [6], while attenuation of the K^{*0} and $\bar{K}^{*0}$ states in the hadronic phase of the expanding fireball in HICs has been observed by the NA49 Collaboration [7]. Moreover, the production of K^{*0} in p+p, p+Pb and Pb+Pb collisions has been studied by the ALICE Collaboration, also observing a centrality-dependent and multiplicity-dependent suppression of the K^{*0}/K ratio [8].

All such findings indicate that a reliable determination of the interaction of strange mesons with the surrounding hadronic environment is important, since it might help to give answer to several open questions in matter, such as the experimental signatures of in-medium effects coming from the hadronic phase on the final observables in HICs. Therefore, in this paper the properties of strange

⋆e-mail: tolos@ice.csic.es

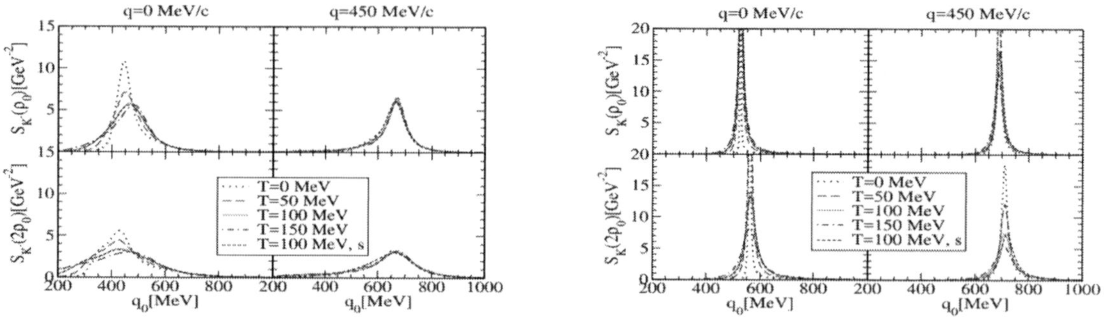

Figure 1. $\bar{K}$ (left) and K (right) spectral functions in hot dense matter. Both plots are taken from [10].

pseudoscalar and vector mesons and some strange baryon resonances in matter are reviewed for the experimental conditions ranging from SIS to RHIC and LHC energies.

2 Strange pseudoscalar mesons

The properties of the strange pseudoscalar mesons (K,$\bar{K}$) in hot dense matter are described by means of their self-energies and, hence, spectral functions. The K and $\bar{K}$ self-energies are obtained from the s- and p-waves in-medium KN and $\bar{K}N$ interactions within a chiral unitary approach [9, 10]. On the one hand, the s-wave amplitude of the $\bar{K}N$ comes from the Weinberg-Tomozawa term of the chiral Lagrangian. Unitarization in coupled channels is imposed on the on-shell amplitudes with a cutoff regularization. The $\Lambda(1405)$ resonance in the $I = 0$ channel is generated dynamically, whereas a satisfactory description of low-energy scattering observables is achieved. On the other hand, the s-wave KN effective interaction is also obtained from unitarization using the same cutoff parameter.

The in-medium s-wave amplitude accounts for Pauli-blocking effects, mean-field binding on the nucleons and hyperons via a $\sigma - \omega$ model, and the dressing of the π and $K/\bar{K}$ propagators. The self-energy is then obtained in a self-consistent manner summing the in-medium on-shell amplitudes for the different isospins over the nucleon Fermi distribution at a given temperature [10]. In the case of the $\bar{K}$ meson, the model includes, in addition, a p-wave contribution to the self-energy from hyperon-hole (Yh) excitations, where Y stands for Λ, Σ and Σ^* states. For the K meson the p-wave self-energy results from YN^{-1} excitations in crossed kinematics.

Once the self-energies are determined, the corresponding K and $\bar{K}$ spectral functions are obtained, as shown in Fig. 1. The $\bar{K}$ spectral function (left) shows a broad peak resulting from a strong mixing between the quasi-particle peak and the $\Lambda(1405)N^{-1}$ and YN^{-1} excitations. These p-wave YN^{-1} subthreshold excitations affect mainly the properties of the $\bar{K}$ at finite momentum. Temperature and density soften the p-wave contributions to the spectral function at the quasi-particle energy. As for the K meson, the spectral function (right) shows a narrow quasi-particle peak which dilutes with temperature and density as the phase space for KN states increases, while shifting it to higher energies with increasing density due to the repulsive character of the KN interaction.

Moreover, the in-medium amplitudes in terms of the transition probabilities give access to the behaviour of $\Lambda(1115)$, $\Lambda(1405)$, $\Sigma(1195)$ and $\Sigma^*(1385)$ in matter. As an example, in Fig. 2 the transition probability for the K^-p elastic reactions is shown as a function of the meson-baryon center-of-mass energy at total vanishing three-momentum. The K^-p state is an admixture of isospin $I = 0, 1$ and, thus, the two Λ states and the Σ show up in the spectrum, as seen in the left panel of Fig. 2. In

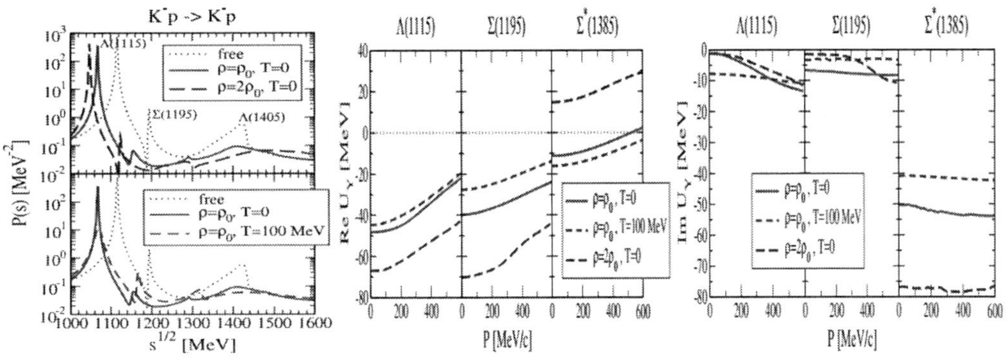

Figure 2. Left: In-medium transition probability $\mathcal{P}$ at zero total three-momentum of the meson-baryon pair for K^-p. Middle and Right: The real and imaginary part of the hyperon potentials. All plots are taken from [11].

the medium, the structure of the $\Lambda(1405)$ is washed out at normal matter density, whereas the p-wave ground states experience moderately attractive mass shifts as density is increased. Temperature broadens the $\Lambda(1115)$ and $\Sigma(1195)$ states as compared to the vacuum case.

In a more quantitative manner, from the analysis of the transition probabilities one can extract information on the single-particle potentials of $\Lambda(1115)$, $\Sigma(1195)$ and $\Sigma^*(1385)$ hyperons at finite momentum, density and temperature (middle and right panels of Fig. 2). At finite density the Λ and the Σ show a potential of roughly -50 and -40 MeV, respectively, at normal matter density ρ_0 and zero temperature. Both hyperons acquire a finite decay width with density and temperature, reflecting the probability to be absorbed by the nuclear medium or have quasi-elastic scattering processes. The Σ^* develops an attractive potential of about -10 MeV at ρ_0 and zero temperature, that turns into a small repulsion for increasing densities, while its decay width is notably enhanced at finite density due to the opening of new absorption channels as pions are dressed. The effect of the temperature in this case is moderate due to the important phase space already available at zero temperature. Also, one finds that the Λ, Σ and Σ^* potentials have a smooth behavior with momentum [11].

At this stage one can address the importance of the in-medium properties of strange hadrons in their production and production in HICs. A systematic study of the experimental results of KaoS collaboration together with a detailed comparison to transport model calculations indicated that, among others, the K^+ feel repulsion while the K^- attraction [2]. However, the conclusions of Leifels and collaborators in SQM2017 conference [12] on recent results from HADES and FOPI indicate that, while the K^+ shows a repulsive interaction in matter, the Φ decay into K^- washes out the effects of the K^- potential in the spectra and flow. Thus, more systematic and high statistic data on K^- production are necessary, while further information in elementary reactions is desirable. The question still remains on the importance of the in-medium effects on the production and propagation of strangeness in HICs.

3 Strange vector mesons

In this section results from different groups on the K^* and $\bar{K}^*$ interactions in free space and in matter are presented as well as some recent conclusions on their production and propagation in low- and high-energy HICs.

The $\bar{K}^*$ self-energy in nuclear matter and, hence, spectral function is obtained within the hidden gauge formalism, taking into account two sources for the modification of the s-wave $\bar{K}^*$ self-energy in nuclear matter [13]. On the one hand, the contribution associated to the decay mode $\bar{K}\pi$ modified by

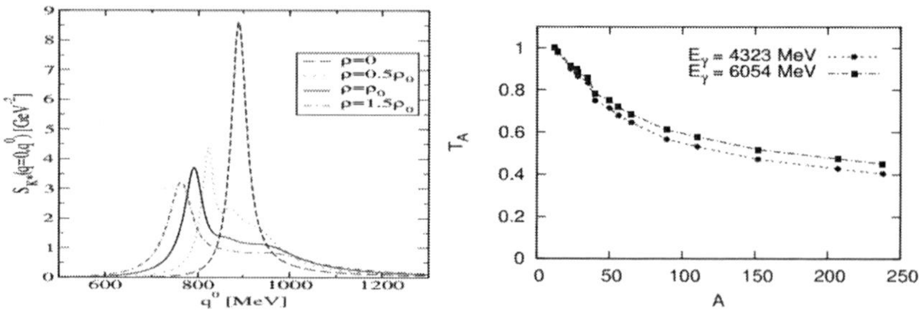

Figure 3. Left: The $\bar{K}^*$ spectral function at zero momentum for different densities. Right: Transparency ratio for $\gamma A \to K^+ K^{*-} A'$. Both plots are taken from [13].

nuclear medium effects on the $\bar{K}$ and π mesons, which accounts for the $\bar{K}^*N \to \bar{K}N, \pi Y, \bar{K}\pi N, \pi\pi Y \ldots$ processes, with $Y = \Lambda, \Sigma$. And, on the other hand, the contribution associated to the interaction of the $\bar{K}^*$ with the nucleons in the medium, which considers the $\bar{K}^*N \to \bar{K}^*N$ process, as well as channels involving other vector mesons, such as $\bar{K}^*N \to \rho Y, \omega Y, \phi Y, \ldots$. As a result, the interaction of $\bar{K}^*N$ and coupled channels in free space gives rise to the appearance of two resonances, $\Lambda(1783)$ and $\Sigma(1830)$, which can be identified with the observed $\Lambda(1800)$ and $\Sigma(1750)$, respectively [14]. These states have been also generated dynamically in subsequent works on $\bar{K}^*N$ interaction that incorporate pseudoscalar meson-baryon states as well as the use of an interaction kernel beyond t-channel [15].

The $\bar{K}^*$ meson spectral function is displayed in the left panel of Fig. 3 as a function of the meson energy q_0, for zero momentum and different densities. The long-dashed line refers to the calculation in free space, where only the $\bar{K}\pi$ decay channel contributes, while the other three lines correspond to the self-consistent calculations, which incorporate $\bar{K}^* \to \bar{K}\pi$ in the medium, as well as the quasielastic $\bar{K}^*N \to \bar{K}^*N$ and other $\bar{K}^*N \to VB$ processes, with V and B denoting vectors and baryons, respectively. The structures above the quasiparticle peak correspond to the dynamically generated $\Lambda(1783)N^{-1}$ and $\Sigma(1830)N^{-1}$ excitations. Density effects result in a dilution and merging of those resonant-hole states, together with a broadening of the spectral function due to the increase of collisional and absorption processes. As a result, the $\bar{K}^*$ spectral function spreads substantially in the medium, becoming five times bigger at ρ_0 than in free space.

In order to test experimentally the broadening of the $\bar{K}^*$ spectral function, one can study the nuclear transparency ratio. The idea is to compare the cross sections of the photoproduction reaction $\gamma A \to K^+ K^{*-} A'$ in different nuclei, and trace them to the in medium K^{*-} width. In the right panel of Fig. 3 the transparency ratio is shown for different nuclei for two different energies of the photon in the lab frame of 4.3 GeV and 6 GeV. A very strong attenuation of the $\bar{K}^*$ survival probability is found due to the decay $\bar{K}^* \to \bar{K}\pi$ or absorption channels $\bar{K}^*N \to \bar{K}N, \pi Y, \bar{K}\pi N, \pi\pi Y, \bar{K}^*N, \rho Y, \omega Y, \phi Y, \ldots$ with increasing nuclear-mass number A. This is due to the larger path that the $\bar{K}^*$ has to follow before it leaves the nucleus, having then more chances to decay or get absorbed.

With regards to K^*, the KN and K^*N interactions are presented using a formalism based on the chiral Lagrangian and the theory of hidden local symmetry. The amplitudes for vector meson-baryon channels are obtained from the s-, t-, u- channel diagrams and a contact interaction, all derived from Lagrangians based on this symmetry. The pseudoscalar meson-baryon interactions are calculated by relying on the Weinberg-Tomozawa theorem. The transition amplitudes between the systems consisting of pseudoscalars and vector mesons are computed by extending the Kroll-Ruderman term for pion photoproduction replacing the photon by a vector meson. In addition, the exchange of light hy-

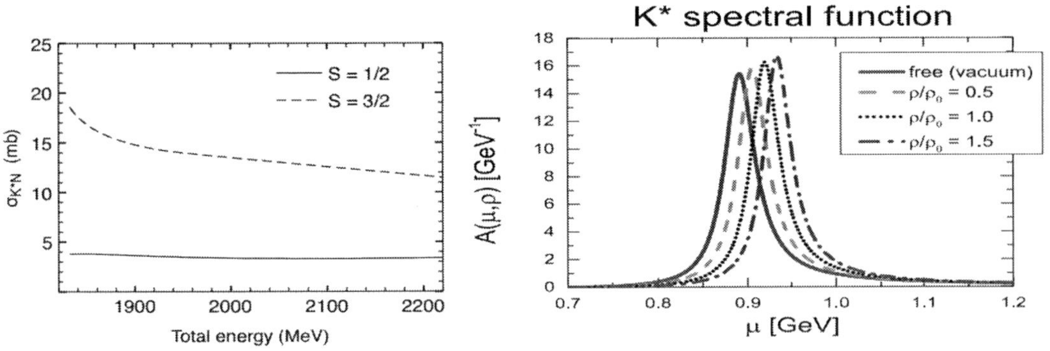

Figure 4. Left: Total cross sections of K^*N in the different spin channels, taken from [16]. Right: K^* spectral function at zero momentum for different densities, taken from [19].

peron resonances, such as $\Lambda(1405)$ and $\Lambda(1670)$ in the u-channel, is included for the $KN \leftrightarrow KN$, $K^*N \leftrightarrow K^*N$ and $K^*N \leftrightarrow KN$ processes [16].

Once the s-wave KN and K^*N interactions are constructed, the Bethe-Salpeter equation is solved in its on-shell factorization form. The subtraction constants required to calculate the loops of the Bethe-Salpeter equation are obtained by fitting the KN amplitudes to the available data for the isospin 0 and 1 s-wave phase shifts, separately [16]. The scattering lengths, $a^{I,S}$, at threshold for the KN system in the different isospin sectors I and $S = 1/2$ can be then calculated, i.e., $a^{0,1/2}_{KN} = -0.16$ fm and $a^{1,1/2}_{KN} = -0.29$ fm. The values found by different partial wave analysis groups for the KN scattering lengths range from -0.105 ± 0.01 fm [17] to -0.23 ± 0.18 fm [18], for isospin 0, and between -0.286 ± 0.06 fm to -0.308 ± 0.003 fm [17], for the isospin 1 case. Thus, the results are compatible with the available data. Moreover, one can obtain the cross sections for K^*N (left panel of Fig. 4) and the scattering lengths for different isospin-spin sectors, i.e., $a^{0,1/2}_{K^*N}(\text{fm}) = (0.2, 0.03)$, $a^{0,3/2}_{K^*N}(\text{fm}) = (-0.08, 0.04)$, $a^{1,1/2}_{K^*N}(\text{fm}) = (0.1, 0.0)$ and $a^{1,3/2}_{K^*N}(\text{fm}) = (-0.31, 0.03)$.

As for K^* in nuclear matter, two sources of medium effects that contribute to the self-energy in a nuclear medium have been considered in [19]: (i) the modification of its dominant decay mode, $K^* \to K\pi$, induced by medium effects on the light pseudoscalars; and (ii) the quasi-elastic interaction of the strange meson with nucleons, $K^*N \to K^*N$ and related absorptive channels with vector mesons. The K^* self-energy has been evaluated straightforwardly in a $T\rho$ approximation, due to the lack of resonant states in the $s = 1$ channel [19].

With all these ingredients, the dynamics of K^* and $\bar{K}^*$ in HICs has been investigated using the PHSD transport model that implements in-medium effects on K^* and $\bar{K}^*$ coming from their production from quark-gluon plasma as well as from the hadronic phase [20]. The PHSD calculations were performed for Au+Au at $\sqrt{s_{NN}} = 200$ GeV (STAR/RHIC) and Pb+Pb $\sqrt{s_{NN}} = 2.76$ TeV (ALICE/LHC) as well as Au+Au for $\sqrt{s_{NN}} = 5$-60 GeV (CBM/FAIR or BMN/NICA or BESII/RHIC) by using the off-shell behaviour of K^* and $\bar{K}^*$. Some of the conclusions are: (i) at LHC/RHIC the main production channel is resonant annihilation of $\pi + K$ $(\bar{K})$ in the final hadronic phase, (ii) the baryon densities tested are rather low at LHC/RHIC, so in-medium effects do not play a role, while at lower energies (CBM/BMN/BESII) the in-medium hadronic effects might be relevant due to the expected longer reaction time and higher densities; and (iii) there are difficulties to extract the in-medium properties of strange vector mesons due to the rescattering and absorption of their decay channels.

4 Summary

The properties of strange pseudoscalar and vector mesons as well as strange baryon resonances are reviewed in hadronic matter. The in-medium modified properties of strange mesons and baryon resonances are being implemented in transport models to analyze the experimental data from low to high-energy HICs, in order to address some fundamental open question, such as the importance of the hadronic phase (and associated in-medium effects) in their production and propagation.

Acknowledgements

The author warmly thanks her collaborators Joerg Aichelin, Elena Bratkovskaya, Daniel Cabrera, Kanchan Khemchandani, Alberto Martinez-Torres, Raquel Molina, Eulogio Oset and Angels Ramos for their comments and contributions. The author also acknowledges support from the Ramón y Cajal research programme, FPA2013-43425-P and FPA2016-81114-P Grants from MINECO, NewCompstar COST Action MP1304 and THOR COST Action CA15213.

References

[1] C. Hartnack, H. Oeschler, Y. Leifels, E. L. Bratkovskaya, and J. Aichelin, Phys. Rept. **510**, 119 (2012)

[2] A. Forster *et al.*, Phys. Rev. C **75**, 024906 (2007)

[3] X. Lopez *et al.*, Phys. Rev. C **81**, 061902 (2010); V. Zinyuk *et al.*, Phys. Rev. C **90**, 025210 (2014); P. Gasik *et al.*, Eur. Phys. J. A **52**, 177 (2016)

[4] G. Agakishiev *et al.*, Phys. Rev. C **90**, 015202 (2014); G. Agakishiev *et al.*, Phys. Rev. C **90**, 054906 (2014)

[5] G. Agakishiev *et al.*, Eur. Phys. J. A **49**, 34 (2013); T. Galatyuk, PoS INPC **2016**, 354 (2017)

[6] J. Adams *et al.*, Phys. Rev. C **71** 064902 (2005); M.M. Aggarwal *et al.*, Phys. Rev. C **84** 034909 (2011); L. Kumar, EPJ Web Conf. **97**, 00017 (2015)

[7] T. Anticic *et al.*, Phys. Rev. C **84**, 064909 (2011)

[8] B.B. Abelev *et al.*, Phys. Rev. C **91**, 024609 (2015); A. Badala, EPJ Web Conf. **142**, 01004 (2017)

[9] L. Tolos, A. Ramos, and E. Oset, Phys. Rev. C **74** 015203 (2006)

[10] L. Tolos, D. Cabrera, and A. Ramos, Phys. Rev. C **78**, 045205 (2008)

[11] D. Cabrera, L. Tolos, J. Aichelin, and E. Bratkovskaya, Phys. Rev. C **90**, 055207 (2014)

[12] Y. Leifels, plenary talk at SQM2017 conference

[13] L. Tolos, R. Molina, E. Oset and A. Ramos, Phys. Rev. C **82**, 045210 (2010)

[14] E. Oset and A. Ramos, Eur. Phys. J. A **44**, 445 (2010)

[15] K.P. Khemchandani, H. Kaneko, H. Nagahiro and A. Hosaka, Phys. Rev. D **83**, 114041 (2011); K.P. Khemchandani, A. Martinez Torres, H. Kaneko, H. Nagahiro and A. Hosaka, Phys. Rev. D **84**, 094018 (2011); K.P. Khemchandani, A. Martinez Torres, H. Nagahiro and A. Hosaka, Phys. Rev. D **85**, 114020 (2012)

[16] K.P. Khemchandani, A. Martinez Torres, F.S. Navarra, M. Nielsen, and L. Tolos, Phys. Rev. D **91**, 094008 (2015)

[17] W.R. Gibbs and R. Arceo, Phys. Rev. C **75**, 035204 (2007)

[18] B.R. Martin, J. Phys. G **4**, 335 (1978)

[19] A. Ilner, D. Cabrera, P. Srisawad and E. Bratkovskaya, Nucl. Phys. A **927**, 249 (2014)

[20] A. Ilner, D. Cabrera, C. Markert and E. Bratkovskaya, Phys. Rev. C **95**, 014903 (2017); A. Ilner, J. Blair, D. Cabrera, C. Markert and E. Bratkovskaya, arXiv:1707.00060 [hep-ph]

EPJ Web of Conferences **171**, 10001 (2018) https://doi.org/10.1051/epjconf/201817110001
SQM 2017

Physics perspectives with AFTER@LHC (A Fixed Target ExpeRiment at LHC)

L. Massacrier[1,*], *M.* Anselmino[2], *R.* Arnaldi[2], *S.J.* Brodsky[3], *V.* Chambert[1], *C.* Da Silva[4], *J.P.* Didelez[1], *M.G.* Echevarria[5], *E.G.* Ferreiro[6], *F.* Fleuret[7], *Y.* Gao[8], *B.* Genolini[1], *C.* Hadjidakis[1], *I.* Hřivnáčová[1], *D.* Kikola[9], *A.* Klein[4], *A.* Kurepin[10], *A.* Kusina[11], *J.P.* Lansberg[1], *C.* Lorcé[12], *F.* Lyonnet[13], *G.* Martinez[14], *A.* Nass[15], *C.* Pisano[16], *P.* Robbe[17], *I.* Schienbein[18], *M.* Schlegel[19], *E.* Scomparin[2], *J.* Seixas[20], *H.S.* Shao[21], *A.* Signori[22], *E.* Steffens[23], *L.* Szymanowski[24], *N.* Topilskaya[10], *B.* Trzeciak[25], *U.I.* Uggerhøj[26], *A.* Uras[27], *R.* Ulrich[28], *J.* Wagner[24], *N.* Yamanaka[1], and *Z.* Yang[8]

[1] *IPNO, Univ. Paris-Sud, CNRS/IN2P3, Université Paris-Saclay, Orsay, France*
[2] *Dip. di Fisica and INFN Sez. Torino, Via P. Giuria 1, Torino, Italy*
[3] *SLAC National Accelerator Laboratory, Stanford University, Menlo Park, USA*
[4] *LANL, P-25, Los Alamos National Laboratory, Los Alamos, NM 87545, USA*
[5] *INFN Sez. Pavia, Via Bassi 6, 27100 Pavia, Italy*
[6] *Dept. de Física de Partículas, USC, Santiago de Compostella, Spain*
[7] *LLR, École Polytechnique, CNRS/IN2P3, Palaiseau, France*
[8] *CHEP, Department of Engineering Physics, Tsinghua University, Beijing, China*
[9] *Faculty of Physics, Warsaw University of Technology, Warsaw, Poland*
[10] *Institute for Nuclear Research, Russian Academy of Sciences, Moscow, Russia*
[11] *Institute of Nuclear Physics Polish Academy of Sciences, PL-31342 Krakow, Poland*
[12] *CPhT, Ecole Polytechnique, CNRS, Université Paris-Saclay, Palaiseau, France*
[13] *Southern Methodist University, Dallas, TX 75275, USA*
[14] *SUBATECH, IMT Atlantique, Université de Nantes, CNRS-IN2P3, Nantes, France*
[15] *Institut für Kernphysik, Forschungszentrum Jülich, Jülich, Germany*
[16] *Dipartimento di Fisica, Universita degli Studi di Pavia, Pavia, Italy*
[17] *LAL, Univ. Paris-Sud, CNRS/IN2P3, Université Paris-Saclay, Orsay, France*
[18] *LPSC, Université Grenoble-Alpes, CNRS/IN2P3, 38026 Grenoble, France*
[19] *Institute for Theoretical Physics, Tübingen U., Tübingen, Germany*
[20] *LIP and IST, Lisbon, Portugal*
[21] *Sorbonne Universités, UPMC Univ. Paris 06, UMR 7589, LPTHE, F-75005 Paris, France CNRS, UMR 7589, LPTHE, F-75005 Paris, France*
[22] *Theory Center, Thomas Jefferson National Accelerator Facility, 12000 Jefferson Avenue, Newport News, VA 23606, USA*
[23] *Physics Institute, Friedrich-Alexander University Erlangen-Nürnberg, Erlangen, Germany*
[24] *National Centre for Nuclear Research (NCBJ), Hoza 69, 00-681, Warsaw, Poland*
[25] *Institute for Subatomic Physics, Utrecht University, Utrecht, The Netherlands*
[26] *Department of Physics and Astronomy, University of Aarhus, Denmark*
[27] *IPNL, Université Claude Bernard Lyon-I, CNRS/IN2P3, Villeurbanne, France*
[28] *Institut für Kernphysik, Karlsruhe Institute of Technology (KIT), Karlsruhe, Germany*

Abstract. AFTER@LHC is an ambitious fixed-target project in order to address open questions in the domain of proton and neutron spins, Quark Gluon Plasma and high-x physics, at the highest energy ever reached in the fixed-target mode. Indeed, thanks to the highly energetic 7 TeV proton and 2.76 A.TeV lead LHC beams, center-of-mass energies as large as $\sqrt{s_{NN}}$ = 115 GeV in pp/pA and $\sqrt{s_{NN}}$ = 72 GeV in AA can be reached, corresponding to an uncharted energy domain between SPS and RHIC. We report two main ways of performing fixed-target collisions at the LHC, both allowing for the usage of one of the existing LHC experiments. In these proceedings, after discussing the projected luminosities considered for one year of data taking at the LHC, we will present a selection of projections for light and heavy-flavour production.

*e-mail: massacrier@ipno.in2p3.fr

1 Introduction

The AFTER@LHC project [1] is a proposal to conduct a multi-purpose fixed-target experiment at the LHC by using its highly energetic proton and lead beams. The energy domain which can be probed ranges from $\sqrt{s_{NN}}$ = 72 GeV in AA collisions up to $\sqrt{s_{NN}}$ = 115 GeV in pp/pA collisions, i.e. the largest energies achieved in the fixed-target mode. This mode offers unique opportunities to access the high Feynman-x_F domain with high luminosities thanks to the large density of the target. The versatility of the target also allows one to perform studies as a function of the nucleus atomic mass. Finally, depending on the chosen technology, it may be possible to polarise the target. The physics programme of AFTER@LHC has extensively been discussed in Refs [2–5] and proposes to address open questions in the domain of high-x, Spin and Quark Gluon Plasma (QGP) physics. In the high-x programme, the aim is to advance our understanding of the high-x gluon, antiquark and heavy-quark content in the nucleon and in the nucleus: by, for instance, constraining the quark Parton Distribution Functions (PDF) and nuclear PDFs (in the EMC region) with Drell-Yan measurements; by searching for the existence of a possible non-perturbative source of c or b quarks in the proton, which is an important input for high-energy-neutrino and cosmic-ray physics; by looking for W boson production near threshold to constrain the light quark sea PDFs at large-x.

In the Spin physics programme (not discussed here), the goal is to advance our understanding of the dynamics and spin of quarks and gluons inside polarised (and unpolarised) nucleons, in particular the Orbital Angular Momentum of quarks and gluons in the proton.

The final part of the AFTER@LHC physics programme concerns the study of heavy-ion collisions toward large rapidities. In Pb-A collisions, at $\sqrt{s_{NN}}$ = 72 GeV, AFTER@LHC is probing the region of high temperature ($\sim$ 1.5 T_c) and low baryon chemical potential, where QGP formation is expected to occur. At such a temperature, the $\Upsilon(2S, 3S)$ excited states are expected to be suppressed in the QGP [6], thus allowing the calibration of its temperature. Conducting measurements of various quarkonium states (together with open heavy-flavours (HF)) as a function of rapidity and system size would permit to scan the phase-transition region. Moreover, AFTER@LHC can study the transport properties of the QGP accounting for its longitudinal expansion. By measuring particle yields and flow coefficients v_N as a function of rapidity, AFTER@LHC can access the temperature dependence of the medium shear viscosity while probing different energy densities. A proper interpretation of the AA data probably requires a complete set of v_N coefficient measurements in smaller systems (pp, pA) to study the collectivity with new observables, like heavy-flavour hadrons, which will be abundantly produced at AFTER@LHC. Other items of interest also include the study of the heavy-quark energy-loss mechanisms in the QGP via D meson measurements as a function of the rapidity and p_T, and the universality of the initial-state effects from pA to AA collisions with the Drell-Yan probe. In these proceedings, we will focus on the performance of AFTER@LHC for light and heavy-flavour production.

2 Possible technical implementations at the LHC and luminosities

In the following, we will discuss the usage of an internal (solid or gaseous) target inside one of the existing LHC experiments (LHCb or ALICE) as a main way to achieve the AFTER@LHC physics goals. Such kinds of solutions enable to conduct a fixed-target programme at a limited cost, with a limited civil engineering, and on a shorter timescale. The LHCb experiment has demonstrated the feasibility of injecting low-density noble gases inside the vacuum chamber of its Vertex Locator detector using the SMOG system [7]. This setup has proven that a parasitic operation of a fixed-target programme with a collider programme is possible, without inducing a decrease of the beam lifetime. However, the physics reach with such a system is limited by: the low gas pressure; the choice of gas species; the limited running time due to the absence of dedicated pumping systems close to the interaction point; the absence of target polarisation. Typical luminosities collected in 2015 in pAr collisions, during about 17 hours of data taking amount to few nb^{-1} [8]. To further increase the gas pressure, two setups are under study. An internal gas-jet is currently used at the RHIC collider to measure proton beam polarisation [9], while a storage-cell gas target has been used by the Hermes experiment at the Hera collider [10]. With both systems, target areal density as high as 10^{15} - 10^{16} $H_2.cm^{-2}$ could be reached with unpolarised H_2, i.e. an increase of the target areal density by several orders of magnitude with respect to the SMOG system. Thanks to a powerful differential pumping system maintaining a high local gas density in the target region, polarised hydrogen target areal density about two orders of magnitude larger than the internal gas-jet one can be obtained with the storage-cell gas target. Finally, the usage of an internal solid target in combination with a bent crystal deflecting the beam halo is also under study [11]. Beam fluxes on the order of 5×10^8 p/s and 10^5 Pb/s can be extracted by means of a bent crystal. An unpolarised solid target, with a length of 5 mm along the beam direction, is currently considered. Such a length should allow an installation inside the LHC beam pipe and limited multiple-scattering inside the target. Table 1 shows the target areal density, instantaneous and integrated luminosities over a year for the gas-jet target, storage-cell target and a bent crystal coupled to a solid-target. Only a selection of beam and target type combinations relevant for the discussions and results in these proceedings is shown. The largest integrated luminosities per year in $pH^{\uparrow}$ ($\sim 10\ fb^{-1}$) and PbXe ($\sim 30\ nb^{-1}$) collisions are obtained with the storage-cell. These luminosities should be considered as maximal and can be further decreased because of the detector data acquision rate capabilities (e.g. the ALICE detector case) or a decrease of the beam lifetime (a 15% beam consumption over a fill has been assumed for PbXe collisions).

Table 1. Summary table of the target areal density, instantaneous and integrated luminosities over a year, for the various technical solutions described in the text, and a selection of beam and target types. The solid target is considered to be 5mm-thick along the beam direction. The proton (lead) LHC year is assumed to last 10^7 s (10^6 s) respectively. The proton (lead) beam flux is considered to be $3.63 \times 10^{18}\ s^{-1}$ ($4.66 \times 10^{14}\ s^{-1}$) for the gas-jet and storage-cell solutions. The symbol ↑ indicates that the target is polarised.

Technical Solution	Beam type	Target type	θ_{target} (cm^{-2})	$\mathcal{L}$ ($cm^{-2}.s^{-1}$)	$\mathcal{L}_{int}$ (pb^{-1}/year)
Gas-Jet Target	p	H↑	1.2×10^{12}	4.3×10^{30}	43
	p	H_2	10^{15} - 10^{16}	3.6×10^{33} - 3.6×10^{34}	36×10^3 - 36×10^4
	Pb	H↑	1.2×10^{12}	5.6×10^{26}	0.56×10^{-3}
	Pb	H_2	10^{15} - 10^{16}	4.7×10^{29} - 4.7×10^{30}	0.47 - 4.7
Storage-Cell Target	p	H↑	2.5×10^{14}	9.2×10^{32}	9200
	p	Xe	6.4×10^{13}	2.3×10^{32}	2300
	Pb	H↑	2.5×10^{14}	1.2×10^{29}	0.120
	Pb	Xe	6.4×10^{13}	3.0×10^{28}	0.030
Bent Crystal + Solid Target	p	Pb	1.6×10^{22}	8.2×10^{30}	82
	Pb	Pb	1.6×10^{22}	1.6×10^{27}	1.6×10^{-3}

The ALICE and LHCb detectors are both well suited to conduct a fixed-target programme of AFTER@LHC. The LHCb detector is fully instrumented in the forward region, with excellent particle identification (PID) performance. It has a mid- to backward coverage in the center-of-mass frame in the fixed-target mode, which reaches large negative Feynman x_F. The ALICE detector also provides a similar coverage in the mid- to backward region thanks to its current muon arm and future Muon Forward Tracker. The long absorber upstream the muon tracking stations is a key feature to reduce the combinatorial background for Drell-Yan studies in pA/AA. In addition, the central barrel of ALICE provides a very backward coverage in the fixed-target mode with good PID capabilities, allowing one to reach the end of the phase space for several identified soft probes.

https://doi.org/10.1051/epjconf/201817110001

3 Projected performance for light and heavy-flavours

The measurement of the flow coefficients at mid- and backward rapidity in the c.m.s frame is interesting to test hydrodynamic calculations accounting for the longitudinal expansion of the formed medium. Indeed, AFTER@LHC would give us insight on the QGP behaviour at low energy, in a rapidity region seldom probed. Figure 1 (left) shows the multiplicity distribution of identified particles (pion, kaon, proton, antiproton) as a function of pseudo-rapidity in the lab ($\eta_{\rm lab}$), in Pb-Pb collisions at $\sqrt{s_{NN}}$ = 72 GeV, in the centrality range 20-30%, generated with the EPOS@LHC Monte Carlo (MC) generator [12, 13]. Identified light particles can be studied in a wide rapidity range with the LHCb and ALICE detectors, and down to their phase space limit with the ALICE central barrel. Based on the identified-particle yields from the figure 1 (left), the projections of the statistical uncertainties on the elliptic flow measurement v_2, as a function of $\eta_{\rm lab}$ for ALICE and LHCb detector acceptances, are shown in figure 1 (right). A minimum particle $p_{\rm T}$ of 0.2 GeV/c (ALICE) and 0.5 GeV/c (LHCb) is required for the PID. The elliptic-flow of identified-particles can be accurately measured in LHCb, after only few hours of Pb-Pb data taking with a solid target (10^8 minimum bias events). In the ALICE central-barrel acceptance, despite the lower yields, an absolute statistical uncertainty on v_2 of 4% for antiprotons, 2% for kaons and better than 1% for pions and protons can already be reached.

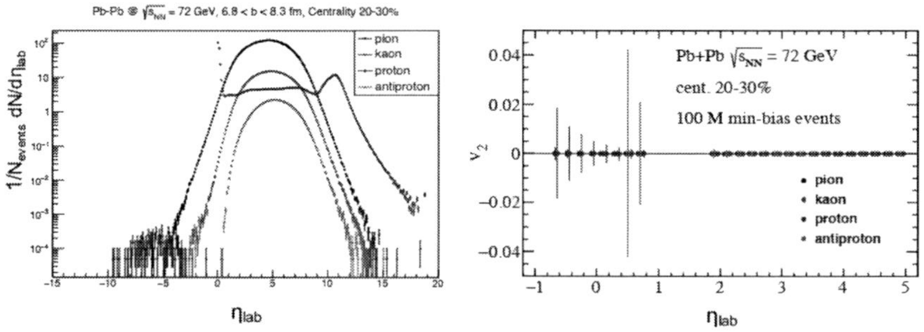

Figure 1. Light-particle multiplicity distributions (left) and elliptic flow v_2 (right) as a function of $\eta_{\rm lab}$ for fixed-target Pb-Pb collisions, in the centrality range 20-30%. Pions (kaons, protons and antiprotons) are represented as black (red, blue, pink) dots, respectively. In the right panel, the statistical uncertainties on the v_2 coefficients (vertical uncertainty bars) have been computed in the ALICE central-barrel and LHCb detector acceptances, assuming the collection of 10^8 minimum bias events.

AFTER@LHC will also be well suited to measure several quarkonium states in order to assess the thermodynamical properties of the QGP in AA collisions. In particular, precise measurements of bottomonium states in pp, pA and AA collisions as a function $p_{\rm T}$, rapidity and system size will provide a range of densities and temperatures to identify the conditions for deconfinement and the quantification of the Cold Nuclear Matter (CNM) effects. Figure 2 (left) shows the dimuon invariant mass distribution for Υ(nS) states after combinatorial background subtraction (using dimuon-like sign pairs), in Pb-Xe collisions at $\sqrt{s_{NN}}$ = 72 GeV, generated with Pythia8 [14] and HELAC-Onia [15, 16] MC generators[1]. The performance of an LHCb-like detector with a storage-cell target have been assumed ($\mathcal{L}_{\rm int}$ = 30 nb^{-1}). Figure 2 (left) is an example of the Υ(nS) yields which can be collected in the rapidity range $3 < y_{\rm lab} < 5$. The three upsilon states are clearly visible (about 450 Υ(3S) are expected per year if no nuclear effects are considered). Figure 2 (right) represents the corresponding projection of the statistical precision on the nuclear modification factor in pXe collisions ($R_{\rm pXe}$, black squares) and Pb-Xe collisions ($R_{\rm PbXe}$, blue dots) for the three upsilon states, in the rapidity range

[1] pp simulations were scaled to PbXe collisions assuming no nuclear effects ($N_{\rm coll}$ scaling for the hard probes).

$3 < y_{\rm lab} < 5$. A total sampled luminosity of $\mathcal{L}_{\rm int} = 250$ pb^{-1} in pXe, and $\mathcal{L}_{\rm int} = 2$ pb^{-1} in pp collisions have been assumed. A 7% (30%) statistical uncertainty on the $R_{\rm PbXe}$ measurement for Υ(1S) (Υ(3S)) respectively is within reach, allowing for the study of Υ excited state suppression in the QGP in a completely new energy domain. The statistical uncertainty of 5% (15%) on $R_{\rm pXe}$ for Υ(1S) (Υ(3S)) respectively will further constrain CNM effects and help to interpret the AA data.

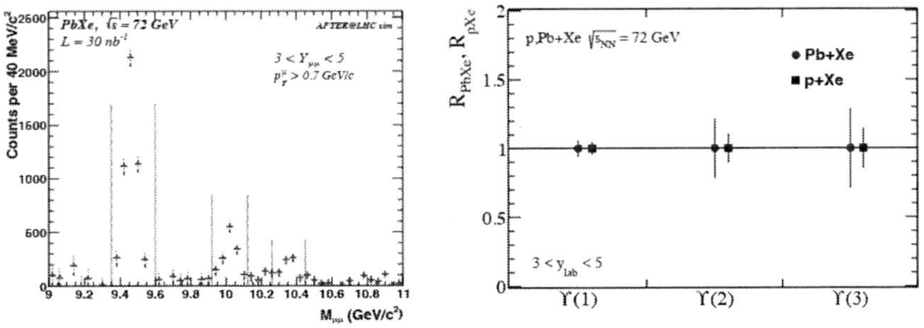

Figure 2. Left panel: Dimuon invariant-mass distribution in the $\Upsilon(nS)$ region after combinatorial background subtraction, in Pb-Xe collisions at $\sqrt{s_{NN}} = 72$ GeV ($\mathcal{L}_{\rm int} = 30$ nb^{-1}). Single muons are required to have $p_T > 0.7$ GeV/c and dimuons to be within $3 < y_{\rm lab} < 5$. Right panel: Projection of the statistical uncertainties (vertical bars) on the nuclear modification factor in pXe collisions (black square), Pb-Xe collisions (blue dot) for $\Upsilon(1S)$, $\Upsilon(2S)$ and $\Upsilon(3S)$, in the rapidity range $3 < y_{\rm lab} < 5$.

In order to better understand the yield of neutrinos originating from charmed hadrons produced during the collision of ultra-high energy cosmic rays with the earth atmosphere, the charm-hadroproduction cross section needs to be accurately determined. In particular, the presence of a non-perturbative intrinsic charm (IC) component in the proton can alter the charm hadron yields. AFTER@LHC, covering the large negative Feynman-x_F region down to very low p_T for charmed hadrons is ideal to put constraints on the IC component of the proton and therefore on the flux of neutrinos from charm at high energy. Figure 3 (left) represents the D^0 meson yield as a function of p_T, in pp collisions at $\sqrt{s} = 115$ GeV, for one year of data taking assuming a LHCb-like detector with a storage-cell target. Large yields up to about 10 GeV/c are expected in the 3 rapidity ranges considered. In figure 3 right, the impact of the IC on the relative D^0 yield uncertainty is shown as a function of p_T in the rapidity range $2 < y_{lab} < 3$ and compared with the projected uncertainties on the D^0 yields predicted for AFTER@LHC (black lines). The red (green) band assumes a fraction of 0.57% (2%) of IC and are derived from theoretical cross sections from Ref [17]. Even considering a 5% systematical uncertainty on the D^0 yield measurement at AFTER@LHC, the precision reached for p_T <12 GeV/c will permit to set strong constraints on IC models.

4 Conclusion

The AFTER@LHC study group is currently investigating the possibilities offered by a future multi-purpose fixed-target experiment at the LHC, in the sector of high-x, spin and heavy-ion physics. Several possible implementations of a fixed-target setup inside an existing LHC experiment (ALICE or LHCb) are currently explored and the expected integrated yearly luminosities have been derived for each setup. In these proceedings, we have presented a selection of projected performance for light and heavy-flavour production at AFTER@LHC. It has been shown that the elliptic flow of identified particles can be measured accurately with an ALICE- or LHCb-like detector, after only few hours of Pb-Pb data taking at $\sqrt{s_{NN}} = 72$ GeV, opening the path for a study of the temperature dependence

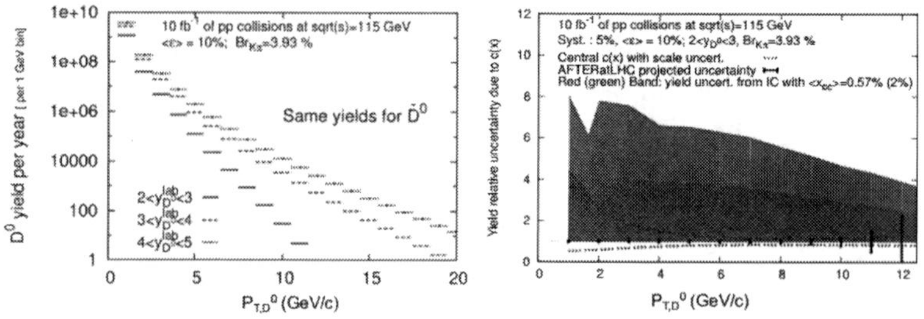

Figure 3. Left panel: D^0 meson yield per year as a function of p_T in fixed-target pp collisions at $\sqrt{s} = 115$ GeV, for three D^0 rapidity ranges: $2 < y_{lab} < 3$ (red line), $3 < y_{lab} < 4$ (green dashed line), $4 < y_{lab} < 5$ (blue dotted line). Right panel: Relative uncertainty on the D^0 meson yield due to the IC content of the proton as a function of p_T. The vertical black lines correspond to the statistical projections of the AFTER@LHC yearly D^0 meson yield in pp collisions at $\sqrt{s} = 115$ GeV, in $2 < y_{lab} < 3$.

of the medium shear viscosity. Both in pXe, and PbXe collisions, the $\Upsilon(nS)$ yields can already be measured after one year of data taking. It will allow to calibrate the QGP thermometer in an energy domain between the ones probed at SPS and RHIC using Υ excited states suppression in the QGP. Accurate measurements in pXe collisions will complement the PbXe studies, for a proper interpretation of the CNM effects. Finally, the large D^0 meson yields expected to be recorded in pp collisions at $\sqrt{s} = 115$ GeV at large-x, will be decisive to constrain the proton IC content and to provide valuable inputs for cosmic-ray physics. The study of open heavy-flavour hadrons in pA/AA will permit to study collectivity in small systems with new probes and to study heavy-quark energy-loss mechanisms in a dense medium.

Acknowledgement

This research was supported by the French P2IO Excellence Laboratory, the French CNRS via the grants FCPPL-Quarkonium4AFTER & Défi Inphyniti–Théorie LHC France, the P2I department of the Paris-Saclay University and by the COPIN-IN2P3 Agreement. AS acknowledges support from U.S. Department of Energy contract DE-AC05-06OR23177, under which Jefferson Science Associates, LLC, manages and operates Jefferson Lab.

References

[1] http://after.in2p3.fr
[2] S.J. Brodsky et al., Phys. Rept. **522**, 239 (2013), arXiv:1202.6585
[3] L. Massacrier, others., Adv. High Energy Phys. **2015**, 986348 (2015), arXiv:1504.05145
[4] D. Kikola et al., Few Body Syst. **58**, 139 (2017), arXiv:1702.01546
[5] B. Trzeciak et al., Few Body Syst. **58**, 148 (2017), arXiv:1703.03726
[6] A. Mocsy, Eur. Phys. J. **C61**, 705 (2009), arXiv:0811.0337
[7] R. Aaij et al. (LHCb), JINST **9**, P12005 (2014), arXiv:1410.0149
[8] R. Aaij et al. (LHCb), LHCb-CONF-2017-001, CERN-LHCb-CONF-2017-001 (2017)
[9] A. Zelenski et al., Nucl. Instrum. Meth. **A536**, 248 (2005)
[10] A. Airapetian et al. (HERMES), Nucl. Instrum. Meth. **A540**, 68 (2005), arXiv:physics/0408137
[11] W. Scandale et al., Phys. Lett. **B692**, 78 (2010)
[12] T. Pierog et al., Phys. Rev. **C92**, 034906 (2015), arXiv:1306.0121
[13] K. Werner, F.M. Liu, T. Pierog, Phys. Rev. **C74**, 044902 (2006), arXiv:hep-ph/0506232
[14] T. Sjostrand, S. Mrenna, P.Z. Skands, Comput. Phys. Commun. **178**, 852 (2008), arXiv:0710.3820
[15] H.S. Shao, Comput. Phys. Commun. **184**, 2562 (2013), arXiv:1212.5293
[16] H.S. Shao, Comput. Phys. Commun. **198**, 238 (2016), arXiv:1507.03435
[17] S.J. Brodsky et al., Adv. High Energy Phys. **2015**, 231547 (2015), arXiv:1504.06287

The sPHENIX Experiment

Carlos E. Pérez Lara[1,⋆], for the sPHENIX Collaboration

[1]*Stony Brook University, Physics and Astronomy Department*

Abstract. Our understanding of QCD under extreme conditions has advanced tremendously in the last 20 years with the discovery of the Quark Gluon Plasma and its characterisation in heavy ion collisions at RHIC and LHC. The sPHENIX detector planned at RHIC is designed to further study the microscopic nature of the QGP through precision measurements of jet, upsilon and open heavy flavor probes over a broad p_T range. The multi-year sPHENIX physics program will commence in early 2023, using state-of-the art detector technologies to fully exploit the highest RHIC luminosities.
The experiment incorporates the 1.4 T former BaBar solenoid magnet, and will feature high precision tracking and vertexing capabilities, provided by a compact TPC, Si-strip intermediate tracker and MAPS vertex detector. This is complemented by highly granular electromagnetic and hadronic calorimetry with full azimuthal coverage.
In this document I describe the sPHENIX detector design and physics program, with particular emphasis on the comprehensive open heavy flavour program enabled by the experiment's large coverage, high rate capability and precision vertexing.

1 Introduction

The ultimate exploration of the properties of the hot QCD matter formed in heavy ion collisions requires a complementary study of the observables produced at both RHIC and LHC energies. Since both accelerators collide ions at different center of mass energy, the initial temperature of the QGP will be different and so will the in-medium evolution of hard probes. The main observables for a full characterization of the QGP dynamics include: jet and heavy flavor energy loss, since jets and heavy flavor are produced early in the collision and can be used to probe the whole evolution of the system; sequential suppression of quarkonia, since they are sensitive to the temperature of the system; among others.

The sPHENIX detector [2] is a new detector at RHIC that replaces the PHENIX experiment, which concluded its final data taking run in 2016. The detector system has been designed to cope with the challenging physics goals that will be pursued at RHIC during its high intensity era in the 2020s. In these proceedings, I will briefly described the main elements of the current detector design, its performance and the running scenario for a possible five year plan.

2 The sPHENIX Experiment Timeline

Since its formation in 2015, the sPHENIX collaboration has been guiding the experiment realization through different stages. The conceptual detector design was approved in late 2016 and currently the

⋆e-mail: carlos.perezlara@stonybrook.edu

Table 1. Running scenario in a five-years plan

Year	System	Weeks	Samp. Lum (all Z)
1	Au+Au	16	34 nb^{-1}
2	p+p	11.5	267 pb^{-1}
2	p+Au	11.5	1.46 pb^{-1}
3	Au+Au	23.5	88 nb^{-1}
4	p+p	23.5	783 pb^{-1}
5	Au+Au	23.5	92 nb^{-1}

design for the proposed detector is being finalized. The detector construction is scheduled to begin in late 2018 and its installation at RHIC will start in late 2020. Data taking is envisioned to start in the middle of 2023 which overlaps greatly with LHC Run 3 timeline and may continue past the LHC Long Shutdown 3 (LS3).

sPHENIX will record data during the RHIC's highest intensity era for three different systems p+p, p+Au and Au+Au at RHIC's top energy $\sqrt{s_{NN}} = 200$ GeV . The proposal for a five-year running plan starts with six weeks of Au+Au the first year followed by 23 weeks of both p+p and p+Au in the next year. The next three years sPHENIX will be recording Au+Au and p+p events at even much higher rates. During the five year running plan, summarized in table 1, sPHENIX will record around 240 billion (1.5 trillion) minimum bias events for Au+Au with a collision vertex reconstructed in $|Z| < 10$ cm (all Z). Furthermore the usage of high level triggers in Au+Au collisions for b-tag or photon-tag events will increase the number of events by about a factor 2.

3 Detector Description

The detector features full azimuthal coverage and a wide pseudorapidity acceptance $|\eta| < 1.1$ and its final configuration is currently under intense R&D. The central detector consists of a charge particle tracking system provided by three subdetectors with complementary technology and a calorimetry system with both electromagnetic and hadronic components. At the time of this conference the following technologies and specifications are part of the detector configuration:

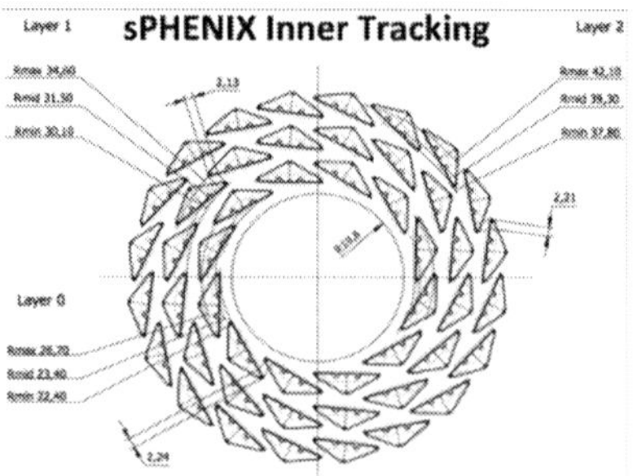

	Layer 0	Layer 1	Layer 2
Min. Radius (mm)	22.4	30.1	37.8
Max. Radius (mm)	26.7	34.6	42.1
Sensitive Length (mm)	271	271	271
Active area (cm^2)	421	562	702
N. of pixel chips	108	144	180
N. of staves	12	16	20

Figure 1. The MVTX detector consists of three layers of silicon pixel detector (MAPS) located very close to the beam pipe and featuring fast response and accurate position measurements for tracking and vertexing.

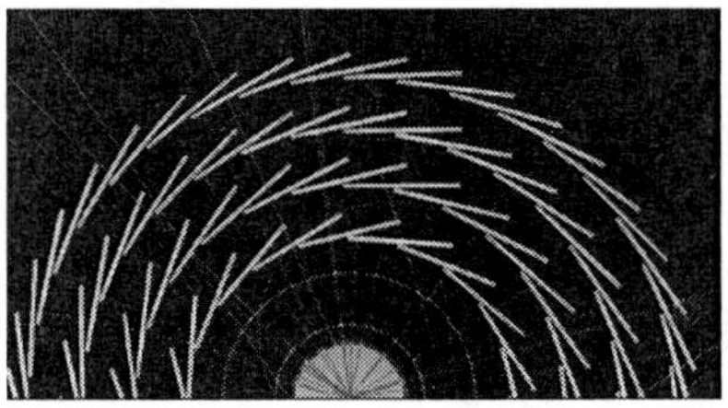

	Radius (cm)	Number of Ladders	Strip Size ($\phi \times$ z, mm)
0	6	20	0.078×18 (18)
1	8	26	0.086×20 (20)
2	10	32	0.086×20 (20)
3	12	38	0.086×20 (20)

Figure 2. The INTT detector has four layers of silicon strip technology situated in between the MVTX and TPC detectors. It aids the pattern recognition and momentum reconstruction for the central tracking system.

3.1 Charged Particle Tracking

The sPHENIX detector is designed around the 1.4 T superconducting solenoid magnet used previously in the BaBar experiment. The magnet is 3.8 m long with an inner radius of 140 cm and a thickness of 33 cm. The charge particle tracking detectors will be housed inside the magnet and is based on three different detector technologies: a silicon pixel detector (MVTX), for accurate measurements of displaced vertices; a silicon strip detector (INTT), for complementary high momentum resolution, and a compact time projection chamber (TPC) for high resolution up to low transverse momentum ($\sim$ 0.2 GeV/c).

The MVTX detector consists of three layers of silicon pixel detectors deployed radially as indicated in figure 1. The technology is based on that used by the ALICE ITS upgrade program: the MAPS sensors. It features a fine pitch of 28×28 μm^2 together with good time resolution of 5 μs. The detector has a DCA resolution better than 25 μm for charged particles with transverse momentum higher than 1 GeV/c.

The INTermediate Tracker (INTT) detector consists of four layers of silicon strip technology used previously in the FVTX detector at PHENIX. The INTT not only improves the momentum resolution for charged particles with the high transverse momentum, but also aids in the pattern recognition and event synchronization. The geometrical configuration of the detector can be seen in figure 2.

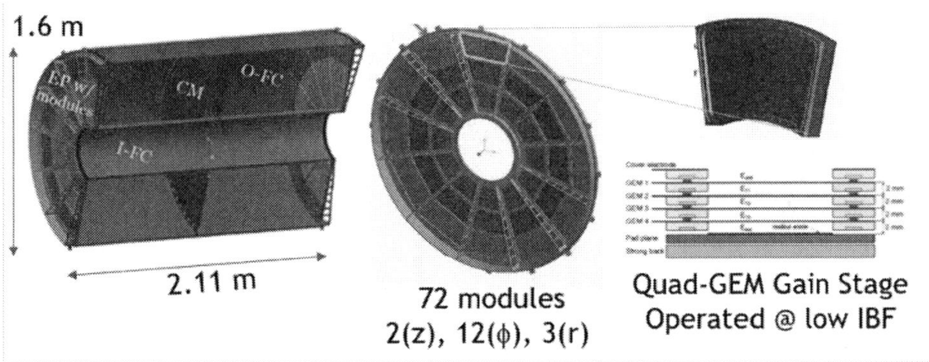

Figure 3. The compact TPC is a small Ne-based detector working in continuos readout using micropattern gas detectors. The high pad segmentation in the cathode plates provide optimal spatial resolution span in 40 radial layers of a mixed zigzag and rectangular pad configuration.

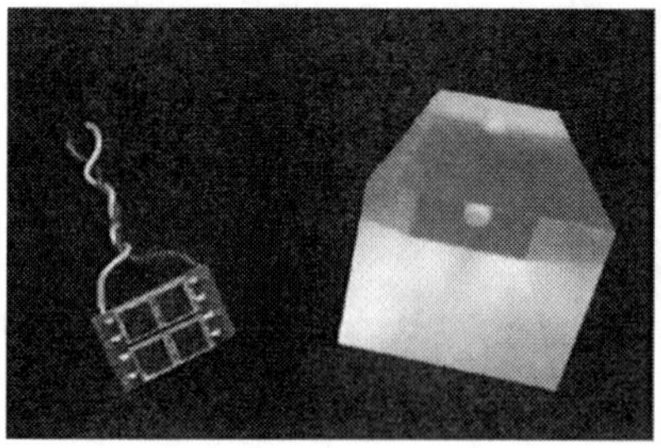

Figure 4. A drawing of part of the EMCal detector is shown on the left, where the block segmentation is also displayed. An individual tower and a SiPMT (resistent to hard magnetic field enviroment) is shown in the right.

The compact Time Projection Chamber (TPC) is the main tracking element of the sPHENIX experiment. The TPC field cage spans a volume from 20 cm to 78 cm in radius and 211 cm in longitude. It will run on continuous readout mode using micro pattern gas detectors at the end plates. The readout will be segmented into 72 modules: $2 \times 12 \times 3$ ($Z \times \phi \times R$) using both rectangular and zigzag pad geometries. The TPC will use a Ne-based gas for high ion mobility and low transverse diffusion in order to minimize spatial distortions from space charge build-up without compromising the desired position resolution. Its geometrical configuration is detailed in figure 3.

3.2 Calorimetry Detectors

The ElectroMagnetic Calorimeter (EMCal), figure 4, is the innermost calorimeter and consists of absorber blocks made from tungsten powder with embedded scintillating fibers. The design is similar to the SPACAL configuration used successfully in the past by various experiments. It features a sampling fraction of 2.3% and highly granular segmentation $\Delta\eta \times \Delta\phi = 0.024 \times 0.024$ (approx. 25k readout channels) while providing an energy resolution better than $15\%/\sqrt{E}$.

The Hadronic Calorimeter (HCal), figure 5, consists of steel and scintillating tiles with wavelength shifter fibers. The HCal system will have two subdetectors: one installed in between the EMCal and

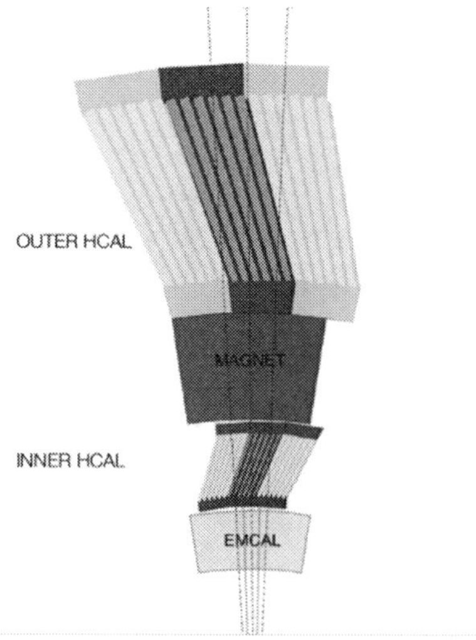

Figure 5. Traverse view of one sector of the hadronic calorimeter system (both inner and outer) together with the EMCal for scale comparison. The steel part of the HCal is also used as a return yoke for magnetic field containment. The geometrical orientation of the wavelength fibers is oposite between the inner and outter calorimeters.

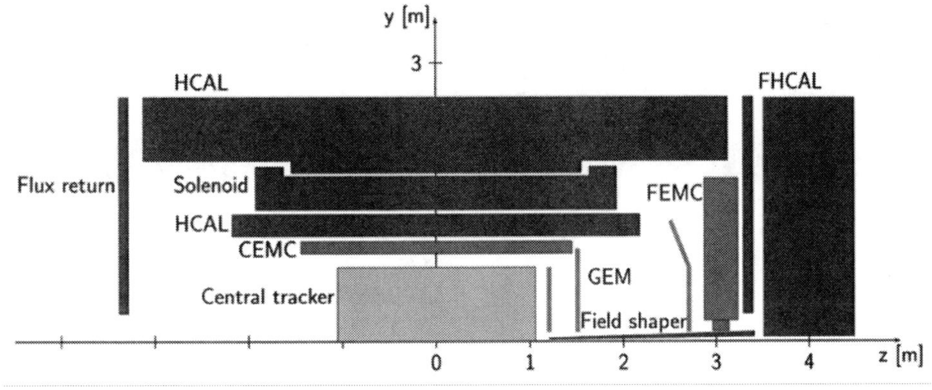

Figure 6. Sided-view drawing of the sPHENIX detector including the proposal for an upgrade in the forward p-going direction. The forward subsystem features a GEM tracker and both electromagnetic and hadronic calorimeters.

the magnet (inner HCal) and one installed outside of the magnet (outer HCal). The later will be used also as a return yoke for the magnetic field. The inner (outer) HCal thickness amounts to 1.0 (3.5) interaction lengths λ_I. Both subdetectors feature a granular segmentation of $\Delta\eta \times \Delta\phi = 0.1 \times 0.1$.

Both calorimeter systems were designed to have full azimuth coverage and complement each other in the reconstruction of jets. The performance of the calorimeter prototype has recently been assessed at a testbeam in Fermilab and the results of this study can be found in [3].

3.3 Detector Upgrade in the Forward Region

The sPHENIX collaboration has also prepared a comprehensive detector upgrade in the forward rapidity region [4]. The forward detector, see figure 6, consists of a forward tracker and calorimetry stations with a pseudorapidity coverage of $-1 < \eta < 4$. The forward upgrade allows for a broad physics program. It will not only allow for the study of nuclear structure and production at low X, but also complement the phase space of the sPHENIX central barrel in the pursue of photon-hadron correlations with a wide rapidity gap.

4 Simulation of Combined Performance

The sPHENIX collaboration has been working to put together a comprehensive software framework for the detector description and the different analyses being performed. There are many different topics studied in the different working groups in the collaboration. The following paragraphs are brief descriptions of a couple of selected topics.

4.1 Suppresion of b-jets

The analysis of b-jets is one of the main corner stones of the sPHENIX program. Since b-quarks, due to their large mass, are mainly created in hard scattering processes early in the collision, they sense the entire evolution of the medium and can be used to scan the properties of the QGP at different

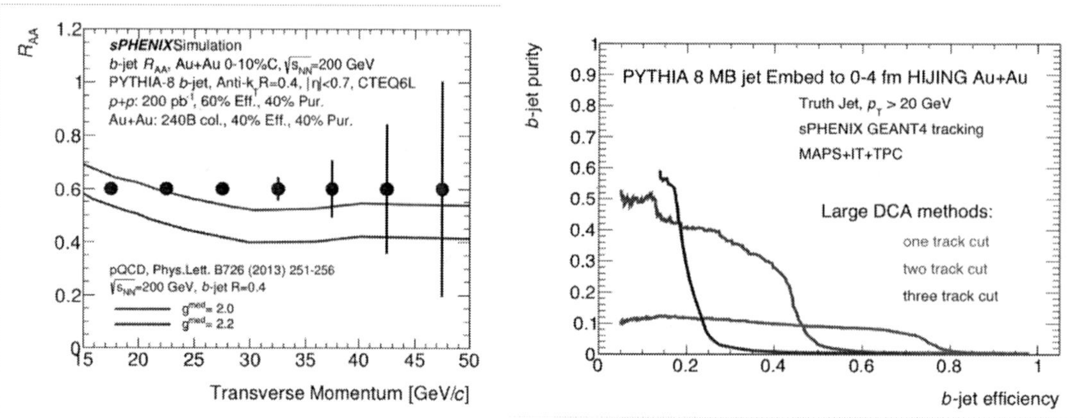

Figure 7. Left: b-jet R_{AA} as a function of transverse momentum for most central Au+Au collisions at $\sqrt{s} = 200$ GeV. The data sample consists of 240 billion events which would be the amount recorded based on the five-years run plan where no L1 trigger was used. Right: Efficiency vs purity of the L1 b-jet trigger performance for HIJING simulation of most central Au+Au collisions using one of the methods being studed in sPHENIX. Respective plots for other methods/conditions can be found in [5]

virtualities. In sPHENIX, jets produced by b quark showers can be reconstructed down to $p_T = 15$ GeV/c which allows for the exploration of the relevant mechanisms for their in-medium energy loss.

Figure 7 shows the accuracy expected in the nuclear modification factor of b-jets for 240 billion most central Au+Au collisions. In order to increase accuracy, specially at high p_T, studies of several L1 trigger algorithms have been performed. One of them requires that tracks from the candidate jet do not originate from the main vertex but are rather displaced. Figure 7 shows the performance of such an algorithm using the sPHENIX tracking system to tag one, two and three tracks with displaced vertex. There is currently work in progress [5] to refine this selection as our tracking system is being

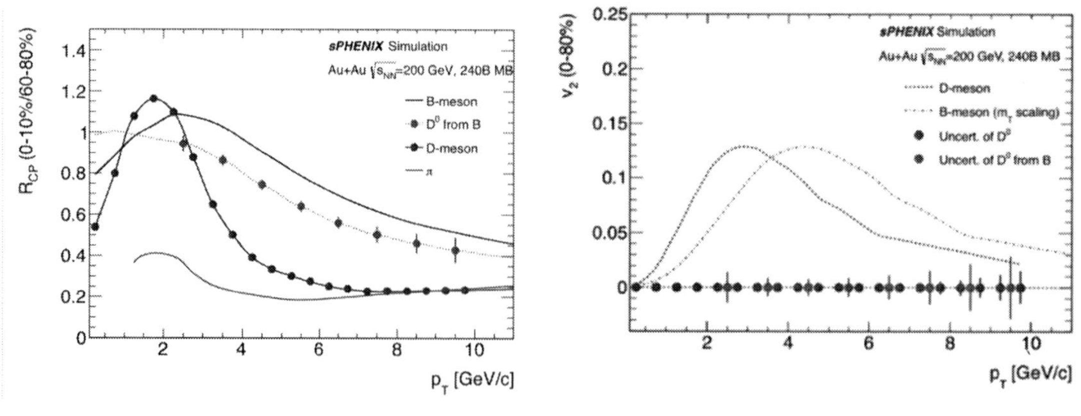

Figure 8. Simulation of the performance of D-meson R_{CP} and v_2 for MB Au+Au collisions at $\sqrt{s} = 200$ GeV. The sample corresponds to 240 billion Hijing Events, which amounts to the expected statistics collected during the full experiment lifetime.

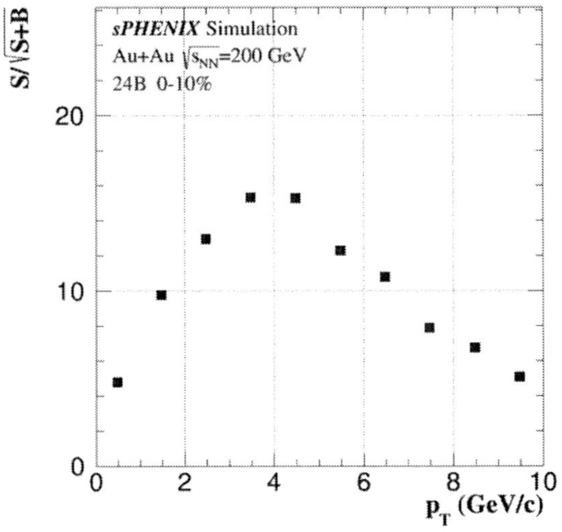

Figure 9. Significance of fully reconstructed B^+ mesons as a function of transverse momentum for 240 billion most central Au+Au collisions at $\sqrt{s}$ = 200 GeV. The peak at 4 GeV/c is due to the large combinatorial background present for low p_T candidates.

finalized, however one can see in this plot the rough interplay of efficiency and purity depending on the severity of the cut used for triggering.

4.2 Fully reconstructed open heavy flavor

The dynamics of the parton energy loss in the QGP can be addressed via the study of suppression and angular correlations of fully reconstructed charm and bottom mesons. These measurements are possible due to the high spatial resolution of the MVTX for reconstruction of displaced vertices and the high momentum resolution for charged particles down to low p_T provided by the TPC and INTT detectors. The decay channel explored is the decay of open charm (one decay vertex) and bottom (two decay vertexes) into hadrons (i.e. $D_0 \rightarrow K^-\pi^+$ and $B^\pm \rightarrow \bar{D}_0(K\pi)\pi^\pm$). Figure 8 shows the accuracy estimation for central to peripheral yield ratio in inclusive D-mesons and D^0 from B using 240 billion MB events. Figure 9 shows the significance of the B^+ meson fully reconstructed via double decay vertex for most central Au+Au collisions at $\sqrt{s}$ = 200 GeV. For both D- and B-mesons, the selection is done via topological multivariate cuts. A feature of sPHENIX is its low transverse momentum reach for heavy flavor measurements, which is important since low p_T hadrons are most sensitive to the medium as it expands and cools down, i.e. collective flow and heavy flavor transport.

5 Summary

In these proceedings, I have summarized the sPHENIX detector's current design. R&D project is ongoing in order to finalize the detector layout by 2018. The reader can go to https://www.sphenix.bnl.gov/ for the most up-to-date detector description. The detector technologies have been chosen to provide a broad acceptance coverage, high momentum resolution down to 0.2 GeV/c for charged particles, highly granular electromagnetic and hadronic calorimetry, and quick readout time in order to cope with the high luminosity RHIC era. sPHENIX is suitable for the study of jets, fully reconstructed open heavy flavor, fully reconstructed upsilons and correlations, thus providing highly accurate measurements for a full characterization of the hot and dense QCD matter formed in heavy ion collisions.

References

[1] A. Aprahamian *et al.*, "Reaching for the horizon: The 2015 long range plan for nuclear science", INSPIRE-1398831

[2] Adare, A. and others, An Upgrade Proposal from the PHENIX Collaboration, ARXIV:1501.06197

[3] Aidala, C. A. and others, Design and Beam Test Results for the sPHENIX Electromagnetic and Hadronic Calorimeter Prototypes, ARXIV:1704.01461

[4] sPHENIX Forward Instrumentation. sPHENIX analysis note: sPH-cQCD-2017-001

[5] Heavy Flavor Jet Simulation and Analysis. sPHENIX simulation analysis note: sPH-HF-2017-001

[6] D^0-meson and B^+-meson production in Au+Au Collisions at $\sqrt{s_{NN}} = 200$ GeV for sPHENIX. sPHENIX simulation analysis note: sPH-HF-2017-002

The Small Acceptance Vertex Detector of NA61/SHINE

M. Deveaux[1,★], *A.* Aduszkiewicz[2], *Y.* Ali[3], *M.* Baszczyk[4], *W.* Brylinski[5], *P.* Dorosz[4], *S.* Di Luise[7], *G.* Feofilov[6], *M.* Gazdzicki[1], *S.* Igolkin[6], *M.* Jablonski[3], *V.* Kovalenko[6], *M.* Koziel[1], *W.* Kucewicz[4], *D.* Larsen[3], *T.* Lazareva[6], *P.* Martinengo[8], *A.* Merzlaya[3,6], *L.* Mik[4], *R.* Planeta[3], *A.* Snoch[1], *V.* Vechernin[6], *D.* Tefelski[5], *M.* Suljic[9], *P.* Staszel[3], and *the NA61/SHINE collaboration*

[1] *Goethe University Frankfurt, Germany*
[2] *University of Warsaw, Warsaw, Poland*
[3] *Jagiellonian University, Cracow, Poland*
[4] *AGH University of Science and Technology, Cracow, Poland*
[5] *Warsaw University of Technology, Warsaw, Poland*
[6] *St. Petersburg State University, St. Petersburg, Russia*
[7] *Eidgenössische Technische Hochschule Zürich, Zürich, Switzerland*
[8] *European Organization for Nuclear Research (CERN), Geneva, Switzerland*
[9] *University & INFN Trieste, Italy*

Abstract. Charmonium production in heavy ion collisions is considered as an important diagnostic probe for studying the phase diagram of strongly interacting matter for potential phase transitions. The interpretation of existing data from the CERN SPS is hampered by a lack of knowledge on the properties of open charm particle production in the fireball. Moreover, open charm production in heavy ion collisions by itself is poorly understood. To overcome this obstacle, the NA61/SHINE was equipped with a Small Acceptance Vertex Detector (SAVD), which is predicted to make the experiment sensitive to open charm mesons produced in A-A collisions at the SPS top energy. This paper will introduce the concept and the hardware of the SAVD. Moreover, first running experience as obtained in a commissioning run with a 150 *A*GeV/*c* Pb+Pb collision system will be reported.

1 Introduction

The NA61/SHINE [1] experiment is a fixed target hadron spectrometer located at the CERN SPS. The experiment can be operated with hadron (π, K, p) and heavy ion beams (Be, Ar, Xe, Pb). The beam energy amounts 13 – 158 *A*GeV/*c* for ions and up to 400 GeV/*c* for protons. Since 2009, NA61/SHINE is undertaking a systematic scan of the nuclear phase diagram, which relies on measurements with different beam energies and system sizes. This phase scan aims for obtaining an understanding of a potential first order phase transition between hadronic matter and QGP and the identification of the predicted critical endpoint of this phase transition.

J/ψ-melting has been proposed as a suited diagnostic probe for identifying this phase transition [2] and was soon experimentally observed at the CERN-SPS. However, a precise and quantitative measurement of the yield and properties of open charm particles is required to discriminate between

★e-mail: deveaux@physik.uni.frankfurt.de

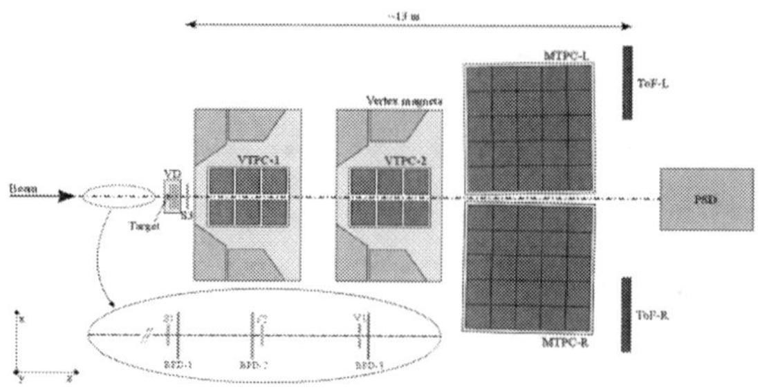

Figure 1. Sketch of the NA61/SHINE detector setup used during the Pb+Pb 2016 data taking. The position of the SAVD is shown.

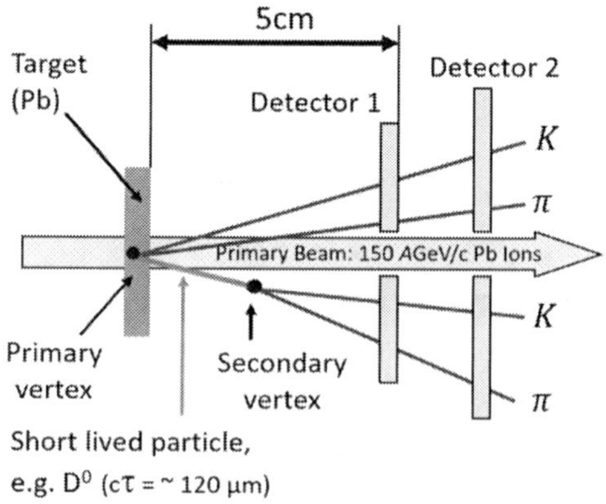

Figure 2. Reconstruction strategy for open charm (not to scale). For details, see section 2.

the onset of phase transition and alternative explanations of the findings [3]. The Small Acceptance Vertex Detector (SAVD) of the NA61/SHINE experiment was build to provide the necessary data by means of doing direct measurements of open charm mesons via their hadronic decay channel (e.g. $D^0 \rightarrow K^- + \pi^+$) at the SPS top energies.

2 Particle reconstruction strategy and detector requirements

Open charm mesons show a typical life-time of $c\tau \approx 100$ μm. Their daughter particles, namely K and π, are produced in large numbers by other sources in heavy ion collisions. To distinguish the daughter particles of open charm particles from those primary hadrons, one aims to accept only hadron pairs generated in a secondary decay vertex (see Figure 2). This requires to separate the latter from the primary vertex of the heavy ion collision, which comes with the need to measure the position of both vertices with excellent precision. This measurement is done by extrapolating the trajectories back to the production target and identifying intersection points. The primary vertex will typically appear as intersection point of multiple tracks while the number of tracks leaving a decay vertex remains small and well-defined.

The uncertainties of a vertex measurement are dominated by the multiple coulomb scattering of the particle in the detector material and by the limited position accuracy of the detector. To minimize both uncertainties, the detector has to provide an excellent spatial resolution. Moreover, it must be build from a minimal amount of material and located the closest possible to the primary vertex.

3 The SAVD

3.1 Detector concept

The SAVD of NA61/SHINE was designed to perform a vertex reconstruction with ~70 μm accuracy. It consists of two spectrometer arms composed each from four detector stations, which are located 5, 10, 15 and 20 cm downstream the target in the stray field of the vertex magnet of NA61/SHINE (see Figure 1). The number of stations was motivated by the wish for a limited stand-alone tracking capability, which eases to match the SAVD tracks with the tracks found in the TPCs. The position of the first station forms a compromise between the need to handle the high radiation load and track

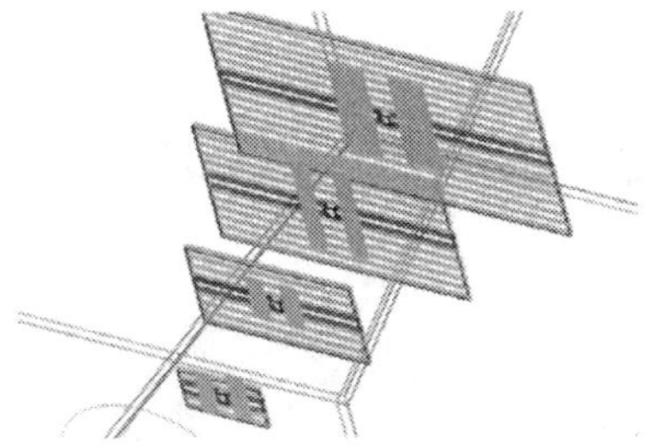

Figure 3. Active surface of the SAVD (orange) vs. the nominal geometrical acceptance of NA61/SHINE in the vertex detector region (green).

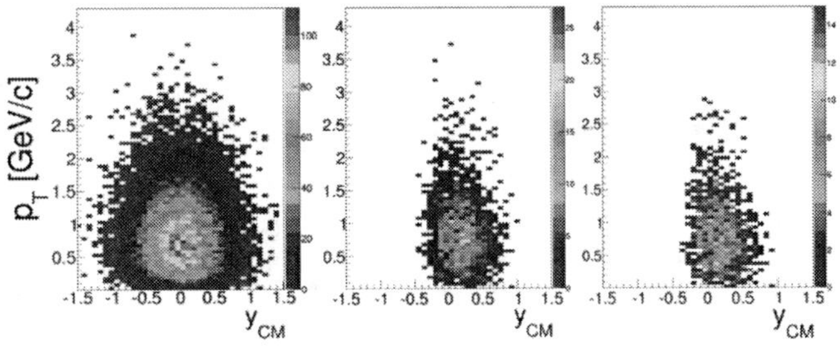

Figure 4. Simulated acceptance of the SAVD for $D^0 \to K^- + \pi^+$ for 150 *A*GeV/*c* Pb+Pb collision systems. Left: Anticipated 4π-emission, middle: full NA61/SHINE acceptance, right: SAVD acceptance.

density found nearby the collision point (which is eased by increasing the distance) and the need to measure the vertex position with good precision (which calls to reduce the distance).

The stations are equipped with the ultra-light, only 50 μm thin, MIMOSA-26AHR CMOS sensors [4, 5] developed by the IPHC Strasbourg. Those silicon sensors provide a very fine pixel pitch of $18.4 \times 18.4\ \mu m^2$ resulting in an outstanding spatial resolution of 3.5 μm. The $1.06 \times 2.13\ cm^2$ active surface of the sensor is covered by 1156 columns made each from 576 pixels. The readout of the pixel is done by means of a rolling shutter and the readout time is 115.2 μs. MIMOSA-26AHR tolerates an integrated non-ionizing dose of $> 10^{13}\ n_{eq}/cm^2$ [5] and an integrated ionizing dose of > 150 krad [6].

The SAVD had to be realized with limited resources, which required to minimize rigorously its size and complexity. As illustrated in Figure 3, each of the first two layers of the SAVD hold only one sensor per arm. The third station and the last station hold two and four sensors per arm respectively. This turns into a total of 16 sensors with a surface of 32 cm^2 and 10^7 pixels, which covers only a small fraction of the geometrical acceptance of NA61/SHINE. However, thanks to the forward boost of the center-of-mass at 150 *A*GeV/*c*, it is sufficient to cover mid rapidity and a substantial part of the forward rapidity (see Figure 4). Simulations [7] suggest that the SAVD accepts geometrically 35% of all D^0 emitted into the full NA61/SHINE acceptance. During the simulations, we assumed the mean multiplicity of $D^0 + \bar{D}^0$ mesons per central (20%) 150 *A*GeV/*c* Pb+Pb collision to be 0.1. Then, the number of reconstructed $D^0 + \bar{D}^0$ mesons from 4×10^6 events amounts 1500 [8]. The S/B was estimated to amount 5 (0.5) assuming perfect (no) particle identification respectively.

3.2 Running conditions

To reduce the beam-gas interactions, the SAVD is operated in a detector vessel flooded with helium, which also contains the target. The magnetic field in the detector region is not homogeneous and amounts $B < 0.25$ T. The SAVD is exposed to radiation from three major sources, which are hadrons emitted from the primary collisions, delta-electrons kicked out of the target by the primary beam and direct beam ions. It was estimated that the two first radiation sources generate a radiation dose of $< 10^{12}\ n_{eq}/cm^2$ and 25 krad during a run of 40 days. The radiation damage caused by the direct beam ions depends decisively on the beam tuning. The radiation load was estimated based on historic data from the NA61/SHINE beam counters. It was found that the most exposed point of the detector might receive $\sim 10^9$ ions per cm^2 and run. According to the Bethe-Bloch equation, this corresponds to about 230 krad. This slightly exceeds the guaranteed capabilities of the sensor, but was considered as acceptable because only few mm^2 detector surface are affected. Moreover, radiation tolerance studies

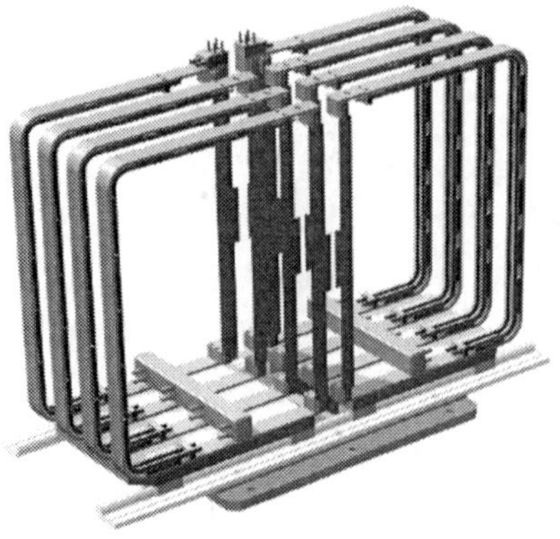

Figure 5. C-frame structure of the SAVD. The ladders, extension plates and cooling pipes are visible.

Figure 6. Photograph of the integrated detector. The sensors, extension plates, ladders and cables can be seen.

presented in [9] suggest that the reaction of MIMOSA-26AHR to modest local overdoses restricts itself likely to local losses of efficiency. The integrated non-ionizing dose was estimated to amount $4 \times 10^{11}\ n_{eq}/cm^2$ by extrapolating theoretical results found in [10].

3.3 Detector implementation

The construction of the detector became possible thanks to the availability of high performance building blocks, which were kindly made available by the PICSEL group of IPHC Strasbourg and the ALICE- and CBM-collaboration. The MIMOSA-26AHR sensors were integrated on 30 cm long, ultra-light, actively cooled carbon fibre ladders developed for the upgrade of the ALICE Inner Barrel [11]. The width of the ladders in the sensor region was extended by means of a thin carbon fibre plate. The sensors and the dedicated, 20 cm long single layer flex print cables used for steering them were first glued to this plate and wedge bonded hereafter. In a next step, the integrated plate was glued to ladder. This integration step relied on tools and experiences obtained from the prototype of the CBM Micro Vertex Detector (MVD) [12].

The design of the flex print cables between the sensors and the related FEE-boards was optimized to minimize the material budget of the SAVD within the acceptance of NA61/SHINE. This was done to preserve the sensitivity of the experiment outside of the SAVD acceptance. The decision was taken knowing that, unlike to newer sensors of the MIMOSA-series, MIMOSA-26AHR is somewhat sensitive to pick-up noise injected by long cables. External filter capacitors were added to alleviate this effect. The front end boards biasing the sensors, and the TRB-data processing boards [13] providing their fast and slow control, were copied from the updated [14] CBM MVD prototype. The related firmware and software [15] was also used but had to be adapted to the needs of NA61/SHINE. Among others, the intrinsically free streaming local DAQ of the MVD prototype had to be interfaced with the triggered DAQ of NA61/SHINE.

The ladders are held by dedicated C-frames made from aluminium. During the December 2016 Pb+Pb run reported in this work, they were complemented by one native ALICE-ITS ladder equipped with ALPIDE sensors. This was done in order to test the sensors to the high particle fluxes found in fixed target experiments. The C-frames of each arm are mounted on a magnet field tolerant moving table, which allows to move the sensors into a safe position during beam tuning or in case of a beam loss. This aims to avoid a direct impact of the beam into the sensors. Moreover, the moving engines allow to extend the only 6 mm wide beam gap in case necessary beam focusing is not reached. We found that the tables available on the market are too slow to react in a reasonable way to an unexpected beam displacement. Fortunately, we also observed that the sensors tolerate direct beam impacts without significant loss of performance (see next section). The target holder was mounted on a separate moving table.

A light and gas tight detector box was realized with metallized Plexiglas with beam entrance and exit windows made from aluminium foil. The cooling system relied on an industrial chiller using demineralized water, which was kept at modest underpressue. This underpressure and the reduced conductivity of the coolant was to reduce the consequences of a potential leakage in the cooling pipes on the nearby electronics.

4 First operation experience

4.1 Immunity to heavy ions

In 2015, the SAVD and its components were tested in beam times with a 30 *A*GeV/*c* Pb beam. The first beam time focused on testing the immunity of the MIMOSA-26AHR sensors to direct beam hits. This included irradiating unpowered MIMOSA-34 test sensors with up to 1.2×10^{10} 30 *A*GeV/*c* Pb-ions per cm^2 and testing them hereafter. As discussed in more detail in [16], the test sensors survived the illumination without significant radiation damage, which is is compatible with the extrapolated predictions of [10].

Independently, an individual thick MIMOSA-26 sensor was operated directly in the heavy ion beam. This test aimed to observe the response of the sensor to heavy ions and to spot the probability of latch-ups and destructive radiation damage on the sensor. Due to missing reference detectors, the experiment was not suited to observe more subtle radiation effects like moderate losses of detection efficiency. It was found that heavy ion impacts generate clusters with about 200 fired pixels. A ring shaped after-glow is sometimes observed. As the sensor was found to be encouragingly immune to ions, it was moved to the beam spot and operated at an ion rate of about 1 kHz (during spill) for 24 hours. We observed no hard error (e.g. stop of operation, reset, data corruption) of the sensor. Three latch-up events causing a modest over-current below the threshold of the related protection system were seen. They were recovered by manual power cycling.

4.2 First observations from the commissioning run

A first commissioning beam time for the SAVD with 150 *A*GeV/*c* Pb+Pb collision system was carried out in December 2016. During this beam time, the SAVD was operated together with the full NA61/SHINE detector. The trigger of NA61/SHINE was tuned to record 20% central collisions, which was not optimal for open charm search but allowed for carrying out other detector tests in parallel. A total of 1.3×10^5 central collisions (after quality cuts, preliminary) were collected.

15 of the nominal 16 sensors were found operational and the detector operated without failure or significant damage during the full beam time. One detector arm generated about 10% of corrupted events, which was associated to a version conflict within the readout and steering software.

The data obtained from the beam test is being analysed. A track finding based on a Hough transform [17] is being used for the stand alone tracking in the SAVD. Identified tracks are extrapolated to a matching plane and merged with TPC-tracks based on position and inclination. Preliminary results show that the data obtained from the SAVD and the remaining NA61/SHINE-detector can be synchronized and that track matching is possible. After a preliminary alignment of the SAVD, a spatial resolution of $\sim 6\ \mu m$ (hit to track residuals) was observed. This very satisfactory result allowed to measure the primary vertex with an about 30 μm precision along beam direction, which matches once more the expectations.

5 Summary and outlook

The SAVD of NA61/SHINE is to reconstruct open charm particles generated in Pb+Pb collisions at highest SPS energies. Its for stations integrate MIMOSA-26AHR CMOS sensors as provided by the IPHC Strasbourg and the ultra-light carbon fibre ladders developed for the ALICE ITS. The DAQ components and integration techniques were developed for the prototype of the CBM-detector. Despite its small active surface of 32 cm^2, the SAVD covers the forward rapidity of the fireball at the nominal beam energy of 150 *A*GeV/*c*. The detector was constructed, and tested among others in a beam time with 150 *A*GeV/*c* Pb+Pb collision system. The results confirm the feasibility of using CMOS sensor based vertex detectors in a fixed target heavy ion environment and can also be considered as a successful test of the building blocks used. Among the positive findings is the unexpectedly high immunity of the sensors to direct heavy ion hits. Moreover, a preliminary analysis of the data confirms the anticipated good spatial and vertexing resolution of the device, which provides an optimistic prospective on the success of future open charm measurements within a production run. Remaining issues like the e.g. the above mentioned software version issue are being eliminated. The improved SAVD will be used and re-tested in a Xe-La beam time end 2017 before performing a high statistics measurement in a 150 *A*GeV Pb+Pb collision system (scheduled for end 2018). Hereafter, it is planned to decommission the SAVD and to replace it by a full-size vertex detector (Large Acceptance Vertex Detector, LAVD) during the long shut down two of the CERN SPS.

Acknowledgements

This work became possible thanks to know-how and support from the ALICE-collaboration, the CBM-collaboration, the TRB-collaboration, and the PICSEL group of IPHC Strasbourg. We wish to thank P. Klaus, J. Michel and M. Wiebusch for valuable help and training on the CBM - MVD prototype DAQ. This work was supported by the Polish National Center for Science (2014/15/B/ST2/02537). M. Deveaux is supported by HIC for FAIR.

References

[1] N. Abgrall et al. (NA61/SHINE), JINST **9**, P06005 (2014)
[2] H. Satz, Nucl.Phys. **B326**, 613 (1989)
[3] H. Satz, Adv. High Energy Phys. **2013**, 242918 (2013)
[4] A. Dorokhov et al., Nucl. Instr. and Meth. **A650**, 174 (2011)
[5] M. Deveaux et al., Journal of Instrumentation **6**, C0 (2011)
[6] M. Koziel et al., Nucl. Instr. and Meth. **A624**, 437 (2010)
[7] Y. Ali, P. Staszel et al., Journal of Physics: Conference Series **509**, 012083 (2014)
[8] NA61/SHINE, CERN-SPSC-2015-038 / SPSC-P-330-ADD-8 (2015)
[9] D. Deoring, Ph.D. thesis, Goethe University Frankfurt (2014)
[10] M. Xapsos et al., IEEE TNS **51**, 3250 (2004)
[11] B. Abelev et al. (ALICE), J. Phys. **G41**, 087002 (2014)
[12] M. Koziel et al., Nucl. Instr. and Meth. **732**, 515 (2013)
[13] J. Michel et al., IEEE TNS **58**, 1745 (2011)
[14] P. Klaus et al., JINST **11**, C03046 (2016)
[15] B. Milanovic et al., PoS **TIPP2014**, 404 (2014)
[16] NA61/SHINE, CERN-SPSC-2016-038 / SPSC-SR-197 (2016)
[17] A. Merzlaya et al., Journal of Physics: Conf. Series **798**, 012072 (2017)

Small collision systems: Theory overview on cold nuclear matter effects

Néstor Armesto[1,⋆]

[1] *Departamento de Física de Partículas and IGFAE, Universidade de Santiago de Compostela, 15782 Santiago de Compostela, Galicia-Spain*

Abstract. Many observables measured at the Relativistic Heavy Ion Collider and the Large Hadron Collider show a smooth transition between proton-proton and proton-nucleus collisions (small systems), and nucleus-nucleus collisions (large systems), when represented versus some variable like the multiplicity in the event. In this contribution I review some of the physics mechanisms, named cold nuclear matter effects, that may lead to a collective-like behaviour in small systems beyond the macroscopic description provided by relativistic hydrodynamics. I focus on the nuclear modification of parton densities, single inclusive particle production and correlations.

1 Motivation

In recent years much attention has been devoted in the high-energy physics community to the experimental findings that indicate a smooth transition between "small" collisions systems (proton-proton, pp, and proton-nucleus, pA) and "large" collision systems (nucleus-nucleus, AA). These findings at both the Large Hadron Collider (LHC) and the Relativistic Heavy Ion Collider (RHIC), are particularly striking as some of the involved observables have been traditionally considered as signatures of the creation of deconfined partonic matter close to thermodynamical equilibrium - Quark Gluon Plasma - in high-energy AA collisions. They include (see the review [1]) azimuthal asymmetries of apparent collective nature, thermal-like spectra and baryon/strangeness enhancement at small and intermediate transverse momentum respectively, hadrochemistry, interferometry, quarkonium suppression,. . . , the only exception being jet quenching.

While this behaviour was first observed in AA and high-multiplicity pA and pp collisions, the interest in this subject has been increased recently by the observation that azimuthal asymmetries and the ridge structure - two particle correlations whose strength extends through several units of pseudorapidity and is maximum at zero (near side region) and π (back-to-back region) azimuthal angle - can also be observed in pp collisions with multiplicities close to minimum bias [2–4]. From the theoretical side, two facts augment the importance of these findings even further: the fact that viscous relativistic hydrodynamics is able to provide a description of such data [5] even if the conditions of large opacity and approximate momentum isotropy seem difficult to fulfil in pp; and the applicability of hydrodynamics in far-from-equilibrium situations that was obtained both at strong [6, 7] and weak coupling [8]. Thus, the traditional statement that the success of hydrodynamics clearly indicated the

⋆e-mail: nestor.armesto@usc.es

existence of an equilibrated partonic medium with vanishing viscosity has moved to the questions [9]:
• What does this apparent continuity in the applicability of collective ideas from small to large systems tell us about the dynamics of a non-equilibrium partonic system in Quantum Chromodynamics (QCD)?;
• How does the macroscopic description that hydrodynamics provides emerge from the microscopic QCD dynamics?

Several reviews exist on the subject, see e.g. [10, 11] and also the numerous presentations at the recent Initial Stages conference [12]. The experimental aspects have been presented in the talks by R. Preghenella and N. Novitzky at this conference. The aim of this contribution is to review effects that may cause or contribute to the collectivity found in small systems, whose origin lies in the initial state or the very initial stages of the collision. Therefore, they may provide an alternative or complementary explanation to that of final state rescattering leading to hydrodynamics. And, in any case, such initial state effects, called cold nuclear matter effects (CNM), must be anyway addressed both from a fundamental point of view - long range rapidity correlations in the ridge must come from the early stages of the collision, and from an instrumental one - hydrodynamics must be provided with initial conditions and the effects on the final observables of the dynamics prior to the hydrodynamic behaviour constrained. Due to limitations of space, I will focus on a few aspects: nuclear modification of parton densities, single inclusive particle production and correlations in small systems. References will not be exhaustive. I apologise in advance to those who may find their work underrepresented.

2 Nuclear modification of parton densities

The fact that structure functions in bound nucleons are different from those in free nucleons has been known for decades [13]. In collinear factorisation that is the standard tool for the calculation of observables with a perturbative scale in hadronic collisions, this fact reflects in nuclear modifications of parton densities (nPDFs), usually studied through the nuclear modification factor for parton density f_i, $R_i(x, Q^2) = f_i^A(x, Q^2)/[Af_i^p(x, Q^2)]$. The most recent analysis are shown in Fig. 1.

Focusing on the two most recent ones [14, 15], the difference between them are mainly the degree of freedom for the nuclear effects on the different quark flavours, applied only to valence in [14] and to both valence and sea quarks in [15], and the inclusion of neutrino and pPb LHC data in the second. The difference can be seen in the left and middle plots in Fig. 2. On the other hand, the effect of neutrino and pPb LHC data can be seen in the plots of the right, through a comparison of [15] with analysis that do not include LHC data [16] or neither neutrino nor LHC data [17]. While parton densities in the different analysis coincide in the regions constrained by experimental data, some comments are in order:
• The effect of the form of the initial parametrisation for DGLAP evolution, widely different in [14, 15], reflects in the fact that the uncertainties obtained by the standard Hessian analysis are very different, and also in the behaviours in regions far from those constrained by experimental data e.g. $x < 0.01$;
• The effect of including more sets of data, given by the comparison of the results of [15] with those of previous analysis [16, 17], seems small and even counterintuitive. This is partly due to the small constraining power of new data except in specific regions (e.g. for the glue at $x \sim 0.1$) and by the fact that a new, more flexible initial parametrisation produces larger uncertainties even if the same sets of data are employed.
Both items emphasise the role of the initial parametrisations on the uncertainties and extrapolations of nPDFs, an aspect well known in the community of proton parton densities and that will play an important role in the analyses of nPDFs in the future.

SET		HKN07 PRC76 (2007) 065207	EPS09 JHEP 0904 (2009) 065	DSSZ PRD85 (2012) 074028	nCTEQ15 PRD93 (2016) 085037	KA15 PRD93 (2016) 014036	EPPS16 EPJC C77 (2017)163
data	eDIS	✔	✔	✔	✔	✔	✔
	DY	✔	✔	✔	✔	✔	✔
	π^0	✘	✔	✔	✔	✘	✔
	νDIS	✘	✘	✔	✘	✘	✔
	pPb	✘	✘	✘	✘	✘	✔
# data		1241	929	1579	740	1479	1811
order		NLO	NLO	NLO	NLO	NNLO	NLO
proton PDF		MRST98	CTEQ6.1	MSTW2008	~CTEQ6.1	JR09	CT14NLO
mass scheme		ZM-VFNS	ZM-VFNS	GM-VFNS	GM-VFNS	ZM-VFNS	GM-VFNS
comments		$\Delta\chi^2$=13.7, ratios, no EMC for gluons	$\Delta\chi^2$=50, ratios, huge shadowing-antishadowing	$\Delta\chi^2$=30, ratios, medium-modified FFs for π^0	$\Delta\chi^2$=35, PDFs, flavour sep., not enough sensitivity	PDFs, deuteron data included	$\Delta\chi^2$=52, ratios, LHC pPb data

Figure 1. Summary of the most recent global analyses of nPDFs indicating the acronym and reference, the sets of employed data, the order of the perturbative evolution, the reference proton parton density used in the analysis, the scheme for treating heavy flavours and some comments like the employed tolerance and peculiarities of the analysis. See the relevant reference for details.

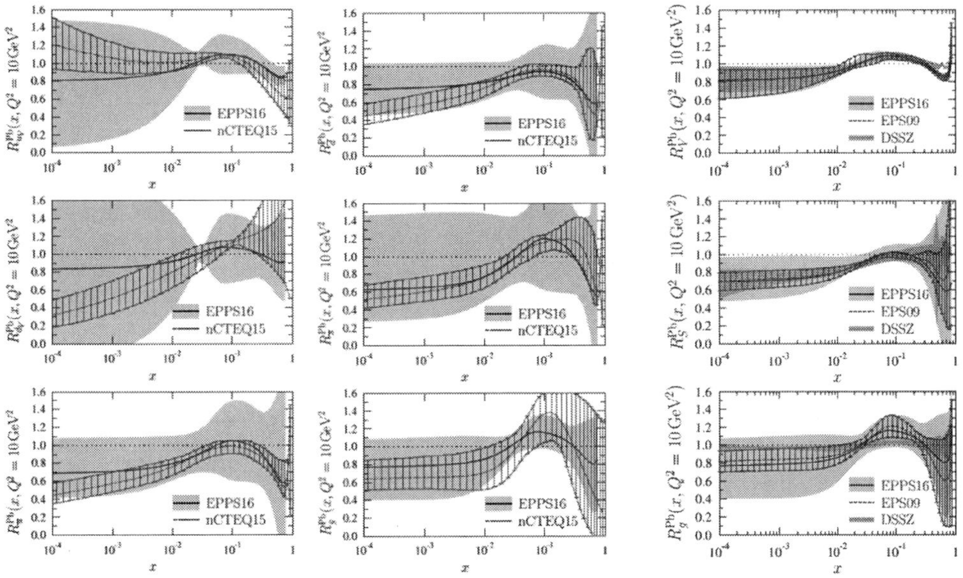

Figure 2. Left and middle plots: comparison between the nuclear modification factors nPDFs from the nCTEQ15 and EPPS16 sets. Plots on the right: id. from the EPPS16, EPS09 and DSSZ sets.

Item	Order	Theory	Pheno-menology	Comments
Evolution eqns.	NLO	✔	~	rcBK and resummations; dilute-dense approx.
DIS impact factor	NLO	✔	✘ at NLO	dilute-dense approx.
Hadrons at y~0	LO	✔	✔	q and Q , dilute-dense approx.
Forward hadrons	NLO	✔	✔	q and Q, hybrid formalism
Quarkonium at y~0	LO	✔	✔	dilute-dense approx.+NRQCD
Forward quarkonium	LO	✔	✔	hybrid formalism
$\gamma^{(*)}$ at y~0	NLO	✔	✘ at NLO	dilute-dense approx., not yet DY at NLO
Forward $\gamma^{(*)}$	LO	✔	✔	hybrid formalism
Dijets at y~0	LO	✔	✔	dilute-dense approx., partial NLO
Forward dijets	LO	✔	✔	hybrid formalism and high-energy factorisation, partial NLO
Diffractive dijets	NLO	✔	✘ at NLO	dilute-dense approx.
g/q/γ-g/q/γ correlations	LO	✔	✔/✘	glasma graph approx. + some density corrections; hybrid formalism

Figure 3. Summary of processes studied in the CGC with the perturbative order of the calculation, their availability for theory and phenomenology, and some comments.

RHIC and the LHC have also provided data with constraining power on nPDFs through nuclear ultraperipheral collisions (UPCs), see the recent review [18] and refs. therein. While data have clearly established the existence of nuclear gluon shadowing in Pb [19], uncertainties in the theory of quarkonium production makes it questionable to introduce such data in a global analysis of nPDFs. On the other hand, UPCs offer a huge potential for constraining the transverse partonic structure of the proton, see [20, 21].

As a final comment, it should be noted that azimuthal asymmetries and other signals of collectivity cannot be presently described within collinear factorisation. Therefore, the use of low transverse momentum observables from dAu or pPb collisions to constrain nPDFs should be carefully examined.

3 Single inclusive particle production

Recently, a huge progress has been developed on calculations in the Color Glass Condensate (CGC) [22], the weak coupling but non perturbative effective theory of dense partonic systems. In Fig. 3 I summarise the status of recent calculations. One point that should be stressed is that most calculations have been done in a dilute-dense regime suitable for the study of pp or pA collisions in the forward (proton) rapidity region.

Concerning single inclusive particle production, see Fig. 4, much progress has been made lately to understand and control the next-to-leading corrections (NLO) computed some years ago [24, 25] in the hybrid formalism[1]. This is a complex calculation where collinear and soft divergencies appear. The collinear ones are absorbed in the DGLAP evolution of standard collinear parton densities and fragmentation functions, while the soft ones produce the small-x JIMWLK evolution of the target colour ensemble, both evolutions appearing at leading order (LO). After realising that the NLO results

[1]The hybrid formalism considers the scattering of a dilute projectile, whose partons can be treated as collinear, on a dense target that is considered an ensemble of colour fields in the CGC framework.

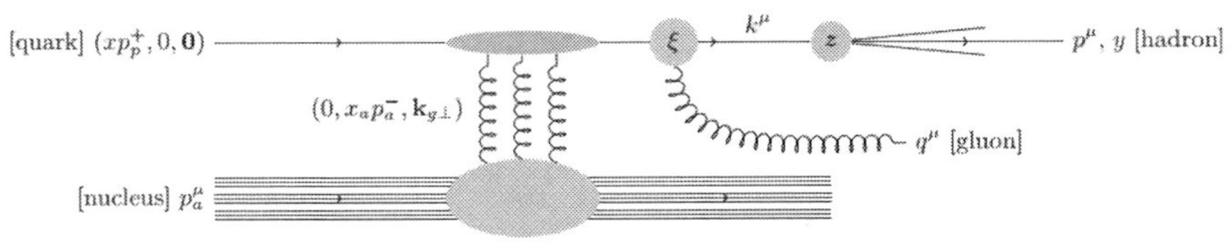

Figure 4. Diagram showing the next-to-leading contribution to single hadron production at forward rapidities in the CGC. Taken from [23].

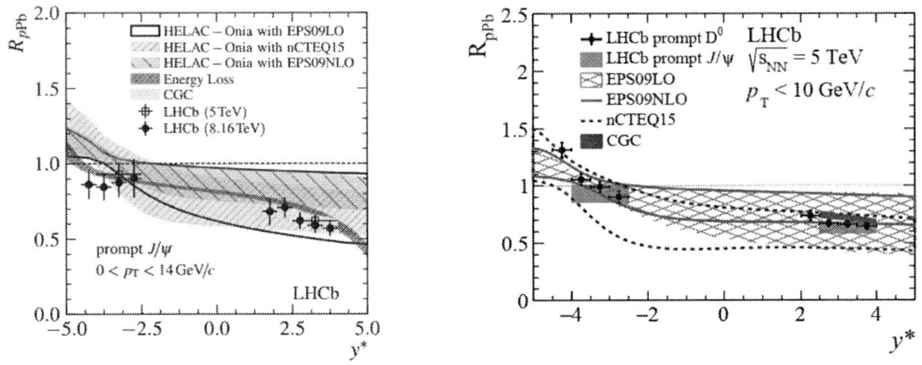

Figure 5. Nuclear modification factor for prompt J/ψ's (left plot, from [33]) and D's (right plot, from [34]) compared to several models. Those labeled 'CGC' are predictions from the hybrid formalism, those labeled 'Energy Loss' contain final state effects for J/ψ, and the rest are purely collinear approaches using different sets of nPDFs. See the experimental papers for details and references.

turned the spectrum negative at moderate transverse momentum (how large this transverse momentum is depends strongly on the employed parametrisation for the dipole-nucleus scattering amplitude), several solution were proposed. Those fluctuations of the projectile too short lived to be resolved by the target had to be removed [26], which gave a natural way to regularise the soft divergencies and produced additional terms in the production cross section that shifted the negativity problem to larger values of transverse momenta.

While a complete understanding is still under progress, at present the key idea lies is devising a procedure for subtraction of the soft divergencies that is able to maintain the positivity of the cross section while resumming leading higher order terms, see [27] and refs. therein. The problem is related to the large negative corrections found in the NLO JIMWLK evolution equations [28, 29]. Several methods have been proposed for this purpose [30, 31]. Meanwhile, the comparison of measurements of the nuclear modification factors in pPb collisions at the LHC for charged particles at midrapidity [32], and of J/ψ's [33] and D's [34] at forward rapidities, are not conclusive as both CGC predictions and collinear calculations (within their large uncertainties) are compatible with existing data.

As in the previous Section, a word of caution is needed in such comparisons, as there may be other effects at play that are included neither in collinear approaches nor in the CGC calculations. Besides the collective effects mentioned there, there may be a contribution from multiple partonic interactions (MPIs), see below, and our understanding of D meson fragmentation and of J/ψ production still contains sizeable uncertainties.

4 Correlations in small systems

The ridge phenomenon lies at the heart of the present discussions in small systems. In this Section I will review its explanation in the CGC and attempts to model it in the MPI framework, see Section 1.

4.1 Color Glass Condensate

All explanations to the ridge phenomenon in the CGC assume that the final state carries the imprint of the correlations in the initial state, and that the CGC wave function is invariant over long rapidity intervals $\Delta y \sim 1/\alpha_s$. Three type of explanations have been proposed:
• "Glasma graphs" [35] that have given rise to a successful phenomenology [36, 37];
• Local anisotropy of target fields [38];
• Spatial variations of partonic density [39].

Concerning the first item, it has been shown that the correlations are due to the Bose enhancement of gluons in the CGC wave function [40], which also give rise to Hanbury-Brown–Twiss correlations [41, 42]. Therefore, they contain information both about the 'source' size (of order of the inverse of the saturation scale of the target) and about the distributions of 'sources', of order of the target radius. Correspondingly, quarks experience Pauli blocking [43]. Its limitations are the fact that the rescattering with the target, modelled through eikonal Wilson lines, is taken at lowest order, and that the colour correlations in the target are modelled as Gaussian and isotropic, which leads to the fact that two particle correlations are subleading in the number of colours N_c, to the absence of odd Fourier harmonics and to the fact that the cumulant $c_2\{4\} > 0$. Work to include high density corrections and overcome these limitations can be found in [44–46], with the appearance of odd harmonics and correlations no longer subleading in N_c. Concerning the second item, models considering non isotropic colour correlations in the target and domains with oriented chromoelectric fields have been developed, see [47] and refs. therein, with the result that $c_2\{4\}$ becomes leading in N_c and negative depending on the size of the anisotropy and on the number of domains.

4.2 Multiple parton interactions

MPIs - the fact that multiple parton-parton and nucleon-nucleon collisions may be at play in pp, pA and AA collisions at high energies - underlie current models for pp interactions, see [48] and refs. therein. They can be traced back to dual unitarisation models, see [49] for a review. Most Monte Carlo simulators for AA collisions [50, 51] contain them in one form or another.

Besides multiple scattering itself, the colour rearragement (CR) between the partons taking part in different parton-parton collisions and the interaction between strings produced in the collision, has proved crucial for reproducing collectivity-like features in small systems. First, CR have been found to describe the baryon/strangeness enhancement at intermediate transverse momentum [52]. Second, string interactions predicted the collectivity features observed in pp collisions at the LHC [53]. Recently, sophisticated models containing both an enhancement of the string tension due to the overlap of the string colour fields, and repulsion among the strings inspired by lattice results, are able to describe the ridge azimuthal distributions [54], see Fig. 6 left. Besides, final state mechanisms (parton rescattering) also produce the azimuthal ridge structure provided a spatially anisotropic system (two separated strings) is considered for initialising the parton transport [55], see Fig. 6 right.

Summarising, small systems have evolved from benchmark to signatures. The experimental finding that they show many of the features found in AA and interpreted as signatures of QGP plus the success of hydrodynamics to describe them, implies that in order to fully interpret the results in AA, we must understand how collectivity is built with increasing system size. Besides nPDFs, I have

https://doi.org/10.1051/epjconf/201817111001

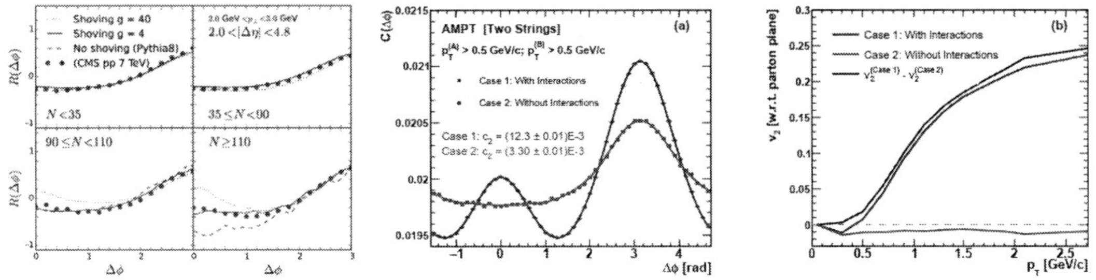

Figure 6. Left plots: two particle azimuthal distributions from the DIPSY model [54] with and without shoving (string repulsion), compared to CMS data. Right plots: two particle azimuthal distributions and second Fourier coefficient in the AMPT model initialised with two strings, with and without parton rescattering [55].

shown some of the CNM alternatives to a hydrodynamical description: the CGC approach to particle production and MPIs. Both contain approximations (e.g. dilute-dense in the CGC) and require non-perturbative input: multiPDFs, models for the colour ensembles in the medium,... The importance of this problem cannot be overstressed as we are addressing a central issue both in QCD and in many other fields: the onset of the transition from the microscopic dynamics to a macroscopic description.

Acknowledgements: I thank the organisers for their invitation and Tolga Altinoluk for a reading of the manuscript. I acknowledge financial support by the ERC grant HotLHC ERC-2011-StG-279579, MiCinn of Spain under project FPA2014-58293-C2-1-P and Unidad de Excelencia María de Maetzu under project MDM-2016-0692, Xunta de Galicia (Consellería de Educación) within the Strategic Unit AGRUP2015/11, and FEDER. This work has been performed in the framework of the COST Action CA15213 THOR.

References

[1] C. Loizides, Nucl. Phys. A **956** (2016) 200.
[2] M. Aaboud *et al.* [ATLAS Collaboration], Phys. Rev. C **96** (2017) no.2, 024908.
[3] V. Khachatryan *et al.* [CMS Collaboration], Phys. Lett. B **765** (2017) 193.
[4] M. Aaboud *et al.* [ATLAS Collaboration], Eur. Phys. J. C **77** (2017) no.6, 428.
[5] R. D. Weller and P. Romatschke, Phys. Lett. B **774** (2017) 351.
[6] P. M. Chesler and L. G. Yaffe, Phys. Rev. D **82** (2010) 026006.
[7] M. P. Heller, R. A. Janik and P. Witaszczyk, Phys. Rev. Lett. **108** (2012) 201602.
[8] A. Kurkela and Y. Zhu, Phys. Rev. Lett. **115** (2015) no.18, 182301.
[9] P. Romatschke, Eur. Phys. J. C **77** (2017) no.1, 21.
[10] K. Dusling, W. Li and B. Schenke, Int. J. Mod. Phys. E **25** (2016) no.01, 1630002.
[11] S. Schlichting and P. Tribedy, Adv. High Energy Phys. **2016** (2016) 8460349.
[12] 4th International Conference on the Initial Stages in High-Energy Nuclear Collisions, Crakow, Poland, September 18th-22nd 2017, `https://indico.cern.ch/event/578078/`.
[13] M. Arneodo, Phys. Rept. **240** (1994) 301.
[14] K. Kovarik *et al.*, Phys. Rev. D **93** (2016) no.8, 085037.
[15] K. J. Eskola, P. Paakkinen, H. Paukkunen and C. A. Salgado, Eur. Phys. J. C **77** (2017) no.3, 163.
[16] D. de Florian, R. Sassot, P. Zurita and M. Stratmann, Phys. Rev. D **85** (2012) 074028.
[17] K. J. Eskola, H. Paukkunen and C. A. Salgado, JHEP **0904** (2009) 065.

[18] S. Klein and J. Nystrand, Phys. Today **70** (2017) no.10, 40.
[19] V. Guzey and M. Zhalov, JHEP **1310** (2013) 207.
[20] H. Mäntysaari and B. Schenke, Phys. Lett. B **772** (2017) 832.
[21] J. Cepila, J. G. Contreras and J. D. Tapia Takaki, Phys. Lett. B **766** (2017) 186.
[22] F. Gelis, E. Iancu, J. Jalilian-Marian and R. Venugopalan, Ann. Rev. Nucl. Part. Sci. **60** (2010) 463.
[23] A. M. Staśto, B. W. Xiao, F. Yuan and D. Zaslavsky, Phys. Rev. D **90** (2014) no.1, 014047.
[24] T. Altinoluk and A. Kovner, Phys. Rev. D **83** (2011) 105004.
[25] G. A. Chirilli, B. W. Xiao and F. Yuan, Phys. Rev. D **86** (2012) 054005.
[26] T. Altinoluk, N. Armesto, G. Beuf, A. Kovner and M. Lublinsky, Phys. Rev. D **91** (2015) no.9, 094016.
[27] B. Ducloué, T. Lappi and Y. Zhu, Phys. Rev. D **95** (2017) no.11, 114007.
[28] I. Balitsky and G. A. Chirilli, Phys. Rev. D **77** (2008) 014019.
[29] A. Kovner, M. Lublinsky and Y. Mulian, Phys. Rev. D **89** (2014) no.6, 061704.
[30] E. Iancu, J. D. Madrigal, A. H. Mueller, G. Soyez and D. N. Triantafyllopoulos, Phys. Lett. B **744** (2015) 293.
[31] G. Beuf, Phys. Rev. D **85** (2012) 034039.
[32] B. B. Abelev *et al.* [ALICE Collaboration], Eur. Phys. J. C **74** (2014) no.9, 3054.
[33] R. Aaij *et al.* [LHCb Collaboration], Phys. Lett. B **774** (2017) 159.
[34] R. Aaij *et al.* [LHCb Collaboration], JHEP **1710** (2017) 090.
[35] A. Dumitru, F. Gelis, L. McLerran and R. Venugopalan, Nucl. Phys. A **810** (2008) 91.
[36] K. Dusling and R. Venugopalan, Phys. Rev. D **87** (2013) no.9, 094034.
[37] K. Dusling, P. Tribedy and R. Venugopalan, Phys. Rev. D **93** (2016) no.1, 014034.
[38] A. Kovner and M. Lublinsky, Int. J. Mod. Phys. E **22** (2013) 133000.
[39] E. Levin and A. H. Rezaeian, Phys. Rev. D **84** (2011) 034031.
[40] T. Altinoluk, N. Armesto, G. Beuf, A. Kovner and M. Lublinsky, Phys. Lett. B **751** (2015) 448.
[41] Y. V. Kovchegov and D. E. Wertepny, Nucl. Phys. A **906** (2013) 50
[42] T. Altinoluk, N. Armesto, G. Beuf, A. Kovner and M. Lublinsky, Phys. Lett. B **752** (2016) 113.
[43] T. Altinoluk, N. Armesto, G. Beuf, A. Kovner and M. Lublinsky, Phys. Rev. D **95** (2017) no.3, 034025.
[44] T. Lappi, B. Schenke, S. Schlichting and R. Venugopalan, JHEP **1601** (2016) 061.
[45] A. Kovner, M. Lublinsky and V. Skokov, Phys. Rev. D **96** (2017) no.1, 016010.
[46] K. Dusling, M. Mace and R. Venugopalan, arXiv:1706.06260 [hep-ph].
[47] A. Dumitru, A. V. Giannini and V. Skokov, arXiv:1503.03897 [hep-ph].
[48] T. Sjöstrand, arXiv:1706.02166 [hep-ph].
[49] A. Capella, U. Sukhatme, C. I. Tan and J. Tran Thanh Van, Phys. Rept. **236** (1994) 225.
[50] X. N. Wang and M. Gyulassy, Phys. Rev. D **44** (1991) 3501.
[51] H. J. Drescher, M. Hladik, S. Ostapchenko, T. Pierog and K. Werner, Phys. Rept. **350** (2001) 93.
[52] A. Ortiz Velasquez, P. Christiansen, E. Cuautle Flores, I. Maldonado Cervantes and G. Paić, Phys. Rev. Lett. **111** (2013) no.4, 042001.
[53] L. Cunqueiro, J. Dias de Deus and C. Pajares, Eur. Phys. J. C **65** (2010) 423.
[54] C. Bierlich, G. Gustafson and L. Lönnblad, arXiv:1710.09725 [hep-ph].
[55] J. L. Nagle, R. Belmont, K. Hill, J. Orjuela Koop, D. V. Perepelitsa, P. Yin, Z. W. Lin and D. McGlinchey, arXiv:1707.02307 [nucl-th].

Small Collision Systems at RHIC

Norbert Novitzky[1,⋆]

[1] *Department of Physics and Astronomy, Stony Brook University, Stony Brook, NY 11790, USA*

Abstract. The observation of long range correlations in highly asymmetric systems, as in p+Pb and d+Au collisions, suggests a creation of a medium with collective behavior. It is still an open question if the quark-gluon plasma is formed in these collision. Hence, the RHIC collider invested time to study the small systems in different collision systems and energies. Here we discuss the recent results from the PHENIX and STAR collaborations in four different collision systems p+Al, p+Au, d+Au and ^{3}He+Au at $\sqrt{s_{NN}}$ = 200 GeV, and also for the energy scan in d+Au collisions between $\sqrt{s_{NN}}$ = 19.6 – 200 GeV.

1 Introduction

The study of non-perturbative aspects of QCD is one of the main goals of high energy nuclear collisions. In such collisions the created colored medium provides a unique way to study the QCD phase transition. Experiments at RHIC and later at LHC established the creation of such medium - called the Quark-Gluon Plasma (QGP).

The recent measurements of flow-like behavior in small-on-large collision systems at LHC [1] and RHIC energies [2] prompted a large interest in studying the possible creation of a quark-gluon droplet in these systems. The azimuthal anisotropy (v_n) can be understood as arising from initial geometry propagated to final-state momentum correlations via interactions between medium constituents. The RHIC collaborations proposed a systematic study of the medium effects in highly asymmetric collisions. The first approach is to change the initial geometry of the collision, therefore we examined the p+Au, d+Au and ^{3}He+Au collisions at $\sqrt{s_{NN}}$ = 200 GeV. In addition, we also investigated the dependence on collision energy and studied the d+Au collisions at four different energies from 19.6 to 200 GeV.

The non-zero azimuthal anisotropy is the first observed signature of possible formation of a collective medium in small collision systems. However, the v_n measurement alone is ambiguous as there could be other effects which would cause the measured azimuthal anisotropy. Therefore it is very important to study also the other signatures of the QGP in order to confirm the medium creation in these small systems. This proceedings focuses not only on the collective motion, but also on the jet quenching phenomena and the strangeness production in small systems.

2 Azimuthal anisotropy measurements

At the beginning of a heavy ion collision, the spatial distribution of the overlapping matter resembles an ellipsoid due to the incomplete overlap of the two colliding nuclei. The strong scattering in this

⋆e-mail: Norbert.Novitzky@cern.ch

stage converts the spatial anisotropy to a momentum anisotropy which is observable as an elliptic flow of the emitted hadrons. The azimuthal anisotropy of the spectra can be characterized in terms of Fourier coefficients. We know that the initial state fluctuations of the nuclei within the nucleus can cause a more complicated overlap area than an ellipsoid and thus result in the higher order components.

It is natural to ask how the study of collective effects in the small system will contribute to the understanding of the flow evolution as a function of the initial geometry and the multiplicity of the events. At RHIC both of the detectors collected excellent data in various collision systems to explore the initial geometry effects. In addition, RHIC provided also an energy scan in d+Au collisions in order to study the onset of the observed small system collective motion as a function of the energy.

2.1 Study of the initial geometry

It is expected that the different collision systems will provide a different initial geometry [3]. For example, the expectation from the d+Au collisions is a natural ellipticity due to the two nucleons in the deuteron, while in the ^{3}He+Au collisions the initial expectation is more triangular. However, one has to be careful with the initial fluctuations of the geometrical orientation also, as the deuteron and the ^{3}He nucleus are not distributed in only the transverse plane but rather 3-dimensionally. This effect can cause various initial geometrical fluctuation which is complicating the simple picture of number of hotspots in the transverse plane. The PHENIX strategy is to utilize the high-multiplicity triggers using the beam-beam counters in order to focus on events only with large number of particles. The STAR collaboration approach is to focus on all (small and large) collision systems at fixed charge particle multiplicities.

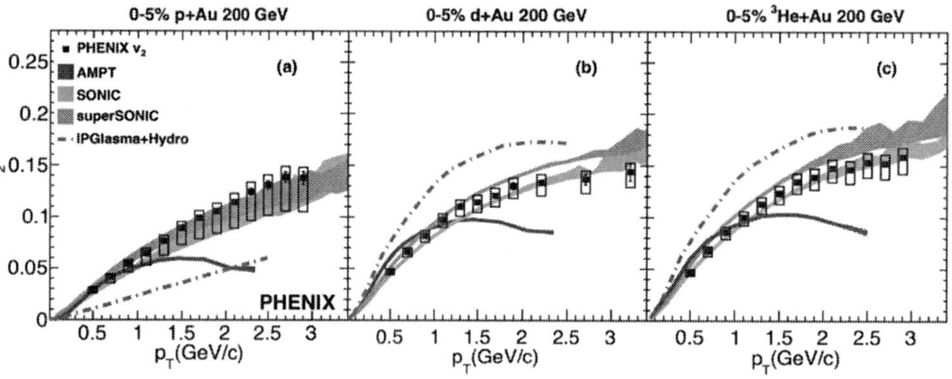

Figure 1. The measured azimuthal anisotropy by PHENIX in the p+Au , d+Au and ^{3}He+Au collisions at $\sqrt{s_{NN}}$ = 200 GeV. The data are from the new high-multiplicity trigger and restricted to 0-5% centrality.

Figure 1 shows the measured v_2 by PHENIX in $p/d/^3$He+Au collisions at $\sqrt{s_{NN}}$ = 200 GeV from the high-multiplicity triggered data which is then restricted to $0-5\%$ centrality categorization. The value for v_2 is very similar in d+Au and ^{3}He+Au collisions. The results for p+Au collisions suggest a smaller v_2 component in the p+Au collisions at low-$p_{\rm T}$ region. The data are first compared with the AMPT [4] event generator calculation which shows good agreement with the data only in the $p_{\rm T}$ < 1 GeV/c region. The AMPT does not contain a hydrodynamical evolution of the system, the predicted v_2 is resulting from the initial partonic and final hadronic interactions and the transition between

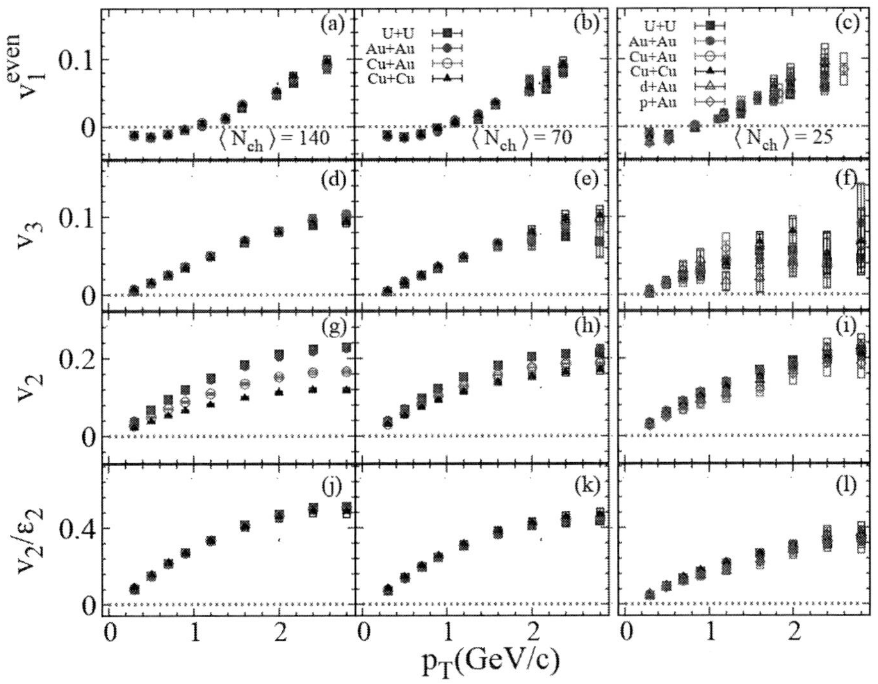

Figure 2. The measured odd (upper panels) and even (lower panels) azimuthal correlations by STAR in various collisions systems (from p+Au to U + U) at fixed charge hadron multiplicity between $\langle N_{ch} \rangle = 140 - 25$.

these two phases of matter. The data shows excellent agreement with the hydro models SONIC and superSONIC [5], where the latter one includes the pre-equilibrium flow also. The agreement is better with the SONIC model, but the uncertainties on the data do not allow yet to invalidate the existence of the pre-equilibrium flow in these systems. The model with IPGlasma initial state and then hydrodynamical evolution [6] largely underestimates the data in p+Au collisions, while overestimates the measured flow in the d+Au and ^{3}He+Au collisions.

Figure 2 shows the measured preliminary STAR data of $v_1^{even\star}$, v_3 and v_2 measurements in various collision systems from very large (U + U and Au + Au) to moderately large (Cu + Au and Cu + Cu) to small systems (d+Au and p+Au). The measurements are done at three fixed charged hadron multiplicities, $\langle N_{ch} \rangle = 140, 70$ and 25 where the smallest multiplicity includes also the p+Au and d+Au results. The non-flow contributions are characterized as short-range and long-range effects. The short-range effect includes HBT and resonance decay contribution and they can be suppressed with a large rapidity (η) gap. The long-range effects enter from the momentum-conservation and di-jets and their contribution. In order to suppress the non-flow contributions, the results are obtained with the multi-particle correlations (sub-event cumulant method). It is important to note that these

*In the absence of fluctuations, the directed flow v_1 develops along the direction of the impact parameter and it is an odd function of the pseudo-rapidity. However, initial-state fluctuations, acting in agreement with hydrodynamic-like expansion, gives an additional rapidity-even component v_1^{even}.

studies were done only on the STAR results and thus the contributions from the non-flow effects on the data are different between the two experiments.

The v_1^{even} and v_3 shows similar trend and magnitudes independent of the collision system only depending on the multiplicity of the event, see Figure 2. While the even harmonics, v_2, shows similar trend in different collision systems, the magnitudes depend on the collision system. However, the values can be divided by the second order eccentricities (ε_2) of the different collision systems and then all the data scale to a single curve. The observation supports that the odd harmonics are only sensitive to fluctuations and not to the average geometries, while the second order harmonics shows significant sensitivity to the initial geometry of the collision systems [7]. In conclusion, the data suggest that at same multiplicities the initial geometry of the collision is different, although the collectivity of the system is very similar from large to small collision systems.

2.2 Energy scan measurements

In 2016 RHIC also delivered a beam energy scan of d+Au collisions at four different collision energies from 19.6 to 200 GeV. The PHENIX measurement on these collision energies are shown in Figure 3 [8]. The data show very similar trend at all four collision energies, while the magnitude of the measured v_2 at $p_T > 1$ GeV$/c$ is slightly increasing towards smaller collision energies. This effect is likely due to an increase in the non-flow contribution, which is not subtracted from the data.

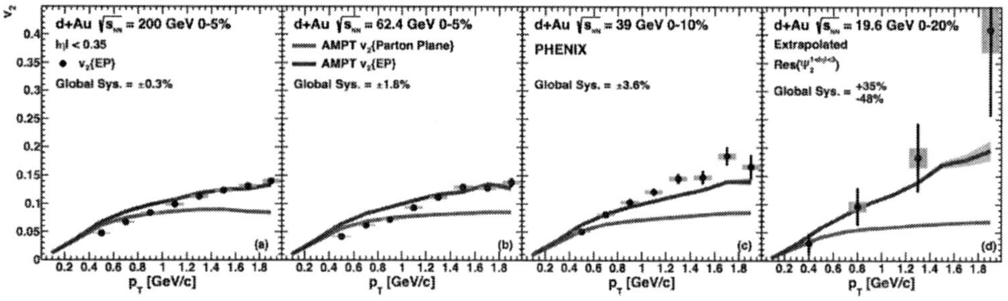

Figure 3. The measured azimuthal anisotropy by PHENIX in d+Au collisions with the high-multiplicity trigger in various collision energies $\sqrt{s_{NN}} = 200 - 19.6$ GeV. The lines represent the AMPT model calculation with two different determination of the reaction plane.

The measured v_2 is compared to the AMPT calculation [4] with two different setups. The green curve shows the v_2 relative to the participant plane, so the result contains only the v_2 relative to the initial geometry. The blue curve shows the calculation of v_2 relative to the event plane. The event plane method is more consistent to the experimental extraction of the v_2 measurement, and the calculation is also in better agreement with the measured data.

3 Searching for energy loss in small systems

The parton propagating through the hot and dense strongly interacting medium loses part of its energy. The energy loss is characterized with the ratio of the yield in heavy ion collisions to the scaled yield in $p + p$ collisions, otherwise known as the nuclear modification factor (R_{AA}). The high-p_T suppression of particle production in heavy-ion collision is a clear indication of the jet energy loss [9]. The direct photons are not participating in the strong interaction with the created medium and after their creation they leave the medium without further interaction. The high-p_T photon measurements confirmed that there is no nuclear modification ($R_{AA} = 1$), which further supports the jet energy-loss picture [10].

The energy loss is one of the key signatures in the discovery of the created plasma. Therefore, if a similar medium would be created in the small systems, one would naturally expect such nuclear modification, too. However, the picture is not so simple in the small system collisions. First, the created medium is expected to be very small, which could mean that the outgoing parton which originated from the hard scattering already left the medium before it is formed. On other hand the outgoing parton could lose energy in the cold nuclear matter. Furthermore, initial state effects can strongly affect the observed nuclear modifications. Thus it is crucial to distinguish between initial state (IS) and final state (FS) effects.

The measurement of neutral pion is a very good analysis to investigate the nuclear modification factor of the leading hadrons of jets in the small system collisions. PHENIX measured the neutral pion production in p+Au , d+Au and ^{3}He+Au minimum bias collisions at $\sqrt{s_{NN}}$ = 200 GeV, shown in Figure 4. There is a clear enhancement observed at p_T ~ 5 GeV/c in the p+Au collisions, while it disappears in the d+Au and ^{3}He+Au collisions. At high-p_T the data show some suppression (R_{AA} < 1), but the uncertainties on the data are still very large.

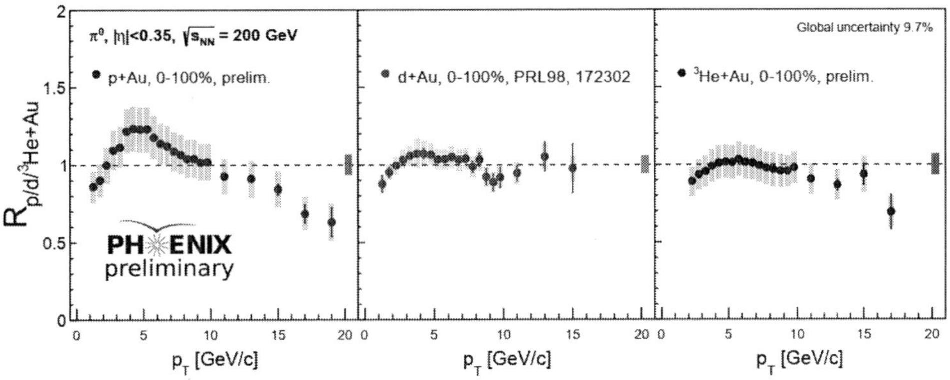

Figure 4. The nuclear modification factor of π^0 in p+Au , d+Au and ^{3}He+Au minimum bias collisions at $\sqrt{s_{NN}}$ = 200 GeV.

The nuclear modification factor was also measured in bins of centrality in p+Au , d+Au and ^{3}He+Au collisions. It is important to note that PHENIX uses the exact same procedure of the centrality selections using the South Beam-Beam Counter ($-3 < \eta < -4$) detector as in heavy ion collisions. Figure 5 shows the measured results in the different centrality selections in p+Au , d+Au and ^{3}He+Au collisions at $\sqrt{s_{NN}}$ = 200 GeV. The enhancement in the mid-p_T region is more pronounced in the most central collisions while it disappears in the more peripheral collisions. The high-p_T part shows suppression at most central collisions, while in more peripheral collisions some enhancement cannot be excluded, but the data is consistent with unity within the experimental uncertainties. The nuclear modification at high-p_T is also consistent with being flat in the two most peripheral collisions.

The interpretation of the results are still ongoing as the data are being finalized. The enhancement at mid-p_T is often called the "Cronin-peak" [11] and it is traditionally attributed to the multiple scattering effect. However, the enhancement occurs at p_T ~ 5 GeV/c higher than the expected contribution from multiple scattering. The high-p_T nuclear modification is very comparable within the different collisions system and the data suggest it is independent on the projectile (p, d or ^{3}He). The data

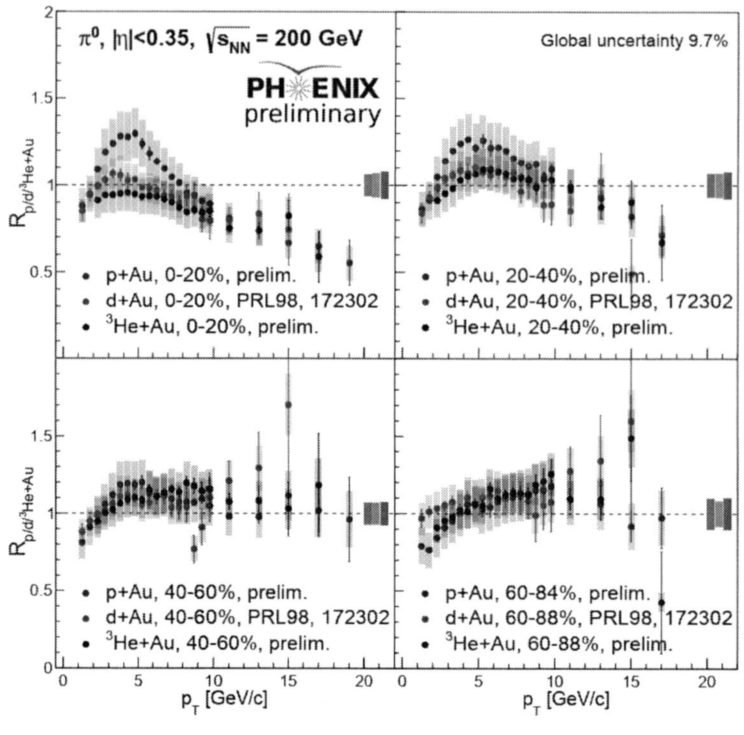

Figure 5. The nuclear modification factor of π^0 as a function of p_T in different centrality classes of p+Au, d+Au and ^{3}He+Au collisions at $\sqrt{s_{NN}}$ = 200 GeV.

could suggest some (cold/hot) energy loss in the most central collisions [12]. The jets are suffering both IS and FS effects, while the direct photons would be sensitive only to the IS effects. In order to correctly interpret the data one needs to compare the neutral pion modification to the direct photon measurements, that are not available yet.

4 Strangeness production in small systems

In this section we present the recent measurements on the ϕ production in small system collisions from PHENIX. The ϕ meson is a pure strange-antistrange ($s\bar{s}$) state and its production mechanism is similar to those of J/ψ and Υ. The ϕ meson has a small inelastic cross section with non-strange hadrons and as such represent a unique probe as they are more sensitive to the initial evolution of the system.

The forward and backward rapidity production of ϕ mesons was measured in p+Al, p+Au and ^{3}He+Au minimum bias collisions at $\sqrt{s_{NN}}$ = 200 GeV, shown in Figure 6. The data show very comparable results in all three different collision systems within the experimental uncertainties. The ϕ production in backward region shows enhancement, while the forward region is suppressed. The results are also very consistent with the previously measured ϕ production in d+Au collisions.

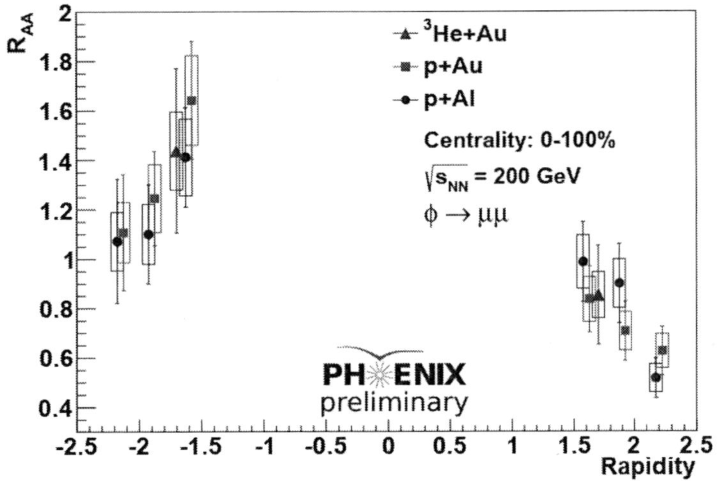

Figure 6. The nuclear modification of ϕ as a function of rapidity in p+Al , p+Au and ^{3}He+Au minimum bias collisions at $\sqrt{s_{NN}}$ = 200 GeV.

5 Summary

This proceedings present the recent results from the RHIC collider experiment on the small collision systems. The measurement of the v_n in the small collision systems are presented to study the initial geometrical effect as well as the energy dependence. The measurements from both PHENIX and STAR suggest strong correlation to the initial geometry of the collision system and a similarity of the observed v_n measurements to the larger heavy-ion collision systems. Therefore the data suggest a smooth transition from $p + p$ to heavy ion collisions. The energy scan of the d+Au collision shows that the long-range correlations are still present at the lowest 19.6 GeVcollision energies.

A systematic study of the neutral pion production was performed in $p/d/^3$He+Au collisions at $\sqrt{s_{NN}}$ = 200 GeV. We observe a small enhancement in the mid-p_T region peaking around $\sim$ 5 GeV/c, where the data suggest an ordering between the different collision systems. The enhancement is traditionally contributed to the multiple scattering, but the peak is at much higher p_T 's than any predictions from theoretical models. At high-p_T the nuclear modification factor is very similar between the different collision systems which suggest no dependency on the size of the projectile. The R_{AA} values for p_T > 10 GeV/c show no strong p_T dependence and fall bellow unity in the most central collisions as well as in the minimum bias collisions in this p_T region. However, jet quenching is not confirmed in these collisions and the suppression could be due to multiple different mechanisms due to initial or final state effects. The direct photon measurement is needed in order to correctly interpret the results in these collision systems.

The ϕ meson production was measured in the forward and backward region in minimum bias p+Al , p+Au and ^{3}He+Au collisions at $\sqrt{s_{NN}}$ = 200 GeV. The results shows small suppression in the forward region, while it is enhanced in the backward region. We also observe no significant difference between the different collision systems within the experimental uncertainties. The current results will help us study the nuclear effects in the forward and backward region in order to tune the theoretical models.

References

[1] S. Chatrchyan et al. (CMS), Phys. Lett. **B718**, 795 (2013), 1210.5482
[2] A. Adare et al. (PHENIX), Phys. Rev. Lett. **111**, 212301 (2013), 1303.1794
[3] A. Bzdak, B. Schenke, P. Tribedy, R. Venugopalan, Phys. Rev. **C87**, 064906 (2013), 1304.3403
[4] Z.W. Lin, C.M. Ko, B.A. Li, B. Zhang, S. Pal, Phys. Rev. **C72**, 064901 (2005), nucl-th/0411110
[5] P. Romatschke, Eur. Phys. J. **C75**, 305 (2015), 1502.04745
[6] B. Schenke, R. Venugopalan, Nucl. Phys. **A931**, 1039 (2014), 1407.7557
[7] B. Schenke, P. Tribedy, R. Venugopalan, Phys. Rev. **C89**, 064908 (2014), 1403.2232
[8] C. Aidala et al. (PHENIX) (2017), 1708.06983
[9] X.N. Wang, M. Gyulassy, Nucl. Phys. **A544**, 559 (1992)
[10] S. Afanasiev et al. (PHENIX), Phys. Rev. Lett. **109**, 152302 (2012), 1205.5759
[11] J.W. Cronin, H.J. Frisch, M.J. Shochet, J.P. Boymond, R. Mermod, P.A. Piroue, R.L. Sumner, Phys. Rev. **D11**, 3105 (1975)
[12] Y.T. Chien, A. Emerman, Z.B. Kang, G. Ovanesyan, I. Vitev, Phys. Rev. **D93**, 74030 (2016), 1509.02936

EPJ Web of Conferences **171**, 11003 (2018) https://doi.org/10.1051/epjconf/201817111003
SQM 2017

Small systems at the LHC

Roberto Preghenella[1,*]

[1]Istituto Nazionale di Fisica Nucleare, Sezione di Bologna

Abstract. In these proceedings, I report on a selection of recent LHC results in small systems from ALICE [1], ATLAS [2] and CMS [3] experiments. Due to the fact that the investigation of QCD in small systems at high multiplicity is becoming an increasingly large subject, interesting the heavy-ion community and more in general the high-energy physics community, not all the related topics can be discussed in this paper. The focus will be given to some of the measurements addressing the physics of collective phenomena in small systems and to the recent results on strangeness enhancement in proton-proton collisions. The reader must be informed that a large number of interesting results did not find space in the discussion reported here.

1 Introduction

There was a time when small collision systems, like proton-proton and proton-nucleus, where almost excusively employed in heavy-ion physics as a reference to compute the nuclear modification factors of particle production and as a proxy to disentangle initial state, cold nuclear matter effects (i.e. due to the nuclear PDF of the target/projectile) from genuine final state, hot nuclear matter effects. As a matter of fact, results based on the analysis of particle production in small systems provided the evidence that the characteristic phenomena observed in heavy-ion collisions, most notably the high-p_T and jet suppression [4, 5], were actually due to the formation of a hot state of QCD matter, the Quark-Gluon Plasma.

With the advent of the LHC, pioneering studies of pp collisions in high-multiplicity events revealed the presence of unexpected phenomena in small systems. The discovery of the "ridge" in pp collisions by the CMS Collaboration was the first of a long list of such results [6]. Long-range, near-side angular correlations in particle production emerged in pp and subsequently in p–Pb collisions [7] and paved the way for a systematic investigation of the existence of collective phenomena, known since long in heavy-ion collisions [8], in the much smaller pp and p–Pb collision systems. A wealth of new, unexpected phenomena have been observed so far with striking similarities to heavy-ion phenomenolgy. In this report, only a few of them can be discussed because of the lack of space. The focus will be given to a selection of recent LHC results aiming at studying collective phenomena and the chemistry of hadron production in small systems.

2 Collective phenomena

The observation of the ridge in pp [6] and subsequently in p–Pb collisions [7] promptly triggered further investigations. Studies of two-particle correlations in p–Pb collisions re-

*on behalf of ALICE, ATLAS and CMS Collaborations (e-mail: preghenella@bo.infn.it)

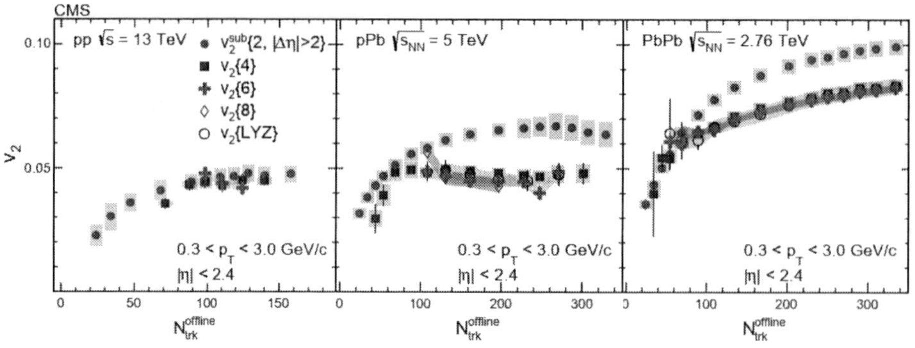

Figure 1. Second-order azimuthal anisotropy Fourier harmonics v_2 measured by CMS in pp, p–Pb and Pb–Pb collisions based on correlations among multiple charged particles [11, 17].

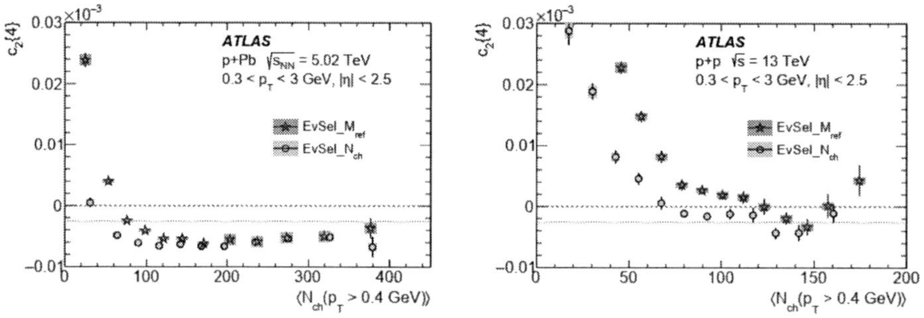

Figure 2. Four-particle cumulants $c_2\{4\}$ measured by ATLAS in p–Pb collisions (left) and in pp collisions (right) using different multiplicity selection methods [18].

ported by the ALICE Collaboration have shown that, when taking the difference between the yields per trigger particle in high-multiplicity and low-multiplicity events, two nearly identical long-range ridge-like excess structures on the near-side and away-side, the so-called "double ridge", arise [9, 10]. Results on two-particle correlations of identified hadrons [11–14] and of Bose-Einstein (HBT) correlations [15, 16] further stressed the initial observations, showing striking similarities between pp, p–Pb and Pb–Pb collisions, consistent with the hydrodynamic picture of a particle-production source expanding more explosively along the collision event plane.

2.1 Multi-particle correlations

A fundamental question is whether the two-particle azimuthal correlation structures observed at large relative pseudorapidity in pp and p–Pb collisions result from correlations exclusively between particle pairs, or if it is a multi-particle, genuine collective effect. A strong hint, originally named as evidence by the CMS Collaboration, for collective multi-particle correlations in pp and p–Pb pp collisions was reported already [11, 17]. Figure 1 shows the second-order azimuthal anisotropy Fourier harmonics v_2 measured in pp, p–Pb and Pb–Pb collisions over a wide pseudorapidity range based on correlations among up to eight particles. The v_2 values obtained with correlations among more than four particles are consistent with four-particle results and with comparable magnitude to those from two-particle corre-

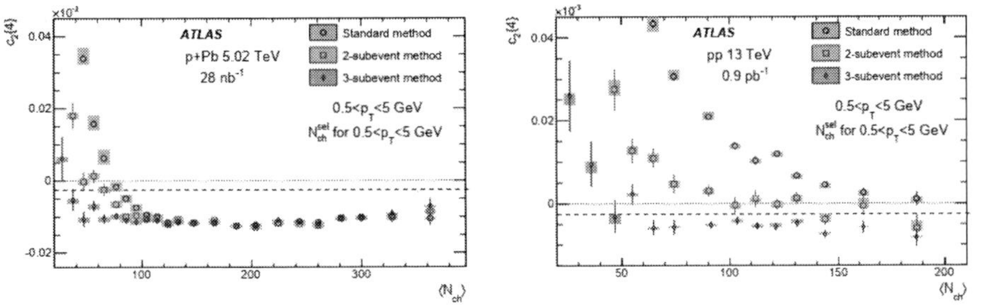

Figure 3. Four-particle cumulants $c_2\{4\}$ measured by ATLAS in p–Pb collisions (left) and in pp collisions (right) with the "standard", "two-subevent" and the "three-subevent" methods [19].

lations. These observations support the interpretation of a collective origin for the observed long-range correlations in high-multiplicity pp and p–Pb collisions.

There is a note of care that must be taken seriously into account. While multi-particle correlation measurements have the advantage of suppressing short-range two-particle correlations such as jets and resonance decays, they are not totally insensitive to such or other so-called "non-flow" effects. This has been stressed in a publication by the ATLAS Collaboration [18], where it is shown how different multiplicity selection methods may yield different four-particle cumulants ($c_2\{4\}$) results. The measurements of $c_2\{4\}$ are shown in Figure 2 for pp and p–Pb collisions. The results confirm the evidence for collective phenomena in p–Pb and low-multiplicity Pb–Pb collisions (not reported in Figure 2). On the other hand, the pp results for $c_2\{4\}$ do not demonstrate collective behaviour, indicating that they may be biased by contributions from non-flow correlations.

Reliably suppressing non-flow correlations in pp collision is a central issue that needs to be solved in order to be able to unveil the actual underlying collectivity. Fortunately, several new methods have been recently developed which aim at tackling this issue. As an example, the ATLAS Collaboration has reported on the measurement of multi-particle correlations in pp and p–Pb collisions with the so-called "two-subevent" and the "three-subevent" methods [20]. Figure 3 shows $c_2\{4\}$ measured in pp and p–Pb collisions with such methods. $c_2\{4\}$ from the standard method is sensitive to the choice of particles used to form the event classes. The sensitivity is greatly reduced in the two-subevent method and is almost fully removed in the three-subevent method, suggesting that the three-subevent method is much more robust against non-flow effects. The three-subevent method shows significant flow in pp collisions in a broad N_{ch} range.

2.2 Identified hadron spectra

The p_{T} distributions of identified hadrons are another tool to probe the collective behaviour of particle production. If collective radial flow develops, this would result in a characteristic dependence of the shape of the transverse momentum distribution on the particle mass. The p_{T} distributions in pp and p–Pb collisions show a clear evolution, becoming harder as the multiplicity increases, with a change which is most pronounced for protons and lambdas. The stronger multiplicity dependence of the spectral shapes of heavier particles is evident when looking at the p_{T}-dependent ratios, such as the Λ/K^0_S ratio measured by the CMS Collaboration [21], shown in Figure 4. The ratios in pp and p–Pb collisions show a significant enhancement at intermediate p_{T} ($\sim$3 GeV/c), quantitatively similar to that measured in Pb–Pb collisions at similar multiplicity. Results from the ALICE Collaboration show that the

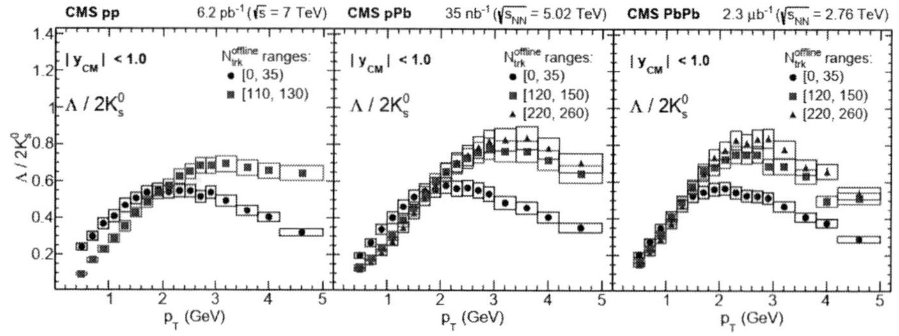

Figure 4. Ratios of p_T spectra of $\Lambda/2K_S^0$ measured by CMS in pp (left), p–Pb (middle), and Pb–Pb collisions (right) [21].

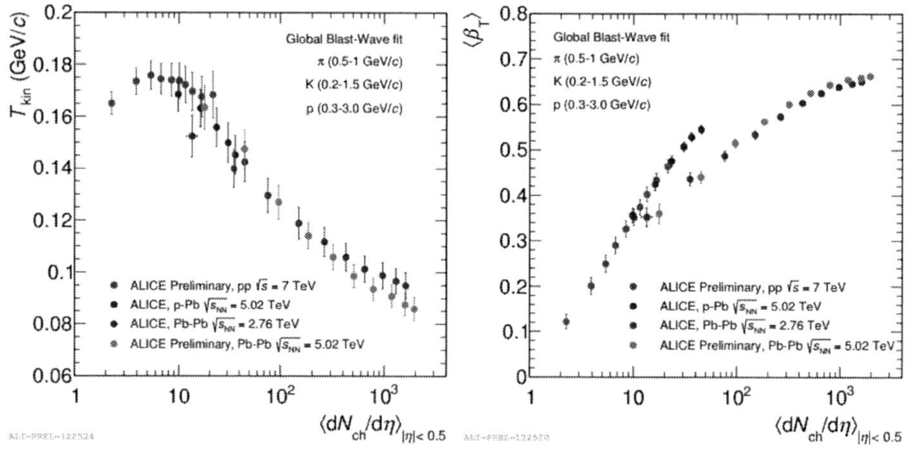

Figure 5. Multiplicity dependence of the $T_{\rm kin}$ and $\beta_{\rm T}$ Blast-Wave [22] parameters fitted to pion, kaon and proton $p_{\rm T}$ spectra measured by ALICE in pp, p–Pb and Pb–Pb collisions.

modification of the $p_{\rm T}$-depentent ratios of p/π and of Λ/K_S^0 follows a rather smooth evolution at low and mid $p_{\rm T}$ across different systems, pointing towards a universal soft mechanism driven by final state $dN_{\rm ch}/d\eta$ [23]. The high-$p_{\rm T}$ part of the ratio, which is dominated by hard fragmentation, is unchanged.

The Blast-Wave model [22] is a useful tool to characterize the spectral shapes of identified hadrons and test data against the radial flow picture. A simultaneous fit to pions, kaons and protons is performed by the ALICE Collaboration following the approach discussed in [23]. The best fit describes the data with good accurancy in the low-$p_{\rm T}$ domain, stressing the consistency of particle production from a thermal source expanding with a common transverse velocity. The results of the fit to pp, p–Pb and Pb–Pb data are shown in Figure 5 as a function of $dN_{\rm ch}/d\eta$. In the context of the Blast-Wave model, when comparing the parameters of different systems at similar $dN_{\rm ch}/d\eta$, $T_{\rm kin}$ values are similar, whereas $\beta_{\rm T}$ is larger for small systems. This might be an indication of a larger radial flow in small systems as consequence of stronger pressure gradients. The ALICE results also show that both $T_{\rm kin}$ and $\beta_{\rm T}$ parameters obtained from pp and p–Pb fits are consistent for similar $dN_{\rm ch}/d\eta$. One has to acknowledge

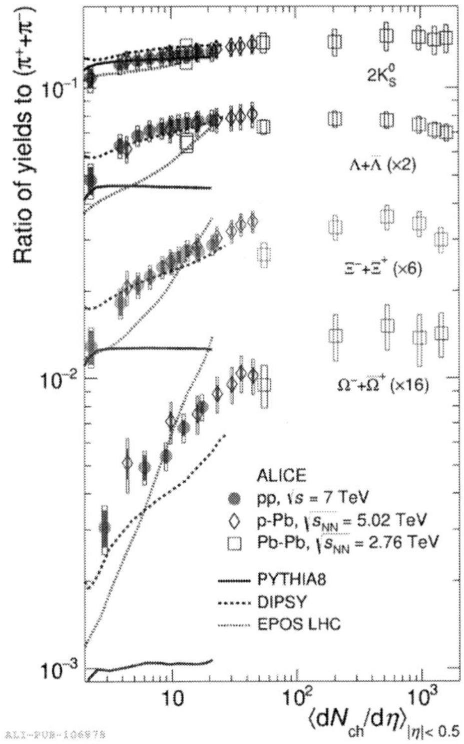

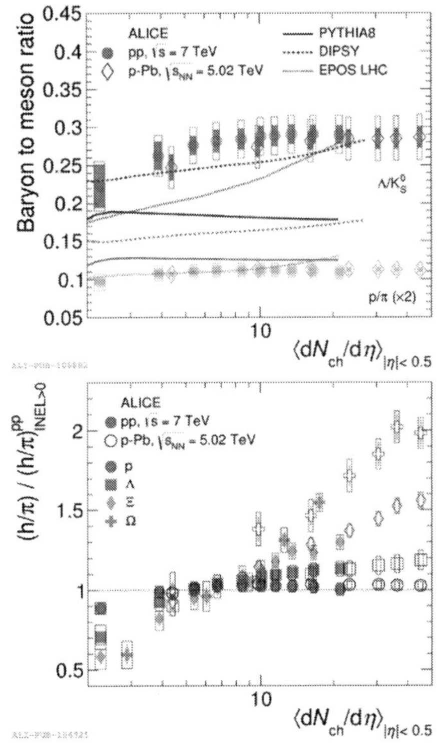

Figure 6. (left) p_T-integrated yield ratios of strange and multi-strange hadrons to pions as a function of $dN_{ch}/d\eta$ measured by ALICE in pp, p–Pb and Pb–Pb collisions [23–26].

Figure 7. (top right) Particle yield ratios Λ/K^0_S and p/π as a function of $dN_{ch}/d\eta$ measured by ALICE in pp and p–Pb collisions. (bottom right) Particle yield ratios to pions normalised to the values measured in the inclusive pp sample.

that a different conclusion emerges from similar fits reported by CMS [21], where pp collisions appear as more explosive than p–Pb collisions, with a larger β_T for similar $dN_{ch}/d\eta$.

3 Strangeness production

The study of the production of strange hadrons in high-energy hadronic interactions provides an important means to investigate the properties of QCD. Unlike up (u) and down (d) quarks, which form ordinary matter, strange (s) quarks are not present as valence quarks in the initial state. Yet they are sufficiently light to be abundantly created by non-perturbative (soft) processes in the course of the collisions. String fragmentation, as an example of non-perturbative QCD process, dominates the production of strange hadrons at low p_T. As a matter of fact, given that the mass of the strange quark is larger than the one of up and down quarks, production of strange hadrons in fragmentation is suppressed relative to hadrons containing only light quarks.

3.1 Strangeness enhancement

An enhanced production of strange hadrons was one of the earliest proposed indicators for the formation of a Quark-Gluon Plasma (QGP) state [27–29], as higher rates for strange

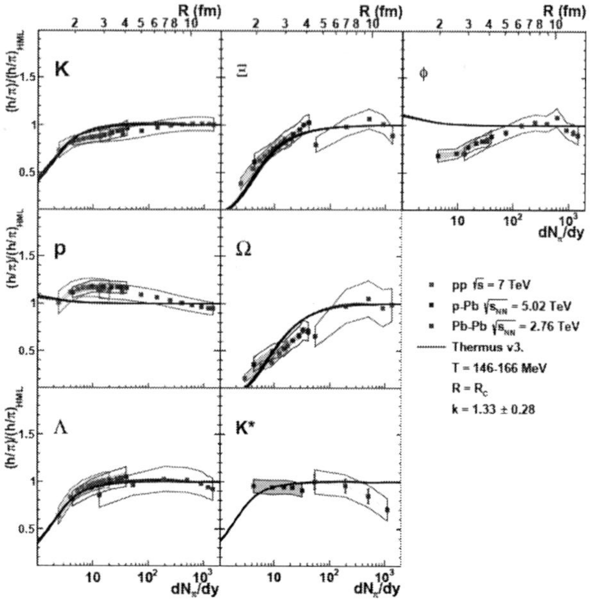

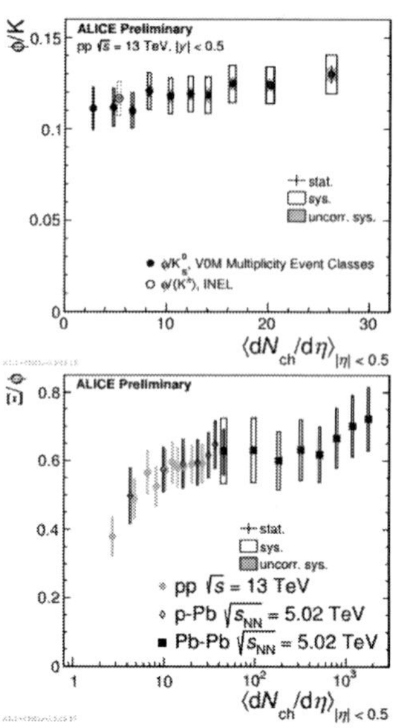

Figure 8. (left) Ratios of several particle species measured by ALICE as a function of the midrapidity pion yields for pp, p–Pb and Pb–Pb collisions compared to the THERMUS strangeness canonical suppression model prediction [45].

Figure 9. (top right) p_T-integrated ϕ/K ratio in pp collisions as a function of $dN_{ch}/d\eta$. (bottom right) p_T-integrated Ξ/ϕ ratio in pp, p–Pb [25, 46] and Pb–Pb collisions as a function of $dN_{ch}/d\eta$.

quark production are expected in a highly-excited state of QCD matter. This strangeness enhancement is expected to be more pronounced for multi-strange baryons, and was indeed observed in collisions of heavy nuclei at the SPS, RHIC and LHC [26, 30–38].

The ALICE Collaboration has recently reported on the multiplicity dependence of the production of primary strange and multi-strange hadrons in pp collisions [24]. Similar measurements have been previously performed by ALICE p–Pb collisions [23, 25] and in Pb–Pb collisions [26, 39]. Figure 6 shows the measured ratios of the p_T-integrated yields of K^0_S, Λ, Ξ and Ω to the pion yield as a function of $dN_{ch}/d\eta$ in pp collisions [24] compared to p–Pb and Pb–Pb results at the LHC [23, 25, 26]. A significant enhancement of strange to non-strange hadron production is observed with increasing charged-particle multiplicity in pp collisions. The behaviour observed in pp collisions resembles that of p–Pb collisions at a slightly lower centre-of-mass energy [25], both in the values of the ratios and in their evolution with $dN_{ch}/d\eta$, suggesting that strangeness production is driven by the event activity of the event rather than by the initial-state collision system or energy. Figure 7 shows that the p_T-integrated yield ratios Λ/K^0_S and p/π do not change significantly with multiplicity, demonstrating that the observed enhanced production rates of strange hadrons with respect to pions is not due to the difference in the hadron masses.

The results in Figures 6 and 7 are compared to calculations from Monte Carlo models commonly used for pp collisions at the LHC: PYTHIA8 [40], EPOS LHC [41] and DIPSY [42–44]. The observation of a multiplicity-dependent enhancement of the production of strange hadrons along with the constant production of protons relative to pions cannot be simultaneously reproduced by any of the Monte Carlo event generators commonly used at the LHC.

3.2 Canonical suppression

The abundances of strange particles in heavy-ion collisions are compatible with those of a hadron gas in thermal and chemical equilibrium and can be described using a grand canonical statistical model [47, 48]. Extensions of the statistical description, like the ones employed in the strangeness canonical suppression [49] and the core-corona superposition [50, 51] models, can effectively produce a suppression of strangeness production in small systems.

The authors in [45] have studied the multiplicity dependence of light flavour hadron production at LHC energies in the strangeness canonical suppression picture using THERMUS [52]. The details can be found therein. As can be observed from Figure 8, the study shows that the model provides a good description of the experimental data, except for the ϕ meson. As a matter of fact, the strangeness canonical suppression model is only sensitive to the strangeness quantum number of the hadron. The observation of an enhancement (suppression) of ϕ meson production with increasing (decreasing) multiplicity clearly signals that the particle production rate is not consistent with the one expected from a hadron with zero strangeness quantum number. This can be observed also in Figure 9, which shows the ratio of ϕ/K and Ξ/ϕ as a function of $\mathrm{d}N_{\mathrm{ch}}/\mathrm{d}\eta$. It is evident that ϕ meson (S = 0) production increases faster than K meson (S = 1) production. When compared to the Ξ, the Ξ/ϕ is constant within uncertainties for $\mathrm{d}N_{\mathrm{ch}}/\mathrm{d}\eta > 10$. This is an indication that the ϕ meson behaviour is closer to the behaviour expected from a hadron formed by two strange quarks. The increase of the relative ϕ meson production with $\mathrm{d}N_{\mathrm{ch}}/\mathrm{d}\eta$ constitues a notable crack in the canonical suppression interpretation of the observed strangeness enhancement in high-multiplicity pp and p–Pb collisions.

4 Conclusions

Several effects, like near-side long-range correlations and mass-dependent hardening of p_{T} distributions, which in nuclear collisions are typically attributed to the formation of a strongly-interacting collectively-expanding quark-gluon medium, have been observed in high-multiplicity pp and p–Pb collisions at the LHC [6, 7, 9–12, 21, 23, 25, 53, 54]. An enhanced production of strange particles as a function of the charged-particle multiplicity density ($\mathrm{d}N_{\mathrm{ch}}/\mathrm{d}\eta$), originally considered to be another signature of QGP formation in nuclear collisions [27–29], has also been recently observed in pp collisions [24].

The study of small collision systems at high multiplicity is undoubtedly of considerable interest. Further studies are essential to assess whether the combination of these observations can be interpreted as signal of the progressive onset of a QGP medium, which starts developing already in small systems.

References

[1] K. Aamodt et al. (ALICE), JINST **3**, S08002 (2008)
[2] G. Aad et al. (ATLAS), JINST **3**, S08003 (2008)
[3] S. Chatrchyan et al. (CMS), JINST **3**, S08004 (2008)
[4] S.S. Adler et al. (PHENIX), Phys. Rev. Lett. **91**, 072303 (2003), `nucl-ex/0306021`
[5] J. Adams et al. (STAR), Phys. Rev. Lett. **91**, 072304 (2003), `nucl-ex/0306024`
[6] V. Khachatryan et al. (CMS), JHEP **09**, 091 (2010), `1009.4122`
[7] S. Chatrchyan et al. (CMS), Phys. Lett. **B718**, 795 (2013), `1210.5482`
[8] B. Abelev et al. (STAR), Phys. Rev. **C80**, 064912 (2009), `0909.0191`
[9] B. Abelev et al. (ALICE), Phys. Lett. **B719**, 29 (2013), `1212.2001`

[10] G. Aad et al. (ATLAS), Phys. Rev. Lett. **110**, 182302 (2013), 1212.5198
[11] V. Khachatryan et al. (CMS), Phys. Lett. **B765**, 193 (2017), 1606.06198
[12] B. Abelev et al. (ALICE), Phys. Lett. **B726**, 164 (2013), 1307.3237
[13] V. Khachatryan et al. (CMS), Phys. Lett. **B742**, 200 (2015), 1409.3392
[14] B. Abelev et al. (ALICE), JHEP **06**, 190 (2015), 1405.4632
[15] ATLAS Collaboration (2017), ATLAS-CONF-2017-008
[16] D. Adamova et al. (ALICE), Phys. Rev. Lett. **118**, 222301 (2017), 1702.01612
[17] V. Khachatryan et al. (CMS), Phys. Rev. Lett. **115**, 012301 (2015), 1502.05382
[18] M. Aaboud et al. (ATLAS), Eur. Phys. J. **C77**, 428 (2017), 1705.04176
[19] M. Aaboud et al. (ATLAS) (2017), 1708.03559
[20] ATLAS Collaboration (2017), ATLAS-CONF-2017-002
[21] V. Khachatryan et al. (CMS), Phys. Lett. **B768**, 103 (2017), 1605.06699
[22] E. Schnedermann et al., Phys. Rev. **C48**, 2462 (1993), nucl-th/9307020
[23] B. Abelev et al. (ALICE), Phys. Lett. **B728**, 25 (2014), 1307.6796
[24] J. Adam et al. (ALICE), Nature Phys. **13**, 535 (2017), 1606.07424
[25] J. Adam et al. (ALICE), Phys. Lett. **B758**, 389 (2016), 1512.07227
[26] B. Abelev et al. (ALICE), Phys. Lett. **B728**, 216 (2014), 1307.5543
[27] J. Rafelski et al., Phys. Rev. Lett. **48**, 1066 (1982)
[28] P. Koch, J. Rafelski, W. Greiner, Phys. Lett. **B123**, 151 (1983)
[29] P. Koch, B. Muller, J. Rafelski, Phys. Rept. **142**, 167 (1986)
[30] E. Andersen et al., Phys. Lett. **B433**, 209 (1998)
[31] E. Andersen et al. (WA97), Phys. Lett. **B449**, 401 (1999)
[32] F. Antinori et al. (NA57), Phys. Lett. **B595**, 68 (2004), nucl-ex/0403022
[33] S.V. Afanasiev et al. (NA49), Phys. Lett. **B538**, 275 (2002), hep-ex/0202037
[34] T. Anticic et al. (NA49), Phys. Rev. Lett. **93**, 022302 (2004), nucl-ex/0311024
[35] J. Adams et al. (STAR), Phys. Rev. Lett. **92**, 182301 (2004), nucl-ex/0307024
[36] J. Adams et al. (STAR), Phys. Rev. Lett. **98**, 062301 (2007), nucl-ex/0606014
[37] B. Abelev et al. (STAR), Phys. Rev. **C77**, 044908 (2008), 0705.2511
[38] F. Antinori et al. (NA57), J. Phys. **G37**, 045105 (2010), 1001.1884
[39] B. Abelev et al. (ALICE), Phys. Rev. Lett. **111**, 222301 (2013), 1307.5530
[40] T. Sjöstrand et al., Comput. Phys. Commun. **178**, 852 (2008), 0710.3820
[41] T. Pierog, I. Karpenko, J. Katzy, E. Yatsenko, K. Werner (2013), 1306.0121
[42] C. Flensburg, G. Gustafson, L. Lonnblad, JHEP **08**, 103 (2011), 1103.4321
[43] C. Bierlich, G. Gustafson, L. Lönnblad, A. Tarasov, JHEP **03**, 148 (2015), 1412.6259
[44] C. Bierlich, J.R. Christiansen, Phys. Rev. **D92**, 094010 (2015), 1507.02091
[45] V. Vislavicius, A. Kalweit (2016), 1610.03001
[46] J. Adam et al. (ALICE), Eur. Phys. J. **C76**, 245 (2016), 1601.07868
[47] J. Cleymans et al., Phys. Rev. **C74**, 034903 (2006), hep-ph/0604237
[48] A. Andronic et al., Phys. Lett. **B673**, 142 (2009), 0812.1186
[49] K. Redlich, A. Tounsi, Eur. Phys. J. **C24**, 589 (2002), hep-ph/0111261
[50] F. Becattini, J. Manninen, J. Phys. **G35**, 104013 (2008), 0805.0098
[51] J. Aichelin et al., Phys. Rev. **C79**, 064907 (2009), 0810.4465
[52] S. Wheaton, J. Cleymans, J. Phys. **G31**, S1069 (2005), hep-ph/0412031
[53] G. Aad et al. (ATLAS), Phys. Lett. **B725**, 60 (2013), 1303.2084
[54] S. Chatrchyan et al. (CMS), Phys. Lett. **B724**, 213 (2013), 1305.0609

BM@N and MPD experiments at NICA

Vladimir Kekelidze[1,*], *Vadim* Kolesnikov[1], and *Alexander* Sorin[1,2]

[1]*Joint Institute for Nuclear Research, 141980 Dubna, Russia*
[2]*National Research Nuclear University (MEPhI), 115409 Moscow, Russia*

Abstract. The project NICA (Nuclotron-based Ion Collider fAcility) aims to study hot and baryon rich QCD matter in heavy ion collisions in the energy range $\sqrt{s_{NN}} = 4 - 11$ GeV. The rich heavy-ion physics program will be performed at two experiments, BM@N (Baryonic Matter at Nuclotron) at beams extracted from the Nuclotron, and at MPD (Multi-Purpose Detector) at the NICA collider. This program covers a variety of phenomena in strongly interacting matter of the highest baryonic density, which includes study of collective effects, production of hyperon and hypernuclei, in-medium modification of meson properties, and event-by-event fluctuations.

1 Introduction

QCD shows the explicit triumph of quantum field theory after the discovery of asymptotic freedom, which makes this theory self-consistent up to the arbitrary small scale at the order of quarks and gluons. In the non-perturbative regime, however, the theory of strong interactions is far from being complete and the study of heavy-ion collisions can shed light on the most fundamental QCD features: confinement and chiral symmetry breaking. Special interest in the study of the QCD phase diagram is related to the region of high baryon density, which is poorly known. In particular, it is of fundamental interest to establish the nature of the deconfinement phase transition and unveil the main features of the QCD diagram, such as the location of the critical endpoint (CEP). One of the main goals of the project Nuclotron-based Ion Collider fAcility (NICA), which is currently under development at the Joint Institute for Nuclear Research (JINR) in Dubna, is to study hot and baryon rich QCD matter in heavy ion collisions in the energy range $\sqrt{s_{NN}} = 4 - 11$ GeV [1–4]. Main objects of the NICA facility are presented in Fig. 1. The Nuclotron accelerator with the circumference of 251 m was put in operation in 1993 and essentially upgraded in 2010-2015. The Nuclotron is designed to provide beams of heavy ions with the kinetic energy up to 4.5 GeV/n and protons with the kinetic energy up to 12 GeV. Beams are extracted to the experimental hall where the first experiment at NICA - Baryonic Matter at Nucloton (BM@N) is allocated. The accelerator development plan includes the construction of the synchrotron Booster and two collider storage rings with the circumference of 503 m, and with two interaction points (IP) for allocation of the Multi-Purpose Detector (MPD) and the Spin Physics Detector (SPD). The NICA collider will provide interactions of gold ions up to center-of-mass energy of 11A GeV with the luminosity of 10^{27} cm^{-2}s^{-1}.

*e-mail: kekelidze@jinr.ru

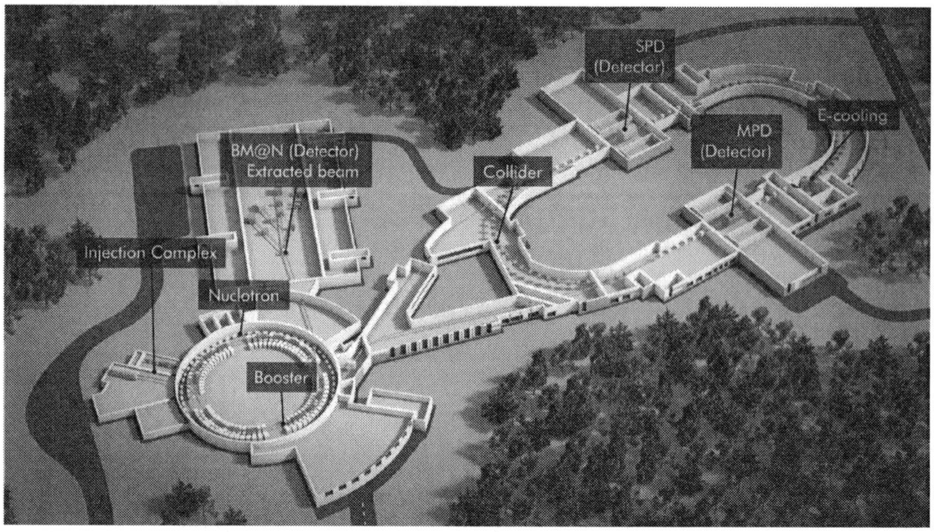

Figure 1. Main objects of the NICA facility; accelerator block: Nuclotron, Booster, Collider; experimental set-ups: BM@N, MPD, and SPD.

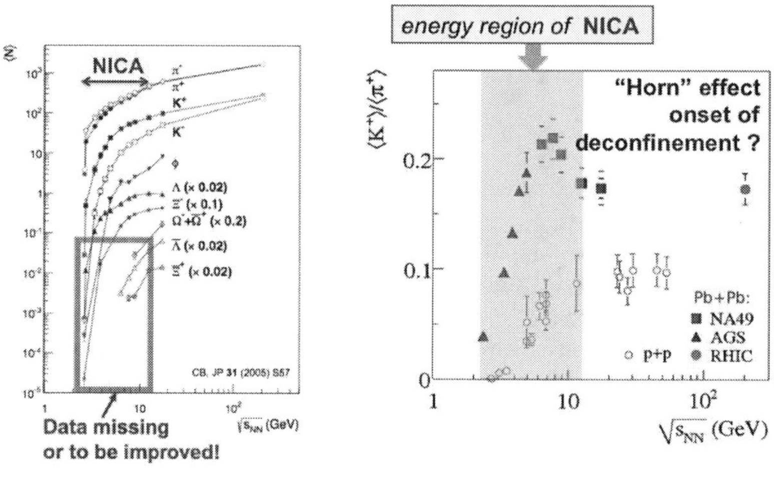

Figure 2. Left: energy dependence of strange particle production; right: excitation function of the K^+/π^+ ratio.

Production of strange particles in A+A collisions is of particular interest because their enhanced yields (relative to pp reactions) have been predicted as the signal of the deconfinement phase transition. Such enhancement, indeed, was observed at SPS energies [5] and then confirmed at the RHIC. However, the data on strangeness production at energy region of the maximum baryonic density are not complete or even missing (see Fig. 2, left). Recently, both the NA49 and the STAR experiments reported that the ratio of K^+ over π^+ has the maximum around 8 GeV/n in the center-of-mass (see Fig. 2, right). The observed non-monotonic excitation function of the strangeness-to-entropy ratio (Horn) was interpreted as the onset of deconfinement at SPS energies [6]. An alternative explanation

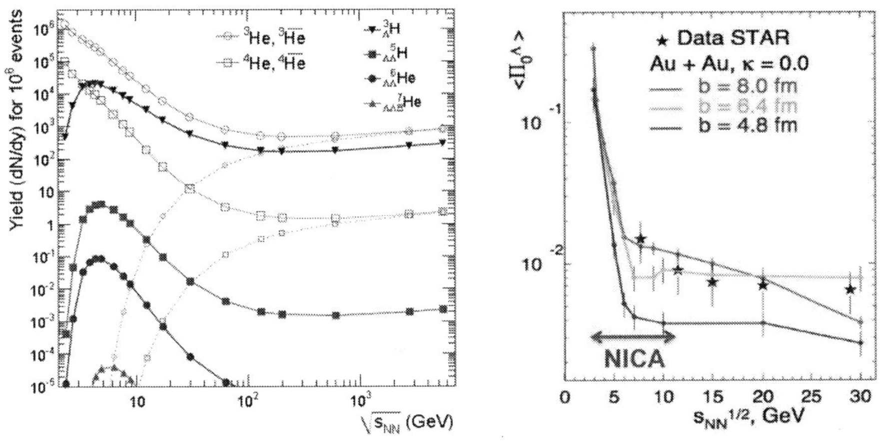

Figure 3. Left: energy dependence of hypernuclei yields in heavy ion collisions calculated in [9]; right: energy dependence of Λ polarization in heavy ion collisions.

of the Horn structure in K/π-ratio was given in Ref. [7] as a result of partial restoration of chiral symmetry (CSR) in hadronic systems at high temperatures and densities. In order to prove or rule out the strangeness enhancement as QGP or CSR indicator more experimental data on strange particle spectra and yields as a function of the system size and centrality are needed. NICA could provide corresponding investigations in the region of the maximum baryonic density, - such maximum for equilibrated states was indeed predicted using thermal model calculations within the NICA energy range [8].

If the deconfinement phase transition in dense hadronic matter is accomplished by partial restoration of the chiral symmetry, the vector meson spectral functions can be modified. Electron-positron pairs (dileptons) from vector meson decays are in this case the best candidates to study such in-medium modifications since they interact only weakly and escape easily having no subsequent strong interactions in the nuclear medium. There is no data on dilepton spectra at center-of-mass energies of several GeV, thus the dilepton program at NICA aims to close this gap.

Hypernuclei provide unique opportunity to study hyperon-nucleus interactions in a many-body environment. Details of the strange sector in the nuclear matter equation-of-state (EOS) are of great importance for astrophysics and should help understanding the mechanism of creation and evolution of super-dense stellar objects - neutron stars. As results of model calculations [9] indicate, the yields of hypernulei in central heavy ion collisions are enhanced within the NICA energy range (see Fig. 3, left), this gives good perspectives for such kind of studies at NICA.

Non-zero polarization of the produced in heavy ion collisions Λ-hyperons due to so-called vorticity effect was predicted in [10, 11]. This polarization is related to the anomalously induced axial current, which is proportional to the square of the chemical potential in the system. Calculations indicate that the polarization of Lambda is rising dramatically with the decreasing energy towards the maximum value within the NICA energy range (see Fig. 3, right). Recent STAR BES results reproduce the predicted trend at high collision energies [12] and the studies at lower energies will be continued at NICA.

A landscape of heavy ion collision experiments in the energy region of the maximum baryonic density, including present and future programs, is presented in Fig. 4. Two NICA experiments - BM@N with a fixed target and MPD at the collider will cover the whole indicated energy region.

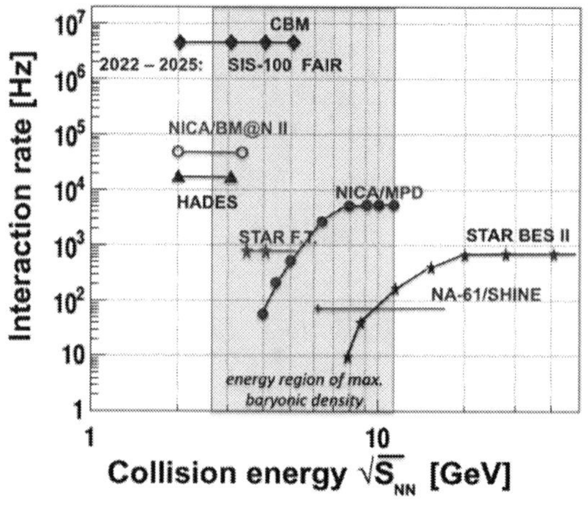

Figure 4. Present and future experiments with heavy ion collisions in the energy range of the maximum baryonic density.

Among the fixed target experiments, CBM at FAIR will provide the largest interaction rate. There are only two collider experiments - NICA and STAR-BES which differ in their interaction rates by 2-3 orders of the magnitude.

Thus, the NICA energy domain has several intriguing features, namely: the transition from baryon dominated matter to meson dominated one; the highest net baryon density - density frontier; it covers the onset of deconfinement region having the maximum in the K^+/π^+ and Λ/π ratios, as well as in the Λ polarization; a hypothetical 1-st order transition and creation of the mixed phase can be accomplished by the presence of the critical endpoint within the range.

2 Baryonic Matter at Nuclotron (BM@N)

A schematic view of the BM@N set-up is shown in Fig. 5. The tracking system is based on 12 planes of GEM detectors installed inside the gap of the dipole magnet of 0.8 T (SP-41) and drift/pad detectors (DCH, CPC) situated outside the magnet. The fully equipped set-up contains two stations of the Time of Flight (TOF) system, a Zero Degree Calorimeter (ZDC), a Recoil detector, and 4 stations of a Silicon Tracker (ST) installed just downstream the target. Several milestones are foreseen to put in operation the fully equipped detector. The first physical run is planed before the end of 2017 with about a half of the GEM chambers installed, and the complete design configuration is foreseen to be operational in 2020. The BM@N research program is aimed at studying elementary and heavy ion collision with the beam kinetic energy covering the region from 3.5 to 4.5 GeV/n. The results of extensive feasibility studies indicate a good performance of the detector in reconstruction of cascades and hypernuclei (see Fig. 6) with a monthly rate of about 10^7 for each species.

3 Multi-Purpose Detector (MPD)

Study of heavy-ion collisions within the energy range of maximum net baryonic density at the NICA collider will be performed with the MPD detector (see Fig. 7) designed as a large uniform acceptance spectrometer for detecting hadrons, electrons and gammas.

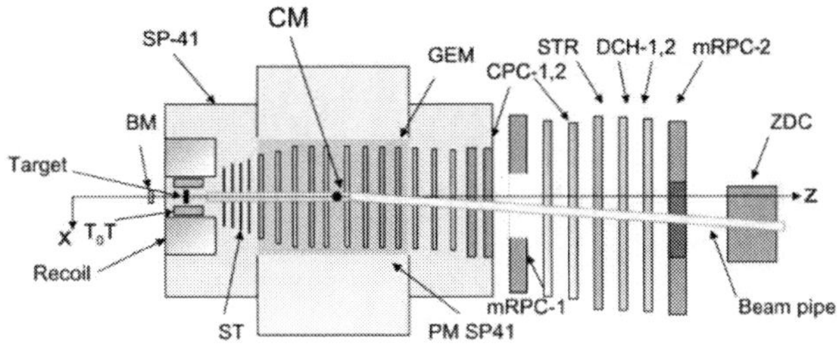

Figure 5. Scheme of the BM@N experiment.

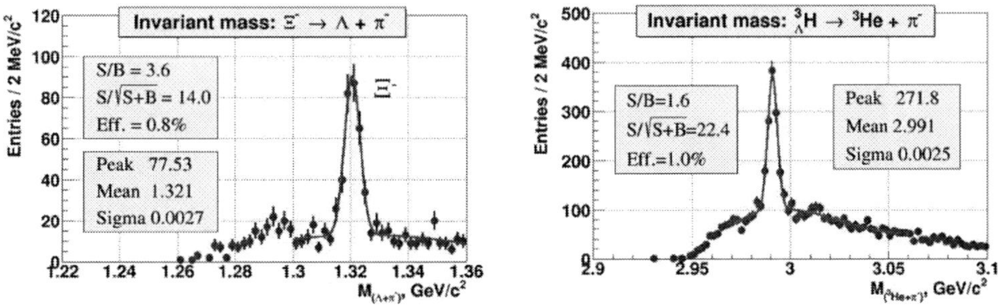

Figure 6. Simulation of reconstructed in BM@N decays of Ξ^- into Λ and π^- (left panel), and decays of hypertriton into ^{3}He and π^- (right panel) for central Au+Au collisions at 4.5A GeV.

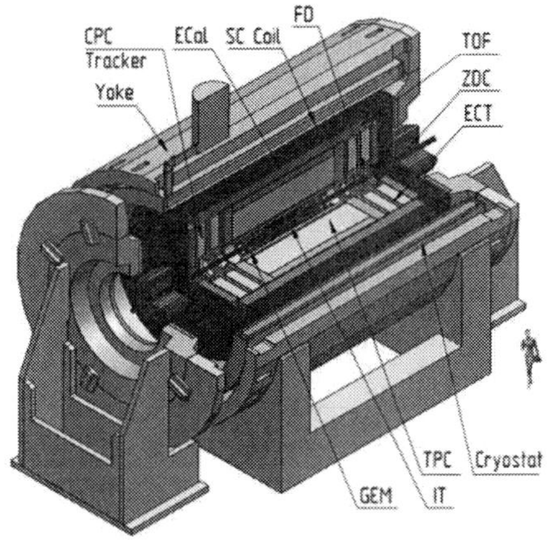

Figure 7. A view of the MPD detector at NICA.

All sub-detectors are placed in a large superconducting solenoid magnet generating the magnetic field of up to 0.5 T. The magnetic field inhomogeneity inside the tracking volume will be less than 10^{-3} in order to ensure a good tracking performance. The main tracking device of the MPD detector - the cylindrical Time-Projection Chamber (TPC) of 2.8 m (diameter) by 3.4 m (length) will provide trajectory reconstruction within the pseudorapidity range $|\eta| < 2$. Signals are readout with wire cham-

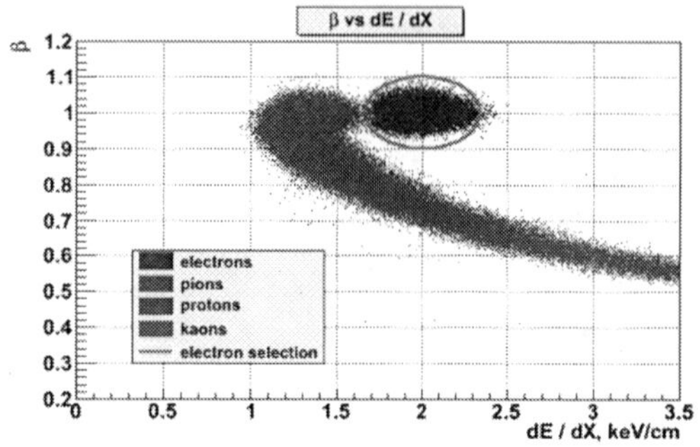

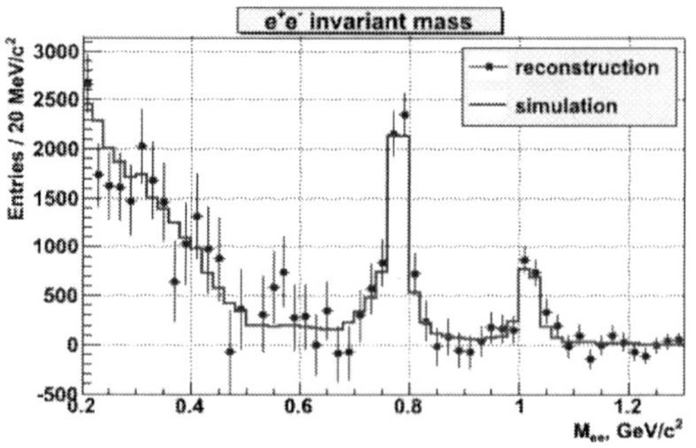

Figure 8. Left: correlation of the dE/dx and TOF information in MPD, the red line indicates the PID ellipse for electrons; Right: background-subtracted invariant mass distributions of electron-positron pairs from central Au+Au collisions at MPD.

bers (98000 channels in total) located at the both end faces. As an alternative read out technology for an upgraded TPC version GEM detectors are considered. For the midrapidity tracks the TPC will provide up to 53 measurements of the specific ionization and thus allow particle identification (PID) via energy loss (dE/dx) measurements with the precision better than 8%. The TOF system, which provides an extra PID power for charged tracks, consists of a barrel and two endcap systems equipped with Multigap Resistive Plate Chambers (MRPC). Having the intrinsic time resolution better than 70 ps the TOF system will allow hadron discrimination up to several GeV/c momenta. Two arrays of quartz counters (FD) located at a distance of 140 cm at each side from the MPD center will provide fast timing and triggering for the experiment. The Electromagnetic Calorimeter (ECAL) for electron and gamma measurements located behind the TOF detector will be built of lead-scintillator sampling modules based on the "shashlyk" technology. ECAL comprises about 43000 modules of 18 radiation length thickness and of about 3 cm^2 cross section forming a barrel shape with projective geometry directing to the IP. The MPD two-arm hadron calorimeter FHCAL, which is made of 44 modules with transverse dimensions of 15x15 cm^2, is planned to be used for event plane reconstruction and definition of the reaction centrality. The FHCAL design covering the pseudorapidity range $2.0 < |\eta| < 4.8$ allows measuring the reaction plane with the accuracy of $\approx 20°$ and to define centrality with a better then 10% precision. The inner tracker (IT) system maximizing the MPD performance for vertex reconstruction will be based on MAPS technology developed by the ALICE experiment.

The MPD detector will be put in operation in two stages. At the first stage MPD will be equipped with the TPC and barrel parts of the TOF and ECAL systems. The FHCal calorimeter and the FD detector will be operational as well. At the second stage MPD will be supplemented with the IT system and the endcap sub-detectors (the Straw tube tracker complemented by the endcap parts of the TOF and ECal systems).

The dedicated Monte-Carlo feasibility studies, including event generation, particle transport and realistic detector simulation, were carried out to investigate MPD performance during the first stage of the NICA physics program realization. Figure 8 (left panel) indicates the reliable electron identification using the combined information from the TOF and TPC detectors. Additional information about the energy deposit from the ECAL calorimeter allows reducing the hadron contamination in the electron sample down to the level of 10^{-5}. That could provide the reliable vector meson reconstruction though their dilepton decays in heavy-ion collisions at MPD (see Fig. 8, right).

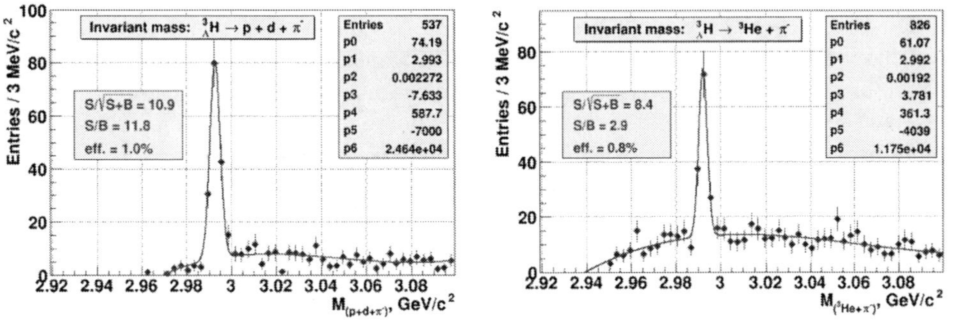

Figure 9. Reconstruction of two modes of hypertriton decays into $p + d + \pi^-$ (left) and ^{3}He + π^- (right).

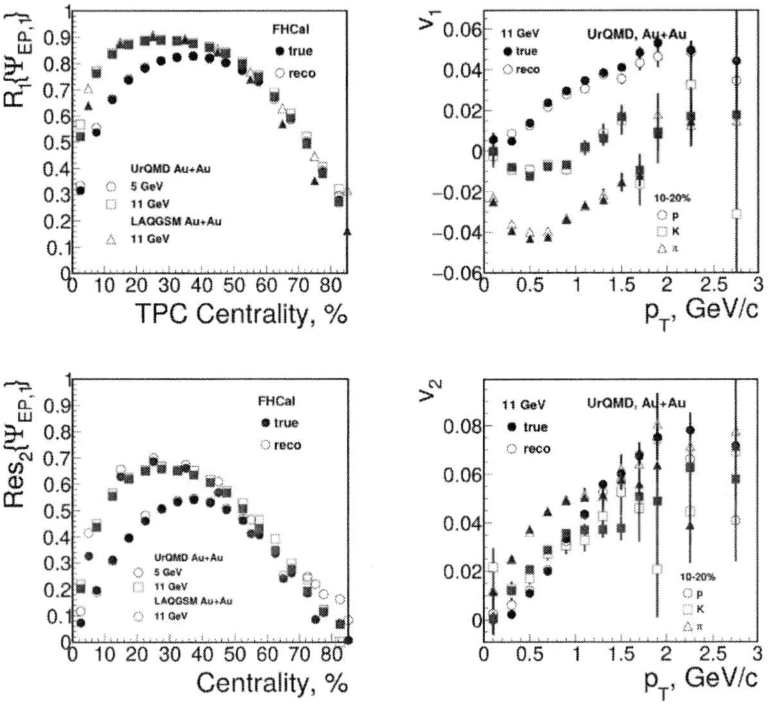

Figure 10. Correction factors of reaction plane reconstructions for the first (upper, left) and second (down, left) harmonics of flow Fourier decompositions; reconstructed first (upper, left) and second (down, right) harmonics of flows versus p_t.

An analysis of the hypernuclei reconstruction has been performed with $5 \cdot 10^5$ central Au+Au events, which corresponds to about 30 minutes of data taking time at NICA. The obtained invariant mass spectra for reconstruction of two modes of simulated hypertriton decays are presented in Fig. 9.

MPD ability to measure azimuthal asymmetry in heavy ion collisions is presented in Fig. 10. Centrality of reaction was defined by means of TPC track multiplicity, and the reaction plane was reconstructed using the FHCal responses. The correction factors $R_{1,2}$ for the first and second harmonics of flow decomposition are presented in the left panels, while the reconstructed first (v_1) and second (v_2) flow coefficients for various particles are presented on the right panels as a function of p_t. The presented results indicate good precisions of reaction plane reconstruction and sufficiently good reproducibility of the generated events in a wide range of transverse momentum.

4 Conclusions and perspectives

The project NICA, which is aimed at the construction of a new accelerator facility for heavy ions and polarized beams, is under development at JINR (Dubna). NICA ambitious goals cover many domains including experimental exploration of the QCD phase diagram with heavy ions, spin program with an emphasis on the nucleon spin structure, accelerator R&D, and intensive infrastructure upgrades. The project realization plan foresees a staged construction of the accelerator complex with the booster commissioning getting started by the end of 2018 as the first milestone. The start-up configuration of the NICA collider will be ready by the end of 2020, and its design high luminosity parameters are foreseen for 2023. The construction of the experimental setups at NICA is going close to the schedule. The BM@N experiment starts taking data in 2017 with a half of the tracking and PID systems installed, while its complete configuration will be ready in 2020. The first stage of the MPD detector for study nucleus-nucleus collisions at midrapidity will be ready in 2020 and further upgraded in 2023 to cover close to the full phase space.

The NICA project is open and attractive to a wide scientific community, its international collaborations currently involve more than 70 institutions from 25 countries. The future scientific infrastructure of the NICA facility will be able to provide research opportunities for about 1000 visitors, scientists and engineers working in accelerator physics, particle and nuclear physics, as well as in applied research areas, including material science and medicine. Beams of the NICA complex will be used not only for fundamental research but also for innovation and technological activities.

Realization of the NICA project, which has been included in the 2016 issue of the road-map of European Strategy Forum of Research Infrastructure (ESFRI), in the list of research infrastructures of global interest of Group of Senior Officials (GSO) on Global Research Infrastructures, and NuPECC Long Range Plan 2017, will provide new opportunities for the experimental exploration of nuclear matter of the highest baryonic density.

References

[1] Kekelidze V.D. et al. (NICA and MPD Collab.), Phys. Atom. Nucl. **75** 542-545 (2012).
[2] Kekelidze V.D. et al. (NICA Collab.), EPJ Web Conf. **95** 01014 (2015).
[3] Kekelidze V.D. et al. (NICA Collab.) Nucl. Part. Phys. Proc. **273-275** 170-174 (2016).
[4] Kekelidze V.D. et al. (NICA Collab.) JINST **12** No.06 C06012 (2017).
[5] Antinori F. et al. (NA57 Collab.) J. Phys. G. **32**, 427 (2006).
[6] Alt C. et al. (NA49 Collab.) Phys. Rev. C **77**, 024903 (2008).
[7] W. Cassing, A. Palmese, P. Moreau, E. Bratkovskaya Phys. Rev. C **93**, 014902 (2016).
[8] Randrup J., Cleymans J. Phys. Rev. C **74**, 047901 (2006).
[9] Andronic A., Braun-Munzinger P., Stachel J., Stocker H., Phys. Lett. B **697**, 203 (2011).
[10] Rogachevsky O., Sorin A., Teryaev O. Phys. Rev. C **82**, 054910 (2010).
[11] Baznat M., Gudima K., Sorin A., Teryaev O. E-print: arXiv:1701.00923.
[12] Adamczyk L. et al. (STAR Collab.), Nature **548**, 62-65 (2017).

The Compressed Baryonic Matter experiment at FAIR

Claudia Höhne for the CBM collaboration[1,*]

[1] *Physics Institute II, Justus-Liebig University Giessen*

Abstract.
The CBM experiment will investigate highly compressed baryonic matter created in A+A collisions at the new FAIR research center. With a beam energy range up to 11 AGeV for the heaviest nuclei at the SIS 100 accelerator, CBM will investigate the QCD phase diagram in the intermediate range, i.e. at moderate temperatures but high net-baryon densities. This intermediate range of the QCD phase diagram is of particular interest, because a first order phase transition ending in a critical point and possibly new high-density phases of strongly interacting matter are expected. In this range of the QCD phase diagram only exploratory measurements have been performed so far. CBM, as a next generation, high-luminosity experiment, will substantially improve our knowledge of matter created in this region of the QCD phase diagram and characterize its properties by measuring rare probes such as multi-strange hyperons, dileptons or charm, but also with event-by-event fluctuations of conserved quantities, and collective flow of identified particles. The experimental preparations with special focus on hadronic observables and strangeness is presented in terms of detector development, feasibility studies and fast track reconstruction. Preparations are progressing well such that CBM will be ready with FAIR start. As quite some detectors are ready before, they will be used as upgrades or extensions of already running experiments allowing for a rich physics program prior to FAIR start.

1 Mapping the high density region of the QCD phase diagram with CBM

Heavy ion experiments at relativistic energies are *the* experimental tool in order to map the conjectured QCD phase diagram, see figure 1 (left), which is of fundamental interest for understanding the strong interaction and strongly interacting matter in particular in the non-perturbative regime. In dependence on the collision energy strongly interacting matter of varying temperature and baryon chemical potential is created. The exploration of matter at high temperatures but low baryon chemical potential in heavy-ion collisions at the highest energies at RHIC and LHC [1, 3] reveals insight into the characteristics of the quark-gluon plasma and the phase transition to hadronic matter. Experimental results can be compared to lattice QCD calculations at $\mu_B = 0$ which predict a crossover from partonic to hadronic matter at temperatures around 160 MeV [4] in accordance to experimental findings. According to lattice QCD, the phase transition is driven by the energy density and occurs at a critical density on the order of 1 GeV/fm^3. At large baryon densities the QCD phase diagram is uncharted territory as lattice calculations are not applicable, and as only few experimental measurements mainly of bulk

*e-mail: claudia.hoehne@physik.uni-giessen.de

probes are available from former AGS experiments. However, this region of the phase diagram is predicted to have rich structures like e.g. new exotic phases as the quarkyonic matter [5] or a first order phase transition and a critical point, see e.g. [6]. The experimental discovery of such structures and the characterization of dense baryonic matter would be a major milestone in our understanding of strongly interacting matter and has triggered a number of experimental campaigns worldwide.

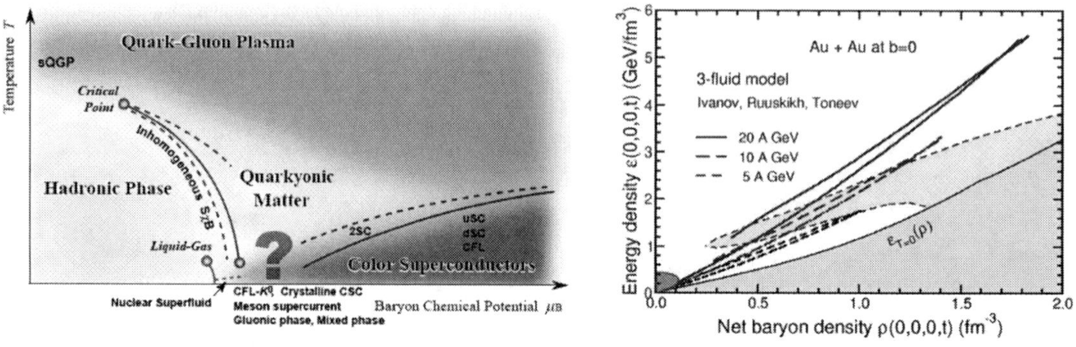

Figure 1. Left: Conjectured QCD phase diagram [2]. Right: Energy density versus baryon density in the central cell of Au+Au collisions at various energies from a 3-fluid model [7].

At FAIR the SIS 100 accelerator will offer heavy-ion beams with energies up to 11 AGeV for Au ions at intensities up to 10^9 ions/s. Light ions and protons will be available at respectively higher energies. Heavy ion collisions in this energy range create matter at energy densities that reach beyond the critical density of about 1 GeV/fm^3. In parallel to the energy density also the baryon density increases in those collisions and can reach up to 10 times normal nuclear matter density in the center of central Au+Au collisions [8], see fig. 1 (right). The CBM experiment is being prepared in order to characterize the dense baryonic matter created in heavy-ion collisions at FAIR by measuring a wide variety of observables including rare probes such as multistrange particles, charm (if the energy is sufficient) and dileptons.

2 Probes of strongly interacting matter at high baryon densities

Unfortunately, to all our knowledge there is no simple key measurement that will reveal the phase structure of dense baryonic matter. The experimental strategy thus relies on systematic measurements of a set of promising observables in dependence on collision energy and system size. In the following, some of the promising observables and their connection to the open physics questions will be discussed briefly, for more details see [9].

Strangeness production

Particles containing strange quarks are important probes of the matter created in heavy-ion collisions. The strange quark plays a special role in investigations that are connected to the question whether the created matter reaches equilibrium. Strange quarks are newly produced quark flavours, however they are still light enough to be produced thermally, i.e. in secondary collisions of produced particles. Whether or not strangeness is in thermal equilibrium may be seen in the yields and phase space distributions of multi-strange particles, in particular the Ω baryon [10]. New HADES measurements show on the other hand [11] that multi-strange particle production may happen through more complicated reaction mechanisms, which may be connected to the equation-of-state at high baryon densities. CBM therefore plans to measure yields, phase space distributions, flow and fluctuations of strange particles including multi-strange baryons and the ϕ-meson in dependence on energy and system-size.

Directed and elliptic flow

The collective flow of hadrons is driven by geometrical anisotropies of the heavy-ion collision and (resulting) pressure gradients. Theoretical calculations compared to experimental results can connect the experimental findings with the equation-of-state of dense matter and in-medium properties of hadrons. However, so far no consistent measurements of in particular also flow of strange hadrons is available in dependence on energy. The various pieces of experimental observations that are available already show very interesting behaviour, and with CBM measurements the aim is to bring all of this together to a consistent picture. For example, measurements from the STAR experiment show an increasing difference in particle and antiparticle flow the lower the collision energy becomes [12]. Measurements from AGS show that Λ and K^0 flow have opposite signs [13] and measurements of kaon flow from FOPI and KAOS [14, 15] reveal a strong dependence on in-medium potentials.

Existence of exotic strange objects

Thermal model calculations predict a maximum of the production of single and double hypernuclei close to top SIS 100 beam energies [16]. This maximum is reached due to a counterplay of increasing hyperon production and decreasing production of light nuclei with higher beam energy. As a result, CBM should have a major discovery potential for the production of light double-Λ hypernuclei. Such measurements would be a breakthrough in hypernuclei physics and substantially improve the knowledge in $\Lambda - \Lambda$ and $\Lambda - N$ interactions. More speculative is even the connection to remnants of dense, chirally restored matter which also should result in the production of more exotic strange objects [17].

Charm production

Charm production at threshold can be investigated with proton beams at SIS 100. Interestingly, in Ref. [18] the authors predict subthreshold charm production in Au+Au collisions at SIS 100 energies. As the CBM experiment is being prepared also for the measurement of open charm and J/ψ, charmed hadrons will be measurable if being produced with the predicted multiplicity. If measured, this signal would open up a new door for the understanding of charm production in dense baryonic matter and for the understanding of the important role of excited baryons in this process.

Event-by-event fluctuations

Event-by-event fluctuations of conserved quantities such as baryon number, electric charge or strangeness can be directly related to thermodynamic susceptibilities and thus provide an ideal tool to investigate the predicted existence of phase transitions, the order of phase transitions or the existence of mixed phases. Event-by-event fluctuations are thus one of the experimental measurements that can rather directly be compared to lattice QCD calculations [19]. Measurements of event-by-event fluctuations have been performed in order to search for the critical point or signs of a first order phase transition, however, always have to fight with the many experimental artefacts that can mimic fluctuations. Currently, the most interesting and promising measurements are an energy scan of net-proton fluctuations from the STAR experiment [20] combined with preliminary results from the HADES experiment [21]. The results clearly call for high-precision measurements at energies in between in order to search for a possible peak in this fluctuation measurement.

Dileptons

Dileptons emitted in heavy-ion collisions offer the unique property of giving insight into the early phase of the created matter. Due to their penetrating nature and the large variety of processes in which they are created, they provide information on the early temperature of the fireball, in-medium properties of vector mesons, lifetime of the fireball, baryon density due to coupling to baryonic resonances and possibly chiral symmetry restauration. In CBM, particular emphasis will be put on measuring dilepton rates in the mass range beyond 1 GeV/c^2 as this allows to directly access the fireball temperature [22] and a contribution from ρ-a_1 chiral mixing [23]. Figure 2 highlights the discovery potential connected to this measurement.

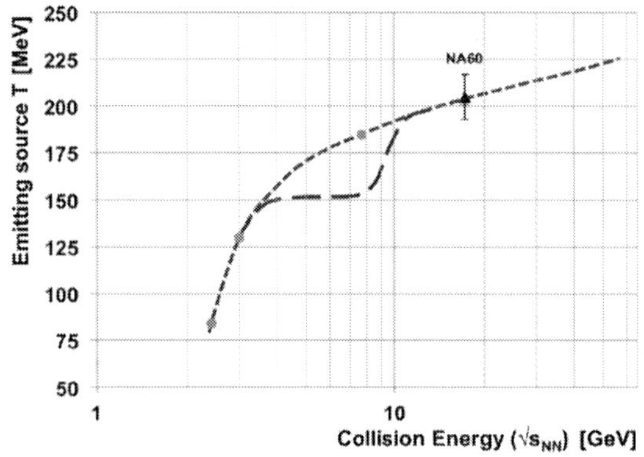

Figure 2. Model calculation (dashed red curve) of the excitation function of the fireball temperature measured with dileptons and speculated shape (violet) with a first order phase transition in the SIS 100 energy range [24].

3 The CBM experiment - challenges and experimental status

CBM aims at systematic measurements of the above introduced observables in order to access their excitation functions and their dependence on centrality and system size. For efficient measurements also of the rare probes, running at highest possible interaction rates is a must for CBM. Figure 3 (left) compares the maximum interaction rates for different experiments in dependence on center of mass energy for Au+Au collisions. The anticipated maximum interaction rates of 10 MHz for CBM outrange all other experiments and open up new and exciting possibilities, however, they come with a challenge: The whole detector, see figure 3 (right), has to be radiation hard and fast, and has to cope with very high particle fluxes due to the fixed target geometry.

In addition to this, the online systems will be as important and challenging as the CBM detector itself [25]. CBM will not employ hardware triggers for event selection but will collect free streaming triggerless data. Each detector hit is sent with a time stamp to a high performance computing farm in the GSI GreenIT cube. A high speed first level event selection (FLES) system [26] will deliver online event reconstruction and selection including secondary vertex finding in real-time [27]. Only this effort will allow to filter out the rare events with, e.g., weakly decaying multi-strange hyperons. In order to cope with the high rates and time-stamped data format, the track reconstruction routines will add time as a fourth dimension in order to disentangle the different events [28]. Reconstructed tracks are then clustered in groups representing the original events (figure 4).

The detector setup of CBM ensures a large acceptance for all interesting observables for the full SIS 100 energy range, and detailed feasibility studies are being performed in order to evaluate the performance, see e.g. [29, 30]. One benchmark observable is the capability to measure Ω-baryons. Assuming one week beamtime each, 1 MHz interaction rate and 10% central Au+Au collisions, CBM

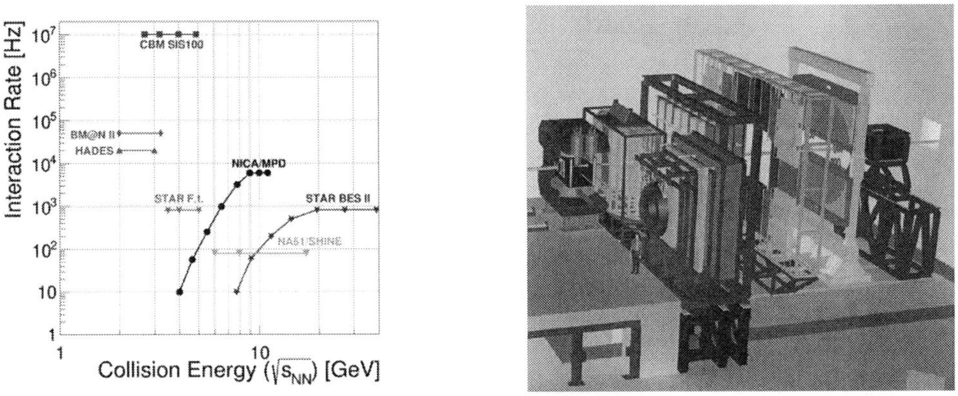

Figure 3. Left: Interaction rates achieved by existing and planned heavy-ion experiments as a function of center-of-mass energy [9]. "STAR F.t." denotes the fixed-target operation of STAR. Right: Setup of the CBM experiment at SIS 100. The beam is entering from left, the sequence of the detectors is: tracking system consisting of silicon pixel (MVD) and micro-strip detectors (STS) in the superconducting dipole magnet, RICH detector, some layers of TRD, TOF, ECAL and a calorimeter for event caracterization (PSD). The muon detector (MuCh) is shown here in the parking position.

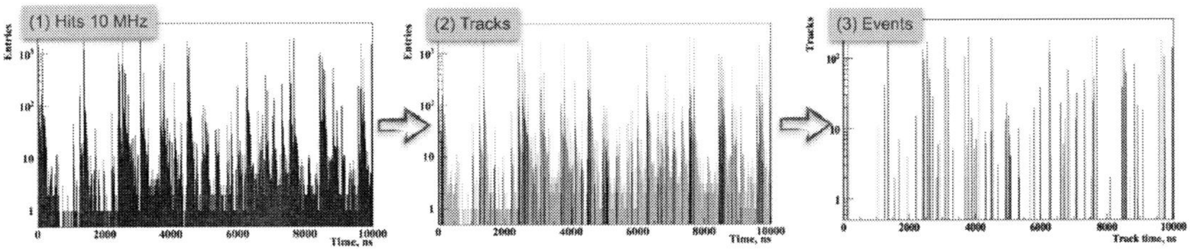

Figure 4. Part of a time-slice with 100 minimum bias Au+Au collisions at 25 AGeV beam energy [28]. Left: time distribution of hits in the selected time-slice. Middle: reconstructed tracks (black) within the time distribution of hits. Right: reconstructed tracks clustered into event groups.

expects to measure 10^6 Ω at 6 AGeV and 10^7 Ω at 10 AGeV beam energy - far enough in order to evaluate yield and phase space distributions. Most notably this performance will allow to reconstruct the elliptic flow of the Ω. Within a week beamtime at 10 AGeV beam energy and running minimum bias collisions with 250 kHz interaction rate only, the Ω flow will be reconstructed with approximately (10-20)% statistical errors at medium p_t [31]. According to new software developments, the high performance silicon tracking system will allow even for Σ^+ and Σ^- baryon reconstruction through their decay topology and a missing mass hypothesis. 4π efficiencies of (1-3)% are expected. This may even allow to address physics questions related to the symmetry energy. Assuming the predicted production rates in Ref. [16] ${}^{5}_{\Lambda\Lambda}H$ double hypernuclei can be discovered in 10^{12} central Au+Au collisions at 10 AGeV beam energy.

The current FAIR schedule foresees the full modularized start version to be operational in 2025, with first detector installations and commisioning as well as first beams 2021-2024. CBM will be the first day experiment of SIS 100 due to its position right behind SIS 100. CBM will be ready until then, even more, several detector components will be ready substantially earlier in time. Therefore,

CBM detector components will be installed prior to the usage at FAIR in existing heavy-ion experiments at other laboratories. This so called *FAIR phase 0* program has many benefits: CBM detector components will substantially improve the detector performance of other heavy-ion experiments and increase their physics output. Furthermore, CBM detectors and their readout chain will be thoroughly tested and calibrated, being fully operational and well understood for the start of CBM at FAIR. In this context several projects are pursued: Four STS stations will improve the tracking of the BM@N experiment in Dubna, 10% of the TOF detector modules will substantially improve the forward particle identification in STAR during beam energy scan II, and 40% of the RICH MAPMTs can be used to upgrade the HADES RICH detector for new beamtimes at SIS 18. This replacement of the old photondetector will improve the efficiency of dilepton reconstruction for close pairs of about a factor 5. Within the next heavy-ion HADES beamtime at SIS 18 we expect a breakthrough in dilepton physics at low beam energies, i.e. the possibility to measure dielectron signal pairs with masses beyond 1 GeV/c^2. Besides the use of CBM detectors in other heavy-ion experiments a proposal for a miniCBM experiment at SIS 18 during FAIR phase 0 has been accepted. MiniCBM will use detector prototypes of all CBM components in order to test and establish the self-triggered readout, data transfer to FLES, time-based event building and the online analysis. A succesful operation of miniCBM will substantially reduce commissioning and setup time at SIS 100.

4 Summary

The CBM experiment at FAIR is part of a world-wide endeavour to experimentally map the QCD phase diagram and characterize strongly interacting matter in dependence on temperature and baryon density. The CBM experiment will be unique for the exploration at high-net baryon densities because CBM will operate at interaction rates several orders of magnitude higher than anywhere else. CBM will thus be able to explore matter at high baryon densities with systematic high-precision measurements of rare observables like multistrange particles, dileptons and charm. This offers the chance to resolve many open fundamental physics questions related to baryon-dense matter such as for the nuclear matter equation of state, phase transitions and possible new phases or in-medium modifications of hadrons. CBM thus has a substantially discovery potential. The CBM detector development is well advanced, and several of the detector components can even be used in other heavy-ion experiments prior to FAIR start, substantially enhancing their physics potential.

References

[1] B. Jacak, B. Müller, Science 337, 310 (2012).
[2] K. Fukushima, T. Hatsuda, Rept. Prog. Phys.74:014001,2011.
[3] P. Foka, M.-A. Janik, Reviews in Physics 1, 154-171 (2016).
[4] Z. Fodor, S.D. Katz, Acta Phys.Polon. B42 (2011) 2791-2810.
[5] L. McLerran and R. Pisarski, Nucl. Phys. A 796 (2007) 83.
[6] C. Fischer et al., *Locating the critical end point of QCD*, Nucl. Phys. A **931** (2014) 774.
[7] Y.B. Ivanov, V.N. Ruuskikh, V.D. Toneev, Phys. Rev. C 73, 44904 (2006).
[8] B. Friman et al. (Eds), The CBM Physics Book: Compressed Baryonic Matter in Laboratory Experiments, Springer Series: Lecture Notes in Physics, Vol. 814 (2011).
[9] T. Ablyazimov et al. [CBM collaboration], Eur. Phys. J. A 53 (2016) 60.
[10] P. Braun-Munzinger, J. Stachel and C. Wetterich, Phys. Lett. B 596 (2004) 61.
[11] G. Agakishiev et al. [HADES collaboration], Phys. Rev. Lett. 103, 132301, (2009).

[12] L. Adamczyk et al. [STAR collaboration], Phys. Rev. Lett. 110, 142301 (2013).
[13] P. Chung et al. [E895 collaboration], Phys. Rev. Lett. 85, 940 (2000).
[14] V. Zinyuk et al. [FOPI collaboration], Phys. Rev. C 90, 025210 (2014).
[15] Y. Shin et al. [KAOS collaboration], Phys. Rev. Lett. 81, 1576 (1998).
[16] A. Andronic, P. Braun-Munzinger, J. Stachel, H. Stöcker, Phys. Lett. B 697, 201 (2011).
[17] H. Stöcker et al., Nucl. Hys. A 827, 624c (2009).
[18] J. Steinheimer, A. Botvina, M. Bleicher, Phys. Rev. C 95, 014911 (2017).
[19] F. Karsch, J. Phys. Conf. Ser. 779, 012015 (2017).
[20] X. Luo et al. [STAR collaboration], PoS CPOD 2014, 019 (2014).
[21] M. Lorenz [HADES collaboration], Nucl. Phys. A 967, 27 (2017).
[22] R. Arnaldi et al. [NA60 Collaboration], Eur. Phys. J C 61 (2009) 711.
[23] P.M. Hohler, R. Rapp, Nucl. Part. Phys. Proc. 267-278, (2016) 253.
[24] T. Galatyuk et al., Eur. Phys. J. A 52 (2016) 131.
[25] V. Friese, Springer Lecture Notes in Computer Science, Volume 7125 (2012) 17.
[26] J. de Cuveland and V. Lindenstruth, J. Phys.: Conf. Ser. 331 (2011) 022006.
[27] V. Friese, J. Phys.: Conf. Ser. 331 (2011) 032008.
[28] I. Kisel, EPJ Web Conf. 108 (2016) 01006.
[29] The CBM collaboration, STS-TDR, GSI-2013-05499, http://repository.gsi.de/record/54798.
[30] The CBM collaboration, TOF-TDR, GSI-2015-01999, http://repository.gsi.de/record/109024.
[31] The CBM collaboration, PSD-TDR, GSI-2015-02020, http://repository.gsi.de/record/109059.

Processes of hypernuclei formation in relativistic ion collisions

Alexander Botvina[1,2] and *Marcus* Bleicher[1]

[1] *Frankfurt Institute for Advanced Studies, J.W.Goethe University, Frankfurt am Main, Germany*
[2] *Institute for Nuclear Research, Russian Academy of Sciences, Moscow, Russia*

Abstract. The study of hypernuclei in relativistic ion collisions open new opportunities for nuclear and particle physics. The main processes leading to the production of hypernuclei in these reactions are the disintegration of large excited hyper-residues (target- and projectile-like), and the coalescence of hyperons with other baryons into light clusters. We use the transport, coalescence and statistical models to describe the whole reaction, and demonstrate the effectiveness of this approach: These reactions lead to the abundant production of multi-strange nuclei and new hypernuclear states. A broad distribution of predicted hypernuclei in masses and isospin allows for investigating properties of exotic hypernuclei, as well as the hypermatter both at high and low temperatures. There is a saturation of the hypernuclei production at high energies, therefore, the optimal way to pursue this experimental research is to use the accelerator facilities of intermediate energies, like FAIR (Darmstadt) and NICA (Dubna).

1 Introduction

Hypernuclei are formed when hyperons ($\Lambda, \Sigma, \Xi, \Omega$) produced in high-energy interactions are captured by nuclei. Presently, the hypernuclear physics is still focused on spectroscopic information and is dominated by a quite limited set of reactions induced by hadrons and leptons [1, 2]. However, many experimental collaborations (e.g., STAR at RHIC [3]; ALICE at LHC [4]; HypHI at GSI [5], FOPI/CBM [6], and Super-FRS/NUSTAR at FAIR [7]; BM@N and MPD at NICA [8]) have started or plan to investigate hypernuclei and their properties in reactions induced by relativistic nucleons and ions. The limits in isospin space, particle unstable states, multiple strange nuclei and precision lifetime measurements are unique topics of these fragmentation reactions. Generally, the investigation of hypernuclei is one of the rapidly progressing fields of nuclear physics, as they provide complementary methods to improve traditional nuclear studies and open new horizons for studying particle physics and nuclear astrophysics.

2 Modelling production of hypernuclei

For description of the hypernuclei production both dynamical and statistical approaches can be employed. We use the Ultrarelativistic Quantum Molecular Dynamics (UrQMD) model, the Hadron String Model (HSD), the Dubna Cascade Model (DCM) as the transport approaches to simulate the initial dynamical reaction stage [9, 10]. Nucleons from overlapping parts of the projectile and target

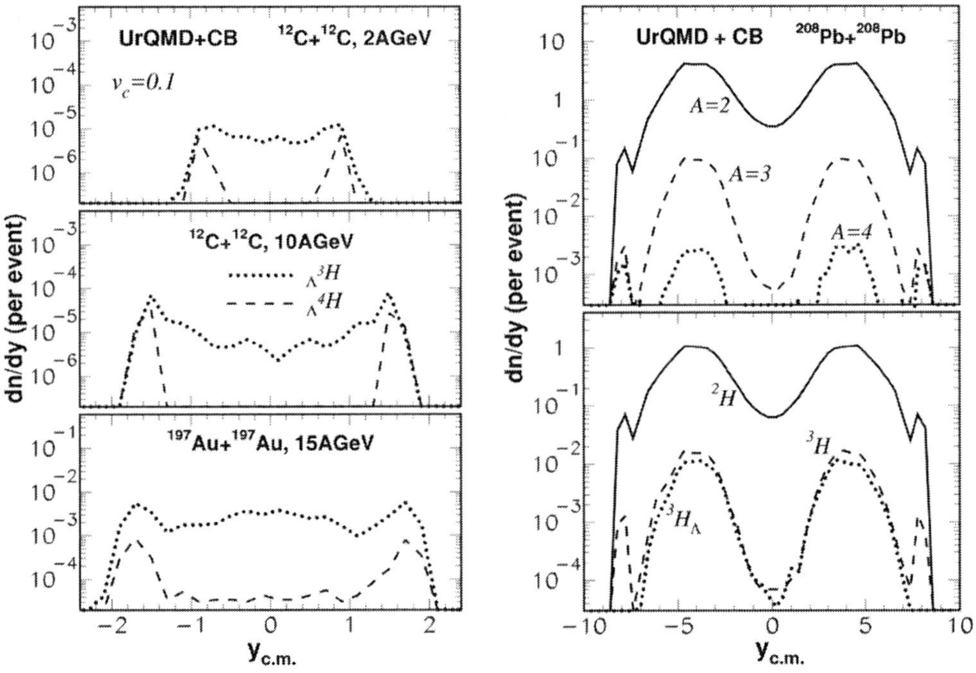

Figure 1. UrQMD and coalescence (CB) calculations of the light nuclei rapidity distributions. Left panels are from Ref. [10]: Relative yields (per inelastic event) of $^3_\Lambda$H and $^4_\Lambda$H hypernuclei in carbon on carbon and gold on gold collisions at intermediate energies shown in the panels. Right panels are from Ref. [18]: Relative yields of nuclei with mass numbers A=2, 3, 4 (top) and ^{2}H, ^{3}H, $^3_\Lambda$H (bottom) in lead on lead collisions at LHC energies ($\sqrt{s_{NN}}$ =2.76 A TeV). Notations for the lines are in the panels.

(participant zone) interact strongly with each other and with hadrons produced in primary and secondary collisions. Nucleons from non-overlapping parts do not interact intensively, and they form residual nuclear systems, which we call spectator residues. The production of hyperons is associated with nucleon-nucleon collisions, e.g., p+n→n+Λ+K$^+$, or collisions of secondary mesons with nucleons, e.g., π^++n→ Λ+K$^+$. In the following the strange particles can re-scatter and undergo secondary interactions. As a result the produced hyperons populate the whole momentum space around the colliding nuclei, including the vicinity of nuclear spectators, and can be captured by the nuclear residues. This can be described both as the capture in a nuclear potential [9] and within a generalized coalescence model [10]. The coalescent mechanism can be applied for the formation of light clusters too. In such a way it is possible to describe the formation of fragments of all sizes, from lightest nuclei to heavy residues, including hyper-fragments, at all rapidities, on the event-by-event basis. The disintegration of excited hyper-fragments can be described by the generalized statistical evaporation and multifragmentation models (for big fragments) [11, 12] and by the Fermi break-up model (for small fragments) [13], which were applied with great success for description of the secondary decays in normal reactions, see e.g. [15–17]. As a new theoretical result we have extended the elaborated previously hybrid description of nuclear fragmentation in hadron and ion relativistic collisions for hyper-nuclear matter. Within these models one can predict the production of all kind of hypernuclei (including multi-strange and the exotic ones) which are not always possible to obtain in traditional hyper-nuclear studies [1, 2].

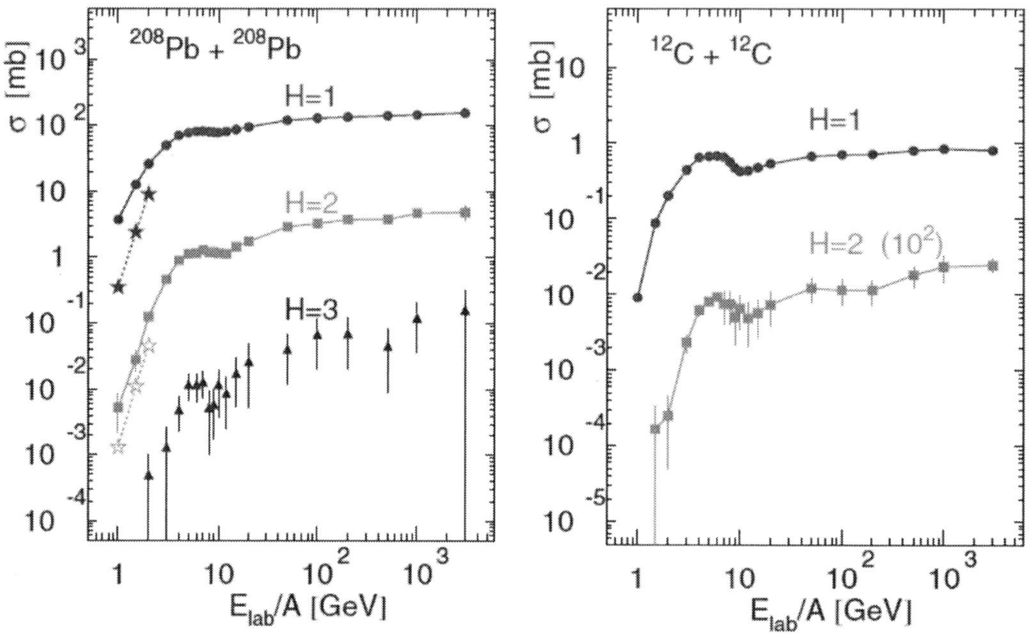

Figure 2. UrQMD (dashed lines, star symbols) and DCM (solid lines, circles, squares, triangle symbols) transport calculations: Absolute yields (millibarn) of spectator hyper-residues with capture of H=1, 2, 3 Λ-hyperons versus the laboratory beam energy in lead on lead (left panel) and carbon on carbon (right panel) collisions [14].

3 Yields of hypernuclei

Within our approach we demonstrate in Fig. 1 the rapidity distributions of light hypernuclei produced at intermediate- and very high-energy collisions. The calculations were performed overall impact parameters (minimal bias). One can see that light hypernuclei ($^{3}_{\Lambda}$H) can be formed at all rapidities by the coalescence mechanism. However, the production of more heavy nuclear species (e.g., $^{4}_{\Lambda}$H) is clearly shifted toward target and projectile rapidities. This is true for both intermediate (left panels) and high (right panels) energies. Moreover, at high energies, besides the expected fragmentation peaks around the target/projectile residues, we have found wide peaks related to so-called the fragmentation regions outside the midrapidity. These regions are produced because of the secondary interactions of strange and normal particles, which effectively decrease the particle's kinetic energy and make their capture in big clusters more probable. A chance to form an exotic state is much larger in such a case.

To evaluate the production of big hypernuclei we have calculated the formation of hyper-nuclear residues in the relativistic collisions with DCM and UrQMD (plus coalescence) models. Fig. 2 demonstrates their yields as function of the incident energy for different numbers of captured Λ hyperons (H). One can see a fast increase of the yields around the threshold energy and then a saturation up to very high energies. The cross-sections are sufficiently large even for multi-strange residues, by taking into account that modern experiments can measure nanobarn yields. The experimental vertex technique [3–5] can be effectively used. We should consider that these big residues are mostly excited and decay afterwards during the time around $10^2 - 10^3$ fm/c. It was demonstrated previously that the disintegration of such excited hyper-nuclear matter leads to a very broad isotope distribution of final hypernuclei [11, 12]. Novel multi-hyperon nuclear states can be formed in this case too.

4 Conclusion and perspectives

Our results demonstrate that the investigation of hypernuclei in relativistic collisions is very promising. In particular, the yields of strange spectator residues and hypernuclei are about $\sim$ 1–100 mb (depending on the colliding nuclei) that is quite a lot in comparison with the traditional hypernuclear methods. An additional advantage is that very exotic hypernuclei (neutron-rich and neutron-poor ones) as well as multi-strange nuclei can be produced. The last ones are specially important for understanding the hyperon-hyperon interactions and the nature of compact astrophysical objects (e.g., neutron stars).

A.S. Botvina acknowledges the support of BMBF (Germany).

References

[1] H. Bando, T. Mottle, and J. Zofka, Int. J. Mod. Phys. **A5**, 4021 (1990).
[2] O. Hashimoto, H. Tamura, Prog. Part. Nucl. Phys. **57**, 564 (2006).
[3] The STAR collaboration, Science **328**, 58 (2010).
[4] B. Dönigus *et al.* (ALICE collaboration), Nucl. Phys. A**904-905**, 547c (2013).
[5] T.R. Saito *et al.* (HypHI collaboration), Nucl. Phys. A **881**, 218 (2012).
[6] I. Vassiliev *et al.* (CBM collaboration). JPS Conf. Proc. **17**, 092001 (2017).
[7] https://indico.gsi.de/event/superfrs3
[8] Topical issue *NICA White Paper*, Edt. D.Blaschke *et al.*, Eur. Phys. J. **A52**, 211-267 (2016)
[9] A.S. Botvina *et al.*, Phys. Rev. C **84**, 064904 (2011)
[10] A.S. Botvina , J. Steinheimer, E. Bratkovskaya *et al.*, Phys. Lett. B **742**, 7 (2015)
[11] A.S. Botvina and J. Pochodzalla, Phys. Rev. C **76**, 024909 (2007)
[12] A.S.Botvina *et al.*, Phys. Rev. **C94**, 054615 (2016)
[13] A.S. Botvina, K.K. Gudima, J. Pochodzalla, Phys. Rev. C **88**, 054605 (2013)
[14] A.S. Botvina *et al.*, Phys. Rev. **C95**, 014902 (2017)
[15] H. Xi *et al.*, Z. Phys. A **359**, 397 (1997)
[16] R.P. Scharenberg *et al.*, Phys. Rev. C **64**, 054602 (2001)
[17] R. Ogul *et al.*, Phys. Rev. C **83**, 024608 (2011)
[18] A.S. Botvina, J. Steinheimer, M. Bleicher. Phys. Rev. **C96**, 014913 (2017)

EPJ Web of Conferences **171**, 13002 (2018) https://doi.org/10.1051/epjconf/201817113002
SQM 2017

Kaon and Phi Production in Pion-Nucleus Reactions at 1.7 GeV/c*

Joana Wirth[1,2,**] for the HADES Collaboration

[1] *Technische Universität München, Fakultät für Physik, E62*
[2] *Excellence Cluster Universe, Technische Universität München*

Abstract. The production and properties of K^+, K^- and ϕ in nuclear reactions $\pi^- + A$ ($A = C, W$) at a beam momentum of 1.7 GeV/c has been studied with the HADES setup at SIS18/GSI. Of particular interest is the K^- absorption in nuclear matter which should be driven by strangeness exchange processes on one ($K^-N \rightarrow Y\pi$) or more nucleons ($K^-NN \rightarrow YN\pi$). In this context, also the ϕ has to be taken into account, since ϕ decays into K^+K^- pairs may substantially affect the measured K^- abundance. A solid reference is needed to evaluate the K^- absorption with in the nucleus and in this work we discuss the determination of this reference. A double ratio of $K^-/K^+(W)/K^-/K^+(C)$ is measured within the acceptance and compared to the reference.

1 Introduction

Pion-nucleus reactions allow for a quantitative study of the strange hadron production at nuclear saturation density. In this context, the FOPI collaboration demonstrated that the total K^0 production cross-section scales with the surface of the target nuclei ($\sigma = \sigma_{eff}\ A^{2/3}$, where A denotes the mass number) [1]. Since kaons (K^+, K^0) do not undergo strong absorption processes within a nucleus, they can provide stringent constraints on the production mechanisms of strange hadrons. Contrary, the K^- meson ($K^- = s\bar{u}$) properties in nuclear matter are more complex and less known. Several effects compete, the antikaon can be rescattered or be absorbed by nucleons and form baryonic resonances leading to the modification of its in-medium properties [2] [3]. This absorption strangeness exchange reaction can occur on one ($K^-N \rightarrow Y\pi$) or more nucleons ($K^-NN \rightarrow YN\pi$). A direct indication for the latter effect can be deduced from the comparison of the K^-/K^+ ratio measured in collisions with heavy targets like Tungsten (W) and lighter ones like Carbon (C). A peculiar role is played by the $\Lambda(1405)$ since this resonance is interpreted often as a molecular state partically composed of $\bar{K}N$ bound state [2].

2 The experiment

The **H**igh-**A**cceptance **Di**-**E**lectron **S**pectrometer (HADES) [4] is a fixed target experiment currently located at the SIS18 accelerator of the GSI Helmholtzzentrum in Darmstadt (Germany). It is composed of six identical sectors surrounding the beam axis covering almost the full azimuthal range and

*supported by the DFG cluster of excellence "Origin and Structure of the Universe" and SFB 1258
**e-mail: joana.wirth@tum.de

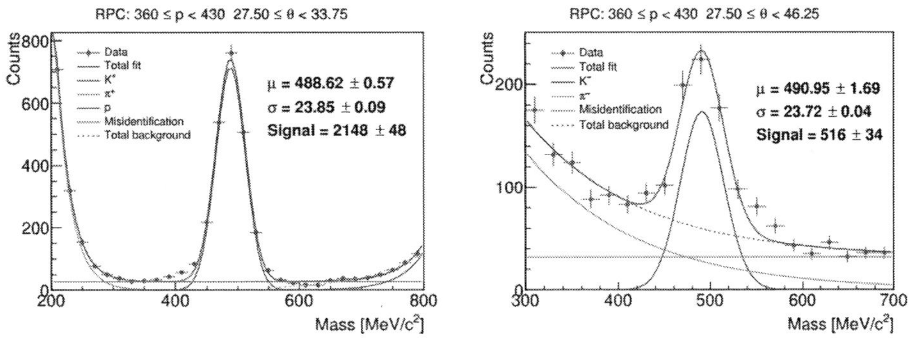

Figure 1. (Color online) Mass spectrum of K^+ (left) and K^- (right) in a specific $p - \theta$ region (see legend) for the RPC in $\pi^- + C$ reactions. The K^+ signal is represented by a Gaussian (magenta line). The background is shown in pink (pions), red (protons) and green (misidentified particles), for details see text. The K^- mass spectra is fitted with a composed function of a Gaussian for the signal (magenta line) and exponential function for the pions (pink line) combined with polynomial (green line).

polar angles from 18° to 85°. Two layers of **M**ultiwire **D**rift **C**hambers (MDC), in front and behind a superconducting magnet, allow for momentum reconstruction with a resolution of $\Delta p/p \approx 3\%$ and particle identification via dE/dx. In this experiment, the LVL1 trigger condition required a T_0 signal in the target detector [5] and a minimum multiplicity of two charged particle hits (M2) in the **M**ultiplicity and **E**lectron **T**rigger **A**rray (META) wall consisting of two time-of-flight detectors, RPC and TOF. Overall, 10×10^7 and 13×10^7 events have been collected in $\pi^- + C$ and $\pi^- + W$ at $p_{\pi^-} = 1.7$ GeV/c, respectively.

3 Antikaon absorption

3.1 (Anti)Kaon abundance measured with HADES

Both charged kaon candidates have been identified with the energy loss information from the MDCs (details in [6]). The final (anti)kaon yield has been extracted by fitting the measured mass distribution, separately for the two time-of-flight detectors, RPC and TOF. High statistics allowed for a two-dimensional analysis in two different kinematic sets of variables (p_T, y) and (p, θ). Figure 1 shows the mass spectrum for K^+ and K^- corresponding to an interval in the RPC of $360 \leq p$ [MeV/c] < 430 and $27.50 \leq \theta$ [°] < 33.75 (K^+)/46.25 (K^-). The (anti)kaon signal is represented by a Gaussian on top of the background composed of polynominal combined with an exponential function for each background particle species. The width of the Gaussian was constrained to vary around the mean value of the width obtained from simulations. Further, in each phase space region the raw yield has been corrected for acceptance and efficiency based on simulations modelling the detector response in combination with GiBUU as an event generator.
Since the phase space coverage of the measured K^+ distribution is more extended compared to K^-, the double ratio $K^-/K^+(W)/K^-/K^+(C)$ was only extracted in the overlap regions within the HADES acceptance[1]. A double ratio of $K^-/K^+(W)/K^-/K^+(C) = 0.442 \pm (0.03)^{stat} \pm \binom{+0.07}{-0.04}^{sys}$ has been extracted.

[1] $220 \leq p$ [MeV/c] < 570, $15 \leq \theta$ [°] < 65

3.2 (Anti)Kaon production based on cross-sections

In order to interpret the extracted double ratio $K^-/K^+(W)/K^-/K^+(C)$ in terms of the K^- absorption, the influence on the double ratio caused by the different number of neutrons and protons in Tungsten ($A = 184, Z = 74, N = 110$) compared to Carbon ($A = 12, Z = 6, N = 6$) with the assumption of pure (anti-)kaon production without absorption has been taken into account. For this purpose all K^+ and K^- production channels that can be accessed are listed in Tab. 1. As not all cross-sections have been measured at the presented beam momentum, a cross-section parametrization proposed by Sibirtsev *et al.* [9] has been adapted to extracted the cross-section for each individual channel at $p_{\pi^-} = 1.7$ GeV/c. It reads as follows:

$$\sigma = a\left(1 - \frac{s_0}{s}\right)^b \left(\frac{s_0}{s}\right)^c, \qquad (1)$$

where σ denotes the cross-section of a specific channel and a, b, c the fit parameters. The channel dependent threshold and center-of -mass energy are given by s_0 and $\sqrt{s}$, respectively. The parametrization has been fitted to the available experimental measurements taken form Landolt-Börnstein [7] as shown for one example channel in Fig. 2. All measured cross-sections are shown as black points, whereas the fit function is represented as red solid curve together with the cross-section parametrization implemented in GiBUU [12] given by the yellow solid curve. Moreover, the blue dashed lines depicted the estimated cross-section for the incident pion beam momentum $p_{\pi^-} = 1.7$ GeV/c. It has to be mentioned that the experimental data for the channels $\pi^- + p \rightarrow \Sigma(1385)^- K^+$ and $\pi^- + n \rightarrow \Sigma^- + \pi^- + K^+$ is scarce, thus the extracted cross-sections are rather an upper limit.
The reference double ratio based on the pure production has been obtained by summing all elementary cross-sections (σ_{fit} listed in Tab. 2) weighted with the relative neutron and proton contents. The scaling by $A^{2/3}$ [1] corresponding the effective number of nucleons on the surface of the nucleus cancels out. A double ratio based on the pure production of $K^-/K^+(W)/K^-/K^+(C) = 0.930 \pm 0.09$ has been extracted. For comparison also the double ratio has been deduced on the basis of the cross-sections ($\sigma_{\rm GiBUU}$ listed in Tab. 2) implemented in GiBUU with an value of $K^-/K^+(W)/K^-/K^+(C) = 0.966$.

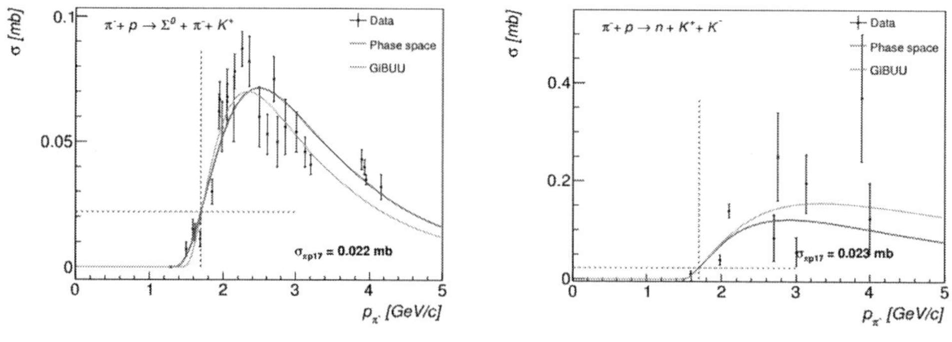

Figure 2. (Color online) Cross-section of $\pi^- + p \rightarrow \Sigma^- + \pi^0 + K^+$ (left) and $\pi^- + p \rightarrow n + K^+ + K^-$ (right) taken from the Landolt-Börnstein compilation [7] as function of the incident pion momentum. The red solid curve indicates the fit corresponding to a cross-section parametrization adapted from Sibirtsev *et al.* [9], while the yellow solid curve pictures the parametrization employed in GiBUU [12]. The blue dashed lines represent the suggested cross-section by the fit at $p_{\pi^-} = 1.7$ GeV/c.

Table 1. Production channels of K^+ and K^- in pion-induced reactions with corresponding threshold momentum for the incident pions. The cross-section σ_{fit} at p_{π^-} = 1.7 GeV/c corresponding to a fit with a cross-section parametrization to experimental cross-sections measured at several energies compared to the cross-section parametrization implemented in GiBUU.

	K^+ production	Threshold p_{π^-} [GeV/c]	σ_{fit} [mb]	σ_{GiBUU} [mb]
$\pi^- + p$	$\Lambda\pi^-K^+$	1.144	0.0793	0.0908 [12]
	$\Lambda\pi^0\pi^-K^+$	1.407	0.0015	
	$\Sigma^+\pi^-\pi^-K^+$	1.568	0.0003	
	$\Sigma^0\pi^-K^+$	1.290	0.0220	0.0212 [12]
	$\Sigma^-\pi^+\pi^-K^+$	1.585	≈ 0	
	$\Sigma^-\pi^0K^+$	1.290	0.0190	0.0152 [12]
	Σ^-K^+	1.035	0.1498	0.1564 [10]
	$\Sigma(1385)^0\pi^-K^+$	1.680	≈ 0	
	$\Sigma(1385)^-\pi^0K^+$	1.680	≈ 0	
	$\Sigma(1385)^-K^+$	1.399	< 0.0693	
$\pi^- + n$	$\Sigma^-\pi^-K^+$	1.296	< 0.0697	0.0247 [12]
	K^+/K^- production			
$\pi^- + p$	nK^+K^-	1.495	0.0231	0.0224 [9]
	$n\phi(K^+K^-,\ BR \approx 48.9\%$ [8])	1.559	0.0274	0.0196 [11]
	K^- production			
$\pi^- + p$	pK^0K^-	1.497	0.0071	0.0110 [9]

3.3 Summary

A comparison of the measured double ratio $K^-/K^+(W)/K^-/K^+(C)$ has been carried out to the reference double ratio based on the pure production. Accordingly, the double ratio extracted on the basis of the cross-section parametrization adapted from Sibirtsev *et al.* [9] and the one implemented GiBUU where averaged leading to a value of $K^-/K^+(W)/K^-/K^+(C) = 0.948 \pm 0.092$. The measured double ratio of $K^-/K^+(W)/K^-/K^+(C) = 0.442 \pm (0.03)^{stat} \pm \binom{+0.07}{-0.04}^{sys}$ is well below the reference directly indicating K^- absorption in the heavy target Tungsten with respect to Carbon.

References

[1] M. L. Benabderrahmane *et al.* [FOPI Collaboration], Phys. Rev. Lett. **102**, 182501 (2009)
[2] Ch. Fuchs, Prog. Part. Nucl. Phys. **56** 1-103 (2006)
[3] E. Y. Paryev *et al.*, J. Phys. G.: Nucl. Part. Phys. **42**, 075107 (2015)
[4] G. Agakishiev *et al.* [HADES Collaboration], Eur. Phys. J. **A41**, 243-277 (2009)
[5] J. Adamczewski-Musch *et al.* [HADES Collaboration], Eur. Phys. J. **A53** 188 (2017)
[6] J. Wirth, Proceeding Bormio 55th Wintermeeting on Nuclear Physics (2017)
[7] Landolt-Börnstein, New Series I/12a.
[8] C. Patrignani *et al.* [PARTICLE DATA GROUP Collaboration], Chin. Phys. **C40**, 100001 (2016)
[9] A. Sibirtsev *et al.*, Z. Phys. **A358**, 101-106 (1997);
A. Sibirtsev *et al.*, Nucl. Phys. **A614**, 415-432 (1997)
[10] Tsushima *et al.*, Phys. Lett. **B337**, 245-253 (1994)
[11] A. Sibirtsev *et al.*, Z. Phys. **A358**, 357 (1997)
[12] 0. Buss *et al.*, Physics Report **512**, 1-124 (2012)

Azimuthal anisotropy of strange hadrons in U+U collisions at $\sqrt{s_{NN}}$ = 193 GeV at RHIC

Vipul Bairathi (For the STAR collaboration)[1,⋆]

[1] *National Institute of Science Education and Research, HBNI, Jatni 752050, India*

Abstract. We present the measurement of the azimuthal anisotropy of strange hadrons (K_s^0, ϕ and Λ) at mid-rapidity ($|y| < 1.0$) in U+U collisions at $\sqrt{s_{NN}}$ = 193 GeV using the STAR detector at RHIC. We present the centrality and transverse momentum dependence of flow coefficients v_n for $n = 2, 3, 4$. A strong centrality dependence of v_2 is observed for the particles K_s^0, ϕ and Λ in U+U collisions at $\sqrt{s_{NN}}$ = 193 GeV similar to Au+Au collisions at $\sqrt{s_{NN}}$ = 200 GeV. We studied the number of constituent quark scaling (NCQ) of the flow coefficients. The NCQ scaling of the flow coefficients holds within uncertainties for the particles studied in the U+U collisions. We also present the comparison of the results to the AMPT transport model.

1 Introduction

A strongly interacting, hot and dense medium named Quark Gluon Plasma (QGP) is believed to be created in high-energy relativistic heavy-ion collisions at Relativistic Heavy Ion Collider (RHIC) [1]. Experimentally, the dynamics and the collective behavior of such strongly interacting medium have been studied by measuring the azimuthal anisotropy of the produced particles relative to the reaction plane [2]. The initial spatial anisotropy is converted into momentum space anisotropy through rescatterings among the constituents in the medium. The initial coordinate space anisotropy diminishes as the system expands. Due to this self-quenching effect, the azimuthal anisotropy is thought to be sensitive to the earliest stages of heavy-ion collisions [3]. The study of azimuthal anisotropy is based on Fourier expansion of the azimuthal angle, ϕ, of the momentum space distribution, $dN/d\phi \propto 1 + \sum_n 2v_n \cos n(\phi - \Psi_n)$. In experiments, the flow harmonics, v_n, can be measured using the equation $v_n = \langle \cos n(\phi - \psi_n) \rangle$, where the v_n and ψ_n are the magnitude and phase of the nth-order of flow harmonics, respectively. The published results for second harmonic (elliptic flow, v_2) from Au+Au collisions at RHIC have demonstrated the collective expansion feature, the mass ordering of elliptic flow among various hadron species [4, 5]. Strange hadrons, especially multistrange hadrons, and the ϕ meson because of their of small hadronic interaction cross section and freeze-out temperature close to the phase transition temperature, are believed to be less sensitive to hadronic rescatterings in the late stage of collisions and thus serve as a good probe for the partonic stage in heavy-ion collision [6]. Recent developments suggest that measurements of higher order flow harmonics can yield information about initial state fluctuations [7], which help to constrain the initial conditions of hydrodynamic simulation for a precise extraction of the QGP transport properties [8].

⋆e-mail: vipul.bairathi@niser.ac.in

2 Analysis Method

In this work, we have used minimum bias dataset from $\sqrt{s_{NN}}$ = 193 GeV U+U collisions collected by the STAR experiment during year 2012. The main detectors used in the analysis are Time Projection Chamber (TPC) and Time Of Flight (TOF) detector for the identification of charged particles at mid-rapidity ($|\eta| < 1.0$). The charged particles with a transverse momentum (p_T) range of $0.2 < p_T < 2.0$ GeV/c are selected for the event plane reconstruction. The event plane angle is calculated in two separated pseudorapidity windows, with a gap of 0.1 units between them to reduce non-flow effects. We reconstruct K_s^0, ϕ and $\Lambda(\bar{\Lambda})$ using invariant mass method through their decay channels: $K_s^0 \rightarrow \pi^+ + \pi^-$, $\phi \rightarrow K^+ + K^-$ and $\Lambda(\bar{\Lambda}) \rightarrow p + \pi^-(\bar{p} + \pi^+)$. Topological and kinematic cuts are applied to reduce the combinatorial background for K_s^0 and $\Lambda(\bar{\Lambda})$. The detailed description of the procedures can be found in the Ref. [9]. We have used standard Event-Plane method for the calculation of flow coefficients v_n relative to the event plane angle ψ_n [2].

3 Results and Discussions

3.1 Flow harmonics (v_n)

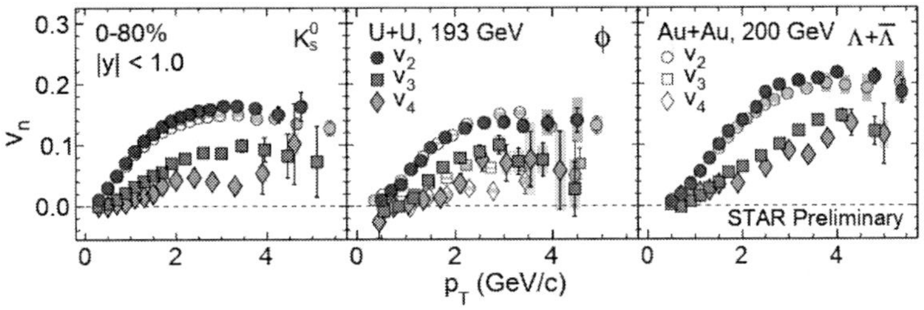

Figure 1. Flow harmonics $v_n(p_T)$ for K_s^0, ϕ and $\Lambda + \bar{\Lambda}$ in 0-80% centrality U+U collisions at $\sqrt{s_{NN}}$ = 193 GeV (solid symbols) compared with those from Au+Au collisions at $\sqrt{s_{NN}}$ = 200 GeV (open symbols) from Ref. [5]. The systematic uncertainties are shown with shaded boxes and the statistical uncertainties by vertical lines.

Figure 1 shows 0-80% centrality $v_n(p_T)$ of K_s^0, ϕ and $\Lambda + \bar{\Lambda}$ at mid-rapidity in U+U collisions at $\sqrt{s_{NN}}$ = 193 GeV. The v_n values increases with p_T for the measured range. The magnitude of flow harmonics $v_2 > v_3 > v_4$. The v_2 values shows similar p_T dependence in U+U and Au+Au collisions.

3.2 Centrality dependence

Figure 2 shows $v_n(p_T)$ of K_s^0, ϕ and $\Lambda+\bar{\Lambda}$ at mid-rapidity for 0-10%, 10-40% and 40-80% most central U+U collisions at $\sqrt{s_{NN}}$ = 193 GeV. A clear centrality dependence of $v_2(p_T)$ is observed for K_s^0, ϕ and $\Lambda + \bar{\Lambda}$ similar to that in Au+Au collisions measured by the STAR experiment [5]. The v_2 values are found to be larger in peripheral collisions (40-80%) compared to the central collisions (0-10%). Higher order flow harmonics v_3 and v_4 shows a much weaker centrality dependence, if any.

3.3 NCQ scaling

Figure 3 shows the v_2, v_3 and v_4 scaled by the power of number of constituent quarks ($n_q^{n/2}$) as a function of $(m_T - m_0)/n_q^{n/2}$ for 0-80% and 10-40% in U+U collisions at $\sqrt{s_{NN}}$ = 193 GeV, where m_T and m_0 are the transverse mass and the rest mass of the particles, respectively. Flow harmonics v_n scaled by the $n_q^{n/2}$ vs. $(m_T - m_0)/n_q^{n/2}$ lie on a single curve for the particles studied within uncertainties in U+U collisions, similar to that is observed for identified hadrons in Au+Au collisions at $\sqrt{s_{NN}}$ = 200 GeV by the PHENIX collaboration in the Ref. [10].

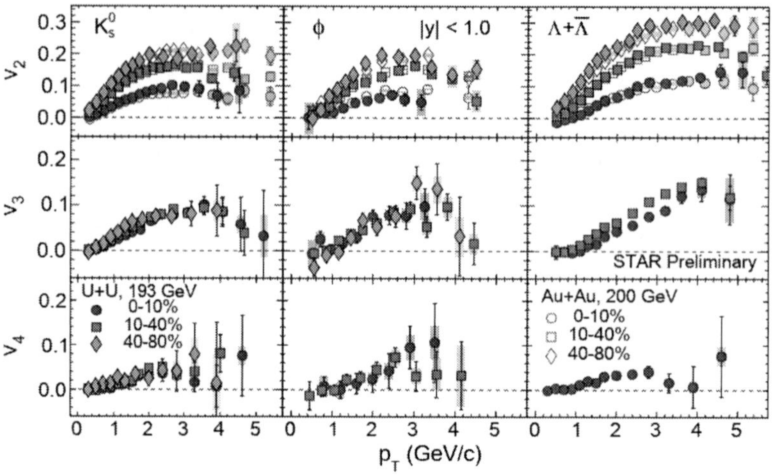

Figure 2. v_2(top panels), v_3(middle panels) and v_4(bottom panels) vs. p_T of K^0_s, ϕ and $\Lambda + \bar{\Lambda}$ for 0-10%, 10-40% and 40-80% centrality in U+U collisions at $\sqrt{s_{NN}}$ = 193 GeV (solid symbols) compared with those from Au+Au collisions at $\sqrt{s_{NN}}$ = 200 GeV (open symbols) from Ref. [5]. The systematic uncertainties are shown with shaded boxes and the statistical uncertainties by vertical lines.

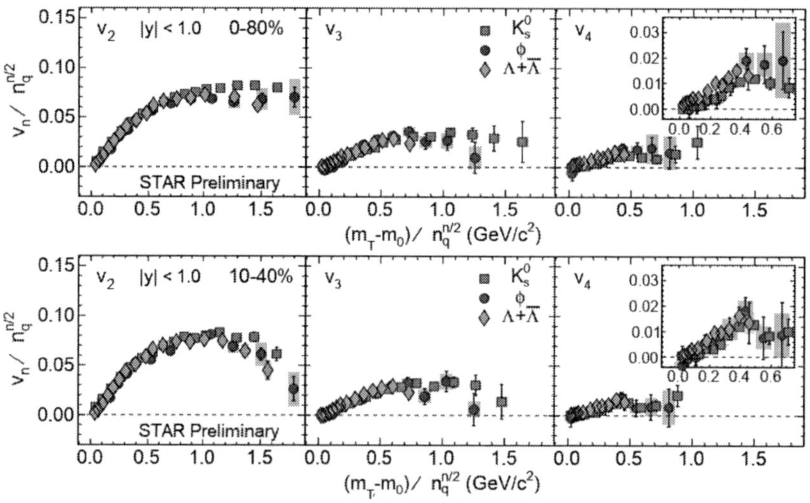

Figure 3. v_n scaled by the power of number of constituent quarks ($n_q^{n/2}$) as function of $(m_T - m_0)/n_q^{n/2}$ for 0-80% (top panels) and 10-40% (bottom panels) in U+U collisions at $\sqrt{s_{NN}}$ = 193 GeV.

3.4 Model comparisons

Figure 4 shows the v_n coefficients of K^0_s, ϕ and $\Lambda + \bar{\Lambda}$ for 0-80%(left panels) and 10-40%(right panels) centrality in U+U collisions at $\sqrt{s_{NN}}$ = 193 GeV. The same v_n analysis method is used for the model study. The AMPT (with the string melting) model [11] with a 3 mb parton cross-section, describes the data at low p_T < 2.0 GeV/c.

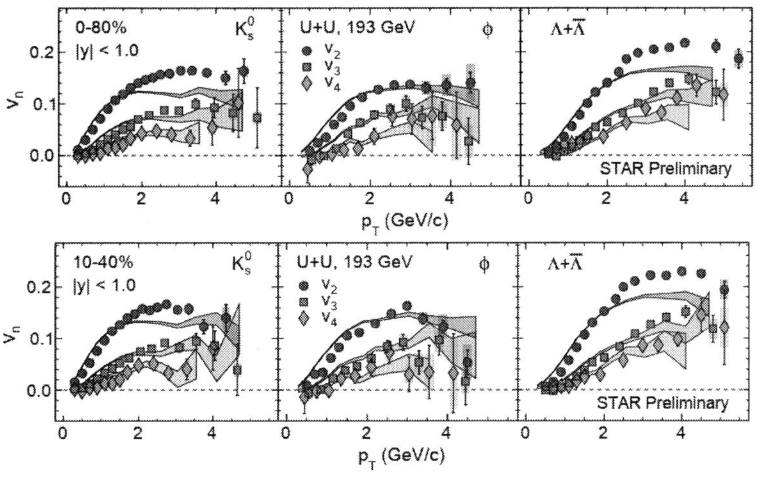

Figure 4. Comparison of the v_n coefficients of K_s^0, ϕ and $\Lambda + \bar{\Lambda}$ for 0-80% (top panels) and 10-40% (bottom panels) centrality in U+U collisions at $\sqrt{s_{NN}}$ = 193 GeV (solid symbols) with AMPT (shaded bands).

4 Summary

We present results on flow coefficients v_n (n = 2,3,4) for strange hadrons in U+U collisions at $\sqrt{s_{NN}}$ = 193 GeV and compared with the corresponding results from Au+Au collisions at $\sqrt{s_{NN}}$ = 200 GeV. The centrality dependence of $v_n(p_T)$ is presented. A clear centrality dependence is observed for elliptic flow v_2 in U+U collisions similar to Au+Au collisions. Higher order flow coefficients (v_3, v_4) are found to be less sensitive to the collision centrality. The constituent quark number scaling $v_n/n_q^{n/2}$ vs. $(m_T - m_0)/n_q^{n/2}$ for different particles scale to a single curve within uncertainties for each harmonic. The results were compared with the transport model AMPT. The AMPT model explains the data at low p_T < 2 GeV/c. This work is supported by the DAE and DST, of the Government of India.

References

[1] I. Arsene *et al.* (BRAHMS Collaboration), Nucl. Phys. A **757**, 1 (2005); B. B. Back *et al.* (PHOBOS Collaboration), Nucl. Phys. A **757**, 28 (2005); J. Adams *et al.* (STAR Collaboration), Nucl. Phys. A **102**, 1 (2005); K. Adcox *et al.* (PHENIX Collaboration), Nucl. Phys. A **757**, 184 (2005).
[2] S. Voloshin and Y. Zhang, Z. Phys. C **70**, 665 (1996); A. Poskanzer and S. Voloshin, Phys. Rev. C **58**, 1671 (1998).
[3] H. Sorge, Phys. Rev. Lett. **78**, 2309 (1997); H. Sorge, Phys. Rev. Lett. **82**, 2048(1999).
[4] J. Adams *et al.* (STAR Collaboration), Phys. Rev. C **72**, 014904 (2005).
[5] B. I. Abelev *et al.* (STAR Collaboration), Phys. Rev. C **77**, 054901 (2008); L. Adamczyk *et al.* (STAR Collaboration), Phys. Rev. Lett. **116**, 062301 (2016).
[6] A. Shor, Phys. Rev. Lett. **54**, 1122 (1985); H. Hecke, *et al.*, Phys. Rev. Lett. **81**, 5764 (1998).
[7] B. Alver and G. Roland, Phys. Rev. C **81**, 054905 (2010).
[8] A. Adare *et al.* (PHENIX Collaboration), Phys. Rev. Lett. **107**, 252301 (2011); L. Adamczyk *et al.* (STAR Collaboration), Phys. Rev. C **88**, 014904 (2013).
[9] C. Adler *et al.* (STAR Collaboration), Phys. Rev. Lett. **89**, 132301 (2002); J. Adams *et al.* (STAR Collaboration), Phys. Rev. Lett. **92**, 052302 (2004).
[10] A. Adare *et al.* (PHENIX Collaboration), Phys. Rev. C **93**, 051902 (2016).
[11] Z. W. Lin, C. M. Ko *et al.*, Phys. Rev. C **65**, 034904 (2002).

EPJ Web of Conferences **171**, 13004 (2018) https://doi.org/10.1051/epjconf/201817113004
SQM 2017

ϕ Meson Measurements at RHIC with the PHENIX Detector

Murad Sarsour[1,*] on behalf of the PHENIX Collaboration

[1] *Georgia State University, Atlanta, Georgia 30303, USA*

Abstract. The measurement of ϕ mesons provides key information on the phase of the hot and dense medium created in the relativistic heavy ion collisions. It has a relatively small hadronic interaction cross section and is sensitive to the increase of strangeness (strangeness enhancement), a phenomenon associated with soft particles in bulk matter. Measurements in the dilepton channels are especially interesting since leptons interact only electromagnetically, thus carrying the information from their production phase directly to the detector. Measurements in different nucleus-nucleus collisions allow us to perform a systematic study of the nuclear medium effects on ϕ meson production. The PHENIX detector provides the capabilities to measure the ϕ meson production in a wide range of transverse momentum and rapidity to study these effects.
In this proceeding, we present measurements of the ϕ mesons in a variety of collision systems at $\sqrt{s_{NN}} = 200$ GeV. In case of small systems, the data are compared with AMPT calculations to study the various cold nuclear medium effects involved in ϕ meson production.

1 Introduction

Ever since RHIC announced the discovery of the hot and dense state of strongly interacting matter called Quark-Gluon Plasma (QGP) [1–3] the main objective has been to quantify its properties. This is accomplished by looking at as many observables as possible, including the nuclear modification of ϕ meson production in the QGP environment. ϕ meson is an excellent probe for studying QGP in heavy ion collisions because it is sensitive to several aspects of the collision, including modifications of strangeness production in bulk matter. Owing to its small inelastic cross section for interaction with nonstrange hadrons, the ϕ meson is less affected by late hadronic rescattering and may reflect the initial conditions of the evolution of the system. Being a nearly pure strange antistrange ($s\bar{s}$) state, the ϕ meson puts additional constraints on models of quark recombination in the QGP. However to gauge the QGP related modifications, we need to know the ϕ meson production in p+p collisions as a baseline. There are additional effects from the nuclear medium itself, cold nuclear matter effects (CNM), and they are accessed by studying ϕ meson production in small system collisions (e.g. d+Au).

The lepton decay channel is of particular interest because of the absence of strong interactions between muons and the surrounding hot hadronic matter. At forward rapidity, it allows us to study the rapidity dependence of ϕ meson production and especially in asymmetric heavy-ion collisions provides the means for accessing different mixtures of initial and final state effects.

*e-mail: msar@gsu.edu

In p+p collisions, the ϕ meson production is important to study because it may have similar production mechanisms to other onia like J/ψ and Υ. In addition, it provides a tool to study effects that scale with mass (e.g. collective effects) since it is the heaviest easily accessible meson made of light quarks.

2 Experimental Setup

The PHENIX detector [4] has a high rate capability utilizing a fast DAQ and specialized triggers, high granularity detectors, and good mass resolution and particle ID. Detection of $e^{\pm}$ utilizes the finely grained electromagnetic calorimeter (EMCal), the drift chamber and ring imaging Cherenkov detector, while the $\mu^{\pm}$ are detected by the forward spectrometers consisting of Muon ID and Muon Tracker. The very forward beam-beam counters (BBC) are used to determine the collision vertex position and time, the beam luminosity and form a minimum bias trigger.

The PHENIX collaboration collected data from a variety of collisions systems provided by the Relativistic Heavy Ion Collider (RHIC) at Brookhaven National Laboratory (BNL). Results from some of these data sets are presented in this proceeding.

3 Results

PHENIX has measured the production of ϕ meson in p+p collisions at $\sqrt{s}$ = 200 GeV over a wide range of p_T and rapidity [5, 6], which is a very important resource for validating the phenomenological models of strangeness production. In addition, these measurements provide baseline for studying cold and hot nuclear modifications.

To gain insight into nuclear medium effects and particle production mechanisms in A+B collisions, the ratio of the ϕ meson yields in A+B collisions to p+p collisions scaled by the number of nucleon-nucleon collisions in the A+B system, N_{coll} [7], is calculated as:

$$R_{AB} = \frac{d^2N_{AB}/dydp_T}{N_{\text{coll}} \times d^2N_{pp}/dydp_T}, \tag{1}$$

where $d^2N_{AB}/dydp_T$ is the per-event yield of particle production in heavy ion collisions and $d^2N_{pp}/dydp_T$ is the per-event yield of the same process in p+p collisions. The p+p invariant yield used in the A+B calculation for the l^+l^- decay channel is the p+p differential cross section [5] divided by the p+p total cross section, 42.2 mb.

Measurements of ϕ meson R_{AB} at midrapidity in heavy ion collisions show that ϕ meson exhibits a different suppression pattern compared to lighter mesons (π^0 and η) and baryons (protons and antiprotons) [8]. For all centralities, ϕ meson is less suppressed than π^0 and η in the intermediate p_T range (2–5 GeV/c), whereas ϕ, π^0 and η show similar suppression at higher p_T. However, ϕ meson R_{AB} at forward and backward rapidities [9] show that ϕ meson yields in most central d+Au collisions are suppressed at low-p_T (< 2 GeV/c) in the d-going direction and strongly enhanced in Au-going direction in the intermediate p_T range (2–3 GeV/c). This modification decreases as we move towards peripheral collisions. No modification of ϕ meson production is observed in the most peripheral d+Au collisions. ϕ meson $R_{d\text{Au}}$ as a function of p_T was measured for different centralities and a nuclear modification very similar to that of the heavy flavor muons is observed [10]. The ϕ meson enhancement in the Au-going direction and the suppression in the d-going direction are consistent with what is observed by ALICE in p+Pb collisions at $\sqrt{s_{NN}}$=5.02 TeV in $-4.46 < y < -2.96$ and $2.03 < y < 3.53$ [11].

The nuclear-modification factor is also studied in Cu+Au collisions, R_{CuAu}, at $\sqrt{s_{NN}}$ = 200 GeV to evaluate the effects of hot and cold nuclear matter on ϕ meson production. Although statistically limited, we observed a dependence of R_{CuAu} on both centrality and rapidity with a similar trend to that observed in d+Au collisions [9].

PHENIX measured the differential cross section of ϕ meson at forward and backward rapidities in p+p collisions at $\sqrt{s}$ = 510 GeV [12]. The integrated cross section over the measured p_T range, 2.79 ± 0.20 (stat) ± 0.17 (syst) ± 0.34 (norm) $\times 10^{-2}$ mb, from this measurement and those from PHENIX at $\sqrt{s}$ = 200 GeV [5] and LHC at $\sqrt{s}$ = 2.76 and 7.0 TeV [11, 13–15] are used to study the energy dependence of ϕ meson production at forward rapidity, $1.2 < y < 2.2$, in p+p collisions, as shown in Fig. 1. Comparisons to PYTHIA6 [16], PYTHIA8 [17], PHOJET [18], AMPT [19], EPOS3 [20] and EPOS-LHC [21] model predictions show that these models exhibit similar energy dependence trend to that of the data. Further studies with EPOS3 model show that the addition of the hydrodynamic evolution of the system induces an enhancement of ϕ meson production at the LHC energies at intermediate p_T, which hints at collective effects, however, no effect is observed at RHIC energies [12].

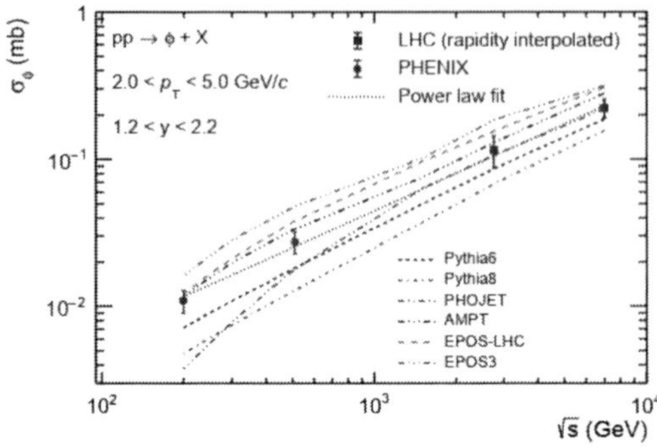

Figure 1. ϕ meson production cross section in $1.2 < y < 2.2$ and $2 < p_T < 5$ GeV/c in p+p collisions versus the center-of-mass energy, $\sqrt{s}$, compared to different model predictions.

PHENIX collected data from ^{3}He+Au, p+Au and p+Al collisions in 2014 and 2015. In addition to d+Au data set collected in 2008 [8], these data allow to carry out a systematic study of the various cold nuclear matter effects included in models like AMPT and EPOS. The left panel of Fig. 2 shows

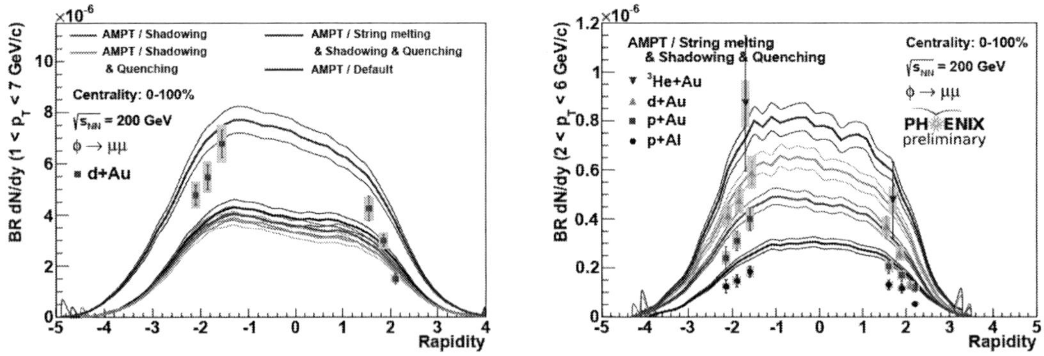

Figure 2. Left: ϕ meson invarient yield compared with AMPT model predictions with different combinations of nuclear medium effects. Right: ϕ meson invarient yields from ^{3}He+Au, d+Au, p+Au and p+Al collisions compared with AMPT model predictions that include string melting, shadowing and quenching [19].

AMPT model predictions with different combination of effects included against ϕ meson invariant yield in d+Au collisions. The right panel shows the AMPT model predictions, with the string melting, shadowing and quenching turned on, compared to the data.

4 Summary

The PHENIX collaboration measured ϕ meson production, over a wide p_T range in the forward and backward rapidities in a variety of collision systems to study cold and hot nuclear matter effects. In d+Au collisions, we observed an enhancement (suppression) of ϕ meson production in the backward (forward) rapidity region in most central collisions. Similar behavior was previously observed for inclusive charged hadrons and open heavy flavor muons which may suggest similar cold nuclear matter effects.

The PHENIX collaboration also measured ϕ meson production in $p+p$ collisions at $\sqrt{s}$ = 510 GeV at backward and forward rapidities, $\sigma_\phi = 2.79 \pm 0.20\,(\text{stat}) \pm 0.17\,(\text{syst}) \pm 0.34\,(\text{norm}) \times 10^{-2}$ mb. Combined with previous PHENIX and LHC measurements, this measurement allowed studying the energy dependence of ϕ meson production in $p+p$ collisions over a wide energy range. Comparisons with EPOS3 model showed a hint of collective effects in $p+p$ at LHC energies and no effect at RHIC energies.

New data sets from p+Au and p+Al collisions collected in 2015 allow ϕ measurement at backward and forward rapidities in less complicated p+Au and p+Al systems. The wealth of small system data sets (^{3}He+Au, d+Au, p+Au and p+Al) along with the introduction of the forward vertex detector in 2015 data sets will allow studying the different CNM effects using models like AMPT and EPOS.

References

[1] K. Adcox *et al.* (PHENIX Collaboration), Nucl. Phys. A **757**, 184 (2005).
[2] J. Adams *et al.* (STAR Collaboration), Nucl. Phys. A **757**, 102 (2005).
[3] R. Nouicer, Eur. Phys. J. Plus **131**, 70 (2016).
[4] K. Adcox *et al.* (PHENIX Collaboration), Nucl. Instrum. Methods Phys. Res. A **499**, 469 (2003).
[5] A. Adare *et al.* (PHENIX Collaboration), Phys. Rev. D **90**, 052002 (2014).
[6] A. Adare *et al.* (PHENIX Collaboration), Phys. Rev. D **83**, 052004 (2011).
[7] A. Adare *et al.* (PHENIX Collaboration), Phys. Rev. C **87**, 034904 (2013).
[8] A. Adare *et al.* (PHENIX Collaboration), Phys. Rev. C **83**, 024909 (2011).
[9] A. Adare *et al.* (PHENIX Collaboration), Phys. Rev. C **92**, 044909 (2015).
[10] A. Adare *et al.* (PHENIX Collaboration), Phys. Rev. Lett. **112**, 252301 (2014).
[11] J. Adam *et al.* (ALICE Collaboration), Phys. Lett. B **768**, 203 (2017).
[12] A. Adare *et al.* (PHENIX Collaboration), arXiv:1710.01656v1 (2017).
[13] B. Abelev *et al.* (ALICE Collaboration), Phys. Lett. B **710**, 557 (2012).
[14] B. Abelev *et al.* (ALICE Collaboration), Eur. Phys. J. C **72**, 2183 (2012).
[15] R. Aaij *et al.* (LHCb Collaboration), Phys. Lett. B **703**, 267 (2011).
[16] T. Sjostrand, S. Mrenna, and P. Z. Skands, J. High Energy Phys. **5**, 26 (2006).
[17] T. Sjostrand, S. Ask, J. R. Christiansen, R. Corke, N. Desai, P. Ilten, S. Mrenna, S. Prestel, C. O. Rasmussen, and P. Z. Skands, Comput. Phys. Commun. **191**, 159 (2015).
[18] F. W. Bopp, R. Engel, and J. Ranft, hep-ph/9803437 (1998).
[19] Z.-W. Lin, C. M. Ko, B.-A. Li, B. Zhang, and S. Pal, Phys. Rev. C **72**, 064901 (2005).
[20] K. Werner, B. Guiot, Iu. Karpenko, and T. Pierog, Phys. Rev. C **89**, 064903 (2014).
[21] T. Pierog, Iu. Karpenko, J. M. Katzy, E. Yatsenko, and K. Werner, Phys. Rev. C **92**, 034906 (2015).

Strange and heavy hadrons production from coalescence plus fragmentation in AA collisions at RHIC and LHC

Salvatore Plumari[1,2,⋆], *Vincenzo* Minissale[1,2], *Santosh K.* Das[1,2], *Francesco* Scardina[1,2], and *Vincenzo* Greco[1,2]

[1] *Department of Physics and Astronomy, University of Catania, Via S. Sofia 64, 1-95125 Catania, Italy*
[2] *Laboratori Nazionali del Sud, INFN-LNS, Via S. Sofia 62, I-95123 Catania, Italy*

Abstract. In a coalescence plus fragmentation approach we study the p_T spectra of charmed hadrons D^0, D_s up to about 10 GeV and the Λ_c^+/D^0 ratio from RHIC to LHC energies. In this study we have included the contribution from decays of heavy hadron resonances and also that due to fragmentation of heavy quarks that are left in the system after coalescence. The p_T dependence of the heavy baryon/meson ratios is found to be sensitive to the heavy quark mass. In particular we found that the Λ_c/D^0 is much flatter than the one for light baryon/meson ratio like p/π and Λ/K.

1 Introduction

Ultra-relativistic Heavy Ion Collisions (uHIC) can be used to probe the properties of the Quark-Gluon Plasma (QGP). In the studies of the QGP created in HIC is necessary to take into account that partonic behavior is not directly projected on the observables measured in uHIC. Thus the choice of the model for hadronization process is a crucial point in order to have a comparison with experimental data. The study of heavy baryons like D mesons or Λ_c baryons and the systematic study of the baryon over meson ratio for different species from light to heavy flavor are important in order to understand the hadronization mechanism. For light and strange hadrons it has been shown an enhancement of the baryon over meson ratio compared to the one for p-p collision. Recent experimental results from STAR collaboration have shown that similar enhancement of the baryon/meson ratio is expected in the heavy flavor sector [1]. The coalescence hadronization process can explain both this features that appears at intermediate p_T. Hadronization via coalescence lead to a modification of the relative abundance of the different heavy hadron species produced. In particular this can manifests in a baryon-to-meson enhancement for charmed hadrons. The Λ_C/D^0 enhancement due to coalescence was first suggested in [2] where based on di-quark or three-quark coalescence mechanism with a full thermalized charm quarks and the predicted Λ_c/D^0 ratio is found to be comparable to the recent experimental data. Other recent calculations on the Λ_c/D^0 can be found in [3, 4].

2 Hadronization via Coalescence and Fragmentation

There are two hadronization mechanisms to produce heavy flavor hadrons: one is the fragmentation where high momentum quarks fragment directly in high momentum hadrons and another one is the

⋆e-mail: salvatore.plumari@hotmail.it

coalescence where heavy quarks hadronize by recombination with light quarks. At increasing p_T the probability to coalescence decreases and the standard fragmentation takes over.

2.1 Fragmentation

The hadron momentum spectra via fragmentation of minijet parton spectra is given by: $dN_{had}/d^2p_T\,dy = \sum_{jet}\int dz \frac{dN_{jet}}{d^2p_T\,dy}\frac{D_{had/jet}(z,Q^2)}{z^2}$ where $z = p_{had}/p_{jet}$ is the fraction of minijet momentum carried by the hadron and $Q^2 = (p_{had}/2z)^2$ is the momentum scale for the fragmentation process. We employ the Peterson fragmentation function $D_{had/jet}(z,Q^2) \propto 1/\left[z\left[1-\frac{1}{z}-\frac{\epsilon_c}{1-z}\right]^2\right]$ [5]. The ϵ_c is a free parameter fixed to $\epsilon_c = 0.06$ by experimental data on D mesons production in $p+p$ collisions [6]. In the absence of the $p+p$ data for the Λ_c the value has been fixed to be a factor two larger than the D meson one [4]. For light quarks we employ the distribution function that at high $p_T > p_0 \sim 2-3$ GeV is evaluated in next-to-leading order (NLO) in a pQCD scheme including the modification due to the jet quenching mechanism. The charm distribution function for both RHIC and LHC have been taken in accordance to the Fixed Order+Next-to-Leading Log (FONLL), as given in Ref.[7, 8].

2.2 Coalescence

The spectrum of hadrons formed by coalescence of quarks as developed in [9, 10] can be written as:

$$\frac{d^2N_H}{dP_T^2} = g_H \int \prod_{i=1}^{n} \frac{d^3p_i}{(2\pi)^3E_i} p_i \cdot d\sigma_i \, f_{q_i}(x_i,p_i) f_H(x_1...x_n, p_1...p_n)\,\delta^{(2)}\left(P_T - \sum_{i=1}^{n} p_{T,i}\right) \quad (1)$$

where $d\sigma_i$ denotes an element of a space-like hypersurface, g_H is the statistical factor to form a colorless hadron from quarks and antiquarks with spin 1/2 while f_{q_i} are the quark (anti-quark) phase-space distribution functions for i-th quark (anti-quark). For $n = 2$ Eq. 1 describes meson formation while for $n = 3$ the baryon one. For D mesons the statistical factors is $g_D = 1/36$ while for baryons i.e. Λ_c the statistical factors is $g_\Lambda = 1/108$. Finally $f_H(x_1...x_n, p_1...p_n)$ is the Wigner function and following Ref.[2, 10] we assume for f_H a Gaussian shape in space and momentum relative coordinates $f_H(x_1...x_n, p_1...p_n) = A_H \exp\left(-\sum_{i=1}^{n-1}\frac{x_{ri}^2}{\sigma_i^2} - \sum_{i=1}^{n-1} p_{ri}^2\sigma_i^2\right)$ with A_H a normalization constant for the Wigner function. While σ_i are the covariant widths and are related to the size of the hadron and they can be evaluated from the charge radius of the hadrons according to quark model [11, 12]. For D meson the widths are fixed by the mean square charge radius of D^+ which is given by $\langle r^2\rangle_{ch} = 0.184\,fm^2$ and it correspond to $\sigma = 0.283\,GeV$ while for Λ_c^+ we have $\langle r^2\rangle_{ch} = 0.15\,fm^2$ and the corresponding widths are $\sigma_1 = 0.18\,GeV$ and $\sigma_2 = 0.342\,GeV$. The overall normalization factor A_H is fixed by requiring that the total recombination probability $P_{coal} = 1$ for a $p = 0$ heavy quark that undergoes to all possible heavy flavor hadron channels. From P_{coal} we can assign a probability of fragmentation as $P_{frag} = 1 - P_{coal}$ and the charm distribution function undergoing to fragmentation is evaluated by convoluting the momentum of heavy quarks which do not undergone to coalescence. For the bulk we assume a thermalized system of gluons and u, d, s quarks and anti-quarks where at $\tau = 4.5 fm/c$ ($\tau = 7.8 fm/c$) at RHIC (LHC) with a temperature of $T_C = 165$ MeV. The longitudinal momentum distribution is assumed to be boost-invariant in $y \in (-0.5, +0.5)$. For the quark-gluon plasma collective flow, we assume a radial flow profile as $\beta_T(r_T) = \beta_{max}\frac{r_T}{R}$, where R is the transverse radius of the fireball. For partons at low transverse momentum, $p_T < 2\,GeV$ we consider a thermal distribution while $p_T > 2\,GeV$, we consider the minijets that have undergone the jet quenching mechanism. The parametrization and the fireball parameters are the same used in [13]. For heavy quarks we use the transverse momentum distribution obtained by solving the relativistic Boltzmann equation that give a good description of R_{AA} and v_2 of D mesons [6].

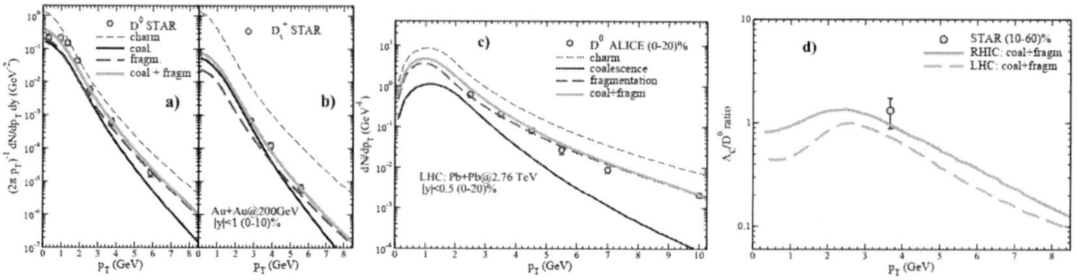

Figure 1. Left panel:p_T spectra for D^0 and D_s for $Au-Au$ collisions at $\sqrt{s} = 200\,GeV$ and for (0−10%) centrality. Green dashed line refers to the charm spectrum. Black solid and red dashed lines refer to the D^0 spectrum via only coalescence and only fragmentation respectively while orange solid line refers to the fragmentation plus coalescence processes. Data from [15]. Middle panel: p_T spectra for D^0 for $Pb-Pb$ collisions at $\sqrt{s} = 2.76\,TeV$ and for (0 − 20%) centrality. Data from [16]. Right panel:Λ_c^+/D^0 ratio as a function of p_T. Solid line refer to $Au-Au$ collisions at $\sqrt{s} = 200\,GeV$ while dashed line to $Pb-Pb$ collisions at $\sqrt{s} = 2.76\,TeV$. Data from [1].

3 Charmed hadrons transverse momentum spectra and ratio

In this section, we show results for the transverse momentum spectra of D^0 and D_s mesons and for Λ_c. In all the spectra the contribution that comes from resonances decay are included with their decay channel and the pertinent branching ratios as given in Particle Data Group [14]. For example for D^0 meson a dominant contribution comes from D^{*+} and D^{*0} where $D^{*+} \to D^0\pi^+$ (68 %) and $D^{*0} \to D^0\pi^0$ (62 %) and $D^{*0} \to D^0\gamma$ (38 %). The inclusion of resonances improves the description at low p_T but does not affect significantly the intermediate p_T. As shown in Fig.1 a) and b) the contribution of both hadronization mechanism (black solid and red dashed lines) is similar for $p_T \lesssim 2.5\,GeV$ while at higher p_T the fragmentation becomes dominant mechanism. The contribution of coalescence plus fragmentation shown by orange lines give a good description of the experimental data. For the spectra of D_s^+ we observe that at low p_T Fig.1 b) the coalescence is the dominant mechanism while fragmentation play a role at $p_T \gtrsim 4GeV$. This is related to the fact that the fragmentation fraction for D_s^+ are small and less than about 8% of produced total heavy hadrons as obtained in [17]. Again the comparison with the experimental data shows that the description by both hadronization mechanisms appear to be quite good. In Fig.1 c) it is shown the spectrum for D^0 meson for (0 − 20%) centrality at LHC energies. As shown the total spectrum (coalescence plus fragmentation) shown by orange line is again in a good agreement with the experimental data. We notice that at LHC energies the fragmentation is the dominant hadronization mechanism to produce the D^0 meson. This is due to the fact that at high energies the coalescence is less significant because the effect of the coalescence depends on the slope of the charm quark momentum distribution. In fact for an harder charm quark distribution like at LHC the gain in momentum reflects in a smaller increase in the slope compared to the one at RHIC energies [6]. In Fig.1 d)it is shown the comparison between RHIC and LHC energies of the Λ_c^+ to D^0 ratio as a function of p_T. As shown the coalescence by itself predicts a rise and fall of the baryon/meson ratio. The inclusion of fragmentation reduce the ratio and we can see that in the peak region we get a quite good agreement with the only experimental data. Notice that in our calculation we obtain similar baryon/meson ratio to the one predicted in [2]. Compared with measured light baryon/meson ratios like $\bar{p}/\pi^-$ and Λ/K_S^0 ratios (see [18–20]), the obtained Λ_c^+/D^0 ratios has a different behavior it is thus much flatter and for $p_T \to 0$ hadronization by coalescence and fragmentation predict $\Lambda_c^+/D^0 \simeq 0.8$ which is much larger respect to the one measured for light baryon/meson ratio with $\bar{p}/\pi^- \simeq 0$ for $p_T \to 0$ [13]. This behavior comes from the large mass of heavy quarks. Finally, we observe that at LHC energies coalescence plus fragmentation predict a smaller Λ_c^+/D^0. This is due to the fact that at LHC the hadronization by fragmentation becomes the dominant hadronization mechanism with the effect to reduce the baryon/meson ratio.

4 Conclusions

We have studied the transverse momentum spectra of charmed hadrons in heavy ion collisions from RHIC to LHC energies. The results obtained for the spectra are in good agreement with the recent experimental data from RHIC to LHC in central collisions.We have also studied the p_T dependence of the Λ_c/D^0 ratio for different energies. The comparison with the light baryon/meson ratio shows that the Λ_c/D^0 ratio have a weaker dependence on the transverse momentum due to the massive charms quarks inside heavy hadrons. We have found that the $\Lambda_c/D^0 \simeq 1.5$ at the peaks at $p_T \simeq 2.5\,GeV$ at RHIC energies and $\Lambda_c/D^0 \simeq 0.8$ in the $p_T \simeq 0$ region. Similar results have been predicted in [2]. We observe that at LHC energies due to the fact that fragmentation start to be the dominant hadronization mechanism for the D meson production this ratio is predicted to be smaller at higher energies.

Acknowledgments

V. Greco, S. Plumari acknowledge the support of the ERC-StG Grant under the QGPDyn project.

References

[1] L. Zhou (STAR) (2017), 1704.04364
[2] Y. Oh, C.M. Ko, S.H. Lee, S. Yasui, Phys. Rev. **C79**, 044905 (2009), 0901.1382
[3] S. Ghosh, S.K. Das, V. Greco, S. Sarkar, J.e. Alam, Phys. Rev. **D90**, 054018 (2014), 1407.5069
[4] S.K. Das, J.M. Torres-Rincon, L. Tolos, V. Minissale, F. Scardina, V. Greco, Phys. Rev. **D94**, 114039 (2016), 1604.05666
[5] C. Peterson, D. Schlatter, I. Schmitt, P.M. Zerwas, Phys. Rev. **D27**, 105 (1983)
[6] F. Scardina, S.K. Das, V. Minissale, S. Plumari, V. Greco (2017), 1707.05452
[7] M. Cacciari, P. Nason, R. Vogt, Phys. Rev. Lett. **95**, 122001 (2005), hep-ph/0502203
[8] M. Cacciari, S. Frixione, N. Houdeau, M.L. Mangano, P. Nason, G. Ridolfi, JHEP **10**, 137 (2012), 1205.6344
[9] V. Greco, C. Ko, P. Levai, Phys.Rev. **C68**, 034904 (2003), nucl-th/0305024
[10] V. Greco, C.M. Ko, R. Rapp, Phys. Lett. **B595**, 202 (2004), nucl-th/0312100
[11] C.W. Hwang, Eur. Phys. J. **C23**, 585 (2002), hep-ph/0112237
[12] C. Albertus, J.E. Amaro, E. Hernandez, J. Nieves, Nucl. Phys. **A740**, 333 (2004), nucl-th/0311100
[13] V. Minissale, F. Scardina, V. Greco, Phys. Rev. **C92**, 054904 (2015), 1502.06213
[14] K.A. Olive et al. (Particle Data Group), Chin. Phys. **C38**, 090001 (2014)
[15] L. Adamczyk et al. (STAR), Phys. Rev. Lett. **113**, 142301 (2014), 1404.6185
[16] B. Abelev et al. (ALICE), JHEP **09**, 112 (2012), 1203.2160
[17] M. Lisovyi, A. Verbytskyi, O. Zenaiev, Eur. Phys. J. **C76**, 397 (2016), 1509.01061
[18] S.S. Adler et al. (PHENIX), Phys. Rev. **C69**, 034909 (2004), nucl-ex/0307022
[19] S.S. Adler et al. (PHENIX), Phys. Rev. **C71**, 051902 (2005), nucl-ex/0408007
[20] B.B. Abelev et al. (ALICE), Phys. Lett. **B736**, 196 (2014), 1401.1250

ϕ meson production at forward rapidity in pp and Pb–Pb collisions with ALICE at the LHC

Alessandro De Falco[1],* for the ALICE Collaboration

[1] *Università and INFN Cagliari*

Abstract. The ALICE Collaboration has measured ϕ meson production in the dimuon channel at the forward rapidity ($2.5 < y < 4$) in pp and Pb–Pb collisions at several center of mass energies. Results in pp collisions at center of mass energies $\sqrt{s}$ = 5.02 and 8 TeV are reported. They complement the previously published results at $\sqrt{s}$ = 2.76 and $\sqrt{s}$ = 7 TeV, providing a solid baseline for Pb–Pb collisions. In Pb–Pb, the preliminary ϕ meson transverse momentum spectra for different centrality classes and the yields as a function of the collision centrality in the transverse momentum range $2 < p_{\mathrm{T}} < 7$ GeV/c are presented.

1 Introduction

Quantum Chromodynamics predicts the occurrence of a phase transition from hadronic matter to a plasma of deconfined quarks and gluons, the Quark-Gluon Plasma (QGP), at extreme conditions of temperature and energy density. Ultrarelativistic heavy-ion collisions provide the means to study this phase of matter in the laboratory. Strangeness production is a key tool to understand the properties of the medium formed in heavy-ion collisions: an enhanced production of strange particles was early proposed as one of the signatures of the QGP [1]. The ϕ meson, due to its $s\bar{s}$ valence quark content, provides insight into strangeness production [2].

ALICE has measured vector meson production in pp and Pb–Pb collisions at the LHC. The measurement at forward rapidity was performed using the muon spectrometer, that covers the pseudorapidity range $-4 < \eta < -2.5$[1]. It consists of a hadron absorber, five tracking stations, each one composed of two cathode pad chambers, for the track reconstruction in a dipole magnetic field, an iron wall acting as a muon filter and two stations of two resistive plate chambers for the muon trigger. The centrality is determined with the V0 detector, that consists of two arrays of plastic scintillators covering the pseudorapidity ranges $2.8 < \eta < 5.1$ and $-3.7 < \eta < -1.7$.

The muon trigger introduces a threshold on the single muon transverse momentum ($p_{\mathrm{T}\mu}$), that was set at ~ 0.5 GeV/c in pp collisions at $\sqrt{s}$ = 5.02 TeV and at ~ 1 GeV/c in pp collisions at $\sqrt{s}$ = 8 TeV and in Pb–Pb collisions at $\sqrt{s_{\mathrm{NN}}}$ = 5.02 TeV. As a consequence, the lowest dimuon p_{T} that can be reached is lower for the data sample with the lowest threshold.

*e-mail: alessandro.de.falco@ca.infn.it

[1] Although in the ALICE reference frame the muon spectrometer covers a negative η range, we chose to present our results in symmetric systems with a positive y notation.

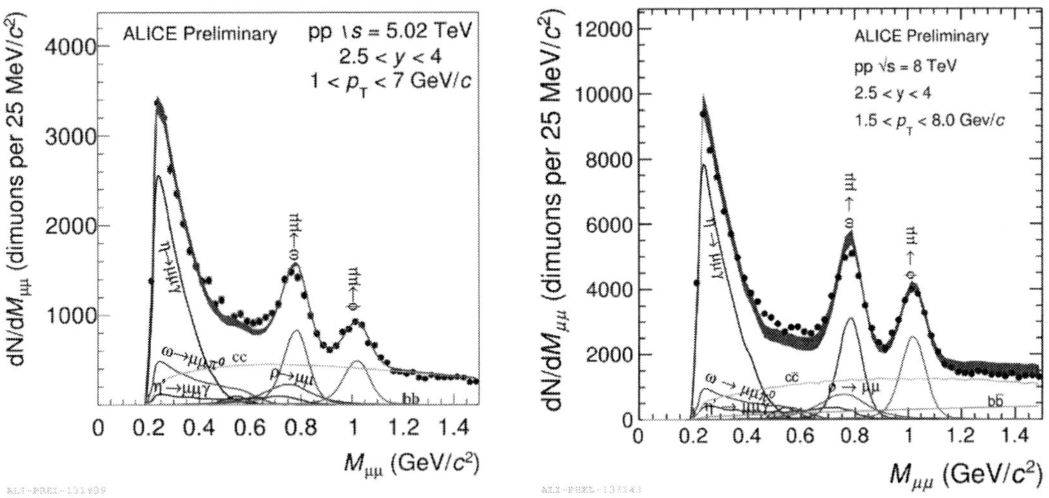

Figure 1. Dimuon invariant mass spectrum for pp collisions at $\sqrt{s}$ = 5.02 TeV (left) and $\sqrt{s}$ = 8 TeV (right) after combinatorial background subtraction. Red band: sum of all simulated contributions. The width of the red band represents the uncertainty on the relative normalization of the sources.

2 Results from pp collisions

New results on ϕ meson production in pp collisions at $\sqrt{s}$ = 5.02 TeV and $\sqrt{s}$ = 8 TeV are presented for the first time, complementing the already published results at $\sqrt{s}$ = 2.76 TeV [3] and $\sqrt{s}$ = 7 TeV [4]. The data samples collected at 5.02 and 8 TeV correspond to the integrated luminosities L_{int} = 106.3 ± 3 nb^{-1} and L_{int} = 2.4 ± 3 pb^{-1}, respectively. Muon tracks were selected requesting that the tracks reconstructed in the tracking stations matched the ones in the trigger chambers and

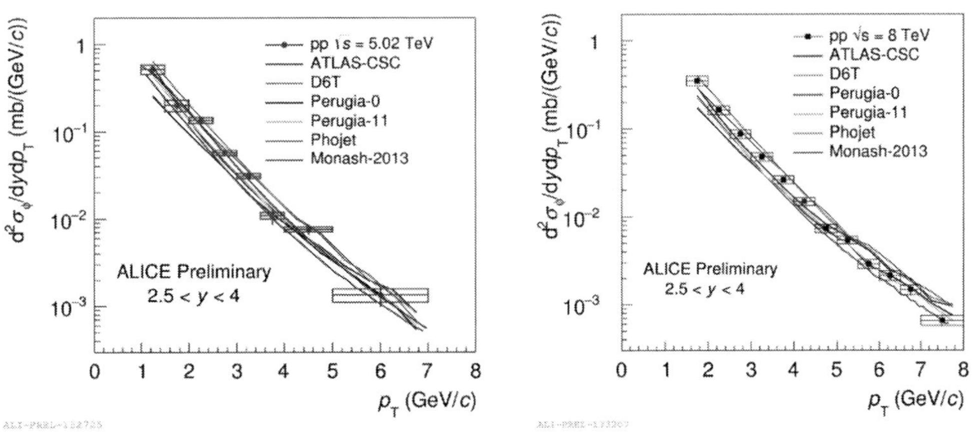

Figure 2. ϕ differential cross section vs p_T for pp collisions at $\sqrt{s}$ = 5.02 TeV (left) and $\sqrt{s}$ = 8 TeV (right) compared with the predictions based on several PYTHIA tunes and PHOJET.

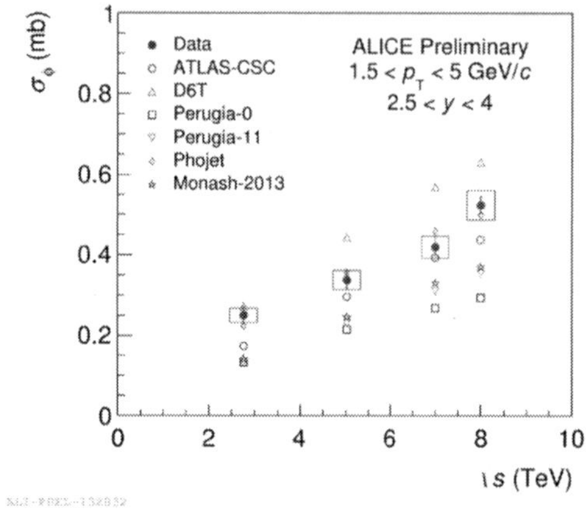

Figure 3. ϕ cross section as a function of $\sqrt{s}$ compared with the predictions based on several PYTHIA tunes and PHOJET.

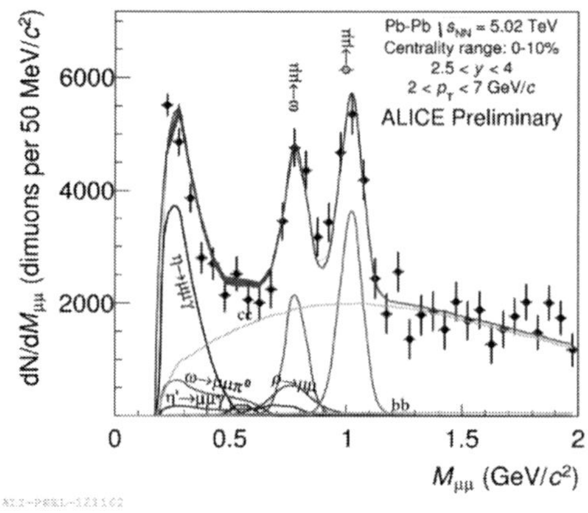

Figure 4. Dimuon invariant mass spectrum for Pb–Pb collisions at $\sqrt{s_{\rm NN}}$ = 5.02 TeV after combinatorial background subtraction.

that their pseudorapidity was in the range $-4 < \eta_\mu < -2.5$. The mass spectra after background subtraction, shown in Fig. 1, were described in terms of the so-called hadronic cocktail, consisting of a superposition of light meson decays into muon pairs, with an additional contribution coming from charm and beauty semi-muonic decays.

The ϕ $p_{\rm T}$-differential cross sections are shown in Fig. 2. In the same figure, the predictions based on several PYTHIA tunes (ATLAS-CSC, D6T, Perugia-0, Perugia-11 and MONASH 2013) and on PHOJET are shown [5–10]. Data are best described by PHOJET at all energies, as also seen when the cross section integrated in the common transverse momentum range, $1.5 < p_{\rm T} < 5$ GeV/c, is plotted as a function of $\sqrt{s}$ (Fig. 3).

3 Results from Pb–Pb collisions

The data from Pb–Pb collisions were collected in 2015 at $\sqrt{s_{\rm NN}}$ = 5.02 TeV and amount to an integrated luminosity of ~225 μb^{-1}. Muon tracks were selected requiring that the tracks reconstructed in the tracking stations matched the ones in the trigger chambers and that their pseudorapidity was in the range $-4 < \eta_\mu < -2.5$. A cut on the single muon $p_{{\rm T}\mu} > 0.85$ GeV/c was also applied. Muon pairs were selected inside the dimuon rapidity interval $2.5 < y < 4$. Due to the fact that the acceptance for low mass dimuons is close to zero for $p_{\rm T} < 2$ GeV/c, only dimuons with $p_{\rm T} > 2$ GeV/c were selected. An example of invariant mass distribution, after the combinatorial background subtraction, is shown in Fig. 4 (right) for $2 < p_{\rm T} < 7$ GeV/c, for the most central bin (0-10%).

The $p_{\rm T}$ spectra at all centralities (Fig. 5 left) are well described by a power-law function ${\rm d}N/{\rm d}p_{\rm T} \propto p_{\rm T}/[1 + (p_{\rm T}/p_0)^2]^n$. Peripheral collisions are characterized by a slightly harder $p_{\rm T}$ tail than central collisions.

The nuclear modification factor is defined as $R_{\rm AA} = \langle\phi\rangle / (\sigma_{\rm pp} \langle T_{\rm AA}\rangle)$, where $\langle\phi\rangle$ is the ϕ multiplicity, $\sigma_{\rm pp}$ is the pp cross section and $T_{\rm AA}$ is the nuclear overlap function. The $R_{\rm AA}$, measured as a function of $p_{\rm T}$, is shown in Fig. 5 (right) for central and semiperipheral collisions. A clear decrease of the $R_{\rm AA}$ at high $p_{\rm T}$, reaching about 0.3 for $p_{\rm T} > 4$ GeV/c in central collisions, shows the onset of the partonic energy loss regime. The decrease is less pronounced for semiperipheral collisions, where the

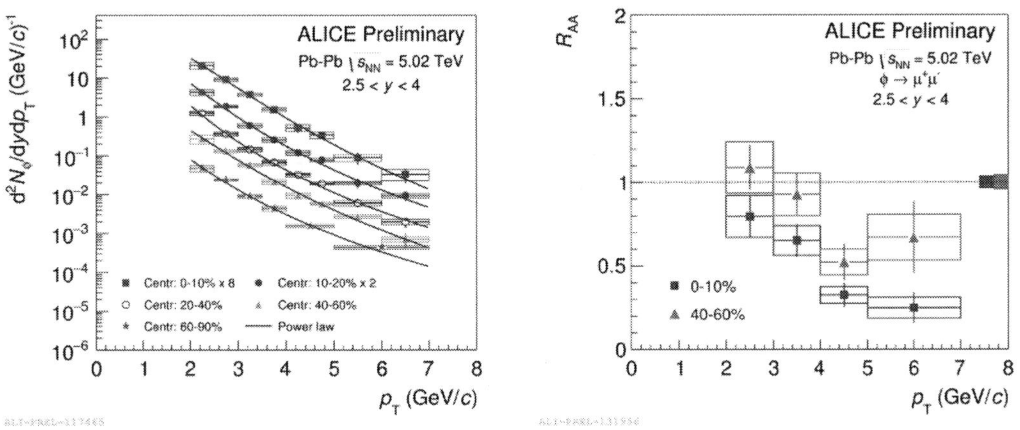

Figure 5. Left: ϕ p_T spectrum for several centrality classes in Pb–Pb collisions at $\sqrt{s_{NN}}$ = 5.02 TeV. Right: ϕ R_{AA} as a function of p_T, for central and semiperipheral collisions.

nuclear modification factor is compatible to unity for p_T < 4 GeV/c and decrease to ~ 0.6 for higher p_T. These results are in qualitative agreement with the ones obtained by PHENIX at RHIC [11].

4 Summary

ϕ meson production was measured in the dimuon channel in the rapidity region $2.5 < y < 4$ in pp and Pb–Pb collisions at several center of mass energies. Results in pp collisions at center of mass energies $\sqrt{s}$ = 5.02 and 8 TeV, together with the already published ones at $\sqrt{s}$ = 2.76 and 7 TeV, give an overall picture of ϕ production at forward rapidity at the LHC energies and allow for a comparison with several calculations. PHOJET best describes the data at all energies. In Pb–Pb, the p_T spectra have been measured in the range $2 < p_T < 7$ GeV/c for several collision centralities. Peripheral collisions are characterized by a slightly harder p_T tail than central ones. The nuclear modification factor was measured as a function of p_T. A clear suppression of R_{AA} for $p_T > 4$ GeV/c is observed at central collisions. The suppression is less pronounced in semiperipheral collisions.

References

[1] J. Rafelski, B. Muller, Phys. Rev. Lett. **48**, 1066 (1982), [Erratum: Phys. Rev. Lett.56,2334(1986)]
[2] B.I. Abelev et al. (STAR), Phys. Lett. **B673**, 183 (2009), 0810.4979
[3] J. Adam et al. (ALICE), Phys. Lett. **B768**, 203 (2017), 1506.09206
[4] B. Abelev et al. (ALICE), Phys. Lett. **B710**, 557 (2012), 1112.2222
[5] T. Sjostrand, S. Mrenna, P.Z. Skands, JHEP **05**, 026 (2006), hep-ph/0603175
[6] C.M. Buttar, D. Clements, I. Dawson, A. Moraes, Acta Phys. Polon. **B35**, 433 (2004)
[7] R. Field, Acta Phys. Polon. **B39**, 2611 (2008)
[8] P.Z. Skands, Phys. Rev. **D82**, 074018 (2010), 1005.3457
[9] P. Skands, S. Carrazza, J. Rojo, Eur. Phys. J. **C74**, 3024 (2014), 1404.5630
[10] R. Engel, J. Ranft, Phys. Rev. **D54**, 4244 (1996), hep-ph/9509373
[11] A. Adare et al. (PHENIX), Phys. Rev. **C83**, 024909 (2011), 1004.3532

EPJ Web of Conferences **171**, 13007 (2018) https://doi.org/10.1051/epjconf/201817113007
SQM 2017

Strangeness production in Pb-Pb collisions at LHC energies with ALICE

Michal Šefčík[1,⋆] for the ALICE Collaboration

[1] *Pavol Jozef Šafárik University, Slovakia*

Abstract. The results on the production of strange and multi-strange hadrons (K^0_S, Λ, Ξ and Ω) measured with ALICE in Pb-Pb collisions at the top LHC energy of $\sqrt{s_{NN}}$ = 5.02 TeV are reported. Thanks to its excellent tracking and particle identification capabilities, ALICE is able to measure weakly decaying particles through the topological reconstruction of the identified hadronic decay products. Results are presented as a function of centrality and include transverse momentum spectra measured at central rapidity, p_T-dependent Λ/K^0_S ratios and integrated yields. A systematic study of strangeness production is of fundamental importance for determining the thermal properties of the system created in ultrarelativistic heavy ion collisions. In order to study strangeness enhancement, the yields of studied particles are normalised to the corresponding measurement of pion production in the various centrality classes. The results are compared to measurements performed at lower energies, as well as to different systems and to predictions from statistical hadronization models.

1 Introduction

Measurement of strangeness production represents an important source of information about the strongly interacting system created in relativistic heavy-ion collisions. LHC Run II data from Pb-Pb collisions constitute a large statistics sample at high collision energy. In these proceedings we describe new results on the production of K^0_S, Λ ($\overline{\Lambda}$), Ξ ($\overline{\Xi}$) and Ω ($\overline{\Omega}$) in Pb-Pb collisions at the energy of $\sqrt{s_{NN}}$ = 5.02 TeV. We first describe the method used for the identification of these particles and then the main results concerning transverse momentum spectra and hadrochemistry.

2 Strange and multi-strange hadron detection in ALICE

The ALICE detector was designed for the study of heavy-ion collisions at the LHC. Its Inner Tracking System (ITS), which consists of six layers of silicon detectors, provides precision tracking and vertexing close to the interaction point. The Time Projection Chamber (TPC) has an excellent tracking performance in high multiplicity collisions, such as central Pb-Pb collisions. It also allows for particle identification through the measurement of the specific energy loss. The V0 detector consists of two scintillator hodoscopes covering the forward pseudo-rapidity regions on both sides of the interaction point. It is used for triggering and multiplicity and centrality estimation. Full description of the ALICE sub-detectors can be found in [1].

⋆e-mail: michal.sefcik@cern.ch

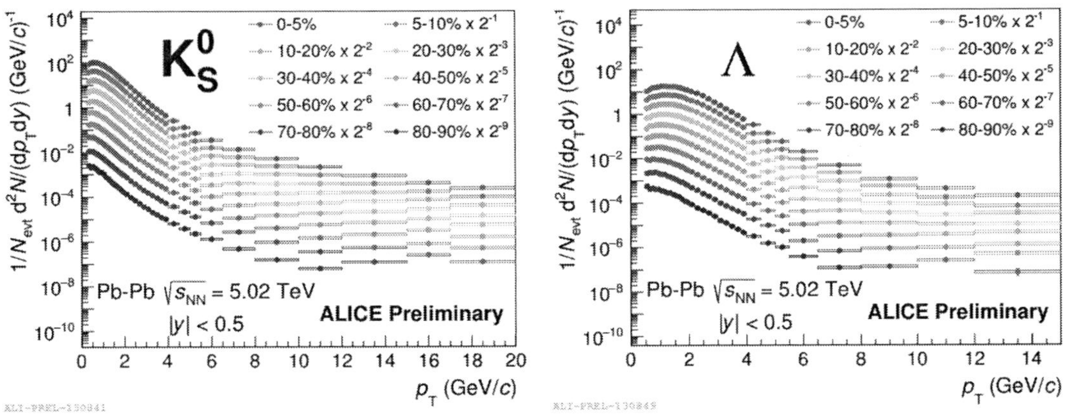

Figure 1. Transverse momentum spectra of K^0_S and Λ for different centrality intervals in Pb-Pb collisions at $\sqrt{s_{NN}}$ = 5.02 TeV.

Strange (K^0_S and Λ) and multi-strange (Ξ and Ω) hadrons are reconstructed via their characteristic weak decay topologies, decaying into two and three charged particles, respectively. Charged particle tracks are reconstructed with ITS and TPC and particle identification is based on a selection on the specific energy loss in the TPC. Charged particle tracks are combined in an invariant mass analysis to reconstruct weak decay candidates. These candidates are required to fulfill selection criteria based on geometric (topological) and kinematic quantities. Yields are extracted by bin counting from the invariant mass distribution in each transverse momentum and centrality bin. Acceptance and efficiency corrections are calculated using Monte Carlo simulations.

3 Transverse momentum spectra

The transverse momentum spectra for K^0_S and Λ are shown in Fig. 1. For all particles under study, going from peripheral to central collisions, a hardening of the spectral shapes is observed. The hardening is more pronounced for heavier particles. This behaviour is qualitatively consistent with expectations from hydrodynamical models. It is also consistent with similar observations for other light-flavour hadrons.

The simultaneous blast-wave model fit to $\pi^{\pm}$, $K^{\pm}$ and p ($\overline{p}$) in Pb-Pb collisions at 5.02 TeV follows the trend from 2.76 TeV with further increase in radial expansion velocity. Strange and multi-strange baryons are not well described by this fit in Pb-Pb collisions, suggesting different kinetic freeze-out conditions for strange and non-strange particles. In contrast, in p-Pb collisions strange and non-strange particles are described by common blast-wave parameters [2].

The Λ to K^0_S ratio in Pb-Pb collisions is shown in Fig. 2. At 5.02 TeV we observe the same features as at the lower energy of 2.76 TeV. From peripheral to central collisions, there is a depletion at low p_T ($\lesssim$ 2 GeV/c), an enhancement at intermediate p_T ($\sim 2-7$ GeV/c) and no change at high p_T ($\gtrsim$ 7 GeV/c). Comparison with models at $\sqrt{s_{NN}}$ = 2.76 TeV shows that the behaviour at low p_T is explained by hydrodynamical models, while the peak at intermediate p_T is described qualitatively by recombination models [3]. EPOS [4], which includes radial flow and jet-medium interactions, describes the dependence over the entire p_T range.

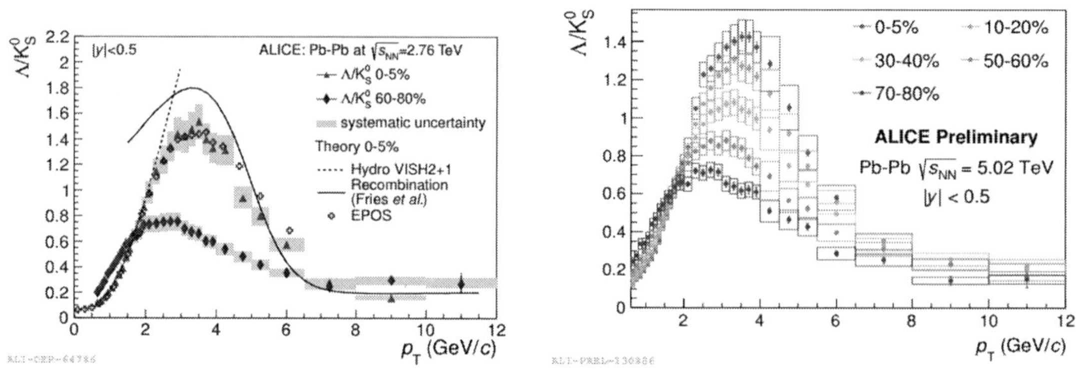

Figure 2. Λ/K^0_S yield ratio as a function of transverse momentum in different centrality classes in Pb-Pb collisions. Left: $\sqrt{s_{NN}}$ = 2.76 TeV [5] Right: $\sqrt{s_{NN}}$ = 5.02 TeV

4 Hadrochemistry

Strangeness enhancement is one of the first proposed signatures of the formation of the quark-gluon plasma (QGP), first suggested by Rafelski and Müller in 1982 [6]. The enhancement was observed at SPS [7], RHIC [8] and LHC [9] and has been found to increase with centrality and strangeness content of the particle and decrease with the energy of the colliding system. At SPS and RHIC the strangeness enhancement was studied by measuring the ratio of the yield in AA collisions and the yield in pp collision at the same energy, normalized to the number of participants. Since the production of charged particles does not scale linearly with the number of participants [10], this is not an ideal quantity to study.

A better observable is the ratio of the yield of the studied particle to the pion yield, which is shown in Fig. 3. We observe a smooth evolution across different colliding systems (pp, p-Pb and Pb-Pb) and no significant dependence on collision energy. For single-strange particles (K^0_S and Λ), there is an indication of an increasing trend with increasing multiplicity in small systems, while for multi-strange particles (Ξ and Ω), the increase is significant. For all particles, there is a flat trend in Pb-Pb collisions at high multiplicity.

Statistical hadronization models based on the grand-canonical approach have been demonstrated to be able to predict particle yield ratios in heavy-ion collisions over a large energy range [11]. Fig. 3 shows that the predictions of statistical hadronization models using a chemical freezeout temperature of 156 MeV are comparable with the measured values in Pb-Pb collisions. In this statistical picture, the lower production of strangeness in small systems (pp collisions) is the consequence of the small phase-space volume available, and it is referred to as canonical suppression [12].

5 Conclusions

The most recent results of the ALICE experiment on the production of strange and multi-strange hadrons in Pb-Pb collisions at 5.02 TeV have been reported. The measured p_T spectra show a mass dependent hardening with centrality, as expected from radial flow. The baryon to meson ratio shows the same behaviour as in 2.76 TeV Pb-Pb collisions. We observe a continuous behaviour of the yield ratio to pions versus multiplicity across colliding systems (pp, p-Pb and Pb-Pb). The new data presented here shows that there is not a significant change of this ratio going from 2.76 TeV to 5.02 TeV.

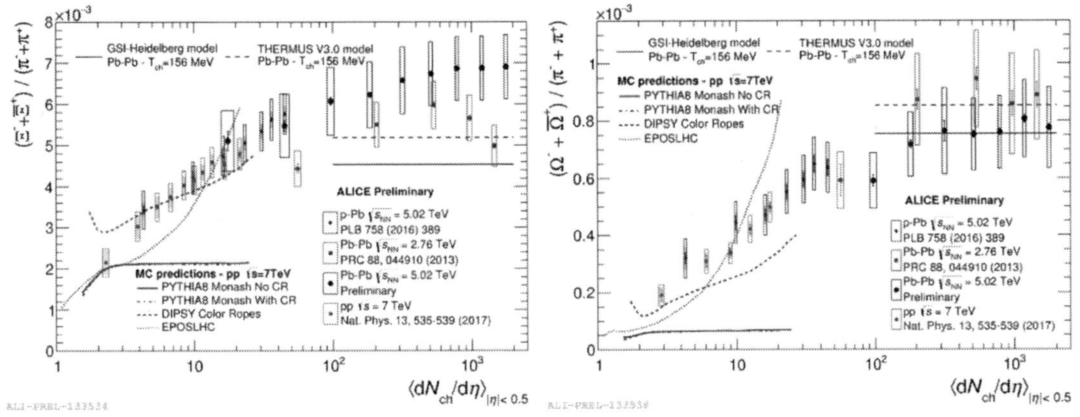

Figure 3. Ratio of Ξ and Ω yields to pion yield as a function of charged particle multiplicity density at medium pseudo-rapidity. Data from pp collisions at 7 TeV (small circles), p-Pb collisions at 5.02 TeV (pluses), Pb-Pb collisions at 2.76 TeV (squares), Pb-Pb collisions at 5.02 TeV (big circles) and comparisons to statistical hadronization models are shown. Open boxes are total systematic uncertainties and shaded boxes show the component that is uncorrelated over multiplicity.

References

[1] Aamodt K. *et al.* (ALICE Colaboration), JINST **3**, S08002 (2008)
[2] Abelev B. *et al.* (ALICE Colaboration), Phys. Lett. B **728**, 25-38 (2014)
[3] Fries R. J., J. Phys. G **30**, S853 (2004)
[4] Werner K., Nucl. Phys. B **175-176**, 81-87 (2008)
[5] Abelev B. *et al.* (ALICE Colaboration), Phys. Rev. Lett. **111**, 222301 (2013)
[6] Rafelski J. and Müller B., Phys. Rev. Lett. **48**, 1066 (1982)
[7] Antinori F. *et al.* (NA57 Colaboration), J. Phys. G **32**, 427-441 (2006)
[8] Abelev B. *et al.* (STAR Colaboration), Phys. Rev. C **77**, 044908 (2008)
[9] Abelev B. *et al.* (ALICE Colaboration), Phys. Lett. B **728**, 216-227 (2014)
[10] Aamodt K. *et al.* (ALICE Colaboration), Phys. Rev. Lett. **106**, 032301 (2011)
[11] Stachel J., Andronic A., Braun-Munzinger P. and Redlich K., J. Phys. Conf. Ser. **509**, 012019 (2014)
[12] Hamieh S., Redlich K. and Tounsi A., Phys. Lett. B **486**, 61-66 (2000)

Strange hadron production in pp, pPb, and PbPb collisions at LHC energies

For the CMS Collaboration

Hong Ni[1,★]

[1] *Department of Physics and Astronomy, Vanderbilt University, Nashville, TN, United States*

Abstract. Identified particle spectra provide an important tool for understanding the particle production mechanism and the dynamical evolution of the medium created in relativistic heavy ion collisions. Studies involving strange and multi-strange hadrons, such as K^0_S, Λ, and Ξ^-, carry additional information since there is no net strangeness content in the initial colliding system. Strangeness enhancement in AA collisions with respect to pp and pA collisions has long been considered as one of the signatures for quark-gluon plasma (QGP) formation. Recent observations of collective effects in high-multiplicity pp and pA collisions raise the question of whether QGP can also be formed in the smaller systems. Systematic studies of strange particle abundance, particle ratios, and nuclear modification factors can shed light on this issue. The CMS experiment has excellent strange-particle reconstruction capabilities over a broad kinematic range, and dedicated high-multiplicity triggers in pp and pPb collisions. The spectra of K^0_S, Λ, and Ξ^- hadrons have been measured in various multiplicity and rapidity regions as a function of p_T in pp, pPb, and PbPb collisions for several collision energies. The spectral shapes and particle ratios are compared in the different collision systems for events that have the same multiplicity and interpreted in the context of hydrodynamics models.

1 Introduction

Recently, the long-range, near-side two-particle correlations in small systems with high-multiplicity [1–4] have brought up an intense debate in the heavy ion field. However, the origin of the observed correlations is still unclear. Among the theories that attempt to interpret these features seen in data, the collective flow scenario of a fluid-like medium provides a natural interpretation [5–7].

In the hydro-dynamic regime, all the particles are supposed to have the same expansion velocity, thus hadrons with different masses will have different p_T behavior. As the strange quark is the most abundant of the heavier quarks, production of strange particles can be a useful tool to test whether a mass effect is present in these systems. Usually, particle correlations are used to study the flow directly. In the results presented here [8], measurements of strange hadrons (K^0_s, $\Lambda + \overline{\Lambda}$(hereafter referred to as Λ), and $\Xi^- + \Xi^+$(hereafter referred to as Ξ^-)) transverse momentum (p_T) spectra in three collision systems (pp, pPb, and PbPb) at different center-of-mass energies ($\sqrt{s_{NN}}$) in several different multiplicity ($N_{trk}^{offline}$) and center-of-mass rapidity (y_{cm}) regions are presented. With the precise

★e-mail: hong.ni@vanderbilt.edu

measurements of the spectra, particle ratio measurements (Λ / K_s^0 and Ξ^- / Λ) are performed. To explore the possible collective behavior of particles in these systems, a blast-wave model is employed to extract the transverse expansion velocity (β_T) and kinetic freeze-out temperature (T_{kin}) in different multiplicity ranges for each system.

2 Results

2.1 Invariant mass peaks

The details of event selection and particle reconstruction can be found in [8]. Figure 1 shows some samples of invariant mass peaks. After applying the reconstruction selection criteria, prominent mass peaks are visible, and the background has been largely reduced. To describe the invariant mass peaks, for all the strange particles studied, two Gaussian functions with a common mean are summed up. For the background component, different functional forms are used for different particle species. For K_s^0, the background is modeled with a quadratic function. For Λ, the analytic form $Aq^{1/2} + Bq^{3/2}$ with $q = m - (m_\pi + m_p)$ is used. For Ξ^-, the form Cq^D with $q = m - (m_\Lambda + m_\pi)$ is applied. A, B, C, and D in above analytic expressions are free parameters used in the fitting.

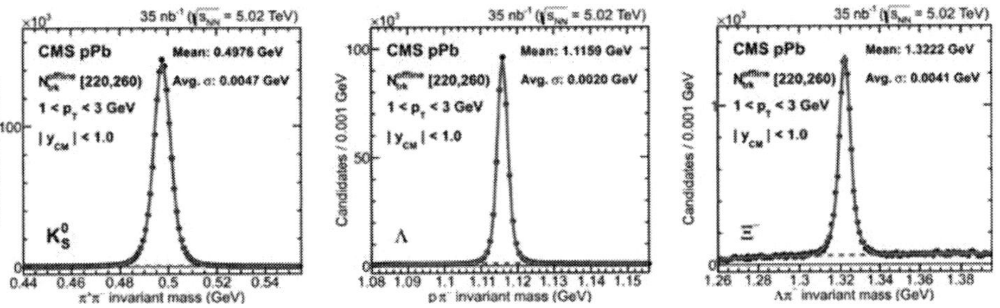

Figure 1. Invariant mass distribution of K_s^0(left), Λ(middle), and Ξ^-(right) candidates in the p_T range 1-3 GeV for $220 \leq N_{trk}^{offline} < 260$ in pPb collisions. Summing up of the charge conjugate states is implied for Λ and Ξ^-. Figure from [8].

2.2 Particle ratios

With the precise measurements of strange particle spectra over different multiplicity ranges, particle ratio measurements can be calculated to check the potential changes of spectra shapes as the multiplicity changes. With the help of a high-multiplicity trigger developed by CMS, particle ratios in similar multiplicity intervals for three different collision systems can be compared. The Λ / K_s^0 and Ξ^- / Λ ratios in a center-of-mass rapidity interval $|y_{CM}| < 1$ are shown in Figure 2. For all multiplicity classes, the Λ / K_s^0 ratio increases with p_T and reaches a maximum at intermediate p_T, then decreases at higher p_T while Ξ^- / Λ ratio increases with p_T and reaches a plateau region at around 3 GeV. For the low p_T region, in each system, at a given p_T, the Λ / K_s^0 ratio is smaller for higher multiplicity events, and the difference between high and low multiplicity events is larger for smaller systems. No significant difference between high and low multiplicity events is shown for Ξ^- / Λ which can be understood as due to the small mass difference between Ξ^- and Λ is small.

Among the three collision systems studied, pPb is the only asymmetric system, and a larger radial flow effect is expected on the Pb-going side, as predicted by hydro-dynamic models. With the large acceptance of the CMS detector, the Λ / K_s^0 ratio in different multiplicity intervals on the Pb-going sides and the p-going sides can be compared, and the results are shown in Figure 3. For both low and high multiplicity events, the Λ / K_s^0 ratio from the Pb-going direction lies above the results from the p-going direction, with the largest difference observed at high p_T in high multiplicity events.

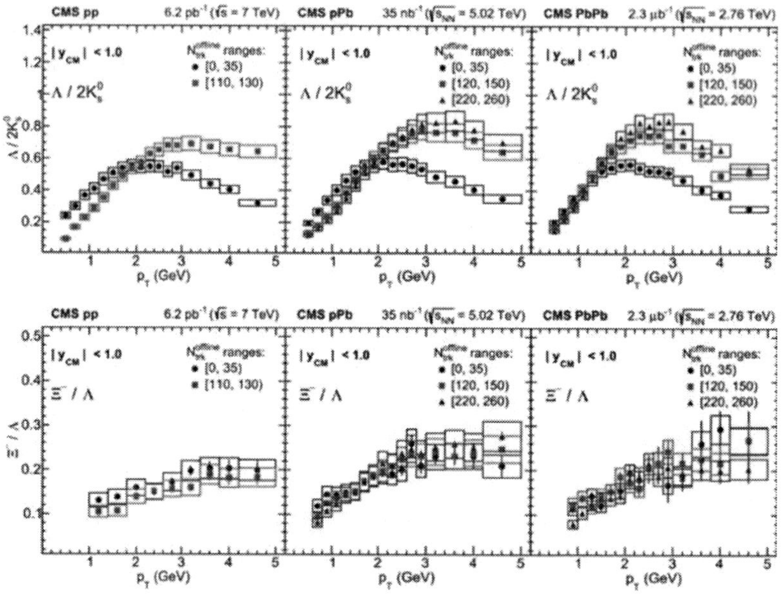

Figure 2. Ratios of p_T spectra for Λ / K_s^0(top) and Ξ^- / Λ(bottom) in the center-of-mass rapidity range $|y_{cm}| < 1.0$ for pp collisions at $\sqrt{s}$ = 7 TeV (left), pPb collisions at $\sqrt{s_{NN}}$ = 5.02 TeV (middle), and PbPb collisions at $\sqrt{s_{NN}}$ = 2.76 TeV (right). The error bars represent the statistical uncertainties, while the boxes indicate the systematic uncertainties. Figure from [8].

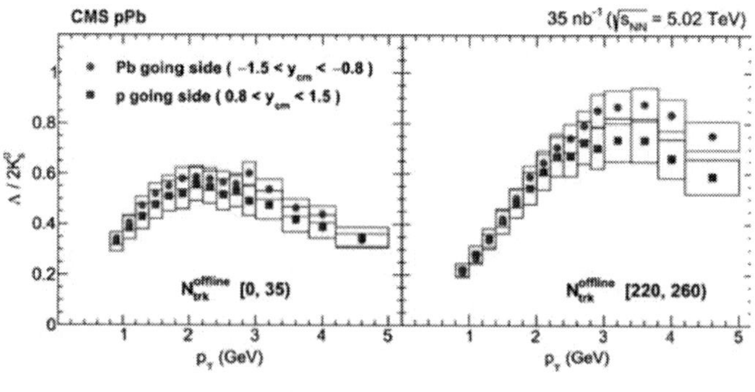

Figure 3. Ratios of p_T spectra for Λ / K_s^0 from the $-1.5 < y_{cm} < -0.8$ (Pb-going) and $0.8 < y_{cm} < 1.5$ (p-going) rapidity regions in pPb collisions at $\sqrt{s_{NN}}$ = 5.02 TeV (middle). Results are presented for two multiplicity ranges $0 \leq N_{trk}^{offline} < 35$ (left) and $220 \leq N_{trk}^{offline} < 260$ (right). The error bars represent the statistical uncertainties, while the boxes indicate the systematic uncertainties. Figure from [8].

2.3 Blast-wave fit

Benefitting from the fact that the strange particles are measured down to very low p_T, the hydrodynamically motived blast-wave fit [9] can be performed with small uncertainties. In the blast-wave model, common values of T_{kin} and average radial-flow velocity $\langle\beta_T\rangle$ are assumed for all particle species. A detailed description of the blast-wave fit can be found in [8]. The extracted values of T_{kin} and $\langle\beta_T\rangle$ are shown in Figure 4. According to the plot, for different systems at similar multiplicities the T_{kin} values are similar, while $\langle\beta_T\rangle$ is larger for the system with smaller size, which indicates a larger expansion velocity for smaller systems, although the exact meaning of the two parameters is model dependent.

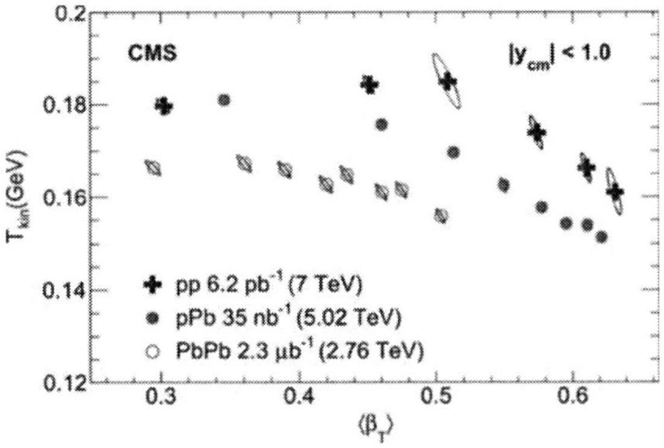

Figure 4. The extracted kinetic freeze-out temperature (T_{kin}) versus the average radial-flow velocity $\langle\beta_T\rangle$ from a simultaneous blast-wave fit to the K_s^0 and Λ p_T spectra at $y_{cm} < 1$ for different multiplicity intervals in pp, pPb, and PbPb collisions. The multiplicity intervals are exactly the same as in Figure 2. For the results in this plot, the multiplicity increases from left to right. The correlation ellipses represent the statistical uncertainties. Systematic uncertainties, which are evaluated to be on the order of a few percent, are not shown. Figure from [8].

3 Summary

In summary, measurements of strange hadron (K_s^0, $\Lambda + \overline{\Lambda}$, and $\Xi^- + \Xi^+$) transverse momentum (p_T) spectra in three collision systems (pp, pPb, and PbPb) at different center-of-mass energies in several different multiplicity and center-of-mass rapidity regions are presented. With these spectra, particle ratios have been calculated and simultaneous blast-wave fits have been performed. The results are in qualitative agreement with hydro-dynamic models.

References

[1] CMS Collaboration, JHEP **09**, 091 (2010)
[2] CMS Collaboration, Phys. Lett. B **718**, 795 (2013)
[3] ATLAS Collaboration, Phys. Rev. Lett. **110**, 182302 (2013)
[4] ALICE Collaboration, Phys. Lett. B **719**, 29 (2013)
[5] P.Bozek, Phys. Rev. C **85**, 014911 (2012)
[6] P.Bozek, W.Broniowski, Phys. Lett. B **718** 1557 (2013)
[7] A.Bzdak, B.Schenke, P.Tribedy, R.Venugopalan, Phys. Rev. C **87** 064906 (2013)
[8] CMS Collaboration, Phys. Lett. B **768** 103 (2017)
[9] E.Schnedermann, J.Sollfrank, U.Heinz, Phys. Rev. C **48** 2462 (1993)

EPJ Web of Conferences **171**, 14001 (2018) https://doi.org/10.1051/epjconf/201817114001
SQM 2017

Role of repulsive interactions in the interplay with missing strange resonances

Paolo Alba[1,★]

[1] *Frankfurt Institute for Advanced Studies, Goethe Universität Frankfurt, D-60438 Frankfurt am Main, Germany*

Abstract. The standard implementation of the HRG model has been shown to be unable to describe all the available data on QCD matter. Here we show the balance of repulsive and attractive hadronic interactions on QCD thermodynamics through observables both calculated by lattice simulations and measured in experiment. Attractive interactions are mediated by resonance formation, which are here implemented through extra states predicted by the Quark Model, while repulsive interactions are modelled by means of Excluded Volume (EV) effects. Informations on flavour dependent effective sizes are extracted. It is found that EV effects are present in lattice QCD thermodynamics, and are essential for a comprehensive description of higher order fluctuations of conserved charges.

1 Introduction

The current abundance of results from lattice simulations [1, 2] allows for a detailed study of QCD thermodynamics. It has been shown how the standard implementation of the Hadron-Resonance Gas (HRG) model is unable to describe all the available data, with quite strong hints for the presence of extra higher-mass resonances [3, 4], being them compatible with a prediction from the Quark-Model (QM) [5, 6]. On the other hand the relevance of repulsive interactions has been pointed out through the S-matrix approach [7], and a clear connection between the experimental phase shift and fluctuations of conserved charges has been established [8]. These interactions can be modelled in the HRG framework through the assumption of hadrons as hard spheres, which then acquire an effective radius r [9]. In the present paper we study the combination of QM states with Excluded Volume (EV) effects, extracting from lattice QCD thermodynamics informations about the effective sizes of hadrons as it was already done for the pure gauge sector [10]. Interesting implications have been found at imaginary chemical potential [11].

2 The HRG model

The HRG model describes a system of interacting hadrons as a system of non interacting hadrons and resonances, where resonances formation mediates attractive hadronic interactions. Thus the partition

★e-mail: alba@fias.uni-frankfurt.de

function can be written as:

$$\ln \mathcal{Z}(T, \{\mu_B, \mu_Q, \mu_S\}) = \sum_{i \in Particles} (-1)^{B_i+1} \frac{d_i}{(2\pi^3)} \int d^3\vec{p} \ln\left[1 + (-1)^{B_i+1} e^{-(\sqrt{\vec{p}^2+m_i{}^2}-\mu_i)/T}\right], \quad (1)$$

where the sum runs over all the particles listed by Particle Data Group (PDG) [12] with their properties. From this, fluctuations of conserved charges can be calculated as:

$$\chi_{lmn}^{BQS} = \frac{\partial^{l+m+n}(\ln \mathcal{Z}/T^3)}{\partial(\mu_B/T)^l \partial(\mu_Q/T)^m \partial(\mu_S/T)^n}. \quad (2)$$

For details please see [13, 14].

2.1 Excluded Volume

The effective radius for hadrons results into a shifted single-particle chemical potential:

$$\mu_i^* = \mu_i - v_i\, p\ , \quad (3)$$

which implies a transcendental equation for pressure p, where $v_i = \frac{16}{3}\pi r_i^3$ is the single particle *eigenvolume*. In order to be consistent with the virial expansion to second order one must use, instead of eigenvolumes, the coefficients $b_{ij} = \frac{2}{3}\pi\left(r_i + r_j\right)^3$ which properly account for two particle interactions (*crossterms*). This leads to a further modification of the single particle chemical potential, and to a system of coupled transcendental equations. For details see [15, 16].

3 Results

We performed fits to continuum extrapolated lattice QCD data at physical values of quark masses, employing different hadronic spectra and different EV schemes. We find that the best description is given by a list inspired by QM predictions where particle eigenvolume increases with the mass, but with smaller values for strange particles with respect to light ones with equal masses (*2b* scheme).
In figure 1 it is shown how EV effects are able to reproduce the characteristic suppression of 4th to 2nd ratios of fluctuations of conserved charges, preserving the agreement with lower order observables like the pressure. The very same happens when crossterms are employed, showing that the standard EV implementation is already sufficient to catch all the relevant features of lattice QCD thermodynamics. For more details see [19].

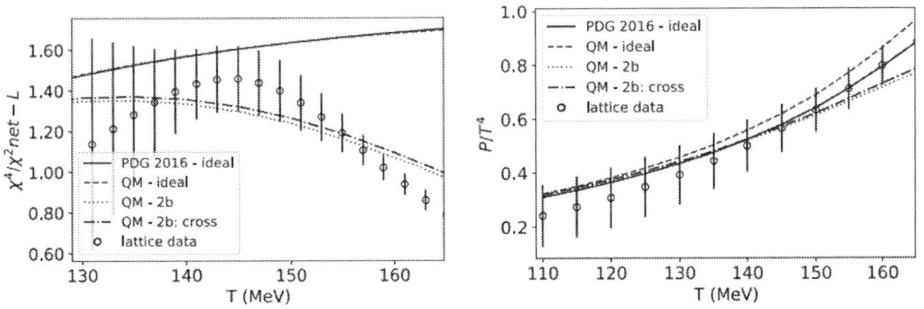

Figure 1. Lattice data for χ^4/χ^2 for net-light number and pressure [17, 18] compared to HRG calculations with different lists and EV schemes.

3.1 Comparison with the experimental results

In figure 2 are shown fits to experimental particle yields measured by the ALICE Collaboration at 2.76 TeV $\sqrt{s_{NN}}$ in 0-5% centrality class [20–23]; fit results are summarised in table 1. It is shown how the parameterisation extracted by lattice QCD essentially improves the fit mainly due to EV effects acting on (anti-)protons and (anti-)Ξs, solving then the so-called proton anomaly; in particular with it there is no presence for the second minimum structure which affects some EV schemes [24]. This effect has already been observed in [25], where however a different scheme was employed. It should be noted that the effectiveness of the 2b scheme found here, is also supported by experimental measurement on the electric radii of a few ground-state hadrons [12].
It is worth to note that accounting for crossterms, without changing EV parameterisation, pushes the agreement with the experimental results. For more details see [19].

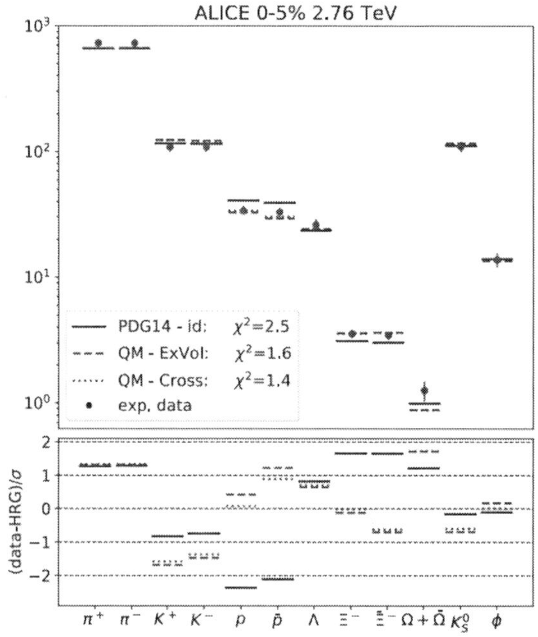

Figure 2. Particle yields measured by the ALICE Collaboration for PbPb collisions at 2.76 TeV in the 0-5% centrality class, [20–23] in comparison to HRG fits with different lists and EV schemes.

Table 1. Parameters resulting from the fit to particle yields measured by the ALICE Collaboration, using different lists and EV schemes.

	χ^2	T (MeV)	μ_B (MeV)	V (fm^3)
PDG14 - id	22.24/9	154.1 ± 2.2	3.5 ± 6.9	4992.9 ± 643.7
QM - ExVol	14.49/9	149.1 ± 1.7	7.7 ± 7.9	7560.9 ± 697.2
QM - Cross	12.95/9	149.4 ± 1.7	7.5 ± 7.5	7274.8 ± 698.6

4 Conclusions

In this proceeding we show how EV effects are manifest both in lattice QCD thermodynamics and in experimentally measured particle yields. This was achieved without the inclusion of additional van-der-Waals attractive terms [26], which may attain rather large values if observables on the lattice are fitted [27]. Furthermore in our approach we can consistently assign a finite radius to mesons too, fact that could have interesting implications on key observables. See [19] for more details.

References

[1] R. Bellwied, S. Borsanyi, Z. Fodor, S.D. Katz, A. Pasztor, C. Ratti, K.K. Szabo, Phys. Rev. **D92**, 114505 (2015), 1507.04627
[2] A. Bazavov et al., Phys. Rev. **D95**, 054504 (2017), 1701.04325
[3] A. Bazavov et al., Phys. Rev. Lett. **113**, 072001 (2014), 1404.6511
[4] P. Alba et al., Phys. Rev. **D96**, 034517 (2017), 1702.01113
[5] D. Ebert, R.N. Faustov, V.O. Galkin, Phys. Rev. **D79**, 114029 (2009), 0903.5183
[6] S. Capstick, N. Isgur, Phys. Rev. **D34**, 2809 (1986), [AIP Conf. Proc.132,267(1985)]
[7] W. Broniowski, F. Giacosa, V. Begun, Phys. Rev. **C92**, 034905 (2015), 1506.01260
[8] P. Huovinen, P. Petreczky (2017), 1708.00879
[9] D.H. Rischke, M.I. Gorenstein, H. Stoecker, W. Greiner, Z. Phys. **C51**, 485 (1991)
[10] P. Alba, W.M. Alberico, A. Nada, M. Panero, H. Stoecker, Phys. Rev. **D95**, 094511 (2017), 1611.05872
[11] V. Vovchenko, A. Pasztor, Z. Fodor, S.D. Katz, H. Stoecker (2017), 1708.02852
[12] C. Patrignani et al. (Particle Data Group), Chin. Phys. **C40**, 100001 (2016)
[13] P. Alba, W. Alberico, R. Bellwied, M. Bluhm, V. Mantovani Sarti, M. Nahrgang, C. Ratti, Phys. Lett. **B738**, 305 (2014), 1403.4903
[14] P. Alba, R. Bellwied, M. Bluhm, V. Mantovani Sarti, M. Nahrgang, C. Ratti, Phys. Rev. **C92**, 064910 (2015), 1504.03262
[15] V. Vovchenko, H. Stoecker, Phys. Rev. **C95**, 044904 (2017), 1606.06218
[16] L.M. Satarov, V. Vovchenko, P. Alba, M.I. Gorenstein, H. Stoecker, Phys. Rev. **C95**, 024902 (2017), 1610.08753
[17] R. Bellwied, S. Borsanyi, Z. Fodor, S.D. Katz, C. Ratti, Phys. Rev. Lett. **111**, 202302 (2013), 1305.6297
[18] S. Borsanyi, Z. Fodor, C. Hoelbling, S.D. Katz, S. Krieg, K.K. Szabo, Phys. Lett. **B730**, 99 (2014), 1309.5258
[19] P. Alba (2017), 1711.02797
[20] B. Abelev et al. (ALICE), Phys. Rev. **C88**, 044910 (2013), 1303.0737
[21] B.B. Abelev et al. (ALICE), Phys. Rev. Lett. **111**, 222301 (2013), 1307.5530
[22] B.B. Abelev et al. (ALICE), Phys. Lett. **B728**, 216 (2014), [Erratum: Phys. Lett.B734,409(2014)], 1307.5543
[23] F. Becattini, E. Grossi, M. Bleicher, J. Steinheimer, R. Stock, Phys. Rev. **C90**, 054907 (2014), 1405.0710
[24] V. Vovchenko, H. Stoecker, J. Phys. **G44**, 055103 (2017), 1512.08046
[25] P. Alba, V. Vovchenko, M.I. Gorenstein, H. Stoecker (2016), 1606.06542
[26] V. Vovchenko, M.I. Gorenstein, H. Stoecker, Phys. Rev. Lett. **118**, 182301 (2017), 1609.03975
[27] S. Samanta, B. Mohanty (2017), 1709.04446

Strangeness freeze-out: role of system size and missing resonances

Sandeep Chatterjee[1,*], *Sabita* Das[2], *Ajay Kumar* Dash[3], *Debadeepti* Mishra[3], *Bedangadas* Mohanty[3], and *Subhasis* Samanta[3]

[1] *AGH University of Science and Technology, al. Mickiewicza 30, 30-059 Krakow, Poland*
[2] *Institute of Particle Physics and Key Laboratory of Quark & Lepton Physics (MOE), Central China Normal University, Wuhan-430079, China*
[3] *National Institute of Science Education and Research, HBNI, Jatni-752050, India*

Abstract. The conventional approach to treat strangeness freezeout has been to consider a unified freezeout scheme where strangeness freezes out along with the non-strange hadrons (1CFO), with or without an additional parameter accounting for out-of-equilibrium strangeness production (γ_S). Several alternate scenarios have been formulated lately. Here, we will focus on flavor dependent freezeout with early freezeout of strangeness (2CFO) in comparison to 1CFO and its variants with respect to the roles played by the system size and missing resonances predicted by different theoretical approaches but yet to be seen in experiments. In contrast to the performance of 1CFO with/without γ_S that is insensitive to system size, 2CFO exhibits a clear system size dependence- while for Pb+Pb the χ^2/NDF is around 0-2, for smaller system size in p+Pb and p+p, the χ^2/NDF> 5 and larger than 1CFO+γ_S. This clearly shows a system size dependence of the preference for the freezeout scheme, while 2CFO is preferred in Pb+Pb, 1CFO+γ_S is preferred in p+Pb and p+p. We have further investigated the role of the missing resonances on strangeness freezeout across SPS to LHC beam energies.

1 Introduction

Freezeout is the latest phenomenon to occur prior to the detection of the produced hadrons in relativistic heavy ion collisions. This makes it imperative to decipher the nitty-gritty of the freezeout stage to probe the properties of the hadronic matter as well as the quark gluon plasma that is produced as a result of the collision. In general, various quantities could freezeout at different times: in this article we will discuss freezeout of the hadron multiplicity. This stage is usually referred to as chemical freezeout (CFO). An ideal gas of all known hadrons and resonances known as the hadron resonance gas (HRG) model has been amazingly successful in providing a description of CFO within a thermodynamic framework. For detailed information on the work presented here as well as references, we request the reader to Refs. [1–3] on which the work below is based upon.

In this article, we discuss the performance of several freezeout schemes with system size as well as the systematic uncertainties due to the missing resonances from the hadron spectrum on the model predictions [1–3]. A grand canonical ensemble (GCE) description of the thermal state of QCD medium

*e-mail: Sandeep.Chatterjee@fis.agh.edu.pl

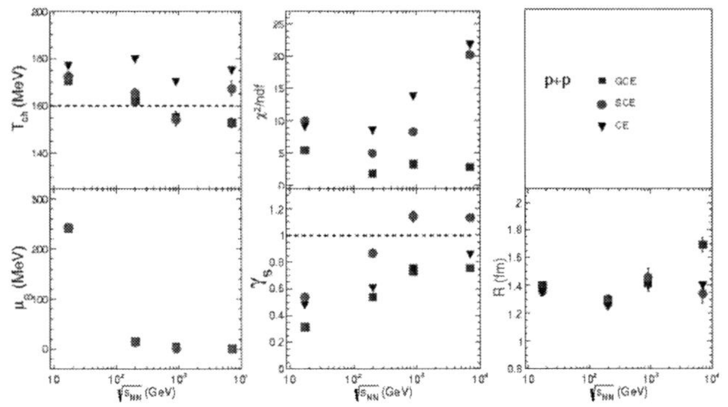

Figure 1. (Color online) A comparison of the performance of the HRG model fits to p+p data in different ensembles [1].

requires four parameters: temperature T and three chemical potentials μ_B, μ_Q and μ_S corresponding to the three conserved charges of QCD namely the baryon number B, electric charge Q and strangeness S respectively. Since phase space integrated hadron multiplicities are fitted, one also requires the volume V. Such a scheme where all hadrons freezeout together in the same thermal state is referred to as 1CFO. One often accounts for possible non-equilibrium production of strange hadrons by intoducing an additional parameter γ_S. We will refer to this as 1CFO+γ_S. Since freezeout is mainly a competition between hadron collisions and fireball expansion rates, it is possible to have flavor-dependent freezeout owing to flavor hierarchy in the hadron-hadron cross sections. The simplest version of such a multiple freezeout scenario is 2CFO where hadrons with strange valence quarks freezeout at a different thermal state as compared to those with only non-strange valence quarks.

An important aspect in the application of thermal models across different system sizes is the choice of the ensemble. It has been a standard practice to treat the mid-rapidity data of heavy-ion collisions in GCE while smaller collision systems like proton-heavy ion and proton-proton collisions have been treated in canonical ensemble (CE). While for 4π acceptance in data, it is clear that one has to work in the CE where the charges are conserved exactly, for analysis of a part of the phase space of the produced particles, e.g mid-rapidity data, the CE choice may not always be appropriate. The other aspect which we also investigate in this article are the systematic uncertainties in the HRG predictions due to our incomplete knowledge of the hadron spectrum.

2 Results

In Fig. 1 we show the comparative performance of the different ensembles in describing the mid-rapidity p+p data. Clearly at the LHC energies, GCE works much better than the other ensembles. Henceforth, we show results only in the GCE. In Fig. 2 we have plotted the system size dependence of the performance of different freezeout schemes as well the extracted best fitted freezeout parameters. It is interesting to note that there is a clear system size dependence of the preferred freezeout scheme-2CFO is preffered in Pb+Pb while 1CFO+γ_S is preferred in smaller systems like p+Pb and p+p.

Finally, in Fig. 3, we show the flavor hierarchy in freezeout parameters T and VT^3 obtained in heavy ion collisions in 2CFO freezeout scheme. We plot the beam energy depedence of the ratio T_S/T_{NS} and $V_S T_S^3/V_{NS} T_{NS}^3$. The plot suggests that flavor hierarchy in freezeout parameters is most prominent between $\sqrt{s_{NN}} \sim 10 - 100$ GeV. The results have been plotted for two different hadron

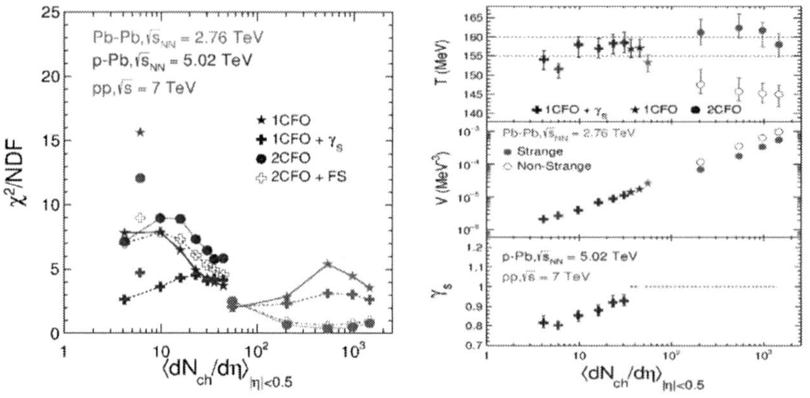

Figure 2. (Color online) Left: System size dependence of the goodness of fit in terms of χ^2/N_{df} compared over three different freezeout schemes- 1CFO, 1CFO+γ_S and 2CFO. Right: The thermal parameters corresponding to the best fits are shown [2].

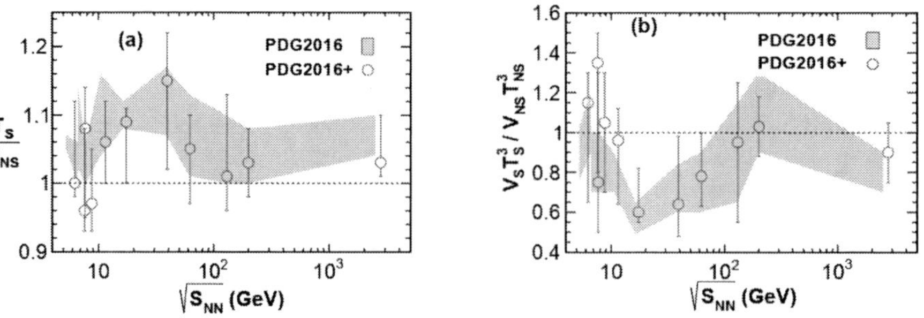

Figure 3. (Color online) The influence of the systematic uncertainties of the hadron spectrum on flavor hierarchy in thermal parameters [3].

spectra: PDG 2016 (that includes only the confirmed resonances from PDG 2016) and PDG 2016+ (that, in addition to PDG2016 also include the other resonacnes that have an unconfirmed status in PDG 2016). We find the extracted flavor hierarchy in T and VT^3 not influenced by such systematic uncertainties of the HRG model.

3 Conclusion

We have presented our recent works on various aspects of strangeness freezeout in relativistic collisions within the framework of the thermal model. GCE performs best in describing the mid-rapidity yields in small p+p collisions at the LHC. We showed that while early freezeout of strangeness is preferred in heavy ion collisions over unified freezeout, in small systems 1CFO with γ_S performs best. This could be due to the difference in hadronic interactions in small versus large systems. Finally, we also showed that the observed flavor hierarchy in heavy-ion collisions is stable under systematic uncertaintites due to the hadron spectrum which is an important input in thermal models.

References

[1] Sabita Das, Debadeepti Mishra, Sandeep Chatterjee and Bedangadas Mohanty, Phys. Rev. **C 95**, 014912 (2017)
[2] Sandeep Chatterjee, Ajay Kumar Dash and Bedangadas Mohanty, J. Phys. **G 44**, 105106 (2017)
[3] Sandeep Chatterjee, Debadeepti Mishra, Bedangadas Mohanty and Subhasis Samanta, Phys. Rev. **C 96**, 054907 (2017)

Rope Hadronization and Strange Particle Production

Christian Bierlich[1,⋆]

[1] *Department of Astronomy and Theoretical Physics, Lund University*
Sölvegatan 14A, Lund, Sweden.

Abstract. Rope Hadronization is a model extending the Lund string hadronization model to describe environments with many overlapping strings, such as high multiplicity pp collisions or AA collisions. Including effects of Rope Hadronization drastically improves description of strange/non-strange hadron ratios as function of event multiplicity in all systems from e^+e^- to AA. Implementation of Rope Hadronization in the MC event generators DIPSY and PYTHIA8 is discussed, as well as future prospects for jet studies and studies of small systems.

1 Introduction

The Lund string model for hadronization [1, 2] is based on the "string" picture of QCD, where the confined colour field between a $q\bar{q}$ pair moving apart is modelled as a classical string, with tension $\kappa \approx 1\text{GeV/fm}$. When it becomes energetically favorable for the string to break into smaller pieces, it breaks according to the Schwinger equation [3] (modified to take transverse degrees of freedom into account) with the probability:

$$\frac{d\mathcal{P}}{dp_\perp} \propto \kappa \exp\left(-\frac{\pi m_\perp^2}{\kappa}\right). \tag{1}$$

Directly from eq. (1), one obtains suppression of hadrons with strange quark content wrt. hadrons without strange quarks. Factorizing the $p_\perp$ dependence out, the suppression of strange quarks in the string breaking becomes:

$$\rho = \frac{\mathcal{P}_s}{\mathcal{P}_{u,d}} = \exp\left(-\frac{\pi(m_s^2 - m_u^2)}{\kappa}\right). \tag{2}$$

Taking eq. (2) at face value, inserting constituent quark masses of $m_u \approx 0.3$ GeV and $m_s \approx 0.5$ GeV, one obtains a suppression factor of $\rho \approx 0.08$, which is too small to describe data. In practise one therefore treats ρ as a free parameter, fitted to data from e^+e^- collisions [4], along with the rest of the parameters of the fragmentation model. Such sets of parameters are usually referred to as a "tune", taken as a hard constraint on parameter values when making predictions for other collision systems such a pp, pA or AA. One example is the Monash tune [5] for PYTHIA8 [6], from which it can be seen that the string model does a good job of described hadron flavours in an e^+e^- environment.

⋆e-mail: christian.bierlich@thep.lu.se. Report numbers: LU-TP 17-31, MCnet-17-15. Work done in collaboration with: J. Christiansen, G. Gustafson, L. Lönnblad and A. Tarasov. Work supported in part by the MCnetITN FP7 Marie Curie Initial Training Network, contract PITN-GA-2012-315877, and the Swedish Research Council (contracts 621-2012-2283, 621-2013-4287 and 2017-00347).

Turning to pp collisions, the picture is somewhat changed. In figure 1 (taken from ref. [7]), it is shown how the unmodified string picture does not hold even for inclusive ratios of identified particles, when emphasis is put on strangeness. The theoretical curves "DIPSY " and "PYTHIA8 def." represent unmodified string hadronization, "PYTHIA8 new" represents a modified colour reconnection scheme in PYTHIA8 described in ref. [8] (which shall not be mentioned further here), while "DIPSY rope" represents the Rope Hadronization model, which is the subject of this contribution to the proceedings.

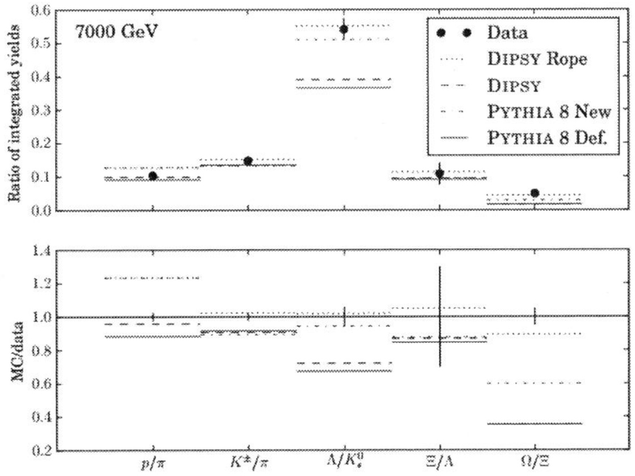

Figure 1. Default string hadronization and rope hadronization compared to inclusive ratios of identified particles in pp collisions at $\sqrt{s} = 7$ TeV.

2 Rope Hadronization

One crucial assumption behind string hadronization, as described in the previous section, is that strings are allowed to fragment independently of each other. It was noted early that in the dense environment of nuclear collisions, strings might act together coherently to form a stronger "rope" [9, 10], which would then hadronize with a larger, effective string tension. This idea has been followed by several authors [11–20], also for pp collisions. In ref. [21], the rope hadronization model was implemented in the parton shower event generator DIPSY [22] (using PYTHIA8 for hadronization), and in ref. [23] as a direct plug-in to the PYTHIA8 MC. The key point of the Rope Hadronization model, is the increase of local string tension. In the DIPSY and PYTHIA8 implementations this is calculated by calculating the transverse-space overlap of string pieces. Denoting the number of string pieces parallel with the breaking string p, and anti-parallel string pieces q, one finds that the effective string tension ($\tilde{\kappa}$) for a break-up becomes:

$$\tilde{\kappa} = \frac{2p + q + 2}{4}\kappa. \quad (3)$$

This result is confirmed by lattice calculations indicating that the string tension of an SU(3) multiplet structure scales with the secondary Casimir operator of the multiplet [24]. A key difference between this Rope Hadronization model and other earlier models, is that the effective string tension in eq. (3) is calculated as the difference of the available energy in the full multiplet and the multiplet reached by breaking one string in the rope (*i.e* replacing $p \to p - 1$). If one instead uses the full multiplet to calculate the effective string tension, effects become much too large. Replacing κ with $\tilde{\kappa}$ in eq. (2), the s-suppression becomes $\tilde{\rho} = \rho^{\kappa/\tilde{\kappa}}$, approaching unity as $\tilde{\kappa}$ grow large, when many strings overlap (in a pp collision at LHC, $\tilde{\rho}$ reaches values of up to $\approx 3\rho$).

One should note in particular the importance of multi-strange particles. Especially the ϕ-meson, consisting of two s-quarks, is important, as its mass is quite similar to the mass of the proton, making it an excellent laboratory for studying whether enhancement with increasing multiplicity is really due to strangeness enhancement, or if it is due to a mass effect, as indicated by some thermodynamical calculations [25].

3 Results

In figure 2, predicted enhancement for several strange and multi-strange hadrons as function of event multiplicity is shown. Results are shown both for the original Dipsy implementation, as well as the Pythia8 implementation (the latter for pp only). The figure can be compared to experimental results obtained by ALICE [26]. Rope Hadronization can be seen to describe strangeness in all systems from e^+e^- to PbPb well. As noted above, the ϕ/π ratio is expected to rise, even in pp. As this can serve as a potential discriminator between Rope Hadronization and thermodynamical calculations, the measurement is theoretically well motivated.

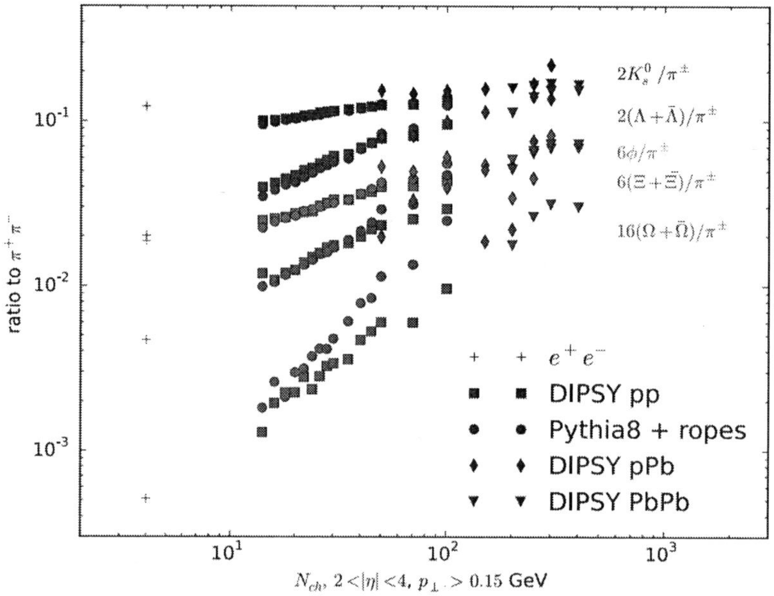

Figure 2. Ratios of strange particles to pions in e^+e^-, pp, pPb and PbPb collisions as function of multiplicity (centrality) with Rope Hadronization as implemented in the Pythia8 and Dipsy MC event generators.

4 Outlook

Strangeness enhancement has historically been seen as an important observable for QGP formation. In this contribution to the proceedings it has been shown that the Rope Hadronization mechanism – a microscopic model assuming neither a deconfined nor thermalized plasma – can describe strangeness in collision systems from e^+e^- to AA. The question which begs to be answered is if such a microscopic model can describe other features usually ascribed to plasma formation, such as flow and jet quenching. Regarding flow, ongoing efforts [27] aim to investigate whether the transverse pressure gradient which is naturally built up due to string overlaps, can be translated into flow. The effects of Rope Hadronization on jets in pp collisions is also worth investigating further [28], though the Rope Hadronization model is develeoped primarily with soft effects in mind. Finally it is worth mentioning the prospects of looking further into effects even smaller systems, such as central diffractive pp collisions or e^+e^- collisions at an FCC-ee machine [29]. The absence of a coloured initial state in the latter makes it a valuable precision tool for determining whether collective effects arise from initial state correlations or final state interactions.

References

[1] B. Andersson, G. Gustafson, B. Söderberg, Z. Phys. **C20**, 317 (1983)
[2] B. Andersson, G. Gustafson, Z. Phys. **C3**, 223 (1980)
[3] J.S. Schwinger, Phys. Rev. **82**, 664 (1951)
[4] K. Hamacher, M. Weierstall (DELPHI) (1995), hep-ex/9511011
[5] P. Skands, S. Carrazza, J. Rojo, Eur. Phys. J. **C74**, 3024 (2014), 1404.5630
[6] T. Sjöstrand, S. Ask, J.R. Christiansen, R. Corke, N. Desai, P. Ilten, S. Mrenna, S. Prestel, C.O. Rasmussen, P.Z. Skands, Comput. Phys. Commun. **191**, 159 (2015), 1410.3012
[7] C. Bierlich, J.R. Christiansen, Phys. Rev. **D92**, 094010 (2015), 1507.02091
[8] J.R. Christiansen, P.Z. Skands, JHEP **08**, 003 (2015), 1505.01681
[9] T.S. Biro, H.B. Nielsen, J. Knoll, Nucl. Phys. **B245**, 449 (1984)
[10] A. Białas, W. Czyz, Phys. Rev. **D31**, 198 (1985)
[11] A.K. Kerman, T. Matsui, B. Svetitsky, Phys. Rev. Lett. **56**, 219 (1986)
[12] M. Gyulassy, A. Iwazaki, Phys. Lett. **165B**, 157 (1985)
[13] B. Andersson, P.A. Henning, Nucl. Phys. **B355**, 82 (1991)
[14] M. Braun, C. Pajares, Nucl. Phys. **B390**, 542 (1993)
[15] M. Braun, C. Pajares, Phys. Rev. **D47**, 114 (1993)
[16] N.S. Amelin, M.A. Braun, C. Pajares, Z. Phys. **C63**, 507 (1994)
[17] N. Armesto, M.A. Braun, E.G. Ferreiro, C. Pajares, Phys. Lett. **B344**, 301 (1995)
[18] H. Sorge, M. Berenguer, H. Stoecker, W. Greiner, Phys. Lett. **B289**, 6 (1992)
[19] M. Bleicher, W. Greiner, H. Stoecker, N. Xu, Phys. Rev. **C62**, 061901 (2000), hep-ph/0007215
[20] S. Soff, J. Randrup, H. Stoecker, N. Xu, Phys. Lett. **B551**, 115 (2003), nucl-th/0209093
[21] C. Bierlich, G. Gustafson, L. Lönnblad, A. Tarasov, JHEP **03**, 148 (2015), 1412.6259
[22] C. Flensburg, G. Gustafson, L. Lönnblad, A. Ster, JHEP **06**, 066 (2011), 1103.4320
[23] C. Bierlich, PoS **DIS2016**, 051 (2016), 1606.09456
[24] G.S. Bali, Phys. Rev. **D62**, 114503 (2000), hep-lat/0006022
[25] V. Vislavicius, A. Kalweit (2016), 1610.03001
[26] J. Adam et al. (ALICE), Nature Phys. **13**, 535 (2017), 1606.07424
[27] C. Bierlich, G. Gustafson, L. Lönnblad (2016), 1612.05132
[28] M.L. Mangano, B. Nachman (2017), 1708.08369
[29] C. Bierlich, *Colour reconnections in pp collisions*, in *Proceedings, Parton Radiation and Fragmentation from LHC to FCC-ee: CERN, Geneva, Switzerland, November 22-23, 2016* (2017), pp. 149–154

Baryon spectra and antiparticle-to-particle ratios from the improved AMPT model

Yuncun He[1,2,⋆] and *Zi-Wei* Lin[2,3,⋆⋆]

[1] *Faculty of Physics and Electronic Technology, Hubei University, Wuhan 430062, China*
[2] *Department of Physics, East Carolina University, Greenville, North Carolina 27858, USA*
[3] *Key Laboratory of Quarks and Lepton Physics (MOE) and Institute of Particle Physics, Central China Normal University, Wuhan 430079, China*

Abstract. The current version of a multi-phase transport (AMPT) model with string melting can reasonably describe the dN/dy yields, p_T spectra and anisotropic flows of pions and kaons at low p_T in heavy ion collisions at RHIC and LHC energies, although it failed to reproduce the dN/dy and p_T spectra of baryons. In this work, we improve the quark coalescence mechanism in AMPT by removing the forced separate number conservations of mesons, baryons and antibaryons in each event. We find that the improved AMPT model can better describe the yields at midrapidity, the p_T spectra and elliptic flow of low-p_T baryons in comparison with the experimental data. Antiparticle-to-particle ratios of strange baryons are also significantly improved.

1 Introduction

The original string melting version of AMPT [1], which converts all the excited strings to partons before parton cascade and hadronizes with the quark coalescence mechanism, could fit the elliptic flow and two-pion interferometry at RHIC very well, but it failed to reproduce the p_T spectra and rapidity distributions of hadrons. With a new set of key parameters, the string melting AMPT model was improved to reasonably reproduce the dN/dy yields, p_T spectra and elliptic flows of pions and kaons at low p_T in heavy ion collisions at $\sqrt{s_{NN}} = 200$ GeV and 2.76 TeV [2]. It has also been used to predict the identified particles dN/dy, p_T spectra, azimuthal anisotropies and longitudinal correlations in the Pb+Pb collisions at $\sqrt{s_{NN}} = 5.02$ TeV [3]. However, it still has some deficiencies on baryon descriptions. It overestimates the proton yield at midrapidity and underestimates the slope of the proton p_T spectra significantly. In addition, the antiparticle-to-particle ratios from the current AMPT model are often above unity for strange baryons, contrary to naive expectations or the data.

The coalescence in the current AMPT model forces the numbers of mesons, baryons and antibaryons in an event to be conserved separately. The model first searches partners for quarks from the melting of mesons to form mesons, then searches partners for the remaining quarks or antiquarks to form baryons or antibaryons. In this study [4], we remove the artificial separate conservation of the numbers of mesons, baryons and antibaryons in each event; while the conservations of the net

⋆e-mail: hyuncun@foxmail.com
⋆⋆e-mail: linz@ecu.edu

baryons and net strangeness are still satisfied. Each (anti)quark now has the freedom to form a meson or a (anti)baryon depending on the distances between the coalescing partners. A new constant parameter r_{BM}, which controls the relative probability of a quark forming a baryon versus a meson, is introduced and its value has been determined by fitting the experimental data [4]. However, note that in principle this parameter should be determined from the hadron wavefunctions. In this new AMPT model, we use the Lund string fragmentation parameters $a = 0.55$ for Au+Au collisions at 200 GeV and $a = 0.2$ for Pb+Pb collisions at 2.76 TeV respectively, and $b = 0.15$ GeV^{-2} for both collisions. The relative production of strange to nonstrange quarks from Lund string fragmentation is still kept 0.4 and the strong coupling constant is set to $\alpha_s = 0.33$ as in earlier studies [2, 3], while the parton cross section is set to 1.5 mb instead of 3 mb in those earlier studies.

2 Results for protons and antiprotons

Figure 1 shows the rapidity distributions (left panel), p_T spectra at midrapidity (middle panel), and v_2 as function of p_T (right panel) of protons and antiprotons in Au+Au collisions at 200 GeV. Figure 2 shows the results for Pb+Pb collisions at 2.76 TeV. The experimental data are from PHENIX [5, 6] and ALICE [7, 8] respectively, and both of them have included the corrections from weak decays. The thin curves represent results produced by the old quark coalescence, which gives higher yields at midrapidity (particularly at LHC) and too soft p_T spectra at both RHIC and LHC. The thick curves represent results produced by the new quark coalescence, which gives lower yields and harder p_T spectra and is thus more consistent with the data. The p_T dependence of v_2 from the new coalescence shows little change; this is expected since the elliptic flow is a normalized variable. We have checked that the new model also reasonably describes the dN/dy, p_T spectra and elliptic flow of low-p_T pions and kaons. Thus the string melting AMPT model with the new quark coalescence provides a much better description of the bulk matter in high-energy heavy ion collisions.

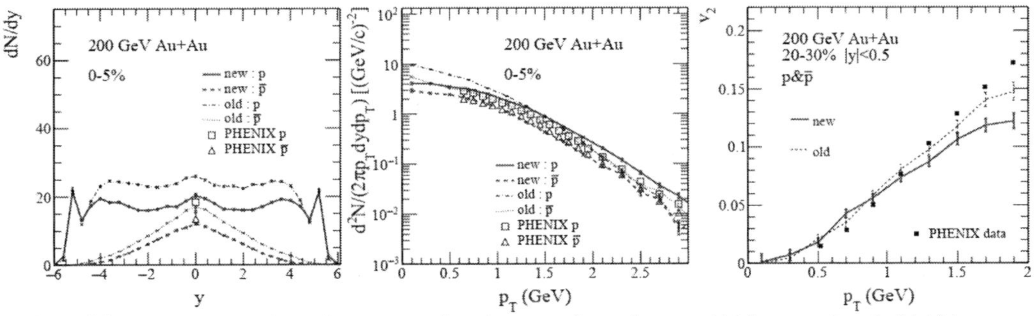

Figure 1. dN/dy, p_T spectra and v_2 of protons and antiprotons from the new (thick curves) and old (thin curves) quark coalescence for Au+Au collisions at 200 GeV in comparison with the PHENIX data.

3 Results for antiparticle-to-particle ratios

Figure 3 shows the antiparticle-to-particle ratios of π^-/π^+, K^-/K^+, $\bar{n}/n$, $\bar{p}/p$, $\bar{\Sigma}^+/\Sigma^-$, $\bar{\Lambda}/\Lambda$, $\bar{\Sigma}^-/\Sigma^+$, $\bar{\Xi}^+/\Xi^-$, $\bar{\Xi}^0/\Xi^0$and $\bar{\Omega}/\Omega$ around midrapidity for central Au+Au collisions at 200 GeV (left panel) and Pb+Pb collisions at 2.76 TeV (right panel) in comparison with the experimental data [7, 9–13]. We see that the results from the new quark coalescence (solid curves) are generally more consistent with the experimental data than results from the old quark coalescence (dashed curves). In particular, the

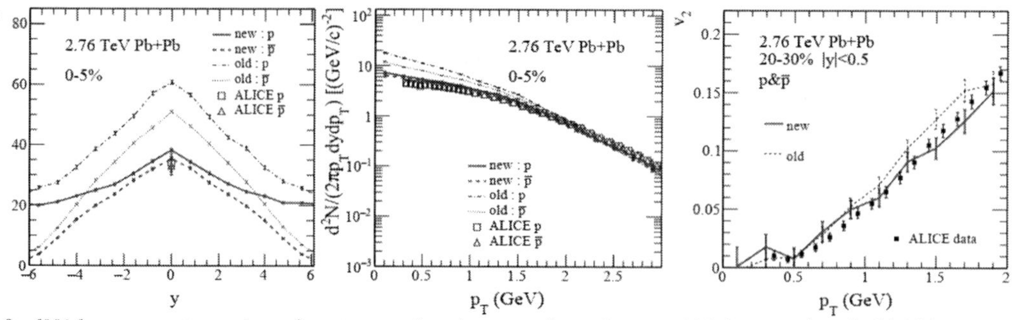

Figure 2. dN/dy, p_T spectra and v_2 of protons and antiprotons from the new (thick curves) and old (thin curves) quark coalescence for Pb+Pb collisions at 2.76 TeV in comparison with the ALICE data.

antiparticle-to-particle ratios of multi-strange baryons $\bar{\Xi}^+/\Xi^-$, $\bar{\Xi}^0/\Xi^0$ and $\bar{\Omega}^+/\Omega^-$ are much lower than before and now agree well with the data. We also see that the new ratios at 2.76 TeV are closer to unity than those at 200 GeV, since baryons and antibaryons are more symmetric in higher energy collisions.

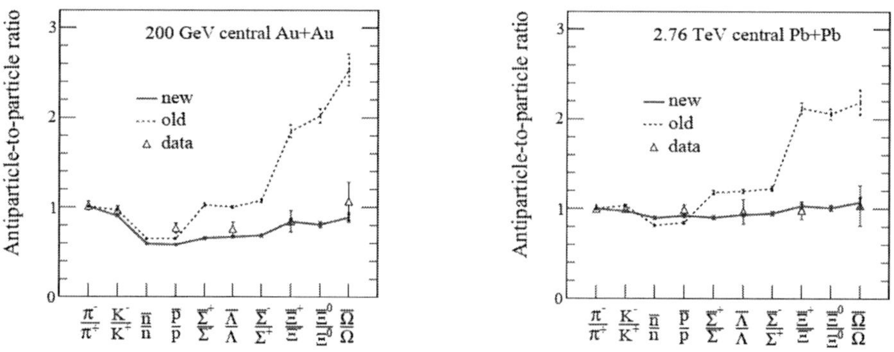

Figure 3. Antiparticle-to-particle ratios around midrapidity from the new (solid curves) and old (dashed curves) quark coalescence for Au+Au collisions at 200 GeV and Pb+Pb collisions at 2.76 TeV in comparison with data.

Figure 4 shows the antibaryon-to-baryon ratios around midrapidity in central collisions as functions of energy, including the results for Au+Au collisions at 62.4 GeV, 130 GeV, 200 GeV and Pb+Pb collisions at 2.76 TeV and 5.02 TeV. Note that we have used the Lund string parameter $a = 3.2$ for 62.4 GeV and $a = 1.65$ for 130 GeV. We see that the results from the new quark coalescence (left panel) are generally more consistent with the experimental data than the old results (right panel), especially for the ratios of multistrange baryons. Overall, the ratios from the new quark coalescence for all these baryons gradually approach unity as the colliding energy increases, as we would expect.

The new quark coalescence is much more physical, however some problems still exist. For example, the antiproton-to-proton ratios are well below the data at RHIC and LHC as shown in Figure 3. Further improvements, such as the implementation of energy-momentum conservation in quark coalescence, may be necessary.

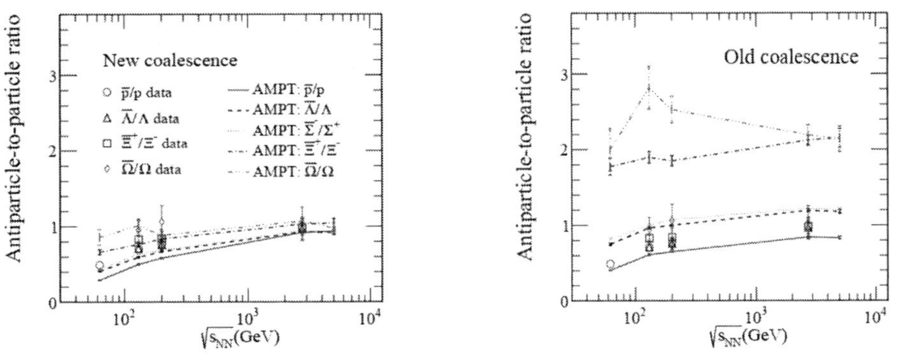

Figure 4. Antibaryon-to-baryon ratios around midrapidity as functions of collision energies from the new (left panel) and old (right panel) quark coalescence in comparison with data.

4 Conclusion

We have improved the quark coalescence component in the string melting version of a multi-phase transport model AMPT. In particular, we have removed the artificial separate conservation of the numbers of mesons, baryons, and antibaryons in an event, where only the conservation of the net-baryon number in an event is needed for the coalescence process. In the new quark coalescence, a quark now has the freedom to form either a baryon or a meson, depending on the distance to its coalescence partner(s). We have compared the results with the experimental data and shown that the new model can reasonably describe the dN/dy, p_T spectra, and v_2 of pions, kaons, and most (anti)baryons. Significant improvements are also made on the antibaryon-to-baryon ratios of Ξ and Ω.

We thank support from the NSFC of China under Grants No. 11628508, No. 11547016 and No. 11447190.

References

[1] Z. W. Lin, C. M. Ko, B. A. Li, B. Zhang and S. Pal, Phys. Rev. C **72**, 064901 (2005)
[2] Z. W. Lin, Phys. Rev. C **90**, no. 1, 014904 (2014)
[3] G. L. Ma and Z. W. Lin, Phys. Rev. C **93**, no. 5, 054911 (2016)
[4] Y. He and Z. W. Lin, Phys. Rev. C **96**, no. 1, 014910 (2017)
[5] S. S. Adler *et al.* [PHENIX Collaboration], Phys. Rev. C **69**, 034909 (2004)
[6] A. Adare *et al.* [PHENIX Collaboration], Phys. Rev. C **93**, no. 5, 051902 (2016)
[7] B. Abelev *et al.* [ALICE Collaboration], Phys. Rev. C **88**, 044910 (2013)
[8] B. B. Abelev *et al.* [ALICE Collaboration], JHEP **1506**, 190 (2015)
[9] J. Adams *et al.* [STAR Collaboration], Phys. Rev. Lett. **98**, 062301 (2007)
[10] C. Suire [STAR Collaboration], Nucl. Phys. A **715**, 470 (2003)
[11] B. I. Abelev *et al.* [STAR Collaboration], Phys. Rev. C **79**, 034909 (2009)
[12] S. Schuchmann, Ph.D. thesis, Goethe University Frankfurt, 2015
[13] B. B. Abelev *et al.* [ALICE Collaboration], Phys. Lett. B **728**, 216 (2014) Erratum: [Phys. Lett. B **734**, 409 (2014)]

Baryons in the plasma: In-medium effects and parity doubling

Gert Aarts[1,⋆], *Chris* Allton[1], *Davide* de Boni[1], *Simon* Hands[1], *Benjamin* Jäger[2], *Chrisanthi* Praki[1], and *Jon-Ivar* Skullerud[3,4]

[1] *Department of Physics, College of Science, Swansea University, Swansea SA2 8PP, United Kingdom*
[2] *Institute for Theoretical Physics, ETH Zürich, CH-8093 Zürich, Switzerland*
[3] *Department of Theoretical Physics, National University of Ireland Maynooth, County Kildare, Ireland 4*
[4] *School of Mathematics, Trinity College Dublin, Dublin 2, Ireland*

Abstract. We investigate the fate of baryons made out of u, d and s quarks in the hadronic gas and the quark-gluon plasma, using nonperturbative lattice simulations, employing the FASTSUM anisotropic $N_f = 2+1$ ensembles. In the confined phase a strong temperature dependence is seen in the masses of the negative-parity groundstates, while the positive-parity groundstate masses are approximately temperature independent, within the error. At high temperature parity doubling emerges. A noticeable effect of the heavier s quark is seen. We give a simple description of the medium-dependent masses for the negative-parity states and speculate on the relevance for heavy-ion phenomenology via the hadron resonance gas.

1 Introduction

The behaviour of hadrons under the extreme conditions of nonzero temperature and/or density is a longstanding question relevant for e.g. heavy-ion collisions and neutrons stars, as well as for the fundamental aspects of chiral symmetry and confinement. Due to the nonperturbative nature of QCD, one has to rely on either models or first-principle studies using the lattice discretisation. While mesons at finite temperature have been fairly well studied on the lattice, in various contexts (see e.g. Ref. [1] for an overview), for baryons this is not the case. In fact, there are only a few lattice studies of baryonic thermal screening [2, 3] and temporal [4] masses. Here we report on our work on baryons [1, 5, 6], using lattice QCD simulations on the FASTSUM anisotropic $N_f = 2 + 1$ ensembles [7, 8].

We are particularly interested in (the absence of) parity doubling: in the presence of unbroken chiral symmetry, it can be shown that positive- and negative-parity baryonic channels are degenerate, and hence there is parity doubling. When chiral symmetry is (spontaneously) broken, this is no longer the case and indeed, in vacuum the positive-parity groundstate is typically lighter than the negative-parity one. For the nucleon, one finds e.g. $m_+ \equiv m_N = 939$ MeV while $m_- \equiv m_{N^*} = 1535$ MeV. Since chiral symmetry is restored around the deconfinement transition, one expects a degeneracy to emerge. Here we address *how* this degeneracy emerges, which is a nonperturbative, dynamical question, not answerable using symmetry considerations alone.

⋆Speaker, e-mail: g.aarts@swan.ac.uk

2 From low to high temperature

The FASTSUM collaboration uses highly anisotropic lattices, with $a_\tau/a_s \ll 1$, which are constructed specifically for spectral studies of QCD at nonzero temperature. The temperature is varied by changing the number of time slices N_τ, via the standard relation, $T = 1/(a_\tau N_\tau)$, i.e., we use a fixed-scale approach. We have generated several ensembles, four below and four above the deconfinement transition, see Table 1. Tuning of the lattice parameters and also the "zero-temperature" ensemble ($N_\tau = 128$) were kindly provided by the HadSpec collaboration [9]. The crossover temperature, denoted with T_c, is determined via the inflection point of the renormalised Polyakov loop and is found to be $T_c = 185(4)$ MeV. This is higher than in nature, due to the light quarks not having their physical masses ($m_\pi = 384(4)$ MeV). The strange quark mass has its physical value, though. For further details, we refer to Refs. [7, 8].

Table 1. Number of time slices N_τ and the corresponding temperature, in units of T_c. Also shown are the number of configurations and sources. The spatial lattice size is 24^3, with $a_s = 0.1227(8)$ fm and $a_s/a_\tau = 3.5$.

N_τ	128	40	36	32	28	24	20	16
T/T_c	0.24	0.76	0.84	0.95	1.09	1.27	1.52	1.90
$N_{\rm cfg}$	139	501	501	1000	1001	1001	1000	1001
$N_{\rm src}$	16	4	4	2	2	2	2	2

We computed all octet (spin 1/2) and decuplet (spin 3/2) baryon correlators, for both positive and negative parity. Results in the N, Δ and Ω channels can be found in Ref. [1]; the others (Λ, Σ, Σ^*, Ξ, Ξ^*) will appear in Ref. [6]. In the confined phase, we find that the correlators can be described by combinations of exponentials, allowing us to determine the groundstate masses $m_\pm$ as a function of the temperature. These results are summarised in Fig. 1, where we show $m_\pm(T)$ in the various channels, normalised with m_+ at the lowest temperature.

Several observations can be made. The positive-parity masses are largely temperature independent. A slight increase and subsequent drop close to T_c can be seen, but it is not significant within current errors. The negative-parity masses on the other hand drop in all channels in a similar way, and

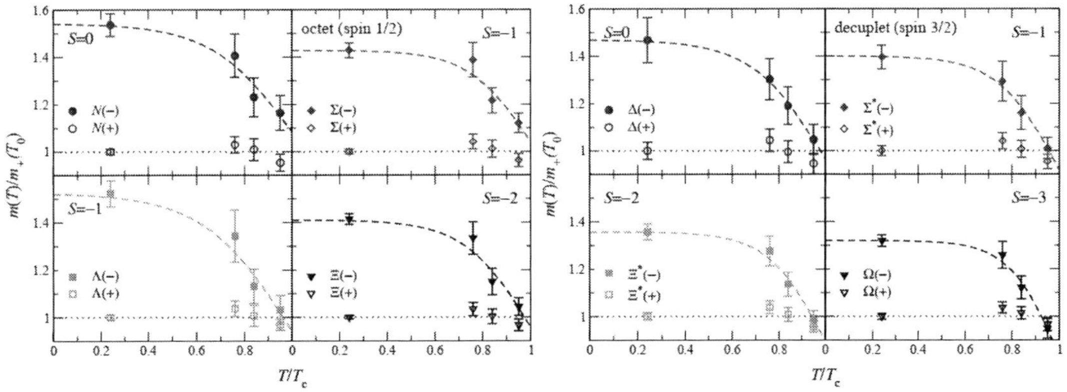

Figure 1. Temperature dependence of the groundstate masses, normalised with m_+ at the lowest temperature, $m_\pm(T)/m_+(T_0)$, in the hadronic phase, for octet (left) and decuplet (right) baryons. Positive- (negative-) parity masses are indicated with open (closed) symbols.

 https://doi.org/10.1051/epjconf/201817114005

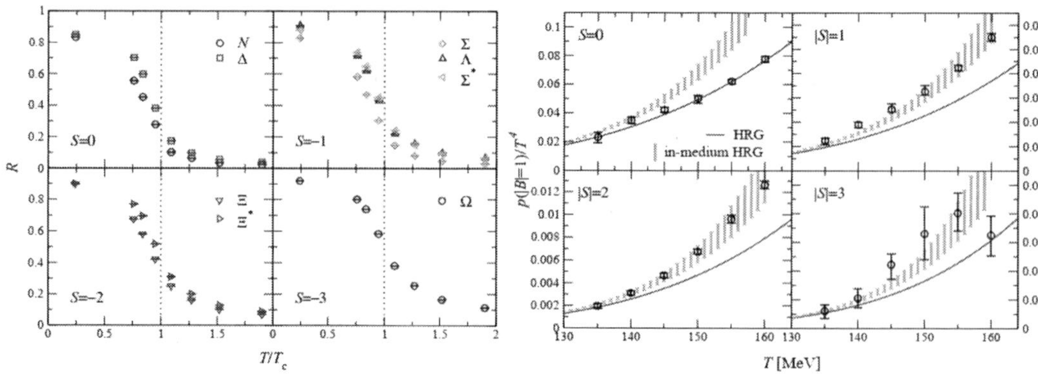

Figure 2. Left: temperature dependence of the R ratio, see Eq. (2). Right: contributions to the normalised pressure p/T^4 from the sectors with baryon number $|B| = 1$ and strangeness $|S| = 0, 1, 2, 3$, according to the lattice data of Ref. [14], the ideal hadron resonance gas (HRG) and the in-medium HRG, using temperature-dependent negative-parity groundstate masses (1), with $\gamma = 0.3$, $1 < m_-(T_c)/m_+(0) < 1.1$ and $T_c = 155$ MeV.

become near-degenerate with the corresponding positive-parity mass close to T_c. The mass splittings at low temperature are around 500~600 MeV, in agreement with nature. The dashed lines show a fit using a simple Ansatz, interpolating between $m_-(0)$ and $m_-(T_c)$,

$$m_-(T) = w(T,\gamma)m_-(0) + [1 - w(T,\gamma)]\, m_-(T_c), \tag{1}$$

with the transition function $w(T,\gamma) = \tanh[(1 - T/T_c)/\gamma] / \tanh(1/\gamma)$, such that $w(0,\gamma) = 1$ and $w(T_c,\gamma) = 0$. Small (large) γ corresponds to a narrow (broad) transition region, which is expected to depend on the masses of the light quarks. We have carried out fits in each of the 8 channels and find $0.22 \lesssim \gamma \lesssim 0.35$ and $0.85 \lesssim m_-(T_c)/m_+(0) \lesssim 1.1$. The largest uncertainty resides in $m_-(T_c)$, since it assumes that the concept of a well-defined groundstate at or close to T_c remains sensible.

To summarise our findings as $T \to T_c$, we have confirmed that parity doubling indeed emerges, as expected, and established that the degeneracy develops from a reduction of the negative-parity masses with temperature, while the positive-parity masses remain approximately constant. We find this to be similar in all the channels studied. Effective parity-doublet models [10] introduce a chirally invariant component to baryon masses to model this type of behaviour and our findings can be used to assess or constrain those models. Some recent work along these lines can be found in Refs. [11–13].

We now turn to the deconfined phase. At our first temperature above T_c, $T/T_c = 1.09$, no clearly identifiable boundstates appear to be present and e.g. exponential fits no longer describe the data [1]. Hence we focus on the signal for parity doubling directly in the correlators $G_\pm$, using the ratio [4, 5]

$$R(\tau) = \frac{G_+(\tau) - G_-(\tau)}{G_+(\tau) + G_-(\tau)}, \qquad\qquad R = \frac{\sum_n R(\tau_n)/\sigma^2(\tau_n)}{\sum_n 1/\sigma^2(\tau_n)}. \tag{2}$$

It should be noted that $G_\pm$ are related via $G_\pm(\tau) = -G_\mp(1/T - \tau)$ and hence $R(1/T - \tau) = -R(\tau)$. In the case of parity doubling, $R(\tau) = 0$, while in absence of parity doubling and with clearly separated groundstates satisfying $m_- \gg m_+$, $R(\tau) = 1$. By summing over the temporal lattice points as in Eq. (2), where σ denotes the error, we can construct the quasi-order parameter R, which equals 0 in the case of parity doubling and 1 in the case of single states satisfying $m_- \gg m_+$.

The results for R as a function of temperature are shown in Fig. 2 (left), organised by strangeness. We note that R indeed changes from 1 to 0 as T increases and that the crossover behaviour coincides with the deconfinement transition. For baryons with larger strangeness, R is still distinct from 0 at the highest temperatures studied. Since the strange quark mass breaks chiral symmmetry explicitly, its effect is expected to disappear eventually as $m_s/T \to 0$. We hence conclude that the interpretation of parity doubling due to chiral symmetry restoration in the baryon sector is indeed valid.

Finally, we consider an application and speculate about the relevance for heavy-ion phenomenology. The distinct medium-dependent masses we found in the confined phase are easily implemented in the hadron resonance gas (HRG). In the usual (ideal) HRG, vacuum masses are used, see e.g. Ref. [14] and references therein. In our in-medium modification, we keep the positive-parity masses fixed, but reduce the negative-parity groundstate masses. Concretely, we use the PDG2016 baryon masses classified with 3 and 4 stars, up to 2.5 GeV, and for the in-medium modification (1) we take $\gamma = 0.3$, vary $1 < m_-(T_c)/m_+(0) < 1.1$ (the effect of varying γ is negligible) and $T_c = 155$ MeV [15]. The results for the baryonic contributions to the pressure are shown in Fig. 2 (right), again organised by strangeness. The ideal and in-medium HRG are compared to the lattice data of Ref. [14]. While in the $S = 0$ sector, the ideal HRG describes the data very well, in the $|S| = 1, 2, 3$ sectors the in-medium HRG leads to an improvement compared to the ideal one. It would hence be interesting to see how this effect compares with other modifications of the HRG that are currently under consideration [14, 16].

Acknowledgments. We thank Claudia Ratti for providing the lattice data shown in Fig. 2 (right). We are grateful for support from STFC, SNF, ICHEC, the Royal Society, the Wolfson Foundation and COST Action CA15213 THOR. Computing resources were made available by HPC Wales and the STFC-funded DiRAC HPC Facility, which is part of the National E-Infrastructure.

References

[1] G. Aarts, C. Allton, D. De Boni, S. Hands, B. Jäger, C. Praki and J. I. Skullerud, JHEP **1706** (2017) 034 [arXiv:1703.09246 [hep-lat]].
[2] C. E. Detar and J. B. Kogut, Phys. Rev. Lett. **59** (1987) 399; Phys. Rev. D **36** (1987) 2828.
[3] I. Pushkina *et al.* [QCD-TARO Collaboration], Phys. Lett. B **609** (2005) 265 [hep-lat/0410017].
[4] S. Datta, S. Gupta, M. Padmanath, J. Maiti and N. Mathur, JHEP **1302** (2013) 145 [arXiv:1212.2927 [hep-lat]].
[5] G. Aarts, C. Allton, S. Hands, B. Jäger, C. Praki and J. I. Skullerud, Phys. Rev. D **92** (2015) no.1, 014503 [arXiv:1502.03603 [hep-lat]].
[6] G. Aarts, C. Allton, D. De Boni, S. Hands, B. Jäger and J. I. Skullerud, in preparation.
[7] G. Aarts, C. Allton, T. Harris, S. Kim, M. P. Lombardo, S. M. Ryan and J. I. Skullerud, JHEP **1407** (2014) 097 [arXiv:1402.6210 [hep-lat]].
[8] G. Aarts, C. Allton, A. Amato, P. Giudice, S. Hands and J. I. Skullerud, JHEP **1502** (2015) 186 [arXiv:1412.6411 [hep-lat]].
[9] R. G. Edwards, B. Joo and H. W. Lin, Phys. Rev. D **78** (2008) 054501 [arXiv:0803.3960 [hep-lat]].
[10] C. E. Detar and T. Kunihiro, Phys. Rev. D **39** (1989) 2805.
[11] Y. Motohiro, Y. Kim and M. Harada, Phys. Rev. C **92** (2015) no.2, 025201 Erratum: [Phys. Rev. C **95** (2017) no.5, 059903] [arXiv:1505.00988 [nucl-th]].
[12] A. Mukherjee, S. Schramm, J. Steinheimer and V. Dexheimer, arXiv:1706.09191 [nucl-th].
[13] C. Sasaki, arXiv:1707.05081 [hep-ph].
[14] P. Alba *et al.*, Phys. Rev. D **96** (2017) no.3, 034517 [arXiv:1702.01113 [hep-lat]].
[15] S. Borsanyi *et al.* [Wuppertal-Budapest Collaboration], JHEP **1009** (2010) 073 [arXiv:1005.3508 [hep-lat]].
[16] V. Vovchenko, M. I. Gorenstein and H. Stoecker, Phys. Rev. Lett. **118** (2017) no.18, 182301 [arXiv:1609.03975 [hep-ph]].

van der Waals Interactions and Hadron Resonance Gas: Role of resonance widths modeling on conserved charges fluctuations

Volodymyr Vovchenko[1,2,⋆], *Paolo* Alba[2], *Mark I.* Gorenstein[2,3], and *Horst* Stoecker[1,2,4]

[1]*Institut für Theoretische Physik, Goethe Universität Frankfurt, D-60438 Frankfurt am Main, Germany*
[2]*Frankfurt Institute for Advanced Studies, Giersch Science Center, D-60438 Frankfurt am Main, Germany*
[3]*Bogolyubov Institute for Theoretical Physics, 03680 Kiev, Ukraine*
[4]*GSI Helmholtzzentrum für Schwerionenforschung GmbH, D-64291 Darmstadt, Germany*

Abstract. The quantum van der Waals (QvdW) extension of the ideal hadron resonance gas (HRG) model which includes the attractive and repulsive interactions between baryons – the QvdW-HRG model – is applied to study the behavior of the baryon number related susceptibilities in the crossover temperature region. Inclusion of the QvdW interactions leads to a qualitatively different behavior of susceptibilities, in many cases resembling lattice QCD simulations. It is shown that for some observables, in particular for χ_{11}^{BQ}/χ_2^B, effects of the QvdW interactions essentially cancel out. It is found that the inclusion of the finite resonance widths leads to an improved description of χ_2^B, but it also leads to a worse description of χ_{11}^{BQ}/χ_2^B, as compared to the lattice data. On the other hand, inclusion of the extra, unconfirmed baryons into the hadron list leads to a simultaneous improvement in the description of both observables.

1 Introduction

Lattice QCD simulations provide the equation of state of strongly interacting matter at zero net baryon density [1, 2]. A smooth crossover-type transition between hadronic and partonic matter is observed [3]. A common model for the hadronic phase – the ideal hadron resonance gas (IHRG) model – successfully describes many lattice observables at lower temperatures, $T \sim 100 - 150$ MeV. Agreement of the IHRG model with lattice data rapidly breaks down at $T \simeq 150 - 160$ MeV for fluctuations and correlations of conserved charges [4]. Sometimes, this breakdown was interpreted as a signal for deconfinement [4, 5].

On the other hand, it has recently been shown that the onset of deviations of IHRG from the lattice data can well be explained by the QvdW-type interactions between baryons [6]. In this work we assess simultaneous effects of QvdW interactions, the modeling of the finite widths of the resonances, and also the HRG hadron list. The importance of the potentially missing hadron states in the hadron list was pointed out in several recent publications, in the context of the IHRG model [7–10] and also in the context of the excluded volume (EV) HRG model [11], while the effects due to the finite widths of the resonances were barely discussed at all. None of the two effects were studied simultaneously with the presence of the QvdW interactions.

⋆e-mail: vovchenko@fias.uni-frankfurt.de

2 Model

QvdW extension. The Quantum van der Waals extension of the HRG model of Ref. [6] – the QvdW-HRG model – is based on the following assumptions:

1. QvdW interactions exist between all pairs of baryons and between all pairs of antibaryons. The QvdW parameters a and b for all (anti)baryons taken to be equal to those of nucleons, as obtained from the fit to the ground state of nuclear matter [12]: $a \simeq 329$ MeV fm^3 and $b \simeq 3.42$ fm^3.

2. The baryon-antibaryon, meson-meson, and meson-(anti)baryon QvdW interactions are neglected. Note that the model still contains, by construction, the meson-related hadronic interactions that lead to the formation of resonances.

The QvdW-HRG model contains basic nuclear matter physics, in contrast to IHRG, and it yields the liquid-gas first-order phase transition in the symmetric nuclear matter with a critical point located at $T_c \simeq 19.7$ MeV and $\mu_c \simeq 908$ MeV ($n_c \simeq 0.07$ fm$^3 = 0.45\, n_0$).

The QvdW-HRG consists of three sub-systems: Ideal gas of mesons, QvdW gas of baryons, and QvdW gas of antibaryons. The total pressure reads $p(T,\boldsymbol{\mu}) = p_M(T,\boldsymbol{\mu}) + p_B(T,\boldsymbol{\mu}) + p_{\bar{B}}(T,\boldsymbol{\mu})$ with

$$p_M(T,\boldsymbol{\mu}) = \sum_{j\in M} p_j^{\rm id}(T,\mu_j), \qquad p_{B(\bar{B})}(T,\boldsymbol{\mu}) = \sum_{j\in B(\bar{B})} p_j^{\rm id}(T,\mu_j^{B(\bar{B})*}) - a\, n_{B(\bar{B})}^2, \tag{1}$$

where M stands for mesons, $B(\bar{B})$ for (anti)baryons, $\boldsymbol{\mu} = (\mu_B, \mu_S, \mu_Q)$ are the chemical potentials for net baryon number B, strangeness S, and electric charge Q, $\mu_j^{B(\bar{B})*} = \mu_j - b\, p_{B(\bar{B})} - a\, b\, n_{B(\bar{B})}^2 + 2\, a\, n_{B(\bar{B})}$, $\mu_j = B_j \mu_B + S_j \mu_S + Q_j \mu_Q$ is the chemical potential for hadron species j, with B_j, S_j, and Q_j being its corresponding quantum numbers. n_B and $n_{\bar{B}}$ are total densities of baryons and antibaryons.

The calculation of mesonic pressure $p_M(T,\boldsymbol{\mu})$ is straightforward. The shifted chemical potentials $\mu_j^{B(\bar{B})*}$ of (anti)baryons depend explicitly on (anti)baryon pressure $p_{B(\bar{B})}$ and on total (anti)baryon density $n_{B(\bar{B})}$. By taking the derivatives of $p_{B(\bar{B})}$ with respect to the baryochemical potential one obtains additional equation for the total (anti)baryon densities: $n_{B(\bar{B})} = (1 - b\, n_{B(\bar{B})}) \sum_{j\in B(\bar{B})} n_j^{\rm id}(T, \mu_j^{B(\bar{B})*})$.

At given T and $\boldsymbol{\mu}$, the above equations are solved numerically, yielding $p_{B(\bar{B})}(T,\boldsymbol{\mu})$ and $n_{B(\bar{B})}(T,\boldsymbol{\mu})$. The entropy density is $s = (\partial p/\partial T)_\mu$, and the energy density is obtained from the Gibbs relation.

Finite widths of the resonances. The ideal Fermi or Bose gas pressures $p_i^{\rm id}(T,\mu_i)$ in Eq. (1) contain the additional integration over hadron's mass:

$$p_i^{\rm id}(T,\mu_i) = \frac{d_i}{6\pi^2} \int_{m_i^{\rm min}}^{m_i+2\Gamma_i} dm\, \rho_i(m) \int dk\, \frac{k^4}{\sqrt{m^2+k^2}} \left[\exp\left(\frac{\sqrt{m^2+k^2}-\mu_i}{T}\right) \pm 1\right]^{-1}. \tag{2}$$

For $n_j^{\rm id}(T,\mu_i)$ the expressions are analogous to (2). The function $\rho_i(m)$ is the properly normalized mass distribution for hadron type i. For stable hadrons, or whenever the zero width approximation is applied, one has $\rho_i(m) = \delta(m - m_i)$. The finite widths of the resonances are taken into account in a simplified way, by the integration over their relativistic Breit-Wigner shapes (see, e.g., Refs. [13, 14]):

$$\rho_i(m) = \frac{2\, m\, m_i\, \Gamma_i}{(m^2 - m_i^2)^2 + m_i^2 \Gamma_i^2} \bigg/ \left(\int_{m_i^{\rm min}}^{m_i+2\Gamma_i} d\tilde{m}\, \frac{2\, \tilde{m}\, m_i\, \Gamma_i}{(\tilde{m}^2 - m_i^2)^2 + m_i^2 \Gamma_i^2}\right), \tag{3}$$

where $m_i^{\rm min} = \max(m_i - 2\Gamma_i, m_i^{\rm thr})$ with $m_i^{\rm thr}$ the minimum decay threshold mass for resonance i.

Hadron list. In the standard scenario, the hadron list includes all established (3- and 4-star) strange and non-strange hadrons which are listed in the Particle Data Tables [15]. This list is denoted as PDG and it contains about 380 different hadron species. In addition, we also consider an extended PDG-based list, which also includes unconfirmed hadron states. This PDG+ list contains about 580 hadron species. Using the PDG and PDG+ lists we test the sensitivity of the results to the input hadron list.

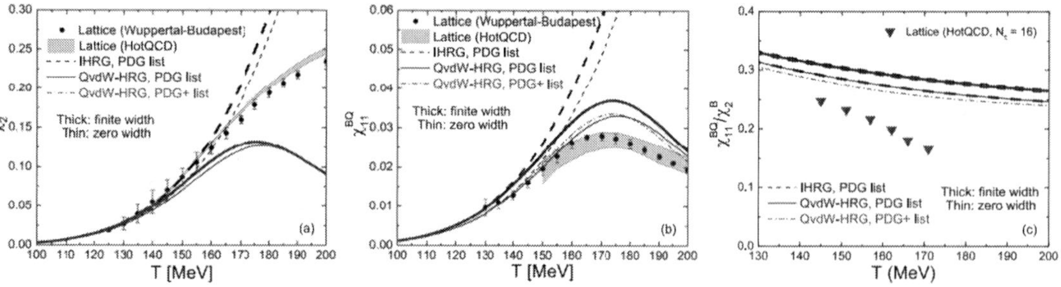

Figure 1. The temperature dependence of (a) net baryon number susceptibility χ_2^B, (b) baryon-electric charge correlator χ_{11}^{BQ}, and (c) χ_{11}^{BQ}/χ_2^B, calculated within IHRG (dashed black lines) and QvdW-HRG (solid blue lines) with PDG hadron list. Calculations with finite (zero) widths of the resonances are depicted by the thick (thin) lines. QvdW-HRG calculations with PDG+ hadron list are exhibited by the dash-dotted red lines. Lattice QCD results of the Wuppertal-Budapest [16, 17] and HotQCD [9, 18, 19] collaborations are shown, respectively, by black symbols and green bands/symbols.

3 Results

We consider temperature dependence of the conserved charges susceptibilities at $\mu = 0$, defined as

$$\chi_{lmn}^{BSQ} = \frac{\partial^{l+m+n} p/T^4}{\partial(\mu_B/T)^l \, \partial(\mu_S/T)^m \, \partial(\mu_Q/T)^n} . \tag{4}$$

We focus on the net baryon susceptibility χ_2^B, baryon-electric charge correlator χ_{11}^{BQ}, and the ratio χ_{11}^{BQ}/χ_2^B. The temperature dependences of these quantities are shown in Fig. 1, along with the lattice data.

The inclusion of QvdW interactions between baryons leads to a qualitatively different behavior of χ_2^B and χ_{11}^{BQ} at high temperatures, as seen from comparison between the dashed black lines (IHRG) and the solid blue lines (QvdW-HRG) in Fig. 1. Some qualitative features seen in lattice simulations, such as the inflection point in the temperature dependence of χ_2^B and the peak in χ_{11}^{BQ} are reproduced by the QvdW-HRG model, as reported previously in Ref. [6].

It is interesting that the ratio χ_{11}^{BQ}/χ_2^B (Fig. 1c) is virtually unaffected by the QvdW interactions. This result can be proved analytically for the Boltzmann approximation. In this case the ratio $n_i/n_i^{\rm id}$ between the density n_i of the baryon species i in the QvdW-HRG model and the corresponding ideal gas density $n_i^{\rm id}$ is the same for all baryons and is a function of the total density of baryons n_B only. At $\mu_B = 0$ one has $n_{\bar{B}} = n_B$, therefore the modification factor is the same for baryons and antibaryons. Therefore, the effects of QvdW interactions between baryons cancel out in the ratio χ_{11}^{BQ}/χ_2^B at $\mu_B = 0$:

$$\left(\chi_{11}^{BQ}/\chi_2^B\right)_{\rm QvdW} = \frac{\sum_{i\in B} Q_i\, n_i^{\rm id}}{\sum_{i\in B} n_i^{\rm id}} = \left(\chi_{11}^{BQ}/\chi_2^B\right)_{\rm id} . \tag{5}$$

This explains the results shown in Fig. 1c. Note that, in general, there is no such cancellation if the QvdW interaction parameters would be assumed to be different for different baryon-baryon pairs [11].

All three observables, including the ratio χ_{11}^{BQ}/χ_2^B, are sensitive to the modeling of the finite widths of the resonances, in particular the Δ and N^* resonances. Applying the prescription given by Eqs. (2) and (3) one obtains an improved description of χ_2^B, but also a worse description of χ_{11}^{BQ} and χ_{11}^{BQ}/χ_2^B, within both the IHRG and QvdW-HRG models. The influence of the finite resonance widths is also the likely source of the discrepancy between the IHRG and QvdW-HRG results for χ_{11}^{BQ}/χ_2^B reported in Ref. [18]. Present results imply a necessity for a more involved modeling of the resonances in a HRG. One possibility is to use the S-matrix approach [20].

The description of all the considered observables is improved when an extended PDG+ hadron list, which contains additional baryons, is used. The improvement is rather modest, and the lattice data for the χ_{11}^{BQ}/χ_2^B ratio are still not described well by all the considered models. A notably improved description can be obtained by using the quark model states [18]. Another interesting possibility, also presented at this conference, are the in-medium mass modifications for the negative-parity states [21, 22]. It would be interesting to compare and combine these modifications with the QvdW approach.

To summarize, the effects of quantum van der Waals interactions in the HRG model on the baryon number susceptibilities are studied simultaneously with the effects of finite resonance widths and input hadron list. QvdW interactions lead to a qualitatively different behavior of χ_2^B and χ_{11}^{BQ}, some features resembling the lattice data, but they cancel out in the ratio χ_{11}^{BQ}/χ_2^B. The inclusion of finite resonance widths via the relativistic Breit-Wigner distribution improves the description for χ_2^B, but also leads to a worse agreement for χ_{11}^{BQ} and χ_{11}^{BQ}/χ_2^B. Inclusion of the extra, unconfirmed hadron states from PDG appears to improve slightly the agreement for all the observables considered.

Acknowledgments. We thank F. Karsch for useful discussions regarding χ_{11}^{BQ}/χ_2^B and for providing the lattice data for this quantity. We acknowledge support from HIC for FAIR, HGS-HIRe, Judah M. Eisenberg Laureatus Chair at Goethe University, and National Academy of Sciences of Ukraine.

References

[1] S. Borsanyi, *et al.* [Wuppertal-Budapest Collaboration], Phys. Lett. B **730**, 99 (2014).
[2] A. Bazavov *et al.* [HotQCD Collaboration], Phys. Rev. D **90**, 094503 (2014).
[3] Y. Aoki, G. Endrodi, Z. Fodor, S. D. Katz and K. K. Szabo, Nature **443**, 675 (2006).
[4] A. Bazavov *et al.*, Phys. Rev. Lett. **111**, 082301 (2013).
[5] A. Bazavov *et al.*, Phys. Rev. D **88**, 094021 (2013).
[6] V. Vovchenko, M. I. Gorenstein and H. Stoecker, Phys. Rev. Lett. **118**, 182301 (2017).
[7] A. Majumder and B. Muller, Phys. Rev. Lett. **105**, 252002 (2010).
[8] A. Bazavov *et al.*, Phys. Rev. Lett. **113**, 072001 (2014).
[9] A. Bazavov *et al.*, Phys. Rev. D **95**, 054504 (2017).
[10] P. Alba *et al.*, Phys. Rev. D **96**, 034517 (2017).
[11] P. Alba, arXiv:1711.02797 [nucl-th].
[12] V. Vovchenko, D. V. Anchishkin and M. I. Gorenstein, Phys. Rev. C **91**, 064314 (2015).
[13] F. Becattini, Z. Phys. C **69**, 485 (1996).
[14] S. Wheaton, J. Cleymans and M. Hauer, Comput. Phys. Commun. **180**, 84 (2009).
[15] K. A. Olive *et al.* [Particle Data Group], Chin. Phys. C **38**, 090001 (2014).
[16] S. Borsanyi, Z. Fodor, S. D. Katz, S. Krieg, C. Ratti and K. Szabo, JHEP **1201**, 138 (2012).
[17] S. Borsányi, PoS **LATTICE2015**, 015 (2016); Wuppertal-Budapest collaboration, preliminary.
[18] F. Karsch, Nucl. Phys. A **967**, 461 (2017).
[19] A. Bazavov *et al.* [HotQCD Collaboration], Phys. Rev. D **86**, 034509 (2012).
[20] P. M. Lo, B. Friman, K. Redlich and C. Sasaki, arXiv:1710.02711 [hep-ph].
[21] G. Aarts, C. Allton, D. De Boni, S. Hands, B. Jäger, C. Praki and J. I. Skullerud, JHEP **1706**, 034 (2017).
[22] G. Aarts, C. Allton, D. de Boni, S. Hands, B. Jäger, C. Praki and J. I. Skullerud, these proceedings [arXiv:1710.00566 [hep-lat]].

Melting and freeze-out conditions of hadrons in a thermal medium

Juan M. Torres-Rincon[1,⋆], *Joerg* Aichelin[1,2], *Hannah* Petersen[1,3,4], *Jean-Bernard* Rose[1,3], and *Joseph* Tindall[5]

[1] *Frankfurt Institute for Advanced Studies, Ruth-Moufang-Str. 1, 60438 Frankfurt am Main, Germany*
[2] *Subatech, UMR 6457, IN2P3/CNRS, Université de Nantes, École de Mines de Nantes, 4 rue Alfred Kastler, 44307 Nantes, France*
[3] *Institute for Theoretical Physics, Goethe University, Max-von-Laue-Str. 1, 60438 Frankfurt a.M., Germany*
[4] *GSI Helmholtzzentrum für Schwerionenforschung, Planck-Str. 1, 64291 Darmstadt, Germany*
[5] *Department of Physics, University of Bath, Claverton Down, Bath, United Kingdom*

Abstract. We describe two independent frameworks which provide unambiguous determinations of the deconfinement and the decoupling conditions of a relativistic gas at finite temperature. First, we use the Polyakov-Nambu-Jona–Lasinio model to compute meson and baryon masses at finite temperature and determine their melting temperature as a function of their strangeness content. Second, we analyze a simple expanding gas within a Friedmann-Robertson-Walker metric, which admits a well-defined decoupling mechanism. We examine the decoupling time as a function of the particle mass and cross section. We find evidences of an inherent dependence of the hadronization and freeze-out conditions on flavor, and on mass and cross section, respectively.

1 Introduction

Statistical thermal models [1] have been very successful in characterizing the thermodynamical properties of the hadronic medium created in heavy-ion collisions at the chemical freeze-out. By simply measuring the total multiplicities of particles, the temperature T_f and the baryochemical potential $\mu_{B,f}$ can be extracted with a satisfactory statistical confidence. However, the assumption of a single T_f entails a tension between multistrange and nonstrange baryons in the thermal fits made to the data of the ALICE collaboration [2]. The Ξ and Ω baryons seems to prefer a larger freeze-out temperature (with a difference of $\sim$16 MeV) than the T_f for protons.

On the other hand, the computation of generalized susceptibilities by lattice-QCD shows that the inflection point of the 2nd-order susceptibility—indicator of the crossover transition—is located for the strange sector at a temperature which is around 15 MeV larger than the one for its light counterpart [3]. Therefore, there exist indications of a flavor hierarchy in (at least) two different observables: the QCD phase transition and the freeze-out temperatures [4]. This issue have created much interest at this SQM2017 conference, e.g. [5].

In this contribution we provide qualitative evidences of such a hierarchy by presenting two separate studies. The QCD deconfinement temperature will be accessed by the melting temperature of

⋆Speaker, e-mail: torres@fias.uni-frankfurt.de

hadrons in the PNJL model of strong interactions. The freeze-out temperature will be represented by the decoupling temperature of an expanding system of particles using a relativistic transport model.

2 Melting temperature hierarchy in the PNJL model

The PNJL model is an effective realization of the low-energy dynamics of QCD [6]. Thanks to its Lagrangian formulation it can be easily extended to finite densities and temperatures. The dynamical degrees of freedom are $N_f = 3$ massive quarks. Nevertheless, colorless states like mesons and baryons can also be addressed in this model with the use of few-body techniques [7]. Mesons are modeled as quark-antiquark bound states which are stable in vacuum. At higher temperatures mesons can melt into pairs. This happens at the so-called Mott (or melting) temperature, which is understood as the effective deconfinement temperature (as the PNJL model lacks true confinement).

Similar considerations apply to bound states of 2 quarks (diquarks), and to bound states of a diquark and a quark. The later are identified with physical baryons when they belong to a singlet color representation. In Ref. [7] we computed the masses and Mott temperatures of mesons and baryons in several flavor/spin channels. A summary of the melting temperatures is shown in Table 1. Notice that the η' is unstable already at $T = 0$ because its vacuum mass ($m_{\eta'} = 987$ MeV, according to our calculation in Ref. [7]) is larger than the sum of its quark plus antiquark masses. In addition, the baryons masses versus temperature (up to the Mott temperature) are plotted in Fig. 1.

Table 1. Mott (or melting) temperatures of several mesons and baryons calculated in the PNJL model [7].

Meson	π	K	η	η'	ρ	K^*	ω	ϕ
T_{Mott} (MeV)	282	286	245	0	253	266	253	382
Baryon	p	Λ	Σ	Ξ	Δ	Σ^*	Ξ^*	Ω
T_{Mott} (MeV)	254	269	195	287	223	231	239	288

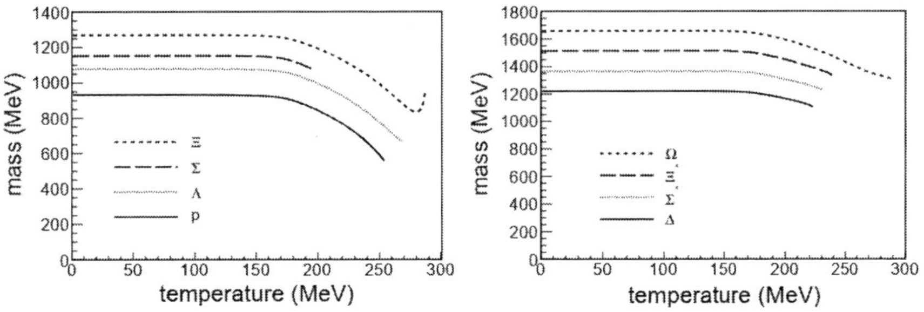

Figure 1. Masses of baryons belonging to the flavor octet (left panel) and decuplet (right panel) representations as functions of the temperature until the melting (or Mott) temperature.

Several conclusions can be extracted from Table 1 and Fig. 1: **1)** The effective deconfinement temperature is not universal. **2)** The melting temperatures of Ξ and Ω are larger (ca. 30 MeV) than the one for the proton. This relative ordering might be responsible for the situation observed in the statistical thermal fits [2]. **3)** Multistrange states generally have larger melting temperatures (except the Ξ^*). In particular, the T_{Mott} of the ϕ meson (strange-antistrange pair) is abnormally large.

These results point to a hierarchy in the deconfinement conditions, with a larger temperature for the strange sector than the light one, as reflected in lattice-QCD results for the susceptibilities [3].

3 Expanding system and its decoupling

In a completely different approach we have focused on the freeze-out mechanism and its dependence on the particle properties. For this case we have used a transport model to describe a system of relativistic particles interacting via constant cross sections. SMASH (Simulating Many Accelerated Strongly-interacting Hadrons) is a newly developed transport model that aims to describe the hadronic stage of heavy-ion collisions with applications from GSI energies to LHC physics [8]. We have adapted SMASH to work with a background Friedmann-Robertson-Walker metric so that the system exhibits a Hubble-like expansion in time. This expansion mimics the so-called "blast-wave model", used to describe the late stages of central heavy-ion collisions. We use a spherical volume filled with a large number of particles, initialized with an isotropic, homogeneous spatial distribution.

For a particular initial condition, we were able to reproduce the exact analytical solution of the massless Boltzmann equation found in Ref. [9]. This represents a nontrivial check for the SMASH code in a far-from-equilibrium scenario [10].

For massive particles, we apply a Hubble expansion whose initial rate is much smaller than the scattering rate. In this regime the system is able to reach and maintain equilibrium during the expansion. When the expansion rate becomes larger than the scattering rate, the equilibrium state cannot be maintained anymore, as the number of collisions is not sufficient to redistribute the momentum.

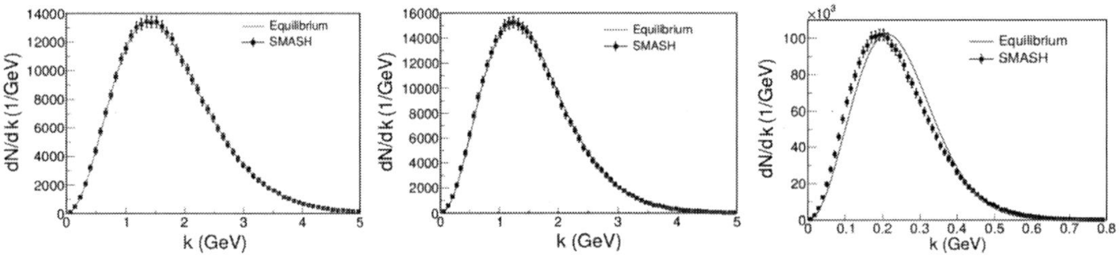

Figure 2. Distribution function at different times of the Hubble expansion as a function of the particle momentum. From left to right $t = 0.1, 5$ and 20 fm. In solid line we show the predicted equilibrium distribution for each time.

From Fig. 2 we can observe that SMASH perfectly reproduces the theoretical distribution function (solid line) until decoupling happens soon after the middle panel. At $t = 20$ fm the distribution from SMASH does not match the equilibrium expectations. Taking into account the absence of collisions and the redshift in momentum, the distribution after the freeze-out (decoupling) reads

$$f_{freeze-out}(t,k) = f_{equilibrium}(t_D, ka(t)/a_D) \propto \exp\left[-T_D{}^{-1}\sqrt{(ka(t)/a_D)^2 + m^2}\right], \tag{1}$$

where t_D is the decoupling time, $a(t)$ is the scale factor, and a_D, T_D are the value of the scale factor and the temperature at the decoupling time, respectively.

The fit to this function provides the value of the decoupling time and temperature. In the example of Fig. 2 it gives $t_D = 5.3 \pm 0.6$ fm. Repeating this procedure for particles with different masses and cross sections, we obtain the results in Fig. 3 for the extracted freeze-out times.

The main conclusions of this study are: **1)** The more massive particles freeze out earlier (at higher temperature). **2)** The larger the cross section is, the later the freeze-out appears (lower temperature). Two caveats are in order. This scenario is closer to the kinetic freeze-out (the absence of interactions) rather than the chemical freeze-out (relevant for statistical thermal fits). So far, inelastic processes are absent in our study. In addition, we have not explicitly introduced strangeness in the system.

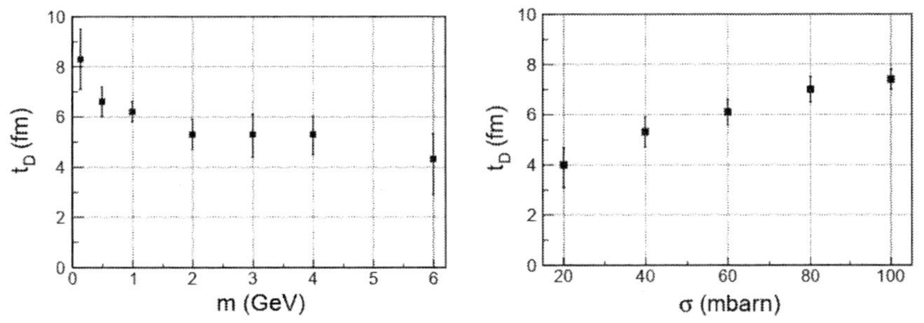

Figure 3. Decoupling time versus the mass of the particles with fixed cross section $\sigma = 40$ mb (left panel), and versus their total cross section with fixed mass $m = 2$ GeV (right panel).

However, one can notice that strange hadrons generally entail larger masses, and smaller cross sections in comparison with their lighter counterparts. According to the results in Fig. 3, the combined effect would result in an earlier freeze-out for the hadrons with strangeness (i.e. higher values of T_f).

4 Conclusion

We provide evidences of a flavor-nonuniversal deconfinement and freeze-out conditions that support previous results in this directions [3–5]. On one hand, using the PNJL model we observe that multistrange hadrons melt at higher temperatures rather systematically. This points towards an earlier confinement transition for the strange sector in heavy-ion collisions. On the other hand, the decoupling mechanism in an expanding spacetime provides direct indications that more massive and less interacting particles freeze out earlier and at higher temperatures. This observation supports the higher freeze-out temperature found for (multi)strange hadrons in statistical thermal fits [2].

Work supported by DAAD (Germany), Helmholtz Young Investigator Group VH-NG-822 (Helmholtz Association and GSI), the HIC for FAIR within the framework of the LOEWE program (Hesse, Germany), and Project No.FPA2013-43425-P (Spain), and Program TOGETHER from Région Pays de la Loire and the I3-Hadron Physics (EU). We acknowledge LOEWE-Center for Scientific Computing for computing resources.

References

[1] J. Cleymans and H. Satz, Z. Phys. **C57**, 135 (1993)
P. Braun-Munzinger *et al.*, Phys. Lett. **B518**, 41 (2001)
[2] R. Preghenella (ALICE Collaboration), Acta Phys. Polon. **B43**, 555 (2012)
[3] R. Bellwied *et al.*, Phys. Rev. Lett. **111**, 202302 (2013)
[4] S. Chaterjee *et al.*, Phys. Lett. **B727**, 554 (2013)
[5] Talks by R. Bellwied, S. Chaterjee, P. Alba at this conference
[6] S,P, Klevansky, Rev. Mod. Phys. **64**, 649 (1992)
M. Buballa, Phys. Rept. **407**, 205 (2005)
C. Ratti, M. A. Thaler and W. Weise, Phys. Rev. **D73**, 014019 (2006)
[7] J.M. Torres-Rincon, B. Sintes, and J. Aichelin, Phys. Rev. **C91**, 065206 (2015)
[8] J. Weil *et al.*, Phys. Rev. **C94**, , 054905 (2016)
[9] D. Bazow *et al.*, Phys. Rev. Lett. **116**, 022301 (2016)
[10] J. Tindall, J.M. Torres-Rincon, J.-B. Rose, and H. Petersen, Phys. Lett. **B770**, 532 (2017)

EPJ Web of Conferences **171**, 14008 (2018) https://doi.org/10.1051/epjconf/201817114008
SQM 2017

Mass hierarchy and energy scaling of the Tsallis – Pareto parameters in hadron productions at RHIC and LHC energies

Gábor Bíró[1,2,⋆], *Gergely Gábor* Barnaföldi[1], *Tamás Sándor* Biró[1], and *Keming* Shen[3]

[1] *Wigner Research Centre for Physics of the H.A.S., P.O. Box, H-1525 Budapest, Hungary*
[2] *Institute of Physics, Eötvös University, 1/A Pázmány P. Sétány, H-1117 Budapest, Hungary*
[3] *Key Laboratory of Quark & Lepton Physics and Institute of Particle Physics, Central China Normal University, Wuhan 430079, China*

Abstract. The latest, high-accuracy identified hadron spectra measurements in high-energy nuclear collisions led us to the investigation of the strongly interacting particles and collective effects in small systems. Since microscopical processes result in a statistical Tsallis – Pareto distribution, the fit parameters q and T are well suited for identifying system size scalings and initial conditions. Moreover, parameter values provide information on the deviation from the extensive, Boltzmann – Gibbs statistics in finite-volumes. We apply here the fit procedure developed in our earlier study for proton-proton collisions [1, 2]. The observed mass and center-of-mass energy trends in the hadron production are compared to RHIC dAu and LHC pPb data in different centrality/multiplicity classes. Here we present new results on mass hierarchy in pp and pA from light to heavy hadrons.

1 Introduction

In high-energy nuclear physics the investigation of small colliding nuclear systems is fundamentally important for understanding collective effects. Strong correlation phenomena typical for heavy-ion collisions were recently observed in pp and pA collisions [3, 4]. While correlations in such small systems are more clearly observable thanks to the lower background, their statistical and thermodynamical description points towards to exceed the classical Gibbs – Boltzmann framework. A modern tool for such studies is provided by the non-extensive statistics. In the application one fits Tsallis – Pareto-type distributions to describe the measured spectra in wide central-of-mass energy and p_T ranges [5–8]. In our recent works [1, 2], we analyzed the spectra of various hadron species in pp collisions from RHIC to LHC energies. We compared the results to theoretical model calculations and explored the mass hierarchy and c.m. energy scaling of the parameters. In this short contribution, we extend our investigations to the latest identified strange- and heavy-flavored hadron spectra measured in dAu and pPb collisions from $\sqrt{s_{NN}} = 200$ GeV RHIC to $\sqrt{s_{NN}} = 5.02$ TeV LHC energies in different centralities. We used data from STAR [11] and ALICE [12, 13] measurements and analyzed them following the procedure described in details in Refs. [1, 2].

⋆e-mail: biro.gabor@wigner.mta.hu

2 Identified hadron spectra from pA collisions

Recent high-precision measurements in dAu and pPb collisions at RHIC and LHC energies present a *Janus-face* behavior, depending on the multiplicity class of the selected data set. While low-multiplicity events are more jet-like, the high multiplicity ones include correlations and present sizeable nuclear modification, especially at the low- and intermediate momentum ranges. The same duality has been already observed in nucleus-nucleus (AA) collisions [14–16]. In smaller systems it is a special challenge to identify these components [17]. We quantify and separate these observations and deduce the parameters of the microscopical interactions. We apply the same formula as in Refs. [1, 2], for all centrality classes (0-20%, 20-40% and 40-100% for the STAR data, 0-5%, 5-10%, 10-20%, 20-40%, 40-60%, 60-80% and 80-100% for the ALICE data):

$$\left.\frac{\mathrm{d}^2N}{N_{ev}2\pi p_T \mathrm{d}p_T \mathrm{d}y}\right|_{y\approx 0} \sim \left[1+(q-1)\frac{m_T}{T}\right]^{-\frac{1}{q-1}} \quad . \tag{1}$$

Here, $m_T = \sqrt{p_T^2 + m^2}$ is the *transverse mass* of the given hadron species. We note that this formula can be derived from the Tsallis q-entropy and, as we pointed out in Refs. [6, 18], the parameters carry a certain physical meaning: the *non-extensivity* parameter, q measures the deviation from the (extensive) Boltzmann – Gibbs state ($q = 1$), while a generalized temperature-like parameter, T provides a temperature value taking into account the finite-size effects and fluctuations. These parameters are interconnected via the multiplicity distribution, e.g. a (negative) binomial distribution [19]. On Fig. 1

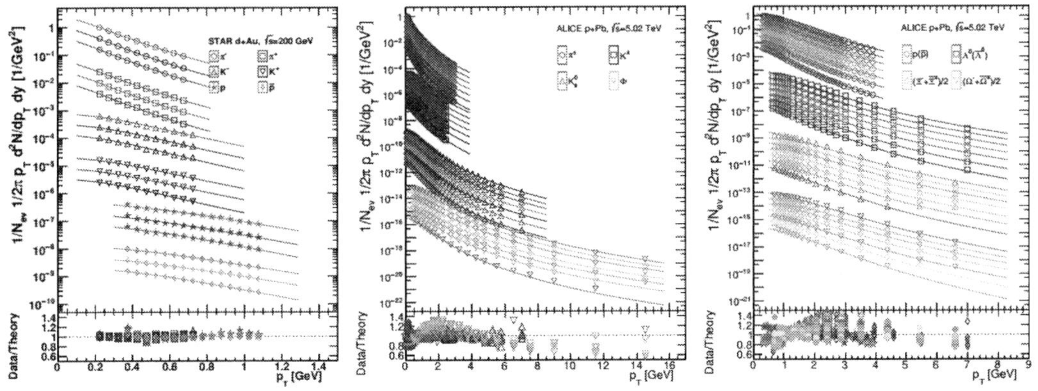

Figure 1. The fitted spectra for various identified hadrons measured in dAu collisions at $\sqrt{s_{NN}} = 200$ GeV at STAR, RHIC (*left*) and pPb collisions at $\sqrt{s_{NN}} = 5.02$ TeV at ALICE, LHC (*middle panel for mesons and right panel for baryons*). Data have been scaled for better visibility. *Bottom panels* present the Data/Fit ratios using the same color as on the spectra.

the analyzed identified hadron spectra measured in small colliding systems along with the fitted distributions were plotted. For both the $\sqrt{s_{NN}} = 200$ GeV dAu at RHIC and $\sqrt{s_{NN}} = 5.02$ TeV pPb at LHC energy collisions we extraced the parameters of Eq. (1) utilizing the ROOT data analysis framework, performing fits based on the minimal χ^2 method. Based on the available identified hadron data published in Refs. [11–13], the fits were performed in various centrality bins on as wide p_T ranges as possible. Lower panels present the data-to-fit ratios with agreement of 20% and 40% – 50% for dAu (RHIC) and pPb (LHC) data respectively.

3 Results: parameters and mass hierarchy, compared to pp

The fit parameters of Eq. (1) are plotted on the parameter-space diagram, Fig. 2. T is plotted against $q-1$, with the latter measuring the deviation from the Boltzmannian case directly. The empty markers and the shaded area show the pp results with different energies (62.4 GeV $\leq \sqrt{s} \leq$ 7 TeV) as it was presented in Ref. [1] along with the STAR dAu data at $\sqrt{s_{NN}}$ = 200 GeV. The full markers show the parameters from the present fit for the pPb spectra at $\sqrt{s_{NN}}$ = 5.02 TeV. The new points that

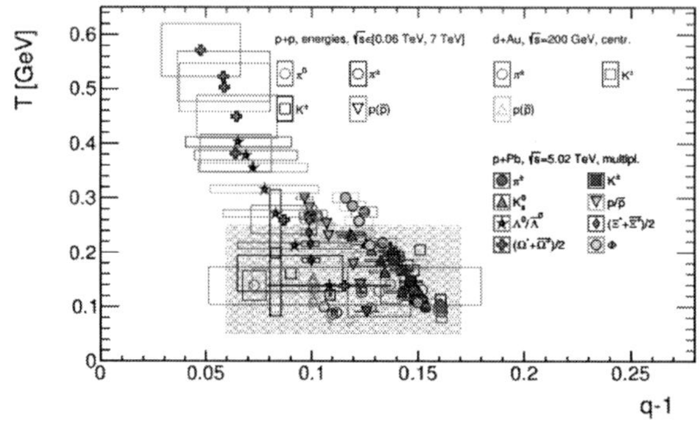

Figure 2. The T vs $(q-1)$ parameter space for identified hadron spectra in pPb at $\sqrt{s_{NN}}$ = 5.02 TeV with different multiplicity classes (*filled markers*) and for the same pp measurements for various $\sqrt{s}$ energies from Ref. [1] including dAu STAR data [11] at $\sqrt{s}$ = 200 GeV (*empty markers and shaded area*).

are consistent with the previous results (i.e. those inside the shaded box) are the most pp-like points, especially for the smallest multiplicity class. We observe deviance only for those points which are extracted from high-multiplicity spectra. Increasing the multiplicity either by increasing the c.m. energy or getting to more central events, lead to both higher q and T. Points in the parameter space originating form datasets with similar multiplicities group together nicely. The dAu data from STAR accompany the lowest energy pp fit parameters. It is either due to the low multiplicity or due to the very limited p_T range of the identified transverse momentum spectra. The absence of strong power-law tail results in a weak determination of the Tsallis – Pareto parameters. Fig. 2 also represents the mass hierarchy: the higher the mass of the identified hadron, the higher the T and the lower the q is. This effect is more visible on Fig. 3, where the q (*left panel*) and T (*right panel*) parameters are plotted for each identified hardon species in order to analyze the mass hierarchy effects. Here we used the same markers as on the figures above. This result extends our earlier analysis in pp collisions in Refs. [1, 2] using only π, K, and $p(\bar{p})$, while adding the hadron species with higher masses, we can see a mass hierarchy for both parameters. Based on the extracted parameters, the hadron species with the highest mass has the highest T, meanwhile the non-extensitivity is reduced as $q \to 1$, especially for the low multiplicity events. Trends with growing s quark content supports the idea of strangeness enhancement [3]. It can be a signature of the quark-gluon plasma like, close-to-thermalized, hot and dense matter in small colliding nuclear systems like pp, dAu, or pPb.

4 Summary

Mass and strangeness hierarchy is analyzed and presented for pp, dAu, and pPb collisions. We analyzed data in a wide $\sqrt{s_{NN}}$ range form RHIC to LHC energies in various centrality/multiplicity ranges. The results are consistent with our previous center-of-mass energy scaling investigations for the most pp-like, peripheral centrality classes, however deviations are observed for central (c.f. high multiplicity) collisions for hadron species with high mass. Our main observation is that the measure of the

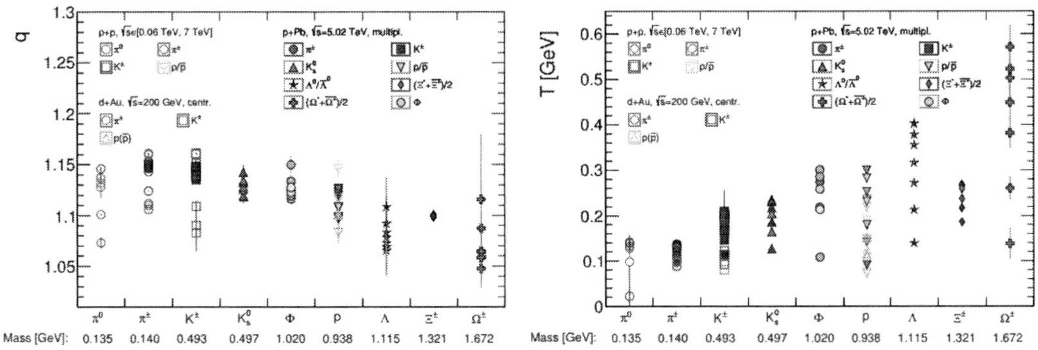

Figure 3. The fitted Tsallis – Pareto parameters q (*left panel*) and T (*right panel*) obtained in pp, dAu, and pPb collisions for all investigated hadron species in increasing mass order.

non-extensitivity, $q-1$, decreases with increasing hadron mass, thus higher-mass hadrons present less deviation from the Boltzmann – Gibbs $q = 1$ case. On the other hand, similar to the exponential fits, higher-mass hadron species require higher T to obtain the best fit results.

Acknowledgement

This work was supported by the Hungarian-Chinese cooperation grant No TéT 12 CN-1-2012-0016 and No. MOST 2014DFG02050, Hungarian National Research Fund (OTKA) grants K120660 and K123815. Author G.G. Barnaföldi also thanks the THOR COST action CA15213. We acknowledge the support of the Wigner GPU Laboratory.

References

[1] G. Biro *et al.*, Entropy **88**, 19(3) (2017)
[2] G. Biro *et al.*, AIP Conf. Proc. **1853** 080001 (2017)
[3] ALICE Collaboration, Nat. Phys. **13**, 9 (2017)
[4] A. Toia, J. Phys. Conf. Ser. **798** 012068 (2017)
[5] C. Tsallis, J. Stat. Phys. **52**, 479 (1988)
[6] Biró, T.S.; Ván, P.; Barnaföldi, G.G.; Ván, P. Eur. Phys. J. **A49**, 110 (2013)
[7] J. Cleymans, J. Phys. Conf. Ser. **779**, 012079 (2017)
[8] M. Rybczynski; G. Wilk; Z. Włodarczyk, EPJ Web Conf. **90** 01002 (2015)
[9] S. Grigoryan, Phys.Rev. **D95**, 056021 (2017)
[10] A.S. Parvan, O.V. Teryaev, J. Cleymans, Eur.Phys.J. **A53**, 102 (2017)
[11] STAR Collaboration, Phys. Rev. C. **79**, 034909 (2009)
[12] ALICE Collaboration, Phys.Lett. B. **728**, 25-38 (2014)
[13] ALICE Collaboration, Phys.Lett. B. **758**, 389-401 (2016)
[14] A. Ortiz; G. Bencédi; H. Bello, J. Phys. G: Nucl. Part. Phys. **44**, 065001 (2017)
[15] G. G. Barnaföldi; K. Ürmössy; G. Bíró, J. Phys. Conf. Ser. **612** 012048 (2015)
[16] X. Yin; L. Zhu; H. Zheng, Adv. High Energy Phys. **2017** 6708581 (2017)
[17] S. Tripathy; T. Bhattacharyya; P. Garg; P. Kumar; R. Sahoo; J. Cleymans, Eur. Phys. J. A **52**, 289 (2016)
[18] T. S. Biró; G. G. Barnaföldi; P. Ván, EPJ A **49**, 110 (2013)
[19] T. S. Biró; G. G. Barnaföldi; G. Bíró; K.M. Shen, J. Phys. Conf. Ser. **779**, 012081 (2017)

Production of (anti-)(hyper-)nuclei at LHC energies with ALICE

Maximiliano Puccio[1,2,⋆] for the ALICE collaboration

[1] *Dipartimento di Fisica, Università degli Studi di Torino*
[2] *Istituto Nazionale di Fisica Nucleare, Sezione di Torino*

Abstract. The ALICE experiment at the LHC has measured a variety of (anti-)(hyper-)nuclei produced in Pb–Pb collisions at $\sqrt{s_{\rm NN}}$ = 5.02 TeV and at 2.76 TeV. In addition, a large sample of high quality data was collected in pp collisions at $\sqrt{s}$ = 7 TeV and 13 TeV and in p-Pb collisions at $\sqrt{s_{\rm NN}}$ = 5.02 TeV. These data are used to study the production of different (anti-)(hyper-)nuclei in the collisions, namely (anti-)deuteron, (anti-)^{3}He, (anti-)alpha and (anti-)$^3_\Lambda$H. The identification of these (anti-)(hyper-)nuclei is based on the energy loss measurement in the Time Projection Chamber and the velocity measurement in the Time-Of-Flight detector. In addition, the Inner Tracking System is used to distinguish secondary vertices originating from weak decays from the primary vertex. New results on deuteron production as a function of multiplicity in pp, p–Pb and Pb–Pb collisions will be presented, as well as the measurement of ^{3}He in p–Pb and Pb–Pb collisions. Special emphasis will be given to the new results of the (anti-)$^3_\Lambda$H in its charged-two-body decay mode. The large variety of measurements at different energies and system sizes constrains the production models of light flavour baryon clusters, in particular those based on coalescence and the statistical hadronisation approaches.

1 Introduction

The nuclei, anti-nuclei and hyper-nuclei formation mechanism in high energy collisions is still unknown. Thanks to the LHC which provides pp, p–Pb and Pb–Pb collisions at the highest energy ever reached in the laboratory, the ALICE experiment is able to shed light on the phenomenology of (anti-)(hyper-)nuclei production. The latest results on the production spectra of (anti-)deuteron and (anti-)^{3}He in pp collisions at $\sqrt{s}$ = 13 TeV and Pb–Pb collisions at $\sqrt{s_{\rm NN}}$ = 5.02 TeV will be discussed here. The results are compared with the expectation from the statistical hadronisation and the hadron coalescence models. Finally the new measurement of the (anti-)$^3_\Lambda$H lifetime by the ALICE experiment, which is the most precise ever performed, will be discussed.

2 Analyses details

The key features that allow ALICE to measure (anti-)(hyper-)nuclei are the precise vertexing, tracking capabilities and the redundancy of particle identification detectors.

Tracking and vertexing are performed using the Inner Tracking Systems (ITS), a silicon tracker featuring 6 cylindrical layers, and the Time Projection Chamber (TPC) [1]. Thanks to the extended

⋆e-mail: maximiliano.puccio@cern.ch

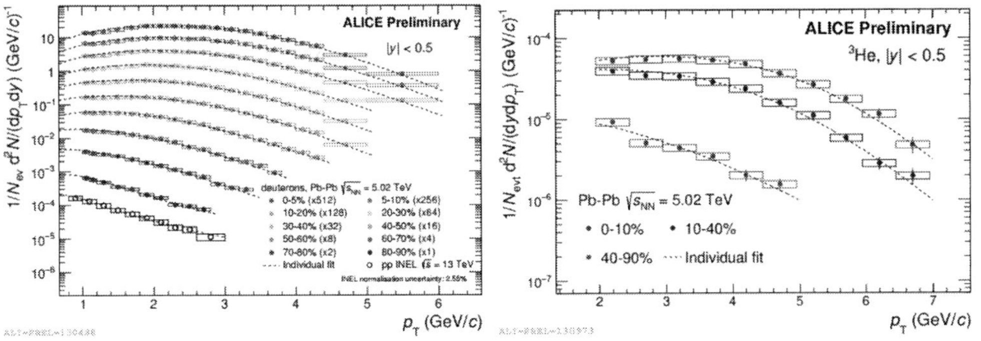

Figure 1. Transverse momentum spectra of deuterons (on the left) and ^{3}He (on the right) in Pb–Pb collisions at $\sqrt{s_{NN}}$ = 5.02 TeV. The p_T production spectrum of deuterons in pp at $\sqrt{s}$ = 13 TeV is also reported in the left panel. Statistical uncertainties are represented as vertical bars whereas boxes represent the systematic ones. The dashed lines represent the fit to the individual spectra (Blast Wave function in Pb–Pb [8] and the Lévy-Tsallis in pp [9]) to extrapolate the production yield in the unmeasured p_T regions.

lever arm and a maximum solenoidal magnetic field of 0.5 T, the momentum resolution is better than 1% for particles with $p \leq 100$ GeV/c [2]. Tracks reconstructed with points in the innermost ITS layer have a pointing resolution better than 300 μm [2]. For (anti-)(hyper-)nuclei identification the TPC specific energy loss signal ($\mathrm{d}E/\mathrm{d}x$) is used to clearly separate nuclei with Z=2 over the full measured momentum range from the bulk of the produced charged particles. Lighter nuclei with Z=1 can be identified clearly by means of the TPC $\mathrm{d}E/\mathrm{d}x$ only at low momenta (e.g. $p \leq 1.4$ GeV/c for (anti-)deuterons), but thanks to the Time Of Flight detector (TOF) it is possible to identify, for instance, (anti-)deuterons up to p_T = 6 GeV/c using a statistical unfolding technique [3]. In order to identify (anti-)$^3_\Lambda$H decaying in (anti-)^{3}He and $\pi^{-(+)}$ it is necessary to reconstruct its decay vertex. For each possible pair of (anti-)^{3}He and $\pi^{-(+)}$ the point of closest approach between the two tracks is computed and a set of topological selection is applied. A pair of (anti-)^{3}He and $\pi^{-(+)}$ that passes all these selections is considered to be a (anti-)$^3_\Lambda$H candidate. Finally a two components fit is performed to extract the yield of (anti-)$^3_\Lambda$H, similarly to what is illustrated in [4].

3 Results

The production spectra of deuterons in Pb–Pb collisions at $\sqrt{s_{NN}}$ = 5.02 TeV and in pp collisions at $\sqrt{s}$ = 13 TeV are shown in Figure 1 (left). The typical hardening of the spectra with increasing centrality, already observed for lighter particles [5, 6] and for deuterons [3, 7] at lower energies, is visible here and in the ^{3}He spectra (right panel of Figure 1): this pattern suggests an increasing radial flow with increasing centrality. Moreover, the ratio between the production spectra of nuclei and their anti–matter partners is compatible with unity [10], like for the other baryon over anti-baryon ratios at the LHC energies. This observation is compatible with expectations from the statistical hadronisation [11] and the hadron coalescence models [12]. In both approaches the abundances of a (anti-)nuclide X with mass number A are coupled to those of (anti-)protons according to the following relation: $\overline{X}/X \approx (\overline{p}/p)^A$. This expectation for the coalescence model holds only if the spectra of (anti-)neutrons are compatible with those of (anti-)protons. Under this assumption it is also possible to compute the coalescence parameter B_A which is defined for a nuclide with mass number A as the

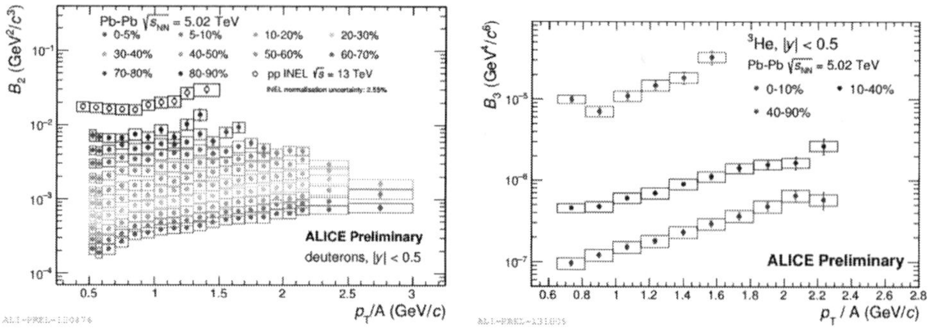

Figure 2. Coalescence parameters of deuterons (B_2, left panel) and ^{3}He (B_3, right panel) measured in Pb–Pb collisions at $\sqrt{s_{NN}} = 5.02$ TeV. The B_2 measured in pp collisions at $\sqrt{s} = 13$ TeV is also shown in the left panel. Statistical uncertainties are represented as vertical bars whereas boxes represent the systematic ones. The ALICE preliminary proton p_T-production spectra [6] are used to compute the coalescence parameters.

ratio between the nuclide invariant production spectrum and the proton spectrum to the power of A. Figure 2 shows the coalescence parameters for deuterons and ^{3}He. In the simple hadron coalescence approach the coalescence parameter is assumed to be momentum independent. However, the presence of correlations in the proton and neutron initial emission, such as jets or common emission radius from the thermal source [13], can introduce a momentum dependence similar to the measured B_A.

Another observable qualitatively described by the simple hadron coalescence is the deuteron over proton ratio as a function of multiplicity in small systems, see Figure 3. As more protons and neutrons are created in events with increasing multiplicities, the probability of finding two nucleons close in momentum space and consequently the probability of forming a deuteron increase in small systems. However, this scaling breaks in semi-central Pb–Pb collisions where the deuteron over proton ratio as a function of multiplicity becomes flatter. A flat nucleus over proton ratio as a function of multiplicity, observed also in the case of ^{3}He shown in Figure 3, is expected if the statistical hadronisation is the underlying production mechanism in nucleus nucleus collisions. At very high multiplicity the measurements of the deuteron over proton ratios in Pb–Pb collisions at $\sqrt{s_{NN}} = 2.76$ TeV and $\sqrt{s_{NN}} =$ 5.02 TeV show a hint of suppression. This kind of behaviour might point to the presence of a hadronic rescattering phase that might break nuclei, similar to what is proposed in [14].

Finally, the ALICE collaboration has been able to perform one of the most precise measurements of the $^3_\Lambda$H lifetime using the Pb–Pb data sample at $\sqrt{s_{NN}} = 5.02$ TeV collected in 2015. The lifetime was measured analysing the $^3_\Lambda\mathrm{H} \rightarrow ^3\mathrm{He} + \pi^-$ and the $^3_{\bar{\Lambda}}\overline{\mathrm{H}} \rightarrow ^3\overline{\mathrm{He}} + \pi^+$ decays in different ct intervals and performing an exponential fit to the obtained $\mathrm{d}N/\mathrm{d}(ct)$ distribution, similarly to what has been done in [4]. The preliminary measured value $\tau = 237^{+33}_{-36}$ (stat.) $\pm$ 17 (syst.) ps is compatible with both the previously computed world average ($\tau = \left(216^{+18}_{-16}\right)$ ps, [4]) and with the free Λ lifetime.

4 Conclusions

The harvest of results from the LHC Run 2 data in the (anti-)nuclei sector just started. The new results at the current LHC top energies confirm the picture depicted by the Run 1 results [3, 7]: at the LHC energies the nuclei production mechanisms in small systems and in Pb–Pb seem to be different. On one hand the simple hadron coalescence describes qualitatively the measurements of deuteron over proton ratio and B_2 up to peripheral Pb–Pb collisions. On the other hand the steady rise of the B_2 with

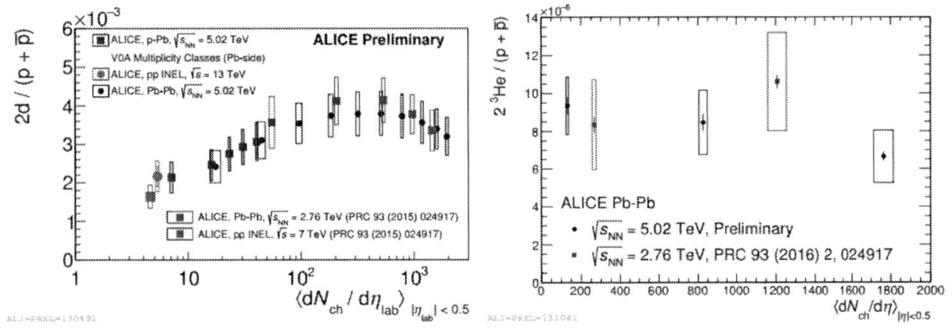

Figure 3. Deuteron (left panel) and ^{3}He (right panel) over proton ratios as a function of the charged particle multiplicity in the laboratory in different collision systems. Statistical uncertainties are represented as vertical lines whereas boxes represent the systematic ones.

momentum and the flattening of the deuteron over proton ratio in mid-central and central Pb–Pb collisions require additional assumptions to reconcile the coalescence expectations with ALICE results. The larger data sample expected in the upcoming 2018 Pb–Pb run will help to refine the results for ^{3}He and possibly extend the measurements to heavier nuclei to test the model predictions with increased precision. However, hadron coalescence might not be the production mechanism of (anti-)nuclei in Pb–Pb collisions: the flat d/p ratio starting from mid-central collisions is in agreement with the expectations of the statistical hadronisation model. Further studies on the systematic uncertainties will help to establish whether the decrease of the d/p ratio in very central Pb–Pb collision is significant.

In the strange nuclear sector, the new ALICE measurement of the $^3_\Lambda$H lifetime is compatible with both the current world average and with the theoretical expectation (i.e. the free Λ lifetime [15]).

References

[1] K. Aamodt et al. (ALICE), JINST **3**, S08002 (2008)
[2] B.B. Abelev et al. (ALICE), Int. J. Mod. Phys. **A29**, 1430044 (2014), 1402.4476
[3] S. Acharya et al. (ALICE) (2017), 1707.07304
[4] J. Adam et al. (ALICE), Phys. Lett. **B754**, 360 (2016), 1506.08453
[5] J. Adam et al. (ALICE), Phys. Rev. **C93**, 034913 (2016), 1506.07287
[6] N. Jacazio (ALICE) (2017), 1704.06030
[7] J. Adam et al. (ALICE), Phys. Rev. **C93**, 024917 (2016), 1506.08951
[8] E. Schnedermann, J. Sollfrank, U.W. Heinz, Phys. Rev. **C48**, 2462 (1993), nucl-th/9307020
[9] C. Tsallis, J. Statist. Phys. **52**, 479 (1988)
[10] ALICE (2017), ALICE-PUBLIC-2017-006, http://cds.cern.ch/record/2272148
[11] A. Andronic, P. Braun-Munzinger, J. Stachel, H. Stöcker, Phys. Lett. B **697**, 203 (2011), 1010.2995
[12] J.I. Kapusta, Phys. Rev. **C21**, 1301 (1980)
[13] R. Scheibl, U.W. Heinz, Phys. Rev. **C59**, 1585 (1999), nucl-th/9809092
[14] F. Becattini, E. Grossi, M. Bleicher, J. Steinheimer, R. Stock, Phys. Rev. **C90**, 054907 (2014), 1405.0710
[15] R.J. Prem, P.H. Steinberg, Phys. Rev. **136**, B1803 (1964)

Estimation of the isothermal compressibility from event-by-event multiplicity fluctuation studies

Maitreyee Mukherjee[1,*], *Sumit* Basu[2] , *Sanchari* Thakur[3], *Souvik* P. Adhya[3], *Arghya* Chatterjee[3], *Sandeep* Chatterjee[4], and *Tapan* K. Nayak[3,5]

[1] Bose Institute, Department of Physics and CAPSS, Kolkata-700091, India
[2] Department of Physics and Astronomy, Wayne State University, Detroit, MI 48201, USA
[3] Variable Energy Cyclotron Centre, HBNI, Kolkata-700064, India
[4] AGH University of Science and Technology, al. Mickiewicza 30, 30-059 Krakow, Poland
[5] CERN, Geneva 23, Switzerland

Abstract. The first estimation of the isothermal compressibility (k_T) of matter is presented for a wide range of collision energies from $\sqrt{s_{NN}}$ = 7.7 GeV to 2.76 TeV. k_T is estimated with the help of event-by-event charged particle multiplicity fluctuations from experiment. Dynamical fluctuations are extracted by removing the statistical fluctuations obtained from the participant model. k_T is also estimated from event generators AMPT, UrQMD, EPOS and a hadron resonance gas model. The values of isothermal compressibility are estimated for the Large Hadron Collider (LHC) energies with the help of the event generators.

1 Introduction

To understand the behaviour of the system formed at high temperature and energy density, it is important to understand the thermodynamic state of matter formed in high-energy heavy-ion collisions. Thermodynamic properties of the system, i.e, a set of response functions, like specific heat, compressibility, and different susceptibilities are directly related to event-by-event fluctuation observables, which are experimentally measurable.

2 Methodology

Isothermal compressibility (k_T) is the measure of relative change in volume with respect to change in pressure, at constant temperature.

In the Grand Canonical Ensemble (GCE), variance of particle multiplicity distribution is related to k_T by [1, 2],

$$\sigma^2 = \frac{k_B T \langle N \rangle^2}{V} k_T \tag{1}$$

where k_B is Boltzmann constant. Charged particle multiplicity fluctuations can be defined by scaled variance (ω_{ch}), which is the variance scaled over the mean charged particle multiplicity ($\mu \equiv \langle N \rangle$). Thus, we get a connection between the multiplicity fluctuations and k_T through the following equation as,

$$\omega_{ch} = \frac{\sigma^2}{\mu} = \frac{k_B T \mu}{V} k_T \tag{2}$$

Thus, applying GCE properties to experimental measurements at mid-rapidity and with the help of N_{ch}, T and V of the system, k_T may be estimated at the chemical freeze-out, where the particle ratios get fixed and no new particle is generated later.

* Corresponding author: maitreyee.mukherjee@cern.ch

3 Analysis

3.1 Extraction of dynamical multiplicity fluctuations

It is very important to extract dynamical multiplicity fluctuations in order to calculate k_T. A detailed study of the charged particle multiplicity fluctuations including centrality as well as beam-energy dependence, with proper acceptance corrections have been presented in [3]. The measured ω_{ch} have contributions from statistical as well as dynamical source; the dynamical part is related to thermodynamics. Here, the statistical fluctuations are estimated with the help of the participant model as described in [4], where ω_{ch} is given by,

$$\omega_{ch} = \omega_n + <n>\omega_{Npart} \tag{3}$$

where n is the number of charged particles per participant, ω_n and ω_{Npart} denote fluctuations in n and number of participants, respectively. From proton-proton collision data, and the formulation given in [5], scaled variance from the participant model is estimated and by subtracting this statistical contribution from the experimentally found multiplicity fluctuations, the dynamical multiplicity fluctuations ($\omega_{ch,dyn}$) are extracted [6]. The results are presented in Fig. 1.

3.2 Multiplicity fluctuations from event generators

To validate the results from the experimental data, the results from three event generators, i.e, UrQMD, EPOS and AMPT (both default (deft) and String Melting (SM) mode) are studied in details, for Au+Au collisions at $\sqrt{s_{NN}}$ = 7.7 to 200 GeV. The centrality is selected using minimum bias distributions within 0.5 < |η| <1, and 0.2 < p_T < 2 GeV/c. Narrow centrality bins are chosen to minimise geometrical fluctuations.

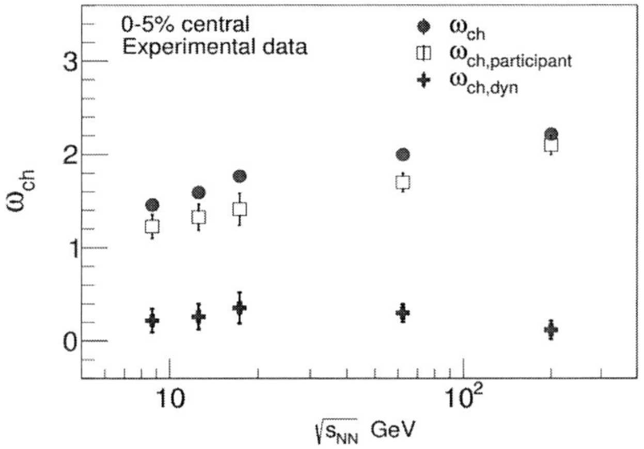

Fig. 1. Beam-energy dependence of multiplicity fluctuations for central (0-5%) Au+Au (Pb+Pb) collisions from experiments. Dynamical multiplicity fluctuation, after subtracting the statistical part from the participant model, is presented too.

The results are estimated for 5% centrality bins, with the application of centrality bin width corrections. For each case, dynamical fluctuations are extracted in the procedure described in section 3.1.

3.3 Estimation of k_T from Hadron Resonance Gas (HRG) model

The calculations of k_T in the HRG model are performed in terms of the species (denoted by subscript 'i') dependence of hadrons instead of the total number of charged particles. Considering the pressure P as a function of T and $\{\mu_i\}$, additionally, with fixed N_i, k_T within HRG can be written as,

$$k_T|_{T,\{<N_i>\}} = \frac{1}{V}\frac{1}{\Sigma_i\left(\frac{\partial P}{\partial \mu_i}\right)^2 / \left(\frac{\partial N_i}{\partial \mu_i}\right)} \quad (4)$$

Following Eq. (4), k_T is estimated within the HRG model in Au+Au collisions as a function of collision energy. Results are compiled with the results from experiment, in Fig. 2.

4 Results and discussions

For the evaluation of k_T from experiment, the values of dynamical-scaled variances are used. The volume and chemical freeze-out temperatures (T_{ch}) are taken from Ref. [7]. The beam-energy dependence of k_T is shown in Fig. 2. From experimental data, we observe that k_T remains almost constant at all energies. Event generators show a decreasing trend from lower to higher energies. For LHC energies, k_T values are presented from AMPT and EPOS.

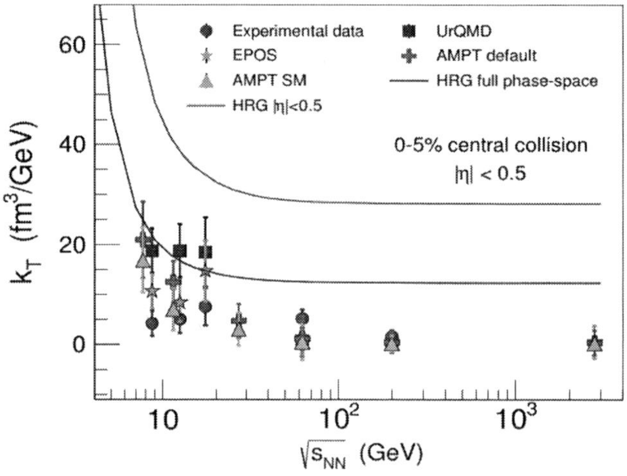

Fig. 2. Isothermal compressibility (k_T) as a function of collision energy for available experimental data for central (0-5%) Au+Au (Pb+Pb) collisions. Estimations of k_T from event generators and HRG model are presented, too.

The errors give the extent of the estimated values. For HRG, k_T is rapidly decreasing from 2 to 20 GeV, and remains constant at higher energies, both for full phase space and for $|\eta| < 0.5$. This observation suggests that collision system is less compressible at higher energies. The estimation of isothermal compressibility (k_T), specific heat (c_v) [8], etc., help to obtain the Equation of State (EOS) of matter, hence, the measurements of these quantities are of immense importance.

This work used resources of the LHC grid computing centres at Variable Energy Cyclotron Centre and Bose Institute, Kolkata.

References

1. S. Mrowczynski, Phys.Lett.B **430** (1998) 9.
2. A. Adare *et al.* (PHENIX Collaboration), Phys. Rev. C **78** (2008) 044902.
3. M. Mukherjee *et al.*, J. Phys. G: Nucl. Part. Phys. **43** (2015) 085102.
4. H. Heiselberg, Phys. Rept. **351** (2001) 161.
5. M. M. Aggarwal *et al.* (WA98 Collaboration), Phys. Rev. C **65** (2002) 054912.
6. M. Mukherjee *et al.*, arxiv : 1708.08692 [nucl-ex].
7. S. Chatterjee *et. al.*, Adv. High Energy Phys. (2015) 349013.
8. S. Basu *et. al.*, Phys. Rev. C **94** (2016) 044901.

Probing the hadronic phase with resonances of different lifetimes in Pb-Pb collisions with ALICE

Neelima Agrawal[1,*] (on behalf of ALICE collaboration)

[1] *Indian Institute of Technology Bombay, India*

Abstract. The ALICE experiment has measured the production of a rich set of hadronic resonances, such as $\rho(770)^0$, $K^*(892)^0$, $\phi(1020)$, $\Sigma^{\pm}(1385)$, $\Lambda(1520)$ and Ξ^{*0} in pp, p-Pb and Pb-Pb collisions at various energies at the LHC. A comprehensive overview and the latest results are presented in this paper. Special focus is given to the role of hadronic resonances for the study of final-state effects in high-energy collisions. In particular, the measurement of resonance production in heavy-ion collisions has the capability to provide insight into the existence of a prolonged hadronic phase after hadronisation. The observation of the suppression of the production of $\Lambda(1520)$ resonance in central Pb-Pb collisions at $\sqrt{s_{\rm NN}}$ =2.76 TeV adds further support to the existence of such a dense hadronic phase, as already evidenced by the ratios $K^*(892)^0/K$ and $\rho(770)^0/\pi$.

1 Introduction

Lattice QCD calculations performed with 2 + 1 flavors of quarks predict that at high enough temperature (T_c ~154 ± 9 MeV) and energy density (> 1 GeV/fm^3), nuclear matter undergoes a phase transition to a deconfined state of quarks and gluons known as the Quark-Gluon Plasma (QGP) [1]. These critical conditions are experimentally achieved by colliding heavy ions at ultra-relativistic energies [2, 3]. In such collisions, a fireball of very hot and dense partonic matter is formed and persists for a very short time (few fm/c). The QGP matter cools down and transitions back to confined hadronic matter, which subsequently undergoes chemical and kinetic freeze-out. Due to their lifetimes shorter or comparable to the lifetime of the fireball (~ 10 fm/c [4]), hadronic resonance are good probes to investigate the properties of the hadronic medium existing in between the chemical and kinetic freeze-out. Resonances are produced at hadronisation and can decay inside the hadronic medium. If the hadronic phase lasts long enough, the decay daughters of very short-lived resonances experience its full evolution and suffer rescattering via elastic collisions in the dense hadronic medium, which could modify their correlations and hence the experimentally measured resonance yields. As a consequence, the reconstructible resonance yield is lower than the original yield. Alternatively, depending upon the density and duration of the hadronic medium, hadrons might interact pseudoelastically to form resonances, which can increase the resonance yield. The competing effects of rescattering and regeneration on the final reconstructible resonance yields is dependent on the time scale between chemical and kinetic freeze-out, the relevant hadronic interaction cross sections and the resonance

*e-mail: neelima.agrawal@cern.ch

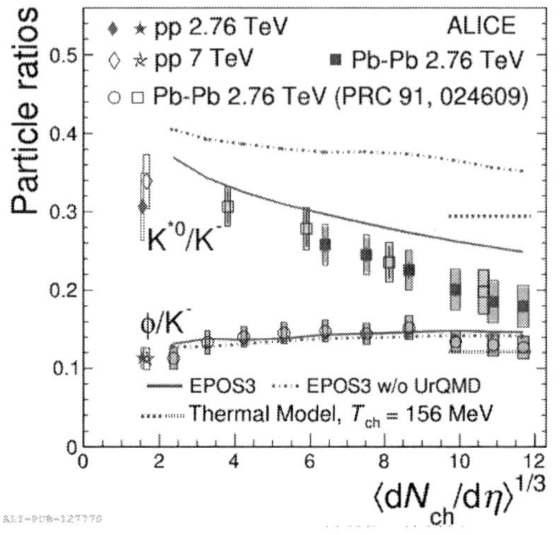

Figure 1. The p_T-integrated ratios of $K^*(892)^0/K^-$ and $\phi(1020)/K^-$ as a function of $\langle dN_{ch}/d\eta\rangle^{1/3}$ measured at mid-rapidity in pp collisions at $\sqrt{s} = 2.76$ TeV and 7 TeV and Pb–Pb collisions at $\sqrt{s_{NN}} = 2.76$ TeV [8]. The statistical, total systematics and centrality uncorrelated systematic uncertainties are represented by the bars, empty boxes and shaded boxes, respectively. The expectations from a SHM calculation with a chemical freeze-out temperature of 156 MeV for the most central collisions and EPOS3 calculations are shown.

lifetime. The Statistical Hadronisation Model (SHM) predicts ratios of resonance yields to stable particle yields at the chemical freeze-out stage [5–7]. Deviation from the model predictions can be useful to identify the possibility of the presence of hadronic phase. Figure 1 shows the ratio of $K^*(892)^0$ (τ = 4.2 fm/c) and ϕ (τ = 46.2 fm/c) to kaons as a function of $\langle dN_{ch}/d\eta\rangle^{1/3}$ [8]. The production of the short-lived $K^*(892)^0$ is suppressed in central Pb–Pb collisions, suggesting that the phenomenon might be due to the rescattering of the $K^*(892)^0$ decay products within the dense hadronic medium before the kinetic freeze-out. No suppression is observed for the $\phi(1020)$ mesons which could indicate that it mostly decays outside the fireball due to its longer lifetime.

2 Analysis Method and Results

Several hadronic resonances with different lifetimes have been measured by ALICE at midrapidity ($|y| < 0.5$) in pp, p–Pb and Pb–Pb collisions at various centre-of-mass energies. The properties of these resonances are listed in Table 1. The resonance signal is reconstructed using an invariant mass technique. The decay products are identified using the information from the TPC and TOF detectors. The invariant mass distribution of the correct-sign pair of decay daughters is constructed for each centrality/multiplicity class and p_T interval. The combinatorial background is estimated from either mixed events or like-charged tracks from the same event and subtracted from the signal distribution; the result is then fitted to extract the resonance yield. The p_T-integrated yields are obtained for all resonances and collision systems and are corrected for the decay channel branching ratio, detector

Table 1. Hadronic resonances with their lifetime, decay width, valance quark content, decay mode exploited for the measurements and corresponding branching ratio (BR) [9].

resonances	τ (fm/c)	Γ (MeV)	valance quark contents	decay mode	BR[%]
$\rho(770)^0$	1.3	149.1 ± 0.8	$(u\bar{u} + d\bar{d})/\sqrt{2}$	$\pi^+ + \pi^-$	100
$K^*(892)^0$	4.2	47.3 ± 0.5	$d\bar{s}$	$K^+ + \pi^-$	66.6
$\phi(1020)$	46	4.3 ± 0.02	$s\bar{s}$	$K^+ + K^-$	48.9
$\Sigma(1385)^+$	5.5	36.0 ± 0.7	uus	$p + \pi^- + \pi^+$	87.0
$\Lambda(1520)$	12.6	15.6 ± 1.0	uds	$K^- + p$	22.5
$\Xi(1530)^0$	21.7	9.1 ± 0.5	uss	$p + 2\pi^- + \pi^+$	66.7

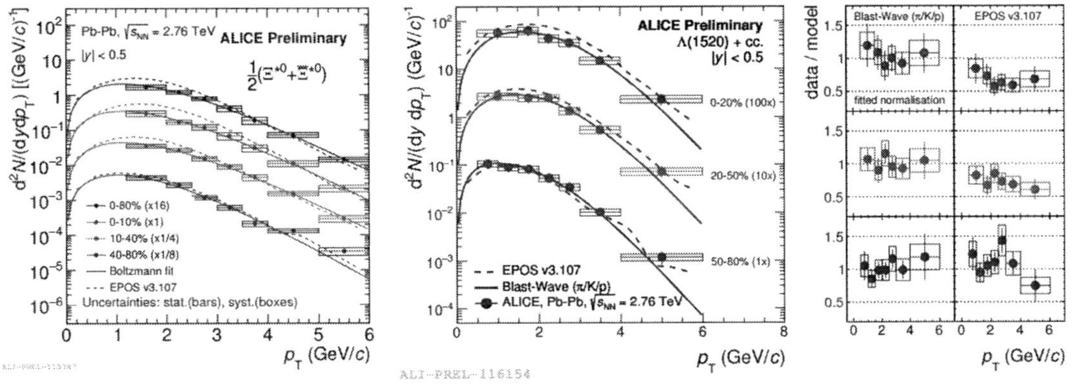

Figure 2. The p_T spectra of Ξ^{*0} (left) and $\Lambda(1520)$ (middle) in Pb–Pb collisions at $\sqrt{s_{NN}}$ = 2.76 TeV. The $\Lambda(1520)$ spectra ratios with respect to Blast-Wave and EPOS3 predictions (right).

acceptance and reconstruction efficiency. For example, Figure 2 shows the p_T-spectrum of Ξ^{*0} (left) (τ = 21.7 fm/c) and $\Lambda(1520)$ (middle) (τ = 12.6 fm/c) in Pb–Pb collisions measured in the p_T range 1.2-6.0 GeV/c and 0.5-6.0 GeV/c, respectively. The p_T spectra are fitted with appropriate parametrisations (Boltzmann for Ξ^{*0} and Blast-Wave for $\Lambda(1520)$) to obtain the particle yield in the unmeasured p_T region. The p_T distribution have also been compared with predictions from the EPOS3 model [10] (Figure 2 middle), which describes the full evolution of heavy ion collisions and embeds the UrQMD model for the description of the hadronic phase. This model reproduces the p_T distribution reasonably well but over-predicts the yield for central Pb–Pb collisions.

The integrated yield ratios of resonances to stable hyperons with the same strangeness content is calculated in order to investigate the impact of the hadronic phase on the reconstructed resonance yield. Figure 3 shows the $\Lambda(1520)/\Lambda$ (left) and Ξ^{*0}/Ξ (right) ratios as a function of $\langle dN_{ch}/d\eta\rangle^{1/3}$ in pp, p–Pb and Pb–Pb collisions at 7, 5.02 and 2.76 TeV, respectively. Predictions from Monte Carlo generators commonly used in pp and p–Pb collisions, EPOS3 and various SHM predictions [5–7] for Pb–Pb collisions are also shown. The $\Lambda(1520)/\Lambda$ shows suppression in central Pb–Pb collisions with respect to peripheral Pb–Pb and p–Pb and pp collisions. The SHM predictions for Pb–Pb collisions overestimate the data. On the other hand, EPOS3 qualitatively reproduces the suppression trend of ratio, however, it overestimates the value. The $\Lambda(1520)/\Lambda$ ratio suppression trend indicates the dominance of rescattering of the decay daughters over regeneration. The Ξ^{*0}/Ξ ratio shows relatively weak suppression from pp, p–Pb, peripheral Pb–Pb to central Pb–Pb collisions. The ratio values are underestimated by PYTHIA8 in pp, DPMJET in p–Pb and overestimate by various SHM calculations in Pb–Pb collisions. In addition, EPOS3 with UrQMD predicts a slight decrease of the Ξ^{*0}/Ξ ratio from peripheral to central Pb-Pb collisions and a yield value which is consistent with the SHM predictions for most central Pb-Pb. The trend is qualitatively consistent with the observed one, given the present uncertainties of the measurement, however the measured yield ratio is lower than the predictions.

3 Conclusion

In this paper, we have presented the first measurement of $\Lambda(1520)$ and Ξ^{*0} in pp, p–Pb and Pb–Pb collisions at $\sqrt{s}$ = 7, $\sqrt{s_{NN}}$ = 5.02 and 2.76 TeV, respectively. The $\Lambda(1520)/\Lambda$ ratio is suppressed in central collisions compared to the values observed in peripheral Pb–Pb, pp and p–Pb collisions. This is similar to the behaviour observed in the ρ/π and $K^*(892)^0/K^-$ ratios. The suppression of $\Lambda(1520)$ in central Pb–Pb collisions suggests the dominance of (pseudo)elastic rescattering of decay daugh-

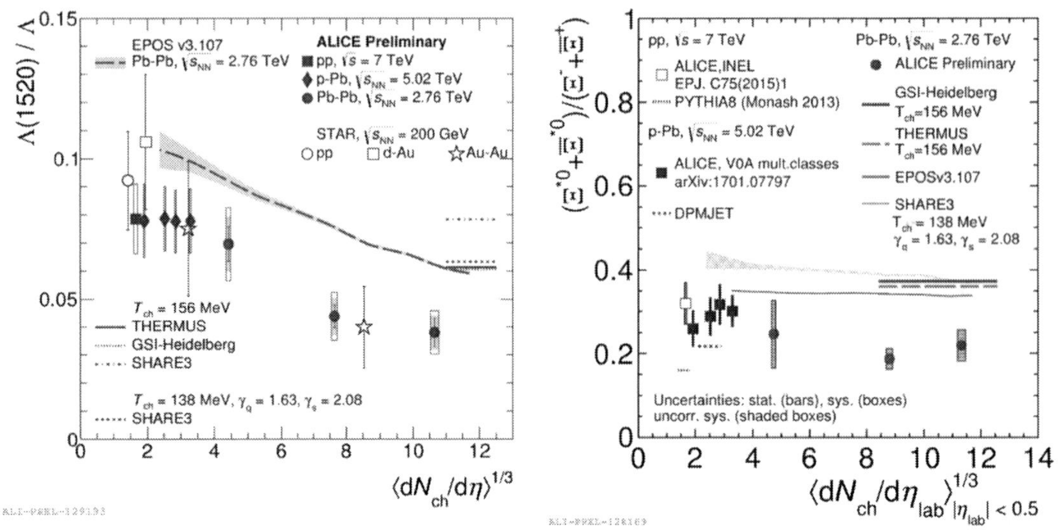

Figure 3. The p_T-integrated ratio of $\Lambda(1520)/\Lambda$ (left) and Ξ^{*0}/Ξ (right) as a function of $\langle dN_{ch}/d\eta\rangle^{1/3}$ measured in Pb–Pb collisions at $\sqrt{s_{NN}} = 2.76$. The statistical errors, systematic uncertainties and centrality uncorrelated systematic uncertainties are represented by the bars, empty boxes and shaded boxes, respectively. Predictions from several SHM models [5–7] as well as the prediction from EPOS3 [10] are compared to the data.

ter particles in the hadronic phase. Moreover, the suppressed value of $\Lambda(1520)/\Lambda$ ratio at highest $\langle dN_{ch}/d\eta\rangle^{1/3}$ could be attributed to a larger fireball size and its life-time ($\tau_{Hadr} \sim 10$ fm/c) accessible at LHC energies. The Ξ^{*0}/Ξ ratio exhibits a very weak suppression from pp, p–Pb, peripheral Pb–Pb to central Pb–Pb collisions. The $\Lambda(1520)/\Lambda$ and Ξ^{*0}/Ξ ratios in central Pb–Pb collisions are compared and found to be lower than the calculations from several implementations of the Statistical Hadronisation Models in Grand Canonical description and one non-equilibrium model. The deviations of these ratios in the data as compared to the SHM models are useful to decode effects related to the interactions in hadronic medium. The use of SHM models in combination with a microscopic description of the interactions in the hadronic phase (UrQMD) allow one to estimate the time span between chemical and kinetic freeze-out. These results further support the existence of a hadronic phase lasting long enough to cause a significant reduction of the reconstructible yield of short-lived resonances.

References

[1] F. Karsch, Lect. Notes Phys. **583**, 209 (2002), hep-lat/0106019
[2] G. Roland, K. Safarik, P. Steinberg, Prog. Part. Nucl. Phys. **77**, 70 (2014)
[3] A. Ohnishi, Prog. Theor. Phys. Suppl. **193**, 1 (2012), 1112.3210
[4] K. Aamodt et al. (ALICE), Phys. Lett. **B696**, 328 (2011), 1012.4035
[5] A. Andronic et al., Nucl. Phys. **A772**, 167 (2006), nucl-th/0511071
[6] S. Wheaton, J. Cleymans, J. Phys. **G31**, S1069 (2005), hep-ph/0412031
[7] M. Petran et al., Comput. Phys. Commun. **185**, 2056 (2014), 1310.5108
[8] B. Abelev et al. (ALICE), Eur. Phys. J. **C72**, 2183 (2012), 1208.5717
[9] C. Patrignani et al. (Particle Data Group), Chin. Phys. **C40**, 100001 (2016)
[10] A.G. Knospe et al., Phys. Rev. **C93**, 014911 (2016), 1509.07895
[11] J. Adam et al. (ALICE), Phys. Rev. **C93**, 024905 (2016), 1507.06842

EPJ Web of Conferences **171**, 15002 (2018) https://doi.org/10.1051/epjconf/201817115002
SQM 2017

Resonances in a sudden chemical freeze-out model

Viktor Begun[1,⋆]

[1] *Warsaw University of Technology, Faculty of Physics, Koszykowa 75, 00-662 Warsaw Poland*

Abstract. The prediction for p_T spectra of various resonances produced in Pb+Pb collisions at 2.76 TeV at the LHC in equilibrium and non-equilibrium models is made. It includes the η, $\rho(770)$, $\Sigma(1385)$, $\Lambda(1520)$, and $\Xi(1530)$. The apparent differences may allow to distinguish between the models.

A chemical non-equilibrium model [1] with a single freeze-out [2] appeared to be rather successful in describing the LHC ALICE data at 2.76 TeV for various particles [3, 4]. In this model the mean *multiplicities* are described with the use of four thermodynamic parameters: temperature T, volume V, and two non-equilibrium parameters - γ_s and γ_q. It fixes the area under the curve for the p_T spectra. The *form* of the spectra is best reproduced by the Hubble-like single freeze-out hyper-surface. Then, the *slopes* of the spectra are described with only one extra parameter - the ratio of the freeze-out time τ_f to the freeze-out radius r_f, because their combination, the cylinder of volume $\pi * r^2 * \tau$, is equal to the volume V, which is determined from multiplicities on the previous step.

It appears that the p_T spectra of pions, kaons, protons, $K^*(892)^0$, and $\phi(1020)$ are described by the same parameters in the single freeze-out model [3, 4]. This is very surprising for the $K^*(892)^0$ and the $\phi(1020)$, because the first one is short living, while the second one is long living. The description of both of them may question the necessity of the long re-scattering phase, which is also successfully used to describe the ALICE data [5]. It may also indicate that the non-equilibrium, as implemented in [3, 4], may effectively include the re-scattering, because γ_q and γ_s are equivalent to non-equilibrium chemical potentials for each particle, see [3, 4]. It is important to differentiate between the equilibrium with the re-scattering, and the single sudden freeze-out in the non-equilibrium, because the non-equilibrium also leads to pion condensation [6, 7].

A good test for the non-equilibrium single freeze-out scenario [3, 4] is the comparison to different resonances, especially strange resonances, because this scenario requires a special relation between the strange and the non-strange chemical potentials, depending on the quark content of a resonance. The heavy Λ, Ξ and Ω can be still described by the non-equilibrium very well, if one assumes a smaller slope for them [4]. This introduces the dependence on the mass of the resonance, but is also supported by smaller flow of heavy particles in other approaches, see e.g. [8]. The parameters obtained in the fit to the 2.76 TeV Pb+Pb LHC data in equilibrium (EQ), non-equilibrium (NEQ) [3, 4], and non-equilibrium with the possibility of pion Bose-Einstein condensation (BEC) on the ground state [6, 7] in hadron-resonance gas, using correspondingly modified SHARE [9] and THERMINATOR [2] codes, are shown in Fig. 1. One can see that the system is closer to the scenario with the condensate in central collisions. However, the uncertainty is rather large, which means that more mean multiplicities are needed to constrain the fit. At large chemical potentials finite size effects should be taken into

⋆e-mail: viktor.begun@gmail.com

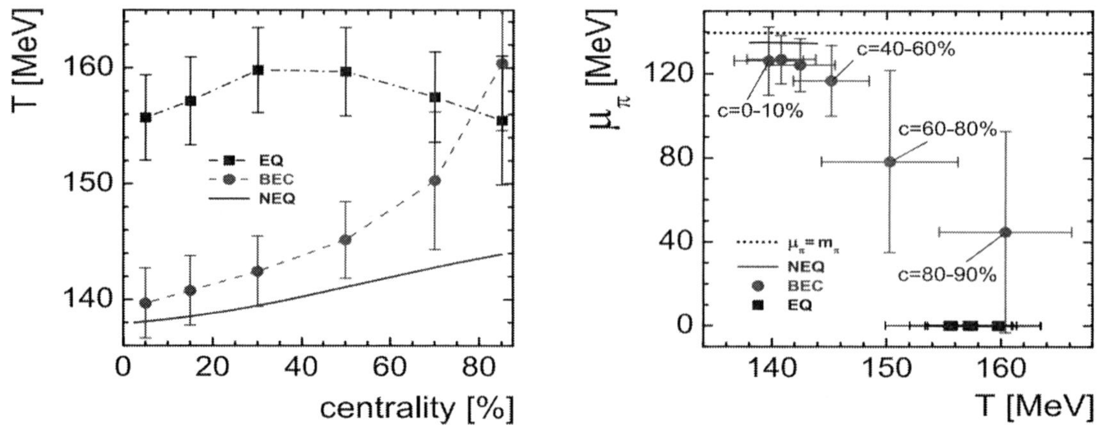

Figure 1. Temperature and non-equilibrium pion chemical potential, $\mu_\pi = 2T \ln \gamma_q$, in Pb+Pb at 2.76 TeV [6, 7].

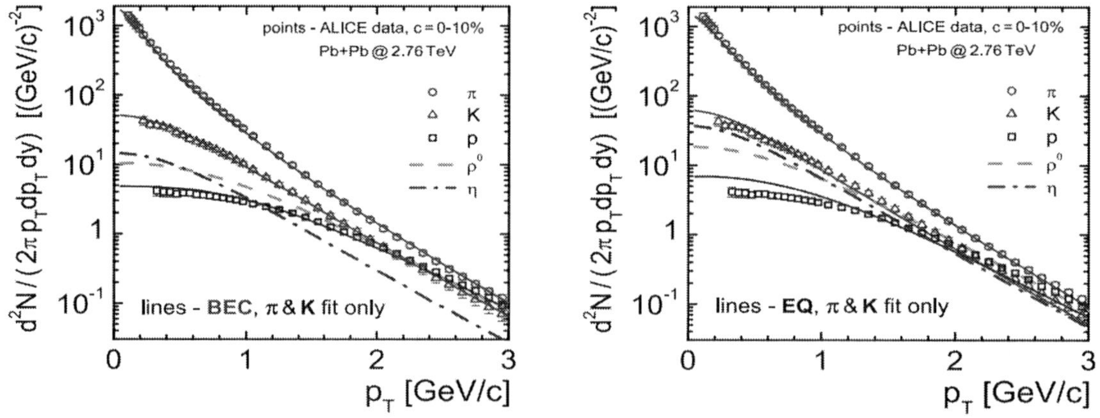

Figure 2. The fit of pions and kaons, and the prediction for protons, ρ and η in 0-10% centrality window.

account. The corresponding BEC fit of pion and kaon spectrum gives a good description of protons, while protons in EQ require a different freeze-out hypersurface. The amount and spectra of ρ^0 and η mesons are significantly different in EQ and BEC, see Fig. 2. Charged pions favor BEC[1], see Fig. 3, while the η/π^0 ratio favors EQ (data from A. Morreale [10]). However, the uncertainty, again, seems to be too large to judge. Both BEC and EQ explain K_S^0 and ϕ spectra similarly good, see Fig. 4. The $K^*(892)^0$ is closer to BEC prediction. Note, that $K^*(892)^0$ was not included neither in the fit of mean multiplicities, nor in the fit of spectra (data from [11]). It means that BEC can be treated as an effective parameterizations of the freeze-out. Strange baryons require different freeze-out hypersurface compared to that one for π, K, p, K_S^0, K^*, and ϕ, see Fig. 5, and also [8, 12]. There is the mass dependence in BEC - the heavier the baryon, the smaller is the slope, i.e. the flow, or, equivalently, smaller radius of the hypersurface. BEC predicts similar multiplicities and spectra of $\Lambda(1520)$ and $\Xi(1530)$, see Fig. 6. EQ predicts larger multiplicity difference between $\Lambda(1520)$ and

[1] According to my best knowledge, there is no other model that explained this low p_T excess of pions together with proton spectra without giving pions and protons extra parameters since the data appearance in 2012.

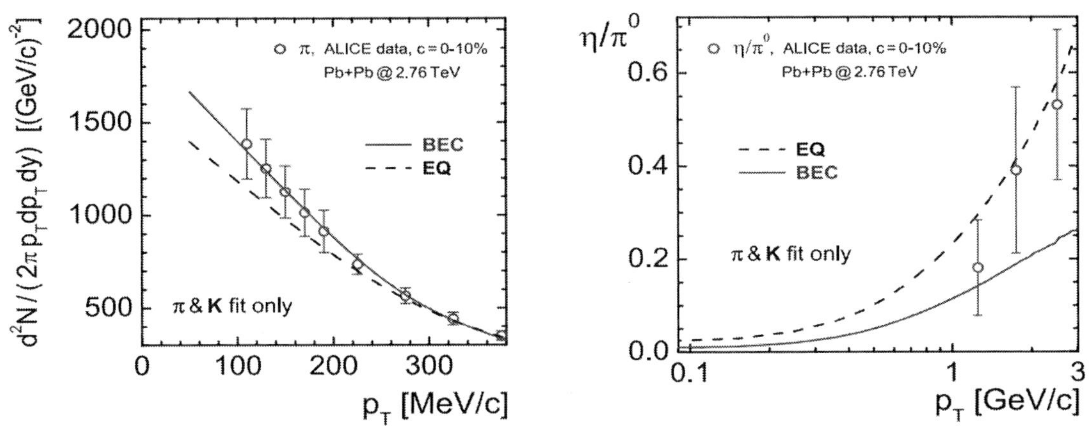

Figure 3. Low p_T charged pions and η/π^0 in EQ and BEC.

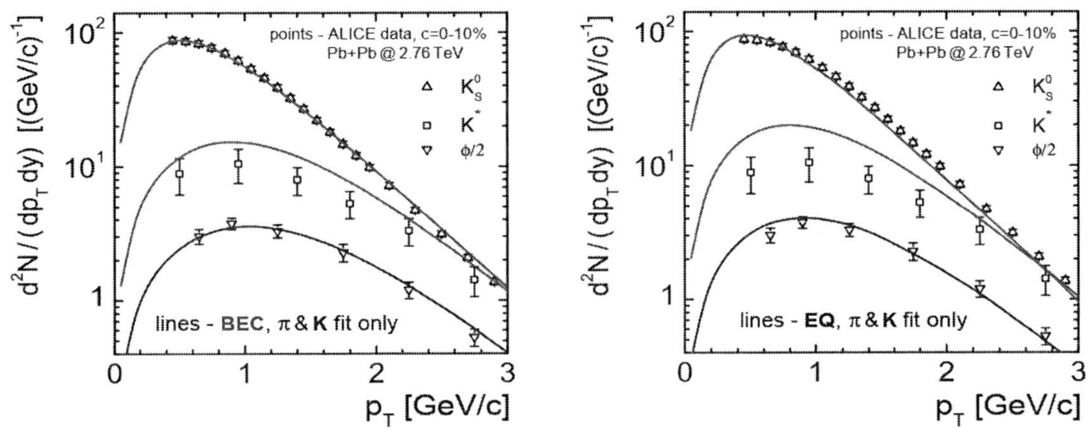

Figure 4. Strange mesons obtained for the hypersurface, which was fitted to pions and kaons only.

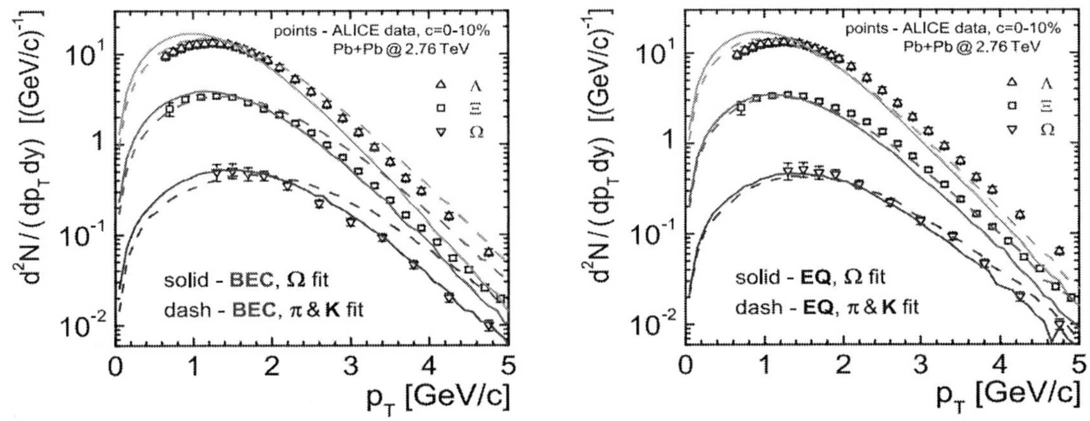

Figure 5. Strange baryons obtained for the hypersurface that was fitted either to pions and kaons, or to Ω baryons.

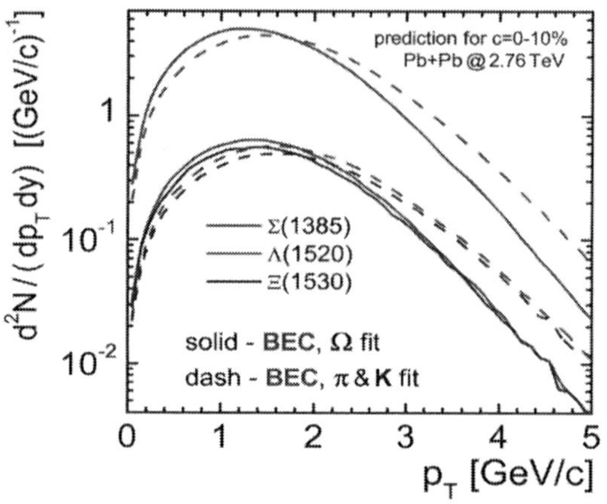

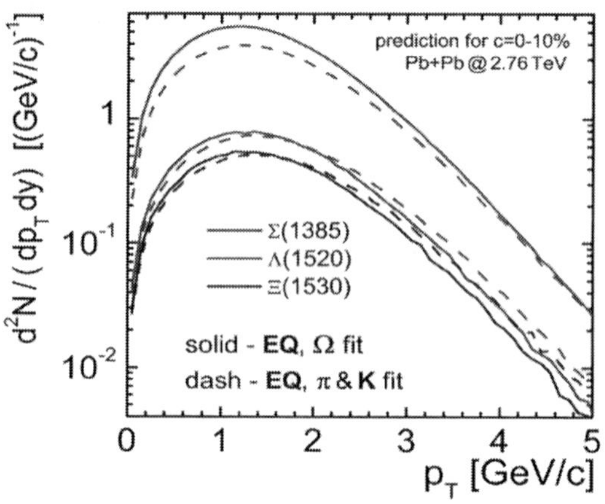

Figure 6. Prediction for the half-sum of $\Sigma^{+}(1385)$, $\Sigma^{0}(1385)$, $\Sigma^{-}(1385)$, and their anti-particles, as well as $(\Lambda^0(1520) + \overline{\Lambda^0}(1520))/2$, and $(\Xi^0(1530) + \overline{\Xi^0}(1530))/2$.

$\Xi(1530)$ than BEC. There is a significant dependence of the spectra on the freeze-out hypersurface for heavy strange baryons in BEC, while in EQ only $\Sigma(1385)$ is sensitive to the hypersurface, see Figs. 5, 6.

Therefore, one may conclude that π, K, K_S^0 and ϕ particles may have a common freeze-out hypersurface in both BEC and EQ models. The BEC additionally allows to explain protons, low p_T pions, and $K^*(892)^0$. Strange baryons require different freeze-out in both models. The predictions for ρ^0, η, $\Sigma(1385)$, $\Lambda(1520)$, and $\Xi(1530)$ are significantly different in BEC and EQ.

Acknowledgments: I thank L. Leardini, A. Marin, C. Markert and A. Morreale for discussions. The research was partly carried out in laboratories created under the project "Development of research base of specialized laboratories of public universities in Swietokrzyskie region" no POIG 02.2.00-26-023/08 dated May 19, 2009.

References

[1] M. Petrán, J. Letessier, V. Petráček and J. Rafelski, Phys. Rev. C **88**, no. 3, 034907 (2013).
[2] M. Chojnacki *et al.*, Comput. Phys. Commun. **183**, 746 (2012).
[3] V. Begun, W. Florkowski and M. Rybczynski, Phys. Rev. C **90**, no.1, 014906 (2014).
[4] V. Begun, W. Florkowski and M. Rybczynski, Phys. Rev. C **90**, no.5, 054912 (2014).
[5] A. G. Knospe et al., Phys. Rev. C **93**, no.1, 014911 (2016).
[6] V. Begun and W. Florkowski, Phys. Rev. C **91**, 054909 (2015).
[7] V. Begun, Phys. Rev. C **94**, no.5, 054904 (2016).
[8] I. Melo and B. Tomasik, J. Phys. G **43**, no.1, 015102 (2016).
[9] M. Petran *et al.*, Comput. Phys. Commun. **185**, 2056 (2014).
[10] A. Morreale [ALICE Collaboration], J. Phys. Conf. Ser. **779**, no. 1, 012051 (2017).
[11] J. Adam *et al.* [ALICE Collaboration], Phys. Rev. C **95**, no. 6, 064606 (2017).
[12] S. Chatterjee, B. Mohanty and R. Singh, Phys. Rev. C **92**, no. 2, 024917 (2015).

Kaon femtoscopy in Au+Au collisions at $\sqrt{s_{\mathrm{NN}}}$ = 200 GeV at the STAR experiment

Jindřich Lidrych[1,*] for the STAR Collaboration

[1] *Faculty of Nuclear Science and Physical Engineering, Czech Technical University in Prague Břehová 7, Prague 1, 115 19, Czech Republic*

Abstract. In this proceedings, the STAR preliminary results on femtoscopic correlations of identical kaons from Au+Au collisions at $\sqrt{s_{\mathrm{NN}}}$=200 GeV are presented. The measured kaon source radii are studied as a function of collision energy as well as centrality and transverse pair mass m_{T}. In addition, extracted kaon blast-wave freeze-out parameters are presented.

1 Introduction

Quantum Chromodynamics calculations predict that under sufficiently high temperature or energy density nuclear matter undergoes a phase transition from hadrons to a state of deconfined quarks and gluons, the Quark-Gluon Plasma. The properties of this novel state of matter have been extensively studied in high-energy nuclear collisions at RHIC. Two-particle interferometry measurement, known as femtoscopy or HBT, is a standard method of studying space-time characteristics of the created system during the heavy-ion collisions [1]. It has been most common to perform the femtoscopic analyses with pions [2, 3] as they are the most abundantly produced particles in heavy-ion collisions. Recently collected high-statistic datasets allow measurements with the heavier but less abundant particles. In this proceedings, we report on the latest results from kaon femtoscopy at the STAR experiment. In comparison with pions, the kaons are less affected by the long-lived resonance decays, contain strange quarks and have a smaller cross-section with the hadronic matter. Hence the kaon femtoscopy can be sensitive to the earlier stage of the collisions evolution and can provide additional information about the created system in the heavy-ion collisions.

2 Two-particle correlation function

Experimental correlation function is defined as a ratio of two-particle spectrum from the same events, $N(\vec{q})$, and two-particle spectrum from the mixed events, $D(\vec{q})$,:

$$C(\vec{q}) = \frac{N(\vec{q})}{D(\vec{q})}, \tag{1}$$

*e-mail: jlidrych@gmail.com

where $\vec{q}$ is the pair 4-momentum difference of the first and the second particle decomposed in the Bertsch-Pratt coordinate system [4, 5] to q_{out}, q_{side}, and q_{long} components. The source radii are extracted from the correlation functions by the standard Bowler-Sinyukov [6, 7] method, which is based on the separation of the Coulomb interaction, $K(q_{\text{inv}}, R_{\text{inv}})$, from the pure Bose-Einstein correlation function. In this analysis, the experimental correlation functions were fitted assuming Gaussian shape of the correlation function

$$C(\vec{q}) = N\left[1 - \lambda + \lambda K(q_{\text{inv}}, R_{\text{inv}})\left(1 + \exp\left(-R_{\text{out}}^2 q_{\text{out}}^2 - R_{\text{side}}^2 q_{\text{side}}^2 - R_{\text{long}}^2 q_{\text{long}}^2\right)\right)\right], \quad (2)$$

where λ is the correlation strength, N is the normalization and R_{out}, R_{side} and R_{long} are the Gaussian source radii.

3 Results

3.1 Kaon source radii

Figure 1 shows the STAR preliminary results on kaon femtoscopic radii from Au+Au collisions at $\sqrt{s_{NN}} = 200$ GeV as a function of the centrality and the average transverse pair momentum $k_T = \left|\vec{p_1} + \vec{p_2}\right|_T/2$, where $\vec{p_1}$ and $\vec{p_2}$ is the 3-momentum of the first and second particle, respectively. The presented analysis was performed for 4 centralities, namely 0-10%, 10-30%, 30-50% and 50-80% most central Au+Au collisions and for 6 k_T bins. The correlation functions were constructed separately for positive and negative kaon pairs. Within errors, no difference between source radii for positive and negative kaons has been found. The extracted kaon source radii follow typical centrality and transverse pair momentum dependence. The expansion of the system and the transverse flow causes the decrease of the R_{out} and R_{side} with the increasing k_T. The longitudinal expansion is encoded in the falling of R_{long} with the increasing k_T.

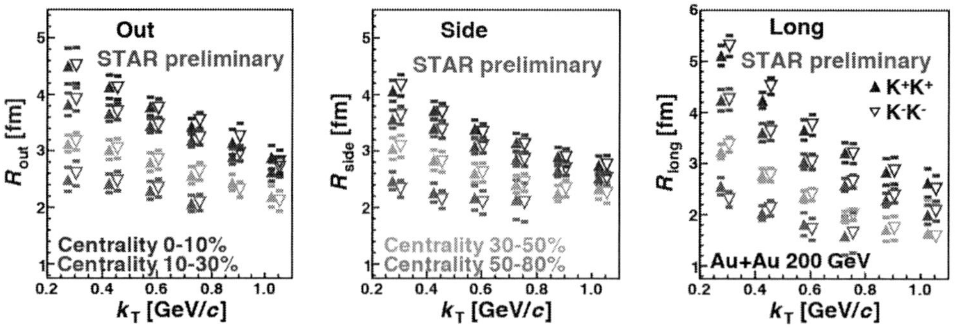

Figure 1. The R_{out}, R_{side} and R_{long} as a function of the transverse pair momenta k_T for 4 centrality (0-10%, 10-30%, 30-50% and 50-80%) for positive kaon pairs (solid triangles) and negative kaon pairs (open triangles) from Au+Au collisions at $\sqrt{s_{NN}} = 200$ GeV. The points are horizontally offsetted for better clarity.

Figure 2 shows a comparison of kaon and pion source radii for the most 0-5% and 30-40% central Au+Au collisions. This analysis extended the previous pion femtoscopy results [3] to higher m_T using the ToF detector, which allows pion identification up to momenta $p \sim 1.55$ GeV/c. The R_{out} values for kaons are systematically larger than those of pions, while the R_{side} values for kaons and pions are comparable, and the m_T-dependence of R_{long} is different between kaons and pions. Similar behaviors have been recently reported by the ALICE collaboration from Pb+Pb collisions at $\sqrt{s_{NN}}$=2.76 TeV [8].

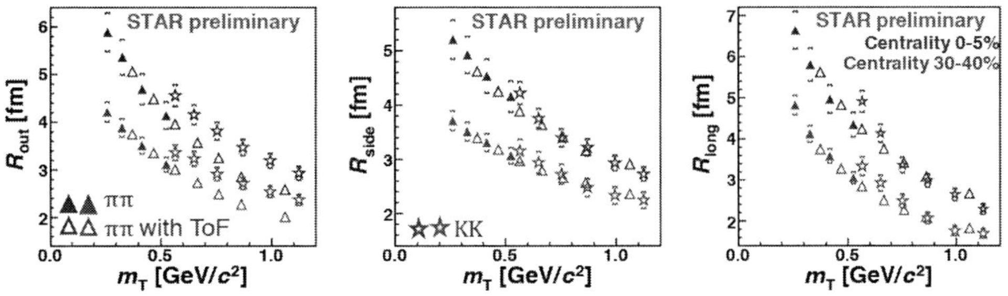

Figure 2. The comparison of the kaon and pion source radii from Au+Au collisions at $\sqrt{s_{NN}}$=200 GeV in STAR. The solid triangles represents the pion source radii taken from [3] .

These behaviors indicate different space-time configuration at the kinetic freeze-out for the pions and kaons. However, deeper understanding of the observed behavior needs a comparison with the full three-dimensional hydrodynamic calculations with rescattering in the hadronic phase.

3.2 Kinetic freeze-out parameters

One of the most commonly used source models in femtoscopy is the blast-wave parametrization, such as in [9], which aims to describe the system at the kinetic freeze-out with a minimal set of parameters. The extracted kaon source radii from Figure 1 and transverse particle spectra ($\pi^{\pm}$, $K^{\pm}$, p, $\bar{p}$) [10] were simultaneously fitted in order to obtain the following 5 parameters of the kinetic freeze-out: temperature T, radius of the system R, maximum transverse rapidity ρ_0, emission duration $\Delta\tau$ and the system proper lifetime τ. Figure 3 shows the extracted kinetic freeze-out parameters of the kaon source as a function of the collision centrality.

While the temperature, T, decreases, the system size, R, and the maximum transverse rapidity, ρ_0, increase with increasing centrality of the collisions. The expansion of the larger system takes longer as shown by the decreasing system evolution time τ and the emission duration last also longer $\Delta\tau$ for more central collisions. Figure 3 also shows that within systematic errors the STAR preliminary results on the blast-wave model kinetic freeze-out parameters for kaons are consistent with the published PHENIX results [12]. At the same time the comparison with STAR pion results from [2] reveals that while the extracted temperature is systematically larger for kaons than for pions, the radius of the system and the system proper time is smaller for kaons. Such behavior can indicate different kinetic freeze-out configuration for pions and kaons.

4 Conclusions

In this proceedings, preliminary results from the STAR experiment on identical kaon femtoscopy from Au+Au collisions at $\sqrt{s_{NN}}$ = 200 GeV have been presented. The kaon source radii R_{out}, R_{side} and R_{long} are extracted from the three-dimensional analysis for different collision centralities and transverse pair momenta. The extracted source radii show centrality and transverse pair momenta dependence typical for collectively expanding source. Parameters of blast-wave model describing the kaon source at freeze-out are extracted by a simultaneous fit of the kaon source radii and particle spectra. The parameters of the kinetic freeze-out configuration are different for kaons and pions within the blast-wave parametrization. Further comparisons to hydrodynamic models are required to reach a quantitative physics conclusion.

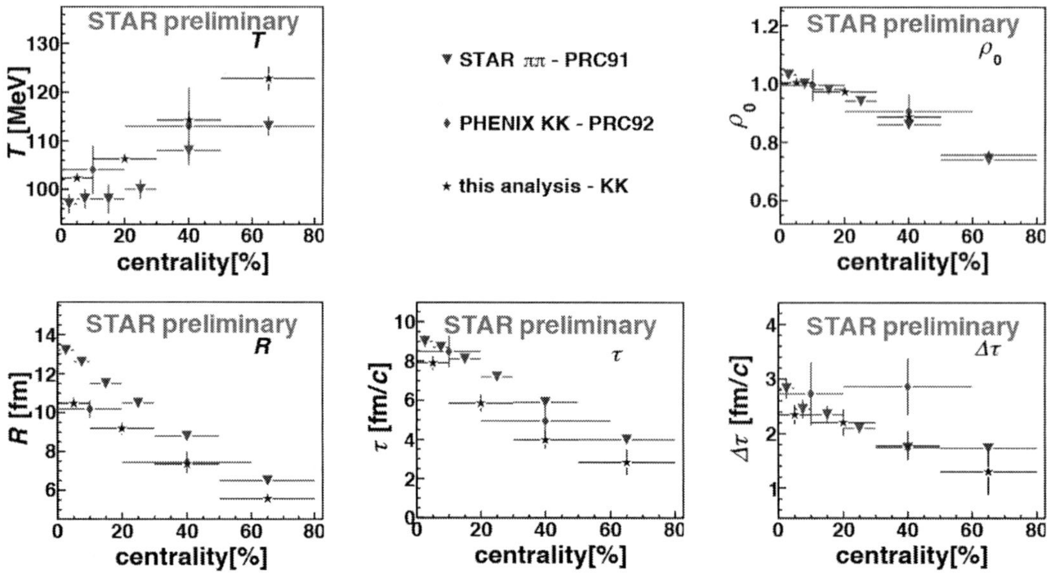

Figure 3. Fit results from kaon source radii: kinetic freeze-out temperature T, radius of the system R, maximum transverse rapidity ρ_0, emission duration $\Delta\tau$ and the system proper time τ as a function of centrality for Au+Au collisions at $\sqrt{s_{NN}} = 200$ GeV. The results from pion femtoscopy are taken from [11] and the PHENIX results are taken from [12].

Acknowledgments

This work was supported by the grant of the Grant Agency of the Czech Technical University in Prague, grant No. SGS16/238/OHK4/3T/14 and by the grant LG15001 of the Ministry of Education, Youth and Sports of the Czech Republic.

References

[1] M.A. Lisa, S. Pratt, R. Soltz, U. Wiedemann, Ann. Rev. Nucl. Part. Sci. **55**, 357 (2005)
[2] J. Adams et al. (STAR), Phys. Rev. **C71**, 044906 (2005)
[3] L. Adamczyk et al. (STAR), Phys. Rev. **C92**, 014904 (2015)
[4] G. Bertsch, M. Gong, M. Tohyama, Phys. Rev. **C37**, 1896 (1988)
[5] S. Pratt, T. Csorgo, J. Zimanyi, Phys. Rev. **C42**, 2646 (1990)
[6] M.G. Bowler, Phys. Lett. **B270**, 69 (1991)
[7] Yu. Sinyukov, R. Lednicky, S.V. Akkelin, J. Pluta, B. Erazmus, Phys. Lett. **B432**, 248 (1998)
[8] S. Acharya et al. (ALICE) (2017), 1709.01731
[9] F. Retiere, M.A. Lisa, Phys. Rev. **C70**, 044907 (2004)
[10] S.S. Adler et al. (PHENIX), Phys. Rev. **C69**, 034909 (2004)
[11] B.I. Abelev et al. (STAR), Phys. Rev. **C80**, 024905 (2009)
[12] A. Adare et al. (PHENIX), Phys. Rev. **C92**, 034914 (2015)

Beam energy and system dependence of rapidity-even dipolar flow

Niseem Magdy (For the STAR Collaboration)[1]*

[1] *Department of Chemistry, Stony Brook University, Stony Brook, NY, 11794-3400, USA*

Abstract. New measurements of rapidity-even dipolar flow, v_1^{even}, are presented for several transverse momenta, p_T, and centrality intervals in Au+Au collisions at $\sqrt{s_{NN}}$ = 200, 39 and 19.6 GeV, U+U collisions at $\sqrt{s_{NN}}$ = 193 GeV, and Cu+Au, Cu+Cu, d+Au and p+Au collisions at $\sqrt{s_{NN}}$ = 200 GeV. The v_1^{even} shows characteristic dependencies on p_T, centrality, collision system and $\sqrt{s_{NN}}$, consistent with the expectation from a hydrodynamic-like expansion to the dipolar fluctuation in the initial state. These measurements could serve as constraints to distinguish between different initial-state models, and aid a more reliable extraction of the specific viscosity η/s.

1 Introduction

Heavy-ion collisions (HIC) at the Relativistic Heavy Ion Collider (RHIC) and the Large Hadron Collider (LHC) are aimed at studying the properties of the strongly interacting quark-gluon plasma (QGP) created in such collisions. Recent studies have emphasized the use of anisotropic flow measurements to study the transport properties of the QGP [1–7]. A crucial question in these studies was the role of initial-state fluctuations and their influence on the uncertainties associated with the extraction of η/s for the QGP produced in HIC [8, 9]. This work emphasizes new measurements for rapidity-even dipolar flow, v_1^{even}, which could aid a distinction between different initial-state models and facilitate the extraction of η/s with better constraints.

Anisotropic flow is characterized by the Fourier coefficients, v_n, obtained from a Fourier expansion of the azimuthal angle (ϕ) distribution of the emitted particles [10]:

$$\frac{dN}{d\phi} \propto 1 + 2\sum_{n=1} v_n \cos(n(\phi - \Psi_n)), \tag{1}$$

where Ψ_n represents the nth-order event plane, the coefficients v_1, v_2 and v_3 are called directed, elliptic and triangular flow, respectively. The flow coefficients v_n are related to the two-particle Fourier coefficients $v_{n,n}$ as:

$$v_{n,n}(p_T^a, p_T^b) = v_n(p_T^a)v_n(p_T^b) + \delta_{NF}, \tag{2}$$

where p_T^a and p_T^b are the transvers momentum of particles (a) and (b), respectively, and δ_{NF} is a so-called non-flow (NF) term, which includes possible contributions from resonance decays, Bose-Einstein correlations, jets, and global momentum conservation (GMC) [11–15]. The directed flow, v_1,

* e mail: niseemm@gmail.com

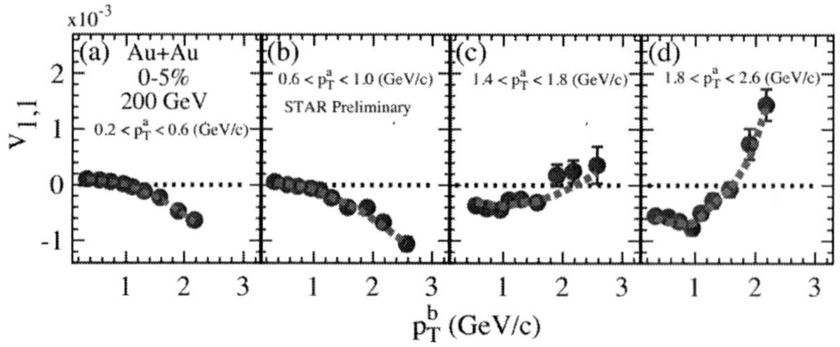

Figure 1. $v_{1,1}$ vs. p_T^b for several selections of p_T^a for 0-5% central Au+Au collisions at $\sqrt{s_{NN}}$ = 200 GeV. The dashed curve shows the result of the simultaneous fit with Eq. 5.

can be separated into an odd function of pseudorapidity (η) [16] which develops along the direction of the impact parameter, and a rapidity-even component [13, 17] which results from the effects of initial-state fluctuations acting in concert with a hydrodynamic-like expansion; $v_1(\eta) = v_1^{even}(\eta) + v_1^{odd}(\eta)$, where Ψ_1^{odd} and Ψ_1^{even} are uncorrelated. The magnitude of v_1^{even} is related to the fluctuations-driven dipole asymmetry ε_1 and η/s [14, 17, 18].

2 Measurements

The correlation function technique was used to generate the two-particle $\Delta\phi$ correlations:

$$C_r(\Delta\phi, \Delta\eta) = \frac{(dN/d\Delta\phi)_{same}}{(dN/d\Delta\phi)_{mixed}}, \tag{3}$$

where $(dN/d\Delta\phi)_{same}$ represent the normalized azimuthal distribution of particle pairs from the same event and $(dN/d\Delta\phi)_{mixed}$ represents the normalized azimuthal distribution for particle pairs in which each member is selected from a different event but with a similar classification for the vertex, centrality, etc. The pseudorapidity requirement $|\Delta\eta| > 0.7$ was also imposed on track pairs to minimize possible non-flow contributions associated with the short-range correlations from resonance decays, Bose-Einstein correlations and jets.

The two-particle Fourier coefficients $v_{n,n}$ are obtained from the correlation function as:

$$v_{n,n} = \frac{\sum_{\Delta\phi} C_r(\Delta\phi, \Delta\eta)\cos(n\Delta\phi)}{\sum_{\Delta\phi} C_r(\Delta\phi, \Delta\eta)}, \tag{4}$$

and then used to extract v_1^{even} via a simultaneous fit of $v_{1,1}$ as a function of p_T^b, for several selections of p_T^a with Eq. 2:

$$v_{1,1}(p_T^a, p_T^b) = v_1^{even}(p_T^a)v_1^{even}(p_T^b) - Cp_T^a p_T^b. \tag{5}$$

Here, $C \propto 1/(\langle Mult\rangle\langle p_T^2\rangle)$ takes into account the non-flow correlations induced by a global momentum conservation [14, 15] and $\langle Mult\rangle$ is the mean multiplicity.

For a given centrality selection, the left hand side of Eq. 5 represents the $N \times N$ matrix which we fit with the right hand side using $N + 1$ parameters; N values of $v_1^{even}(p_T)$ and one additional parameter C, accounting for momentum conservation [19]. Fig. 1 shows a representative result for this fitting procedure for $0-5\%$ central Au+Au collisions at $\sqrt{s_{NN}}$ = 200 GeV. The dashed curve (obtained with Eq. 5) in each panel illustrates the effectiveness of the simultaneous fits, as well as the constraining power of the data. That is, $v_{1,1}(p_T^b)$ evolves from negative to positive values as the selection range for p_T^a is increased.

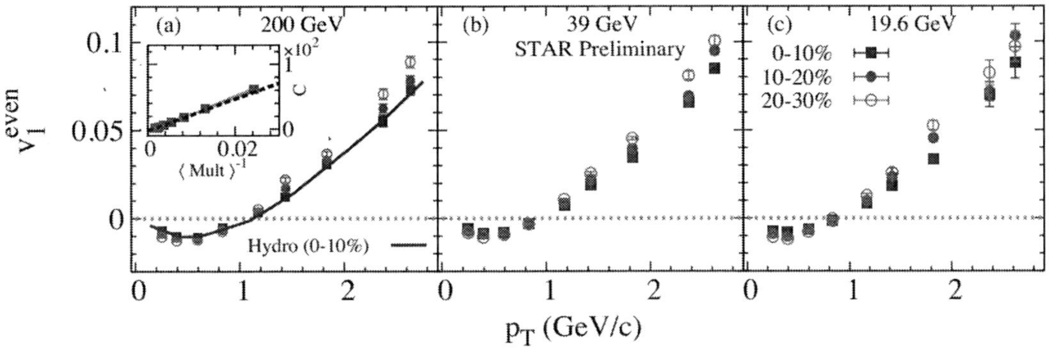

Figure 2. Extracted values of v_1^{even} vs. p_T for different centrality selections (0-10%, 10-20% and 20-30%) Au+Au collisions for several values of $\sqrt{s_{NN}}$ as indicated; the v_1^{even} values are obtained via fits with Eq. (5). The solid line in panel (a) shows the result from a hydrodynamic calculations with $\eta/s = 0.16$ [14]. The inset in panel (a) shows a representative set of the associated values of C vs. $\langle Mult \rangle^{-1}$.

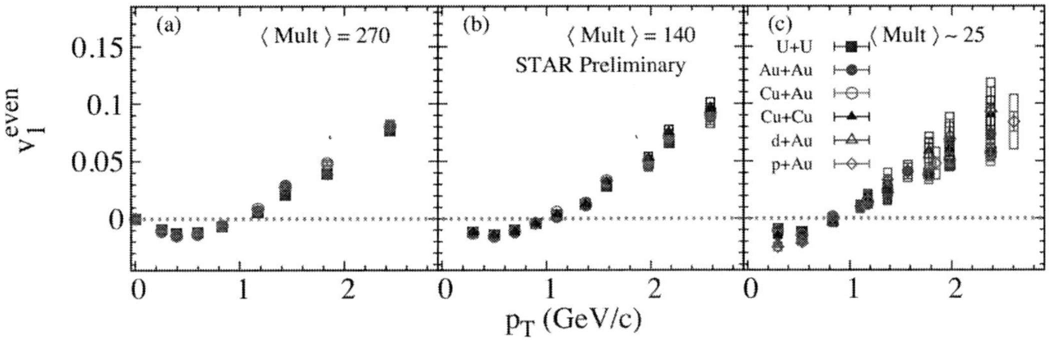

Figure 3. Extracted values of v_1^{even} vs. p_T for different $\langle Mult \rangle$ selections for different collisions system at $\sqrt{s_{NN}} \sim 200$ GeV as indicated; the v_1^{even} values are obtained via fits with Eq. (5).

3 Results

Representative v_1^{even} results for Au+Au collisions at $\sqrt{s_{NN}} = 200$, 39, and 19.6 GeV and for different collision systems U+U at $\sqrt{s_{NN}} = 193$ GeV, and Cu+Au, Cu+Cu, d+Au and p+Au at $\sqrt{s_{NN}} = 200$ GeV are summarized in Figs. 2 and 3. The values of $v_1^{even}(p_T)$ extacted for different centrality selections (0-10%, 10-20% and 20-30%) are shown in Fig. 2; the solid line in panel (a) shows the a hydrodynamic calculations with $\eta/s = 0.16$[14], which in good agreement with our measurements, the inset shows the corresponding results for the associated momentum conservation coefficient, C, extracted for several centralities at $\sqrt{s_{NN}} = 200$ GeV. The $v_1^{even}(p_T)$ values indicate the characteristic pattern of a change from negative $v_1^{even}(p_T)$ at low p_T to positive $v_1^{even}(p_T)$ for $p_T > 1$ GeV/c, with a crossing point that shifts with $\sqrt{s_{NN}}$. They also indicate that v_1^{even} increase as the centrality become more peripheral, as might be expected from the centrality dependence of ε_1.

The extracted values of $v_1^{even}(p_T)$, for different collision systems are compared in Fig. 3 for different values of $\langle Mult \rangle$. Figs. 3(a), 3(b) and 3(c) indicate similar $v_1^{even}(p_T)$ magnitudes for the systems specified at each $\langle Mult \rangle$, as well as the characteristic pattern of a change from negative $v_1^{even}(p_T)$ at low p_T to positive $v_1^{even}(p_T)$ for $p_T > 1$ GeV. This pattern confirms the predicted trends for rapidity-even dipolar flow [13, 14, 17] and further indicates that for the selected values of $\langle Mult \rangle$, $v_1^{even}(p_T)$ does not show a strong dependence on the collision system. This apparent system independence of $v_1^{even}(p_T)$ for the indicated $\langle Mult \rangle$ values suggests that the fluctuations-driven initial state eccentricity

ε_1, is similar for the six collision systems. It also suggests that the viscous effects that are related to η/s are comparable for the matter created in each of these collision systems.

4 Conclusion

In summary, we have used the two-particle correlation method to carry out new differential measurements of rapidity-even dipolar flow, v_1^{even}, in Au+Au collisions at different beam energies, and in U+U, Cu+Au, Cu+Cu, d+Au and p+Au collisions at $\sqrt{s_{NN}} \simeq 200$ GeV. The measurements confirm the characteristic patterns of an evolution from negative $v_1^{even}(p_T)$ for $p_T > 1$ GeV/c to positive $v_1^{even}(p_T)$ for $p_T > 1$ GeV/c, expected when initial-state geometric fluctuations act in concert with the hydrodynamic-like expansion to generate rapidity-even dipolar flow. This measurements provide additional constraints which are important to discern between different initial-state models, and to aid precision extraction of the temperature dependence of the specific shear viscosity.

Acknowledgments

This research is supported by the US Department of Energy under contract DE-FG02-87ER40331.A008.

References

[1] D. Teaney, Phys.Rev. **C68**, 034913 (2003), nucl-th/0301099
[2] R.A. Lacey, A. Taranenko, PoS **CFRNC2006**, 021 (2006), nucl-ex/0610029
[3] B. Schenke, S. Jeon, C. Gale, Phys.Lett. **B702**, 59 (2011), 1102.0575
[4] H. Song, S.A. Bass, U. Heinz, Phys.Rev. **C83**, 054912 (2011), 1103.2380
[5] H. Niemi, G. Denicol, P. Huovinen, E. Molnar, D. Rischke, Phys.Rev. **C86**, 014909 (2012), 1203.2452
[6] G.Y. Qin, H. Petersen, S.A. Bass, B. Muller, Phys.Rev. **C82**, 064903 (2010), 1009.1847
[7] N. Magdy (STAR), J. Phys. Conf. Ser. **779**, 012060 (2017)
[8] B. Alver, G. Roland, Phys. Rev. **C81**, 054905 (2010), [Erratum: Phys. Rev.C82,039903(2010)], 1003.0194
[9] R.A. Lacey, D. Reynolds, A. Taranenko, N.N. Ajitanand, J.M. Alexander, F.H. Liu, Y. Gu, A. Mwai, J. Phys. **G43**, 10LT01 (2016), 1311.1728
[10] A.M. Poskanzer, S.A. Voloshin, Phys. Rev. **C58**, 1671 (1998), nucl-ex/9805001
[11] R.A. Lacey, Nucl. Phys. **A774**, 199 (2006), nucl-ex/0510029
[12] N. Borghini, P.M. Dinh, J.Y. Ollitrault, Phys. Rev. **C62**, 034902 (2000), nucl-th/0004026
[13] M. Luzum, J.Y. Ollitrault, Phys. Rev. Lett. **106**, 102301 (2011), 1011.6361
[14] E. Retinskaya, M. Luzum, J.Y. Ollitrault, Phys. Rev. Lett. **108**, 252302 (2012), 1203.0931
[15] G. Aad et al. (ATLAS), Phys. Rev. **C86**, 014907 (2012), 1203.3087
[16] P. Danielewicz, R. Lacey, W.G. Lynch, Science **298**, 1592 (2002), nucl-th/0208016
[17] D. Teaney, L. Yan, Phys. Rev. **C83**, 064904 (2011), 1010.1876
[18] F.G. Gardim, F. Grassi, Y. Hama, M. Luzum, J.Y. Ollitrault, Phys. Rev. **C83**, 064901 (2011), 1103.4605
[19] J. Jia, S.K. Radhakrishnan, S. Mohapatra, J. Phys. **G40**, 105108 (2013), 1203.3410

EPJ Web of Conferences **171**, 16003 (2018) https://doi.org/10.1051/epjconf/201817116003
SQM 2017

Enhancements of high order cumulants across the 1st order phase transition boundary

Lijia Jiang[1,2,*], *Shanjin* Wu[2,**], and *Huichao* Song[2,3,4,***]

[1] *Frankfurt Institute for Advanced Studies,Ruth Moufang Strasse 1, D-60438, Frankfurt am Main, Germany*
[2] *Department of Physics and State Key Laboratory of Nuclear Physics and Technology, Peking University, Beijing 100871, China*
[3] *Collaborative Innovation Center of Quantum Matter, Beijing 100871, China*
[4] *Center for High Energy Physics, Peking University, Beijing 100871, China*

Abstract. In this proceeding, we investigate the dynamical evolution of the σ field with a trajectory across the 1st order phase transition boundary, using the Langevin equation from the linear sigma model. We find the high order cumulants of the σ field are largely enhanced during the dynamical evolution, compared with the equilibrium values, due to the supercooling effect of the first order phase transition.

1 Introduction

The STAR Beam Energy Scan (BES) program aims at searching and locating the QCD critical point. It was proposed that the fluctuations of net baryon number and net protons are largely enhanced in the critical region due to the increased correlation length [1, 2]. In experiment, large deviations of high order cumulants and cumulants ratios of net protons from the Poisson baselines have been observed at the collision energies below 40 GeV [3–5], indicating the potential of discovering the critical point.

Many past research has studied the static critical fluctuations near the critical point, which could explain the acceptance dependence of the net proton's fluctuations, but could not qualitatively describe all the cumulants measured in experiment [6–8]. Recently, the investigations on the dynamical critical fluctuations based on the Fokker-Plank equation and Langevin dynamics have shown that critical slowing-down effects play important roles in the critical regime, where the signs of high order dynamical cumulants of the σ field are even different from the corresponding equilibrium values [9, 10].

In this proceeding, we focus on studying the dynamical evolution of the σ's cumulants with the 1st order phase transition scenario, using the Langevin dynamics within the framework of the linear sigma model. We will show that the high order cumulants of the σ field are largely enhanced during the dynamical evolution when compared with the equilibrium values, due to the supercooling effect across the first order phase transition boundary.

*jiang@fias.uni-frankfurt.de
**shanjinwu2014@pku.edu.cn
***huichaosong@pku.edu.cn

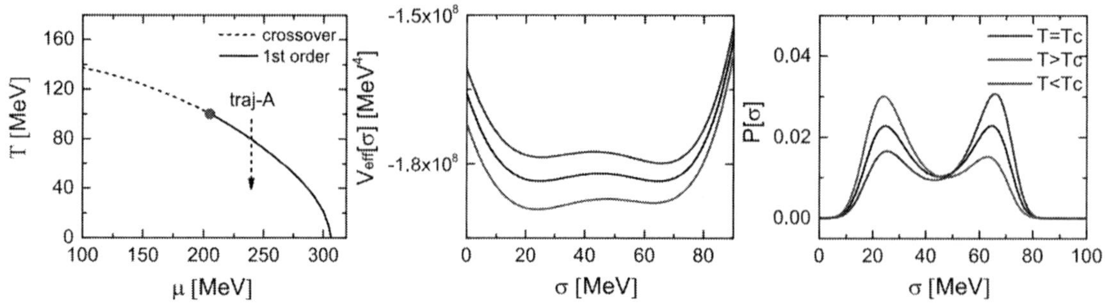

Figure 1. Left panel: the phase diagram of the linear sigma model. The middle and right panel: the effective potential $V_{eff}(\sigma)$ and probability distribution function $P[\sigma]$ at different temperature $T > T_c$, $T = T_c$ and $T < T_c$, along a chosen trajectory with $\mu = 240$ MeV.

2 The formulism

Within the framework of the linear sigma model with constituent quarks, we solve the time evolution of the σ field based on the Langevin equation. The effective potential of this model gives a phase diagram with cross-over, the first order phase transition as well as a critical point [11–13]. The long wavelength mode of the σ field is the order parameter of the chiral phase transition, and its mass vanish at the critical point. According to the classification of [14], the dynamics of the σ field in the critical regime belongs to model A, which only evolves the non-conserved order parameter field. The Langevin equation of the σ field is written as:

$$\partial^{\mu}\partial_{\mu}\sigma(t,x) + \eta\partial_t\sigma(t,x) + \frac{\delta V_{eff}(\sigma)}{\delta\sigma} = \xi(t,x), \tag{1}$$

which is a semi-classical equation. The damping η and noise $\xi(t,x)$ term in the above equation satisfy the fluctuation-dissipation theorem: $\langle\xi(t)\xi(t')\rangle \sim \eta\delta(t-t')$. The equilibrium thermodynamical potential is obtained by integrating out the thermal quarks:

$$V_{eff}(\sigma) = U(\sigma) + \Omega_{\bar{q}q}(\sigma) = \frac{\lambda^2}{4}\left(\sigma^2 - v^2\right)^2 - h_q\sigma - U_0 + \Omega_{\bar{q}q}(\sigma). \tag{2}$$

Here $U(\sigma)$ represents the vacuum potential of the σ field, $\Omega_{\bar{q}q}(\sigma) = -d_q\int\frac{d^3p}{(2\pi)^3}\{\omega + T\ln[1 + e^{-(\omega-\mu)/T}] + T\ln[1 + e^{-(\omega+\mu)/T}]\}$ comes from the contribution of thermal quarks, the particle energy $\omega = \sqrt{p^2+M^2}$, and the effective mass of quarks $M(\sigma) = m_0 + g\sigma$, with the current quark mass $m_0 = 5.5$ MeV, and coupling $g = 3.3$. The parameters h_q, v, λ, U_0 in the effective potential are determined by the properties of vacuum hadrons. Due to the spontaneously broken of chiral symmetry in vacuum, the vacuum expectation of σ field is $\langle\sigma\rangle = f_\pi = 93$ MeV. h_q is the symmetry explicitly broken term, $h_q = f_\pi m_\pi^2$ with $m_\pi = 138$ MeV. The parameter v is determined by $v^2 = f_\pi^2 - m_\pi^2/\lambda^2$, the in vacuum mass of σ is $m_\sigma \sim 600$ MeV by setting $\lambda^2 = 20$. The zero-point energy $U_0 = m_\pi^4/\left(4\lambda^2\right) - f_\pi^2 m_\pi^2$. Note that we have neglected the fluctuations of $\vec{\pi}$, since its mass is finite in the critical regime.

From the thermodynamic potential in eq. (2), we depict the phase diagram on the μ-T plane with a critical point located at $(\mu_c, T_c) = (205, 100.2)$ MeV, as shown in Fig. 1 (left). The middle and right panel of Fig. 1 present the effective potential and probability distribution function of the σ field

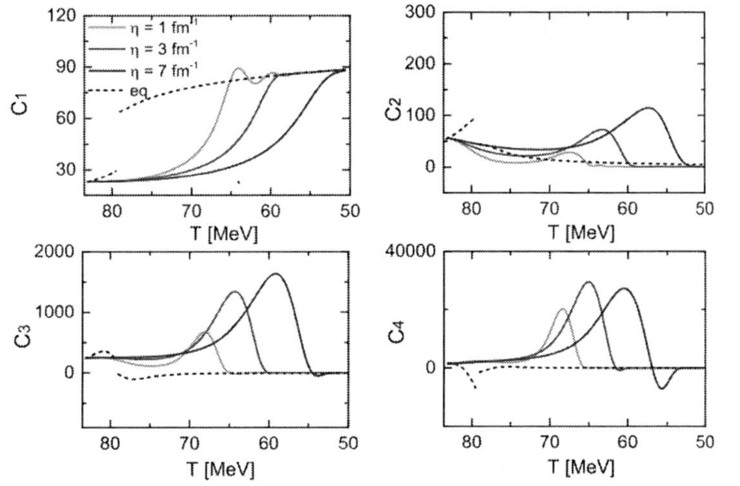

Figure 2. Dynamical evolution of the σ's cumulants along trajectory-A across the first order phase transition with μ = 240 MeV. The colored lines represent the dynamical cumulants with different damping coefficients. The black dotted lines are the corresponding equilibrium cumulants.

$\left(P[\sigma(\mathbf{x})] \sim \exp(-E(\sigma)/T)\right.$ with $\left.E(\sigma) = \int d^3x \left[\frac{1}{2}(\nabla\sigma(x))^2 + V_{eff}(\sigma(x))\right]\right)$ at different temperature $T > T_c$, $T = T_c$ and $T < T_c$, along a chosen trajectory with $\mu = 240$ MeV.

For the numerical implementation of the Langevin dynamics, we need to input the initial configurations of the σ field to start the simulations. Here, we construct the event-by-event initial conditions according to the probability function $P[\sigma(\mathbf{x})]$. In addition, the simulations of the Langevin equation requires to input the changing temperature and chemical potential profiles $T(t, x, y, z)$ and $\mu(t, x, y, z)$ from the heat bath. For simplicity, we assume the system is isothermal and evolves along a trajectory across the first order phase transition boundary with a fixed chemical potential, and the temperature decreases in a Hubble like way: $\frac{T(t)}{T_0} = \left(\frac{t}{t_0}\right)^{-0.45}$.

From the dynamical evolution of the σ field, one can obtain the spatial information of $\sigma(x)$ at each time step for each event, and then sum all the event configurations to obtain the associated moments: $\mu'_n = \langle\sigma^n\rangle = \frac{\int d\sigma\sigma^n P[\sigma]}{\int d\sigma P[\sigma]}$, where $\sigma = \int d^3x\sigma(\mathbf{x})$. The cumulants of the σ field can be calculated from these various moments with:

$$C_1 = \mu'_1,\ C_2 = \mu'_2 - \mu'^2_1,\ C_3 = \mu'_3 - 3\mu'_2\mu'_1 + 2\mu'^3_1,\ C_4 = \mu'_4 - 4\mu'_3\mu'_1 - 3\mu'^2_2 + 12\mu'_2\mu'^2_1 - 6\mu'^4_1. \quad (3)$$

3 Numerical results

In this proceeding, we focus on investigating the evolution of the σ field along a trajectory across the first-order phase transition boundary (as shown in Fig. 1 traj-A). Fig. 2 shows the time evolution of dynamical cumulants of the σ field along traj-A, with different damping coefficients (solid lines with different colors). The black dotted lines represent the equilibrium cumulants of the σ field at fixed temperature and chemical potential along the trajectory of the evolving heat bath, obtained from the perturbative calculations around the global minimum [1].

As shown in Fig. 2, the magnitude of damping coefficients directly influence the dynamical evolution of σ field. A larger damping coefficient η leads to a slower evolution of the σ field. Besides, the evolution with a larger damping coefficient tends to develop larger fluctuations at late evolution time. For all three chosen damping coefficients, we find clear memory effects. When the system goes below T_c, the high order cumulants, especially C_3 and C_4 can memorize the sign of the earlier stage during the evolution.

On the other hand, the dynamical cumulants of the σ field are largely enhanced compared with the equilibrium values, after the system evolves across the first order phase transition boundary. As shown in Fig. 1 (middle and right), the effective potential and corresponding probability distribution function shows double well (peak) structures near the first order phase transition above and below T_c. Due to the supercooling effect, the σ field tends to distribute in both minima of the potential, which largely increases the magnitude of high order cumulants.

4 Summary

In this proceeding, we numerically simulate the dynamical evolution of the σ field with a trajectory across the 1st order phase transition boundary, using the event-by-event simulations of the Langevin dynamics. The dynamical cumulants present clear memory effects compared with the equilibrium ones, where C_3 and C_4 can memorize the signs of the early stage during the dynamical evolution. Besides, the dynamical critical fluctuations are largely enhanced after the system across the 1st order phase transition boundary, due to the associated supper cooling effect.

References

[1] M. A. Stephanov, Phys. Rev. Lett. **102**, 032301 (2009).
[2] M. A. Stephanov, Phys. Rev. Lett. **107**, 052301 (2011).
[3] M. M. Aggarwal *et al.* [STAR Collaboration], Phys. Rev. Lett. **105**, 022302 (2010).
[4] L. Adamczyk *et al.* [STAR Collaboration], Phys. Rev. Lett. **112**, 032302 (2014).
[5] X. Luo [STAR Collaboration], PoS CPOD **2014**, 019 (2014).
[6] B. Ling and M. A. Stephanov, Phys. Rev. C **93**, no. 3, 034915 (2016)
[7] L. Jiang, P. Li and H. Song, Phys. Rev. C **94**, no. 2, 024918 (2016)
[8] L. Jiang, P. Li and H. Song, Nucl. Phys. A **956**, 360 (2016)
[9] S. Mukherjee, R. Venugopalan and Y. Yin, Phys. Rev. C **92**, no. 3, 034912 (2015).
[10] L. Jiang, S. Wu and H. Song, Nucl. Phys. A **967**, 441 (2017), and paper in preparation.
[11] D. U. Jungnickel and C. Wetterich, Phys. Rev. D **53**, 5142 (1996).
[12] V. Skokov, B. Friman, E. Nakano, K. Redlich and B.-J. Schaefer, Phys. Rev. D **82**, 034029 (2010).
[13] M. Nahrgang, S. Leupold, C. Herold and M. Bleicher, Phys. Rev. C **84**, 024912 (2011).
[14] P. C. Hohenberg, B. I. Halperin, Rev. Mod. Phys. **49**, 435-479 (1977).

Fluctuating fluid dynamics for the QGP in the LHC and BES era

Marcus Bluhm[1,2,⋆], *Marlene* Nahrgang[2], *Thomas* Schäfer[3], and *Steffen A.* Bass[4]

[1] *Institute of Theoretical Physics, University of Wrocław, 50204 Wrocław, Poland*
[2] *SUBATECH UMR 6457 (IMT Atlantique, Université de Nantes, IN2P3/CNRS), 44307 Nantes, France*
[3] *Physics Department, North Carolina State University, Raleigh, NC 27695, USA*
[4] *Department of Physics, Duke University, Durham, NC 27708-0305, USA*

Abstract. In an era of high-precision determinations of QGP properties a full incorporation of fluid dynamical fluctuations into our models has become crucial, in particular, when describing the dynamics of small systems or near the conjectured QCD critical point. In this talk we discuss some effects of the propagation of these fluctuations. For LHC physics we focus on fluctuations in the energy-momentum tensor, while the impact of fluctuations in the diffusive net-baryon density is studied to improve our knowledge on the formation of critical fluctuations being searched in current and future BES programs.

1 Introduction

The bulk expansion of the quark-gluon plasma (QGP) formed in relativistic heavy-ion collisions [1, 2] can successfully be described by means of conventional dissipative fluid dynamics [3]. In conventional fluid dynamics thermal averages of energy, momentum and conserved charges, such as net-baryon number, are propagated. Including dissipative effects into the fluid dynamical framework demands, however, to also incorporate fluctuations. This is because the role of the dissipation, in accordance with the fluctuation-dissipation theorem, is to counteract intrinsic fluid dynamical fluctuations which drive systems constantly out of equilibrium. These depend on the number of particles or the volume and, thus, become essential for small systems or systems undergoing a phase transition where long-range correlations develop.

In the vicinity of the conjectured QCD critical point fluctuation signals, as searched in the beam energy scan (BES) programs, are expected to be significantly enhanced [4–7]. The diffusive net-baryon density n_B becomes the slowest critical mode [8, 9], and fluctuations in the net-baryon number and related observables promise therefore indicative signatures. Nonequilibrium effects caused by the fast expansion of the matter can, however, quantitatively affect the fluctuation signals [10–17]. This talk reports, in continuation of [18], on our recent progress in developing dynamical models capable of fully propagating intrinsic fluctuations in fluid dynamics as relevant both for vanishing n_B (Sec. 2) and for the dynamics of critical fluctuations (Sec. 3).

2 The role of fluid dynamical fluctuations

Fluid dynamical fluctuations are embedded into the full evolution equations of a relativistic fluid via

$$\partial_\mu T^{\mu\nu} = \partial_\mu \left(T^{\mu\nu}_{\rm eq} + \Delta T^{\mu\nu}_{\rm visc} + \Xi^{\mu\nu}\right) = 0 \tag{1}$$

⋆e-mail: marcus.bluhm@uwr.edu.pl

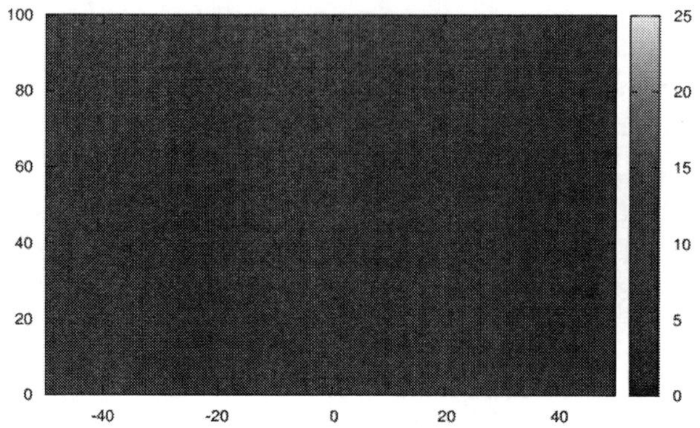

Figure 1. Spatio-temporal evolution of the energy density in units of GeV/fm^3 shown in one spatial direction within the first 100 fm/c for a fluid confined in a box of length 100 fm. The fluid has a specific shear viscosity $\eta/s = 0.2$ and initially a homogeneous energy density $e_0 = 10$ GeV/fm^3. This result is obtained for a resolution of $1/\Delta x = 1/\text{fm}$.

by adding the noise field tensor $\Xi^{\mu\nu}$ to the conventional dissipative energy-momentum tensor $T^{\mu\nu}$. The correlators of the noise components are obtained in linear response theory and read for a conformal fluid in 3 + 1 dimensions in Gaussian white noise approximation

$$\langle \Xi^{\mu\nu}(x)\,\Xi^{\alpha\beta}(x')\rangle = 2T\eta\left(\Delta^{\mu\alpha}\Delta^{\nu\beta} + \Delta^{\mu\beta}\Delta^{\nu\alpha} - \frac{2}{3}\Delta^{\mu\nu}\Delta^{\alpha\beta}\right)\delta^{(4)}(x-x') \tag{2}$$

while their mean values vanish. For relativistic fluids the dissipative corrections $\Delta T^{\mu\nu}_{\rm visc}$ are dynamical quantities whose evolution is subject to Israel-Stewart type relaxation equations. It makes sense to evolve the noise fields in a similar fashion via $u^\gamma\partial_\gamma\Xi^{\langle\mu\nu\rangle} = -(\Xi^{\mu\nu} - \xi^{\mu\nu}_{\rm Gauss})/\tau_\xi$, where the components $\xi^{\mu\nu}_{\rm Gauss}$ are correlated as in Eq. (2). As a consequence, the noise is correlated in time within the relaxation time τ_ξ.

We have incorporated fluctuations in $T^{\mu\nu}$ into the 3 + 1d relativistic viscous fluid dynamics code vHLLE [19]. For exemplary numerical studies we choose the noise relaxation time to be the same as that of the shear stresses and consider a fluid in a static box with periodic boundary conditions. With decreasing lattice spacing $\Delta x < 1$ fm the handling of large gradients caused by the increasing variance of the noise becomes challenging for the present algorithm such that we opt for smoothing the noise fields over lengths of 1 fm.

Figure 1 shows the evolution of the energy density, which we initialize at a homogeneous value e_0, along one spatial direction. Fluctuations induce local deviations from the average energy and momentum densities which are propagated by the fluid dynamical equations. As time progresses, a non-zero local variance of the energy density builds up and saturates. Moreover, the average energy density in the box is reduced by about 3% at late times. This is because the nonlinearities in fluid dynamics modify thermal averages in a nontrivial way: the interactions of the fluctuations lead, for example, to cutoff-dependent corrections in the equation of state and the transport coefficients [20, 21]. A diligent numerical implementation of fluctuating fluid dynamics needs to correct such effects. This will be a crucial aspect of future work on small systems as created in p-p(A) collisions [22].

3 Stochastic diffusion of critical fluctuations

Similarly, fluctuations can be included into the conservation equation of the net-baryon number $\partial_\mu N^\mu = \partial_\mu\left(N^\mu_{\rm eq} + \Delta N^\mu_{\rm visc} + I^\mu\right) = 0$ via a noise vector I^μ. Considering, for simplicity, a nonrelativistic fluid with space-time independent fluid velocity at a homogeneous temperature T, the conservation equation reduces to a stochastic diffusion of the net-baryon density,

$$\partial_t n_B(t,x) = \Gamma\,\nabla^2\mathcal{F}'[n_B] + \nabla J(t,x)\,, \tag{3}$$

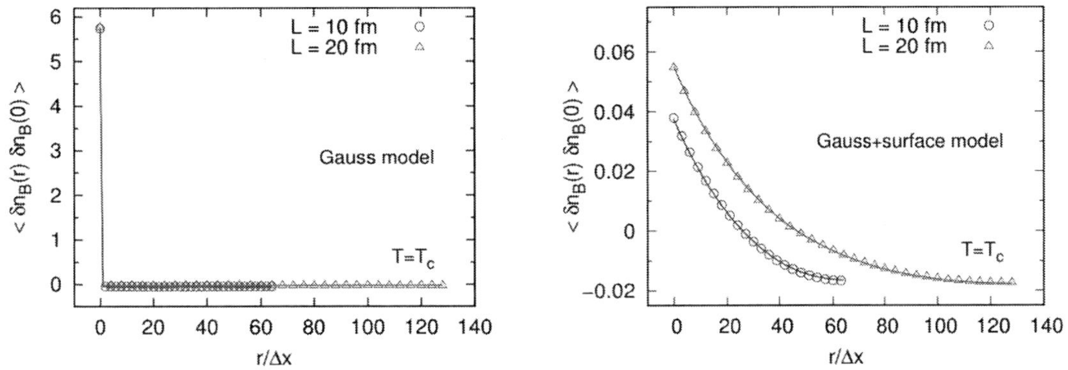

Figure 2. Equal-time correlation functions as a function of distance r in units of fixed $\Delta x = L/N_x$ for $T = T_c$ and different L in the Gauss ($K = 0$, left panel) and Gauss+surface ($K = 1/\xi_0$, right panel) model. The number of grid sites in the numerics (symbols) is $N_x = 128$ for $L = 10$ fm and $N_x = 256$ for $L = 20$ fm. The solid curves depict analytic expectations for the employed implicit scheme.

which follows the minimization of the free energy functional

$$\mathcal{F}[n_B] = T \int \mathrm{d}^3x \left(\frac{m^2}{2n_c^2}\delta n_B^2 + \frac{K}{2n_c^2}(\nabla\delta n_B)^2 \right). \tag{4}$$

Here, Γ is the mobility coefficient, $\delta n_B = n_B - n_c$ with critical density n_c, and $J(t,x) = \sqrt{2T\Gamma}\zeta(t,x)$ is a stochastic current with Gaussian white noise $\zeta(t,x)$, which itself has unit variance. We include criticality via the coupling $m^2 = 1/(\xi_0\xi^2)$ which is a function of the equilibrium correlation length ξ whose temperature-dependence can be obtained from the 3d Ising model equation of state [23, 24]. The kinetic term in Eq. (4) depends on the surface tension K.

The free energy functional contains only Gaussian terms such that Eq. (3) is linear in n_B. This allows us to study the stochastic equation numerically with a simple implicit scheme. Moreover, we confine the diffusion to a one-dimensional system of length L with periodic boundaries which implies that cutoff-effects are absent and the results approach continuum expectations with decreasing lattice spacing. In the numerics we initialize n_B homogeneously at $n_c = 1/(3\text{fm}^3)$, choose $\Gamma = 1/(3T\text{fm}^2)$ and $\xi_0 = 0.479$ fm, fix the pseudo-critical temperature $T_c = 0.15$ GeV at which $\xi(T_c) \simeq 6.4\,\xi_0$, and analyze equilibrium observables only after a sufficient equilibration time.

In Fig. 2 we show the equal-time correlation functions $\langle\delta n_B(r)\,\delta n_B(0)\rangle$ for the cases $K = 0$ and $K = 1/\xi_0$. For $K = 0$, fluctuations are uncorrelated over distances larger than Δx. Instead, a non-zero surface tension term permits the development of real long-range correlations: for $L = 20$ fm, we find a numerically realized correlation length of about $5.8\,\xi_0$ which goes asymptotically to $\xi(T_c)$ with increasing L. In both cases, we observe a non-zero local variance $\langle\delta n_B(0)^2\rangle$ emerging from the purely white noise. While $\langle\delta n_B(0)^2\rangle_{K=0}$ depends on $1/\Delta x$ in line with the thermodynamic expectation $\langle\delta n_B(r)^2\rangle_{K=0} = (n_c^2/m^2)\,\delta(r)$, for $K \neq 0$ the numerics approaches the grandcanonical continuum expectation $\langle\delta n_B(r)\,\delta n_B(0)\rangle = (n_c^2/\sqrt{4m^2K})\,e^{-r/\xi}$ as $L \to \infty$. For a finite L, however, one finds in both cases negative correlations at large r as a consequence of exact charge conservation $\int_L \langle\delta n_B(r)\,\delta n_B(0)\rangle\,dr = 0$ in the finite size system. Note that the non-Gaussian fluctuations remain zero for the considered Gaussian models.

4 Conclusions

The full implementation of intrinsic fluctuations into fluid dynamics presents a challenge for current theoretical models, but is also a crucial step toward a realistic description of the QCD dynamics probed in heavy-ion collisions. The accurate treatment of nonlinear interactions connected with the necessity of a careful renormalization of cutoff-effects requires a systematic effort. We discussed a stochastic diffusion of the net-baryon density which is able to capture the critical phenomena of the Gaussian fluctuations and highlights, in particular, the importance of charge conservation in finite-size systems. In future work, we will report on the statics and real-time dynamics of non-Gaussian fluctuations [25].

Acknowledgments

M.B. is funded by the European Union's Horizon 2020 research and innovation programme under the Marie Skłodowska Curie grant agreement No 665778 via the National Science Center, Poland, under grant Polonez UMO-2016/21/P/ST2/04035. This work was financed in parts by the U.S. Department of Energy with grants DE-FG02-03ER41260 and DE-FG02-05ER41367. The authors thank Y. Karpenko for providing the vHLLE code and acknowledge fruitful discussions within the Beam Energy Scan Theory (BEST) Topical Collaboration.

References

[1] E. V. Shuryak, Phys. Rept. **61** (1980) 71.
[2] U. W. Heinz, nucl-th/0512051.
[3] P. Romatschke, Int. J. Mod. Phys. E **19** (2010) 1.
[4] M. A. Stephanov, K. Rajagopal and E. V. Shuryak, Phys. Rev. Lett. **81** (1998) 4816.
[5] M. A. Stephanov, K. Rajagopal and E. V. Shuryak, Phys. Rev. D **60** (1999) 114028.
[6] M. A. Stephanov, Phys. Rev. Lett. **102** (2009) 032301.
[7] M. Asakawa, S. Ejiri and M. Kitazawa, Phys. Rev. Lett. **103** (2009) 262301.
[8] D. T. Son and M. A. Stephanov, Phys. Rev. D **70** (2004) 056001.
[9] H. Fujii and M. Ohtani, Phys. Rev. D **70**, 014016 (2004).
[10] B. Berdnikov and K. Rajagopal, Phys. Rev. D **61** (2000) 105017.
[11] M. Nahrgang, S. Leupold, C. Herold and M. Bleicher, Phys. Rev. C **84** (2011) 024912.
[12] M. Nahrgang, S. Leupold and M. Bleicher, Phys. Lett. B **711** (2012) 109.
[13] M. Nahrgang, C. Herold, S. Leupold, I. Mishustin and M. Bleicher, J. Phys. G **40** (2013) 055108.
[14] C. Herold, M. Nahrgang, I. Mishustin and M. Bleicher, Phys. Rev. C **87** (2013) no.1, 014907.
[15] M. Kitazawa, M. Asakawa and H. Ono, Phys. Lett. B **728** (2014) 386.
[16] S. Mukherjee, R. Venugopalan and Y. Yin, Phys. Rev. C **92** (2015) no.3, 034912.
[17] C. Herold, M. Nahrgang, Y. Yan and C. Kobdaj, Phys. Rev. C **93** (2016) no.2, 021902.
[18] M. Nahrgang, M. Bluhm, T. Schäfer and S. Bass, Acta Phys. Polon. Supp. **10** (3) (2017) 687.
[19] I. Karpenko, P. Huovinen and M. Bleicher, Comput. Phys. Commun. **185** (2014) 3016.
[20] P. Kovtun, G. D. Moore and P. Romatschke, Phys. Rev. D **84** (2011) 025006.
[21] C. Chafin and T. Schäfer, Phys. Rev. A **87** (2013) no.2, 023629.
[22] L. Yan and H. Grönqvist, JHEP **1603** (2016) 121.
[23] M. Bluhm, M. Nahrgang, S. A. Bass and T. Schäfer, Eur. Phys. J. C **77** (2017) no.4, 210.
[24] M. Nahrgang, M. Bluhm, T. Schäfer and S. A. Bass, Nucl. Phys. A **967** (2017) 824.
[25] M. Nahrgang, M. Bluhm, T. Schäfer and S. A. Bass to be published soon.

EPJ Web of Conferences **171**, 16005 (2018) https://doi.org/10.1051/epjconf/201817116005
SQM 2017

Identifying QCD Transition Using Deep Learning

Kai Zhou[1,2,⋆], *Long-gang* Pang[1,3,4,⋆⋆], *Nan* Su[1,⋆⋆⋆], *Hannah* Petersen[1,2,5,⋆⋆⋆⋆], *Horst* Stoecker[1,2,5,†], and *Xin-Nian* Wang[3,6,‡]

[1] *Frankfurt Institute for Advanced Studies, 60438 Frankfurt am Main, Germany*
[2] *Institut für Theoretische Physik, Goethe Universität, 60438 Frankfurt am Main, Germany*
[3] *Department of Physics, University of California, Berkeley, CA 94720, USA*
[4] *Nuclear Science Division, Lawrence Berkeley National Laboratory, Berkeley, CA 94720, USA*
[5] *GSI Helmholtzzentrum für Schwerionenforschung, 64291 Darmstadt, Germany*
[6] *Key Laboratory of Quark and Lepton Physics (MOE) and Institute of Particle Physics, Central China Normal University, Wuhan, 430079, China*

Abstract. In this proceeding we review our recent work using supervised learning with a deep convolutional neural network (CNN) to identify the QCD equation of state (EoS) employed in hydrodynamic modeling of heavy-ion collisions given only final-state particle spectra $\rho(p_T, \Phi)$. We showed that there is a traceable encoder of the dynamical information from phase structure (EoS) that survives the evolution and exists in the final snapshot, which enables the trained CNN to act as an effective "EoS-meter" in detecting the nature of the QCD transition.

1 Introduction

The QCD phase structure and the search for the critical end point are the central and primary motivations for high energy heavy ion collisions, which allows to study the birth of the universe in the laboratory on earth as well. Large international collaborations, both in theory and in experiment, have been searching for signals of this phase structure at huge accelerator centres worldwide which are constructed specifically for this exciting view into this distant past. The forthcoming program at FAIR (GSI) and the current beam energy scan project at RHIC (BNL) aim at locating the critical end point in the QCD phase diagram. This critical end point separates the crossover transition and the conjectured first order phase transition from hadrons to deconfined quark-gluon matter [1, 2]. Critical fluctuations [3, 4] are used in experiments to locate this critical end point usually. However, currently the observed signals from experiments are too weak to pin down its location. Moreover, it is rather involved to disentangle different physical factors (like initial states fluctuation, transport coefficients, freeze-out and further hadronic cascade) in a heavy ion collision evolution given only the final states

⋆e-mail: zhou@fias.uni-frankfurt.de
⋆⋆e-mail: pang@fias.uni-frankfurt.de
⋆⋆⋆e-mail: nansu@fias.uni-frankfurt.de
⋆⋆⋆⋆e-mail: petersen@fias.uni-frankfurt.de
†e-mail: stoecker@fias.uni-frankfurt.de
‡e-mail: xnwang@lbl.gov

information. Thus, it is difficult to clearly extract physics about bulk properties of the QCD matter from the experimental raw data. We are thereby lacking of a direct and reliable bridge between the bulk properties of the matter produced during the collisions and the raw experimental observables.

Deep Learning (DL) is a branch of machine learning, which aims at exploring high-level representations of data using a deeper structure of multiple processing layers. Recently, the application of DL to physics research is rapidly growing, such as in particle physics [9–12], nuclear physics [13], and condensed matter physics [14–16]. DL is shown to be very powerful in exploring pertinent hidden features especially for complex non-linear systems with high-level correlations beyond conventional technique's capability. This suggests that DL could be adopted to help uncovering hidden physical information from the highly implicit heavy-ion collision experimental raw data.

In a recent work [17], we give an exploratory study in directly connecting QCD bulk properties and raw data from heavy-ion collisions using state-of-the-art deep-learning techniques. The relativistic hydrodynamic models are utilized to generate raw data of final state pion's spectra $\rho(p_T, \Phi)$ in heavy ion collisions, where different QCD transition types embedded in EoS can be applied directly. Then supervised learning using convolutional neural networks (CNN) is performed with labeled spectra, through which we reveal unique and exclusive encoders of the bulk EoS inside $\rho(p_T, \Phi)$. Here in this proceeding we will review this exploratory study.

2 Training and testing datasets

The evolution of strongly coupled QCD matter in heavy-ion collisions can be well described by 2nd-order dissipative hydrodynamics. The EoS of the medium is a crucial ingredient in solving the hydrodynamic equations, via which the nature of the QCD transition (1st order or crossover) strongly affects the hydrodynamic evolution. The input of our CNN training is set to be final charged pion's spectra $\rho(p_T, \Phi)$ at mid-rapidity, which can be obtained from the Cooper-Frye formula in hydrodynamic simulation:

$$\rho(p_T, \Phi) \equiv \frac{dN_i}{dY p_T dp_T d\Phi} = g_i \int_\sigma p^\mu d\sigma_\mu f_i \,, \tag{1}$$

Here N_i is the particle number density, Y is the rapidity, g_i is the degeneracy, $d\sigma_\mu$ is the freeze-out hypersurface element, f_i is the thermal distribution. In the following, we employ the lattice-EoS parametrization [18] (dubbed as EOSL) for the crossover transition and Maxwell construction [19] (dubbed as EOSQ) for the first-order phase transition.

The training dataset of $\rho(p_T, \Phi)$ (labelled with EOSL or EOSQ) is generated by event-by-event hydrodynamic package CLVisc [20, 21] with fluctuating AMPT initial conditions [22]. The simulation generated about 22000 $\rho(p_T, \Phi)$ for different types of collisions. Details can be seen Tab.1 in Ref. [17].

The testing dataset contains two groups of samples. In the first group, we generate 7343 $\rho(p_T, \Phi)$ events using the second-order event-by-event hydrodynamic package iEBE-VISHNU [23] with MC-Glauber initial condition. In the second group, we generate 8917 $\rho(p_T, \Phi)$ events using the CLVisc package with the IP-Glasma-like initial condition [24]. The testing datasets are constructed to explore very different regions of parameters (different set up for η/s, τ_0 and freeze-out temperature) as compared to training dataset. The details are listed in Tab.2 in Ref. [17]. Note that all the training and testing $\rho(p_T, \Phi)$ are preprocessed by $\rho' = \rho/\rho_{max} - 0.5$ to normalize the input data, and each being accompanied with its label of EoS type — EOSQ is labeled by $(0, 1)$ and EOSL is labeled by $(1, 0)$.

3 Convolutional Neural Network

Being inspired by the excellent performance of CNN [26, 27] in tasks such as image and video recognition, here for our purpose we construct the CNN with an architecture shown in Fig.3 in Ref. [17]

Table 1. Tesing accuracies for three groups (CLVisc with AMPT initial state, iEBE-VISHNU and CLVisc with IP-Glasma-like initial condition) of the testing dataset.

TESTING DATA	GROUP 0	GROUP 1	GROUP 2
Number of events	4000	7343	10953
Accuracy	99.88±0.04%	93.46±1.35%	93.91±3.92%

to handle the QCD transition classification. The input $\rho(p_T, \Phi)$ consists of 15 p_T-bins and 48 Φ-bins. We use two convolutional layers each followed by batch normalization, dropout and PReLU activation. Brief introduction information about these technical terms can be found in the supplementary materials in Ref. [17]. In a convolutional layer, each neuron only locally connects to a small chunk of neurons in the previous layer by a convolution operation – this is a key reason for the success of the CNN architecture. Such an architecture works efficiently to prevent overfitting that may generate model-dependent features from the training dataset and thus hinder the generalizability of the method. The final output layer is a fully connected layer with softmax activation and 2 neurons to indicate the type of the EoS.

Supervised learning with the above CNN structure is performed on the targeting binary classification problem here — EOSQ $(0, 1)$ or EOSL $(1, 0)$. The difference between the true label and the predicted label from the two output neurons, quantified by cross entropy [28], serves as the loss function $l(\theta)$, where θ are the trainable parameters of the neural network. Training attempts to minimize the loss function by updating $\theta \to \theta - \delta\theta$. Here $\delta\theta = \alpha\ \partial l(\theta)/\partial\theta$ where α is the learning rate with initial value 0.0001 and adaptively changed in AdaMax method [29].

4 Results and Conclusion

After training and validating the network, it is tested on the testing dataset of $\rho(p_T, \Phi)$ events. The percentage during test that deep CNN can correctly identify the input EoS is usually defined as accuracy here to indicate the learning performance. As shown in Tab. 1, high prediction accuracies – on average larger than 95% – are achieved for these three groups of testing datasets, which indicates that our method is highly independent of initial conditions. The network is robust against shear viscosity and τ_0 due to the inclusion of events with different η/s and τ_0 in the training. In the testing stage the neural network identifies the type of the QCD transition solely from the spectra of each single event. Furthermore, in the training only one freeze-out temperature is used, while the network is tolerant to a wide range of freeze-out temperatures during the testing. For simplicity, the exploratory study has not included pions from resonance decays (the hadronic transport module UrQMD is switched off in iEBE-VISHNU to exclude contributions from resonance decays in testing data).

The present method yields a novel perspective on identifying the nature of the QCD transition in heavy-ion collisions. By applying state-of-the-art deep CNNs, we firmly demonstrate that, there do exist discriminative and traceable encoder of the dynamical information from phase structure (EoS) inside the collision evolution's final snapshot—final-state $\rho(p_T, \Phi)$ in heavy-ion collisions, which can survive even though may not be intuitive and thus not captured by conventional observables well. Meanwhile the deep CNN can exclusively and efficiently decode these EoS information directly from the implicit $\rho(p_T, \Phi)$ after the hydrodynamic evolution. In this way, high-level representations, which help identifying the QCD transition inside EoS in the present method, act as an "EoS-meter" for the QCD matter created in heavy-ion collisions. Our study might provide a key to the success of the experimental determination of QCD EoS and the search for the critical end point. Another intriguing application of our framework is to extract the QGP transport coefficients from heavy-ion collisions. The present method can be further improved by including hadronic rescattering and detector efficiency corrections.

References

[1] H. Stöcker and W. Greiner, Phys. Rept. **137**, 277 (1986).
[2] M. A. Stephanov, PoS LAT **2006** (2006) 024.
[3] M. A. Stephanov, K. Rajagopal and E. V. Shuryak, Phys. Rev. Lett. **81** (1998) 4816
[4] M. Nahrgang, T. Schuster, M. Mitrovski, R. Stock and M. Bleicher, Eur. Phys. J. C **72** (2012) 2143
[5] M. Luzum and P. Romatschke, Phys. Rev. C **78**, 034915 (2008).
[6] S. Pratt, E. Sangaline, P. Sorensen, and H. Wang, Phys. Rev. Lett. **114**, 202301 (2015)
[7] J. Schmidhuber, "Deep learning in neural networks: An overview," Neural Netw. **61**, 85 (2015).
[8] Y. LeCun, Y. Bengio, and G. Hinton, Nature **521**, 436 (2015).
[9] P. Baldi, P. Sadowski, and D. Whiteson, Nature Commun. **5**, 4308 (2014).
[10] P. Baldi, P. Sadowski, and D. Whiteson, Phys. Rev. Lett. **114**, 111801 (2015).
[11] J. Barnard, E. N. Dawe, M. J. Dolan, and N. Rajcic, Phys. Rev. D **95**, 014018 (2017).
[12] I. Moult, L. Necib, and J. Thaler, J. High Energy Phys. **12**, 153 (2016).
[13] R. Utama, W. C. Chen, and J. Piekarewicz, J. Phys. G **43**, 114002 (2016).
[14] J. Carrasquilla and R. G. Melko, Nat. Phys. http://dx.doi.org/10.1038/nphys4035 (2017).
[15] G. Carleo and M. Troyer, Science **355**, 602 (2017).
[16] G. Torlai and R. G. Melko, Phys. Rev. B **94**, 165134 (2016).
[17] L. G. Pang, K. Zhou, N. Su, H. Petersen, H. Stöcker and X. N. Wang, Nature Communications, in preprint.
[18] P. Huovinen and P. Petreczky, Nucl. Phys. A **837**, 26 (2010).
[19] J. Sollfrank, P. Huovinen, M. Kataja, P. V. Ruuskanen, M. Prakash, and R. Venugopalan, Phys. Rev. C **55**, 392 (1997).
[20] L. G. Pang, Q. Wang, and X. N. Wang, Phys. Rev. C **86**, 024911 (2012).
[21] L. G. Pang, Y. Hatta, X. N. Wang, and B. W. Xiao, Phys. Rev. D **91**, 074027 (2015).
[22] Z. W. Lin, C. M. Ko, B. A. Li, B. Zhang, and S. Pal, Phys. Rev. C **72**, 064901 (2005).
[23] C. Shen, Z. Qiu, H. Song, J. Bernhard, S. Bass, and U. Heinz, Comput. Phys. Commun. **199**, 61 (2016).
[24] C. Gale, S. Jeon, B. Schenke, P. Tribedy, and R. Venugopalan, Phys. Rev. Lett. **110**, 012302 (2013).
[25] J. E. Bernhard, J. S. Moreland, S. A. Bass, J. Liu, and U. Heinz, Phys. Rev. C **94**, 024907 (2016)
[26] A. Krizhevsky, I. Sutskever, and G. E. Hinton, Advances in Neural Information Processing Systems 25 (NIPS 2012).
[27] K. Simonyan and A. Zisserman, arXiv:1409.1556 [cs.CV].
[28] S. Kullback and R. A. Leibler, Ann. Math. Statist. **22**, 79 (1951).
[29] D. Kingma and J. Ba, arXiv:1412.6980 [cs.LG].
[30] F. Chollet, https://github.com/fchollet/keras.
[31] M. Abadi, *et al.*, arXiv:1603.04467 [cs.DC], http://tensorflow.org/.
[32] G. E. Hinton, N. Srivastava, A. Krizhevsky, I. Sutskever, and R. R. Salakhutdinov, arXiv:1207.0580 [cs.NE].

EPJ Web of Conferences **171**, 16006 (2018) https://doi.org/10.1051/epjconf/201817116006
SQM 2017

News on mean pion multiplicity from NA61/SHINE

Michał Naskręt[1]* for the NA61/SHINE Collaboration

[1] *Institute of Theoretical Physics, University of Wroclaw*

Abstract. NA61/SHINE is a large acceptance fixed target experiment at the CERN SPS which studies final hadronic states in interactions between various particles and nuclei [1]. The main topic of this contribution are preliminary results for mean negatively charged pion multiplicities $\langle\pi^-\rangle$ from central Ar+Sc and Be+Be collisions. The data were taken recently by the NA61/SHINE collaboration for a wide range of beam momenta. Measured rapidity distributions $\frac{dn}{dy}$ were extrapolated to unmeasured regions to obtain total multiplicities $\langle\pi^-\rangle$. A new scheme to calculate the mean number of wounded nucleons $\langle W\rangle$ utilizing the EPOS MC model is described. Using data from other experiments, a comparison of $\frac{\langle\pi\rangle}{\langle W\rangle}$ for different collisions and beam momenta is discussed.

1 π^- rapidity distributions

The starting point of the analysis described herein are double differential spectra $\frac{d^2n}{dy dp_{\rm T}}$ of negatively charged hadrons, where y and $p_{\rm T}$ are rapidity and transverse momentum of partcles, respectively. To identify negatively charged pions, the "h$^-$" method (see details in Ref. [2]) is used. Centrality was determined by selecting the 5% of collisions with the smallest forward going energy as measured by the Projectile Spectator Detector (PSD) [1]. The acceptance is as high as 2 GeV/c in p_T and 3.5 in y.

The spectra were obtained from reconstructed tracks applying a series of quality cuts. In order to correct for trigger and reconstruction inefficiencies, one needs to apply a Monte Carlo correction. To this end, the EPOS MC [3] is used in NA61/SHINE. A large statistics of ion collisions is generated and particles are accumulated in bins $n^{i,j}_{\rm gen}$ in transverse momentum $p_{\rm T}$ versus rapidity y. The generated data undergo the regular reconstruction procedure. Selecting negatively charged pions results in the distribution $n^{i,j}_{\rm sel}$. The correction factor $c^{i,j}$ is then calculated as the ratio of the two Monte-Carlo generated spectra $c^{i,j} = n^{i,j}_{\rm gen}/n^{i,j}_{\rm sel}$. The final experimental spectra are obtained as $n^{i,j} = n^{i,j}_{\rm data}c^{i,j}$.

In order to estimate the mean π^- multiplicity in the full acceptance, one needs to extrapolate the experimental data to unmeasured regions. The extrapolation process consisted of two steps - extrapolation in transverse momentum $p_{\rm T}$ for each bin of rapidity y and extrapolation of dn/dy in rapidity. For the latter a sum of two Gaussian functions was fitted, $g(y) = g_{\rm T}(y) + g_{\rm P}(y)$, where

$$g_{\rm T}(y) = \frac{A_0 A_{rel}}{\sigma_0\sqrt{2\pi}}\exp\left(-\frac{(y-y_0)^2}{2\sigma_0^2}\right), g_{\rm P}(y) = \frac{A_0}{\sigma_0\sqrt{2\pi}}\exp\left(-\frac{(y+y_0)^2}{2\sigma_0^2}\right) \quad (1)$$

In order to calculate the mean negatively charged pion multiplicity $\langle\pi^-\rangle$, the following formula was utilized:

*e-mail: mnaskret@cern.ch

$$\langle \pi^- \rangle = \int_{-4}^{y_{\min}} g(y)dy + \sum_{y_{\min}}^{y_{\max}} dy \left(\frac{dn}{dy}\right)_{\text{extrapolated in} p_T} + \int_{y_{\max}}^{4} g(y)dy \tag{2}$$

Thus the final result is the sum over measured values of dn/dy in the acceptance region and the integral over the Gaussian fits outside. Statistical uncertainties were calculated and systematic uncertainties were assumed to be 5% based on the previous NA61 analysis of p+p collisions [4].

2 The mean number of wounded nucleons

The number of wounded nucleons can not be measured experimentally in NA61/SHINE. It has to be calculated using Monte Carlo models. Two models were used to perform calculations - Glissando 2.73 [5] based on the Glauber model and EPOS 1.99 (version CRMC 1.5.3) [3] using a parton ladder model. Glissando provides values that are consistent with previous measurements and applicable to the wounded nucleon model [6]. EPOS, on the other hand, allows for more detailed centrality analysis and event selection. It is possible to reproduce Glauber-based $\langle W \rangle$ values in EPOS and they are in good agreement with Glissando as shown in Fig. 1, where both - statistical and systematic uncertainties are calculated and combined. Therefore, Glauber-based EPOS values are used in later considerations.

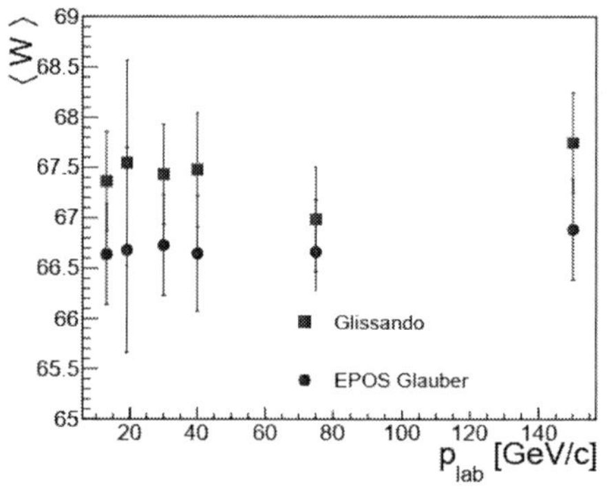

Figure 1. Average number of wounded nucleons $\langle W \rangle$ calculated by Glissando and EPOS "a la Glauber".

3 Results

Preliminary π^- rapidity spectra for Ar+Sc and Be+Be collisions are presented in Fig. 2

Preliminary results on $\langle \pi^- \rangle$ and $\langle W \rangle$ were calculated according to procedures described in sections 1 and 2. These are presented in Table 1 for Ar+Sc and Be+Be collisions.

In order to compare results obtained for different systems, an isospin correction should be taken into account. To this end phenomenological formulas are used

$$\langle \pi^- \rangle_{\mathrm{N+N}} = \langle \pi^- \rangle_{\mathrm{p+p}} + \frac{1}{3}, \langle \pi^- \rangle^{\mathrm{I}}_{\mathrm{Au+Au}} = (\langle \pi^- \rangle_{\mathrm{Au+Au}} + \langle \pi^+ \rangle_{\mathrm{Au+Au}})/2 \tag{3}$$

The correction is only applied to measurements where its effect is the strongest. This assumption is based on the compilation of the world data presented in [7] and the model presented therein. Where needed, the data is corrected for slight differences in beam momentum. Applying this correction one

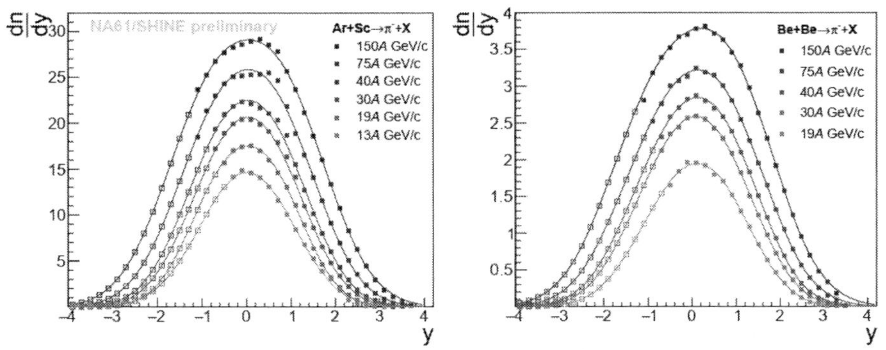

Figure 2. Preliminary π^- rapidity spectra for the 5% most central Ar+Sc and Be+Be collisions.

Ar+Sc			Be+Be		
p_{lab} [A GeV/c]	$\langle\pi^-\rangle$	$\langle W\rangle$	p_{lab} [A GeV/c]	$\langle\pi^-\rangle$	$\langle W\rangle$
13	38.46 ± 1.92	66.63 ± 0.50	20	5.32 ± 0.54	10.99 ± 1.02
19	48.03 ± 2.40	66.68 ± 1.02	30	7.61 ± 0.76	10.86 ± 0.50
30	59.72 ± 2.98	66.72 ± 0.50	40	8.75 ± 0.44	10.86 ± 0.57
40	66.28 ± 3.31	66.64 ± 0.57	75	10.98 ± 0.55	10.83 ± 0.52
75	86.12 ± 4.30	66.66 ± 0.52	158	14.32 ± 0.72	10.79 ± 0.50
150	108.92 ± 5.44	66.88 ± 0.50			

Table 1. Preliminary reslults on $\langle\pi^-\rangle$ and $\langle W\rangle$ for the 5% most central Ar+Sc and Be+Be collisions.

can plot the $\langle\pi^-\rangle/\langle W\rangle$ ratios for different systems for 30A and 150A GeV/c, see Fig. 3. Comparison of the preliminary Be+Be and Ar+Sc results with those from other systems suggest a monotonic increase with $\langle W\rangle$ at 150A GeV/c and a constant value at 30A GeV/c.

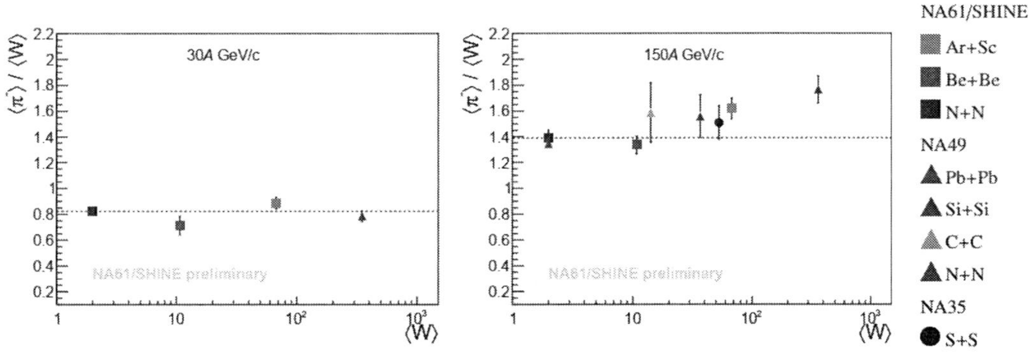

Figure 3. Measurements of the $\langle\pi^-\rangle/\langle W\rangle$ ratio in nucleon-nucleon and central nucleus-nucleus collisions.

The Fermi statistical model predicts a linear increase of $\langle\pi\rangle/\langle W\rangle$ with the Fermi energy measure $F = \left[(\sqrt{s_{\text{NN}}} - 2m_{\text{N}})^3/\sqrt{s_{\text{NN}}}\right]^{1/4}$. An increase of the slope of $\langle\pi\rangle/\langle W\rangle$ versus F – the kink – at the

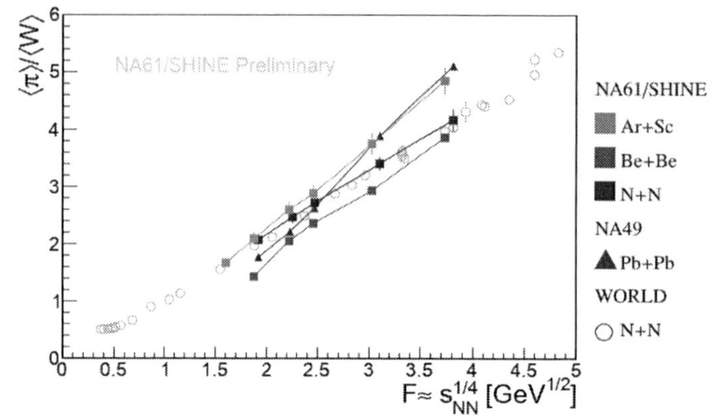

Figure 4. Energy dependence of the $\langle\pi\rangle/\langle W\rangle$ ratio - the kink plot.

onset of deconfinement is predicted by the SMES [8] model due to the larger number of effective degrees of freedom in comparison to the hadron resonance gas (HRG) model.

As for the NA61 Ar+Sc, Be+Be and p+p data only the value of $\langle\pi^-\rangle$ was obtained, the multiplicities of $\langle\pi^+\rangle$ and $\langle\pi^0\rangle$ are approximated by multiplying the previously isospin asymmetry corrected π^- multiplicities by a factor 3: $\langle\pi\rangle = 3\langle\pi^-\rangle$. This approach is motivated by the fact that the NA61/SHINE acceptance is the largest for π^-. The energy dependence of the $\langle\pi\rangle/\langle W\rangle$ ratio is presented in Fig. 4. The Ar+Sc results follow p+p for low energies whereas for high energies they follow Pb+Pb. No simple systematics are observed at low SPS energies. This might be caused by different physics processes and/or systematic bias in the estimate of $\langle W\rangle$. Full simulation of the fragmentation process and PSD response is needed in order to reduce the latter uncertainty.

4 Conclusions

Preliminary results on π^- rapidity spectra and multiplicities in central Be+Be and Ar+Sc collisions at the CERN SPS are presented. Total multiplicities as function of collision energy and ratios with the average number of wounded nucleons were obtained. These new results are compared to previous measurements on nucleon-nucleon and nucleus-nucleus interactions. Results from Ar+Sc follow those obtained in Pb+Pb reactions at higher SPS energies whereas for low energies they follow p+p.

This work was partially supported by the National Science Centre, Poland (grant 2015/18/M/ST2/00125).

References

[1] N. Abgrall et al. (NA61), JINST **9**, P06005 (2014), 1401.4699
[2] A. Aduszkiewicz, Ph.D. thesis, University of Warsaw (2015)
[3] K. Werner, F.M. Liu, T. Pierog, Phys. Rev. **C74**, 044902 (2006), hep-ph/0506232
[4] N. Abgrall et al. (NA61/SHINE), Eur. Phys. J. **C74**, 2794 (2014), 1310.2417
[5] M. Rybczynski et al., Comput. Phys. Commun. **185**, 1759 (2014), 1310.5475
[6] A. Białas, M. Bleszyński, W. Czyż, Nuclear Physics B **111**, 461 (1976)
[7] A.I. Golokhvastov, Physics of Atomic Nuclei **64**, 84 (2001)
[8] M. Gazdzicki, M.I. Gorenstein, Acta Phys. Polon. **B30**, 2705 (1999), hep-ph/9803462

ϕ Meson Spin Alignment and the Azimuthal Angle Dependence of Λ ($\bar{\Lambda}$) Polarization in Au+Au collisions at RHIC

Biao Tu[1,2,*] (For the STAR collaboration)

[1] *Key Laboratory of Quark and Lepton Physics (MOE) and Institute of Particle Physics, Central China Normal University, Wuhan 430079, China*
[2] *Brookhaven National Laboratory, Upton, NY 11973, USA*

Abstract. Initial large global angular momentum in non-central relativistic heavy-ion collisions can produce strong vorticity, and through the spin-orbit coupling, causes the spin of particles to align with the system's global angular momentum. We present the azimuthal angle dependent (relative to the reaction plane) polarization for Λ and $\bar{\Lambda}$ in mid-central Au+Au collisions at $\sqrt{s_{NN}}$ = 200 GeV. We also present the ϕ meson spin alignment parameter, ρ_{00} in Au+Au collisions at $\sqrt{s_{NN}}$ = 19.6, 27, 39, 62.4 and 200 GeV. The implications of the results are discussed.

1 Introduction

High-energy relativistic heavy-ion collisions at Relativistic Heavy Ion Collider (RHIC) produce a strongly interacting, hot and dense medium known as Quark Gluon Plasma (QGP) [1]. The initial orbital angular momentum, associated with the receding spectators, is large ($\sim 1000\hbar$) in non-central collisions may be transferred to quarks through spin-orbit coupling [2–4], which is then transmitted to final-state hadrons and is detectable through the $\Lambda(\overline{\Lambda})$ polarization and ϕ meson spin alignment. Therefore, measurements of the polarization of the particles produced in heavy-ion collisions can provide new insights into the initial conditions and evolution of the QGP [5, 6].

The STAR experiment at RHIC has observed for the first time a significant alignment between the angular momentum of the medium produced in non-central collision and the spin of $\Lambda(\overline{\Lambda})$ hyperons (J=1/2), revealing that the matter produced in heavy-ion collisions is by far the most vortical system ever observed [7]. Such vorticity is expected to be maximal at the equator and due to the low viscosity of the system the vorticity may not be efficiently propagated to the poles. This can lead to a larger in-plane than out-of-plane polarization for hyperons. The study of the azimuthal angle dependence of hyperon polarization can help us in understanding transport properties of the system and shed light on dynamics in a highly vortical, low viscous environment.

The strong vorticity, when acting together with particle production mechanisms (e.g. coalescence and hadronization), may also influence the spin alignment of ϕ mesons (J=1). The magnitude and the transverse momentum (p_T) dependence of the spin alignment are expected to be sensitive to different hadronization scenarios [4]. Thus the ϕ meson spin alignment also probes the particle production mechanisms.

*e-mail: tubiao@mails.ccnu.edu.cn

2 Method

The global polarization of spin-$\frac{1}{2}$ hyperons can be determined from the angular distribution of hyperon decay products relative to the system orbital momentum **L** [8]:

$$\frac{dN}{d\cos\theta^*} = \frac{1}{2}(1 + \alpha_H P_H \cos\theta^*) \tag{1}$$

where P_H is the hyperon global polarization, α_H is the hyperon decay parameter (α_Λ = -$\alpha_{\overline{\Lambda}}$ = 0.642), and θ^* is the angle in the hyperon rest frame between the system orbital momentum L and the three-momentum of the baryon daughter from the hyperon decay. Averaged over all phase space, we extract the average projection of the polarization on **L**. It is shown that [9]:

$$\overline{P}_H = \frac{8}{\pi\alpha_H} \frac{< \sin(\phi^* - \Psi_{EP}) >}{R_{EP}} \tag{2}$$

where Ψ_{EP} is the angle of the first-order event plane that estimates the reaction plane angle Ψ_{RP}, ϕ^* azimuthal angle of the daughter proton (antiproton) in the $\Lambda(\overline{\Lambda})$ frame, $R_{EP} = < \cos(\Psi_{RP} - \Psi_{EP}) >$ is the event plane resolution.

The spin alignment for a spin-1 vector meson is described by a spin-density matrix ρ, 3×3 Hermitian matrix with a unit trace. A deviation of the diagonal elements ρ_{mm} (m= -1,0,1) from 1/3 signals a net spin alignment. Because vector mesons decay strongly, the diagonal elements ρ_{-1-1} and ρ_{11} are degenerate and ρ_{00} is the only independent observable. It can be determined from the angular distribution of the decay products that [10]:

$$\frac{dN}{d\cos\theta^*} = N_0 \times [(1 - \rho_{00}) + (3\rho_{00} - 1)\cos^2\theta^*] \tag{3}$$

where N_0 is the normalization and θ^* is the angle between the system orbital momentum **L** and the momentum direction of a daughter particle in the rest frame of the parent vector meson.

This analysis use charged particles reconstructed by the Time Projection Chamber (TPC) and matched to the Time Of Flight (TOF) detector near mid-rapidity ($|\eta| < 1.0$). We reconstruct $\Lambda(\overline{\Lambda})$ and ϕ meson's invariant masses through their decay channels: $\Lambda(\overline{\Lambda}) \to p + \overline{\pi}(\overline{p} + \pi)$ and $\phi \to K^+ + K^-$, respectively. Topological and kinematic cuts are applied to reduce the combinatorial background. In the study of $\Lambda(\overline{\Lambda})$ polarization, the direction of **L** is determined by the first-order event plane reconstructed with information from the Shower Maximum Detectors at Zero Degree Calorimeter. In the study of ϕ spin alignment, the direction of **L** is determinted by the second-order event plane reconstructed with TPC tracks. For $\Lambda(\overline{\Lambda})$ analysis, the results are based on 440 M and 1000 M minimum bias events taken in years 2011 and 2014. For ϕ spin alignment analysis, the results are based on 118 M(39 GeV), 55 M(62.4 GeV) minimum bias events taken in 2010 and 29 M(19.6 GeV), 58 M(27 GeV), 375 M(200 GeV) minimum bias events taken in 2011.

3 Results

3.1 Azimuthal angle dependence of $\Lambda(\overline{\Lambda})$ polarization

The left panel of Figure 1 shows the Λ and $\overline{\Lambda}$ polarization(P_H) as a function of $\phi - \Psi_{obs}$ at midrapidity in 20-50% central Au+Au collisions at $\sqrt{s_{NN}}$ = 200 GeV. The ϕ here is the azimuthal angle of $\Lambda(\overline{\Lambda})$. The P_H from the off-peak for Λ and $\overline{\Lambda}$ are consistent with zero, which is as expected and could be served as a consistency check. The P_H from the mass peak for Λ and $\overline{\Lambda}$ decreases with increasing $\phi - \Psi_{obs}$. This feature is the same for both Λ and $\overline{\Lambda}$, and there is no significant difference in values

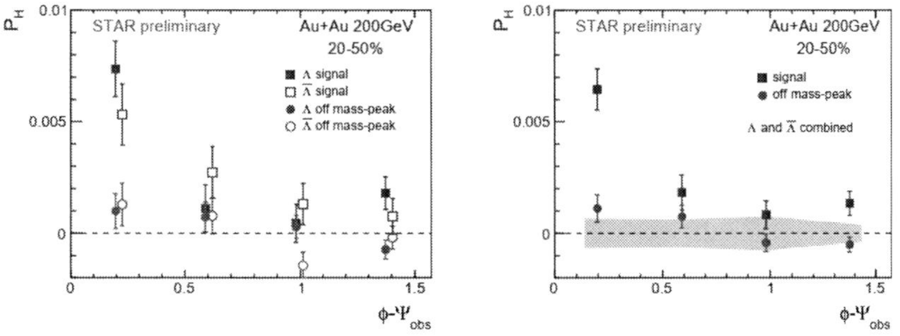

Figure 1: (Color online) Left panel: P_H for Λ and $\overline{\Lambda}$ are plotted as a function of $\phi - \Psi_{obs}$ in 20-50% central Au+Au collisions at $\sqrt{s_{NN}}$ = 200 GeV. Only the statistical errors are plotted. Right panel: Combined P_H between Λ and $\overline{\Lambda}$, are plotted as a function of $\phi - \Psi_{obs}$ in 20-50% central Au+Au collisions at $\sqrt{s_{NN}}$ = 200 GeV. The systematic uncertainties are shown with shadowed band.

between Λ and $\overline{\Lambda}$. Because Λ and $\overline{\Lambda}$ have opposite signs in intrinsic magnetic moments, it is expected that a magnetic field will enhance (reduce) the polarization for $\overline{\Lambda}$ (Λ). Within statistics our data shows that such effect is not visible for Au+Au collisions at $\sqrt{s_{NN}}$ = 200 GeV. This may be due to the short lifetime of magnetic field which gives little time for particles to align their spin, and/or the late hyperon production time at which moment the magnetic field diminished. The finite averaged P_H over four bins is ~ 0.2% which is consistent with STAR's previous published result [7].

The right panel shows the combined P_H between Λ and $\overline{\Lambda}$ as a function of $\phi - \Psi_{obs}$ at mid-rapidity in 20-50% central Au+Au collisions at $\sqrt{s_{NN}}$ = 200 GeV. The significance of ΔP_H, for Λ and $\overline{\Lambda}$ combined, between $\phi - \Psi_{obs}$ bin of [0, π/8] and [3π/8, π/2] is 4.7σ. The larger in-plane than out-of-plane polarization is consistent with the picture of maximum vorticity in the equator and low viscosity of the system. Please note that although P_H has been corrected for the event plane resolution, when presented as a function of $\phi - \Psi_{obs}$. The smearing correction of $\phi - \Psi_{obs}$ bins is not applied yet.

3.2 ϕ meson spin alignment

The left panel of Figure 2 shows ρ_{00} as a function of p_T in 20-60% central Au+Au collisions at $\sqrt{s_{NN}}$ = 19.6, 27, 39, 62.4 and 200 GeV. No clear p_T dependence has been found within currently large systematic uncertainties. The systematic uncertainties are dominated by the residual background estimation and ρ_{00} extraction procedure, and are under further investigation.

The right panel shows ρ_{00} for ϕ meson within $0.4<p_T<3.0$ GeV as a function of beam energy for 20-60% central Au+Au collisions [11]. The central values of ρ_{00} are slightly larger than 1/3 while the large systematic uncertainties prevents us from making definite conclusions. Note that the efficiency corrections due to kinematic cuts have not been applied to the data.

4 Summay

The measurement of the Λ and $\overline{\Lambda}$ polarization as a function of azimuthal angle relative to the reaction plane is presented. The difference of P_H, for Λ and $\overline{\Lambda}$ combined, between the most in-plane bin [0, π/8] and the most out-of-plane bin [3π/8, π/2] is 4.7σ. The data are consistent with the picture of a low

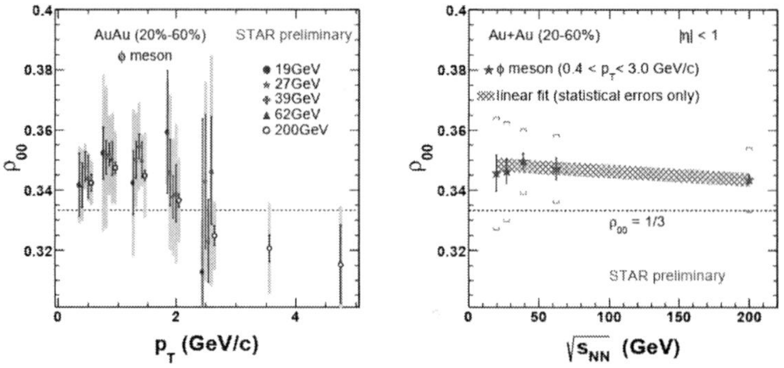

Figure 2: (Color online) Left panel: The ρ_{00} for ϕ meson are plotted as a function of p_T in 20-60% central Au+Au collisions at $\sqrt{s_{NN}}$ = 19.6, 27, 39, 62.4 and 200 GeV. The gray shadow indicate the systematic uncertainties. Right panel: The ρ_{00} for ϕ meson within 0.4<p_T<3.0 GeV are plotted as a function of beam energy for 20-60% central Au+Au collisions. The square brackets indicate the systematic uncertainties. The shadow band is a linear fit for these points. The acceptance and efficiency corrections have not been applied.

viscosity system with maximum vorticity at the equator. The measurement of ϕ meson spin alignment at $\sqrt{s_{NN}}$ = 19.6-200 GeV is shown. Within current large systematics errors, ϕ spin alignment signal ρ_{00} shows little beam energy dependence and non-significant p_T dependence.

Acknowledgments

We thanks the RHIC Operations Group and RCF at BNL, and the NERSC Center at LBNL. B. Tu is supported by the China Scholarship Council (CSC). This work was supported in part by the MoST of China 973-Project No. 2015CB856901, NSFC under Grants No. 11521064.

References

[1] I. Arsene *et al.* (BRAHMS Collaboration), Nucl. Phys. A **757**, 1 (2005); B.B. Back *et al.* (PHOBOS Collaboration), Nucl. Phys. A **757**, 28 (2005); J. Adams *et al.* (STAR Collaboration), Nuclear Physics A **757**, 102 (2005); K. Adcox *et al.* (PHENIX Collaboration), Nucl. Phys. A **757**, 184 (2005)
[2] Z.T. Liang, X.N. Wang, Physical Review Letters **94**, 102301 (2005)
[3] S.A. Voloshin, arXiv preprint nucl-th/0410089 (2004)
[4] Z.T. Liang, X.N. Wang, Physics Letters B **629**, 20 (2005)
[5] B. Betz, M. Gyulassy, G. Torrieri, Physical Review C **76**, 044901 (2007)
[6] F. Becattini, F. Piccinini, J. Rizzo, Physical Review C **77**, 024906 (2008)
[7] L. Adamczyk, *et al.* (STAR Collaboration), Nature **548**, 62 (2017)
[8] G. Bunce, R. Handler, R. March, P. Martin, L. Pondrom, M. Sheaff, K. Heller, O. Overseth, P. Skubic, T. Devlin et al., Physical Review Letters **36**, 1113 (1976)
[9] B. Abelev, *et al.* (STAR Collaboration), Physical Review C **76**, 024915 (2007)
[10] K. Schilling, P. Seyboth, G. Wolf, Nuclear Physics B **15**, 397 (1970)
[11] X. Sun, Quark Matter 2017 Poster Presentation

Measurements of spin alignment of vector mesons and global polarization of hyperons with ALICE at the LHC

Bedangadas Mohanty (for the ALICE collaboration)[1,*]

[1] *National Institute of Science Education and Research, HBNI, Jatni 752050, India*

Abstract. We present the measurements related to global polarization of Λ hyperons and spin alignment of K^{*0} vector mesons at mid-rapidity for Pb-Pb collisions at $\sqrt{s_{NN}} = 2.76$ TeV using the ALICE detector at the LHC. The global polarization measurements are carried out with respect to the first order event plane while the spin alignment measurements are carried out with respect to the production plane. No global polarization signal for Λ is observed for 5-15% and 15-50% central Pb-Pb collisions. The spin density matrix element ρ_{00} is found to have values slightly below 1/3 at low transverse momentum (p_T) for K^{*0} mesons, while it is consistent with 1/3 (no spin alignment) at higher p_T. No spin alignment is observed for K^{*0} in pp collisions at $\sqrt{s} = 13$ TeV and for the spin zero hadron K^0_S in 20-40% Pb-Pb collisions at $\sqrt{s_{NN}} = 2.76$ TeV.

1 Introduction

Large magnetic field [1] and large angular momentum [2] are expected to be present in the initial stages of high energy heavy-ion collisions. One of the physics interests of the heavy-ion program using the ALICE detector at the LHC is to look for signatures of these effects. This can be achieved by studying the angular distributions of the decay daughters of hyperons and vector mesons [3–5]. The angular distributions are measured with respect to a quantization axis, which can either be perpendicular to the production plane of the hadron (for the spin alignment measurement), or normal to the reaction plane of the system (for the global polarization measurement). The production plane is defined by the momentum of the hadron under study and the beam direction, whereas the reaction plane is defined by the impact parameter and beam direction. The angle denoted as θ^* is that made by one of the decay daughters of the hadron in the rest frame of the hadron with respect to the quantization axis. In general, the angular distribution for vector mesons is expressed as [6],

$$\frac{\mathrm{d}N}{\mathrm{d}\cos\theta^*} \propto [1 - \rho_{00} + \cos^2\theta^*(3\rho_{00} - 1)], \tag{1}$$

while that for hyperons as [3],

$$\frac{\mathrm{d}N}{\mathrm{d}\cos\theta^*} \propto [1 + \alpha_H P_H \cos\theta^*]. \tag{2}$$

ρ_{00} is the zeroth element of the spin density matrix, α_H is the decay parameter and P_H is the polarization. If initial conditions or the final hadronization process cause polarization effects in heavy-ion

* e-mail: bedanga@niser.ac.in

collisions, then the angular distributions as defined in Eq.1 and Eq.2 would become non uniform. This would lead to ρ_{00} values being different from 1/3 and/or non-zero values for P_{H}.

In this work we present the first results at LHC energies related to the spin alignment of K^{*0} vector mesons through the measurement of ρ_{00} in pp and Pb-Pb collisions with respect to the production plane. In addition, we report on the global polarization of Λ hyperons through measurement of P_{H} in Pb-Pb collisions with respect to the event plane using the ALICE detector [7].

2 Global polarization of Λ hyperons

The global polarization studies of Λ hyperons are carried out using 49 million minimum bias Pb-Pb events at $\sqrt{s_{\mathrm{NN}}} = 2.76$ TeV. The analysis is performed at mid-rapidity ($-0.5 < y < 0.5$) for 5-15% and 15-50% central collisions. The trigger and centrality selection use the V0 detectors [7]. Flow based techniques are used as discussed in [4, 8]. The information from the Time Projection Chamber (TPC) is used to reconstruct the Λ hyperons from their $p\pi$ decay channel. The measurements are done using the first order event plane reconstructed using the zero degree calorimeter [8]. The p_{T} integrated values of P_{H} for Λ hyperons in $-0.5 < y < 0.5$ are measured to be $P_{\mathrm{H}}(\Lambda) = -0.01 \pm 0.13$ (stat) $\pm$ 0.04 (syst) and $P_{\mathrm{H}}(\Lambda) = -0.08 \pm 0.10$ (stat) $\pm$ 0.04 (syst) for 5-15% and 15-50% central Pb–Pb collisions at $\sqrt{s_{\mathrm{NN}}} = 2.76$ TeV, respectively [8]. The corresponding values for $\bar{\Lambda}$ are $P_{\mathrm{H}}(\bar{\Lambda}) = -0.09 \pm 0.13$ (stat) $\pm$ 0.08 (syst) for 5-15% and $P_{\mathrm{H}}(\bar{\Lambda}) = 0.05 \pm 0.10$ (stat) $\pm$ 0.03 (syst) for 15-50% central Pb–Pb collisions [8]. The P_{H} values are consistent with zero.

3 Spin alignment of K^{*0} vector mesons

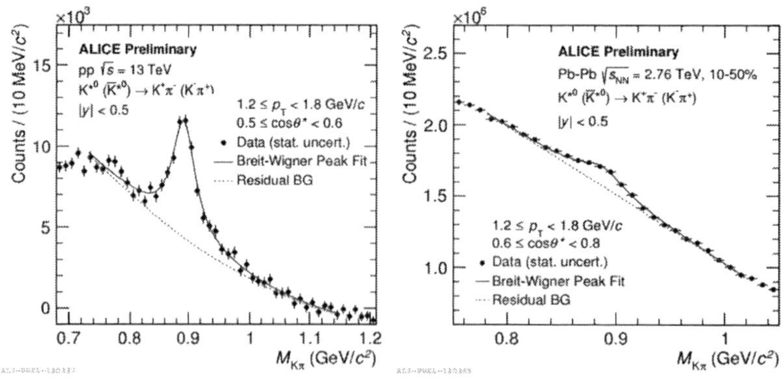

Figure 1. (Color online) Left Panel: Kaon-Pion invariant mass ($M_{\mathrm{K}\pi}$) distribution at mid-rapidity after mixed background subtraction in minimum bias pp collisions at $\sqrt{s} = 13$ TeV. The data are fitted with a Breit-Wigner function and a 2nd order polynomial function in $M_{\mathrm{K}\pi}$ reflecting the residual background. Right Panel: Same as left panel but for 10-50% central Pb–Pb collisions at $\sqrt{s_{\mathrm{NN}}} = 2.76$ TeV.

The spin alignment studies for the spin one meson K^{*0} are carried out using 43 million minimum bias events in pp collisions at $\sqrt{s} = 13$ TeV and 14 million minimum bias events in Pb–Pb collisions at $\sqrt{s_{\mathrm{NN}}} = 2.76$ TeV. The K^{*0} analysis in Pb–Pb collisions uses the data taken in year 2010 whereas the global polarization studies of Λ hyperons use both the data sets taken in 2010 and 2011. The measurements are performed at mid-rapidity for both pp and Pb–Pb collisions. Results for 10-50%

 https://doi.org/10.1051/epjconf/201817116008

collision centrality in the Pb–Pb system are presented here. In order to study the null hypothesis, a similar analysis for 20-40% collision centrality using the spin zero meson K^0_S is also carried out. The K^{*0} are reconstructed via the decay into charged K and π while the K^0_S is reconstructed via its decay into $\pi\pi$. The ALICE TPC and TOF are used for identifying the decay products of the above hadrons and the signals are extracted via the invariant mass technique [9, 10]. The background, dominantly coming from combinatorics, is removed by constructing the invariant mass distributions using mixed event techniques [9, 10]. The mixed event background subtracted invariant mass $M_{K\pi}$ distributions for pp and Pb–Pb collisions are shown in Fig. 1. The results are shown for a particular p_T and $\cos\theta^*$ bin as a representative case, but the same procedure is followed for each p_T and $\cos\theta^*$ bin of the analysis. The distributions are then fitted with a Breit-Wigner function for the signal and 2nd order polynomial for the residual background to extract the yields, which are then corrected for the corresponding reconstruction efficiency and acceptance in each $\cos\theta^*$ and p_T bin using Monte Carlo simulations of the ALICE detector and response [9, 10]. The efficiency and acceptance corrected $dN/d\cos\theta^*$ distributions at mid-rapidity for $1.2 \leq p_T < 1.8$ GeV/c in minimum bias pp collisions and for $0.4 \leq p_T < 1.2$ GeV/c in 10-50% central Pb-Pb collisions are shown in Fig. 2. These distributions are fitted with the functional form shown in Eq. 1 to extract the ρ_{00} values for each p_T bin. Figure 3

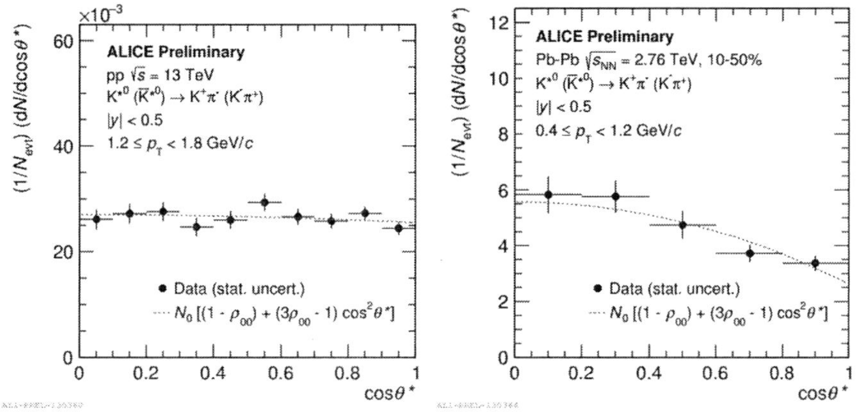

Figure 2. (Color online) Left Panel: $dN/d\cos\theta^*$ versus $\cos\theta^*$ at mid-rapidity for minimum bias pp collisions at $\sqrt{s}$ = 13 TeV. Right Panel: Same but for 10-50% central Pb–Pb collisions at $\sqrt{s_{NN}}$ = 2.76 TeV. The error bars shown are statistical only. The angular distributions are obtained with respect to production plane.

shows the ρ_{00} values as a function of p_T for K^{*0} and K^0_S. The K^{*0} results for pp collisions at $\sqrt{s}$ = 13 TeV are consistent with no spin alignment ($\rho_{00} \sim 1/3$). The ρ_{00} values for K^{*0} at $0.4 \leq p_T < 1.2$ GeV/c is about 2.5 standard deviations (σ) lower than 1/3, about 1.4σ lower than 1/3 for the p_T bin 1.2 - 1.8 GeV/c and consistent with 1/3 for higher p_T values in 10-50% central Pb–Pb collisions at $\sqrt{s_{NN}}$ = 2.76 TeV. The σ is calculated by adding the statistical and systematic uncertainties on ρ_{00} in quadrature. The ρ_{00} values are consistent with 1/3 for the whole p_T range studied for K^0_S in 20-40% central Pb–Pb collisions at $\sqrt{s_{NN}}$ = 2.76 TeV. The sources of systematic uncertainties on ρ_{00} include: reconstruction efficiency and acceptance, track selection conditions, particle identification procedure, signal extraction procedure, and material budget estimates [9, 10].

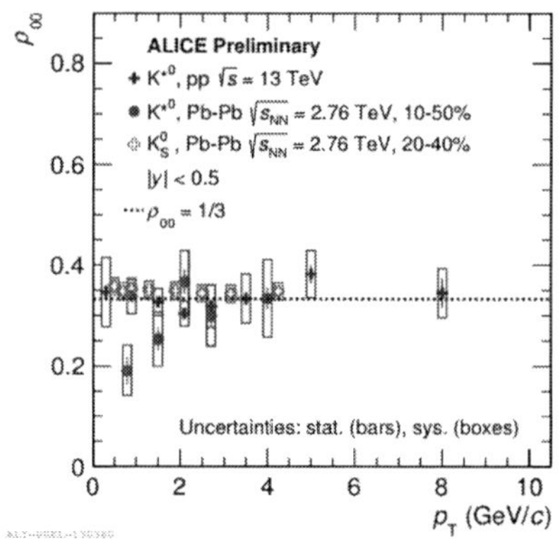

Figure 3. (Color online) ρ_{00} versus p_T extracted from the $\cos\theta^*$ distributions at mid-rapidity for K^{*0} vector meson in minimum bias pp collisions at $\sqrt{s}$ = 13 TeV and 10-50% central Pb–Pb collisions at $\sqrt{s_{NN}}$ = 2.76 TeV. Also shown are the corresponding results for spin zero K^0_S meson for 20-40% Pb–Pb collisions at $\sqrt{s_{NN}}$ = 2.76 TeV. The vertical lines on the data points are statistical errors and the boxes reflect the systematic uncertainties on ρ_{00}.

4 Summary and outlook

We have presented results related to the global polarization of Λ hyperons measured with respect to the reaction plane and those related to spin alignment of K^{*0} with respect to the production plane in Pb-Pb collisions at $\sqrt{s_{NN}}$ = 2.76 TeV. No polarization effect was observed for the Λ produced at mid-rapidity for 5-15% and 15-50% central collisions. The spin density matrix element ρ_{00} extracted from the angular distributions of the decay daughter of K^{*0} at mid-rapidity in 10-50% central Pb-Pb collisions was found to be slightly lower than 1/3 for p_T below 1.8 GeV/c. The ρ_{00} values are consistent with 1/3, namely indicating no spin alignment, for high p_T K^{*0} in Pb-Pb collisions, for the full p_T range studied in pp collisions and for spin zero K^0_S produced in 20-40% central Pb-Pb collisions.

The analysis of the centrality dependence of spin alignment for vector mesons with respect to both production and reaction planes is ongoing. Results with increased statistical precision are expected for both global polarization and spin alignment studies with the Pb-Pb data set at $\sqrt{s_{NN}}$ = 5.02 TeV. This will help in clarifying the significance of the observed effects.

References

[1] D. E. Kharzeev, L. D. McLerran and H. J. Warringa, Nucl. Phys. A **803**, 227 (2008).
[2] F. Becattini, F. Piccinini and J. Rizzo, Phys. Rev. C **77**, 024906 (2008).
[3] B. I. Abelev *et al.* [STAR Collaboration], Phys. Rev. C **76**, 024915 (2007).
[4] L. Adamczyk *et al.* [STAR Collaboration], Nature **548**, 62 (2017).
[5] B. I. Abelev *et al.* [STAR Collaboration], Phys. Rev. C **77**, 061902 (2008).
[6] K. Schilling, P. Seyboth and G. E. Wolf, Nucl. Phys. B **15**, 397 (1970).
[7] B. Abelev *et al.* [ALICE Collaboration], Int. J. Mod. Phys. A **29**, 1430044 (2014).
[8] A. Timmins [ALICE Collaboration], Nucl. Phys. A **967**, 43 (2017) and M. Konyushikhin (ALICE Collaboration), QM2017 poster presentation.
[9] J. Adam *et al.* [ALICE Collaboration], Phys. Rev. C **95**, 064606 (2017).
[10] B. Abelev *et al.* [ALICE Collaboration], Phys. Rev. C **91**, 024609 (2015).

Study of Lambda polarization at RHIC BES and LHC energies

Iurii Karpenko[1,2,⋆] and *Francesco* Becattini[1,3,⋆⋆]

[1] *INFN - Sezione di Firenze, Via G. Sansone 1, I-50019 Sesto Fiorentino (FI), Italy*
[2] *SUBATECH, University of Nantes – IN2P3/CNRS – IMT Atlantique, Nantes, France*
[3] *Universitá di Firenze, Via G. Sansone 1, I-50019 Sesto Fiorentino (FI), Italy*

Abstract. In hydrodynamic approach to relativistic heavy ion collisions, hadrons with nonzero spin, produced out of the hydrodynamic medium, can acquire polarization via spin-vorticity thermodynamic coupling mechanism. The hydrodynamical quantity steering the polarization is the thermal vorticity, that is minus the antisymmetric part of the gradient of four-temperature field. Based on this mechanism there have been several calculations in hydrodynamic and non-hydrodynamic models for non-central heavy ion collisions in the RHIC Beam Energy Scan energy range, showing that the amount of polarization of produced Λ hyperons ranges from few percents to few permille, and decreases with collision energy. We report on an extension of our existing calculation of global Λ polarization in UrQMD+vHLLE model to full RHIC and LHC energies, and discuss the component of polarization along the beam direction, which is the dominant one at high energies.

1 Introduction

Recently, STAR collaboration has reported on the first observation of significantly nonzero global polarization of $\Lambda(\bar{\Lambda})$ hyperons produced in non-central Au-Au collisions in the RHIC Beam Energy Scan (BES) Program [1]. The magnitude of measured polarization is in general agreement with calculations from different hydrodynamic models [2, 3] and AMPT [4, 5]. In the hydrodynamic models the Λ hyperons produced at particlization (fluid to particle transition) hypersurface acquire polarization via spin-vorticity coupling mechanism. In case of AMPT the spin-vorticity coupling mechanism is introduced with the help of coarse-graining procedure. The magnitude of polarization depends on local *thermal vorticity* of the fluid, $\varpi_{\mu\nu} = (\partial_\nu\beta_\mu - \partial_\mu\beta_\nu)$ at the points of Λ production, $\beta_\mu = u_\mu/T$ being the inverse four-temperature field. In this proceeding we report on an extension of our existing calculations of global Λ polarization in RHIC BES [2] and discuss the polarization component along the beam direction.

2 Model and results

We study production and polarization of Λ in a hybrid transport model, which consists of initial state UrQMD hadron/string cascade, followed by a 3 dimensional viscous hydrodynamic expansion. At a

⋆e-mail: yu.karpenko@gmail.com
⋆⋆e-mail: becattini@fi.infn.it

hypersurface where the local rest frame energy density reaches a critical value $\epsilon_{sw} = 0.5$ GeV/fm^3, the particlization (i.e. transition from fluid to particle description) is performed with the help of Cooper-Frye formula. Along with the rest of hadrons, Λ as well as other heavy resonances which decay to it, are produced and acquire their polarization at this hypersurface. Details of the cascade+viscous hydrodynamic model used are thoroughly described in [6].

We have reported global polarization of Λ hyperons calculated with this model in the energy range $\sqrt{s_{\rm NN}} = 7.7, \ldots, 200$ GeV in [2]. It was concluded that, in line with the experimental results, in 20-50% central Au-Au collisions the global Λ polarization decreases with increasing collision energy from about 2% at $\sqrt{s_{\rm NN}} = 7.7$ GeV to 0.2% at $\sqrt{s_{\rm NN}} = 200$ GeV. Only about one quarter of Λ are the ones produced directly at particlization, the rest are products of resonance decays, including cascade processes $X \to \Sigma^0 \to \Lambda$. We have estimated [7] that such resonance feed-down results in 15% reduction of the polarization of all Λ at all collision energies in consideration.

We have extended this result with a 3+1 dimensional viscous hydrodynamic calculation for $\sqrt{s_{\rm NN}} = 200$ GeV RHIC Au-Au and 2760 GeV Pb-Pb collisions using a Monte Carlo Glauber initial state with parametrized rapidity dependence [8] tuned to describe dN/dy and directed flow of charged hadrons at these energies. The combined results are shown on Fig. 1, left. For the Glauber IS, we show two lines (connecting $\sqrt{s_{\rm NN}}$ =200 and 2760 GeV points), the solid one corresponding to longitudinally boost invariant initial flow, and the dashed one corresponding to a small amount of initial shear longitudinal flow as described in [9]. The latter results in higher initial thermal vorticity which in turn leads to a larger magnitude of Λ polarization. Also, for the RHIC BES range we show the curves for both averaged initial state and event-by-event hydro with fluctuating initial state. From the plot one can conclude that 1) the global polarization signal in the model is relatively insensitive to whether the initial state is fluctuating or not, and 2) the global Λ polarization further decreases with energy from RHIC to LHC, and although there are no published results yet from the LHC, we estimate that at the LHC the global polarization possibly goes below the experimental limit for being detected.

One expects that the global (*i.e.* integrated over all Λ in an event) polarization vector is parallel to the global angular momentum vector, which is confirmed by our calculations. However doing a p_T differential analysis we have found that, at high energies, for mid-rapidity Λ the dominant component of polarization is the one along the beam direction [10]. In a simplified scenario with no initial state fluctuations, this component has a quadrupole structure in the transverse momentum plane, and can be decomposed into Fourier series with only sine terms of even multiples of the azimuthal angle ϕ of the transverse momentum vector:

$$P^z(\mathbf{p}_T, Y=0) = \sum_{k=1}^{\infty} f_{2k}(p_T) \sin 2k\varphi. \tag{1}$$

In this expansion, the dominant component is f_2, which we show on the right panel of Fig. 1. Contrary to global polarization (left panel of the same figure) which decreases by almost 1 order of magnitude between $\sqrt{s_{\rm NN}}$=7.7 and 200 GeV, f_2 decreases by only 35%. Even at 2.76 TeV LHC energy, it reaches 1% for $p_T = 3$ GeV Λ, making it feasible to detect it in the experiment.

Another useful feature of the f_{2k} components of the Fourier expansion is that, different from the global polarization vector, they are independent on the orientation (sign) of the event plane, which simplifies the experimental measurements.

The structure of the polarization component along the beam direction can also be accessed via correlations of the corresponding polarization components of Λ pairs. One can show with a simple derivation that, keeping the second Fourier harmonic only:

$$P^z = f_2 \sin 2(\phi - \Psi) \quad \Rightarrow \quad \langle P^z(\phi) P^z(\phi + \Delta\phi) \rangle_\phi = \frac{f_2^2}{2} \cos 2\Delta\phi, \tag{2}$$

where $\langle A\rangle_\phi = \frac{1}{2\pi}\int A(\phi)d\phi$. Note that such covariance function is independent of the event plane angle Ψ.

The resulting covariance functions for P^z are plotted on Fig. 2. The left panel corresponds to averaged initial state and a single hydrodynamic simulation for given collision energy, where due to the geometrical symmetry and the above argument we obtain the expected $\cos(2\Delta\phi)$ shape, with collision energy dependent amplitude. The right panel shows the covariance function calculated in event-by-event hydrodynamic case for $\sqrt{s_{\rm NN}} = 200$ GeV RHIC scenario. For the event-by-event case, we observe a relative enhancement of the covariance function at $\Delta\phi = 0$, caused by vorticity patterns formed by hot spots in the initial state. The overall shape and magnitude of the covariance function are qualitatively consistent with the results in [11].

Note that the quadrupole pattern of the longitudinal component of polarization is present in case of average (non-fluctuating) initial state. Even more, this effect is still present and has a similar magnitude when the initial state has a complete longitudinal boost invariance. The reason is that the longitudinal component of polarization is linked to the anisotropies in the transverse expansion, therefore the only requisite for it is anisotropic transverse flow, which also drives the elliptic flow. Indeed we have shown that in a simple Blast-Wave model there is a linear relation between the p_T dependent f_2 harmonic and elliptic flow:

$$f_2(p_T) = 2\frac{dT}{d\tau}\frac{1}{mT}v_2(p_T) \tag{3}$$

where $dT/d\tau$ is assumed time derivative of temperature T (which in Bjorken picture depends only on proper time τ) at the freezeout hypersurface, and m is mass of particle.

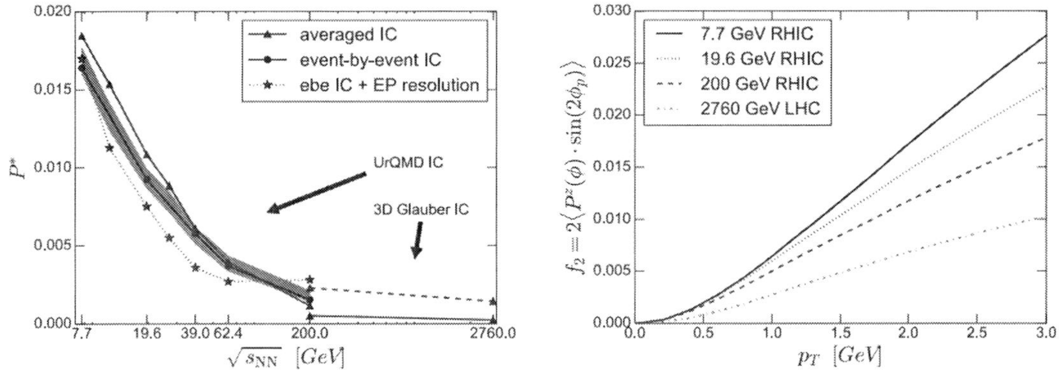

Figure 1. Left panel: collision energy dependence of global Λ polarization along global angular momentum of the fireball, calculated with averaged initial state from UrQMD ($\sqrt{s_{\rm NN}}$=7.7,...,200 GeV) and Monte Carlo Glauber ($\sqrt{s_{\rm NN}}$=200,2760 GeV). Right panel: second order Fourier harmonic coefficient of polarization component along beam direction, calculated as a function of p_T for different collision energies; 200 and 2760 GeV points correspond to Monte Carlo Glauber initial state.

3 Conclusions

We have extended the results for global Λ polarization in RHIC Beam Energy Scan [2] by using Monte Carlo Glauber initial state with parametrized rapidity dependence, tuned to reproduce basic

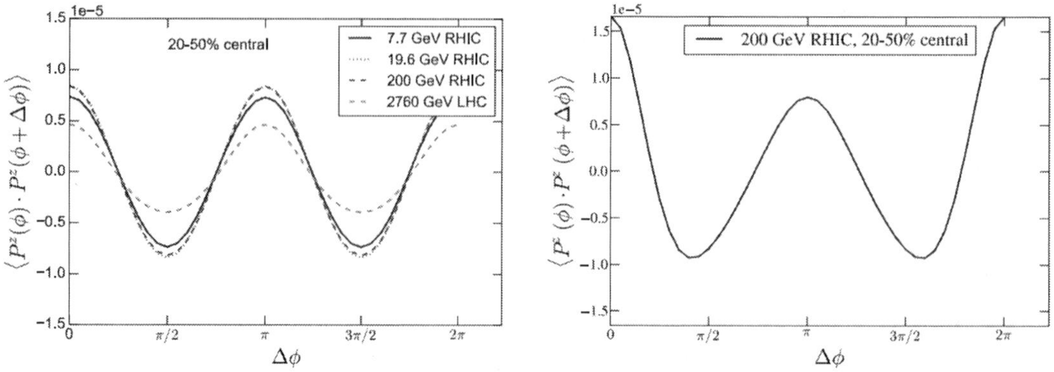

Figure 2. Covariance of polarization components in beam direction, for pairs of midrapidity Λ separated by azimuthal angle $\Delta\phi$ in transverse momentum space. Left panel: results with averaged initial state from UrQMD ($\sqrt{s_{NN}}$=7.7,...,200 GeV Au-Au) and Monte Carlo Glauber ($\sqrt{s_{NN}}$=200 Au-Au, 2760 GeV Pb-Pb). Right panel: event-by-event hydrodynamic calculation with UrQMD initial state for Au-Au collisions at $\sqrt{s_{NN}}$ = 200 GeV.

observables at $\sqrt{s_{NN}}$ =200 GeV RHIC and 2760 GeV LHC energies. We find that, as expected, the global polarization of Λ hyperons further decreases from the full RHIC to the LHC energy. However, at such high energies the dominant component of Λ polarization vector in p_T differential analysis is the one along the beam direction. This component of polarization has a quadrupole structure in the transverse momentum plane, and shows a mild decrease with collision energy. Its second order Fourier harmonic for p_T = 3 GeV Λ reaches 1% at 2760 GeV LHC energy. Such quadrupole structure also results in correlations of polarization components of Λ pairs. We stress that the longitudinal component of polarization is a generic effect which emerges even in a simple longitudinal boost invariant non-fluctuating hydrodynamic picture of heavy ion collisions.

This work was partly supported by the University of Florence grant *Fisica dei plasmi relativistici: teoria e applicazioni moderne*.

References

[1] L. Adamczyk *et al.* [STAR Collaboration], Nature **548**, 62 (2017).
[2] I. Karpenko and F. Becattini, Eur. Phys. J. C **77**, no. 4, 213 (2017).
[3] Y. Xie, D. Wang and L. P. Csernai, Phys. Rev. C **95**, no. 3, 031901 (2017).
[4] H. Li, L. G. Pang, Q. Wang and X. L. Xia, arXiv:1704.01507 [nucl-th].
[5] Y. Sun and C. M. Ko, Phys. Rev. C **96**, no. 2, 024906 (2017).
[6] I. A. Karpenko, P. Huovinen, H. Petersen and M. Bleicher, Phys. Rev. C **91**, no. 6, 064901 (2015).
[7] F. Becattini, I. Karpenko, M. A. Lisa, I. Upsal and S. A. Voloshin, Phys. Rev. C **95**, no. 5, 054902 (2017).
[8] P. Bozek and W. Broniowski, Phys. Rev. C **85**, 044910 (2012).
[9] F. Becattini *et al.*, Eur. Phys. J. C **75**, no. 9, 406 (2015).
[10] F. Becattini and I. Karpenko, arXiv:1707.07984 [nucl-th].
[11] L. G. Pang, H. Petersen, Q. Wang and X. N. Wang, Phys. Rev. Lett. **117**, no. 19, 192301 (2016).

Anisotropic flow of identified particles in Pb–Pb collisions at $\sqrt{s_{\mathrm{NN}}} = 5.02$ TeV

Redmer Alexander Bertens[1,★]

[1] *University of Tennessee (Knoxville, USA)*

Abstract. Anisotropic flow is sensitive to the shear (η/s) and bulk (ζ/s) viscosity of the quark-gluon plasma created in heavy-ion collisions, as well as the initial state of such collisions and hadronization mechanisms. In these proceedings, elliptic (v_2) and higher harmonic (v_3, v_4) flow coefficients of $\pi^{\pm}$, $K^{\pm}$, p($\overline{\mathrm{p}}$) and the ϕ-meson, are presented for Pb—Pb collisions at the highest-ever center-of-mass energy of $\sqrt{s_{\mathrm{NN}}} = 5.02$ TeV. Comparisons to hydrodynamic calculations (IP-Glasma, MUSIC, UrQMD) are shown to constrain the initial conditions and viscosity of the medium.

1 Introduction

Ultra-relativistic heavy-ion collisions are used to study the properties of the quark-gluon plasma (QGP), a state of deconfined quarks and gluons that is created at high energy densities and temperatures. Past measurements of azimuthal anisotropies in particle production relative to collision symmetry plane angles Ψ_n have shown that the QGP behaves as a nearly perfect fluid [1]. These anisotropies in particle production arise when initial spatial anisotropies, resulting from the approximately elliptic overlap region of the colliding nuclei in non-central collisions, combined with the initial inhomogeneities of the system density, are transformed, through multiple interactions between the produced particles, into an anisotropy in momentum space. The efficiency of this process depends on the system's transport coefficients such as shear (η/s) and bulk (ζ/s) viscosity.

Momentum anisotropy in particle production is quantified as harmonic coefficients v_n of a Fourier series of the azimuthal angle (φ) distribution relative to the system's symmetry plane angles Ψ_n [2]

$$\frac{\mathrm{d}N}{\mathrm{d}(\varphi - \Psi_n)} \propto 1 + \sum_{n=1}^{\infty} 2v_n \cos\left(n\left[\varphi - \Psi_n\right]\right). \quad (1)$$

Harmonic coefficients v_n – commonly called *flow coefficients* – are, in addition to being a probe for η/s and ζ/s, also sensitive to the initial state of the system, freeze-out conditions, hadronization mechanisms, and the lifetime of the system.

2 Data analysis

A sample of Pb–Pb collisions used for this work were recorded with the ALICE [3] detector at a center of mass energy per nucleon pair of $\sqrt{s_{\mathrm{NN}}} = 5.02$ TeV. The data set comprises $\approx 6 \times 10^7$ collisions with

★e-mail: rbertens@cern.ch

a primary vertex reconstructed within ±10 cm along the beam pipe. The Inner Tracking System (ITS) and Time Projection Chamber (TPC) are used to reconstruct charged particles at $|\eta| < 0.9$ and $|y| < 0.5$. The V0 scintillator detectors, located at $2.8 < \eta < 5.1$ (V0A) and $-3.7 < \eta < -1.7$ (V0C), are used for centrality determination, and reconstruction of the $\mathbf{Q}_n^{\mathrm{V0}}$ vectors (see Eq. 2). Particle identification is performed using ionization energy loss measured in the TPC, combined with the arrival time of particles in the Time of Flight (TOF) detector. The ϕ-meson is reconstructed in the $\phi \to K^+K^-$ channel, using the analysis method outlined in [4].

The scalar product method [5] is used to measure flow coefficients v_n, and is defined as

$$v_n\{\mathrm{SP, V0C}\} = \langle\langle \mathbf{u}_n \cdot \mathbf{Q}_n^{\mathrm{V0C}*}\rangle\rangle \Bigg/ \sqrt{\frac{\langle \mathbf{Q}_n^{\mathrm{V0C}} \cdot \mathbf{Q}_n^{\mathrm{TPC}*}\rangle\langle \mathbf{Q}_n^{\mathrm{V0C}} \cdot \mathbf{Q}_n^{\mathrm{V0A}*}\rangle}{\langle \mathbf{Q}_n^{\mathrm{TPC}} \cdot \mathbf{Q}_n^{\mathrm{V0A}*}\rangle}}, \quad (2)$$

where $\mathbf{u}_n = \exp(in\varphi)$, in which φ represents azimuthal angle, is the unit vector of a particle of which v_n is measured. Reference flow vectors $\mathbf{Q}_n = \sum_j \exp(in\varphi_j)$ (the sum runs over all j tracks and * denotes the complex conjugate) are measured in the TPC and V0 detectors, where V0C is chosen as reference flow detector as it still has a high event plane resolution. The large rapidity gap between $\mathbf{u}_n$ and $\mathbf{Q}_n^{\mathrm{V0C}*}$ ensures that contributions of short-range correlations that are unrelated to the initial geometry – *'non-flow'* – are suppressed.. Brackets $\langle \ldots \rangle$ in Eq. 2 represent an all-event average; the double brackets in the numerator indicate that prior to the all-event average, an average over all $\mathbf{u}_n \cdot \mathbf{Q}_n^{\mathrm{V0C}*}$ within the single event is taken.

3 Results

Figure 1 shows p_{T}-differential v_2 (top panel) of $\pi^\pm$, $K^\pm$, p($\overline{\mathrm{p}}$) and the ϕ-meson for 10-20% (left) and 40-50% (right) collision centralities. For $p_{\mathrm{T}} < 4$ GeV/c, only $\overline{\mathrm{p}}$ are considered to exclude a contamination from detector material. A mass splitting of v_2 of different particle species is found for $p_{\mathrm{T}} < 2$ GeV/c, which is indicative of strong radial flow [6]. For $3 < p_{\mathrm{T}} < 8$ GeV/c the curves for different particles group according to their constituent quark number, supporting the concept of particle production via quark coalescence [7]. Flow of the ϕ-meson most directly illustrates mass splitting and quark number scaling since the ϕ is a meson with a mass close to proton mass. The ϕ-meson v_2 indeed follows proton v_2 at low p_{T}, but pion v_2 at intermediate p_{T}. The fact that p($\overline{\mathrm{p}}$) v_2 is larger than $\pi^\pm$ v_2 for $3 \lesssim p_{\mathrm{T}} \lesssim 10$ GeV/c, after which the v_2 converge, suggests that partonic energy loss is flavor independent at high p_{T}.

Inhomogeneities in the initial nucleon distribution produce higher harmonic flow coefficients ($n > 2$). These coefficients are thought to be more sensitive to transport coefficients than v_2 [8]. Non-zero v_3 is observed for $\pi^\pm, K^\pm$, p($\overline{\mathrm{p}}$) up to $p_{\mathrm{T}} \approx 8$ GeV/c, as is shown in the middle panel of Fig. 1; v_4 is non-zero in the entire measured range ($p_{\mathrm{T}} < 4$GeV/c, lower panel). The aforementioned mass splitting is visible for v_3 and v_4 up to $p_{\mathrm{T}} \approx 2.5$ GeV/c.

Model predictions from [9] are shown to test the validity of the hydrodynamic description of the QGP in Fig. 2. The curves are based on an IP-Glasma initial state and use a viscous hydrodynamic medium evolution ($\eta/s = 0.095$ with a temperature-dependent ζ/s), followed by a hadronic cascade procedure for hadronization. Mass splitting is violated (ϕ-meson v_2 > p($\overline{\mathrm{p}}$) v_2) in the model, likely resulting from an under-prediction of the hadronic cross section of the ϕ-meson. The model is in agreement with the data for $p_{\mathrm{T}} < 1$ GeV/c in central collisions, but overestimates v_2 already at lower p_{T} for more peripheral collisions.

To test the hypothesis of particle production via quark coalescence, the axes of Fig. 1 are scaled by the number of constituent quarks n_q independently for each species [7] in Fig. 3. From $p_{\mathrm{T}}/n_{\mathrm{q}} > 1.5$ GeV/c particles group approximately according to their type (baryon, meson). Similar behavior is

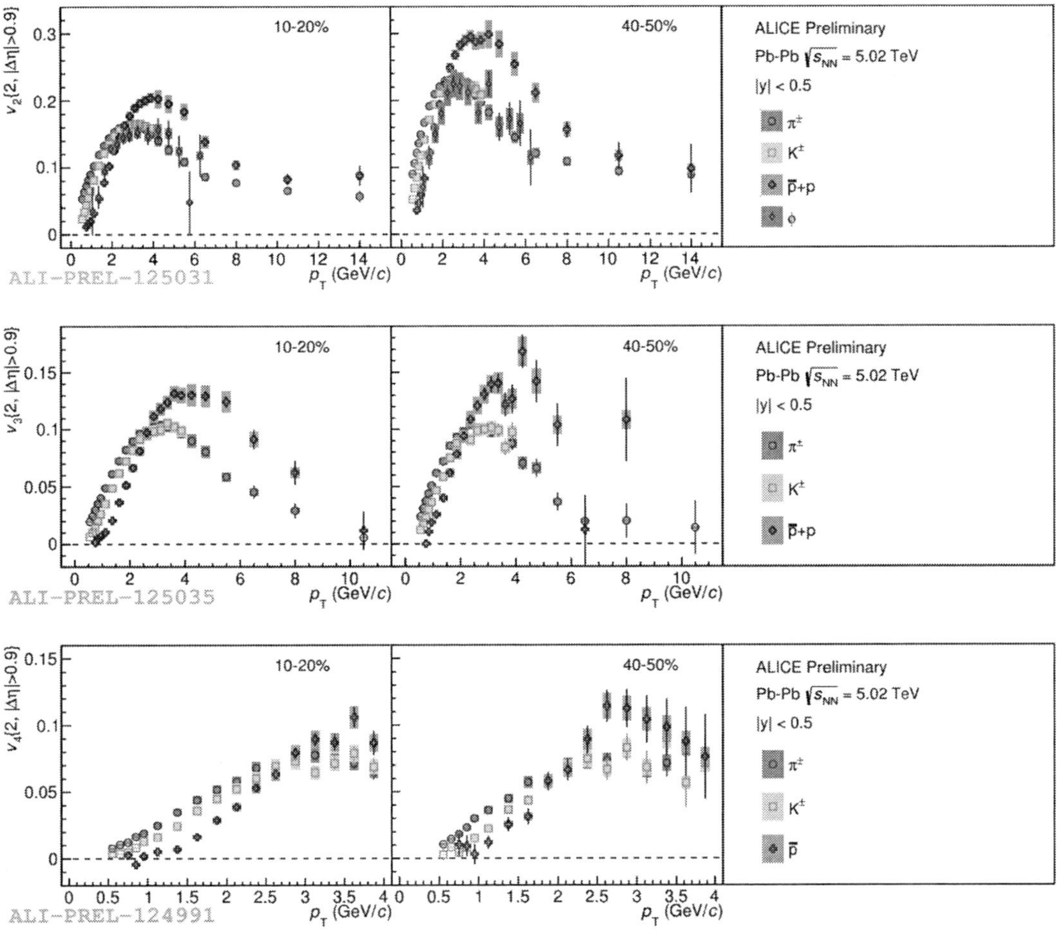

Figure 1. Flow coefficient v_2 (top), v_3 (middle) and v_4 (bottom) of $\pi^{\pm}$, $K^{\pm}$, p($\overline{\text{p}}$) and the ϕ-meson for 10-20% (left) and 40-50% (right) collision centrality as function of p_T. Statistical uncertainties are shown as bars and systematic uncertainties as boxes.

observed for v_3 and v_4 (not shown) but it is stressed that the observed scaling only holds approximately, as was also seen in [4].

4 Summary

Elliptic (v_2) and higher harmonic (v_3, v_4) flow coefficients of $\pi^{\pm}$, $K^{\pm}$, p($\overline{\text{p}}$) and the ϕ-meson, measured in Pb—Pb collisions at $\sqrt{s_{NN}}$ = 5.02 TeV, have been measured. Mass splitting is observed for $p_T <$ 2 GeV/c, as well as approximate quark number scaling for $p_T >$ 2.5 GeV/c. These high-precision measurements can be used to put new constraints on model calculations, furthering the understanding of the initial state of heavy-ion collisions, as well as of the transport coefficients and lifetime of the QGP.

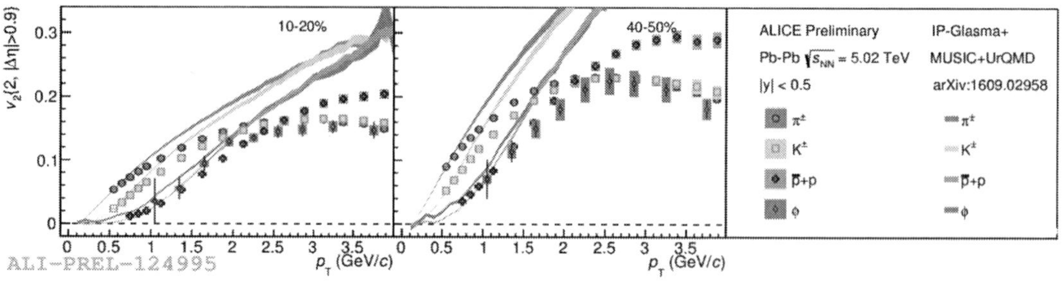

Figure 2. Flow coefficient v_2 of $\pi^{\pm}, K^{\pm}$, p($\overline{\mathrm{p}}$) and the ϕ-meson for 10-20% (left) and 40-50% (right) collision centrality compared to predictions from relativistic hydrodynamic calculations [9]. Statistical uncertainties are shown as bars and systematic uncertainties as boxes.

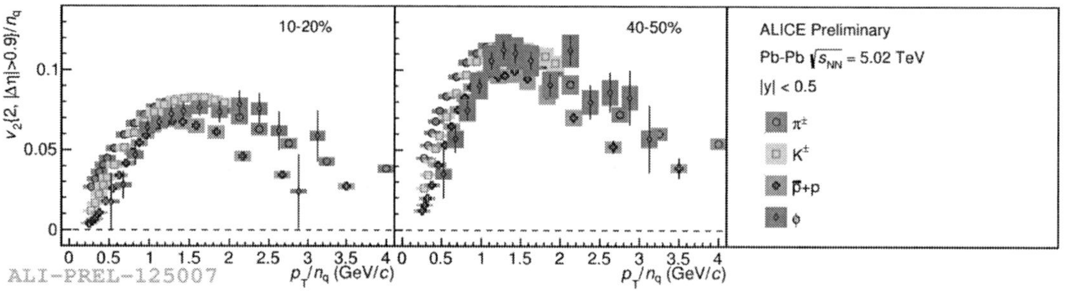

Figure 3. Scaling test for p_T-differential v_2 of $\pi^{\pm}, K^{\pm}$, p($\overline{\mathrm{p}}$) and the ϕ-meson for 10-20% (left) and 40-50% (right) collision centrality, The $x-$ and $y-$axes are scaled by the number of constituent quarks n_q for each species independently. Statistical uncertainties are shown as bars, systematic uncertainties as boxes.

References

[1] J. Adams et al. (STAR), Nucl. Phys. **A757**, 102 (2005)
[2] J.Y. Ollitrault, Phys. Rev. **D46**, 229 (1992)
[3] B. Abelev et al. (ALICE), Int. J. Mod. Phys. **A29**, 1430044 (2014)
[4] B. Abelev et al. (ALICE), JHEP **06**, 190 (2015)
[5] S.A. Voloshin, A.M. Poskanzer, R. Snellings (2008)
[6] P. Huovinen, P.F. Kolb, U.W. Heinz, P.V. Ruuskanen, S.A. Voloshin, Phys. Lett. **B503**, 58 (2001)
[7] D. Molnar, S.A. Voloshin, Phys.Rev.Lett. **91**, 092301 (2003)
[8] G.Y. Qin, H. Petersen, S.A. Bass, B. Muller, Phys. Rev. **C82**, 064903 (2010)
[9] S. McDonald, C. Shen, F. Fillion-Gourdeau, S. Jeon, C. Gale (2016)

Production of D^0 meson in pp and PbPb Collisions at $\sqrt{s_{\rm NN}} = 5.02$ TeV with CMS

Yen-Jie Lee[1,⋆] on behalf of the CMS Collaboration

[1] *Massachusetts Institute of Technology*

Abstract. Heavy flavour mesons are used as powerful tools for the study of the strongly interacting medium in heavy ion collisions as heavy quarks are sensitive to the transport properties of the medium. In these proceedings, D^0 nuclear modification factors, comparing the yields in PbPb and pp collisions, and azimuthal anisotropies in PbPb collisions are reported. Prompt D^0 mesons and their antiparticles have been measured with the CMS detector via the hadronic decay channels $D^0 \to K^-\pi^+$ and $\overline{D}^0 \to K^+\pi^-$ in PbPb and pp collisions at a centre-of-mass energy of 5.02 TeV. Nonprompt D^0 from b quark decays are subtracted. The D^0 results are compared to inclusive charged particles, nonprompt J/ψ mesons from b decays and B^+mesons in order to reveal possible meson mass dependence of the observables.

Heavy quarks are effective probes to study the properties of the deconfined medium created in heavy ion collisions. These quarks are mostly produced in primary hard QCD scatterings with a production timescale that is shorter than the formation time of the Quark Gluon Plasma (QGP) [1, 2]. Therefore, they carry information about the early stages of the QGP. During their propagation through the medium, heavy quarks lose energy via radiative and collisional interactions with the medium constituents. Quarks are expected to lose less energy than gluons as a consequence of their smaller colour factor. In addition, the so-called "dead-cone effect" is expected to reduce small-angle gluon radiation of heavy quarks when compared to both gluons and light quarks [3–5]. Energy loss can be studied using the nuclear modification factor ($R_{\rm AA}$), defined as the ratio of the PbPb yield to the pp cross-section scaled by the nuclear overlap function [6]. Moreover, the azimuthal anisotropy of produced D^0 mesons can be characterized by the Fourier coefficients v_n in the azimuthal angle (ϕ) distribution of the hadron yield, $dN/d\phi \propto 1 + 2\sum_n v_n \cos[n(\phi - \Psi_n)]$, where Ψ_n is the azimuthal angle of the direction of the maximum particle density of the nth harmonic in the transverse plane [7]. At low transverse momentum ($p_{\rm T}$), the charm hadron v_n coefficient can help quantify the extent to which charm quarks flow with the medium, which is a good measure of their interaction strength. The measurements of $R_{\rm AA}$ and v_n can also help explore the coalescence production mechanism for charm hadrons where charm quarks recombine with light quarks from the medium, which could also lead to positive charm hadron v_n [8, 9]. At high $p_{\rm T}$, the charm hadron v_n coefficient can constrain the path length dependence of charm quark energy loss [10, 11], complementary to the $R_{\rm AA}$ measurements. Precise measurements of the $R_{\rm AA}$ and the azimithal anisotropy of particles containing both light and heavy quarks can thus provide important tests of QCD predictions at extreme densities and temperatures. In these proceedings, the production of prompt D^0 mesons in PbPb collisions at 5.02 TeV is measured for the first time

⋆e-mail: yenjie@mit.edu

up to a p_T of 100 GeV/c, allowing one to study the properties of the in-medium energy loss in a new kinematic regime. The D^0 meson and its antiparticle are studied in the central rapidity region ($|y| < 1$) of the CMS detector via the hadronic decay channels $D^0 \to K^-\pi^+$ and $\overline{D}^0 \to K^+\pi^-$.

A detailed description of the CMS experiment can be found in Ref. [12]. The D^0 candidates are reconstructed by combining pairs of oppositely charged particle tracks with an invariant mass within 0.2 GeV/c^2 of the world-average D^0 mass [13]. Each track is required to have $p_T > 1$ GeV/c in order to reduce the combinatorial background. For high-p_T D^0 mesons (above 20 GeV/c) in PbPb data, the single track cut is raised to $p_T > 8.5$ GeV/c to account for the selection ($p_T > 8$ GeV/c) performed with a software-based high level trigger. All tracks are also required to be within $|\eta| < 1.5$. For each pair of selected tracks, two D^0 candidates are created by assuming that one of the particles has the mass of the pion while the other has the mass of the kaon, and vice-versa. The D^0 mesons are required to be within $|y| < 1$, optimized in conjunction to the track pseudorapidity selection to give the best signal to background ratio over the whole range of D^0 p_T studied. In order to further reduce the combinatorial background, the D^0 candidates are selected based on three topological criteria: on the three-dimensional (3D) decay length L_{xyz} normalized to its uncertainty, on the pointing angle θ_p (defined as the angle between the total momentum vector of the tracks and the vector connecting the primary and the secondary vertices), and on the χ^2 probability of the D^0 vertex fit. Details of the D^0 selection criteria and the signal extraction procedure for the R_{AA} and v_n analyses are documented in Refs. [14, 15].

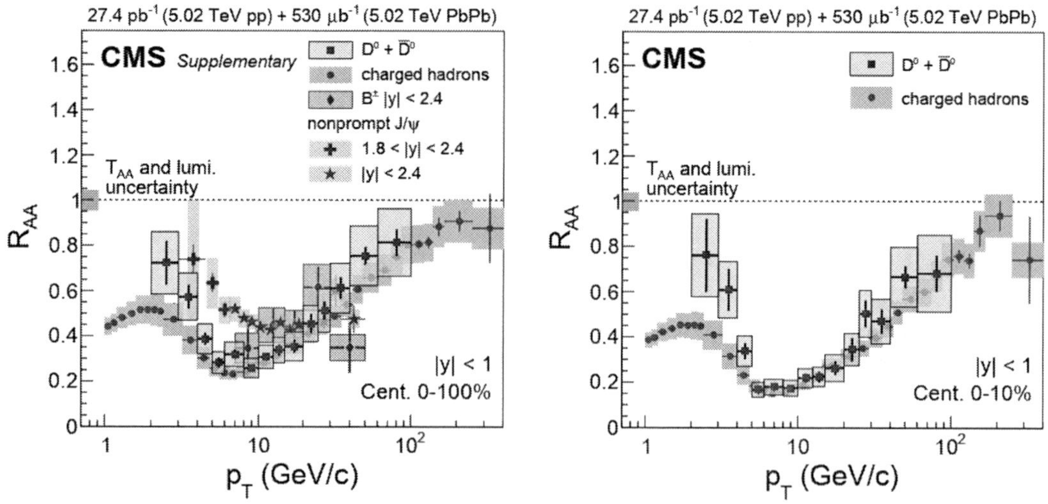

Figure 1. (Left) Nuclear modification factor R_{AA} as a function of p_T in the centrality range 0–100% (green squares) compared to the R_{AA} of charged particles (red circles) [16], $B^{\pm}$ mesons (blue triangles) [17] and to the preliminary results of R_{AA} of nonprompt J/ψ meson (purple crosses and stars) [18] in the same centrality range. (Right) Nuclear modification factor R_{AA} as a function of p_T in the centrality range 0–10% (green squares) compared to the R_{AA} of charged particles (red circles) [16] in the same centrality range.

Figure 1 shows the D^0 R_{AA} as a function of p_T in two centrality classes. The D^0 meson yield is found to be strongly suppressed in PbPb collisions when compared to the measured pp reference data scaled by the number of binary nucleon-nucleon collisions. These measurements are consistent with the R_{AA} of charged hadrons in both centrality intervals for $p_T > 4$ GeV/c. A hint of a smaller suppression of D^0 R_{AA} with respect to charged particle R_{AA} is observed for $p_T < 4$ GeV/c. The D^0 R_{AA} was found to be compatible with the $B^{\pm}$ R_{AA} in the intermediate p_T region and significantly lower than the nonprompt J/ψ meson R_{AA} for $p_T < 10$ GeV/c.

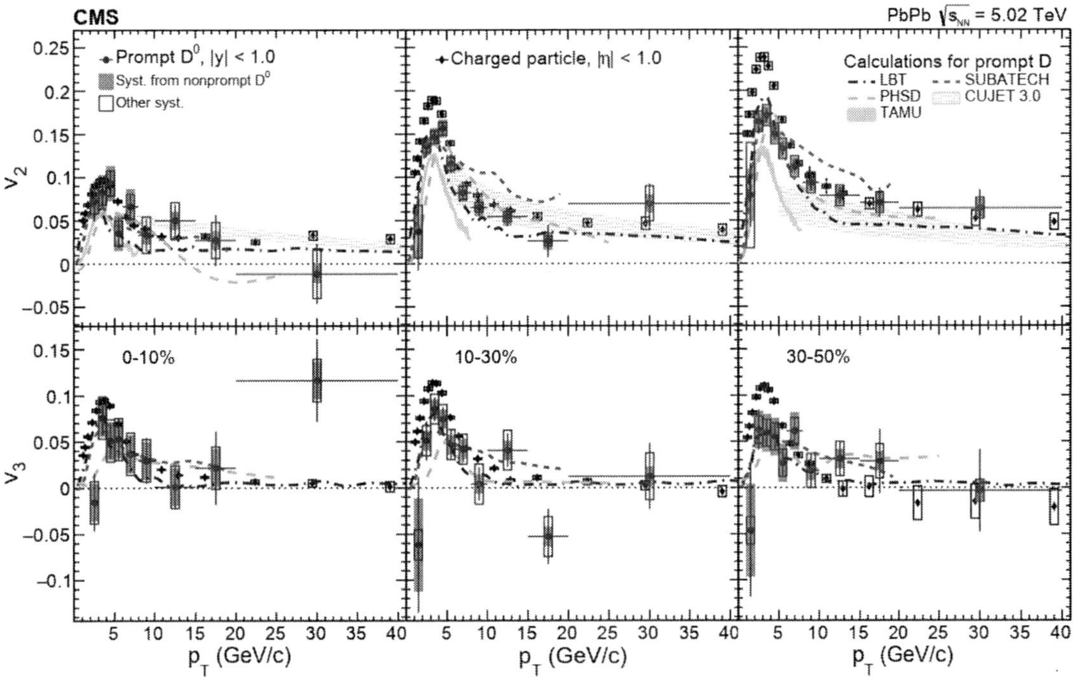

Figure 2. Prompt D^0 meson v_2 (upper) and v_3 (lower) coefficients at midrapidity ($|y| < 1.0$) for the centrality classes 0–10% (left), 10–30% (middle), and 30–50% (right). The vertical bars represent statistical uncertainties, grey bands represent systematic uncertainties from nonprompt D^0 mesons and open boxes represent other systematic uncertainties. The measured v_n coefficient of charged particles at midpseudorapidity ($|\eta| < 1.0$) [19] and theoretical calculations for prompt D meson v_n coefficient [20–24] are also plotted for comparison.

Figure 2 shows the prompt D^0 meson v_2 and v_3 coefficients at midrapidity ($|y| < 1.0$) in different centrality classes and compares them to those of charged particles (dominated by light flavor hadrons) at midpseudorapidity ($|\eta| < 1.0$) [19]. The D^0 meson v_2 and v_3 coefficients increase with p_T to significant positive values in the low-p_T region, and then decrease for higher p_T. Compared to those of charged particles, the D^0 meson v_2 and v_3 coefficients exhibit a similar p_T dependence. As has been observed for charged particles, the D^0 meson v_2 coefficient increases with decreasing centrality in the 0–50% centrality range, while the v_3 coefficient shows little centrality dependence. For $p_T < 6$ GeV/c, the magnitudes of D^0 meson v_2 and v_3 coefficients are smaller than those for charged particles in the centrality classes 10–30% and 30–50%. The comparison between the D^0 meson results and theoretical calculations in this low-p_T region suggests a collective motion of charm quarks. For $p_T > 6$ GeV/c, the D^0 meson v_2 values remain positive, suggesting a path length dependence of the charm quark energy loss; the D^0 meson v_3 precision is limited by the available data. The D^0 meson v_2 values are consistent with those of charged particles, suggesting that the path length dependence of charm quark energy loss is similar to that of light quarks.

In summary, the D^0 meson nuclear modification factor and azimuthal anisotropy coefficients in the central rapidity region ($|y| < 1$) at $\sqrt{s_{_{NN}}} = 5.02\,\text{TeV}$ from CMS are presented. The results are compared to those from inclusive charged particles, nonprompt J/ψ and B^+ mesons. Those high precision D^0 R_{AA}, v_2 and v_3 data provide new constraints on models of the interactions between charm quarks and the quark-gluon plasma, and the charm quark energy loss mechanisms.

References

[1] P. Braun-Munzinger, J. Phys. G **34**, S471 (2007), nucl-th/0701093
[2] F.M. Liu, S.X. Liu, Phys. Rev. C **89**, 034906 (2014), 1212.6587
[3] Y.L. Dokshitzer, D.E. Kharzeev, Phys. Lett. B **519**, 199 (2001), hep-ph/0106202
[4] N. Armesto, C.A. Salgado, U.A. Wiedemann, Phys. Rev. D **69**, 114003 (2004), hep-ph/0312106
[5] A. Andronic et al., Eur. Phys. J. C **76**, 107 (2016), 1506.03981
[6] M.L. Miller, K. Reygers, S.J. Sanders, P. Steinberg, Ann. Rev. Nucl. Part. Sci. **57**, 205 (2007), nucl-ex/0701025
[7] A.M. Poskanzer, S.A. Voloshin, Phys. Rev. C **58**, 1671 (1998), nucl-ex/9805001
[8] D. Molnar, S.A. Voloshin, Phys. Rev. Lett. **91**, 092301 (2003), nucl-th/0302014
[9] V. Greco, C.M. Ko, P. Levai, Phys. Rev. Lett. **90**, 202302 (2003), nucl-th/0301093
[10] M. Gyulassy, I. Vitev, X.N. Wang, Phys. Rev. Lett. **86**, 2537 (2001), nucl-th/0012092
[11] E.V. Shuryak, Phys. Rev. C **66**, 027902 (2002), nucl-th/0112042
[12] S. Chatrchyan et al. (CMS), JINST **3**, S08004 (2008)
[13] Particle Data Group, C. Patrignani et al., Chin. Phys. C **40**, 100001 (2016)
[14] A.M. Sirunyan et al. (CMS) (2017), 1708.04962
[15] A.M. Sirunyan et al. (CMS) (2017), 1708.03497
[16] V. Khachatryan et al. (CMS), JHEP **04**, 039 (2017), 1611.01664
[17] A.M. Sirunyan et al. (CMS), Phys. Rev. Lett. **119**, 152301 (2017), 1705.04727
[18] CMS Collaboration (CMS), CMS Physics Analysis Summary CMS-PAS-HIN-16-025 (2016), https://cds.cern.ch/record/2235781
[19] A.M. Sirunyan et al. (CMS) (2017), 1702.00630
[20] S. Cao, T. Luo, G.Y. Qin, X.N. Wang, Phys. Rev. C **94**, 014909 (2016), 1605.06447
[21] M. Nahrgang, J. Aichelin, S. Bass, P.B. Gossiaux, K. Werner, Phys. Rev. C **91**, 014904 (2015), 1410.5396
[22] T. Song, H. Berrehrah, D. Cabrera, W. Cassing, E. Bratkovskaya, Phys. Rev. C **93**, 034906 (2016), 1512.00891
[23] M. He, R.J. Fries, R. Rapp, Phys. Lett. B **735**, 445 (2014), 1401.3817
[24] J. Xu, J. Liao, M. Gyulassy, JHEP **02**, 169 (2016), 1508.00552

EPJ Web of Conferences **171**, 18001 (2018) https://doi.org/10.1051/epjconf/201817118001
SQM 2017

A data-drive analysis for heavy quark diffusion coefficient

Yingru Xu[1,*], *Marlene* Nahrgang[2], *Shanshan* Cao[3], *Jonah* E. Bernhard[1], and *Steffen* A. Bass[1]

[1] *Department of Physics, Duke University, Durham, NC 27708, USA*
[2] *SUBATECH, UMR 6457, IMT Atlantique, Universit de Nantes, IN2P3/CNRS, Nantes, France*
[3] *Department of Physics and Astronomy, Wayne State University, Detroit, MI, 48201*

Abstract. We apply a Bayesian model-to-data analysis on an improved Langevin framework to estimate the temperature and momentum dependence of the heavy quark diffusion coefficient in the quark-gluon plasma (QGP). The spatial diffusion coefficient is found to have a minimum around 1-3 near T_c in the zero momentum limit, and has a non-trivial momentum dependence. With the estimated diffusion coefficient, our improved Langevin model is able to simultaneously describe the D-meson R_{AA} and v_2 in three different systems at RHIC and the LHC.

1 Introduction

Over the past few decades, facilities at RHIC and the LHC have been exploring the strongly coupled QGP medium by conducting heavy-ion collision experiments. Since the QGP is not directly observable, the study of its properties relies on the comparison between experimental measurements and theoretical modeling. Open heavy flavor, as one of the hard probes of the QGP medium, is valuable yet difficult to study, since it involves measuring 'rare process', as well as modeling the complex evolution of heavy quarks during different stages of the collision. Nevertheless, significant progress has been made in recent years: a number of theoretical models are able to describe some of the heavy quark observables, and provide qualitative agreement on the heavy quark diffusion coefficient [1]. However, the simultaneous description of different heavy quark observables in different collision systems still poses challenges for some of the models. Moreover, the current studies of the heavy quark often rely on 'eye-ball fit' or 'χ^2-goodness fit' comparisons between the model calculations and the experimental measurements, leading to very limited distinguishability between the performance of different models, and possible bias on the estimate of the heavy quark diffusion coefficient.

A rigorous and systematic approach to a model-to-data comparison requires optimizing the model and simultaneously describing different experimental observables (which are often highly interdependent and have correlated uncertainties). A Markov chain Monte Carlo (MCMC) algorithm and Bayesian statistics can be used to optimize the process, and perform the comparison and constrain the parameters in a data-driven manner. A Bayesian analysis has already been applied to the study of soft medium properties with great success, for example, the determination of the temperature dependence of shear and bulk viscosities $\eta/s(T), \zeta/s(T)$ [2], or the constraint of the QCD matter equation of state purely from experimental measurements [3]. In this work, we shall extend this type of analysis to the heavy quark sector.

*e-mail: yx59@phy.duke.edu

2 Heavy quark in-medium evolution

Our study of the full space-time evolution of heavy quarks utilizes the well-established improved Langevin framework developed by the Duke QCD group [4]: The initial entropy density and the heavy quark positions are generated by an effective parametric initial condition model T$_R$ENTo – mimicking the scaling behavior of IP-Glasma [5]. The heavy quark initial transverse momentum distribution is provided by FONLL [6]; After being produced, heavy quarks propagate in the QGP medium according to an improved Langevin equation [4], which takes into account both collisional and radiative energy loss; at the critical temperature T_c = 154 MeV, heavy quarks hadronize into heavy mesons via a hybrid model of recombination and fragmentation. After hadronization, the heavy mesons continuously experience hadronic re-scattering with light hadrons – all hadronic interactions are modeled by UrQMD [8].

The improved Langevin equation that describes the dynamics of heavy quarks inside the QGP medium is:

$$\frac{d\vec{p}}{dt} = -\eta_D(p)\vec{p} + \vec{\xi} + \vec{f_g}. \tag{1}$$

where $-\eta_D(p)\vec{p}$ and $\vec{\xi}$ are the drag and thermal random forces. They are responsible for the heavy quark collisional energy loss and related to the transport coefficient $\hat{q}$ via $\eta_D(p) = \hat{q}/(4TE)$ and $\left\langle \xi_i(t)\xi_j(t')\right\rangle = 1/2\hat{q}\delta_{ij}\delta(t-t')$. The third term $\vec{f_g} = -d\vec{p}_g/dt$ is the recoil force when the heavy quark emits bremsstrahlung gluons, with $\vec{p}_g$ being the gluon momentum. We adopt the higher twist formalism for the medium-induced gluon radiation [7] and the recoil force is dependent on $\hat{q}$. With sufficient large gluon energy limit, the system will equilibrate when the Einstein relationship is satisfied [4]. Under this construction, the drag force, the thermal random force and the recoil force are dependent on the heavy quark transport coefficient $\hat{q}$. In literature, the spatial diffusion coefficient $D_s = 4T^2/\hat{q}$ is often used to characterize the interacting strength between heavy quarks and the medium.

At high temperature or large momentum, the heavy quark transport coefficient can be calculated using the perturbative QCD approach (pQCD). However, it has been found that the pQCD calculation is not sufficient to explain the suppression experienced by heavy quarks when propagating in the medium, nor able to simultaneously describe both the heavy flavor R_{AA} and v_2. In order to introduce a non-perturbative effect, we parameterize the spatial diffusion coefficient as a combination of linear temperature dependent component and a pQCD component:

$$D_s 2\pi T(T, \boldsymbol{p}; \alpha, \beta, \gamma) = \frac{1}{1+(\gamma^2 p)^2}(D_s 2\pi T)^{\text{soft}} + \frac{(\gamma^2 p)^2}{1+(\gamma^2 p)^2}(D_s 2\pi T)^{\text{pQCD}}. \tag{2}$$

The soft component $(D_s 2\pi T)^{\text{soft}} = \alpha * (1 + \beta * (T/T_c - 1))$ is the heavy quark diffusion coefficient in the $p = 0$ GeV/c limit. It accounts for the non-perturbative effects. The $(D_s 2\pi T)^{\text{pQCD}} = 8\pi T^3/\hat{q}^{\text{pQCD}}$ is calculated by summing over the leading-order $2 \to 2$ scattering between heavy quarks (M_c = 1.3 GeV) and massless light partons at fixed coupling α_s = 0.3. The value of heavy quark diffusion coefficient is determined by three parameters $\boldsymbol{x} = (\alpha, \beta, \gamma)$, which can be interpreted as following: Parameter α is $D_s 2\pi T(p = 0)$ around T_c, parameter β is its slope above T_c. Parameter γ controls the ratio between the linear component and the pQCD component. A small value of γ indicates the non-perturbative effects are non-negligible towards the high momentum region, and a large value of γ indicates a quick conversion to the pQCD dominated region.

The evolution of heavy quarks depends on the surrounding matter. In this study, the QGP medium is simulated by an event-by-event (2+1)-dimensional viscous hydrodynamical model VISHNEW [9], which also includes shear and bulk viscous corrections. All parameters related to the medium evolution (such as $\eta/s, \zeta/s$) as well as the initial entropy deposition have been calibrated to experimental data of charged particles by an independent Bayesian analysis [2].

3 Statistical analysis and calibration

According to Bayes' theorem, the posterior probability distribution of the model parameters ($P(\boldsymbol{x}|\boldsymbol{y}_{\rm exp})$) is proportional to the product of the parameters' prior distribution ($P(\boldsymbol{x})$), and the likelihood of observing experimental results given parameter ($L(\boldsymbol{y}_{\rm exp}|\boldsymbol{x})$). In this case, we assume the prior parameter have a uniform distribution , which implies that we don't have any initial constraint/knowledge on the parameters. Thus the posterior distribution of the parameters is proportional to the likelihood:

$$L(\boldsymbol{y}_{\rm exp}|\boldsymbol{x}) \propto \exp\left[-\frac{1}{2}(\boldsymbol{y}-\boldsymbol{y}_{\rm exp})\Sigma^{-1}(\boldsymbol{y}-\boldsymbol{y}_{\rm exp})^T\right] \quad (3)$$

where $\boldsymbol{y}_{\rm exp}$ is the experimental data (D-meson $R_{\rm AA}, v_2$ at different centralities at AuAu collisions (200 GeV) and PbPb collisions (2.76 and 5.02 TeV)) [10], $\boldsymbol{y}$ is the model calculations at parameter $\boldsymbol{x} = (\alpha, \beta, \gamma)$, Σ is the covariance matrix which includes quantified uncertainties (experimental and theoretical).

In practice, we explore the parameter space and sample the Bayesian posterior distribution by performing a MCMC random walk, which requires evaluating the model at thousands or even millions of parameter values. However, the simulation of heavy quark evolution in heavy-ion collision is computationally highly expensive. Therefore, efficient techniques such as Gaussian process emulators are applied to solve the problem – these act as a fast surrogate model of the full Langevin simulation. A detailed description of the construction of the emulators and performance of the MCMC algorithm can be found in Ref. [2].

4 Posterior results and discussion

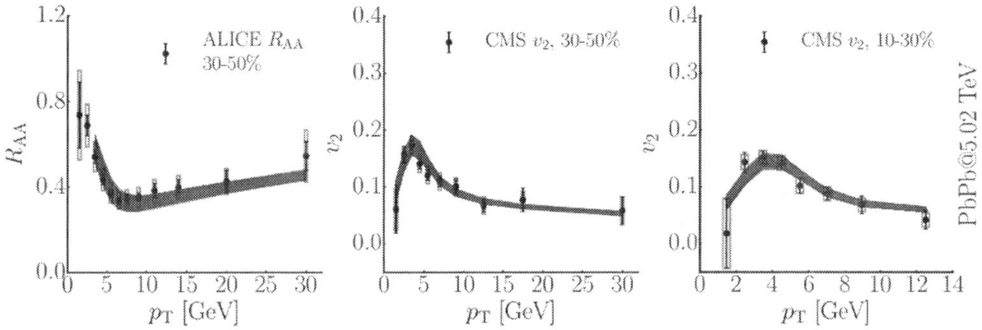

Figure 1: (Color online) 200 random parameters $\boldsymbol{x} = (\alpha, \beta, \gamma)$ sampled from the posterior distribution and evaluated by emulators. A selections of the output – D-meson $R_{\rm AA}$ and v_2 at PbPb collisions at 5.02 TeV are compared with experimental data.

To verify our analysis, 200 sets of parameters are randomly selected from the posterior distributions and evaluated by the GP emulators. Figure 1 compares a selection of observables with experimental data. Two different colors correspond to two independent Bayesian analysis: the blue lines correspond to the results obtained by calibrating on a single collision system – PbPb collision at 5.02 TeV, the red lines correspond to the results obtained by calibrating on the three collision systems at RHIC and the LHC. After the calibration, our improved Langevin approach is capable of simultaneously describing the experimental data. The spreads of the posterior $R_{\rm AA}$ and v_2 visualize the uncertainties kept in our analysis, which come from the experimental statistic and uncorrelated systematic uncertainties, as well as the uncertainty arising from the prediction made by the GP emulators.

The posterior distributions of parameters (α, β, γ) are plotted in Fig. 2, where the diagonal histograms are the marginal distribution of each parameter with the others integrated out. The off-

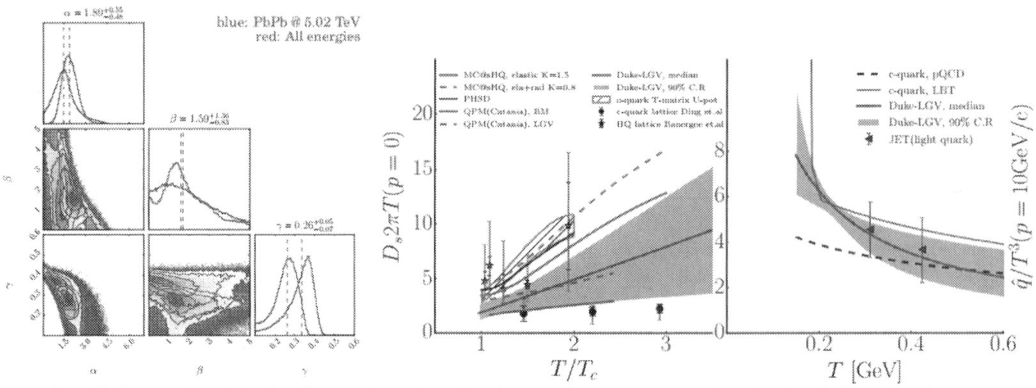

Figure 2: (Color online) **left:** Posterior distributions of parameters (α, β, γ); **right:** the heavy quark diffusion coefficients estimated by taking the posterior samples, compared with other models' calculations.

diagonal plots show the correlation between each pair. We observe a narrow distribution for parameters α and γ, implying those two are constraint with current experimental data and uncertainties. Although parameter β is poorly constraint, the strong negative correlation between α and β indicates some collinear interdependency and should not be ignored.

In Fig. 2 we also estimate the diffusion coefficient by taking the posterior samples from the combined analysis (with 90% credibility). Our estimate of $D_s 2\pi T(p = 0)$ shows a minimum between (1-3) around T_c, and has a positive temperature dependent. Comparing our estimate with the coefficient used/calculated by a number of other models in the market [1, 11], as well as the lattice QCD calculations [12], our estimate is consistent with lattice QCD calculations within the large uncertainties current inherent in lattice QCD calculations. Although the diffusion coefficient used in different models are rather different, they all have a minimum near T_c with value range $D_s 2\pi T(p = 0) \sim (1-7)$. Nevertheless, we should note that in order to make an apple-to-apple comparison among different models, it is essentially important for heavy quarks to propagate in the same QGP medium, especially for those transport models, which is not the case for the current comparison. A detailed investigation is needed, but beyond the scope of this work.

References

[1] P. B. Gossiaux and J. Aichelin, Phys. Rev. C **78**, 014904; M. He, R. J. Fries and R. Rapp, Phys. Rev. Lett. **110**, 112301; T. Song, H. Berrehrah, D. Cabrera, W. Cassing and E. Bratkovskaya, Phys. Rev. Lett. **100**, 192301; F. Scardina, S. K. Das, V. Minissale, S. Plumari and V. Greco, arXiv:1707.05452; S. Cao, T. Luo, G. Y. Qin and X. N. Wang, Phys. Rev. C **94**, 014909.

[2] J. E. Bernhard, J. S. Moreland, S. A. Bass, J. Liu and U. Heinz, Phys. Rev. C **94**, 024907;

[3] S. Pratt, E. Sangaline, P. Sorensen and H. Wang, Phys. Rev. Lett. **114**, 202301 (2015)

[4] S. Cao, G. Y. Qin and S. A. Bass, Phys. Rev. C **88**, 044907

[5] J. S. Moreland, J. E. Bernhard and S. A. Bass, Phys. Rev. C **92**, no. 1, 011901 (2015)

[6] M. Cacciari, M. Greco and P. Nason, JHEP **9805** (1998) 007 ;M. Cacciari, S. Frixione and P. Nason, JHEP **0103** (2001) 006;

[7] X. f. Guo and X. N. Wang, Phys. Rev. Lett. **85**, 3591; B. W. Zhang, E. Wang and X. N. Wang, Phys. Rev. Lett. **93**, 072301

[8] S. A. Bass *et al.*, Prog. Part. Nucl. Phys. **41**, 225 ;M. Bleicher *et al.*, J. Phys. **G25**, 1859.

[9] H. Song and U. W. Heinz, Phys. Rev. C **77**, 064901; C. Shen, Z. Qiu, H. Song, J. Bernhard, S. Bass and U. Heinz, Comput. Phys. Commun. **199**, 61; U. W. Heinz and J. Liu, Nucl. Phys. A **956**, 549.

[10] STAR, G. Xie, Nucl. Phys. **A956**, 473 ;STAR, L. Adamczyk *et al.*, Phys. Rev. Lett. **118**, 212301; ALICE, J. Adam *et al.*, JHEP **11**, 205; ALICE, B. B. Abelev *et al.*, Phys. Rev. **C90**, 034904; ALICE, A. M. Barbano *et al.*, Quark Matter 2017 CMS, C. Collaboration, CMS-PAS-HIN-16-007

[11] JET, K. M. Burke *et al.*, Phys.Rev. **C90**, 014909 (2014), arXiv:1312.5003.

[12] H. T. Ding, A. Francis, O. Kaczmarek, F. Karsch, H. Satz and W. Soeldner, Phys. Rev. D **86**, 014509; D. Banerjee, S. Datta, R. Gavai and P. Majumdar, Phys. Rev. D **85**, 014510.

Heavy flavour energy loss from AdS/CFT: A novel diffusion coefficient

R. Hambrock[1,*] and *W. A.* Horowitz[1,**]

[1] *Department of Physics, University of Cape Town, Private Bag X3, Rondebosch 7701, South Africa*

Abstract. Two AdS/CFT based energy loss models are used to compute the suppression and azimuthal correlations of heavy quarks in heavy ion collisions. The model with a velocity independent diffusion coefficient is in good agreement with B and D meson data up to high p_T. The partonic azimuthal correlations we calculate exhibit an order of magnitude difference in low momentum correlations to pQCD calculations [1]. We thus propose heavy flavour momentum correlations as a distinguishing observable of weakly- and strongly-coupled energy loss mechanisms.

1 Introduction

The quark gluon plasma is of great interest since it represents our first case study of the emergent physics of the non-abelian gauge theory QCD. A key step in understanding this state of matter is identifying its relevant coupling strength. The perturbative techniques of QCD are only adequate in a weakly coupled plasma, with calculations for strongly coupled plasmas constrained to methods like AdS/CFT-based approaches or Resonance Scattering [2].

Both strong- and weak-coupling show qualitative agreement with R^D_{AA} [3], suggesting that they have attained sufficient maturity to compare them with more differential observables. We will argue that the momentum correlations of bottom quarks constitute a promising candidate as a differentiator between weakly and strongly coupled plasmas.

We will use two different AdS/CFT based energy loss models, one [4] with a velocity dependent diffusion coefficient and the other [5] with a diffusion coefficient independent of the heavy quark's velocity. Futhermore, we will probe the spectrum of their possible predictions with two plausible [6] 't Hooft coupling constants ($\lambda_1 = 5.5$ and $\lambda_2 = 12\pi\alpha_s \approx 11.3$ with $\alpha_s = 0.3$) where for the former, temperature and the Yang-Mills coupling are equated, while for the latter constant, energy density and the coupling are equated.

We will compare these with the pQCD based azimuthal correlations calculated in [1]. These will provide a secondary indicator for the momentum correlations.

Finally, we will compute the suppression of B and D mesons and compare with measurements from CMS [7] and ALICE [8] respectively.

*e-mail: roberthambrock@gmail.com
**e mail: wahorowitz@gmail.com

2 Energy Loss Model and Suppression

2.1 Overview

Subsequent to initializing the momenta of heavy quark pairs either to leading order with FONLL [9] or to next-to-leading order with aMC@NLO [10] using Herwig++ [11] for the showering and hadronization, the production points of the heavy quarks are weigthed by the Glauber binary distribution [4]. The particles are propagated through the plasma via the energy loss mechanism described in 2.2 until the temperature in their local fluid cell drops below the Tc threshold and hadronization is presumed to occur or 8.6fm had passed, being the maximum time of the VISHNU background [12].

2.2 Langevin Energy Loss

The stochastic equation of motion for a heavy quark in the fluid's rest frame is [13]

$$\frac{dp_i}{dt} = -\mu p_i + F_i^L + F_i^T \tag{1}$$

where F_i^L and F_i^T are longitudinal and transverse momentum kicks with respect to the quark's direction of propagation and with μ, the drag loss coefficient, being given by $\mu = \pi\sqrt{\lambda}T^2/2M_Q$ [14] where M_Q is the mass of a heavy quark in a plasma of temperature T with 't Hooft coupling constant λ. The correlations of momentum kicks are given by

$$\langle F_i^T(t_1)F_j^T(t_2)\rangle = \kappa_T(\delta ij - \frac{\vec{p}_i\vec{p}_j}{|p|^2})g(t_2 - t_1) \tag{2}$$

$$\langle F_i^L(t_1)F_j^L(t_2)\rangle = \kappa_L\frac{p_ip_j}{|p|^2}g(t_2 - t_1) \tag{3}$$

where g is only known numerically [4] and with

$$\kappa_T = \pi\sqrt{\lambda}T^3\gamma^{1/2} \tag{4}$$

$$\kappa_L = \gamma^2\kappa_T = \pi\sqrt{\lambda}T^3\gamma^{5/2} \tag{5}$$

It should be noted that this construction does not obey the fluctuation-dissipation theorem [4]. The computations based on this model will be labeled *D(p)*.

2.3 Development on energy loss model

The problem with the energy loss mechanism described in 2.2 is that since the longitudinal momentum fluctuations grow as $\gamma^{\frac{5}{2}}$, our setup breaks down for high momenta, where in a perturbative QCD setting, Brehmstrahlung would restrict the momentum growth of the quark. Via a novel calculation presented in [5, 15, 16], we instead consider a stationary string in AdS_d hanging into a black hole horizon and calculate $s^2(t, a, d)$ of the free endpoint. For the the $d = 3$ result, the average squared distance travelled can be determined analytically for small string lengths, which is identical to the asymptotically late time behavior of a string with arbitrary initial length. We thus find the asymptotically late time behavior of a string in d dimensions by

$$s^2(t \gg \beta, a, d) = s^2_{\text{small}}(t \gg \beta, a, d) = (\frac{d-1}{2})^2 s^2_{\text{small}}(t \gg \beta, a, d = 3) = \frac{(d-1)^2}{8\pi\sqrt{\lambda}}\beta(1 - \frac{a}{2}) \tag{6}$$

where $\beta = T^{-1}$ and a parametrizes between a heavy quark for $a = 0$ and a light quark for $a = 1$. At late times, the motion is diffusive, thus we can extract the diffusion coefficient

$$D(a, d) \sim \frac{1}{2}s^2(t \gg \beta, a, d) \tag{7}$$

which in AdS_5 for a heavy quark reads $2\beta/\pi\sqrt{\lambda}$. From this, we obtain

$$\kappa_T = 2T^2/D = \pi\sqrt{\lambda}T^2/\beta = \pi\sqrt{\lambda}T^3 \tag{8}$$

$$\hat{q} = \langle p_\perp(t)^2\rangle\lambda \approx \kappa_T t/\lambda = (\pi T^3 t)/\sqrt{\lambda} \tag{9}$$

Requiring these fluctuations to obey the fluctuation-dissipation theorem, we attain $\mu = \pi\sqrt{\lambda}T^2/2E$. The computations based on this model will be labeled *D=const*.

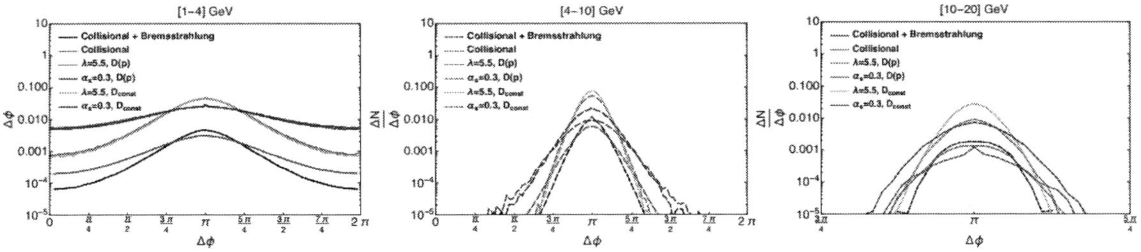

Figure 1: Bottom quark $\frac{dN}{d\phi}$ correlations for the specified classes.

3 Azimuthal correlations and R_{AA}

In [1], at leading order, the weak coupling based computations exhibited very efficient broadening of initial azimuthal correlations for low p_T $b\bar{b}$ pairs ([4 – 10] GeV), which were washed out once NLO production processes were taken into consideration.

Both for mid- and high-p_T ([4 – 10] GeV and [10 – 20] GeV respectively), the initial correlations survive to a large degree, both at leading order and at next-to-leading order, suggesting that they may still be observable in an experimental context.

We compare our strong coupling azimuthal correlations to the weak coupling ones in Fig. 1. For [10 – 20] GeV, our correlations are significant more peaked at their initial back-to-back correspondence. At [4 – 10] GeV, this observation still holds for the upper bound of our parameters with $\lambda_1 = 5.5$, while the $\lambda_2 = 11.3$ bounded result is of similar magnitude but looser angular correlation than either the collisional or the collisional + Bremsstrahlung based results. In the [1 – 4] GeV range, the azimuthal correlations are almost entirely washed out for $\lambda_2 = 11.3$, while for $\lambda_1 = 5.5$, they are broadened with similiar efficiency to the weak coupling results.

Having compared our strong coupling results with pQCD calculations at leading order, we now turn to a more realistic simulation that we can compare with data. The results shown in Fig. 2 are from the next-to-leading order procedure described in 2.1.

In Fig. 2, we compare our calculations for the suppression of B mesons with measurements from CMS. At low p_T, both *D(p)* and *D=const* models are consistent with data. At high p_T, the agreement is inconclusive.

Conversely, for the calculation of averaged D^0, D^+, and D^-, the *D(p)* model diverges from ALICE data [8]. This divergence is due to the momentum fluctuations going as $\gamma^{\frac{5}{2}}$. The *D=const* model remains consistent with data even for high-p_T.

4 Conclusion & Outlook

We have compared the azimuthal correlations predicted by pQCD and AdS/CFT based computations and found that, while the azimuthal correlations are qualitatively similar, the momentum correlations tell a different tale. In particular, the surprise of our findings is the large dissimilarity in low momentum correlations of the pQCD and AdS/CFT based simulations; see Fig. 1 (left). Thus, bottom quark momentum correlations present an opportunity to distinguish between the energy loss mechanisms of the two frameworks.

Although stronger momentum fluctuations in the weakly coupled plasma have been identified as a plausible explanation for this disparity, it can only be verified by computing initial correlations for the weak coupling based correlations as well. Furthermore, whether this order of magnitude difference in predictions for low p_T correlations of bottom quarks exposes weaknesses in either or both

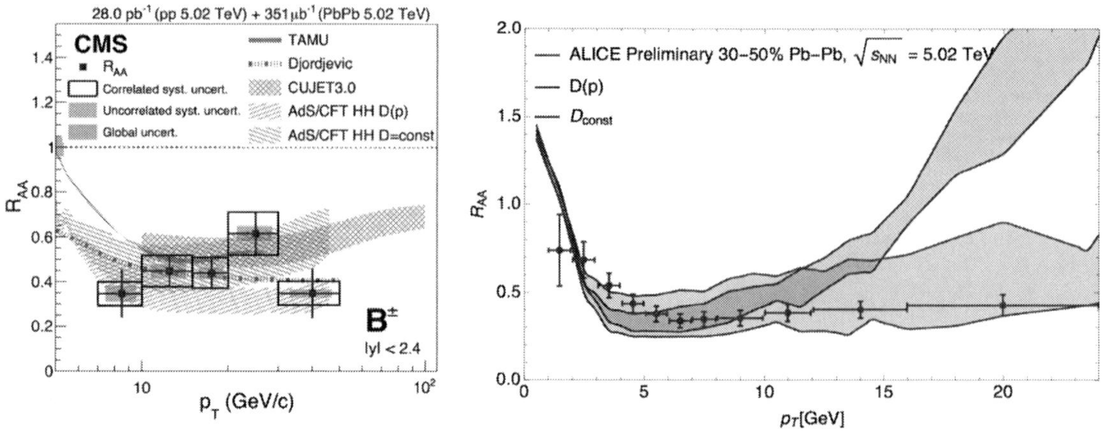

Figure 2: (Left) Comparison with R^B_{AA} data from CMS [7] with $\sqrt{s_{NN}} = 5.02$ TeV, $|y| < 2.4$. (Right) Comparison with R^D_{AA} data from ALICE [8] with $\sqrt{s_{NN}} = 5.02$ TeV, $|y| < 0.5$.

of the frameworks cannot be declared until experimental data of bottom quark momentum correlations emerge.

While the agreement with CMS data for B meson suppression is comparable between the *D(p)* and *D=const* models, the comparison with ALICE data for D mesons shows the limited validity range of the *D(p)* model, whereas the *D=const* model remains consistent with data even for high-p_T. More fundamentally, for the *D(p)* model, the AdS/CFT picture naturally breaks down at $p_T \sim 100$ GeV [4]. For the *D=const* model, there is no such natural breakdown. Only for asymptotically large p_T *and T* is one guaranteed that the physics is perturbative.

Acknowledgements

The authors wish to thank the South African National Research Foundation and SA-CERN for their generous financial support.

References

[1] M. Nahrgang et al., Phys. Rev. **C90**, 024907 (2014), 1305.3823
[2] J. Casalderrey-Solana, H. Liu, D. Mateos, K. Rajagopal, U.A. Wiedemann (2011), 1101.0618
[3] W.A. Horowitz, Nucl. Phys. **A904-905**, 186c (2013), 1210.8330
[4] W.A. Horowitz, Phys. Rev. **D91**, 085019 (2015), 1501.04693
[5] R.W. Moerman, W.A. Horowitz (2016), 1605.09285
[6] S.S. Gubser, Nucl. Phys. **B790**, 175 (2008), hep-th/0612143
[7] A.M. Sirunyan et al. (CMS), Phys. Rev. Lett. **119**, 152301 (2017), 1705.04727
[8] ALICE Collaboration (2017), ALI-PREL-128542
[9] M. Cacciari et al., JHEP **10**, 137 (2012), 1205.6344
[10] J. Alwall et al., JHEP **07**, 079 (2014), 1405.0301
[11] M. Bahr et al., Eur. Phys. J. **C58**, 639 (2008), 0803.0883
[12] C. Shen, U. Heinz, P. Huovinen, H. Song, Phys. Rev. **C84**, 044903 (2011), 1105.3226
[13] G.D. Moore, D. Teaney, Phys. Rev. **C71**, 064904 (2005), hep-ph/0412346
[14] C.P. Herzog et al., JHEP **07**, 013 (2006), hep-th/0605158
[15] R. Hambrock, W.A. Horowitz, J. Phys. Conf. Ser. **889**, 012015 (2017)
[16] R. Hambrock, W.A. Horowitz (2017), 1703.05845

Correlation between heavy flavour production and multiplicity in pp and p-Pb collisions at high energy in the multi-pomeron exchange model

Grigory Feofilov[1], *Vladimir* Kovalenko[1,*], and *Andrei* Puchkov[1]

[1] *Saint Petersburg State University, Russia*

Abstract. The multiplicity dependence of heavy flavour production in pp-collisions at LHC energies is studied in the framework of the multi-pomeron exchange model. The model is introducing the string-string interaction collectivity effects in pp collisions, which modifies multiplicity and transverse momenta, leading to the non-trivial mean p_t vs. multiplicity ($\langle p_t \rangle_{N_{\rm ch}}$ – $N_{\rm ch}$). correlation. The string collectivity strength parameter is fixed by experimental data on multiplicity and transverse momentum correlation in a wide energy range (from ISR to LHC). The particles discrimination is implemented according to Schwinger mechanism taking into account the strong decays of hadron resonances. We demonstrate, that the faster-than-linear growth of the open charm production with the event charged particle multiplicity, observed in experimental pp high energy collisions, can be explained by the modification of the string tension due to the increasing overlap and interaction of quark-gluon strings. The model is extended for p-A interactions and the calculations for p-Pb collisions are performed.

1 Introduction

Measurements of open charm production are traditionally considered as a probe for hot and dense matter formed in heavy ion collisions at high energy. Charm quarks, produced in hard collisions, can interact with QCD medium, and the final distribution of the hadrons with open and hidden charm reflects the transport properties of the QGP. However, the mechanism of the heavy quark production is not fully understood. Perturbation QCD results have large uncertainties leaving a room for alternative mechanisms of charm production. In this connection proton-proton interaction is considered as a benchmark for the analysis of proton-nucleus and heavy ion collisions.

The recent measurements on multiplicity dependence of the heavy flavour hadron production, performed by the ALICE Collaboration [1], demonstrated quite unexpected results. It was obtained that the D meson yield $N_{\rm D}$, plotted as a function of charged multiplicity $N_{\rm ch}$ in relative variables, shows faster-than-linear behaviour. This data have been interpreted in the simple string percolation model [2] where it was assumed that the bulk multiplicity decreases due to strings interaction and fusion. Later it was checked [3] in more realistic Monte Carlo model with string fusion [4, 5], where the string configurations are simulated event-by-event. Similar effects appear in EPOS 3 model [6] with the hydrodynamical stage of the evolution.

In this paper, we calculate the yield of D mesons as a function of multiplicity in the Extended Multi-Pomeron Exchange Model [7–9]. The parameters of the model are fixed by experimental data on total (unidentified) charged multiplicity and correlation between transverse momentum with $N_{\rm ch}$. The particle differentiation is implemented according to modified Schwinger mechanism [10].

*e-mail: v.kovalenko@spbu.ru

2 Model description

According to the Regge-Gribov multi-pomeron exchange approach, several pomeron exchanges in one pp-collision can happen, and the probability w_n of n pomeron exchanges [11, 12] is the following:

$$w_n=\sigma_n/\sum_{n'}\sigma_{n'}, \qquad \text{where} \qquad \sigma_n=\frac{\sigma_P}{nz}\left(1-e^{-z}\sum_{l=0}^{n-1}\frac{z^l}{l!}\right), \qquad z=\frac{2C\gamma s^{\Delta}}{R^2+\alpha'\log(s)} \tag{1}$$

We used the following numerical values of the parameters, which were tuned to describe total and elastic cross sections in high energy range [13]:

$$\Delta=0.139, \quad \alpha'=0.21\ \text{GeV}^{-2}, \quad \gamma=1.77\ \text{GeV}^{-2}, \quad R_0^2=3.18\ \text{GeV}^{-2}, \quad C=1.5\,. \tag{2}$$

Each pomeron corresponds to a pair of strings [11, 12]. The probability of the production of N_{ch} particles from an event with n pomerons is given by Poisson distribution. The distribution of multiplicity and the transverse momentum of the hadrons of different types appears as a linear combination of the corresponding one-particle functions.

Accoring to the modified Schwinger mechanism [7, 10] in case of exchange of n pomerons the probability of the production of a primary hadron of the type ν with transverse momentum p_t is proportional to the value

$$\exp\left(-\frac{\pi\left(p_t{}^2+m_\nu{}^2\right)}{n^\beta t}\right), \tag{3}$$

where m_ν is a mass of the particle, t is the tension of a single string, and β is the parameter, responsible for the collectivity. Thus in $n^\beta t$ it takes into account the increase of string tension due to the string fusion mechanism [14]. The parameter is fixed as a function of the collision energy from the data on $\langle p_t\rangle_{N_{\text{ch}}}-N_{\text{ch}}$ correlations in pp and p$\bar{\text{p}}$ collisions. Parameter $\beta=0$ means no collectivity, and no correlation between transverse momentum and multiplicity. At $\beta>0$ ($\beta<0$) the positive (negative) $\langle p_t\rangle_{N_{\text{ch}}}-N_{\text{ch}}$ correlation appears. The procedure for fixing the parameters is in detail described in the [8, 9]. Particularly, parameter t is taken to be a constant: t = 0.566 GeV^2, and for parameter β a smooth behavior with energy has been obtained and approximated by $\beta=1.16\,(1-\log(\sqrt{s}/(1\ \text{GeV}))-2.52)^{0.19}$. No extra parameter tuning has been made for particle differentiation.

The final hadron spectrum is modified due to the cascade decay of resonances by the strong interaction. They are effectively taken into account by a cascade branching matrix $M_{\mu\nu}$, extracted from the particle decayer built into THERMINATOR 2 Monte Carlo generator [15].

Finally, the relative yields of particles corrected for the decays of hadrons have the form:

$$Y_\nu\sim\sum_\mu M_{\mu\nu}\cdot\left(2S_\mu+1\right)\cdot\exp\left(-\frac{\pi\left(p_t^2+m_\mu^2\right)}{n^\beta t}\right), \tag{4}$$

where S_μ is spin of a particle of the type μ, $M_{\mu\nu}$ is effective branching matrix. The yields of particles are normalized keeping the total multiplicity N_{ch}.

2.1 Extension of the model for p-A collisions

The present model can be extended to the case of proton-nucleus collisions. Taking into account the fact that from the proton side in all inelastic events we have one participating nucleon, the total distribution of the number of pomerons is determined by the distribution of the participating nucleons of the nucleus. From the experimental data [16] we know, that in p-A and d-A collisions the charged multiplicity is approximately proportional to the number of participating nucleons, so that $\frac{dN/d\eta}{\langle N_{\text{part}}\rangle}|_{p\text{-}A}\simeq\frac{dN/d\eta}{2}|_{pp}$. In the present model, for p-A collisions we assume that in an event for each participant pair one emits n pomerons with probability w_n. Consequently, the probability of the production of n pomerons in p-A collision appears as a convolution of the distribution of the number of participating nucleons with the distribution w_n obtained for pp interaction (see figure 1). The distribution of N_{part} was calculated in Glauber model [17].

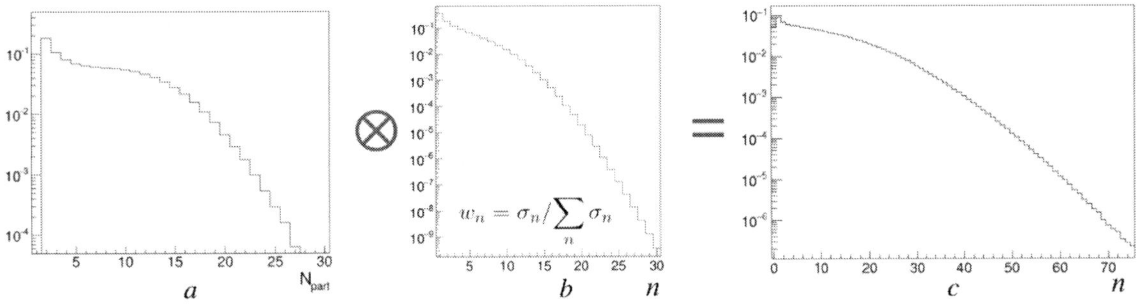

Figure 1. Probability of production of n pomerons (c) obtained as a convolution of $N_{\rm part}$ distribution (a) and the distribution of n pomerons from pp interaction (b).

3 Results

In figure 2 the relative yields of strange and multi-strange hadrons as a function of the charged multiplicity are shown for pp collisions at 7 TeV [18]. The results demonstrate the fast growth of relative strangeness, similarly to what was recently observed in the ALICE experiment [19]. We note that this increase of strange hadrons is obtained due to modification of string tension, similarly to the predictions of string fusion model [14].

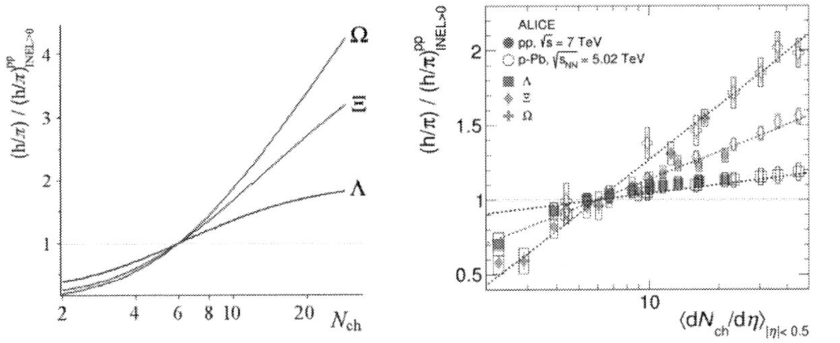

Figure 2. Multiplicity dependence of the relative yields of Ω, Ξ and Λ baryons, normalized to pion multiplicity, in pp-collisions at $\sqrt{s}$ =7 TeV. Left plot shows the model calculation [18], right – experimental data [19]. Linear growth would correspond to unity, which is marked by dotted line.

Figure 3 shows the multiplicity dependence of open charm production in pp and p-Pb collisions at $\sqrt{s}$= 7 TeV. The results show, that the present model reproduces the faster-than-linear growth of D meson yield with multiplicity of pp collisions. The fact that the enhancement in the multi-pomeron model is even stronger than in the experiment [1], may be because the model collective parameter β effectively takes into account all the collective processes leading to a correlation between the multiplicity and the transverse momentum, although the other contributions due to hard scattering processes, which are not related to the modification of the string tension [20], also exist. The results of the model calculation for p-Pb collisions show the dependence similar to pp case. This behavior is in agreement with the experimental data [21].

4 Conclusions

A generalization of the multi-pomeron exchange model with effective account of interaction between strings is proposed allowing for the production of charm particles in the Schwinger mechanism. The model parameters are determined by experimental data on $\langle p_t \rangle_{N_{\rm ch}} - N_{\rm ch}$ correlation. No additional parameters for particle differentiation are introduced. The multiplicity dependence of the strange baryons and charmed meson yields calculated in the model is in qualitative agreement with experimental data. The model is extended for the description of p-Pb collisions and it predicts a similar yield of the open charm as in the case of pp interaction at the same multiplicity, which is consistent with the experimental data.

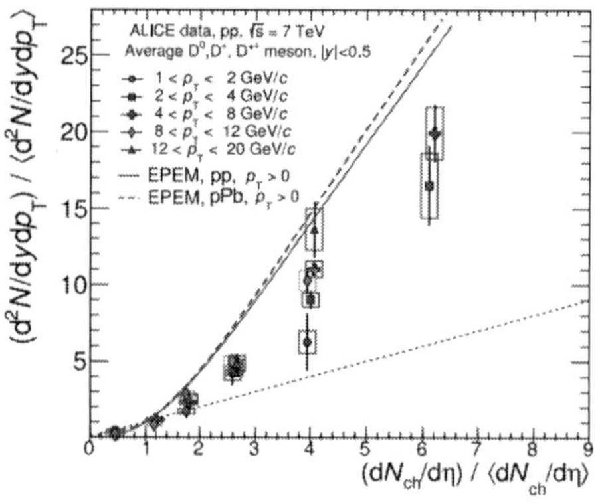

Figure 3. The relative open charm yield in pp collisions at $\sqrt{s}$=7 TeV as a function of normalized charged multiplicity, calculated in extended multi-pomeron exchange model (EPEM), compared with the experimental data [1]. The dotted line shows linear dependence of the yield (proportional to the total charged multiplicity). The solid line shows the model results for pp collisions, dashed – for p-Pb.

5 Acknowledgements

The research was supported by the grant of the Russian Science Foundation (project 16-12-10176).

References

[1] J. Adam, et al. (ALICE Collaboration), JHEP **09**, 148 (2015), arXiv:1505.00664 [nucl-ex].
[2] E. G. Ferreiro, C. Pajares, arXiv:1501.03381 [hep-ph] (2015).
[3] V. Kovalenko, V. Vechernin, ICNFP2016, EPJ Web of Conferences (2017, in press).
[4] V. N. Kovalenko, Phys. Atom. Nucl. **76**, 1189 (2013), arXiv:1211.6209 [hep-ph].
[5] V. Kovalenko, V. Vechernin, PoS (Baldin ISHEPP XXI) 077, arXiv:1212.2590 [nucl-th], (2012).
[6] K. Werner, et al, DESY Conf. Proc. 2016-01, p. 66-70, arXiv:1602.03414 [nucl-th] (2016).
[7] N. Armesto, D. A. Derkach, G. A. Feofilov, Phys. Atom. Nucl. **71**, 2087 (2008).
[8] E. Bodnia, D. Derkach, G. Feofilov, V. Kovalenko, A. Puchkov, PoS (QFTHEP 2013) 060 (2013), arXiv:1310.1627 [hep-ph].
[9] E. O. Bodnia, V. N. Kovalenko, A. M. Puchkov, G. A. Feofilov, AIP Conf. Proc. **1606**, 273-282 (2014), arXiv:1401.7534 [hep-ph].
[10] J. Schwinger, Phys. Rev. **82**, 664 (1951); T. S. Biro, et al, Nucl. Phys. B. **245**, 449 (1984).
[11] A. B. Kaidalov, Phys. Lett. B **116** (1982) 459.
[12] A. B. Kaidalov, K. A. Ter-Martirosyan, Phys. Lett. B **117**, 247 (1982).
[13] I. A. Lakomov, V. V. Vechernin, PoS (Baldin ISHEPP XXI) 072 (2012).
[14] M. A. Braun, C. Pajares, Nucl. Phys. B **390**, 542 (1993).
[15] M. Chojnacki, et al, Comput. Phys. Commun. **183**, 746-773 (2012), arXiv:1102.0273 [nucl-th].
[16] B. Abelev, et al (ALICE Collaboration) Phys. Rev. Lett. **110** (2013) no.3, 032301.
[17] M. L. Miller, et al, Ann. Rev. Nucl. Part. Sci. **57**, 205-243 (2007).
[18] G. Feofilov, V. Kovalenko, A. Puchkov, arXiv:1710.08895 [hep-ph] (2017).
[19] J. Adam, et al (ALICE Collaboration), Eur. Phys. J. C **755**, 226 (2015).
[20] V. N. Kovalenko, Phys. Part. Nucl. **48**, 945 (2017).
[21] R. Russo (for the ALICE Collaboration), Nucl. Phys. A **931**, 552-557 (2014).

Coupled dynamics of heavy and light flavor flow harmonics from EPOSHQ

P.B. Gossiaux[1,*], *J.* Aichelin[1], *M.* Nahrgang[1], *V.* Ozvenchuk[1,2], and *K.* Werner[1]

[1] *SUBATECH, UMR 6457, IMT Atlantique, Université de Nantes, IN2P3/CNRS. 4 rue Alfred Kastler, 44307 Nantes cedex 3, France*
[2] *Cracow, INP, ul. Radzikowskiego 152, 31-342 Kraków, Poland*

Abstract. We pursue the study of event by event correlations between low-mass particles and heavy mesons flow harmonics in ultrarelativistic heavy ion collisions and clarify some ambiguities found in one of our previous work.

1 Introduction and method

The joint study of light and heavy particles produced in ultrarelativistic heavy ion collisions offers new insights into several physical aspects, in particular into the coupling of heavy quarks with the quark gluon plasma (QGP) formed in these collisions. In [1], we have focused on the comparison between flow harmonics of light and heavy mesons and conjectured that heavy mesons would benefit less from the bulk flow due to their inertia. Such an effect could f.i. manifest itself in event by event (EBE) correlations between those flow harmonics, as studied recently in [2] and [3], where anomalous fluctuations where found.

In this study, we follow the same method as in [3], using the EPOSHQ model which allows to describe and predict both, the light and the heavy, sectors in ultrarelativistic pp, pA and AA collisions. Heavy quarks strongly interact with the QGP and thus benefit from the collective flows of the bulk. Later, D mesons are allowed to interact with light hadrons in the hadronic phase, leading to some moderate extra flow. For the sake of simplicity, D mesons are assumed to be produced exclusively through fragmentation mechanism while the role of coalescence will be investigated in an upcoming study. The analysis is performed EBE using the event/participant plane method, which does not suffer from the usual caveats as the participant plane is known with a very good accuracy from the initial state conditions. In [3], we have artificially enhanced the c quark production by a factor 50 in order to reduce the v_n fluctuations resulting from the limited number of c quarks [4]. This, however, seems to lead to abnormally large fluctuations and even led us to conjecture the existence of large EBE fluctuations specific to heavy flavor production and propagation in QGP. In this work, we do not resort to such an oversampling.

*e-mail: gossiaux@subatech.in2p3.fr

2 Results and analysis

In the following, "^" symbols will be used for EBE quantities. In fig. 1, we show the distribution of the $\hat{v}_{n=2,3}$ obtained[1] for π, p, c-quarks (before hadronization) and D mesons in EPOSHQ events, taking $\hat{\Psi}_n^{\rm PP}$ as the EBE reference plane for all species. The large variances observed for the flows of HF mesons imply that their individual EBE flows are often found with opposite sign as compared to the average. Following [4], we expect the variance of $\hat{v}_{n=2,3}$ to be $\mathrm{var}(\hat{v}_n) \approx \frac{1}{2N} + \mathrm{var}_{\mathrm{EBE},n}$. In table 1, we provide the numbers associated to the various cases shown in fig. 1. We observe a clear transition from π (dominance of EBE fluctuations) $\rightarrow$ N $\rightarrow$ D, for which the fluctuations are quite well explained by the usual $1/2N$ term, in contrast with our previous observation [3]. We have checked that this also holds for other centrality classes, other energies, larger HF coupling to the QGP bulk, finite p_T cut off, etc.

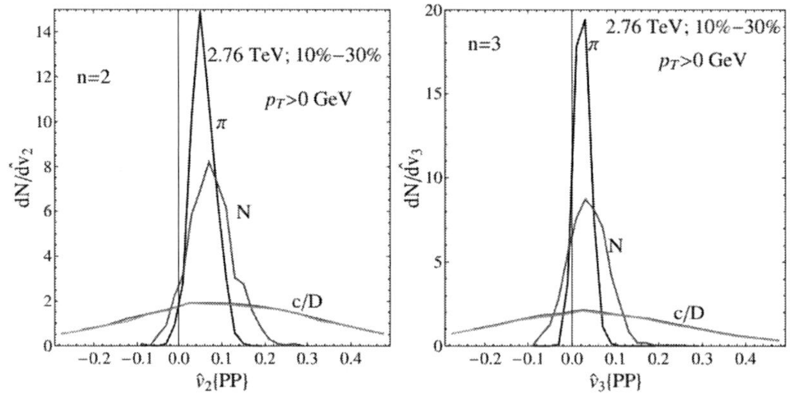

Figure 1. The distributions of the elliptic (left) and triangular (right) flow of final pions, nucleons (N), charmed quarks and D mesons produced in Pb-Pb collisions at $\sqrt{s}$ = 2.76 TeV in the 10%-30% centrality class. Distributions of c quarks and D mesons are quite similar.

Table 1. Decomposition of the flow variance for the 10%-30% centrality class.

	$\mathrm{var}(\hat{v}_{2/3})$	$\langle\frac{1}{2N}\rangle$	residual $(\mathrm{var}(\hat{v}_{2/3}) - \langle\frac{1}{2N}\rangle)$	%age of fluct. due to $\langle\frac{1}{2N}\rangle$
π	8.2E-4/3.3E-4	1.3E-4	6.9E-4/2E-4	16%/39%
N	2.9E-3/2.2E-3	1.8E-3	1.1E-3/4E-4	62%/82%
D	8.5E-2/8E-2	7.9E-2	5.7E-3/1.6E-3	93%/99%

In fig. 2, we show the elliptic and triangular flow of D mesons as a function of the spatial eccentricities $\hat{\epsilon}_n$, in a similar way as our QM2014 approach [5]. A linear increase of $\langle\hat{v}_n\rangle$ with $\langle\hat{\epsilon}_n\rangle$ in the bulk part of the event distribution implies $\mathrm{var}_{\mathrm{EBE},n} \approx (\langle\hat{v}_n\rangle/\langle\hat{\epsilon}_n\rangle)^2\,\mathrm{var}(\hat{\epsilon}_n)$, i.e. $\mathrm{var}_{\mathrm{EBE},2} \approx$ 6.E-4/1.2E-3/6E-4 for π/N/D mesons. These numbers should be compared with the 3rd column of table 1, with excellent agreement for π and nucleons. For HF, an extra source of fluctuations cannot be excluded, but it is definitively much smaller that what was advocated in [3] and is superseded by the $1/2N_D$ contribution for practical purposes.

[1] According to $\hat{v}_n\{\mathrm{PP}\} = \frac{1}{N}\sum_{i=1}^{N}\cos(n(\phi_i - \hat{\Psi}_n^{\rm PP}))$, where N is the number of selected particles in the sample and ϕ_i the azimuthal angles of individual particles.

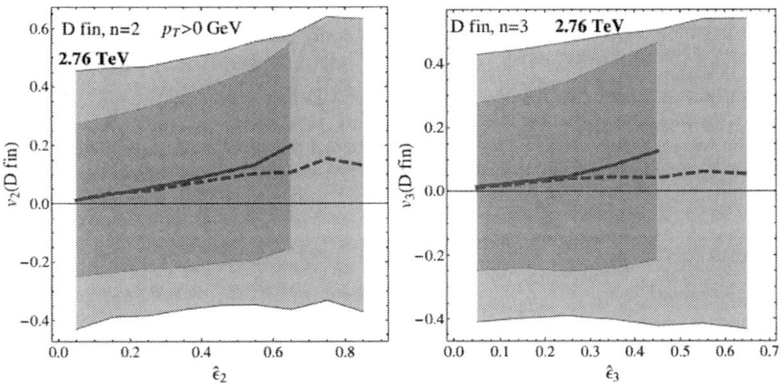

Figure 2. The elliptic (left) and triangular (right) flow of final D mesons produced in Pb-Pb collisions at $\sqrt{s}$ = 2.76 TeV as a function of spatial eccentricity; the solid lines correspond to average values (plain for 10%-30% centrality class and dashed for 30%-50% centrality class). The red (30%-50%) and darker (10%-30%) bands represent the average $\pm 1\ \sigma$ interval.

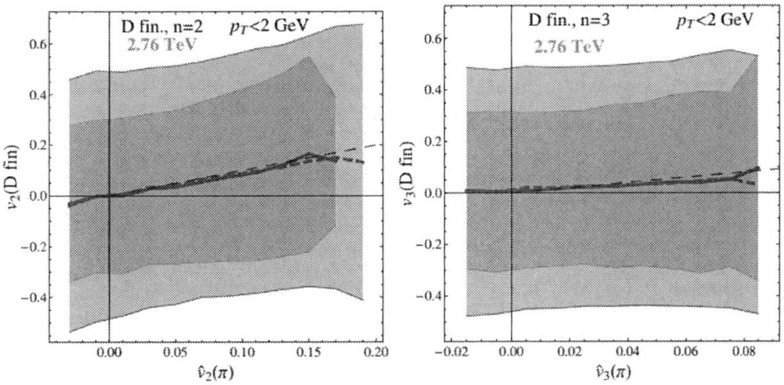

Figure 3. The elliptic (left) and triangular (right) flow of final p_T < 2 GeV D mesons produced in Pb-Pb collisions at $\sqrt{s}$ = 2.76 TeV as a function of equivalent quantity for π; same conventions as for fig. 2. The thin dashed line represents the $\langle \hat{v}_n(D) \rangle = \langle \hat{v}_n(\pi) \rangle$ case.

We have performed a refined analysis to identify the origin of the large fluctuations found previously with the oversampling method. In practice such an oversampling is achieved by performing 50 so-called "HQ events" – i.e. evolution of the HQ in the bulk – for a single EPOS3 global event with the *same* initial heavy quarks. This is an unavoidable feature as the production of heavy-quarks is deeply rooted in the semi-hard pomeron approach of EPOS3. The 50 replications of the HQ $\hat{v}_n$ are therefore totally correlated at initial time and gradually decorrelate with the time evolution. During our previous analysis, we had taken care to check that a very good level of decorrelation is achieved at the end of the evolution. However, we have realized since then that a small persisting correlation generates, in turn, an extra contribution in var($\hat{v}_n$) that scales like the number of replications and thus appears as a contribution to $\mathrm{var}_{\mathrm{EBE},n}$. This explains our misinterpretation in [3].

Having the fluctuations under control, we focus on the averages $\langle \hat{v}_n \rangle$ in fig. 2. Our results are compatible with the conclusions of our study [1] performed with EPOS2 as a description of the bulk:

The higher the eccentricities, the harmonics (and the mass) the less heavy quarks benefit from the bulk flow. In fig. 3 and 4, we present the correlation between the flow of low p_T final pions[2] and the equivalent flow for low and intermediate p_T D-mesons. In the low p_T case – where one expects nearly perfect thermalization of heavy quarks –, one finds a very good correlation between the light and the heavy sector. For intermediate p_T D-mesons, the correlation is still quite stong but less pronounced, which is an indication that c quarks with finite p_T are slightly less coupled with the bulk. Equivalent studies with b quarks are underway.

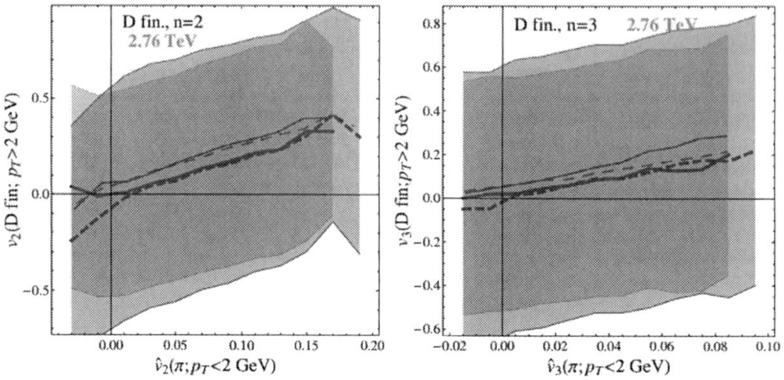

Figure 4. Same as fig. 3 for D mesons with $p_T > 2$ GeV; thin lines show the corresponding flows for pions with $p_T > 2$ GeV for comparison.

3 Conclusions

We have refined our analysis of heavy-light flow correlations initiated in [3] in two major aspects: First, we have understood that the HQ oversampling procedure leads to spurious extra-fluctuations in the EPOSHQ framework and thus clarified the ambiguities raised in [3]. Although some specific EBE fluctuations of the HF harmonic-flow cannot be ruled out, they appear to be subdominant. Second, we have increased our samples and present more robust predictions for the correlations between D mesons and pions harmonic flows. These could be measured experimentally by event engineering techniques, triggering on the $\hat{v}_n$ of pions either in minimum bias events or for a given centrality class. Going towards larger $\sqrt{s}$ facilitates this study, as the fluctuations are reduced due to the more abundant HQ production.

References

[1] M. Nahrgang et al., Phys. Rev. **C91**, 014904 (2015)
[2] C.A.G. Prado et al, arxiv:1611.02965
[3] P.B. Gossiaux et al., Nucl. Phys. A**967**, 672 (2017)
[4] A. Poskanzer and S. Voloshin, Phys. Rev. C**58**, 1671 (1998)
[5] https://indico.cern.ch/event/219436/contributions/1523417/

[2]Taken as the best proxy to probe the initial eccentricities of the Pb-Pb collisions.

Heavy Flavored Jets with CMS

Kurt Jung (for the CMS Collaboration)[1,★]

[1] *University of Illinois at Chicago, 845 W. Taylor St., Chicago, IL 60607*

Abstract. The energy loss of jets in heavy-ion collisions is expected to depend on the mass and flavor of the initiating parton. Thus, measurements of jet quenching with tagged partons place powerful constraints on the thermodynamic and transport properties of the hot and dense medium. Furthermore, recent results that constrain the jet production mechanism will shed additional light on the contributions of leading and next-to-leading order heavy flavor jet production with regard to the global energy loss picture. To this end, we present recent results measuring spectra and nuclear modification factors of jets associated to charm and bottom quarks in both pPb and PbPb collisions, as well as measurements of dijet asymmetry of pairs of b-jets in PbPb collisions.

1 Introduction

The energy loss of highly energetic partons in large collision systems is generally attributed to a process called "jet quenching." Interactions of the jet with the hot and dense medium known as Quark Gluon Plasma, thought to be created in collisions of large ions, induce constituent particle energy loss and jet broadening, where constituent particles are pushed far from the jet axis. Both features contribute significantly to the measured jet energy modification. Measurements of jet quenching are expected to depend on parton flavor primarily through a modification of two jet quenching mechanisms: radiative and collisional energy loss [1, 2], as both these processes are thought to include mass-dependent effects [3, 4]. A process known as the dead cone effect suppresses forward gluon radiation in an angle proportional to the mass of the parton, and collisional energy transfer to the medium is mitigated somewhat by the heavy parton mass. In addition to these effects, a typical inclusive-jet measurement contains a large fraction of jets seeded by high-p_T gluons, while a measurement of tagged b-jets, for example, will contain significantly fewer gluon-seeded jets. Under the assumption that radiative energy loss is the dominant component of jet quenching, gluon jets are expected to quench more strongly than quark jets, due to the larger color factor for gluon emission from gluons than from quarks [5].

The heavy flavored jet measurements summarized in these conference proceedings include finalized measurements of charm-tagged jets [6] in proton-lead (pPb) collisions, as well as measurements of di-b-jets [7] in lead-lead (PbPb) collisions. Measurements of jet quenching in pPb collisions are not expected to include effects from the QGP and are made primarily to test effects of initial-state "cold nuclear matter" processes on jet production. The presented inclusive charm quark jet cross-section measurement is the first of its kind and will shed light on the nature of the charm jet production and

★e mail: kurtcjung@gmail.com

alllow further exploration of the charm-tagged sector at high energies. Measurements of di-b-jets in PbPb attempt to clarify the magnitude of heavy flavored jet energy loss at leading-order, as opposed to inclusive b-jet studies, which are dominated by next-to-leading order production graphs [8]. These next-to-leading order processes like gluon splitting, where a highly energetic gluon splits into a $b\bar{b}$ pair, are expected to contribute significantly to an inclusive b-jet measurement. While interesting in their own right, the back-to-back selection of b-jets will preferentially select leading-order b-jet production to study b quarks that interact directly with the medium.

2 Identification of Flavored Jets

Jets containing heavy quarks are tagged via combinations of quantities that relate to the relatively long lifetimes of B and D mesons. For jet spectra measurements, reconstructed secondary vertices are used to identify both b- and c-jets. The b-jets are tagged through associating the jet to a secondary vertex that has at least two associated tracks and a flight distance significance greater than 2 standard deviations away from the primary vertex. The c-jets are tagged similarly, associating the jets to a secondary vertices with a looser flight distance significance of least 1.68 standard deviations away from the primary vertex, though these vertices are required to have at least three tracks. Once the jets are tagged, the tagging purity is calculated using template fits of both tagged and untagged jets to distributions of the secondary vertex mass. The template fits for the c-jet analysis use a modification of the secondary vertex mass that includes missing energy from neutral particles and neutrinos. If the secondary vertex displacement vector is not parallel to the momentum vector of the vertex, an additional energy is added to the vertex mass to account for the undetected momentum. The correction for missing particles allows enough discrimination power between b- and c-jets such in the template fits so that all three flavor contributions (light, c, and b) are accurately calculated simultaneously in a sample. The maximum b-tagging purity achieved is roughly 60%, while the maximum c-tagging purity is roughly 30%. For further details, see the b-jet [9] and c-jet [6] publications.

The b-dijet analysis uses a more advanced b-jet identification strategy, using machine learning algorithms that effectively combine information from secondary vertices, leptons, and charged particles. A boosted decision tree is used to optimize combinations of more than 29 different parameters such that the b-jet sample purity is upwards of 90%. An additional light jet rejection is imposed by calculating the yields of dijet pairs close together in azimuthal angle ($\Delta\phi < \pi/3$) and removing this yield across the entire $\Delta\phi$ range. This subtraction exploits the fact that while the b-jet $\Delta\phi$ distributions peak around $\Delta\phi = \pi$, the light jet pairs are generally combinatorial in nature and have a flat $\Delta\phi$ distribution. Further corrections are applied to account for situations where light jets are mistagged as the away-side partner of a leading b-jet. These inefficiencies are highly p_T and centrality dependent, ranging from about 50% at 40 GeV to very small values (< 1%) at 100 GeV. This correction is derived by inverting the b-tagging selection and subtracting the inverted yields from the signal, where the inverted yield is weighted by the mistagging fraction, as derived in Monte Carlo simulations. More details can be found in the preliminary document [7].

3 Results

The subleading-to-leading-jet p_T ratio (x_j) ranges from 0 to 1 and depends on the jet kinematic selection. The di-b-jet measurement selects leading jets of at least 100 GeV and subleading jets of at least 40 GeV, separated in azimuthal angle by more than $\Delta\phi > 2\pi/3$. This kinematic selection leads to an average $\langle x_j \rangle$ of around 0.7 for b-jets in pp collisions and between 0.6 and 0.7 for b-jets in PbPb collisions, depending on collision centrality. These $\langle x_j \rangle$ values can be compared to inclusive jet measurements, where similar values are found for all collision centralities, indicating no significant quenching differences between leading-order inclusive dijets and di-b-jets, shown in Fig. 1 (left). We observe a significant centrality dependent trend for $\langle x_j \rangle$, indicating the presence of jet quenching, however the magnitude and shape of this trend is very similar for both the dijet samples studied. This is directly compared in Fig. 1 (right), which shows the ratios of the x_j values for b-jets to inclusive jets. The largest separation is in the 10-30% centrality bin, which shows a separation of about 1.3 standard deviations. The similarity of the inclusive dijet and di-b-jet measurements strengthen the conclusions that high-p_T b-jets behave very similarly to high-p_T inclusive jets, first observed by single b-jet measurements at CMS [9, 10].

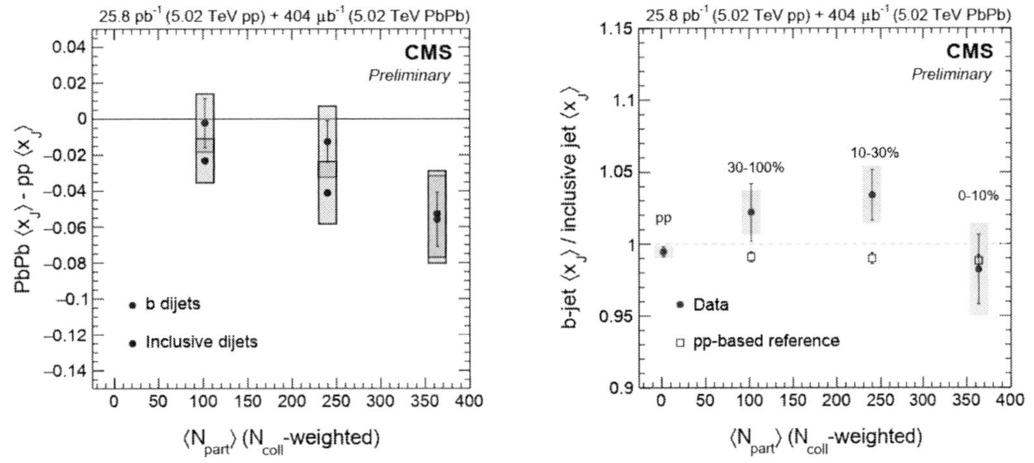

Figure 1. The left figure shows the difference of the average x_j values of dijets between PbPb and pp collisions for inclusive-jets (blue) and for b-jets (red). The right figure shows the ratio of the average x_j values between b-jets and inclusive-jets. Both figures show distributions as a function of the average number of collision participants, weighted by the average number of binary collisions [7].

Smaller system measurements are also presented in these proceedings. Measurements of jets in collision systems like pPb are expected to probe nuclear effects in the initial-state (pre-collision), as a QGP medium is not expected to be produced. To study these initial-state effects, a quantity known as R_{pA} is measured, comparing the yields of jets in pPb collisions to pp collisions. The R_{pA} value is defined as:

$$R_{pA} = \frac{1}{A}\frac{d\sigma_{PbPb}/dp_T}{d\sigma_{pp}/dp_T}, \tag{1}$$

where A is the nuclear mass number of Pb (208), and is included in order to account for the geometrical enhancement of jet production based on the additional number of hard scatterings in pPb collisions relative to pp collisions. The R_{pA} is measured for both b-jets and c-jets in Fig. 2 on the left and lower right panels, respectively. Both measurements show consistency with unity, which indicates that the nuclear initial-state has only a minor effect on jet production at high-p_T. Note especially that the b-jet measurement is convoluted somewhat by the use of a Pythia 6 simulation [11] instead of true pp data, due to the fact that 5 TeV pp data was not yet available when the measurement was published. To account for this, a 22% global uncertainty was assigned to the measurement to account for any differences between the pp data and the PYTHIA simulation. This factor was derived by interpolating the adherence of the PYTHIA calculation to data at both 2.76 and 7 TeV. When these uncertainties are taken into account, both the b-jet and c-jet R_{pA} values are consistent with both unity and each other. Finally, these measurements can be compared to the inclusive-jet R_{pA}, as measured by the CMS Collaboration [12]. The R_{pA} of all three jet flavors are consistent with one another across the entire kinematic range, indicating the lack of sensitivity required to probe the flavor dependence of initial-state jet energy modification.

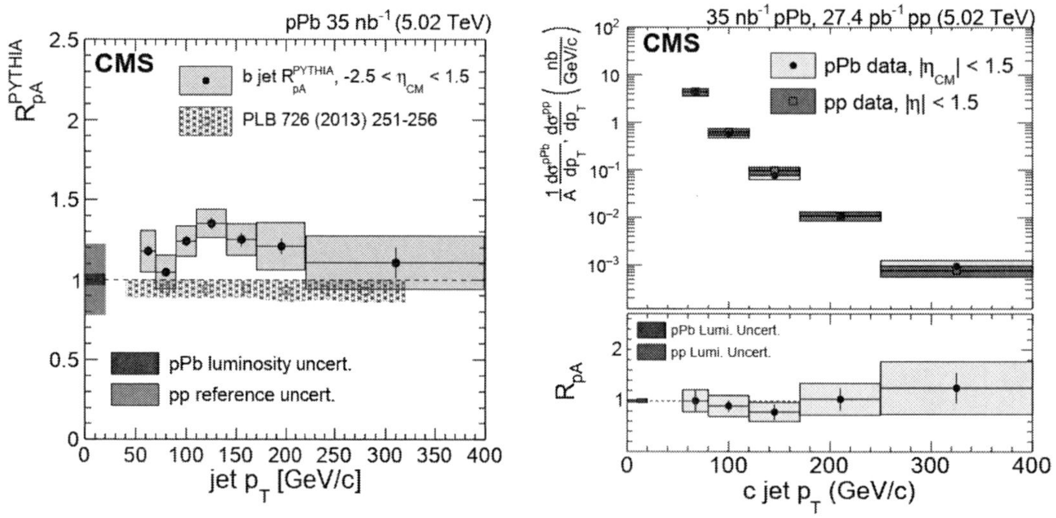

Figure 2. The R_{pA} values for b-jets (left [9]) and c-jets (right, lower panel [6]) are shown as a function of jet p_T. The b-jet measurements are compared to a prediction including only energy-loss effects from Vitev., et. al. [3]. The upper panel of the right figure shows the c-jet cross-sections in pp and pPb, where the pPb spectrum is scaled by the mass number of Pb. Also shown are uncertainties from luminosity on both figures, as well as a pp reference uncertainty on the b-jet measurement.

References

[1] H. van Hees, R. Rapp, Phys. Rev. C **71**, 034907 (2005), nucl-th/0412015
[2] G.D. Moore, D. Teaney, Phys. Rev. C **71**, 064904 (2005), hep-ph/0412346
[3] J. Huang, Z.B. Kang, I. Vitev, Phys. Lett. B **726**, 251 (2013), 1306.0909
[4] D.K. Yu.L. Dokshitzer, Phys. Lett. B **519**, 199 (2001), hep-ph/0106202
[5] D. d'Enterria, *Jet quenching* (Springer Materials, 2010), Vol. 23, chap. 6.4, 0902.2011
[6] V. Khachatryan et al. (CMS), Phys. Lett. B **772**, 306 (2017), 1612.08972
[7] CMS Collaboration (CMS) (2016), http://cds.cern.ch/record/2202805
[8] A. Banfi, G.P. Salam, G. Zanderighi, JHEP **07**, 026 (2007), 0704.2999
[9] V. Khachatryan et al. (CMS), Phys. Lett. B **754**, 59 (2016), 1510.03373
[10] S. Chatrchyan et al. (CMS), Phys. Rev. Lett. **113**, 132301 (2014)
[11] R. Field, Acta Phys. Polon. B **42**, 2631 (2011), 1110.5530
[12] V. Khachatryan et al. (CMS), Eur. Phys. J. C **76**, 372 (2016), 1601.02001

EPJ Web of Conferences **171**, 18006 (2018) https://doi.org/10.1051/epjconf/201817118006
SQM 2017

Measurements of charm hadron production and anisotropic flow in Au+Au collisions at 200 GeV with the STAR experiment at RHIC

Sooraj Radhakrishnan[1,*] for the STAR Collaboration

[1] *Lawrence Berkeley National Laboratory, Berkeley, CA 94720. USA.*

Abstract. Heavy flavor quarks, owing to their large masses, are predominantly produced through initial hard parton scatterings in heavy-ion collisions, and thus are excellent probes to study properties of the strongly coupled Quark Gluon Plasma (sQGP) medium produced in these collisions. Measurements of anisotropic flow harmonics of heavy flavor hadrons can provide information on the properties of the medium, including the heavy flavor transport coefficient. Charm quark hadronization mechanism in the sQGP medium can be studied through measurements of yields of different charm hadrons. In these proceedings we report on the measurements of elliptic and triangular flow harmonics of D^0 mesons as well as the yield ratios of $D_s^{\pm}/D^0$ and $\Lambda_c^{\pm}/D^0$, in Au+Au collisions at $\sqrt{s_{NN}}$ = 200 GeV at RHIC with the STAR detector. These measurements use the STAR Heavy Flavor Tracker (HFT) to reconstruct charm hadrons via their hadronic decay channels. Results are compared to model calculations and the implications on the understanding of charm quark dynamics in the medium are discussed.

1 Introduction

Relativistic heavy-ion collisions at the RHIC and LHC create a hot dense medium of deconfined quarks and gluons. Experimental evidences, particularly the strong collective anisotropic flow of light flavor hadrons [1], suggest that the medium produced is strongly interacting and is in a local thermal equilibrium. It is therefore referred to as the strongly coupled Quark Gluon Plasma (sQGP) [2, 3]. In heavy-ion collisions, heavy flavor quarks (charm and bottom) are predominantly produced through initial hard parton scatterings and therefore are valuable probes to study the properties of the sQGP. In particular, measurements of the anisotropic flow (quantified using the coefficients v_n in the Fourier decomposition of the azimuthal distribution of particles [4]) of heavy flavor hadrons can help understand the interaction of heavy flavor quarks with the medium and constrain the values of the spatial diffusion coefficient D_s for the heavy flavor quarks in sQGP [5]. In these proceedings we present measurements of the elliptic (v_2) and triangular (v_3) flow of D^0 mesons in Au+Au collisions at $\sqrt{s_{NN}}$ = 200 GeV at RHIC with the STAR detector.

We also present measurements of the yield ratios of $D_s^{\pm}/D^0$ and $\Lambda_c^{\pm}/D^0$ in $\sqrt{s_{NN}}$ = 200 GeV Au+Au collisions at RHIC. These measurements can help understand the hadronization mechanism of charm quarks in the sQGP medium. Enhancements of these ratios, relative to the values in p+p collisions, are expected in the intermediate p_T region (3 - 6 GeV/c) if charm quarks hadronize via

[*] e-mail: skradhakrishnan@lbl.gov

the coalescence mechanism in sQGP [6–9]. The magnitude of this enhancement is also sensitive to strangeness enhancement in sQGP for $D_s^{\pm}/D^0$ [6] and the presence of diquarks in the medium for $\Lambda_c^{\pm}/D^0$ [9].

2 Experiment and Analysis

The STAR detector at RHIC has a full azimuthal acceptance and a pseudorapidity (η) coverage of $|\eta| < 1$ [10]. The Heavy Flavor Tracker (HFT) [11] is a high-resolution silicon detector installed close to the beam pipe and provides excellent track pointing resolution, e.g. the track pointing resolution for kaons with $p_T = 1$ GeV/c is less than 40 μm. This allows to place cuts on the quantities describing the decay topology of heavy flavor hadrons, which can greatly enhance the signal significance. Particle identification (PID) at STAR is provided by the ionization energy loss (dE/dx) measured in the Time Projection Chamber (TPC) and the velocity measured using the Time of Flight (TOF) detector [10]. Results presented in these proceedings use about 900 M minimum-bias Au+Au events from the 2014 run at RHIC.

The $D^0(\overline{D^0})$ candidates are reconstructed through the $K^-\pi^+(K^+\pi^-)$ decay channel. Anisotropic flow measurements are made for D^0 (results presented in these proceedings are for D^0 and $\overline{D^0}$ hadrons combined) using the event plane method [4]. The D^0 candidates are correlated with the event plane constructed using TPC tracks in the opposite η hemisphere of the D^0 candidate and outside an additional η-gap of $|\Delta\eta| > 0.05$, to reduce non-flow contributions. More details on the measurement of elliptic flow of D^0 can be found in [12]. Reconstruction of the $D_s^{\pm}$ is done using the $D_s^{\pm}$ –> $\pi^{\pm}\phi(1020)$ –> $\pi^{\pm}K^{\pm}K^{\mp}$ decay channel and $\Lambda_c^{\pm}$ in the $\Lambda_c^{\pm}$ –> $p^{\pm}K^{\mp}\pi^{\pm}$ channel. The TPC track reconstruction efficiency, acceptance corrections and PID efficiency are evaluated using embedding and data-driven methods [13], while the HFT track matching and topological cut efficiencies are evaluated through a fast simulation which takes the HFT to TPC track matching ratios and track pointing resolutions from data.

3 Results

Figure 1 shows the v_2 of D^0 mesons as a function of the D^0 p_T in the 0-80% centrality interval. It can be seen that the D^0 mesons have significantly non-zero values of v_2 for $p_T > 1.5$ GeV/c. The measured values are compared to the calculations from different models. The hydrodynamic model [14] tuned to reproduce the light flavor hadron v_2 is able to describe the D_0 v_2 for $p_T < 4$ GeV/c, suggesting that charm quarks may have achieved thermal equilibrium in these collisions. The TAMU model with no charm quark diffusion and the DUKE model with a fixed value for the charm diffusion coefficient, $2\pi TD_s = 7$ are inconsistent with the data, while transport models with a temperature-dependent diffusion coefficient in the range of $2 < 2\pi TD_s < 12$ are able to describe the measured D_0 v_2 [12]. The left panel of figure 2 shows the v_2 values divided by the number of constituent quarks (n_q) for D^0 mesons and a number of light flavor hadrons, as a function of $(m_T - m_0)/n_q$, where $m_T = \sqrt{m_0^2 + p_T^2}$ is the transverse mass and m_0 is the rest mass. The n_q-scaled v_2 values for all the species, including D^0, fall on the same curve and hence suggest that charm quarks have gained significant flow from their interactions with the sQGP medium. The right panel of figure 2 shows a similar plot for v_3. The scaled v_3 values for D^0 also agree with those of light flavor hadrons, albeit with larger statistical uncertainties.

The left panel of figure 3 shows the D_s/D^0 yield ratio (D_s here denotes both D_s^+ and D_s^-), as a function of p_T for the 0-10% and 10-40% centrality bins. The D_s/D^0 ratios are significantly enhanced in Au+Au collisions compared to PYTHIA [15] predictions for p+p collisions and to the world-data average of the fragmentation ratio from the measurements in p+p, e+e, and e+p collisions. The model

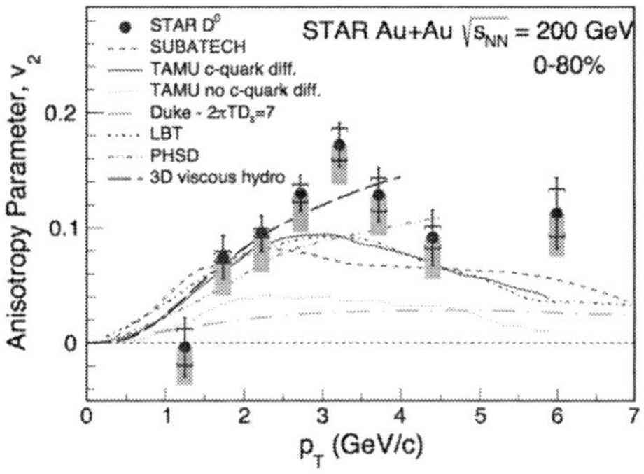

Figure 1. v_2 as a function of p_T for D^0 mesons in 0-80% centrality bin. The vertical bars and brackets represent statistical and systematic uncertainties, respectively. The grey band represents the estimated non-flow contribution [12].

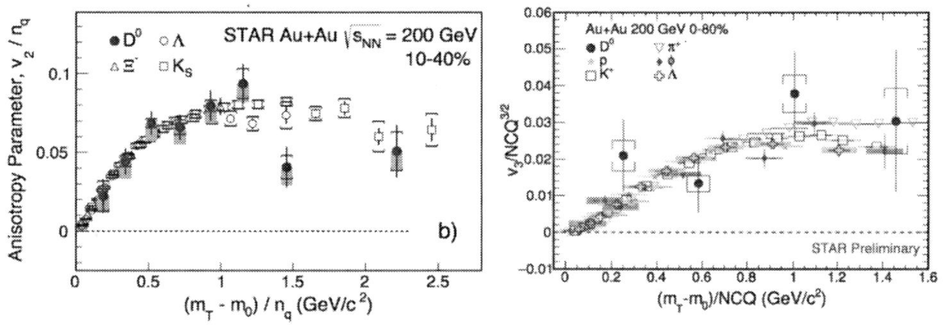

Figure 2. v_2/n_q in 10-40% centrality [12] bin (left) and $v_3/(n_q)^{3/2}$ in 0-80% centrality bin (right), as a function of $(m_T - m_0)/n_q$ for D^0 (solid circles) and light flavor hadrons.

prediction from the TAMU group [6], which includes coalescence of charm quarks, underpredicts the data, while the SHM calculation [16] is comparable to data in the low p_T region. The right panel of figure 3 shows the Λ_c/D^0 (Λ_c here denotes both Λ_c^+ and Λ_c^-) ratio measured in $3 < p_T < 6$ GeV/c. The data shows a significant enhancement of the ratio compared to the PYTHIA prediction for p+p collisions. The SHM model [16] also underpredicts the data. The Ko model (0-5%) [9] calculations include coalescence of thermalized charm quarks and are closer to the measured values. However, both calculations, with and without di-quarks in the sQGP medium, are consistent with the data in the measured p_T region. The Greco model [8] also includes coalescence of charm quarks, but the values shown are for the Λ_c yield over the total D meson yield.

4 Summary and Outlook

Measurements of the elliptic and triangular flow harmonics for D^0 mesons in Au+Au collisions at $\sqrt{s_{NN}} = 200$ GeV at RHIC are presented. The number-of-constituent-quark scaled v_2 for D^0 mesons agrees with those for light flavor hadrons suggesting that charm quarks acquire significant flow from their interactions with the medium. Hydrodynamic calculation can describe the data for $p_T < 4$

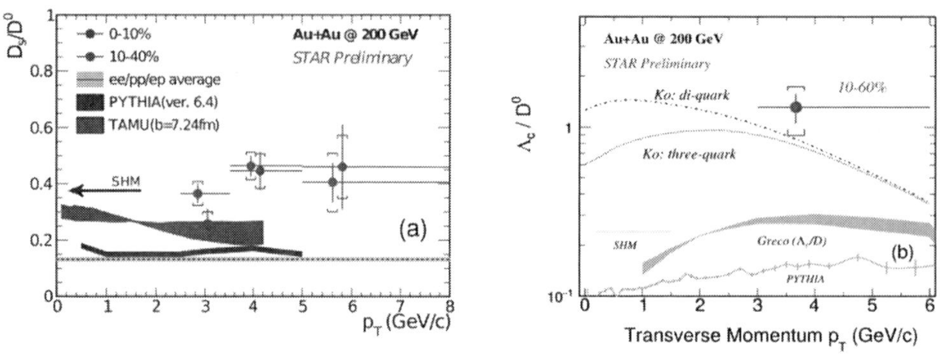

Figure 3. Yield ratios of D_s/D^0 in 0-10% and 10-40% Au+Au collisions (left) and of Λ_c/D^0 in 10-60% central Au+Au collisions (right) as a function of p_T. Vertical bars and brackets represent statistical and systematic uncertainties respectively

GeV/c, suggesting that charm quarks achieve as strong collectivity as light quarks in these collisions. Transport model calculations with a temperature-dependent charm diffusion coefficient in the range of $2 < 2\pi T D_s < 12$ for $T_c < T < 2T_c$ can also describe the data. The measurements of the D_s/D^0 and Λ_c/D^0 yield ratios in Au+Au collisions at $\sqrt{s_{NN}}$ = 200 GeV are also presented. Both ratios show significant enhancements compared to the values in p+p collisions predicted by PYTHIA. For the D_s/D^0 ratio, the TAMU model calculation with coalescence of charm quarks underpredicts the measured values, while for the Λ_c/D^0 ratio, model calculations with coalescence of thermalized charm quarks in the sQGP medium are comparable with the data. STAR has collected two times more data in 2016 than in 2014, which will allow for measurements with better precision, particularly for the triangular flow and Λ_c/D^0 ratio, and extend the kinematic reach of the measurements.

References

[1] C. Gale, S. Jeon, B. Schenke, P *et al.*, Phys. Rev. Lett.**110**, 012302 (2013).
[2] STAR Collaboration, Nucl. Phys.A**757**, 102 (2005).
[3] B. Muller, J. Schukraft, and B. Wyslouch, Ann. Rev. Nucl. Part. Sci. **62**, 361 (2012).
[4] A. M. Poskanzer and S. A. Voloshin, Phys. Rev.C**58**, 1671 (1998).
[5] A. Beraudo, A. De Pace, M. Monteno, M. Nardi, and F. Prino, Eur. Phys. J.C**75**, 121 (2015).
[6] M. He, R. J. Fries and R. Rapp, Phys. Rev. Lett.**110**, 112301 (2013).
[7] Y. Oh, C. M. Ko, S. H. Lee and S. Yasui, Phys. Rev. C**79**, 044905 (2009).
[8] S. Ghosh, S. K. Das, V. Greco, S. Sarkar and J. E. Alam, Physical Review D**90(5)**, 2-7 (2014).
[9] S. H. Lee, K. Ohnishi, S. Yasui, I.-K. Yoo and C. M. Ko Phys. Rev. Lett.**100**, 222301 (2008).
[10] STAR Collaboration, Nucl. Inst. Meth. A**499**, 624–632 (2003).
[11] H. Qiu (STAR Collaboration), Nucl. Phys.A**931**, 1141 (2014).
[12] STAR Collaboration, Phys. Rev. Lett.**118**, 212301 (2017).
[13] STAR Collaboration, Phys. Rev. Lett.**113**, 142301 (2014).
[14] L. G. Pang, Y. Hatta, X. N. Wang, and B. W. Xiao, Phys. Rev.D**91**, 074027 (2015).
[15] T. Sjöstrand, S. Mrenna and P. Skands, Journal of High Energy Physics**2006**, 026 (2006).
[16] I. Kuznetsova and J. Rafelski, The European Physical Journal **C51** 113–133 (2007).

EPJ Web of Conferences **171**, 18007 (2018) https://doi.org/10.1051/epjconf/201817118007
SQM 2017

Measurement of the D-meson nuclear modification factor and elliptic flow in Pb–Pb collisions at $\sqrt{s_{\rm NN}}$ = 5.02 TeV with ALICE at the LHC

Fabrizio Grosa[1,2,⋆] *for the ALICE Collaboration*

[1] *Politecnico di Torino, Corso Duca degli Abruzzi 24, 10129 Torino Italy*
[2] *INFN sez. Torino, via Pietro Giuria 1, 10125 Torino Italy*

Abstract. Heavy-flavour hadrons are recognised as a powerful probe for the characterisation of the deconfined medium created in heavy-ion collisions, the Quark-Gluon Plasma (QGP). The ALICE Collaboration measured the production of D^0, D^+, D^{*+} and D_s^+ mesons in Pb–Pb collisions at $\sqrt{s_{\rm NN}}$ = 5.02 TeV. The measurement of the nuclear modification factor ($R_{\rm AA}$) provides a strong evidence of the in-medium parton energy loss. The comparison between the D_s^+ and the non-strange D-meson $R_{\rm AA}$ can help to study the hadronisation mechanism of the charm quark in the QGP. In mid-central collisions, the measurement of the D-meson elliptic flow v_2 at low transverse momentum ($p_{\rm T}$) gives insight into the participation of the charm quark into the collective motion of the system, while at high $p_{\rm T}$ it constrains the path-length dependence of the energy loss. The D_s^+ v_2, measured for the first time at the LHC, is found to be compatible to that of non-strange D mesons and positive with a significance of about 2.6 σ. The coupling of the charm quark to the light quarks in the underlying medium is further investigated for the first time with the application of the Event-Shape Engineering (ESE) technique to D-meson elliptic flow.

1 Introduction

Heavy quarks (i.e. charm and beauty) are excellent probes for the characterisation of the deconfined medium created in ultra-relativistic heavy-ion collisions, the Quark-Gluon Plasma (QGP). Because of their large mass, heavy quarks are predominantly produced in hard-scattering processes, before the formation of the QGP [1]. Therefore, they experience the whole system evolution, interacting with the medium constituents via collisional and radiative processes [2–6]. The observation of a large suppression of the heavy-flavour hadron yields at intermediate/high $p_{\rm T}$ in central Pb–Pb collisions with respect to those in pp collisions indicated by the measurement of nuclear modification factor $R_{\rm AA}(p_{\rm T}) = ({\rm d}N_{\rm AA}/{\rm d}p_{\rm T})/(\langle T_{\rm AA}\rangle \cdot {\rm d}\sigma_{\rm pp}/{\rm d}p_{\rm T})$ significantly smaller than unity, provides a strong evidence of the in-medium parton energy loss [7]. The comparison between heavy-flavour and light-flavour hadrons gives insight into the colour-charge and quark-mass dependence of the energy loss [8]. Moreover, it is predicted that a fraction of heavy quarks could hadronise via coalescence in the medium and, therefore, could be sensitive to the enhanced production of strange quarks in high-energy

⋆e-mail: fabrizio.grosa@cern.ch

heavy-ion collisions [9]. In this scenario, the measurement of heavy-flavour hadrons with strange-quark content (e.g. the D_s^+ meson) is crucial to understand the modification of the charm-quark hadronisation in the deconfined medium [10]. Complementary information on the interaction of heavy quarks with the QGP is provided by the measurement of the azimuthal anisotropy in the momentum distribution of heavy-flavour hadrons and, in particular, by the elliptic flow $v_2 = \langle \cos(2(\varphi - \Psi_2)) \rangle$, which is defined as the second-order harmonic coefficient of the Fourier expansion of the azimuthal distribution with respect to the reaction plane angle Ψ_2. The measurement of the v_2 at low p_T helps to quantify to which extent heavy quarks are influenced by the collective dynamics of the underlying medium, while at high p_T has the potential to constrain the path-length dependence of the parton energy loss in the QGP [11–13].

2 D-meson reconstruction

Open-charm production in Pb–Pb collisions at $\sqrt{s_{NN}} = 5.02$ TeV was measured by ALICE via the exclusive reconstruction of D mesons at mid-rapidity ($|y| < 0.8$), in the hadronic decay channels $D^0 \to K^-\pi^+$ ($c\tau \simeq 123\,\mu$m, BR = 3.93%), $D^+ \to K^-\pi^+\pi^+$ ($c\tau \simeq 312\,\mu$m, BR = 9.46%), $D^{*+} \to D^0\pi^+$ (strong decay, BR = 67.7%) and $D_s^+ \to \phi\pi^+ \to K^-K^+\pi^+$ ($c\tau \simeq 150\,\mu$m, BR = 2.67%) [17]. The decay topologies were reconstructed exploiting the excellent vertex-reconstruction capabilities of the Inner Tracking System (ITS). Kaons and pions were identified with the Time Projection Chamber (TPC) via their specific energy loss and with the Time-Of-Flight detector (TOF). The raw D-meson yields were extracted via an invariant-mass analysis after having applied topological selections to enhance the signal over background ratio. The efficiency times acceptance corrections were obtained from MC simulations based on HIJING [18] and PYTHIA 6 [19] event generators. The fraction of prompt D mesons was estimated with a FONLL-based approach [20, 21]. The centrality and the Event-Plane angle (estimator of Ψ_2) were provided by the V0 scintillators, which cover the pseudorapidity regions $-3.7 < \eta < -1.7$ and $2.8 < \eta < 5.1$.

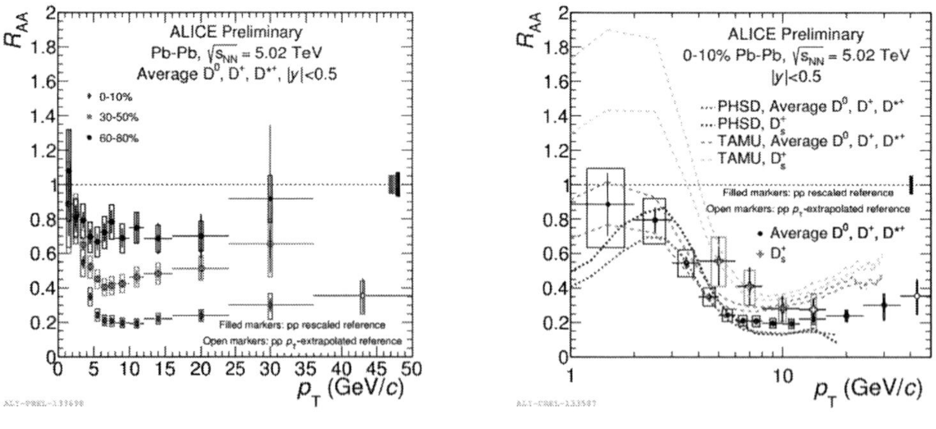

Figure 1. Left: average prompt D^0, D^+, D^{*+} p_T-differential R_{AA} in central 0–10% (diamonds), mid-central 30–50% (squares) and peripheral 60–80% (circles) Pb–Pb collisions at $\sqrt{s_{NN}} = 5.02$ TeV [14]. Right: comparison between prompt D_s^+ (crosses) and non-strange D-meson (circles) R_{AA} in Pb–Pb collisions at $\sqrt{s_{NN}} = 5.02$ TeV in the 0–10% centrality class [14]. The data are compared to models that include charm hadronisation via coalescence in the QGP [15, 16].

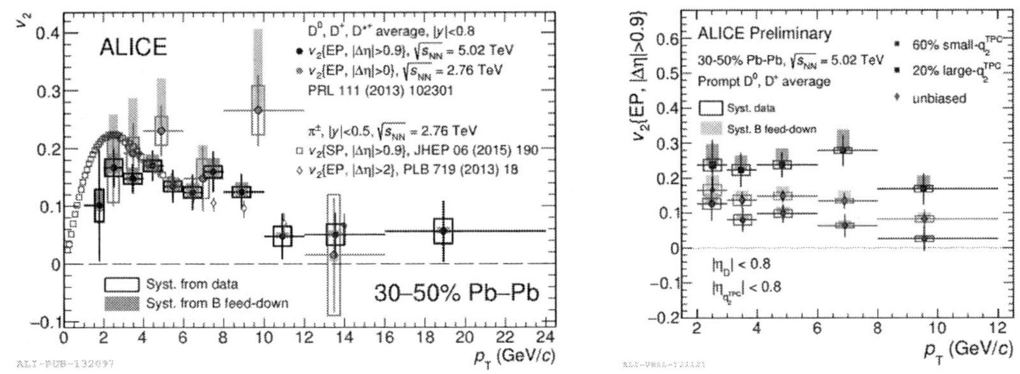

Figure 2. Left: average prompt D^0, D^+, D^{*+} v_2 as a function of p_T in Pb–Pb collisions at $\sqrt{s_{NN}}$ = 5.02 TeV in the 30–50% centrality class compared to the same measurement and the $\pi^{\pm}$ v_2 at $\sqrt{s_{NN}}$ = 2.76 TeV [21]. Right: average prompt D^0 and D^+ v_2 in 30–50% Pb–Pb collisions at $\sqrt{s_{NN}}$ = 5.02 TeV for the 60% of events with smallest q_2 and the 20% of events with largest q_2 compared to the unbiased result.

3 Prompt D-meson nuclear modification factor and elliptic flow

The p_T-differential R_{AA} of prompt D^0, D^+, D^{*+} and D_s^+ was measured in central 0–10%, mid-central 30–50% and peripheral 60–80% Pb–Pb collisions at $\sqrt{s_{NN}}$ = 5.02 TeV [14]. The proton-proton reference was obtained from the measured production cross section at $\sqrt{s}$ = 7 TeV [22], scaled to $\sqrt{s}$ = 5.02 TeV using the FONLL prediction. The average non-strange D-meson R_{AA} shows an increasing suppression from peripheral to central events up to a factor about 5 for $p_T > 5$ GeV/c, as reported in the left panel of Figure 1. The prompt D_s^+ nuclear modification factor is systematically higher, although compatible within uncertainties, than that of non-strange D mesons in the full p_T range measured, for the three centrality classes. In the right panel of Figure 1 the D_s^+ and the non-strange D-meson R_{AA} in the 0–10% centrality class are compared to models that include the hadronisation of charm quark via coalescence in a strangeness-enhanced QGP [15, 16]. These models predict a smaller suppression of the D_s^+ with respect to the non-strange D mesons. The left panel of Figure 2 shows the prompt D-meson v_2 as a function of p_T measured in Pb–Pb collisions at $\sqrt{s_{NN}}$ = 5.02 TeV in the 30–50% centrality class [21], using the Event-Plane method [23]. The pseudorapidity gap ($|\Delta\eta| > 0.9$) between the reconstructed D mesons and the particles used to estimate the Event-Plane angle suppresses the non-flow contributions in the v_2 arising from the correlation in azimuth of particles from decays or from jets. The D^0, D^+ and D^{*+} v_2 are consistent with each other and larger than zero in $2 < p_T < 10$ GeV/c, indicating that low- and intermediate-p_T charm quarks participate to the collective expansion of the medium. For $p_T > 10$ GeV/c the D-meson v_2 is still positive, though comparable with zero. Furthermore, the D-meson v_2 at $\sqrt{s_{NN}}$ = 5.02 TeV is found to be compatible to that measured at $\sqrt{s_{NN}}$ = 2.76 TeV and similar in magnitude to that of charged pions in the same centrality class, as shown in the left panel of Figure 2. The elliptic flow coefficient of prompt D_s^+ is also positive in $2 < p_T < 8$ GeV/c with a significance of about 2.6 σ and is compatible within uncertainties to that of non-strange D mesons [21]. The D^0 and D^+ v_2 were also investigated for the first time with an Event-Shape Engineering (ESE) technique. This method, consisting in measuring the D-meson v_2 for classes of events with different average elliptic flow, provides more insight into the coupling of the charm quark with the bulk of light quarks. The average ellipticity can be quantified by the magnitude of the second-harmonic reduced flow vector $q_2 = |\boldsymbol{Q}_2|/\sqrt{M}$ [24], where M is the multiplicity and $\boldsymbol{Q}_2$ is the second-harmonic flow vector, whose components are given by $Q_{2,x} = \sum_{i=1}^{M} \cos 2\varphi_i$ and $Q_{2,y} = \sum_{i=1}^{M} \sin 2\varphi_i$. The events in the 30–50% centrality class were divided in the 60% of events with smallest q_2 and the 20% of the events with largest q_2. The average elliptic flow of prompt D^0 and D^+ mesons for the two q_2 classes of events compared to the unbiased one is presented in the right panel of Figure 2. The observation of a significant separation between the D-meson v_2 in the two classes of event shapes suggests that the charm quark is sensitive to the light-quark collectivity and to the event-by-event initial-state fluctuations. However, the effect could be slightly enlarged by autocorrelations between D mesons and q_2, since they are measured in the same pseudorapidity region $|\eta| < 0.8$.

4 Conclusions

The ALICE Collaboration measured the R_{AA} and the v_2 of prompt D^0, D^+, D^{*+} and D_s^+ mesons in Pb–Pb collisions at $\sqrt{s_{NN}} = 5.02$ TeV. The D-meson R_{AA} shows an increasing suppression from peripheral to central events. A hint of charm hadronisation via coalescence is provided by the observation of a higher D_s^+ R_{AA} with respect to that of non-strange D mesons. A significantly non-zero D-meson v_2 is measured for $2 < p_T < 10$ GeV/c in mid-central collisions, confirming the participation of the c-quark to the expanding dynamics of the hot medium observed at $\sqrt{s_{NN}} = 2.76$ TeV. For the first time at the LHC, the D_s^+ v_2 was measured and the ESE technique was applied to the non-strange D-meson elliptic flow.

References

[1] F.M. Liu, S.X. Liu, Phys. Rev. **C89**, 034906 (2014), 1212.6587
[2] E. Braaten, M.H. Thoma, Phys. Rev. **D44**, R2625 (1991)
[3] M. Gyulassy, M. Plumer, Phys. Lett. **B243**, 432 (1990)
[4] R. Baier, Y.L. Dokshitzer, A.H. Mueller, S. Peigne, D. Schiff, Nucl. Phys. **B484**, 265 (1997), hep-ph/9608322
[5] B.W. Zhang, E. Wang, X.N. Wang, Phys. Rev. Lett. **93**, 072301 (2004), nucl-th/0309040
[6] Y.L. Dokshitzer, D.E. Kharzeev, Phys. Lett. **B519**, 199 (2001), hep-ph/0106202
[7] J. Adam et al. (ALICE), JHEP **03**, 081 (2016), 1509.06888
[8] J. Adam et al. (ALICE), JHEP **11**, 205 (2017), 1506.06604
[9] I. Kuznetsova, J. Rafelski, Eur. Phys. J. **C51**, 113 (2007), hep-ph/0607203
[10] J. Adam et al. (ALICE), JHEP **03**, 082 (2016), 1509.07287
[11] S. Batsouli, S. Kelly, M. Gyulassy, J.L. Nagle, Phys. Lett. **B557**, 26 (2003), nucl-th/0212068
[12] M. Gyulassy, I. Vitev, X.N. Wang, Phys. Rev. Lett. **86**, 2537 (2001), nucl-th/0012092
[13] B.B. Abelev et al. (ALICE), Phys. Rev. **C90**, 034904 (2014), 1405.2001
[14] ALICE (2017), ALICE-PUBLIC-2017-003, https://cds.cern.ch/record/2265109
[15] M. He, R.J. Fries, R. Rapp, Phys. Lett. **B735**, 445 (2014), 1401.3817
[16] T. Song, H. Berrehrah, D. Cabrera, J.M. Torres-Rincon, L. Tolos, W. Cassing, E. Bratkovskaya, Phys. Rev. **C92**, 014910 (2015), 1503.03039
[17] C. Patrignani et al. (Particle Data Group), Chin. Phys. **C40**, 100001 (2016)
[18] X.N. Wang, M. Gyulassy, Phys. Rev. **D44**, 3501 (1991)
[19] T. Sjostrand, S. Mrenna, P.Z. Skands, JHEP **05**, 026 (2006), hep-ph/0603175
[20] M. Cacciari, M. Greco, P. Nason, JHEP **05**, 007 (1998), hep-ph/9803400
[21] S. Acharya et al. (ALICE) (2017), 1707.01005
[22] S. Acharya et al. (ALICE), Eur. Phys. J. **C77**, 550 (2017), 1702.00766
[23] A.M. Poskanzer, S.A. Voloshin, Phys. Rev. **C58**, 1671 (1998), nucl-ex/9805001
[24] S.A. Voloshin, A.M. Poskanzer, R. Snellings (2008), 0809.2949

Measurement of heavy-flavour production, correlations and jets with ALICE

Shingo Sakai [a1] for the ALICE collaboration
[1] *Univ. of Tsukuba, Faculty of Pure and Applied Sciences, 1-1-1 Tennodai Tsukuba Ibaraki 305-8571 Japan*

Abstract. In this article, recent ALICE measurements of the R_{AA} and v_2 of electrons at mid-rapidity ($|y|<0.6$) and muons at forward rapidity ($2.5<y<4$), from semi-leptonic decays of heavy-flavour hadrons in Pb–Pb collisions at $\sqrt{s_{NN}}$ = 5.02 TeV are shown. The results of D-tagged jet production cross-section in pp collisions at $\sqrt{s}$ =7 TeV are presented. In addition, the azimuthal correlation of D mesons with charged particles in p–Pb collisions at $\sqrt{s_{NN}}$ = 5.02 TeV, and electrons from semi-leptonic decays of heavy-flavour hadrons with charged particles in p–Pb and Pb–Pb collisions at $\sqrt{s_{NN}}$ = 5.02 TeV are reported.

1 Introduction

Heavy-flavour (charm and beauty) production provides relevant information to investigate the strongly interacting medium, the Quark–Gluon Plasma (QGP), formed in heavy-ion collisions at ultra-relativistic energies. Heavy quarks are produced primarily in the initial hard partonic interactions, and they propagate through the QGP. During that propagation they are expected to interact with the constituents of the QGP. The key observables for understanding the interaction are the nuclear modification factor (R_{AA}) and azimuthal anisotropy (v_2). The R_{AA} is sensitive to the energy loss of partons in the hot and dense QCD matter which occurs via radiative [1] and collisional [2] processes. The energy loss has been predicted to depend on mass and colour charge of partons. In this picture charm and beauty quarks are expected to suffer a smaller energy loss than light flavour quarks (u, d and s). The v_2 is sensitive to the transport properties of the medium. And it reflects the collective motion of heavy quarks in the medium at low p_T as well as the path-length dependence of the energy loss at high p_T in the non-central collisions. Further understanding of properties of heavy quarks in the QGP can be obtained by measurements of jets and the azimuthal correlation of particles originating from heavy flavours, which are sensitive to the possible modification of the fragmentation of heavy quarks. Such measurements in pp and p–Pb collisions allow studies of jet properties, and address cold nuclear matter and collective effects of heavy quarks.

[a] sakai.shingo.gw@u.tsukuba.ac.jp

2 Results

2.1 Heavy-flavour production in pp and p–Pb collisions

Figure 1 shows the transverse-momentum-differential cross section of charged jets containing a D meson in pp collisions at $\sqrt{s} = 7$ TeV. The D^0 cross section is well described by calculations based on perturbative QCD (pQCD) [3]. The measurement of charm-jet production allows one to further test pQCD calculations for heavy-flavour production. The measured cross section is compared with the expectation from POWHEG + PYTHIA in the figure. The plot in the bottom panel shows the ratio of data and the calculation, and it was found that the cross section from the data and the calculations are in good agreement in the range $5 < p_T < 30$ GeV/c. In ALICE, heavy-flavour production has also been studied by measuring the azimuthal correlation between heavy flavours and charged particles. Figure 2 shows the azimuthal correlation of D mesons with charged particles in p–Pb collisions at $\sqrt{s_{NN}} = 5.02$ TeV. The correlation was measured in various intervals of p_T ranges for the D meson and its associated charged particles. In the azimuthal correlation, the near-side ($\Delta\varphi = 0$) correlation is sensitive to the fragmentation function of charm quarks. On the other hand, the away-side correlation ($\Delta\varphi = \pi$) is sensitive to the properties of the recoil jet. In p–Pb collisions, possible modifications of the jet properties could arise from cold nuclear matter effects, like parton-density shadowing or saturation [4], a Cronin-like enhancement [5] and energy loss [6] in the initial state. The correlation was compared with PYTHIA calculations with different tunings. The PYTHIA calculations are in good agreement with the measured azimuthal correlation of D mesons with charged partcles in various p_T regions. The agreement indicates that the modification of jets from charm is not significantly affected by the cold nuclear matter effects as well as from possible final-state effects.

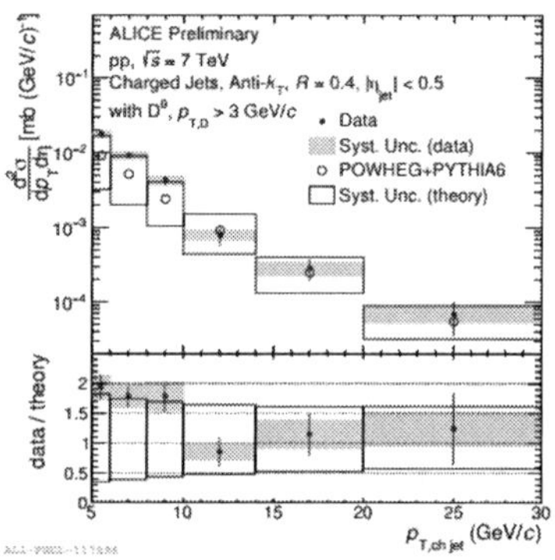

Figure 1: Transverse-momentum-differential cross section of charged jets containing a D meson in pp collisions at $\sqrt{s} = 7$ TeV.

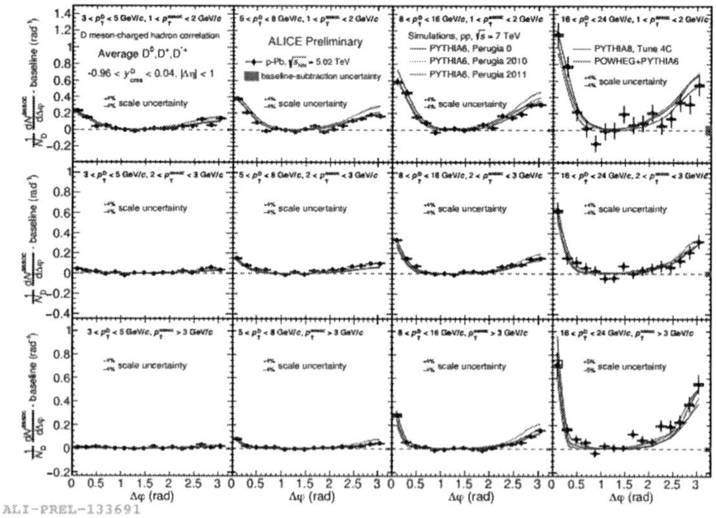

Figure 2: The azimuthal correlation of D mesons with charged particles in p–Pb collisions at $\sqrt{s_{NN}} = 5.02$ TeV.

2.2 Heavy-flavour production in Pb–Pb collisions

The left panel of Fig. 3 shows R_{AA} of electrons from semi-leptonic decays of heavy-flavour hadrons (e^{HF}) at |y|<0.6 in 0–10% centrality in Pb–Pb collisions at $\sqrt{s_{NN}} = 5.02$ TeV. The result shows that

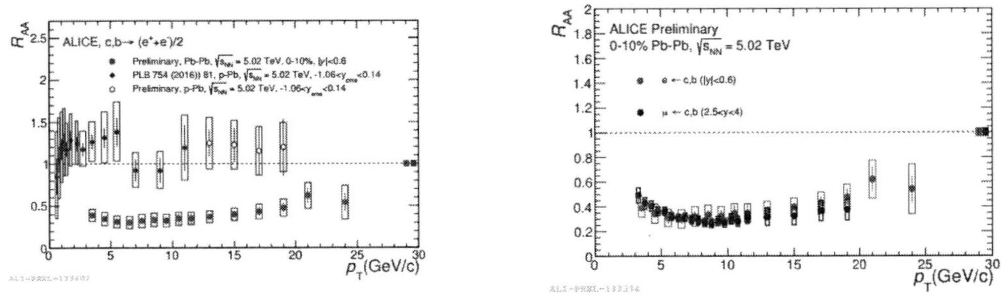

Figure 3: The nuclear modification factor for e^{HF} in Pb–Pb and in p–Pb collisions at $\sqrt{s_{\mathrm{NN}}}$ =5.02 TeV (left), and of e^{HF} ($|y|$<0.6) and μ^{HF} (2.5<y<4) in 0-10% Pb–Pb collisions (right).

the R_{AA} is less than unity (R_{AA} <1) and it suggests that e^{HF} production is significantly suppressed up to 25 GeV/c. On the other hand, such strong suppression of e^{HF} is not observed in p–Pb collisions at $\sqrt{s_{\mathrm{NN}}}$ = 5.02 TeV. The nuclear modification factor of e^{HF} in p–Pb collisions at $\sqrt{s_{\mathrm{NN}}}$ = 5.02 TeV is also shown in the Fig 3. It is consistent with unity within uncertainties, indicating that strong suppression observed in Pb–Pb collisions is mainly due to energy loss of charm and beauty quarks in the hot medium formed in the collisions. A similar strong suppression is also observed in muons from heavy-flavour decays (μ^{HF}) at forward rapidity (2.5<y<4) and the R_{AA} for e^{HF} and μ^{HF} are consistent in the range 3< p_{T} < 20 GeV/c (Fig. 3, right). Further insights on heavy-flavour production in the hot and dense QCD matter can be obtained by measuring the per-trigger azimuthal correlation between e^{HF} and hadrons. Figure 4 shows the associated yield of charged particle in the near-side correlation peak in 0–20% and 20–50% central in Pb–Pb collisions divided by the yield in p–Pb collisions as a function of associated charged particles p_{T}. The yield in Pb–Pb collisions tends to increase at lower p_{T}. Such an enhancement of associated charged hadrons at low p_{T} has been observed in the $\pi^0 - h$ and $h - h$ correlations in Pb–Pb collisions [7, 8]. In addition, a positive v_2 of e^{HF} has been observed

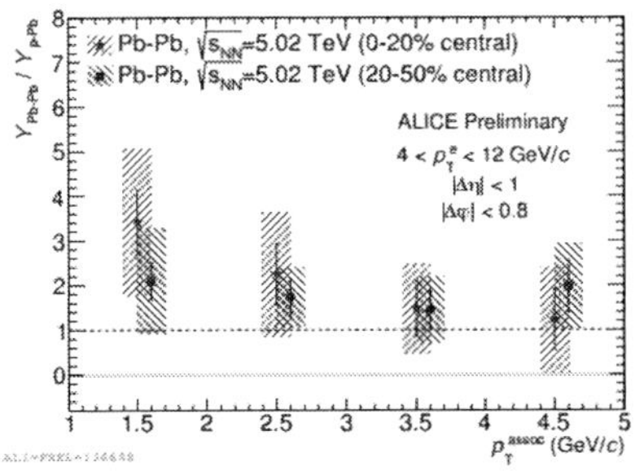

Figure 4: Ratio of near-side associated yields in Pb–Pb (0–20% and 20–50%) and in p–Pb collisions as a function of associated particles p_{T}.

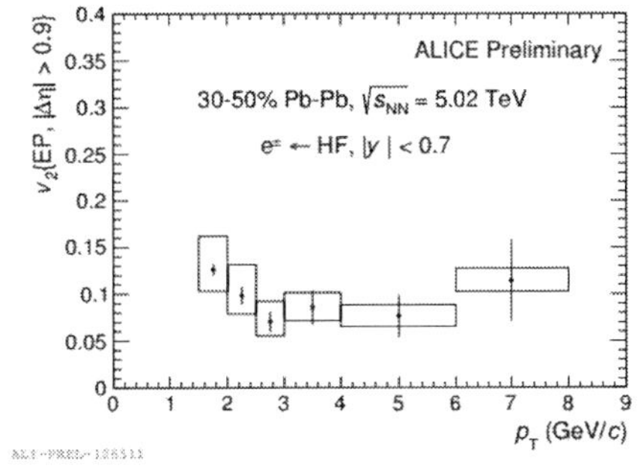

Figure 5: Elliptic flow (v_2) for e^{HF} in semi-central (30-50%) Pb–Pb collisions at $\sqrt{s_{\mathrm{NN}}}$ = 5.02 TeV.

in semi-central (30–50%) in Pb–Pb collisions at $\sqrt{s_{\mathrm{NN}}}$ = 5.02 TeV [9] (Fig. 5). The positive v_2 of e^{HF} suggests that heavy-flavour quarks participate to the collective motion of the system. The R_{AA}

and v_2 for e^{HF} are reproduced by pQCD–based energy loss and transport models [10–15] although the simultaneous description of the R_{AA} and v_2 is challenging as shown in Fig. 6.

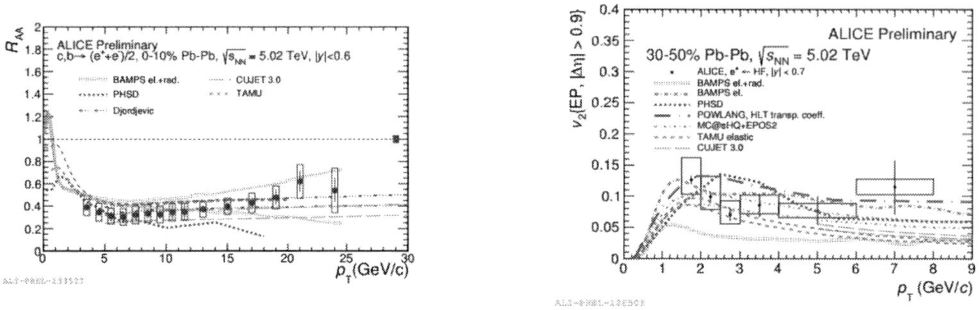

Figure 6: Transverse momentum dependence of R_{AA} in 0–10% (left) and v_2 in 30–50% (right) of e^{HF} compared with expectations from theoretical models [10–15].

3 Summary

The ALICE experiment measured several differential observables in various collision systems and energies, with the goal of understanding the interaction of heavy quarks with the constituents of the QGP formed in Pb–Pb collisions. The production of D-tagged jet in pp collisions and the azimuthal correlation of D mesons with charged particles in p–Pb collisions are consistent with expectations from PYTHIA and POWHEG+PYTHIA event generators. A strong suppression and positive v_2 of leptons from semi-leptonic decays of heavy-flavour hadrons is observed in Pb–Pb collisions at $\sqrt{s_{NN}}$ = 5.02 TeV. Those results suggest significant energy loss of charm and beauty quarks in the hot and dense QCD matter and their participation to the collective motion of the system as a result of strong interaction with medium constituents.

References

[1] M. Djordjevic *et al.* Phys. Lett. B, 632 (2006) 81
[2] A. Adil, I. Vitev, Phys. Lett. B, 649 (2007) 139
[3] B. Abelev *et al.* (ALICE collaboration), JHEP 01 (2012) 128
[4] N. Armesto, J. Phys. G, 32 (2006), pp. 367–394
[5] J. W. Cronin *et al.* Phys. Rev. D, 11 (1975), p. 3105
[6] Z.-B. Kang, I. Vitev, E. Wang, H. Xing, C. Zhang, Phys. Lett. B, 740 (2015), pp. 23–29
[7] K. Aamodt *et al.* (ALICE collaboration), Phys. Rev. Lett. 108, 092301
[8] J. Adam *et al.* (ALICE collaboration), Phys. Lett. B763, pages238–250
[9] D. Godoy for the Alice collaboration, Nuclear Physics, Section A 967C (2017) pp. 636-639
[10] A. Beraudo, A. De Pace, M. Monteno, M. Nardi, F. Prino, Eur. Phys. J. C75 (2015) 121.
[11] M. Nahrgang, J. Aichelin, P. B. Gossiaux, K. Werner, Phys. Rev. C89 (2014) 014905.
[12] J. Upho, F. Senzel, O. Fochler, C. Wesp, Z. Xu, C. Greiner, Phys. Rev. Lett. 114 (2015) 112301.
[13] M. He, R. J. Fries, R. Rapp, Physics Letters B 735 (2014) 445–450.
[14] M. Djordjevic, M. Djordjevic, Phys. Rev. C92 (2015) 024918.
[15] T. Song, H. Berrehrah, D. Cabrera, W. Cassing, E. Bratkovskaya, Phys. Rev. C93 (2016) 034906.

Charmonium production in pPb and PbPb collisions at 5.02 TeV with CMS

Javier Martín Blanco on behalf of the CMS Collaboration[1,*]

[1] *Centre National de la Recherche Scientifique, Laboratoire Leprince-Ringuet, École polytechnique, France*

Abstract. Charmonium states, such as the J/ψ and $\psi(2S)$ mesons, are excellent probes of the deconfined state of matter, the Quark-Gluon Plasma (QGP) created in heavy ion collisions. In addition, the measurements in pPb collisions allow to study the cold nuclear matter effects, being crucial to disentangle these from the QGP-related effects in PbPb collisions. In this talk the new nuclear modification factor R_{AA} of prompt and nonprompt J/ψ in PbPb collisions at $\sqrt{s_{NN}}$ = 5.02 TeV were presented over a wide kinematic range ($3 < p_T < 50$ GeV/c, $|y| < 2.4$), and fine event-centrality intervals. The results were compared to those at 2.76 TeV over a similar kinematic range. In addition, new prompt $\psi(2S)$ R_{AA} results at 5.02 TeV were reported. Finally the final prompt and nonprompt J/ψ results, as well as preliminary $\psi(2S)$ results, in pPb collisions at 5.02 TeV, were discussed.

1 Introduction

Many phenomena, such as dissociation in the QGP and statistical recombination [1–4], on top of cold nuclear matter effects (modifications of nPDFs, initial-state energy loss and nuclear break-up) [5, 6], are required for a complete understanding of charmonium production in PbPb collisions . Therefore measurements in pPb collisions are also crucial for a quantitative determination of the QGP-related effects in PbPb collisions. The recombination mechanism in the QGP is only expected to affect charmonium production at low transverse momentum (p_T), while at high p_T heavy quarkonium is likely to be produced by parton fragmentation, being sensitive to the parton energy loss in the QGP [7, 8]. In addition, it is interesting to separate the prompt and nonprompt (from b hadron decays) charmonium components, since nonprompt charmonium should reflect medium effects on b hadron production, while the prompt component is affected by dissociation in the QGP and statistical recombination.

In this talk, the prompt and nonprompt J/ψ and prompt $\psi(2S)$ nuclear modification factors (R_{AA}) in PbPb data taken by the LHC CMS experiment at $\sqrt{s_{NN}}$ = 5.02 TeV were presented [9]. Also, the results of the nuclear modification factor of prompt and nonprompt J/ψ [10], and prompt $\psi(2S)$ [11] in pPb collisions at the same energy were briefly discussed. The integrated luminosities were 28 pb^{-1}, 34.6 nb^{-1} and 0.5 nb^{-1} for pp, pPb and PbPb data samples respectively.

2 The CMS detector and analysis procedure

The central feature of the CMS detector is a superconducting solenoid of 6 m internal diameter, providing a magnetic field of 3.8 T. Within the solenoid volume are a silicon pixel and strip tracker,

* e-mail: javier.martin.blanco@cern.ch

a lead tungstate crystal electromagnetic calorimeter, and a brass and scintillator hadron calorimeter, each composed of a barrel and two endcap sections. Forward hadron calorimeters (HF) extend the coverage to $2.9 < |\eta| < 5.2$. Muons are measured in the pseudorapidity range $|\eta| < 2.4$ in gas-ionisation detectors embedded in the steel flux-return yoke outside the solenoid, with detection planes made using three technologies: drift tubes, cathode strip chambers, and resistive plate chambers. A more detailed description of the CMS detector can be found in Ref. [12].

In the results presented here, charmonia were identified via their dimuon decay. The separation of the prompt and nonprompt components was based on the the pseudo-proper decay length $l_{J/\psi} = L_{xyz}m_{J/\psi}/|p_{\mu^+\mu^-}|$, where L_{xyz} is the distance between the primary and dimuon vertex, $m_{J/\psi}$ is the world average value of the J/ψ meson mass (assumed for all dimuons), and $p_{\mu\mu}$ is the dimuon momentum. The invariant-mass and the $l_{J/\psi}$ distributions of $\mu^+\mu^-$ pairs were fitted in an extended bidimensional unbinned maximum likelihood fit. The invariant mass component in the fits was parameterised with the sum of two Crystal Ball functions for the signal, and with a polynomial function for the underlying background. The $l_{J/\psi}$ component was parameterised in collision data and Monte Carlo simulated events, using templates for the per-event $l_{J/\psi}$ uncertainty distributions, a sum of Gaussian functions to describe the $l_{J/\psi}$ resolution, and an empirical combination of exponential functions for the background.

3 Results

In Fig. 1, the R_{AA} of prompt (upper) and nonprompt (lower) J/ψ mesons as a function of p_T (left) and centrality (right) are shown in PbPb collisions $\sqrt{s_{NN}}$ = 5.02 TeV [9]. The results as a function of p_T are compared to those obtained at 2.76 TeV [14], and found to be in good agreement. No significant p_T trend is observed in the prompt R_{AA} results in the range of 5–20 GeV/c, but a lower suppression is observed at higher p_T for the first time. A similar raise, attributed to parton energy loss, is observed for charged hadrons at the same energy [13]. A lower suppression is also observed at low p_T. In the prompt R_{AA} results in the $1.8 < |y| < 2.4$ range as a function of centrality, we observe that the suppression is stronger for higher p_T in the most central range. A similar observation by the ALICE Collaboration is attributed to a regeneration contribution [4]. The p_T dependence of the nonprompt J/ψ R_{AA} shows hints for a smaller suppression at low p_T. Similarly, a stronger suppression is observed for $1.8 < |y| < 2.4$ and $p_T > 6.5$ GeV/c at all centralities.

The prompt ψ(2S) meson R_{AA} is derived from the product of the double ratio results of the ψ(2S) and J/ψ meson yields in pp and PbPb collisions [15], and the prompt J/ψ R_{AA} computed in the wider bins of Ref. [15]. The results are presented in Fig. 2 as a function of dimuon p_T and centrality, at central rapidity ($|y| < 1.6$). In the bins where the double ratio is not significant, 95% CL intervals on the prompt ψ(2S) R_{AA} are given. The ψ(2S) meson R_{AA} shows no clear dependence with p_T, and hints of an increasing suppression with collision centrality. In the entire measured range, we observe that the ψ(2S) meson production is more suppressed than that of J/ψ mesons, showing that the excited states are more strongly affected by the medium created in PbPb collisions.

In Ref. [10], the prompt and nonprompt J/ψ meson R_{pPb} is measured in pPb collisions at $\sqrt{s_{NN}}$ = 5.02 TeV. The results show that the prompt J/ψ meson R_{pPb} is slightly above unity without a clear dependence on p_T, except in the most forward bin where suppression at low p_T is observed. The nonprompt J/ψ meson R_{pPb} is consistent with unity for all rapidities, and there is no clear dependence on p_T or rapidity within the uncertainties. The prompt ψ(2S) meson R_{pPb} results in pPb collisions at 5.02 TeV [11], show a stronger suppression than that observed for J/ψ mesons at the same energy, which is unexpected from the current theoretical models including CNM effects, suggesting the presence of final-state effects affecting ψ(2S) meson production. The relatively small modification of the prompt J/ψ and ψ(2S) meson yields in pPb collisions at $\sqrt{s_{NN}}$ = 5.02 TeV indicates that the strong suppression observed in PbPb collisions at the same energy cannot be ascribed to CNM effects alone.

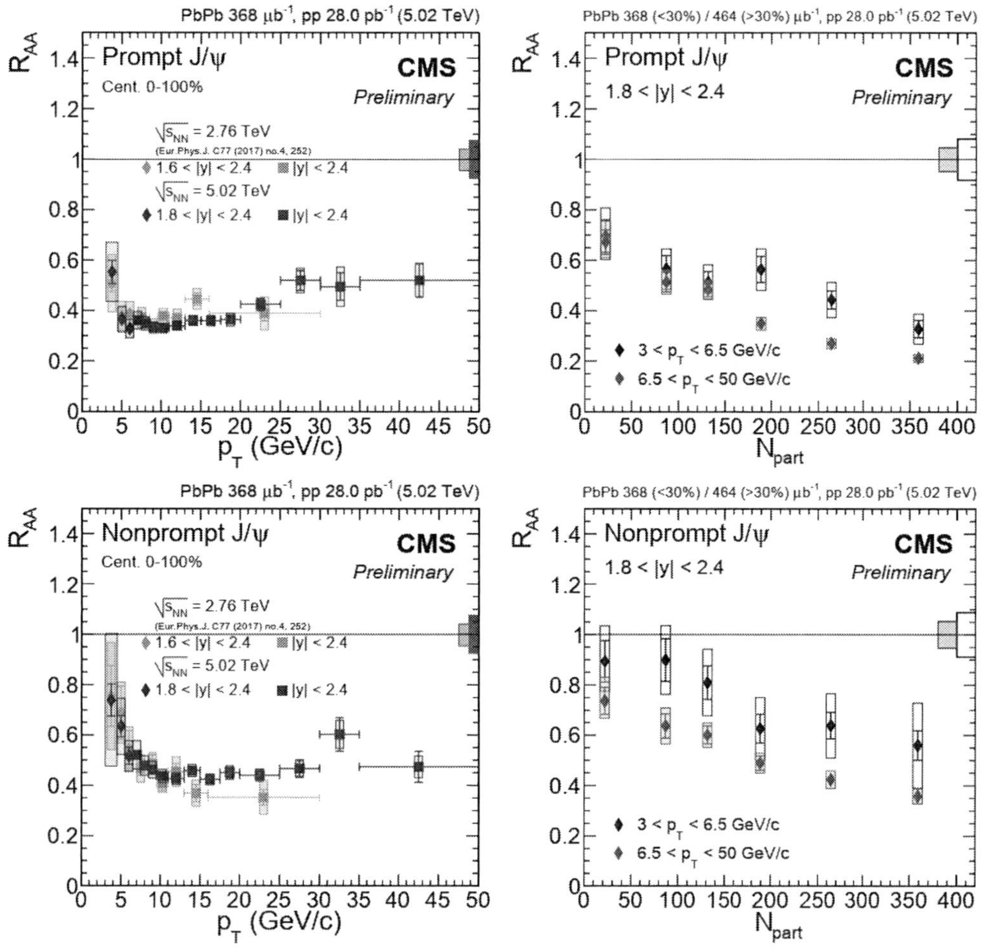

Figure 1. Nuclear modification factor of prompt (upper) and nonprompt (lower) J/ψ mesons as a function of dimuon p_T (left) and centrality (right) [9]. The bars (boxes) represent statistical (systematic) point-by-point uncertainties. The boxes plotted at $R_{AA} = 1$ indicate the size of the global relative uncertainties.

4 Summary

The results of the prompt and nonprompt J/ψ meson R_{AA} have been presented in PbPb collisions at $\sqrt{s_{NN}} = 5.02$ TeV, as a function of transverse momentum (p_T), and collision centrality in different kinematic ranges. The results show a strong centrality dependence, with an increasing suppression towards central events. A smaller suppression at the lowest p_T ranges is observed for both prompt and nonprompt J/ψ mesons. The measurements are consistent with previous results at 2.76 TeV. Also a smaller suppression at high p_T is observed for prompt J/ψ mesons. The prompt $\psi(2S)$ meson R_{AA} has also been measured in PbPb collisions at 5.02 TeV, as a function of p_T and collision centrality. The results show that the $\psi(2S)$ is more suppressed than the J/ψ meson for all the kinematical ranges studied. No p_T dependence is observed within the current uncertainties. Hints of an increasing suppression with collision centrality are also observed.

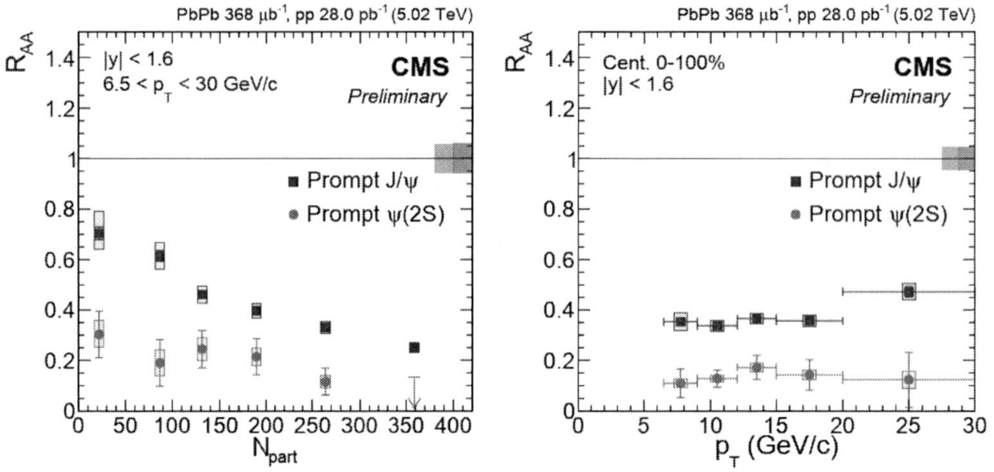

Figure 2. Nuclear modification factor of prompt J/ψ and ψ(2S) mesons as a function of centrality (left) and dimuon p_T (right), at central rapidity [9]. The vertical arrows represent 95% confidence intervals in the bins where the double ratio measurement is consistent with 0 (see text). The bars (boxes) represent statistical (systematic) point-by-point uncertainties. The boxes plotted at $R_{AA} = 1$ indicate the size of the global relative uncertainties.

Finally the prompt and nonprompt J/ψ R_{pPb} results in pPb collisions at 5.02 TeV show small deviations from unity, indicating that the effects observed in PbPb collisions cannot be explained by the presence of CNM effects alone. The prompt ψ(2S) R_{pPb} results show a bigger suppression than for J/ψ mesons, suggesting effects beyond CNM affecting excited states.

References

[1] T. Matsui, H. Satz, Phys. Lett. B178 (1986) 416.
[2] P. Braun-Munzinger, J. Stachel, Phys. Lett. B490 (2000) 196.
[3] R. L. Thews, M. Schroedter, J. Rafelski, Phys. Rev. C63 (2001) 054905.
[4] ALICE Collaboration, Phys. Rev. Lett. 109 (2012) 072301.
[5] R. Vogt et al., Int.J.Mod.Phys. E22 (2013) 1330007.
[6] F. Arleo, S. Peigné, JHEP 1303 (2013) 122.
[7] M. Spousta, Phys. Lett. B767 (2017) 10.
[8] F. Arleo, Phys. Rev. Lett. 119 (2017) 062302.
[9] CMS Collaboration, CMS PAS HIN-16-025 (2017), CDS Record 2274286.
[10] CMS Collaboration, Eur. Phys. J C77. (2017) 269.
[11] CMS Collaboration, CMS-PAS-HIN-16-015 (2017), CDS Record 2244643.
[12] CMS Collaboration, JINST 3 (2008) S08004.
[13] CMS Collaboration, JHEP 04 (2017) 039.
[14] CMS Collaboration, Eur. Phys. J. C77 (2017) 252.
[15] CMS Collaboration, Phys. Rev. Lett. 118 (2017) 162301.

Production cross-sections of prompt Λ_c^+ in pPb collisions at $\sqrt{s_{\rm NN}} = 5$ TeV with the LHCb detector

Jiayin Sun on behalf of the LHCb collaboration[1,*]

[1] *Center for High Energy Physics, Tsinghua University, Beijing, China*

Abstract. The study of the nuclear modification factor and the forward/backward asymmetry in pPb collisions at $\sqrt{s_{\rm NN}} = 5$ TeV is extended to the Λ_c^+ baryon, providing the first measurement of charmed baryon production in pA collisions. The result is compared to the analogous measurement on charmed mesons, providing an insight on the production mechanism of charmed hadrons.

1 Introduction

Heavy quarks are sensitive probes to explore the properties of the Quark-Gluon-Plasma (QGP), a state of matter consisting of deconfined quarks and gluons. They are created in pairs in the early stage of the space-time evolution of heavy-ion collisions. Multiple experimental measurements of D^0 meson production in heavy-ion collisions at RHIC [1] and the LHC [2] indicate strong interactions between the c quark and the medium. To fully intepret these results, one needs to first disentangle quantitatively the cold nuclear matter (CNM) effects, which can be studied in proton-nucleus collisions, from hot medium effects. Measurement of prompt Λ_c^+ production in pPb collisions extends the study of CNM effects to charmed baryons. Furthermore, the baryon-to-meson production ratio for charmed hadrons can be obtained, with possible relevance to the hadronisation mechanism at the charm sector [3, 4]. Measurements of baryon-to-meson ratio for light and strange hadrons have shown a significant baryon enhancement at intermediate transverse momentum ($p_{\rm T}$) in central heavy-ion collisions [5, 6]. Recently, the STAR experiment has reported a significant Λ_c^+ enhancement above theoretical expectations in semi-central AuAu collisions at 200 GeV [7]. The measurement of Λ_c^+ in pPb will provide an important baseline to fully understand the implication of this observation.

2 LHCb detector and heavy-ion data taking

The LHCb detector[8, 9] is a single-arm spectrometer covering the forward pseudorapidity range $2 < \eta < 5$. It is designed for precision measurements of heavy flavour hadrons with unique capabilities, which include hadron reconstruction down to very low $p_{\rm T}$, precise decay vertex reconstruction to distinguish prompt and secondary decay particles, and excellent particle identification based on the RICH detectors.

*e-mail: jiayin.sun@cern.ch

The experiment took its first dataset of heavy-ion collisions in 2013, with pPb collisions at $\sqrt{s_{\mathrm{NN}}} = 5$ TeV. It successfully collected data of $3 - 5\mu\mathrm{b}^{-1}$ PbPb collisions at $\sqrt{s_{\mathrm{NN}}} = 5$ TeV in 2015, and $34\,\mathrm{nb}^{-1}$ pPb collisions at $\sqrt{s_{\mathrm{NN}}} = 8.16$ TeV in 2016. This analysis uses the first data sample of pPb collisions at $\sqrt{s_{\mathrm{NN}}} = 5$ TeV, with $E_p = 4$ TeV and $E_{\mathrm{Pb}} = 1.58$ TeV per nucleon, leading to a rapidity shift of $\Delta y = 0.4645$ in the nucleon-nucleon centre-of-mass system with respect to the laboratory frame. Since the LHCb detector covers only one direction of the full rapidity acceptance, two distinctive beam configurations were used. In the 'forward' ('backward') configuration, the proton (lead) beam enters the LHCb detector from the interaction point. The acceptance is $1.5 < y^* < 4.0$ for the forward configuration and $2.5 < y^* < 5.0$ for the backward configuration, where y^* denotes the rapidity in the nucleon-nucleon centre-of-mass system. The integrated luminosity is $1.06 \pm 0.02\,\mathrm{nb}^{-1}$ and $0.52 \pm 0.01\,\mathrm{nb}^{-1}$ for the forward and backward collisions, respectively.

3 Prompt Λ_c^+ production in pPb collisions at $\sqrt{s_{\mathrm{NN}}} = 5$ TeV

The Λ_c^+ baryons are reconstructed through the $\Lambda_c^+ \to pK^-\pi^+$ decay channel. Tight particle identification requirements based on the RICH detectors are applied to suppress the combinatorial background. The impact parameter with respect to the collision vertex is used to separate the prompt Λ_c^+ and non-prompt Λ_c^+ originating from b-hadron decays. The number of prompt Λ_c^+ signals are determined from fits to the $m(pK^-\pi^+)$ invariant mass distribution and the Λ_c^+ impact parameter distribution. A full detector simulation based on Geant4 [10, 11] is used to calculate the reconstruction and selection efficiency, which is corrected to account for the effects of detector occupancy. The particle identification efficiency is estimated from a calibration data sample. The differential production cross-sections of Λ_c^+ baryons are measured as a function of the Λ_c^+ transverse momentum and rapidity in the kinematic range $2 < p_{\mathrm{T}} < 10\,\mathrm{GeV}/c$ and $1.5 < y^* < 4.0$ for the forward sample and $-4.5 < y^* < -2.5$ for the backward sample. Fig. 1 shows the corresponding one-dimensional differential Λ_c^+ cross-section as a function of p_{T} and y^*.

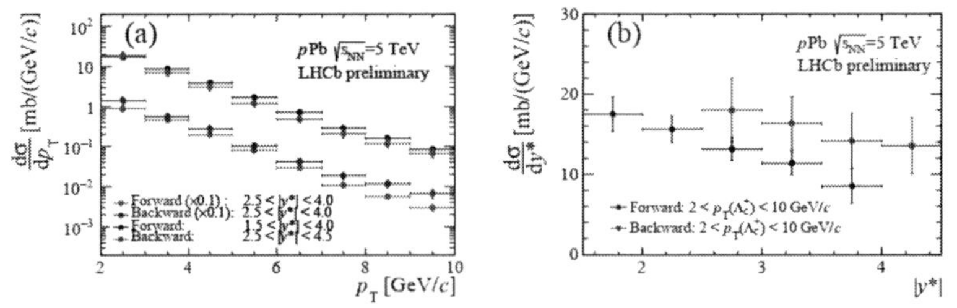

Figure 1. Differential cross-section of prompt Λ_c^+ baryons in pPb collisions as a function of (a) p_{T} and (b) y^* in the forward and backward samples. The error bars represent the quadratic sum of the statistical and the systematic uncertainties.

The forward-backward ratio R_{FB} measures the Λ_c^+ production asymmetry in the forward and backward rapidity regions. It is defined as the ratio of the cross-sections in the two rapidity regions, $R_{\mathrm{FB}} \equiv \sigma(+|y^*|, p_{\mathrm{T}})/\sigma(-|y^*|, p_{\mathrm{T}})$, where $\sigma(+|y^*|, p_{\mathrm{T}})$ and $\sigma(-|y^*|, p_{\mathrm{T}})$ correspond to the cross-sections of the forward and backward rapidity regions, respectively. The left and right panels of Fig. 2 display the prompt Λ_c^+ R_{FB} ratio as a function of p_{T} in the common rapidity region $2.5 < |y^*| < 4.0$, and the R_{FB} ratio as a function of y^* in the region $2 < p_{\mathrm{T}} < 10\,\mathrm{GeV}/c$, respectively. The points are compared to calculations [12] by HELAC-Onia [13, 14]. Constrained by measured prompt Λ_c^+ cross section in pp collisions, the calculations incorporate the parton distribution functions of EPS09LO, EPS09NLO [15] and nCTEQ15 [16]. Good consistency between the data and the calculations is observed.

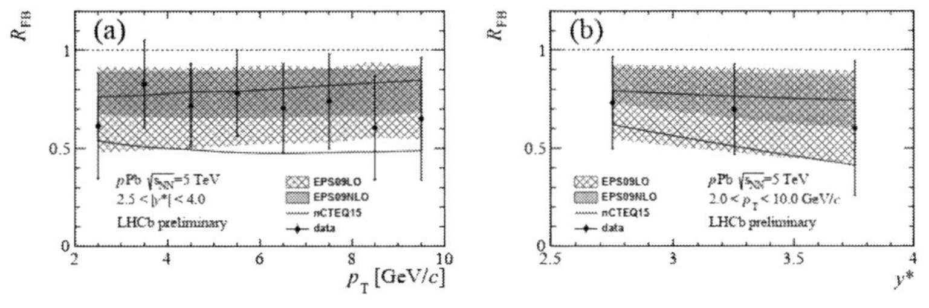

Figure 2. Forward-backward production ratios R_{FB} as function of (a) p_T integrated over $2.5 < y^* < 4.0$ and (b) y^* integrated over $2 < p_T < 10\,\text{GeV}/c$.

The LHCb measurement of D^0 cross-sections in pPb collisions at 5 TeV [17] allows the calculation of the baryon-to-meson production ratio, $R_{\Lambda_c^+/D^0} \equiv \sigma(\Lambda_c^+)/\sigma(D^0)$. Fig. 3 and Fig. 4 show the p_T and y^* dependence of the measured $R_{\Lambda_c^+/D^0}$ ratio. The coloured curves illustrate the HELAC-Onia calculations [12–14]. The data points are in general consistent with the calculations within uncertainty. At higher p_T the data fall below the calculations at forward rapidity. Compared to the relatively flat theoretical curves, the $R_{\Lambda_c^+/D^0}$ ratios tend to increase with increasing $|y^*|$. The trend difference is more pronounced in the backward data. The result is generally below the ALICE measurement of $R_{\Lambda_c^+/D^0}$ ratio in midrapidity in pPb collisions at 5 TeV [18].

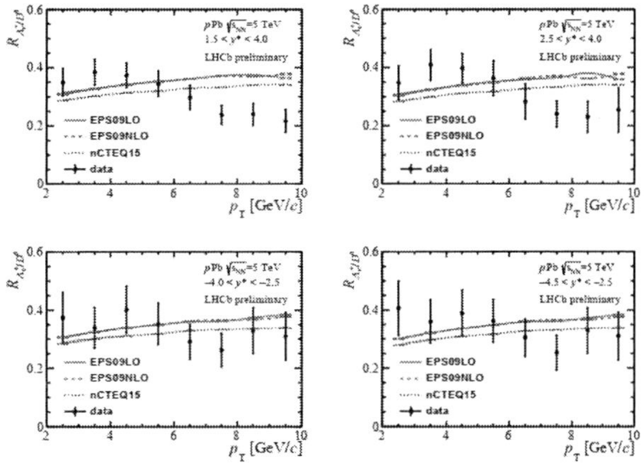

Figure 3. The baryon-to-meson production ratio $R_{\Lambda_c^+/D^0}$ as a function of p_T integrated over four different rapidity regions.

4 Conclusion and prospects

In this proceeding, preliminary results of prompt Λ_c^+ production cross-sections are obtained with pPb collision data at $\sqrt{s_{NN}} = 5$ TeV collected by the LHCb experiment. The prompt Λ_c^+ forward-

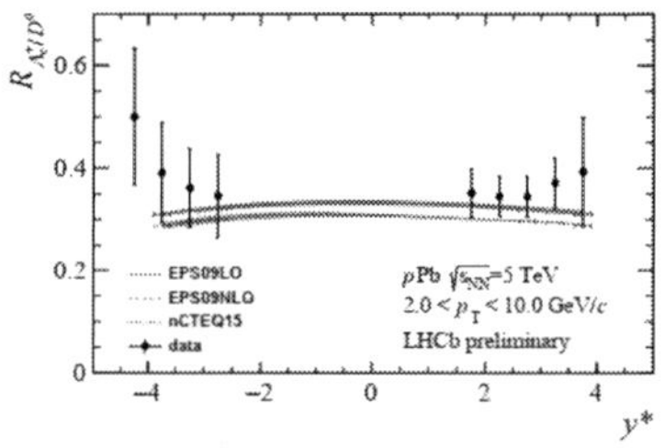

Figure 4. The baryon-to-meson production ratio $R_{\Lambda_c^+/D^0}$ as a function of y^* integrated over $2 < p_T < 10\,\text{GeV}/c$.

backward production ratio R_{FB} and the production ratio $R_{\Lambda_c^+/D^0}$ between Λ_c^+ baryons and D^0 mesons are presented. The results show reasonable agreement with theoretical calculations incorporating nuclear parton distribution functions within uncertainties. The results demonstrate the LHCb experiment's ability to make a significant contribution to precision measurements of open heavy flavour in heavy-ion collisions. With the significantly larger pPb dataset at $\sqrt{s_{NN}}$ = 8.16 TeV, more heavy flavour measurements with improved precision in the forward rapidity are expected in the near future to provide information on the cold nuclear matter effects.

References

[1] L. Adamczyk et al. (STAR collaboration), Phys. Rev. Lett. **113**, 142301 (2014), 1404.6185
[2] B.B. Abelev et al. (ALICE collaboration), Phys. Rev. **C90**, 034904 (2014), 1405.2001
[3] Y. Oh, C.M. Ko, S.H. Lee, S. Yasui, Phys. Rev. **C79**, 044905 (2009), 0901.1382
[4] S.H. Lee, K. Ohnishi, S. Yasui, I.K. Yoo, C.M. Ko, Phys. Rev. Lett. **100**, 222301 (2008), 0709.3637
[5] M.A.C. Lamont (STAR), J. Phys. Conf. Ser. **50**, 192 (2006), [J. Phys.G32,S105(2006)], nucl-ex/0608017
[6] B.B. Abelev et al. (ALICE), Phys. Rev. Lett. **111**, 222301 (2013), 1307.5530
[7] G. Xie (STAR), Nucl. Phys. **A967**, 928 (2017), 1704.04353
[8] A.A. Alves Jr. et al. (LHCb collaboration), JINST **3**, S08005 (2008)
[9] R. Aaij et al. (LHCb collaboration), Int. J. Mod. Phys. **A30**, 1530022 (2015), 1412.6352
[10] S. Agostinelli et al. (Geant4 collaboration), Nucl. Instrum. Meth. **A506**, 250 (2003)
[11] J. Allison, K. Amako, J. Apostolakis, H. Araujo, P. Dubois et al. (Geant4 collaboration), IEEE Trans.Nucl.Sci. **53**, 270 (2006)
[12] J.P. Lansberg, H.S. Shao, Eur. Phys. J. **C77**, 1 (2017), 1610.05382
[13] H.S. Shao, Comput. Phys. Commun. **184**, 2562 (2013), 1212.5293
[14] H.S. Shao, Comput. Phys. Commun. **198**, 238 (2016), 1507.03435
[15] K.J. Eskola, H. Paukkunen, C.A. Salgado, JHEP **04**, 065 (2009), 0902.4154
[16] K. Kovarik et al., Phys. Rev. **D93**, 085037 (2016), 1509.00792
[17] R. Aaij et al. (LHCb collaboration) (2017), submitted to JHEP, 1707.02750
[18] J. Norman (ALICE collaboration), contribution to SQM2017

https://doi.org/10.1051/epjconf/201817118011

Low-mass dielectrons in pp, p–Pb and Pb–Pb collisions measured by the ALICE Experiment

Ivan Vorobyev[1,2,*] on behalf of the ALICE collaboration

[1] *Physik Department, Technische Universität München, Munich, Germany*
[2] *Excellence Cluster Universe, Technische Universität München, Munich, Germany*

Abstract. Dielectrons produced in ultra-relativistic heavy-ion collisions at the LHC provide a unique probe of the system evolution as they are unperturbed by final-state interactions. The dielectron continuum is extremely rich in physics sources: on top of ordinary Dalitz and resonance decays of pseudoscalar and vector mesons, thermal black-body radiation is of particular interest as it carries information about the temperature of the hot and dense system created in such collisions. The dielectron invariant-mass distribution is furthermore sensitive to medium modifications of the spectral function of short-lived vector mesons that are linked to the potential restoration of chiral symmetry at high temperatures. Correlated electron pairs from semi-leptonic charm and beauty decays provide complementary information about the heavy-quark energy loss.

1 Introduction

In ultra-relativistic heavy-ion collisions, hadronic matter is believed to form a new state of matter comprising deconfined quarks and gluons—the quark-gluon plasma (QGP). Electron-positron pairs are produced during all stages of such collisions and carry information to the detector unperturbed by strong final-state interactions, thus allowing us to study the whole space-time evolution of the system. In the mass region below 1 GeV/c^2 the dielectron spectrum is dominated by Dalitz and resonance decays of pseudo-scalar and vector mesons, whereas e^+e^- pairs in the mass region between 1 and 3 GeV/c^2 originate mainly from the semi-leptonic decays of correlated charm and beauty hadrons. In heavy-ion collisions thermal radiation from QGP and hadron gas contribute to the dielectron spectrum over a broad mass range. The measurements of dielectron production in minimum bias pp collisions serve as an important vacuum reference to quantify any observed modifications in heavy-ion collisions, and the studies in p–Pb collisions are used to investigate cold nuclear matter effects.

We present a summary of the results from the ALICE experiment [1] in all three collisions systems at the LHC: pp, p–Pb and Pb–Pb, including the latest analysis of Run-2 pp collisions at $\sqrt{s}$ = 13 TeV collected with a trigger on high charged-particle multiplicities. Event triggering is based on the information from the Silicon Pixel Detector (SPD) and/or from the V0 scintillators, the latter are also used for centrality estimation in Pb–Pb collisions. Charged particle tracks are reconstructed with the help of central barrel detectors ($|\eta| < 0.8$): the Inner Tracking System (ITS) and the Time Projection Chamber (TPC). Specific energy loss in TPC and in ITS and time-of-flight information from the TOF detector are used for electron identification.

*e-mail: ivan.vorobyev@cern.ch

2 Results in pp collisions at $\sqrt{s}$ = 7 TeV and at $\sqrt{s}$ = 13 TeV

The dielectron invariant mass spectrum has been measured in pp collisions at $\sqrt{s} = 7$ TeV and is shown in Fig. 1 (left). The results are found to be consistent within uncertainties with the expectation from the known hadronic sources (cocktail calculations). The contribution from the semi-leptonic decays of correlated D and B mesons has been estimated with the help of PYTHIA 6 simulations and scaled to the measured total charm and beauty cross sections [2]. The dielectron production has been also measured as a function of the transverse pair distance of closest approach to the primary vertex (DCA_{ee}), defined as the quadratic sum of single track DCA divided by the estimated DCA resolution. The DCA_{ee} distribution is shown in Fig. 1 (right) for the mass range of $0.2 < m_{ee} < 1.1$ GeV/c^2. The results are compared to the DCA_{ee} templates obtained from the Monte Carlo simulations of the ALICE detector, and the relative contributions of different sources are normalised according to the hadronic cocktail calculations. The addition of the heavy-flavour contribution is necessary for proper description of the tail of the distribution, so the DCA_{ee} variable allows one to separate prompt and non-prompt dielectron sources.

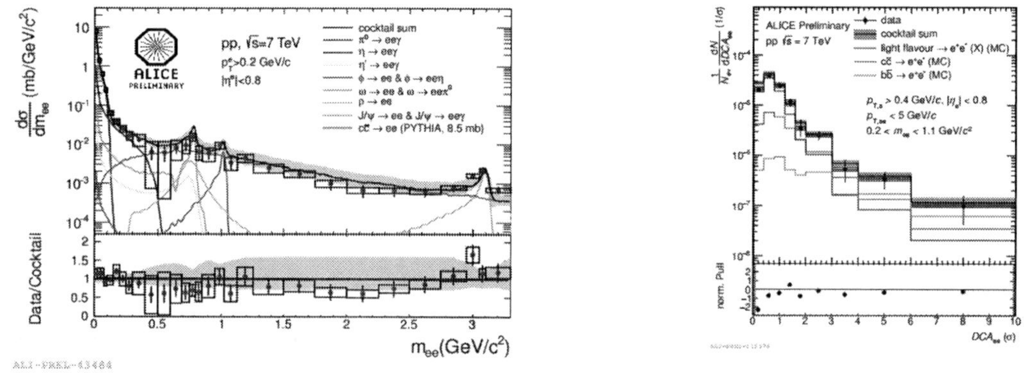

Figure 1. Dielectron invariant mass spectrum in pp collisions at $\sqrt{s} = 7$ TeV compared to the expectation from hadronic cocktail calculations (left), and the DCA_{ee} spectrum in the mass range $0.2 < m_{ee} < 1.1$ GeV/c^2 (right)

The first dielectron studies of pp events with high charged-particle multiplicity have been performed in pp collisions at $\sqrt{s} = 13$ TeV collected in 2016. The goal is to reveal new or heavy-ion like phenomena, e.g. production of ρ meson versus multiplicity or presence of thermal radiation, by comparing the dielectron yield in events with high multiplicity (HM) with average minimum bias (MB) pp collisions (Fig. 2 left). Deviations from multiplicity scaling can be caused by modifications of the charged hadron p_T spectra [3] and by the unknown hadron chemistry in high multiplicity events. Furthermore, the production of open charm mesons is known to increase faster than linearly with multiplicity for $p_T > 1$ GeV/c [4], so the dielectrons from correlated semileptonic charm and bottom decays offer here a unique window at the production of low-p_T D and B mesons. To this end, the ratio of dielectron spectra in high multiplicity over minimum bias events is calculated and is normalised by the charged-particle multiplicity ratio of $\langle N_{ch}^{acc}(\mathrm{HM})\rangle / \langle \mathrm{N}_{\mathrm{ch}}^{\mathrm{acc}}(\mathrm{MB})\rangle = 4.36$. The results are found to be in agreement with the cocktail expectations (Fig. 2 right) which takes into account the modification of the charged hadron p_T spectra and the faster-than-linear production of D mesons. About 5 times more data from 2016 are being analysed in order to investigate the spectrum modification in more details.

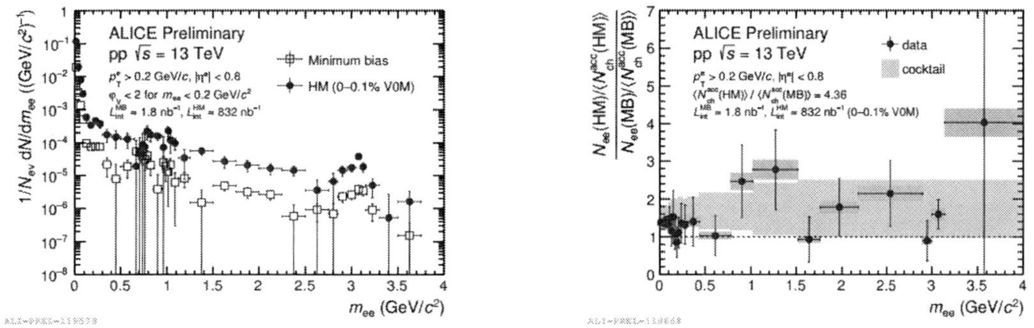

Figure 2. Raw dielectron spectra normalised by number of triggered events (left), and the ratio of dielectron spectra scaled by charged particle multiplicity (right) in pp collisions at $\sqrt{s}$ = 13 TeV

3 Results in p–Pb collisions at $\sqrt{s_{NN}}$ = 5.02 TeV and in Pb–Pb collisions at $\sqrt{s_{NN}}$ = 2.76 TeV

The dielectron production in p–Pb collisions at $\sqrt{s_{NN}}$ = 5.02 TeV has been measured as a function of invariant mass and pair transverse momentum (Fig. 3). The total heavy flavour cross-section (extrapolated from measurements in pp collisions at different energies) has been normalised by the average number of binary nucleon-nucleon interactions in p–Pb collisions. The data are found to be in agreement with the cocktail calculations, with a hint for possible lower charm production, which could be caused by cold nuclear matter effects. A larger p–Pb sample was recorded in 2016 and will help to perform detailed differential analysis as a function of invariant mass, pair p_T and DCA_{ee} to conclude on possible cold nuclear matter effects in the future.

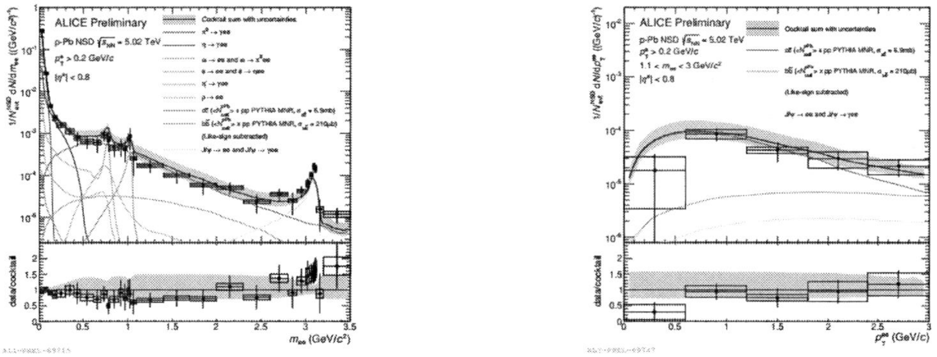

Figure 3. Dielectron spectrum as a function of invariant mass (left) and pair transverse momentum (in the mass region 1.1 < m_{ee} < 3 GeV/c^2, right) in p–Pb collisions at $\sqrt{s_{NN}}$ = 5.02 TeV

In Pb–Pb collisions at $\sqrt{s_{NN}}$ = 2.76 TeV the dielectron production has been measured in the 10% most central collisions as a function of invariant mass (Fig. 4 left). The cocktail calculations are shown for two cases: without the contribution from vacuum ρ meson and with contributions from QGP and hadron gas radiation including modified ρ and ω spectral functions [5, 6]. The data points are consistent with both results within uncertainties, so the current measurement is not yet sensitive to the potential contribution from thermal radiation and in-medium modifications. In the quasi-real photon region ($m_{ee} < 0.3$ GeV/c^2 and ${p_T}^{ee} > 1$ GeV/c) the extraction of the virtual photon fraction (Fig. 4 right) has shown an agreement with the results on real direct photon measurements [7]. Substantial improvements of dielectron measurements are expected following the upcoming ALICE detector upgrades [8], which will allow higher data-taking rates and more effective background rejection.

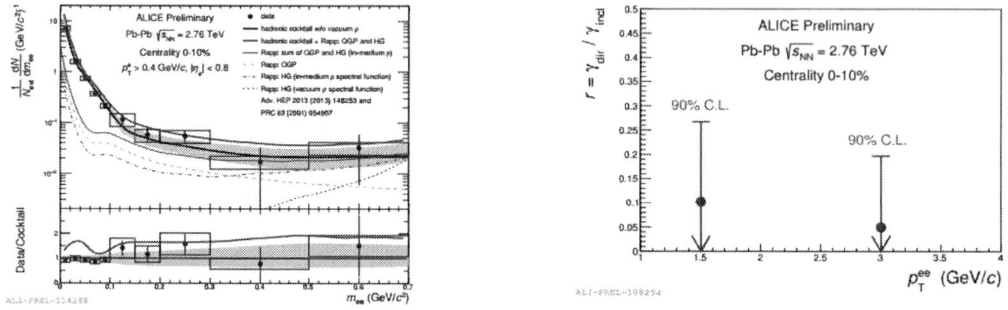

Figure 4. The dielectron mass spectrum compared to the cocktail calculations in central Pb–Pb collisions at $\sqrt{s_{NN}} = 2.76$ TeV (left), and the fraction of direct over inclusive photons (right)

4 Conclusion and outlook

The results of the low-mass dielectron analysis performed with the ALICE experiment are presented for different collision systems. In pp collisions at $\sqrt{s} = 7$ TeV the dielectron production is described with the cocktail of known hadronic sources as a function of invariant mass and pair DCA_{ee}. The first analysis of dielectron production in high-multiplicity pp collisions at $\sqrt{s} = 13$ TeV has shown an agreement with the expectations from the modification of the p_T spectrum of charged hadrons and production of D mesons versus multiplicity. The results in p–Pb collisions at $\sqrt{s_{NN}} = 5.02$ TeV are compatible with the hadronic cocktail expectations within uncertainties. In central Pb–Pb collisions at $\sqrt{s_{NN}} = 2.76$ TeV the measurement is not yet sensitive to the possible contribution from thermal radiation, and the results on virtual direct photons are in agreement with the real photon measurements.

Finally, novel multi-variative analysis techniques (MVA) have been developed on the Monte Carlo data in order to improve the electron identification and to suppress dielectron pairs from conversions of real photons. Usage of such methods is foreseen in future dielectron analyses.

References

[1] The ALICE Collaboration, JINST **3**, S08002 (2008)
[2] The ALICE Collaboration, arXiv:1702.00766 [hep-ex]
[3] The ALICE Collaboration, Phys. Lett. B **753**, 319 (2016)
[4] The ALICE Collaboration, JHEP **09**, 148 (2015)
[5] R. Rapp, Adv. High Energy Phys. **2013**, 148253 (2013)
[6] R. Rapp, Phys. Rev. C **63**, 054907 (2001)
[7] The ALICE Collaboration, Phys. Lett. B **754**, 235-248 (2016)
[8] The ALICE Collaboration, J. Phys. G **41**, 087002 (2014)

EPJ Web of Conferences **171**, 18012 (2018) https://doi.org/10.1051/epjconf/201817118012
SQM 2017

Charmed meson and baryon measurements in pp and p–Pb collisions with ALICE at the LHC

Jaime Norman[1,*] for the ALICE collaboration

[1] *The University of Liverpool, Liverpool L69 7ZE, United Kingdom*

Abstract. We present here recent open heavy-flavour results from the ALICE experiment, including measurements of D-meson, Λ_c baryon and Ξ_c^0 baryon production in pp collisions at $\sqrt{s}$ = 7 TeV and p–Pb collisions at $\sqrt{s_{NN}}$ = 5.02 TeV.

1 Introduction

The measurement of charm production in pp collisions is an important test of perturbative QCD, and in p–Pb collisions the study of charm production can help disentangle cold nuclear matter effects from the effects modifying the p_T spectrum of charmed hadrons in Pb–Pb collisions due to the high-temperature and high energy-density medium formed. The relative abundance of baryons and mesons can shed light on the process of fragmentation - a non-perturbative process - and deviations from measurements made at e^+e^- colliders may hint at specific processes occuring in the higher partonic density environment of pp and p–Pb collisions. In addition, these measurements also provide a reference for future measurements in nucleus-nucleus collisions, where the baryon-to-meson ratio is expected to be sensitive to modified hadronisation mechanisms such as coalescence [1]. Preliminary results from STAR [2] hint at an enhanced Λ_c^+/D^0 ratio in Au–Au collisions.

2 Heavy-flavour reconstruction at ALICE

Heavy-flavour decays are reconstructed at ALICE in the central barrel, which consists of the Inner Tracking System (ITS), the Time Projection Chamber (TPC) and the Time-of-Flight detector (TOF), covering the entire azimuthal range and designed to track and identify charged particles over a wide momentum range. Charmed hadrons are reconstructed at mid-rapidity ($|y| < 0.8$) via their hadronic decays including $D^0 \to K^-\pi^+$, $D^+ \to K^-\pi^+\pi^+$, $D^{*+} \to D^0\pi^+$, $D_s^+ \to \phi\pi^+$, $\Lambda_c^+ \to pK^-\pi^+$ and $\Lambda_c^+ \to pK_S^0$. Selection is made on the hadron decay topology, the signal is extracted via an invariant mass analysis, and corrections are made for the efficiency, acceptance, and the fraction of non-prompt hadrons in the signal sample. The semileptonic decays of Λ_c^+ and Ξ_c^0 baryons are also reconstructed via the decay channels $\Lambda_c^+ \to e^+\nu_e\Lambda$ and $(\Xi_b^0 \to)\Xi_c^0 \to e^+\nu_e\Xi^-$. Here the analyses are not based on topological selections, and are instead based on subtracting the 'wrong-sign' $e^-\Lambda(e^-\Xi^-)$ pair spectra from the 'right-sign' $e^+\Lambda(e^+\Xi^-)$ spectra. Additional corrections include correcting for contributions to the wrong-sign spectra from $\Lambda_b(\Xi_b^0)$, and unfolding the $e^-\Lambda(e^-\Xi^-)$ p_T spectra to obtain the $\Lambda_c^+(\Xi_c^0)$ p_T spectrum. A correction for feed-down from $\Xi_c^{+,0}$ is also included for the Λ_c^+ measurement.

*e-mail: jaime.norman@cern.ch

3 Results

The nuclear modification factor (R_{pPb}) of D^0, D^+, D^{*+} and D_s^+ mesons was measured in p–Pb collisions at $\sqrt{s_{\mathrm{NN}}} = 5.02$ TeV [3]. Figure 1 (left) shows the p_{T}-differential averaged D^0, D^+ and D^{*+} R_{pPb} in comparison with models that include cold nuclear matter effects and models that assume a Quark-Gluon Plasma is formed and include hydrodynamical effects [3]. The statistical precision of the measurements has been improved by approximately a factor of 2 with respect to the previous measurement [4] due to an increased integrated luminosity. The models describe the data well, although a suppression larger than 15-20% for $p_{\mathrm{T}} > 5$ GeV/c, expected from the POWLANG(HTL) and Duke models, is slightly disfavoured by the data. D-meson production has also been measured as a function of centrality in p–Pb collisions at $\sqrt{s_{\mathrm{NN}}} = 5.02$ TeV [3]. Figure 1 (right) shows the D^0 Q_{CP} calculated as the ratio of the D^0-meson nuclear modification factor in central (0-10%) and peripheral (60-100%) centrality intervals. The Q_{CP} tends to increases in the interval $1 < p_{\mathrm{T}} < 4$ GeV/c and reaches about 1.25, and then decreases in the interval $7 < p_{\mathrm{T}} < 24$ GeV/c. The average value of the D^0 Q_{CP} is larger than unity in the interval $3 < p_{\mathrm{T}} < 8$ GeV/c by 1.7 standard deviations of the statistical and systematic uncertainty.

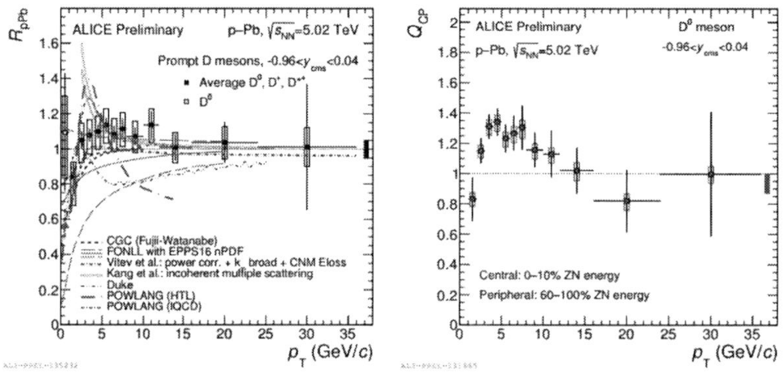

Figure 1. Left: averaged D^0, D^+ and D^{*+} R_{pPb} in p–Pb collisons at $\sqrt{s_{\mathrm{NN}}} = 5.02$ TeV compared to model calculations. Right: prompt D^0 Q_{CP} in p–Pb collisions at $\sqrt{s_{\mathrm{NN}}} = 5.02$ TeV.

The p_{T}-differential cross section of the Λ_c^+ baryon was measured in pp collisions at $\sqrt{s} = 7$ TeV and in p–Pb collisions at $\sqrt{s_{\mathrm{NN}}} = 5.02$ TeV, and is reported in figure 2. The cross section is compared to perturbative calculations at NLO using the GM-VFNS [5, 6] scheme in pp collisions, and at NLO with POWHEG [7] matched with PYTHIA 6.4.25 [8] in pp and p–Pb collisions. For the p–Pb POWHEG the EPS09 nuclear PDF was used [9]. GM-VFNS underpredicts the data by a factor of 2.5 on average, and POWHEG underpredicts the data by a factor of 18(4) at low(high) p_{T} in pp collisions, and by a similar amount in p–Pb collisions. Figure 3 (left) shows the baryon-to-meson ratio Λ_c^+/D^0 measured using the Λ_c^+ cross sections presented in this paper and the D^0 cross sections measured by ALICE [4, 10]. In the same figure theoretical predictions in pp collisions are shown including PYTHIA8 with and without a tune including enhanced colour reconnection [11], DIPSY [12] with rope hadronisation and HERWIG7 [13] with hadronisation via clusters. The Λ_c^+/D^0 ratio in pp collisions is compatible with the same ratio in p–Pb collisions within uncertainties. While all models underpredict the data, PYTHIA8 with enhanced colour reconnection brings the prediction closer to data. Figure 3 (right) shows the nuclear modification factor R_{pPb} for the Λ_c^+ baryon in p–Pb collisions at $\sqrt{s_{\mathrm{NN}}} = 5.02$ TeV, in comparison to the averaged D-meson (D^0, D^+, D^{*+}) R_{pPb} [4], and predictions including POWHEG+PYTHIA with EPS09 nuclear PDF [9] and a prediction for charmed hadrons which assumes a small-size QGP is formed [14]. The R_{pPb} of Λ_c^+ is consistent with unity and with the D-meson R_{pPb}, and does not allow to distinguish between the models presented within the current experimental uncertainties.

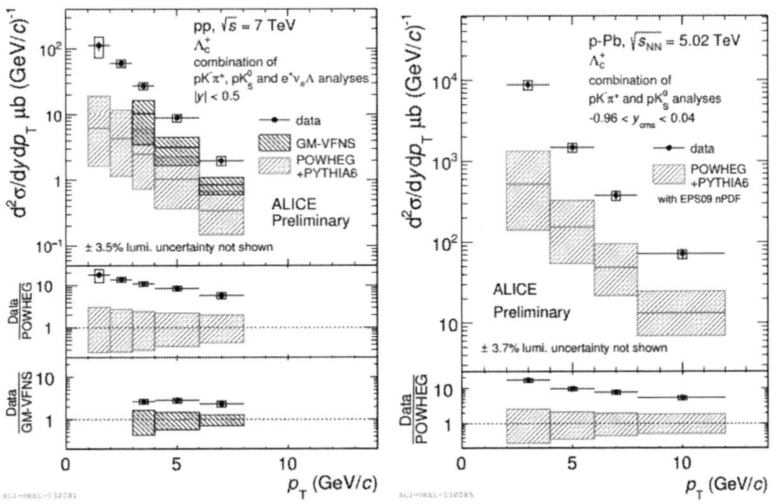

Figure 2. Left: the p_T-differential cross section of prompt Λ_c^+ in pp collisions at $\sqrt{s}$ = 7 TeV. Right: the p_T-differential cross section of prompt Λ_c^+ in p–Pb collisions at $\sqrt{s_{NN}}$ = 5.02 TeV. Both measurements are compared with theoretical predictions.

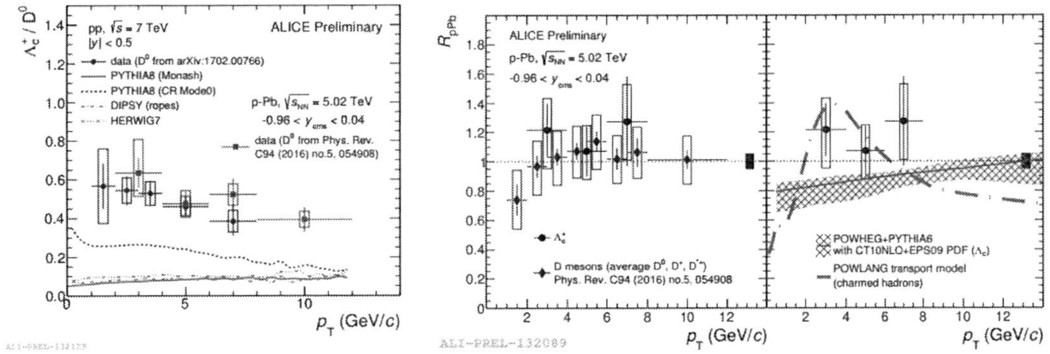

Figure 3. Left: the Λ_c^+/D^0 ratio in pp collisions at $\sqrt{s}$ = 7 TeV, and in p–Pb collisions at $\sqrt{s_{NN}}$ = 5.02 TeV, compared to model predictions. Right: the nuclear modification factor R_{pPb} of Λ_c^+ compared to the R_{pPb} of D mesons, and compared with model predictions.

Figure 4 shows the p_T-differential cross section times branching ratio of the Ξ_c^0 baryon (including prompt and non-prompt contributions), and the baryon-to-meson ratio $\Xi_c^0 \to e^+\nu_e\Xi^-/D^0$ in comparison to predictions from PYTHIA with the aforementioned tunes. The shaded band for the models spans the range of theoretical predictions for the $\Xi_c^0 \to e^+\nu_e\Xi^-$ branching ratio [15–17]. As for the Λ_c^+/D^0 ratio, all predictions significantly underestimate the data.

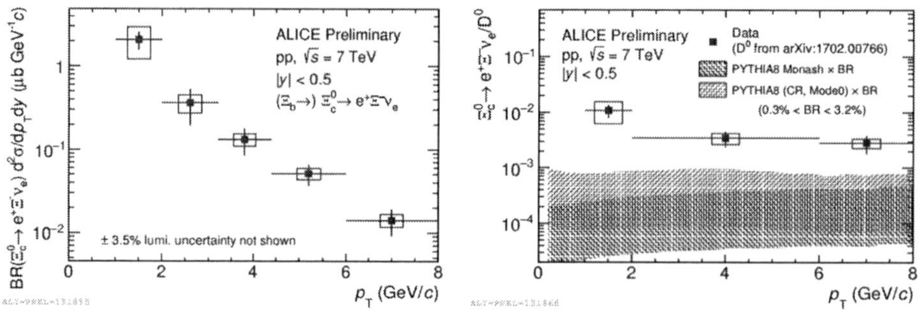

Figure 4. Left: The p_{T}-differential cross section times branching ratio of the Ξ_{c}^{0} baryon in pp collisions at $\sqrt{s}$ = 7 TeV. Right: The $\Xi_{\mathrm{c}}^{0} \rightarrow \mathrm{e}^{+}\nu_{\mathrm{e}}\Xi^{-}/\mathrm{D}^{0}$ ratio compared to theoretical predictions.

4 Summary

Recent charmed meson and baryon measurements by the ALICE collaboration have been presented. The R_{pPb} of D mesons is found to be consistent with unity. The Q_{CP} of D^0 mesons shows a hint of D^0 enhancement at $3 < p_{\mathrm{T}} < 8$ GeV/c in central p–Pb collisions. The cross section of the Λ_{c}^{+} baryon and the $\Lambda_{\mathrm{c}}^{+}/\mathrm{D}^0$ and $\Xi_{\mathrm{c}}^{0} \rightarrow \mathrm{e}^{+}\nu_{\mathrm{e}}\Xi^{-}/\mathrm{D}^{0}$ ratios are found to be underpredicted by theoretical calculations. Finally the R_{pPb} of Λ_{c}^{+} baryons is found to be consistent with unity, with the D-meson R_{pPb} and with theoretical predictions within current uncertainties.

References

[1] Y. Oh, C.M. Ko, S.H. Lee, S. Yasui, Phys. Rev. **C79**, 044905 (2009), 0901.1382
[2] G. Xie (STAR), Nucl. Phys. **A967**, 928 (2017), 1704.04353
[3] ALICE collaboration (2017), ALICE-PUBLIC-2017-008
[4] J. Adam et al. (ALICE), Phys. Rev. **C94**, 054908 (2016), 1605.07569
[5] B.A. Kniehl, G. Kramer, I. Schienbein, H. Spiesberger, Eur. Phys. J. **C41**, 199 (2005), hep-ph/0502194
[6] B.A. Kniehl, G. Kramer, I. Schienbein, H. Spiesberger, Eur. Phys. J. **C72**, 2082 (2012), 1202.0439
[7] S. Frixione, P. Nason, G. Ridolfi, JHEP **09**, 126 (2007), 0707.3088
[8] T. Sjostrand, S. Mrenna, P.Z. Skands, JHEP **05**, 026 (2006), hep-ph/0603175
[9] K.J. Eskola, H. Paukkunen, C.A. Salgado, JHEP **04**, 065 (2009), 0902.4154
[10] S. Acharya et al. (ALICE), Eur. Phys. J. **C77**, 550 (2017), 1702.00766
[11] J.R. Christiansen, P.Z. Skands, JHEP **08**, 003 (2015), 1505.01681
[12] C. Flensburg, G. Gustafson, L. Lonnblad, JHEP **08**, 103 (2011), 1103.4321
[13] M. Bahr et al., Eur. Phys. J. **C58**, 639 (2008), 0803.0883
[14] A. Beraudo, A. De Pace, M. Monteno, M. Nardi, F. Prino, JHEP **03**, 123 (2016), 1512.05186
[15] R. Perez-Marcial, R. Huerta, A. Garcia, M. Avila-Aoki, Phys. Rev. **D40**, 2955 (1989), [Erratum: Phys. Rev.D44,2203(1991)]
[16] R.L. Singleton, Phys. Rev. **D43**, 2939 (1991)
[17] H.Y. Cheng, B. Tseng, Phys. Rev. **D53**, 1457 (1996), [Erratum: Phys. Rev.D55,1697(1997)], hep-ph/9502391

EPJ Web of Conferences **171**, 18013 (2018) https://doi.org/10.1051/epjconf/201817118013
SQM 2017

Quarkonium production in Pb-Pb collisions at $\sqrt{s_{\rm NN}} = 5.02$ TeV with ALICE

Audrey Francisco* *for the ALICE Collaboration*

SUBATECH, IMT-Atlantique, Université de Nantes, CNRS-IN2P3, Nantes, France

Abstract. Ultra-relativistic heavy-ion collisions at the Large Hadron Collider provide a unique opportunity to study the properties of matter at extreme energy densities where a phase transition from the hadronic matter to a deconfined medium of quarks and gluons, the Quark-Gluon Plasma (QGP) is predicted. Among the prominent probes of the QGP, heavy quarks play a crucial role since they are created during the initial stages of the collision, before the QGP formation, and their number is conserved throughout the partonic and hadronic phases of the collision. The azimuthal anisotropy of charmonium production, quantified using the second harmonic Fourier coefficient (referred to as elliptic flow), provides important information on the magnitude and dynamics of charmonium production. Measurements of the quarkonium nuclear modification factor at forward rapidity and J/ψ elliptic flow in Pb-Pb collisions as a function of centrality, transverse momentum and rapidity will be presented and compared to different collision energy results and available theoretical calculations.

Keywords: ALICE, heavy-ion collisions, heavy quark, quarkonium, LHC

1 Introduction

Heavy quarks are produced in primary hard-scattering processes and their formation time ($\tau_c \sim 0.08$ fm/c for the charm and $\tau_b \sim 0.02$ fm/c for the bottom quark) is smaller than the QGP formation time ($\tau_f \sim 0.15$ fm/c) [1]. Therefore they experience the medium evolution through interactions with its constituents. Their study is particularly relevant to extract the properties of the QGP since the same number of heavy quarks per binary collision is expected to be produced in proton-proton and nucleus-nucleus collisions [2]. Bound states of heavy quarks — quarkonium (charmonium for $c\bar{c}$ and bottomonium for $b\bar{b}$) — provide remarkable probes of the medium. Quarkonium suppression by color screening was predicted and proposed early as a signature of the QGP formation[3], with different survival probabilities corresponding to the quarkonium binding energies. This suppression scenario framed the studies performed at SPS and RHIC energies. The yield enhancement observed at the LHC at $\sqrt{s_{\rm NN}} = 2.76$ TeV led to the introduction of a second production source[4]. The increased energy density and number of $c\bar{c}$ pairs enable quark (re)combination at a later stage of the collision [5–7]. This effect is expected to be more relevant for charmonium than bottomonium states. These two antagonistic mechanisms are required by theoretical models to reproduce experimental

*e-mail: audrey.francisco@cern.ch

observations. However many unknowns remain and originate large uncertainties in predictions. The most substantial one corresponds to the charm cross-section but many other parameters are not yet ascertained. Moreover the measurements contain a contribution from higher resonance decays as well as a non-prompt contribution from b hadrons which also require proper estimation. Cold nuclear matter effects can also affect quarkonium production in nucleus-nucleus collisions and their size needs to be precisely quantified.

In non-central collisions the nuclei overlap geometry is anisotropic. If the system is interacting, the density anisotropy will be reflected in the particle momentum distribution. This distribution can be decomposed into a Fourier series whose second coefficient is denoted v_2 and called elliptic flow. The measurement of D meson elliptic flow, recently published by ALICE [8], gives a strong hint of charm quark flow in the medium. Therefore the (re)combined $c\bar{c}$ pairs should inherit the charm quark flow and one can expect the J/ψ v_2 to exhibit large values. First hints of a positive J/ψ v_2 were observed at $\sqrt{s_{\mathrm{NN}}} = 2.76$ TeV both by ALICE [9] and CMS [10] experiments. Lower energy measurements with PHENIX [11] or STAR [12] experiments do not exhibit a significant J/ψ elliptic flow, although their systematics are large and the measurements are also compatible with the LHC results.

2 Analysis

The presented results were obtained from ALICE measurements in Pb-Pb collisions at $\sqrt{s_{\mathrm{NN}}} = 5.02$ TeV during the Run 2 of the LHC. The recorded data represent an integrated luminosity of $\sim$ 225 μb^{-1}. A complete description of the ALICE apparatus can be found at [13]. Inclusive quarkonium measurements in the dimuon decay channel are performed at forward rapidity ($2.5 < y < 4$) and down to zero p_{T} in the muon spectrometer. The silicon pixel detector (SPD) provides the primary vertex and the event plane estimation for flow measurements. The raw number of reconstructed quarkonia were obtained by fitting the dimuon mass spectra with a superposition of signal and background functions. Systematic uncertainties on the extracted yield are obtained by varying the functions and fitting ranges and detector acceptances as well as efficiencies are taken into account by MC simulations.

The nuclear modification factor (R_{AA}) represents the ratio of the measured yields in Pb–Pb to pp collisions at the same energy, scaled by the number of binary collisions. The measurement of the J/ψ elliptic flow was performed with the event plane method where the event plane Ψ was estimated from flow vectors obtained with detector multiplicities. Non-uniform acceptance effects of the detectors are corrected through an equalization procedure and the resolution is calculated with the three sub-event technique. The total dimuon $v_2 = \langle \cos 2(\varphi - \Psi) \rangle$ distribution as a function of the invariant mass $v_2(m_{\ell\ell})$ is evaluated. The total v_2 is modelled as a contribution from the signal and the background: $v_2(m_{\ell\ell}) = v_2^{\mathrm{sig}}\alpha(m_{\ell\ell}) + v_2^{\mathrm{bkg}}(m_{\ell\ell})[1 - \alpha(m_{\ell\ell})]$, where the signal fraction $\alpha(m_{\ell\ell})$ is extracted from the invariant mass distribution fit and the background flow v_2^{bkg} is represented with a generic function.

2.1 Results

The J/ψ nuclear modification factor evolution as a function of centrality (Fig.1 left) at $\sqrt{s_{\mathrm{NN}}} = 5.02$ TeV is similar to the previous observations at $\sqrt{s_{\mathrm{NN}}} = 2.76$ TeV: after an initial decrease, the R_{AA} exhibits almost no centrality dependence for $\langle N_{part} \rangle > 50$. The measurement precision was largely increased. A p_{T} cut below 0.3 GeV/c was applied to reduced the contribution of J/ψ photo-production and the brackets give limits for the remaining contamination [14, 15] on Fig.2 left and right. The R_{AA} dependence with transverse momentum was also studied for different centrality ranges (Fig.2 left). The J/ψ suppression is stronger at high p_{T} and in central collisions. A transport

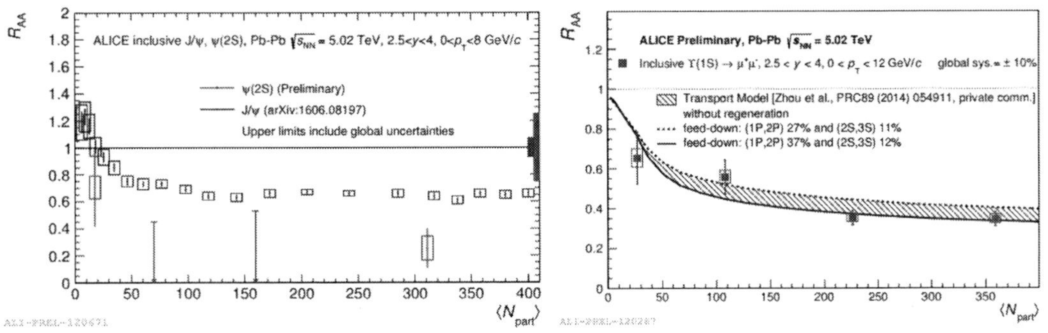

Figure 1. (Left) Comparison of inclusive J/ψ and ψ(2S) $R_{\rm AA}$ and (Right) Υ(1S) $R_{\rm AA}$vs $\langle N_{\rm part} \rangle$ at forward rapidity in Pb–Pb collisions at $\sqrt{s_{\rm NN}}$ = 5.02 TeV

model from Du et al. [16, 17] predicts a similar trend. Coherently, the J/ψ $\langle p_{\rm T} \rangle$ decreases as a function of centrality (Fig.2 right).

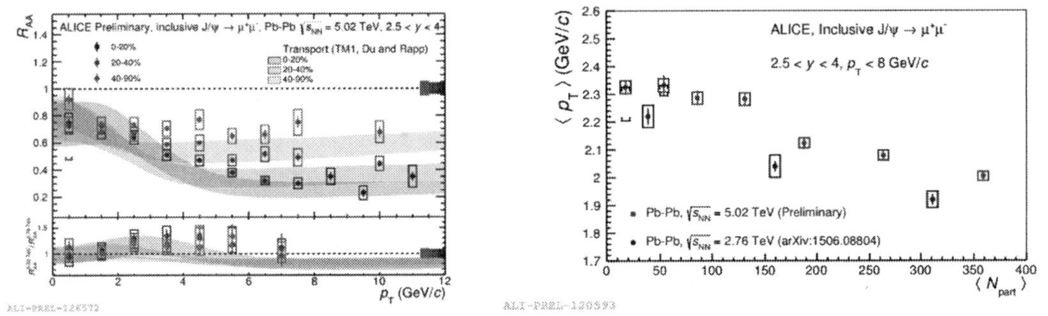

Figure 2. (Left) Differential J/ψ $R_{\rm AA}$ in Pb–Pb collisions at $\sqrt{s_{\rm NN}}$ = 5.02 TeV for 3 centrality classes (0–20%, 20–40% and 40–90%) with comparison to transport model prediction from Du and al. and ratio to $\sqrt{s_{\rm NN}}$ = 2.76 TeV. (Right) J/ψ $\langle p_{\rm T} \rangle$ vs $\langle N_{\rm part} \rangle$ comparison between $\sqrt{s_{\rm NN}}$ = 2.76 TeV and $\sqrt{s_{\rm NN}}$ = 5.02 TeV

The ψ(2S), a losely bound charmonium state, is more suppressed than the J/ψ. Data shows a stronger suppression for semi-central and central collisions. The Run 3 campaign should bring additional statistics and improve the signal over background ratio although the ψ(2S) signal extraction represents a challenging measurement. The Υ(1S) suppression increases as a function of centrality (Fig. 1 right). The observation is compatible with transport models including or not a (re)generation contribution. Precise feed-down measurement from higher resonances (2S and 3S states) are required to further improve the results and discriminate between models.

The J/ψ v_2 as a function of $p_{\rm T}$ is presented on Fig.3. Different centrality classes (5–20%, 20–40% and 40–60%) were studied. Significant values are observed for various centrality and $p_{\rm T}$ bins. The highest J/ψ v_2 is obtained for semi-central collisions (20–40%), with a significance reaching 6.6σ for the 4–6 GeV/c $p_{\rm T}$ class. The comparison to a transport model shows that the v_2 amplitude can only be achieved by including a strong contribution of J/ψ (re)generation. The comparison to open charm flow (Fig.3 right) result is a strong hint of charm (re)combination and thermalization in the medium.

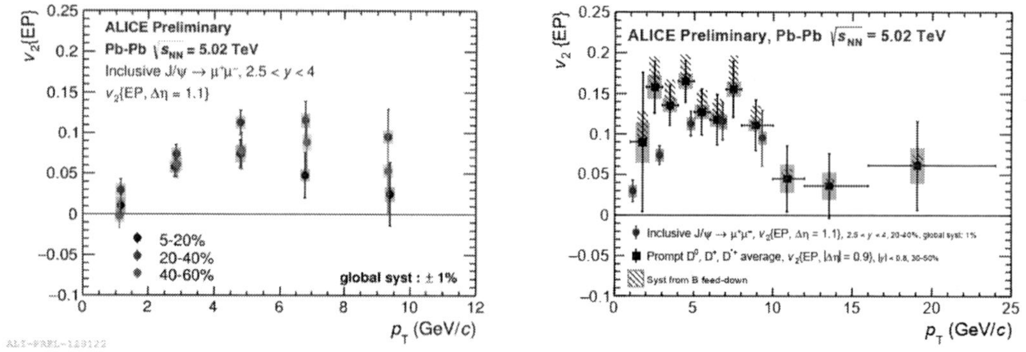

Figure 3. (Left) Inclusive J/ψ v_2 at forward rapidity in Pb–Pb collisions at $\sqrt{s_{\rm NN}}$ = 5.02 TeV for 3 centrality classes. (Right) Comparison of inclusive J/ψ v_2 at forward rapidity (20–40%) with D mesons v_2 (30–50%)

3 Conclusion

The interplay between suppression and (re)generation mechanisms is investigated through quarkonium measurements. Precise results on the J/ψ nuclear modification factor at $\sqrt{s_{\rm NN}}$ = 5.02 TeV, compatible with previous measurements at $\sqrt{s_{\rm NN}}$ = 2.76 TeV, were obtained. Furthermore the J/ψ elliptic flow was asserted with a 6.6 σ significance. Concerning the $\psi(2S)$, a stronger suppression than the J/ψ was observed. The greater centrality dependence of Υ(1S) nuclear modification factor may imply that the suppression plays a dominant role with a minor (re)generation contribution for bottomonium.

References

[1] K. Adcox et al. (PHENIX), Nucl. Phys. **A757**, 184 (2005), nucl-ex/0410003
[2] K. Zhou, Z. Chen, C. Greiner, P. Zhuang, Physics Letters B **758**, 434 (2016)
[3] T. Matsui, H. Satz, Phys. Lett. **B178**, 416 (1986)
[4] B. Abelev et al. (ALICE), Phys. Rev. Lett. **109**, 072301 (2012), 1202.1383
[5] R.L. Thews, M. Schroedter, J. Rafelski, Phys. Rev. C **63**, 054905 (2001)
[6] P. Braun-Munzinger, J. Stachel, Physics Letters B **490**, 196 (2000)
[7] A. Andronic et al., J. Phys. Conf. Ser. **509**, 012019 (2014), 1311.4662
[8] S. Acharya et al. (ALICE) (2017), 1707.01005
[9] E. Abbas et al. (ALICE), Phys. Rev. Lett. **111**, 162301 (2013), 1303.5880
[10] V. Khachatryan et al. (CMS), Eur. Phys. J. **C77**, 252 (2017), 1610.00613
[11] C. Silvestre (for the PHENIX Collaboration), J. Phys. **G35**, 104136 (2008), 0806.0475
[12] L. Adamczyk et al. (STAR), Phys. Rev. Lett. **111**, 052301 (2013), 1212.3304
[13] K. Aamodt et al. (ALICE), JINST **3**, S08002 (2008)
[14] J. Adam et al. (ALICE), JHEP **05**, 179 (2016), 1506.08804
[15] J. Adam et al. (ALICE), Phys. Rev. Lett. **116**, 222301 (2016), 1509.08802
[16] X. Zhao, R. Rapp, Nucl. Phys. **A859**, 114 (2011), 1102.2194
[17] X. Du, R. Rapp, Nucl. Phys. **A943**, 147 (2015), 1504.00670

Heavy Quark Dynamics toward thermalization: R_{AA}, v_1, v_2, v_3

Salvatore Plumari[1,2,⋆], *Santosh K.* Das[1,2], *Francesco* Scardina[1,2], *Vincenzo* Minissale[1,2], and *Vincenzo* Greco[1,2]

[1] *Department of Physics and Astronomy, University of Catania, Via S. Sofia 64, 1-95125 Catania, Italy*
[2] *Laboratori Nazionali del Sud, INFN-LNS, Via S. Sofia 62, I-95123 Catania, Italy*

Abstract. We describe the propagation of Heavy quarks (HQs) in the quark-gluon plasma (QGP) within a relativistic Boltzmann transport (RBT) approach. The interaction between heavy quarks and light quarks is described within quasi-particle approach which is able to catch the main features of non-perturbative interaction as the increasing of the interaction in the region of low temperature near T_C. In our calculations the hadronization of charm quarks in D mesons is described by mean of an hybrid model of coalescence plus fragmentation. We show that the coalescence play a key role to get a good description of the experimental data for the nuclear suppression factor R_{AA} and the elliptic flow $v_2(p_T)$ at both RHIC and LHC energies. Moreover, we show some recent results on the direct flow v_1 and triangular flow v_3 of D meson.

1 Introduction

Charm and bottom quarks are produced at the early stages of a Heavy Ion Collision (HIC) and due to their large masses they are expected to thermalize slower in the QGP. Therefore they represent an ideal probe to study the whole evolution of the QGP. Furthermore, because they are produced out-of-equilibrium they are expected to conserve memory of the history of the plasma evolution. The nuclear suppression factor R_{AA}, which is the ratio between the spectra of heavy flavor hadrons measured in nucleus-nucleus, and the elliptic flow $v_2(p_T)$ which is a measure of the anisotropies in momentum space are the two key observables in heavy flavor sector. Experimental measurements at both RHIC and LHC energies have shown a small value of their R_{AA} and the large values of $v_2(p_T)$ which are almost comparable to those of light hadrons. Several theoretical efforts within different models have been made to study and describe simultaneously the small R_{AA} and large v_2 measured in experiments [1–6] Recently this analysis has been extended to the study of different harmonics showing that also a direct flow v_1 and triangular flow $v_3(p_T)$ of D meson is non vanishing [7, 8].

2 Transport equation for charm quarks in the QGP

We the describe the evolution of the charm quark by solving the RBT equations. In our model the charm quarks scatter in a bulk medium of quarks and gluons as described by the following set of eq.s

$$\begin{aligned} p^\mu \partial_\mu f_Q(x,p) &= C[f_q, f_g, f_Q](x,p) \\ p_k^\mu \partial_\mu f_k(x,p) &= C[f_q, f_g](x_q, p_q) \qquad k = q, g \end{aligned}$$

⋆e-mail: salvatore.plumari@hotmail.it

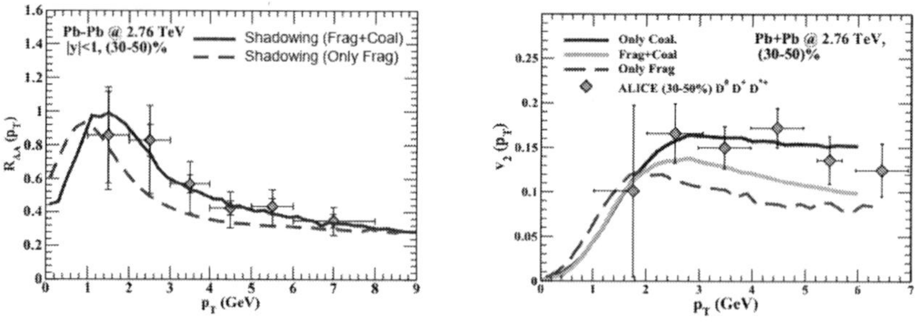

Figure 1. Left and middle panel: D meson $R_{\rm AA}$ and v_2 respectively in $Pb + Pb$ collisions at $\sqrt{s_{\rm NN}} = 2.76\,TeV$ and centrality 30 – 50% compared to ALICE data. Experimental data has been taken from Ref.s [17, 18].

$f_k(x, p)$ indicates the on-shell phase space one-body distribution function of the k parton (quark or gluon) while $C[f_q, f_g, f_Q](x, p)$ is the relativistic Boltzmann-like collision integral. In our calculations, the phase-space distribution function of the bulk medium (quarks and gluons) enters in the evolution equation for charm quarks as an external quantities with $C[f_q, f_g, f_Q]$ and the evolution of f_q and f_g have been assumed to be independent of $f_Q(x, p)$. The evolution of the bulk of quark and gluons is given by the solution of the other two transport equations where the $C[f_q, f_g]$ is tuned to keep fixed the η/s ratio, for a detailed discussion see ref.s [9–11]. We have considered a bulk with massive quarks and gluons that provide a softening of the equation of state with a decreasing speed of sound when the cross over region is approached. Within this approach we describe the evolution of a system that dynamically has approximatively the lQCD equation of state [12, 13]. The quarks and gluons masses are given by the QPM [13]. The hadronization process plays a crucial role in determining the final spectra, $R_{\rm AA}(p_T)$ and $v_2(p_T)$ in comparison to the experimental data. In our calculations we have considered a hybrid model of coalescence plus fragmentation. For a detailed discussion of the hadronization model see [14]. We simulate $Pb + Pb$ collisions at $\sqrt{s_{\rm NN}} = 2.76\,TeV$. The initial conditions for the bulk in the r-space are given by the standard Glauber condition assuming boost invariance along the longitudinal direction. In momentum space we use a Boltzmann-Juttner distribution function up to a transverse momentum $p_T = 2$ GeV while at larger momenta mini-jet distributions as calculated by pQCD at NLO order in [15]. The initial temperature at the center of the fireball is fixed to $T_0 = 490\,MeV$ and the initial time for the simulations is $\tau_0 \simeq 1/T_0 = 0.3\,fm/c$. In the r-space charm quarks are distributed according to N_{coll} while in the p-space the charm quarks are distributed according to the Fixed Order + Next-to-Leading Log (FONLL) calculations, taken from Ref. [16]. In the calculation shown in this paper the dynamical evolution of the bulk is constrained by an $\eta/s = 1/(4\pi)$.

3 Results

In the first part of this section we show the comparison of the results for the nuclear modification factor $R_{\rm AA}$ and for the elliptic flow v_2 with the experimental data. In the second part we discuss the recent results about the direct flow v_1 and triangular flow $v_3(p_T)$ of D meson. In Fig. 1 it is shown the comparison of our results for the $R_{\rm AA}(p_T)$ with the experimental data for $Pb + Pb$ collisions at $\sqrt{s_{\rm NN}} = 2.76\,TeV$ for 30 – 50% centrality. We observe comparing dashed that the effect of the coalescence is to increase the $R_{\rm AA}$ for momenta larger than 1 GeV. This is due to the hadronization mechanism because a D mesons from coalescence, which is composed by one light quark and a charm

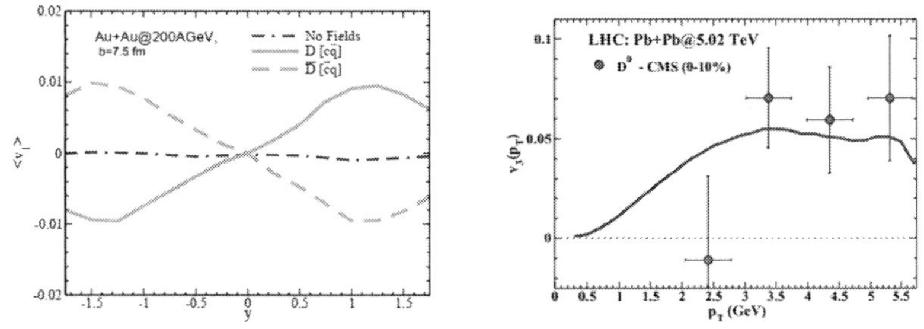

Figure 2. Left panel: Directed flow v_1 as a function of the rapidity for $Au + Au$ at $\sqrt{s_{\rm NN}} = 200\,GeV$ for D and anti-D meson. Right panel: $v_3(p_T)$ for $Pb + Pb$ at $\sqrt{s_{\rm NN}} = 5.02\,TeV$ for $(0-10)\%$ centrality (data from[19]).

quark, it get a larger momentum respect to the D mesons obtained from fragmentation. On the other hand at larger momenta, fragmentation becomes anyway the dominant mechanism of hadronization. In the right panel of Fig. 1, we show the corresponding results for the final $v_2(p_T)$ of D mesons. We show explicitly the different contributions to the $v_2(p_T)$ played by only coalescence (blue solid line) and only fragmentation (red dashed line). As shown, the v_2 developed via only coalescence is larger than the v_2 developed due to fragmentation. This is due mainly to the fact that the D meson is the result of the coalescence of a light quark with a charm quark therefore the final D mesons anisotropy in momentum space reflects both the heavy quark and light quark anisotropies in momentum space and it can even lead to an increase of about a factor two at $p_T > 2\,\text{GeV}$. While, when coalescence plus fragmentation mechanism is included for the hadronization, the v_2 of the D-mesons increases with respect to the v_2 of D meson by about a 30%, see solid orange line. The extracted diffusion coefficient $2\pi D_s$ is in agreement with the data and within the present systematic uncertainties to the lattice QCD calculation for a detailed discussion see [14]. It has been shown that very strong electric and magnetic fields created at early times of uRHICs can affect the charm quarks dynamics [8]. Due to the collision geometry the magnetic field created at early times is dominated by the component along the y axis. The main effect of the magnetic field is the induction of a current in the xz plane while at the same time the electric field is created and it is directed in the x direction. The combined effect of the fields is a current in the xy plane and the effect of which is to generate a finite directed flow v_1. In Fig.2 it is shown the final v_1 of the D mesons as a function of the rapidity for $Au + Au$ at $\sqrt{s_{\rm NN}} = 200\,GeV$ at $b = 7.5\,fm$. In these calculations we have considered for the electromagnetic field the same space-time solution obtained in Ref. [20]. As shown the v_1 at forward rapidity is positive for the D meson and negative for anti D meson which means that the displacement induced by the Faraday current has a stronger effect compared to the Hall drift of the magnetic field. Recently, we have developed an event-by-event transport approach for the bulk in order to study the role of finite η/s on the anisotropic flows $v_n(p_T)$, for details see [10]. In the right panel of Fig.2 we present the $v_3(p_T)$ of D mesons for central collisions for $Pb + Pb$ at $\sqrt{s_{\rm NN}} = 5.02\,TeV$. In our calculation we have used a transport approach to model the soft sector on an event-by-event basis and describe the flow harmonics. Our event-by-event analysis consists of 1000 events for each centrality class. As shown we get a finite $v_3(p_T)$ which is a linearly increasing function at intermediate p_T and as shown we get a good agreement with recent experimental results of CMS collaboration [19]. Notice that the results shown have been obtained including only the hadronization by fragmentation the inclusion of the coalescence will give an enhancement of the final anisotropic flows v_n.

4 Conclusions

We have studied the charm quark propagation in QGP at LHC energies within a relativistic Boltzmann transport approach. The interaction between charm quarks and light quarks of the bulk is described within quasi-particle model. This model is able to catch the main features of non-perturbative interaction and it is possible to describe the evolution of bulk that reproduce the lattice QCD equation of state. Our calculations give a good description for D meson R_{AA} and v_2 both at LHC energies within the experimental uncertainties. In general, we found that the non-perturbative behavior described by the QPM enhances the charm quark bulk interaction near T_c which is essentially the ingredient for the build-up of a large v_2. Moreover, we observe that the effect of the hadronization by coalescence is to increase the R_{AA} and $v_2(p_T)$ for $p_T > 1\,GeV$. We have also studied the charm quark dynamics in the presence of electromagnetic field. We have found that the effect of the initial electromagnetic fields on the charm quarks is to develop a sizable directed flow v_1 which could be measurable at experiments. This suggest that the heavy quark v_1 could be a significant probe to characterize the initial electromagnetic field produced in the HICs.

Acknowledgments

V. Greco, S. Plumari acknowledge the support of the ERC-StG Grant under the QGPDyn project.

References

[1] H. van Hees, M. Mannarelli, V. Greco, R. Rapp, Phys. Rev. Lett. **100**, 192301 (2008).
[2] W.M. Alberico, A. Beraudo, A. De Pace, A. Molinari, M. Monteno, M. Nardi, F. Prino, Eur. Phys. J. **C71**, 1666 (2011), `1101.6008`
[3] S. Cao, G.Y. Qin, S.A. Bass, Phys. Rev. **C92**, 024907 (2015), `1505.01413`
[4] J. Uphoff, O. Fochler, Z. Xu, C. Greiner, Phys. Rev. **C84**, 024908 (2011), `1104.2295`
[5] S.K. Das, F. Scardina, S. Plumari, V. Greco, Phys. Rev. **C90**, 044901 (2014), `1312.6857`
[6] S.K. Das, F. Scardina, S. Plumari, V. Greco, Phys. Lett. **B747**, 260 (2015), `1502.03757`
[7] M. Nahrgang, J. Aichelin, S. Bass, P.B. Gossiaux, K. Werner, Phys. Rev. **C91**, 014904 (2015).
[8] S.K. Das, S. Plumari, S. Chatterjee, J. Alam, F. Scardina, V. Greco, Phys. Lett. **B768**, 260 (2017).
[9] M. Ruggieri, F. Scardina, S. Plumari, V. Greco, Phys.Rev. **C89**, 054914 (2014), `1312.6060`
[10] S. Plumari, G.L. Guardo, F. Scardina, V. Greco, Phys. Rev. **C92**, 054902 (2015), `1507.05540`
[11] S. Plumari, A. Puglisi, M. Colonna, F. Scardina, V. Greco, J.Phys.Conf.Ser. **420**, 012029 (2013), `1209.0601`
[12] S. Borsanyi, G. Endrodi, Z. Fodor, A. Jakovac, S.D. Katz, S. Krieg, C. Ratti, K.K. Szabo, JHEP **11**, 077 (2010), `1007.2580`
[13] S. Plumari, W.M. Alberico, V. Greco, C. Ratti, Phys.Rev. **D84**, 094004 (2011), `1103.5611`
[14] F. Scardina, S.K. Das, V. Minissale, S. Plumari, V. Greco (2017), `1707.05452`
[15] V. Greco, C.M. Ko, P. Levai, Phys. Rev. Lett. **90**, 202302 (2003), `nucl-th/0301093`
[16] M. Cacciari, S. Frixione, N. Houdeau, M.L. Mangano, P. Nason, G. Ridolfi, JHEP **10**, 137 (2012).
[17] B. Abelev et al. (ALICE), Phys. Rev. Lett. **111**, 102301 (2013), `1305.2707`
[18] J. Adam et al. (ALICE), JHEP **03**, 081 (2016), `1509.06888`
[19] A.M. Sirunyan et al. (CMS) (2017), `1708.03497`
[20] U. Gursoy, D. Kharzeev, K. Rajagopal, Phys. Rev. **C89**, 054905 (2014), `1401.3805`

EPJ Web of Conferences **171**, 18015 (2018) https://doi.org/10.1051/epjconf/201817118015
SQM 2017

Quarkonium measurements in heavy-ion collisions with the STAR experiment

Xinjie Huang[1] on behalf of the STAR Collaboration
[1] *Tsinghua University, Beijing 100084, China*

Abstract. In these proceedings, we present the latest measurements of J/ψ and Υ by the STAR experiment. The J/ψ and Υ production measured in p+p collisions provide new baselines for similar measurements in Au+Au collisions, while the measurements in p+Au collisions can help quantify the cold nuclear matter effects. The J/ψ v_2 is measured in both U+U and Au+Au collisions to place constraints on the amount of J/ψ arising from recombination of deconfined charm and anti-charm pairs. Furthermore, the nuclear modification factors for ground and excited Υ states as a function of transverse momentum and centrality are presented, and compared to those measured at the LHC as well as to theoretical calculations.

1 Introduction

Quarkonium suppression in the medium due to the color screening effect has been proposed as a direct signature of the formation of the Quark Gluon Plasma (QGP) [1]. However, other effects, such as cold nuclear matter (CNM) effects and regeneration of quarkonium states from deconfined heavy quark pairs, give rise to additional complications in the interpretation of the observed suppression. Measurements of the J/ψ elliptic flow (v_2) in different collision systems can help disentangle the different sources contributing to the observed J/ψ population. Compared to charmonia, bottomonia receive less regeneration contribution due to the smaller bottom quark cross-section, thus providing a cleaner probe. Furthermore, different bottomonium states with their different binding energies are expected to dissociate at different temperatures. Measurements of this "sequential melting" can help constrain the medium temperature.

In these proceedings, we present the lastest measurements of J/ψ and Υ productions via both the dimuon and dielectron decay channels. The dimuon channel measurements are based on data samples triggered by the Muon Telescope Detector (MTD) corresponding to an integrated luminosity of 14.2 nb^{-1} for Au+Au collisions at $\sqrt{s_{NN}}$ = 200 GeV from the RHIC 2014 run, and integrated luminosities of 122 pb^{-1} for p+p collisions and 409 nb^{-1} for p+Au collisions at $\sqrt{s_{NN}}$ = 200 GeV from the RHIC 2015 run. The dielectron channel measurements are based on data triggered by the Barrel ElectroMagnetic Calorimeter (BEMC) corresponding to an integrated luminosity of 1.1 nb^{-1} for Au+Au collisions at $\sqrt{s_{NN}}$ = 200 GeV from the RHIC 2011 run, and integrated luminosities of 97 pb^{-1} for p+p collisions and 300 nb^{-1} for p+Au collisions at $\sqrt{s_{NN}}$ = 200 GeV from the RHIC 2015 run.

2 Results

In all figures presented in this section, statistical uncertainties are shown as vertical bars while systematic uncertainties are shown as open boxes or brackets around the data points. Filled boxes around unity represent the global uncertainties.

The left panel of Fig. 1 shows the inclusive J/ψ cross-section scaled by the branching ratio B and measured via the dimuon channel in the transverse momentum ($p_{\rm T}$) range of $1 < p_T < 10$ GeV/c (red circles) and via the dielectron channel in $0 < p_T < 14$ GeV/c (blue squares) in p+p collisions at $\sqrt{s_{\rm NN}} = 200$ GeV. The results are consistent in the overlaping p_T range, and can be well discribed by CGC+NRQCD [2] and NLO NRQCD [3] predictions for prompt J/ψ production at low and high p_T, respectively. An improved color evaporation model (ICEM) for direct J/ψ production [4] describes the data in $p_T < 3$ GeV/c, but underestimates the yields at higher p_T.

The right panel of Fig. 1 shows the $R_{\rm pAu}$, which quantifies the CNM effects, as a function of p_T. The $R_{\rm pAu}$ is less than unity at low p_T, and consistent with unity within uncertainties at high p_T. The data is compared with theoretical calculations with the nuclear PDF effect only (color bands) [5–7] and with an additional effect of nuclear absorption (blue dashes) [8]. The calculation with an additional effect of nuclear absorption is favored by the data.

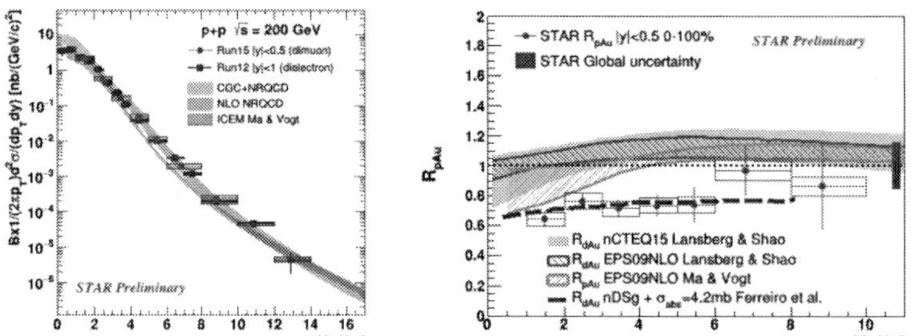

Figure 1. Left: Inclusive J/ψ cross-section in p+p collisions at $\sqrt{s_{\rm NN}} = 200$ GeV measured via the dimuon (red circles) and dielectron channels (blue squares) as a function of p_T. Right: Inclusive J/ψ $R_{\rm pAu}$ as a function of p_T, compared with theoretical calculations [5–8].

Figure 2 shows the J/ψ v_2 as a function of p_T. The new result for U+U collisions at $\sqrt{s_{\rm NN}} = 193$ GeV (red circles) from 2012 data is consistent within error bars with the result for Au+Au collision at $\sqrt{s_{\rm NN}} = 200$ GeV (black circles) from 2010 and 2011 data [9]. For $p_T > 2$ GeV/c, v_2 is consistent with zero within uncertainties, indicating that the contribution from regeneration of fully thermalized charm quarks to J/ψ is likely to be small.

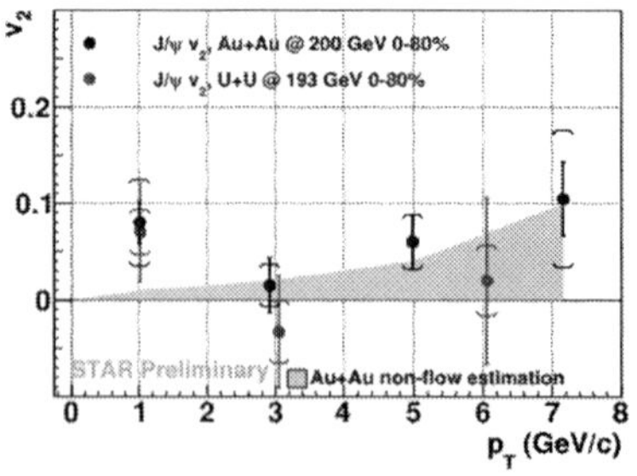

Figure 2. J/ψ v_2 as a function of p_T measured in U+U collisions at $\sqrt{s_{\rm NN}} = 193$ GeV (red circles) and Au+Au collisions at $\sqrt{s_{\rm NN}} = 200$ GeV (black circles) [9]. The grey band shows the estimated non-flow contribution.

With BEMC-triggered data from 2015, the Υ(1S+2S+3S) production cross-section in p+p collisions at $\sqrt{s} = 200$ GeV within $|y| < 0.5$ is measured to be $B \cdot {\rm d}\sigma/{\rm dy} = 81 \pm 5({\rm stat.}) \pm 8({\rm syst.})$ pb, where B is the branching ratio. The new result of better precision is consistent with the published STAR result [10], and follows the trend of world-wide experimental data as well as NLO CEM predictions, as can be seen in the left panel of Fig. 3.

The right panel of Fig. 3 shows the Υ(1S+2S+3S) R_{pAu} as a function of rapidity. The new results measured with BEMC-triggered data from 2015 (red stars) are consistent with the published R_{dAu} (red circles) but with smaller relative uncertainties. The R_{pAu} within $|y| < 0.5$ is $0.82 \pm 0.10(\text{stat.})^{-0.07}_{+0.08}(\text{syst.}) \pm 0.10(\text{norm.})$, hinting at sizable CNM effects.

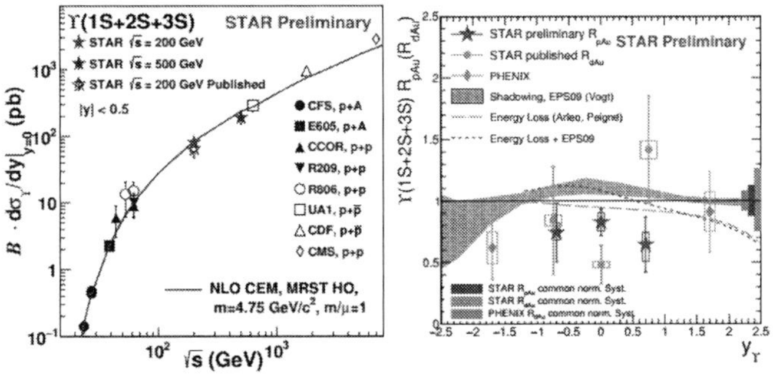

Figure 3. Left: Υ(1S+2S+3S) production cross-section measured in p+p collision at $\sqrt{s}$ = 200 GeV (red star) and $\sqrt{s}$ = 500 GeV(blue star) compared with world-wide data and NLO CEM predictions [11]. Right: the Υ(1S+2S+3S) R_{pAu} (red stars) as a function of rapidity compared to STAR and PHENIX published R_{dAu} results.

The Υ(1S+2S+3S) R_{AA} in Au+Au collisions at $\sqrt{s_{NN}}$ = 200 GeV has been measured via the dimuon channel with MTD-triggered data from 2014 and the dielectron channel with BEMC-triggered data from 2011. Since the nuclear modification factors in both channels are consistent with each other within uncertainlies, they are combined to improve the precision.

The centrality dependence of the combined R_{AA} for Υ(1S) and Υ(2S+3S) is shown in the left and middle panels of Fig. 4, compared with the lastest CMS results [12]. Υ(2S+3S) are more suppressed in central collisions than Υ(1S), which is consistent with the sequential melting. Comparing RHIC and LHC results, the levels of suppression for Υ(1S) are consistent, while Υ(2S+3S) seem to be less suppressed at the RHIC than that at the LHC.

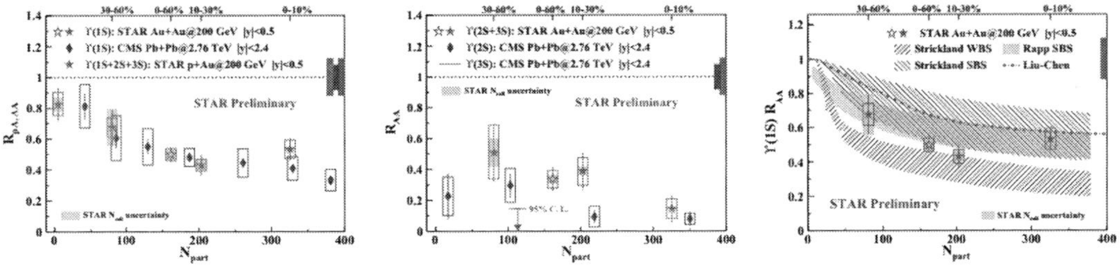

Figure 4. The R_{AA} of Υ(1S) (left panel) and Υ(2S+3S) (middel panel) as a function of N_{part} in Au+Au collisions at $\sqrt{s_{NN}}$ = 200 GeV for $|y| < 0.5$ (red stars), together with the STAR R_{pAu} for Υ(1S+2S+3S) within $|y| < 0.5$ (blue star) and the lastest CMS results in Pb+Pb collisions at $\sqrt{s_{NN}}$ = 2.76 TeV within $|y| < 2.4$ (black diamonds). Right: The combined R_{AA} of Υ(1S) within $|y| < 0.5$ compared to different model calculations [13–15]

The right panel of Fig. 4 shows the centrality dependence of the combined Υ(1S) R_{AA} together with three SBS (strongly bound scenario) model calculations which use internal-energy-based heavy quark potential and a WBS (weakly bound scenario) model calculation which uses free-energy-based heavy quark potential. The Strickland-Bazow model [13], which takes into account the feed down contributions from excited bottomonium states, studies both scenarios but includes neither CNM nor regeneration effect. The Liu-Chen model [14] considers only the dissociation of the excited states. The Emerick-Zhao-Rapp SBS model [15] accounts for CNM effects and the regeneration contribution. The SBS models are favored by data, especially in central collisions.

The p_T dependence of the R_{AA} for Υ(1S) and Υ(2S+3S) measured via the dimuon channel is shown in the left and right panels of Fig. 5, compared with the lastest CMS results [12]. Again for Υ(1S) the suppression is consistent at the RHIC and the LHC, while for Υ(2S+3S) the suppression seems to be stronger at high-p_T at the LHC.

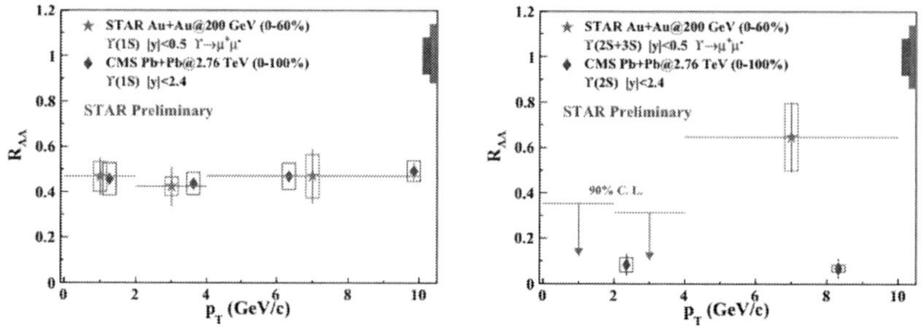

Figure 5. $R_{\rm AA}$ of Υ(1S) (left panel) and Υ(2S+3S) (right panel) at mid-rapidity as a function of $p_{\rm T}$ in Au+Au collisions at $\sqrt{s_{\rm NN}} = 200$ GeV (red stars) compared with CMS results for Pb+Pb collisions at $\sqrt{s_{\rm NN}} = 2.76$ TeV (black diamonds).

3 Summary and Outlook

In summary, we present the latest measurements of J/ψ and Υ productions in p+p, p+Au and Au+Au collisions at $\sqrt{s_{\rm NN}} = 200$ GeV and in U+U collisions at $\sqrt{s_{\rm NN}} = 193$ GeV from the STAR experiment. In p+p collisions, the inclusive J/ψ production cross-section can be described by the CGC+NRQCD at low p_T and NLO NRQCD at high p_T. The new Υ(1S+2S+3S) production cross-section result with better precision compared to previously published results follows the trend of world-wide experimental data as well as NLO CEM predictions. In p+Au collisions, the J/ψ $R_{\rm pAu}$ is below unity at low p_T and consistent with unity at high p_T. The $R_{\rm pAu}$ for Υ within $|y| < 0.5$ is measured to be $0.82 \pm 0.10(\text{stat.})^{-0.07}_{+0.08}(\text{syst.}) \pm 0.10(\text{norm.})$.

In U+U and Au+Au collisions, the J/ψ v_2 results are consistent with each other within uncertainties, and consistent with zero for $p_T > 2$ GeV/c, indicating that the regeneration of fully thermalized charm quarks to J/ψ is unlikely to be the dominant source contributing in this kinematic region. For Υ in Au+Au collisions, the excited Υ states are more suppressed than the ground state in central collisions, which is consistent with the sequential melting. There is also a hint of less suppression at the RHIC than at the LHC for excited Υ states. The statistical precision of quarkonium measurements in Au+Au collisions can be further improved by including the similar amount of additional data on tape.

References

[1] T. Matsui, H. Satz, Phys. Lett. B178 (1986) 416–422.
[2] Ma and Venugopalan,PRL 113 (2014) 192301.
[3] Shao, JHEP 05 (2015) 103.
[4] Ma and Vogt, PRD 94 (2016) 114029.
[5] Ma and Vogt, Private Communication (2017).
[6] Lansberg and Shao, Eur.Phys.J. C77 (2017) no.1, 1.
[7] Shao, Comput. Phys. Commun. 184 (2013) 2562–2570., 198 (2016) 238–259.
[8] Ferreiro*et al.*, Syst. 53 (2012) 27–36.
[9] The STAR Collaboration, PRL 111 (2013) 052301.
[10] The STAR collaboration, Phys. Lett. B 735, 127 (2014).
[11] Frawley, Ullrich, Vogt, Phs. Rev. 462, 125 (2008).
[12] The CMS collaboration, Phys. Lett. B 770, 357 (2017).
[13] Strickland and Bazow, Nucl. Phys. A 879, 25 (2012).
[14] Liu, Chen, Xu, Zhuang, Phys. Lett. B 697, 32 (2011).
[15] Emerick, Zhao, Rapp, Eur. Phys. J. A 48, 72 (2012).

Coherent e^+e^- production at very low transverse momentum at STAR

Chi Yang[1,⋆] For the STAR Collaboration

[1] *Shanda Nanlu 27#, Shandong University, Jinan, Shandong Province, China 250100*

Abstract. We report the measurements of e^+e^- pair production at very low e^+e^- pair transverse momentum (p_T < 0.15 GeV/c) in Au+Au collisions at $\sqrt{s_{NN}}$ = 200 GeV and U+U collisions at $\sqrt{s_{NN}}$ = 193 GeV using the STAR detector at the Relativistic Heavy Ion Collider. In 60-80% centrality, significant excesses are observed with respect to hadronic cocktails in both Au+Au and U+U collisions. These excess yields can not be explained by a theoretical model calculation incorporating in-medium broadened ρ spectral function. Additionally, the p_T^2 distribution for excess yield is shown and found to be exponential at very low p_T.

1 Introduction

The e^+e^- pair production provides a unique observable to study the fundamental properties of the hot and dense medium created in ultra-relativistic heavy-ion collisions. Those pairs are produced during all stages of the collisions and from all forms of the created matter. Due to only electromagnetic interactions with this matter, they can convey information about the dynamics of the entire time evolution of the medium [1]. In the e^+e^- invariant mass region $M_{ll} < M_\phi$ (Low Mass Region, LMR) the production is dominated by in-medium decay of vector mesons (ρ, ω, and ϕ). The broadened ρ model calculation which is linked to chiral symmetry restoration can describe the modified ρ spectra in this region. In intermediate mass region (IMR, $M_\phi < M_{ll} < M_{J/\psi}$) the e^+e^- pair production consists of two parts: thermal radiation of the medium and semi-leptonic decay.

Recently, the ALICE collaboration reported a significant excess at very low p_T in its study of J/ψ yield in peripheral Pb + Pb collisions at $\sqrt{s_{NN}}$ = 2.76 TeV via $\mu^+\mu^-$ channel [2]. The STAR collaboration also observed similar behavior for J/ψ in peripheral Au + Au collisions at $\sqrt{s_{NN}}$ = 200 GeV and U + U collisions at $\sqrt{s_{NN}}$ = 193 GeV at p_T < 0.15 GeV/c via e^+e^- channel [3]. These results via dilepton channels can not be explained by hadronic production mechanisms while coherent photoproduction and photon-photon interaction mechanism in UPC (ultra-peripheral collisions) can describe some observables. In this case, with impact parameter larger than twice of the nucleus radius, virtual photons emitted by one nucleus can fluctuate into quark-antiquark pairs, and then coherently interact with the other nucleus to produce vector mesons. The other process, photon-photon interaction, will also produce dilepton pairs like e^+e^- pairs. It is necessary to measure the e^+e^- production in a wide mass range including vector meson dominant region and beyond.

⋆e-mail: chiyang@sdu.edu.cn

2 Low p_T e^+e^- production

2.1 Analysis

The data used in this analysis are from Au + Au collisions at $\sqrt{s_{NN}} = 200$ GeV and U + U collisions at $\sqrt{s_{NN}} = 193$ GeV. About 720 million Au+Au and 270 million U+U minimum bias (0-80% centrality) events passed data quality assurances and vertex selections in total. The collision vertex was required to be within 30 cm of the mean of the vertex distribution along the beam line, nominally at the center of the Time Projection Chamber (TPC). In the plane perpendicular to the beam line, the collision vertex was selected within 2 cm of the beam line.

The main subsystems used for electron identification were the TPC and the Time-of-Flight (TOF). With the selection requirements on the particle energy loss, dE/dx, measured by the TPC and particle velocity, β, measured by the TOF, high purity electron samples were obtained [4]. The electron purity (fraction of true electrons in the identified electron sample) was about 95% in Au+Au minimum bias collisions on average and was p_T dependent from 0.2 to 2.0 GeV/c [5]. The details for e^+e^- production can be found in Ref. [5].

2.2 Results

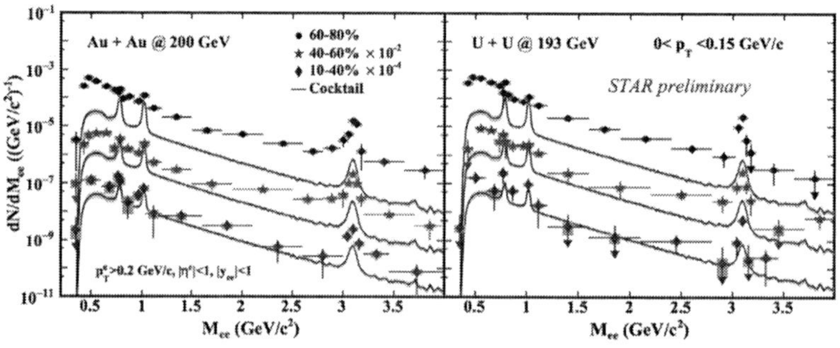

Figure 1. (Color online) The efficiency corrected low p_T (<0.15 GeV/c) e^+e^- pair invariant mass distributions from Au+Au collisions at $\sqrt{s_{NN}}$ = 200 GeV (left) and U+U collisions at $\sqrt{s_{NN}}$ = 193 GeV (right). The distributions for three different centralities 60-80%, 40-60%, and 10-40% are shown. For each centrality, the corresponding hadronic cocktail contribution is shown as a solid black line with the uncertainty as a shaded band. Statistical and systematic uncertainties on the data points are shown as bars and shaded boxes, respectively.

Figure 1 shows the e^+e^- invariant mass, M_{ee}, distribution at $p_T < 0.15$ GeV/c from Au+Au collisions at $\sqrt{s_{NN}} = 200$ GeV and U+U collisions at $\sqrt{s_{NN}} = 193$ GeV. The contributions from ρ meson are excluded from the hadronic cocktail simulations for comparison to model calculations. In 60-80% centrality Au+Au and U+U collisions, there are significant excesses observed with respect to the corresponding hadronic cocktails [5]. Compared to those in 60-80% centrality, the enhancement factors in 40-60% are less significant while in 10-40% the data are consistent with the hadronic cocktail within uncertainties for $M_{ee} > 0.76$ GeV/c^2. The low p_T (<0.15 GeV/c) e^+e^- excess yields with respect to the hadronic cocktail are shown in Fig. 2 from Au+Au and U+U collisions in 60-80% centrality. Theoretical calculations [6] expected from in-medium modified ρ production, QGP thermal radiation, and the sum of the two are shown for comparison. These calculations fail to explain the excess. Additional production mechanisms should be involved in this distribution.

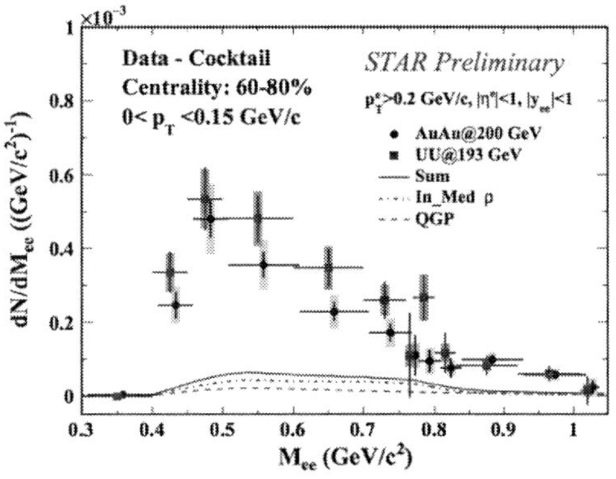

Figure 2. (Color online) The low p_T (<0.15 GeV/c) e^+e^- excess yield with respect to the hadronic cocktail for Au+Au and U+U collisions in 60-80% centrality. Statistical and systematic uncertainties on the data points are shown as bars and shaded boxes, respectively. Theoretical calculations expected from in medium modified ρ production, QGP thermal radiation, and the sum of the two are shown for comparison.

Figure 3 shows the dN/dp_T distribution as a function of p_T for e^+e^- pairs in different mass regions from Au+Au and U+U collisions. Compared to distributions from hadronic cocktail, clear excesses are observed in p_T < 0.15 GeV/c. For the p_T range above 0.15 GeV/c, the hadronic cocktail distributions are consistent with the data within uncertainties. This distribution is similar to that in ultra-peripheral collisions [7].

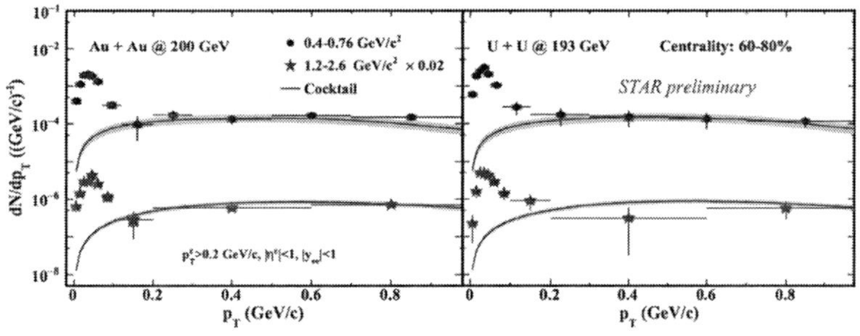

Figure 3. (Color online) The dN/dp_T distribution as a function of p_T for e^+e^- pairs in different mass region for Au+Au collisions at $\sqrt{s_{NN}}$ = 200 GeV (left) and U+U collisions at $\sqrt{s_{NN}}$ = 193 GeV (right). For each mass region, the corresponding hadronic cocktail distribution is shown in solid black line with the uncertainty shown as shaded band. Statistical and systematic uncertainties on the data points are shown as bars and shaded boxes, respectively.

An exponential function ($Ae^{-Bp_T^2}$) is used to fit each p_T^2 distribution in Figure 4. The first data point is excluded from the fit to avoid the shift of the slope because of a predicted dropoff based on quantum electrodynamics calculation [8] (photon-photon process dominant) or a possible destructive interference effect (photon-nucleus process dominant) at very small p_T^2 [9]. The slope parameters from U+U collisions are systematically larger than those from Au + Au collisions in the same mass region. In the ρ-like region, the slope parameter in Au+Au peripheral collisions is consistent with the coherent photoproduction of ρ^0 in ultra-peripheral Au+Au collisions [9]. Since the excess in the ρ-like region is mainly from photon-photon interaction while UPC ρ^0 comes from photon-nucleus process, this consistency might be accidental.

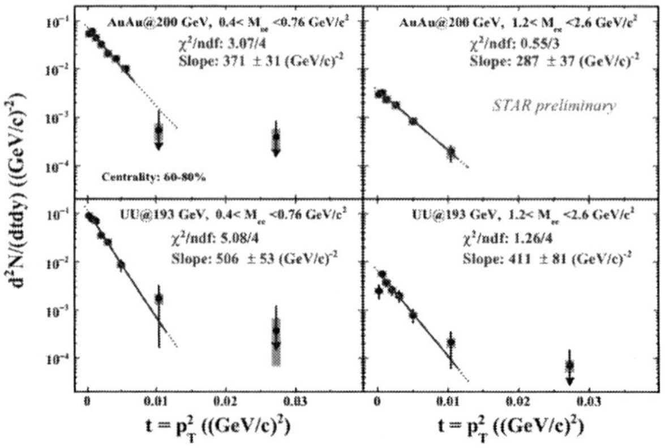

Figure 4. (Color online) The p_T^2 distributions of excess yields in different mass regions for 60-80% centrality Au+Au (upper plots) and U+U (lower plots) collisions. The systematic uncertainties of the data are shown as gray boxes.

3 Summary

We report the measurements of e^+e^- pair production at very low transverse momentum (p_T < 0.15 GeV/c) in Au + Au collisions at $\sqrt{s_{NN}}$ = 200 GeV and U+U collisions at $\sqrt{s_{NN}}$ = 193 GeV at STAR. The e^+e^- pair yields in a wider mass region are significantly enhanced with respect to the hadronic cocktail in 60-80% centrality in both Au+Au and U+U collisions. These excesses are only observed in the very low p_T region (< 0.15 GeV/c).

Acknowledgement

The author is supported partially by the NNSFC, China under grant number 11505181.

References

[1] G. David, R. Rapp, Z. Xu, Phys. Rep. **462**, 176 (2008)
[2] J.Adam, et al.,(ALICE Collaboration), Phys. Rev. Lett. **116**, 222301 (2016)
[3] W.Zha, J. Phys. Conf. Ser. **779**, 12039 (2017)
[4] M. Shao, et al., Nucl. Instrum. Methods A **558**, 419 (2006)
[5] L. Adamczyk, et al., (STAR Collaboration), Phys. Rev. C **92**, 024912 (2015); Phys. Rev. C **(90**, 064904 (2014)
[6] R.Rapp, private communication
[7] L. Adamczyk, et al., (STAR Collaboration), arXiv:1702.07705
[8] J. Adams, et al., (STAR Collaboration), Phys. Rev. C **70**, 031902 (2004)
[9] B. I. Abelev, et al., (STAR Collaboration), Phys. Rev. C **77**, 034910 (2008); Phys. Rev. Lett. **102**, 112301 (2009)

PHENIX results on open heavy flavor production

Takashi Hachiya[1,2,⋆] for the PHENIX collaboration

[1] *Nara Women's University, Department of Physics, Nara, Japan*
[2] *RIKEN BNL Research Center, Brookhaven National Laboratory, NY, USA*

Abstract. PHENIX measures the open heavy flavor productions in $p + p$, Cu+Au, and Au+Au collisions at $\sqrt{s_{NN}} = 200$ and 510 GeV using the silicon tracking detectors for mid- and forward rapidities. In Au+Au collisions, the nuclear modification of single electrons from bottom and charm hadron decays are measured for minimum bias and most central collisions. It is found that bottoms are less suppressed than charms in p_T=3-5 GeV/c and charms in most central collisions are more suppressed than that in minimum bias collisions. In $p + p$ and Cu+Au collisions, J/ψ from B meson decays are measured at forward and backward rapidities. The nuclear modification of B mesons in Cu+Au collisions is consistent with unity.

1 Introduction

Heavy quarks are useful probes to study the property of Quark -Gluon Plasma medium created in high energy heavy ion collisions. Heavy quark production is sensitive to the initial state effects because they are mainly produced via initial hard scattering. Once produced, they don't disappear by the strong interaction and they suffer final state effects such as energy loss and collective flow in QGP. Thus, their momentum and angular distributions are modified.

The PHENIX experiment previously measured a strong suppression of heavy quark production in Au+Au collisions by the measurement of single leptons from the heavy quark decays[1, 2]. The suppression of the charm quark was confirmed with the direct D reconstruction by the STAR experiment[3]. To study the suppression of heavy quarks, especially bottoms, we installed the barrel and forward silicon vertex tracker (VTX and FVTX) in 2011 and 2012. The VTX and FVTX cover $|y| < 1.2$ and $1.2 < y < 2.2$ in rapidity and 2π in azimuth. The VTX at mid-rapidity allows the statistical separation of electrons from bottom and charm decays using the track displacement from the collision vertex. In addition, J/ψ from B decays is a clean channel to measure the open bottom production. The FVTX at forward rapidity enables the extraction of J/ψ from B decays from prompt J/ψ.

In the proceedings, we report the recent results on bottom and charm electrons in central Au+Au collisions and their nuclear modifications, and J/ψ from B decays in $p + p$ and Cu+Au collisions at forward rapidity.

⋆e-mail: hachiya@cc.nara-wu.ac.jp

2 Single electron yields from bottom and charm decays in Au + Au collisions

We analyzed the dataset from Au+Au collisions at $\sqrt{s_{NN}}$ = 200 GeV in 2014. The purpose of the analysis is the separation of electrons from bottom and charm hadron decays using the difference of their decay lengths. B meson has longer decay length than that of D mesons ($c\tau_{B^0} \approx 455\ \mu m$ and $c\tau_{D^0} \approx 123\ \mu m$). Electron identification was first performed using a ring-imaging Čerenkov detector and the energy-momentum matching. The distance of closest approach (DCA) of the electron to the collision vertex was then measured with the VTX. The DCA is proportional to the decay length, thus the shape of the DCA is broader for electrons from bottom decays than that from charm decays. Inclusive electrons contains not only the bottom and charm electrons but a large amount of background electrons. The main sources of these backgrounds are photon conversions, Dalitz decays of π^0 and η as well as the detector effects such as mis-identified hadrons and high multiplicity background. Those backgrounds are estimated by the event driven method and the PHENIX detector simulations.

To separate the bottom and charm components, the unfolding method performs a simultaneous fitting to the DCA distributions and the invariant yield of electrons from inclusive heavy quark decays as described in the article[4]. Figure 1 shows the electron DCA distribution. The bottom and charm components obtained by the unfolding method are also plotted as well as the sum of the all background components.

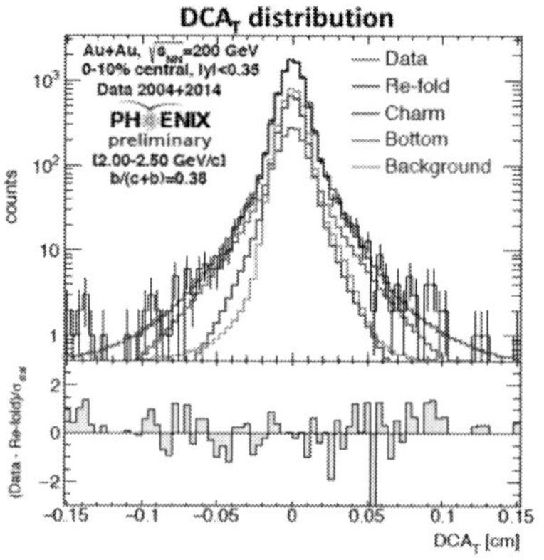

Figure 1. The electron DCA distribution for 0-10% central Au+Au collisions. The bottom and charm components from the unfolding method are also plotted.

The nuclear modification factor (R_{AA}) for bottom electrons and charm electrons are calculated using the bottom electron fractions in Au+Au and $p+p$[5, 6], and R_{AA} of electrons from inclusive heavy quark decays[2]. Figure. 2 show R_{AA}' for minimum bias (MB) (left) and 0-10% central Au+Au collisions (right). We found that bottom electrons are less suppressed than charm electrons decays for $p_T < 4$ GeV/c and similarly suppressed for higher p_T in both 0-10% central and MB collisions. The result is consistent with the expectation of the mass ordering of energy loss, $\Delta E_c > \Delta E_b$. In addition, charms in 0-10% central collisions are more suppressed than that in MB collisions.

3 J/ψ from B decays in $p+p$ and Cu+Au collisions

J/ψ from B meson decays is a clean measurement of open bottom production. The dataset used in the analysis were $p+p$ at $\sqrt{s}$ =200 and 510 GeV and Cu+Au collisions in 2012 and 2015. B production in $p+p$ is good for testing perturbative QCD and provides an important baseline for the measurement in nuclear collisions. The advantage of asymmetric Cu+Au collisions is to study QGP with different initial state effect by measuring at positive and negative rapidities.

PHENIX measures single muons from the decays of $B \to J/\psi(+X) \to \mu^+ + \mu^-$ at forward and backward rapidity. J/ψ is reconstructed from unlike sign di-muon pairs within the invariant mass $2.8 < M_{\mu\mu}$ (GeV/c^2) < 3.5. FVTX measures DCA of single muons from J/ψ decays. The DCA distribution of muons for J/ψ from B decay is broader than that of prompt J/ψ and have longer tail in negative side because the detector acceptance at the negative side is wider, while the DCA for prompt J/ψ is symmetric.

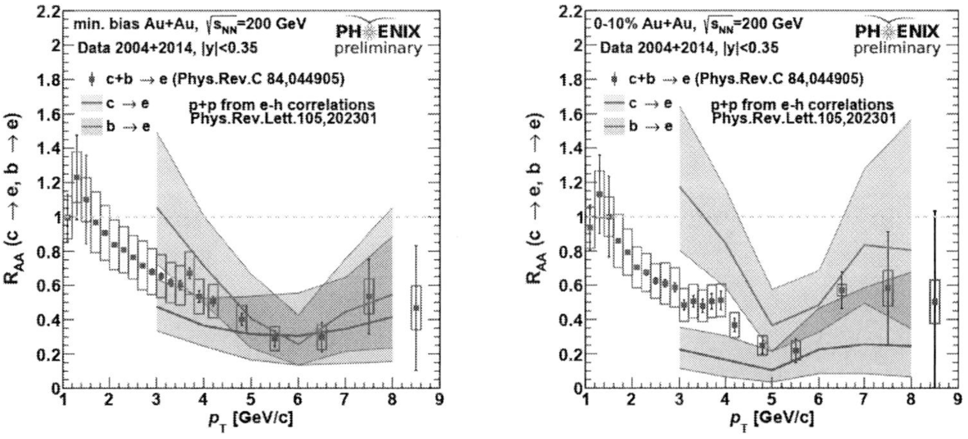

Figure 2. R_{AA} for bottom and charm electrons for MB (left) and 0-10% central Au+Au collisions (right)

Figure 3 shows the single muon DCA distribution from J/ψ decays in Cu+Au collisions. The DCA distribution is fit with the template DCA shapes of J/ψ from B decays and prompt J/ψ to extract the B decay components. The analysis procedure is described in detail[7]. The J/ψ from B decays and prompt J/ψ from the fitting are also plotted.

Figure 4 shows the fraction of J/ψ from B decays relative to inclusive J/ψ in $p + p$ and Cu+Au collisions at $\sqrt{s} = 200$ GeV . By comparing with the FONLL calculation[8] for J/ψ from B decays and Color Evaporation Model[9] for prompt J/ψ, it is found that the calculation is consistent with the data in $p+p$. On the other hand, the fractions in Cu+Au collisions are much larger than that in $p + p$.

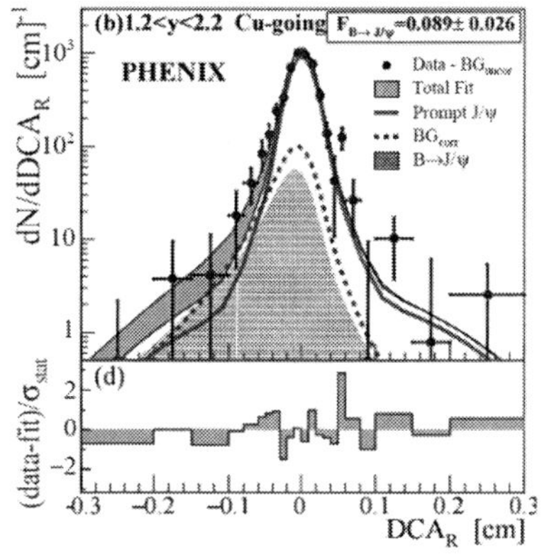

Figure 3. DCA distribution of muons from J/ψ decays in Cu+Au collisions. $B \to J/\psi$ component extracted by the fitting is plotted as well as the prompt J/ψ.

The nuclear modification factor R_{AA} of J/ψ from B decays and prompt J/ψ is calculated using the B $rightarrowJ/\psi$ fractions in Cu+Au and $p + p$, and the R_{AA} of inclusive J/ψ[10]. Figure 5 shows these R_{AA} as a function of rapidity. The R_{AA} of J/ψ of B decays are consistent with unity within uncertainties for both Cu- and Au-going directions even though is also consistent with a small enhancement from the initial state effect predicted by the EPS09 prediction[11].

In 510 GeV $p + p$ collisions, $d\sigma/dy(pp \to b\bar{b})$ is determined by the fraction of J/ψ from B decays and the inclusive J/ψ cross section by the extrapolation of the PHENIX 200 GeV $p + p$ result[12]. The value of $d\sigma/dy(pp \to b\bar{b})$ is $3.57^{+2.38}_{-2.22}$ μb at positive rapidity ($3.68^{+2.08}_{-1.88}$ μb at negative rapidity). These values are comparable with the FONLL calculation within large uncertainties.

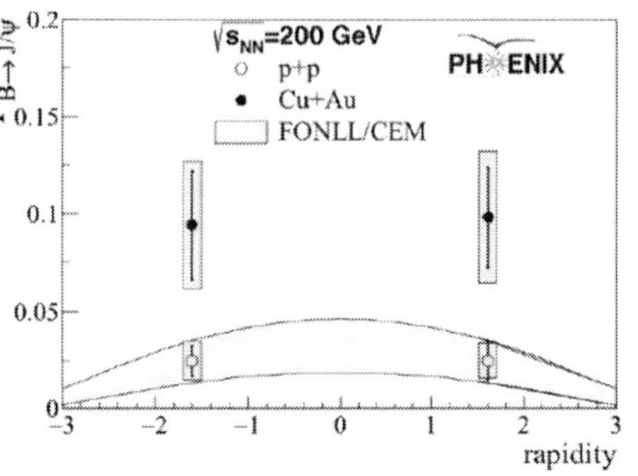

Figure 4. $B \to J/\psi$ fraction as a function of rapidity for $p + p$ and Cu+Au collisions.

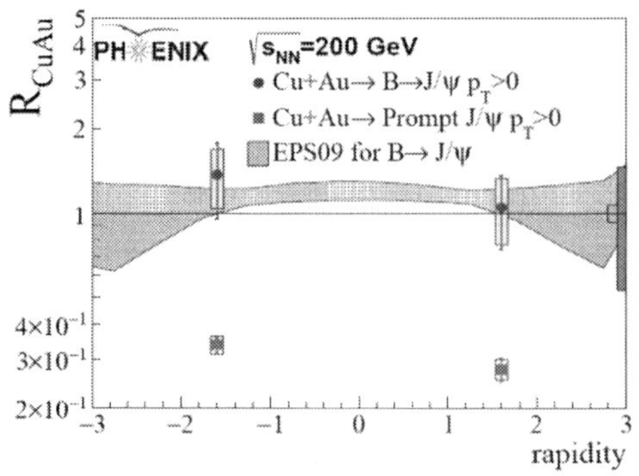

Figure 5. R_{AA} of J/ψ from B decays and prompt J/ψ as a function of rapidity.

4 Summary

PHENIX measured bottom and charm quark productions in $p + p$, Cu+Au, and Au+Au collisions at $\sqrt{s_{NN}}$ = 200 and 510 GeV . The VTX and FVTX enables the statistical separation of the bottom and charm contributions. In Au+*Au*, we studied the nuclear modification of bottoms and charms for 0-10% central collisions and MB collisions. The results show the less suppression of bottoms than that of charms. In addition, charms in 0-10 % central collisions is more suppressed than that in MB collisions. From J/ψ from B decays by the FVTX at forward rapidity, the bottom yield in $p + p$ collisions is consistent with the the FONLL + CEM calculation. In Cu+Au collisions, the nuclear modification of bottoms is consistent with the number of binary collision scaling for both positive and negative rapidities. These results provides important information about the QGP properties from new angle. The analyses with the significantly large amount of Au+Au and $p + p$ data collected in 2014 to 2016 are underway. These new dataset should provide more precise measurement and quantitative constraints on the QGP properties.

References

[1] A. Adare *et al. Phys. Rev. Lett.* **98** 172301 (2007)
[2] A. Adare *et al. Phys. Rev.* C **84**, 044905 (2011)
[3] L. Adamczyk *et al. Phys. Rev. Lett.* **113**, 142301 (2014)
[4] A. Adare *et al. Phys. Rev.* C **93** 034904 (2016)
[5] A. Adare *et al. Phys. Rev. Lett.* 103 082002 (2009)
[6] M. Aggarwal *et al. Phys. Rev. Lett.* 105 202301 (2010)
[7] C. Aidala *et al*, arXiv:1702.01085
[8] M. Cacciari, M. Greco, and P. Nason, JHEP *05, 007* (1998)
[9] A. D. Frawley, T. Ullrich, and R. Vogt, Phys. Rept. 462,790 125 (2008)
[10] C. Aidala *et al., Phys. Rev.* C **90** 064908 (2014)
[11] K. J. Eskola, H. Paukkunen, and C. A. Salgado, JHEP **04**, 065 (2009)
[12] C. Aidala *et al., Phys. Rev.* D **95** 092002 (2017)

Centrality and transverse momentum dependence of J/ψ production in Pb-Pb collisions at $\sqrt{s_{NN}}$ = 5.02 TeV at mid-rapidity with ALICE

Dennis Weiser[1,*] for the ALICE collaboration

[1] *Physikalisches Institut, Im Neuenheimer Feld 226, Heidelberg*

Abstract. We present new results for the nuclear modification factor R_{AA} of J/ψ mesons as a function of centrality and transverse momentum at mid-rapidity. The measurement is carried out with the ALICE central barrel detectors in the acceptance range $|y| < 0.9$ and $p_T > 0$ in the dielectron decay channel.

1 Introduction

The measurement of J/ψ production is a key measurement to study deconfinement in nucleus-nucleus collisions. The initial idea was that the high color density in the medium prevents the formation of charmonia [1]. This would be observable as a suppression of the yield of J/ψ mesons in AA collisions compared to pp collisions scaled to binary nucleon-nucleon collisions. This suppression of the J/ψ yield in nucleus-nucleus collisions has been observed at lower energies at SPS and RHIC [2–5].
At higher center-of-mass energies at the LHC the additional production of J/ψ mesons via the (re)combination of $c\bar{c}$ quarks is possible due to the increasing charm cross section $\sigma_{c\bar{c}}$. There are two main model scenarios describing the production of J/ψ by (re)combination: Thermal models and transport models. The thermal models assume a thermalization of charm in the medium and propose the production of J/ψ through statistical weights at the phase boundary [6]. Transport models describe the production and dissociation of J/ψ as a dynamical process already before hadronization. Throughout the evolution of the medium partial destruction of J/ψ is counterbalanced by continuous production by (re)combination [7]. The data from the LHC Run 1, taken at a collision energy of 2.76 TeV, have indeed shown that a (re)combination process is at work at this energy [8]. However, the discussion on the details of this (re)combination process described by the different model scenarios still did not come to a conclusive picture. The newly taken data at $\sqrt{s_{NN}}$ = 5.02 TeV in the LHC Run 2 provide further insights to the possible production mechanisms, due to more precise and differential measurements enabled by the higher available statistics.

* e-mail: dennis.weiser@cern.ch

2 Analysis

The analysis is based on a Pb-Pb dataset corresponding to an integrated luminosity of $\mathcal{L}_{\rm int} \approx 10\,\mu b^{-1}$ taken at $\sqrt{s_{\rm NN}} = 5.02$ TeV with the ALICE detector [9]. J/ψ mesons are reconstructed at mid-rapidity ($|y| < 0.9$) and down to transverse momentum $p_{\rm T} = 0$ in the dielectron decay channel. The Inner Tracking System (ITS) and the Time Projection Chamber (TPC) are used for track reconstruction. Secondary particles are rejected using the ITS. Electrons are identified using the specific energy loss ($\mathrm{d}E/\mathrm{d}x$) in the TPC. Combinatorial background introduced by electrons stemming from photon conversions is reduced by removing electrons which form pairs with very low invariant mass ($m_{\rm ee} < 50$ MeV) from the pairing. For the description of the background the event mixing technique is used. After subtraction of the background the signal is extracted in the invariant mass window $2.92 < m_{\rm ee} < 3.16$ GeV/c^2 by bin counting. The loss of signal due to the tail stemming from bremsstrahlung is taken into account as part of the efficiency.
The J/ψ proton-proton reference cross-section at $\sqrt{s} = 5.02$ TeV is obtained from an interpolation procedure taking into account measurements at lower and higher collision energies [10–13].

3 Results

The nuclear modification factor $R_{\rm AA}$ for J/ψ mesons is obtained as a function of centrality, rapidity and transverse momentum. For the mid-rapidity measurement the J/ψ candidates with $p_{\rm T} < 150$ MeV have been excluded to avoid a contamination with photoproduced J/ψ's which mostly play a role in peripheral collisions. Figure 1 shows the nuclear modification factor as a function of the mean number of participants $\langle N_{\rm part}\rangle$. The data points show no strong dependence on centrality; for most central collisions a hint of an increase can be noted which is also consistent with a fluctuation. The models contain a significant component of J/ψ production by (re)combination and describe the data rather well within their uncertainties although the transport model TM2 [14] slightly deviates from the data in peripheral collisions. The model uncertainties are dominated by the imprecise knowledge of the total charm cross-section in pp collisions and the modification of the parton distribution functions in the Pb nucleus (shadowing). The global normalization uncertainty for the data points (red box) is dominated by the J/ψ pp reference cross section and amounts to 16.6 %.

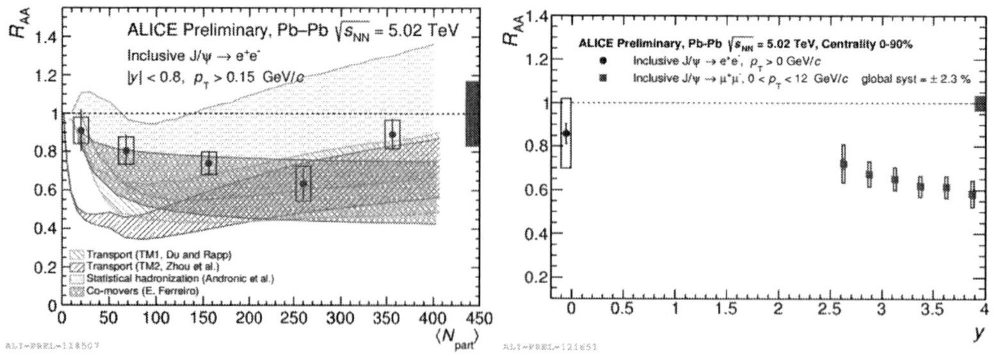

Figure 1. Centrality (left) and rapidity (right) dependent nuclear modification factor. The former is compared to predictions from a thermal model [15], transport models [14, 16, 17] and a comover model [18, 19].

For the rapidity dependent measurement (Fig. 1 right panel) the normalization uncertainty for the mid-rapidity measurement is included in the mid-rapidity uncertainty box. The data show a

trend of an increase in R_{AA} from forward to mid-rapidity. This is in line with the expectation from (re)combination models due to the increase of charm cross-section from forward to mid-rapidity.
The nuclear modification factor as function of p_T is shown in Fig. 2 for the full centrality range. The data show a stronger suppression with increasing p_T. This can be interpreted with the relevance of (re)combination at low p_T while at high p_T color screening or jet-quenching mechanisms should become dominant. The data are confronted with two transport model predictions. The model TM1 [16, 17] slightly undershoots the data, but can describe the data when all uncertainties are taken into account. The model TM2 [14] clearly undershoots the data at high p_T. One might ask whether (re)combination is still relevant at high p_T or whether the hard production and/or B feed-down is modeled accurately enough in this model.

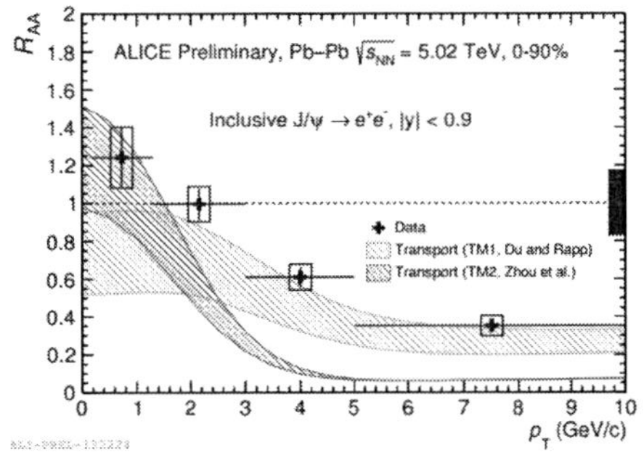

Figure 2. The nuclear modification factor in the 0–90 % centrality range as function of p_T compared to two transport models [14, 16, 17].

The transverse momentum dependent nuclear modification factor measured in the 0–20 % centrality range is shown in Fig. 3 compared to the forward measurement and compared to a transport model prediction. The general evolution of the R_{AA} with p_T is similar as for the centrality integrated case. Special emphasis should be given to the first two dielectron data points which exceed the muon measurement at low p_T. This rapidity dependence at low p_T is also expected from models including (re)combination due to the higher charm cross-section at mid-rapidity. Again, the transport model (TM1) can describe the data when considering all uncertainties.

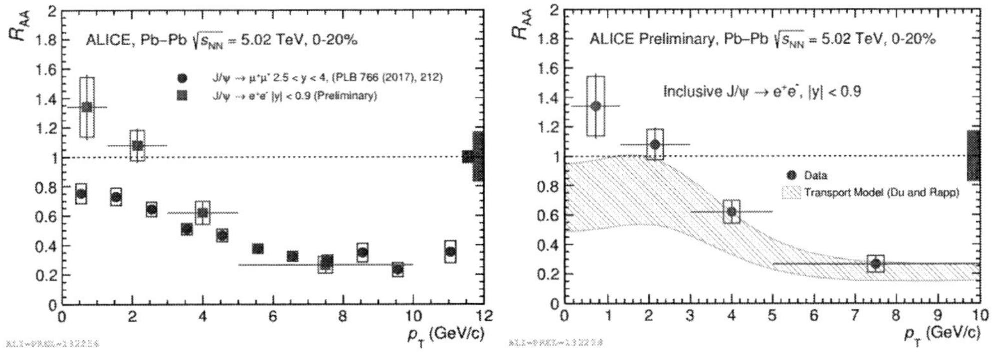

Figure 3. The nuclear modification factor measured in the 0–20 % centrality range compared to the forward measurement (left plot) and compared to a transport model prediction [16, 17] (right plot).

4 Conclusions

The newly taken data by ALICE demonstrate that (re)combination plays an essential role for low p_{T} J/ψ production in Pb-Pb collisions at the highest available collision energies. At low p_{T}, a hint for a J/ψ nuclear modification factor exceeding unity has been observed. This observation is consistent with (re)combination models and thus supports the expectation of deconfinement and thermalization of charm in nucleus-nucleus collisions at LHC energies. Future measurements of the total charm cross-section in Pb-Pb collisions as planned by ALICE for the LHC Run 3 will help to reduce the model uncertainties and will thus allow the models to be put under close scrutiny when confronted with data.

References

[1] T. Matsui and H. Satz, Phys. Lett. **B178**, 416 (1986)
[2] NA50 Collaboration, Eur. Phys. J. **C39**, 335
[3] PHENIX Collaboration, Phys. Rev. Lett. **98**, 232301 (2007)
[4] PHENIX Collaboration, Phys. Rev. **C84**, 054912 (2011)
[5] PHENIX Collaboration, Phys. Rev. Lett. **101**, 122301 (2008)
[6] P. Braun-Munzinger and J. Stachel, Phys. Lett. **B490**, 196 (2000)
[7] R. L. Thews, M. Schroedter and J. Rafelski, Phys. Rev. **C63**, 054905 (2001)
[8] ALICE Collaboration, Phys.Lett. B **734** 314 (2014)
[9] ALICE Collaboration JINST **3**, S08002 (2008)
[10] PHENIX, Phys. Rev. Lett. **98**, 232301 (2007)
[11] CDF, Phys. Rev. D. **71**, 032001 (2005)
[12] ALICE, Phys. Lett. B **718**, 295 (2012)
[13] ALICE, Phys. Lett. B **718**, 692 (2012)
[14] K. Zhou, N. Xu, Z. Xu, and P. Zhuang, Phys. Rev. **C89** 054911 (2014)
[15] A. Andronic et al., Nucl. Phys. **A904-905** 535c (2013)
[16] X. Zhao and R. Rapp, Nucl. Phys. **A859**, 114 (2011)
[17] X. Du and R. Rapp, Nucl. Phys. **A943**, 147 (2015)
[18] E. G. Ferreiro, Phys. Lett. **B731** (2014) 57
[19] E. G. Ferreiro Phys. Lett. **B749** (2015) 98

J/ψ production in proton-lead collisions at 8 TeV with the LHCb detector

Oleksandr Okhrimenko[1,*] on behalf of LHCb Collaboration

[1] *Institute for Nuclear Research NAS of Ukraine, Kiev, Ukraine*

Abstract. The production of J/ψ mesons is studied in proton-lead collisions at a centre-of-mass energy per nucleon pair $\sqrt{s_{NN}} = 8.16\,\text{TeV}$ with the LHCb detector. The double differential cross-sections of prompt and nonprompt J/ψ production are measured as functions of the J/ψ transverse momentum and rapidity. Nuclear modification factors are determined. The results are compared with different theoretical calculations.

1 Introduction

The study of proton-nucleus collisions is important to disentangle the effects of quark-gluon plasma from non-deconfinement ones. The measurements of the production of prompt J/ψ and nonprompt J/ψ mesons are important ingredients for the understanding of the imprints of deconfinement in nucleus-nucleus collisions. They are based on 10 and 40 times larger integrated luminosities than the initial measurements with the 2013 pPb and Pbp data samples, respectively, by the LHCb experiment at $\sqrt{s_{NN}} = 5\,\text{TeV}$ [1]. Furthermore, these results can be used to evaluate models, such as collinear factorisation with different nuclear parton distribution functions (nPDFs) [2], coherent energy loss [3] and the colour glass condensate model (CGC) [4], describing nuclear modifications in proton-nucleus collisions. In addition, the measurements are a natural reference for the measurement of $\phi(2S)$ and χ_c states, which will further clarify the modification of quarkonium production in proton-lead collisions at the TeV scale. In detail this analysis is described in Ref. [5].

2 Analysis Description and Results

The LHCb detector [6, 7] is a single-arm forward spectrometer covering the pseudorapidity range $2 < \eta < 5$, designed for the study of particles containing b or c quarks.

This analysis is based on data acquired during the 2016 LHC heavy-ion run, where protons and ^{208}Pb ions were colliding at a centre-of-mass energy per nucleon pair of $\sqrt{s_{NN}} = 8.16\,\text{TeV}$. Since the energy per nucleon in the proton beam is larger than in the lead beam, the nucleon-nucleon centre-of-mass system has a rapidity in the laboratory frame of 0.465 (−0.465), when the proton (lead) beam travels from the vertex detector towards the muon chambers, which is denoted as pPb (Pbp). Consequently, the LHCb detector covers two different acceptance regions: $1.5 < y^* < 4.0$ for pPb and $-5.0 < y^* < -2.5$ for Pbp, where y^* is the rapidity in the centre-of-mass frame of the colliding

*e-mail: okhrimenko@kinr.kiev.ua

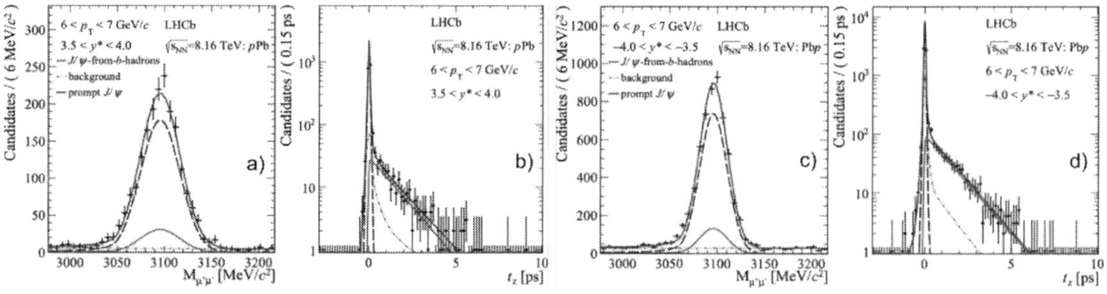

Figure 1. (a, c) Invariant mass and (b, d) pseudo proper time distributions for J/ψ candidates in the bin $6 < p_{\mathrm{T}} < 7\,\mathrm{GeV}/c$ and $3.5 < |y^*| < 4.0$ for the (a, b) pPb and (c, d) Pbp samples respectively. The black circles with error bars represent the LHCb data. The red solid line is the total fit function, the blue dashed line is the prompt J/ψ signal component, the purple solid line is the nonprompt one and the green dashed line is the combinatorial background component.

nucleons, with the proton beam direction defined to be positive y^*. The data samples correspond to an integrated luminosity of $13.6 \pm 0.3\,\mathrm{nb}^{-1}$ of pPb collisions and $20.8 \pm 0.5\,\mathrm{nb}^{-1}$ of Pbp collisions.

Candidate J/ψ are reconstructed through the $J/\psi \to \mu^+\mu^-$ decay channel. Each identified muon track is required to have $p_{\mathrm{T}} > 750\,\mathrm{MeV}/c$, $2 < \eta < 5$ and to have a good-quality track fit. The two muon tracks of the J/ψ candidate must form a good-quality vertex, representing a tighter selection compared to the software trigger requirement. Also, only J/ψ candidates with an invariant mass $M_{\mu^+\mu^-}$ within $120\,\mathrm{MeV}/c^2$ of the known value of the J/ψ mass ($3097\,\mathrm{MeV}/c^2$) are selected. The number of J/ψ candidates is extracted by fitting the invariant mass distribution, where the signal is described by a Crystal Ball function, and the combinatorial background by an exponential function (Fig. 1, a, c plots). Prompt J/ψ can be distinguished from nonprompt J/ψ by exploiting the pseudo proper time distribution (Fig. 1, b, d plots) defined as $t_z \equiv (z_{J/\psi} - z_{\mathrm{PV}}) \cdot M_{J/\psi}/p_z$, where $z_{J/\psi}$ and z_{PV} are the coordinates along the beam axis of the J/ψ decay vertex position and of the primary vertex position, p_z is the z component of the J/ψ momentum and $M_{J/\psi}$ the known J/ψ mass. The t_z distribution of prompt J/ψ is described by a Dirac δ-function $\delta(t_z)$, and that of nonprompt J/ψ by an exponential function for $t_z > 0$. The background t_z distribution is described by an empirical function derived from the shape observed in the J/ψ upper mass sideband, $3200 < M_{\mu^+\mu^-} < 3250\,\mathrm{MeV}/c^2$.

The double differential J/ψ production cross-section in each (p_{T}, y^*) bin is computed as

$$\frac{\mathrm{d}^2\sigma}{\mathrm{d}p_{\mathrm{T}}\mathrm{d}y^*} = \frac{N(J/\psi \to \mu^+\mu^-)}{\mathcal{L} \cdot \epsilon_{tot} \cdot \mathcal{B}(J/\psi \to \mu^+\mu^-) \cdot \Delta p_{\mathrm{T}} \cdot \Delta y^*}, \tag{1}$$

where $N(J/\psi \to \mu^+\mu^-)$ is the number of reconstructed prompt or nonprompt J/ψ signal mesons, $\epsilon_{tot}$is the total detection efficiency in the given kinematic bin, $\mathcal{B}(J/\psi \to \mu^+\mu^-) = (5.961 \pm 0.033)\%$ is the branching fraction of the decay $J/\psi \to \mu^+\mu^-$, $\Delta p_{\mathrm{T}} = 1\,\mathrm{GeV}/c$ and $\Delta y^* = 0.5$ are the bin widths and $\mathcal{L}$ is the integrated luminosity. The total detection efficiency, ϵ_{tot}, is the product of the geometrical acceptance, and the efficiencies for charged track reconstruction, particle identification, candidate and trigger selections. Samples of simulated events are used to evaluate these efficiencies except for the track reconstruction and particle identification, which are determined in a data-driven approach. In the simulation, pPb and Pbp minimum-bias collisions are generated using the Epos LHC [8] and J/ψ candidates from Pythia8 are injected into each event [9]. The measured double-differential cross-sections of prompt and nonprompt J/ψ in the pPb and Pbp data samples are shown in Fig. 2, as a function of p_{T} for the considered y^* bins.

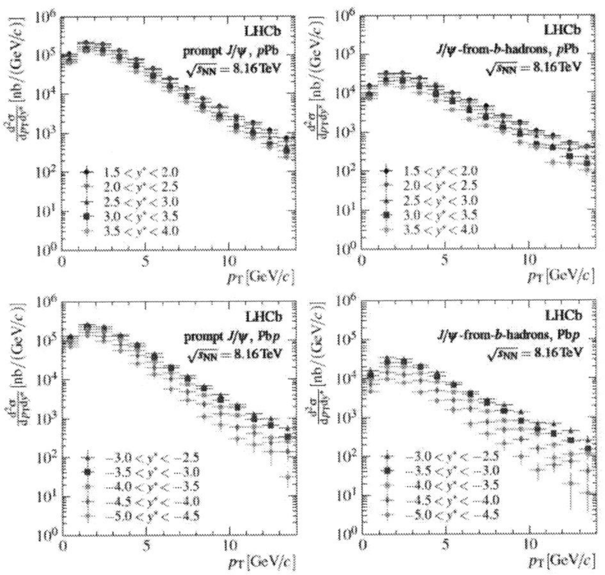

Figure 2. Production cross-section for (left column) prompt J/ψ and (right column) nonprompt J/ψ in (top row) pPb and (bottom row) Pbp collisions as function of p_T for the considered y^* bins.

The nuclear modification factor $R_{p\text{Pb}}$ is computed from the prompt and nonprompt J/ψ production cross-sections in pp [10, 11] and pPb or Pbp collisions using the following definition:

$$R_{p\text{Pb}}(p_T, y^*) \equiv \frac{1}{A}\frac{\mathrm{d}^2\sigma_{p\text{Pb}}(p_T, y^*)/\mathrm{d}p_T\mathrm{d}y^*}{\mathrm{d}^2\sigma_{pp}(p_T, y^*)/\mathrm{d}p_T\mathrm{d}y^*}, \tag{2}$$

where $A = 208$ is the mass number of the Pb ion, $\mathrm{d}^2\sigma_{p\text{Pb}}(p_T, y^*)/\mathrm{d}p_T\mathrm{d}y^*$ the J/ψ production cross-section in pPb or Pbp collisions and $\mathrm{d}^2\sigma_{pp}(p_T, y^*)/\mathrm{d}p_T\mathrm{d}y^*$ the J/ψ reference production cross-section in pp collisions at the same nucleon-nucleon centre-of-mass energy.

The results obtained are shown in Fig. 3 (top and bottom rows denotes to prompt and non-prompt J/ψ, respectively). A suppression of prompt J/ψ production compared to pp collisions of up to 50% (25%) in pPb (Pbp) at the lowest p_T is observed. In both configurations, the nuclear modification factor approaches unity asymptotically at the highest p_T. Theoretical calculations for the nuclear modification factor based on nPDFs, coherent energy loss as well as the CGC model can account for the majority of the observed dependences. For the first time, beauty-hadron production is measured through inclusive decays $b \to J/\psi X$ precisely down to $p_T = 0$ at the LHC in pPb and Pbp collisions. In pPb, a weak suppression at the lowest p_T is observed, whereas in Pbp no significant deviation from unity in the nuclear modification factor is found. This weak modification of beauty production in proton-ion collisions is an important ingredient for the investigation of the modifications of beauty production in heavy-ion collisions.

3 Conclusions

The differential production cross-sections of prompt and nonprompt J/ψ in pPb and Pbp collisions at $\sqrt{s_{NN}} = 8.16\,\text{TeV}$ are measured in the range $0 < p_T < 14\,\text{GeV}/c$. The nuclear modification factors are

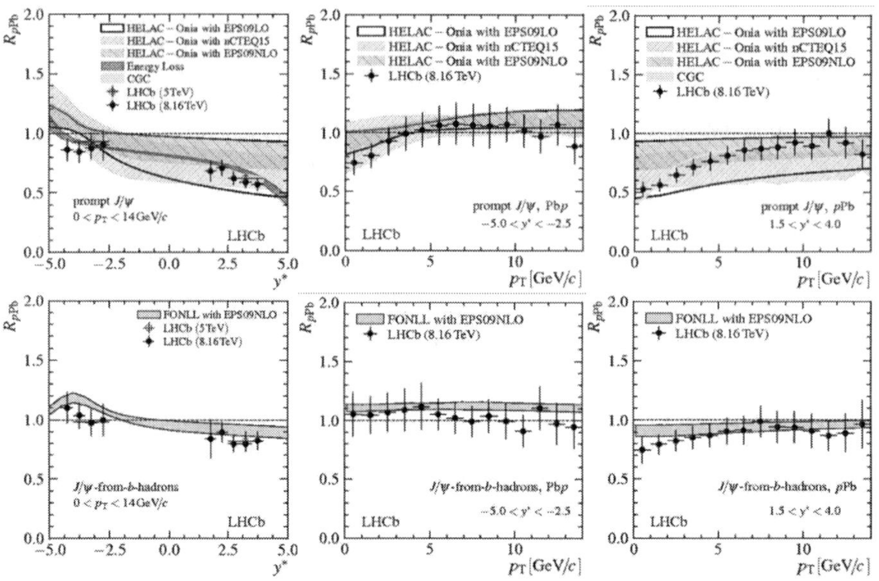

Figure 3. J/ψ nuclear modification factor, $R_{p\mathrm{Pb}}$: (left column) $R_{p\mathrm{Pb}}$ integrated over p_{T} in the range $0 < p_{\mathrm{T}} < 14\,\mathrm{GeV}/c$, as a function of y^*; (middle and right columns) $R_{p\mathrm{Pb}}$ integrated over y^* in the analysis range, as a function of p_{T} in Pbp and pPb collisions, respectively. Top and bottom plots represents prompt and nonprompt J/ψ, respectively. The horizontal error bars indicate the bin widths and vertical ones the total uncertainties. The black circles are the values measured in this letter, the red squares the ones at $\sqrt{s_{NN}} = 5\,\mathrm{TeV}$ from Ref. [1] and the coloured areas the theoretical computations from the models detailed in the text, with their uncertainties.

similar to those found at a collision energy of $\sqrt{s_{NN}} = 5\,\mathrm{TeV}$, but with increased precision. Although the presented measurements have improved precision, it is not possible to single out the main nuclear modification mechanism between different phenomenological models due to large uncertainties of nPDFs ones for charmonium production in proton-lead collisions at the TeV scale. This measurement of J/ψ production is the first step towards measurements of other charmonium states.

References

[1] LHCb Collaboration, JHEP **02**, 072 (2014).
[2] H.-Sh. Shao, Comp. Phys. Com. **184**, 2562 (2013); Comp. Phys. Com. **198**, 238 (2016).
[3] B. Ducloue *et al.*, Phys. Rev. **D91**, 114005 (2015); Phys. Rev. **D94** 074031 (2016).
[4] F. Arleo and S. Peigne, JHEP **03** 122 (2013).
[5] LHCb Collaboration, Phys. Lett. B **774**, 159–178 (2017).
[6] LHCb Collaboration, JINST **3**, S08005 (2008).
[7] LHCb Collaboration, Int. J. Mod. Phys. **A 30**, 15300227 (2015).
[8] T. Pierog *et al.*, Phys. Rev. **C92**, 034906 (2015); arXiv:1306.0121.
[9] T. Sjostrand, S. Mrenna, P. Skands, Comp. Phys. Com. **178**, 852 (2008); arXiv:0710.3820.
[10] ALICE and LHCb Collaborations, LHCb-CONF-2013-013; ALICE-PUBLIC-2013-002.
[11] LHCb Collaboration, EPJ **C 71**, 1645 (2011); JHEP **06**, 064 (2013); JHEP **10**, 172 (2015).

EPJ Web of Conferences **171**, 19002 (2018) https://doi.org/10.1051/epjconf/201817119002
SQM 2017

Quarkonium production in pp and p-A collisions with ALICE at the LHC

Astrid Morreale for the ALICE collaboration[1,*]

[1] *Laboratoire de Physique Subatomique et des Technologies Associees SUBATECH. 4 Rue Alfred Kastler, 44300 Nantes, France*

Abstract. Quarkonia are mesons formed of either a charm and anti-charm quark pair (J/ψ, $\psi(2S)$) or a beauty and anti-beauty quark pair (Υ(1S), (2S) and (3S)). We report on forward rapidity ($2.5 < y < 4$) J/ψ and $\psi(2S)$ production measured in pp collisions at $\sqrt{s} = 13$ TeV, using data collected at the LHC in 2015. The results will be compared with similar measurements performed at $\sqrt{s} = 2.76$, 5.02, 7 and 8 TeV. They will be further compared to NRQCD and FONLL calculations, which describe prompt and non-prompt charmonium production respectively. Results of the J/ψ nuclear modification factor as a function of collision centrality in p–Pb collisions at $\sqrt{s_{\rm NN}} = 8.16$ TeV, at forward and backward rapidities, will also be presented. These measurements will be compared with Run-1 (2009 – 2013) results as well as theoretical calculations and will be interpreted in terms of cold nuclear matter effects.

1 Introduction

Measuring quarkonia in pp collisions may help to understand their production mechanisms and probe the parton distribution functions (PDF), in particular the gluon's PDF down to low x . These measurements also provide a reference to p–Pb and Pb–Pb measurements which in turn are used to quantify the properties of the quark-gluon plasma. In p–Pb these same measurements may probe cold nuclear matter effects such as modification of the PDFs in nuclei, saturation and Cronin enhancement [1, 2].

When we measure quarkonia we have contributions from decays from higher mass resonances[1] and decays from b hadrons. The decay contributions are typically referred to as non prompt J/ψ and ψ(2S). The ALICE quarkonium measurements presented in this document are inclusive which implies that no separation is made between prompt and non-prompt particles.

2 Apparatus

The ALICE detector is described in detail in [3, 4]. The detector subsystems relevant to the present analysis are the muon spectrometer, the Silicon Pixel Detector (SPD) corresponding to the two innermost layers of the Inner Tracking System (ITS), the ZDC detectors, the V0 scintillator hodoscopes and the T0 Cherenkov detectors.

Charmonia are measured in the $\mu^+\mu^-$ decay channel at forward rapidity and down to $p_{\rm T} = 0$ using the ITS for vertex determination, the Muon Trigger (MTR) for triggering, the Muon Chambers

*e-mail: astrid.morreale@cern.ch

[1] For the J/ψ these decay contributions are the ψ(2S) and the χ_c.

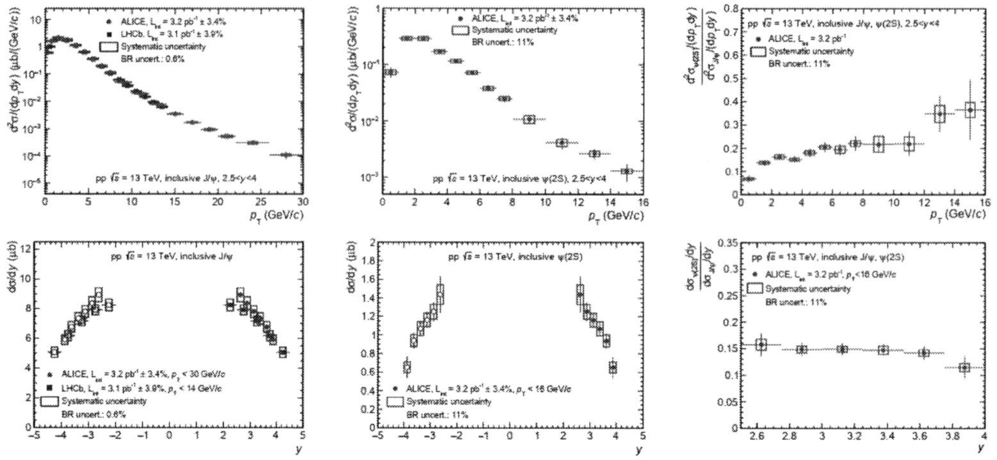

Figure 1. (color online). Inclusive J/ψ cross sections (left), ψ(2S) cross sections (middle) and ψ(2S)-to-J/ψ cross section ratios (right) as a function of p_{T} (top panels) and y (bottom panels) in pp collisions at $\sqrt{s}$ = 13 TeV. J/ψ cross sections are compared to LHCb measurements at the same $\sqrt{s}$ [6]. Open symbols are the reflection of the positive-y measurements with respect to y = 0.

(MCH) for tracking. The V0 detectors are also used for triggering (in coincidence with the MTR), the T0 detectors are used for the luminosity determination and finally the V0 and ZDC detectors are used for the centrality determination in p–Pb together with several charged particle multiplicity estimators defined in [5].

3 J/ψ and ψ(2S) production in pp collisions

Quarkonia are measured using fits to the invariant mass distribution of $\mu^+\mu^-$ pairs detected in the muon system and at forward rapidity. The data sample analyzed for the measurements at $\sqrt{s}$ = 13 TeV correspond to an integrated luminosity $L_{\mathrm{int}} = 3.19 \pm 0.11$ pb^{-1}.

Figure 1 summarizes the inclusive J/ψ and ψ(2S) cross sections as a function of the charmonium transverse momentum p_{T} (top figures) and rapidity y (bottom figures). The left column shows the J/ψ cross sections while the middle and rightmost columns show the ψ(2S) cross sections and the ψ(2S)-to-J/ψ cross section ratios respectively.

The J/ψ production cross sections as a function of p_{T} and y are compared to measurements published by LHCb [6] at the same energy. The measurements of the two experiments are consistent within 1σ of their uncertainties. The ALICE measurement extends the p_{T} reach from 14 GeV/c to 30 GeV/c with respect to the LHCb results. For the ψ(2S) measurement, no comparisons are performed as this is the only measurement available to date at this energy and y range. In Fig. 2, the cross sections presented in the previous section are compared to other forward-y measurements in pp collisions at $\sqrt{s}$ = 2.76 [7], 5.02 [8], 7 [9] and 8 TeV [10].

All ALICE J/ψ and ψ(2S) measurements presented are inclusive and consist of a prompt and a non-prompt contribution. In order to compare model calculations to the data both contributions must be accounted for. This is illustrated in Fig. 3 for the J/ψ production cross section as a function of p_{T} in pp collisions at $\sqrt{s}$ = 13 TeV.

In the left panel of Fig. 3, ALICE data at $\sqrt{s}$ = 13 TeV are compared to three calculations: i) in grey to a prompt J/ψ Next-to-Leading-Order (NLO) NRQCD calculation from Ma, Wang and Chao [11], ii) in blue to a prompt J/ψ Leading Order (LO) NRQCD calculation coupled to a Color Glass Condensate (CGC) description of the low-x gluons in the proton from Ma and Venugopalan [12]

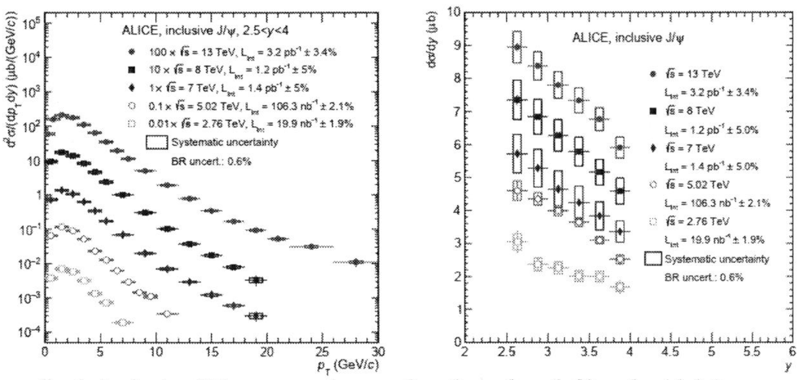

Figure 2. (color online). Inclusive J/ψ cross sections as function of p_T (left) and y (right).

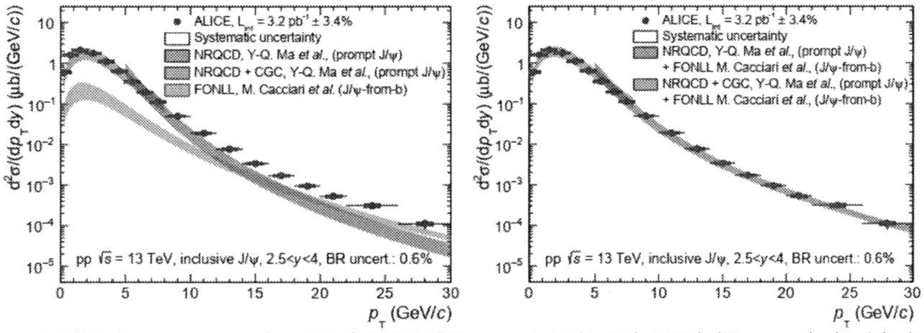

Figure 3. Left: ALICE data compared to NLO NRQCD (grey) [11], LO NRQCD coupled with CGC (blue) [12] and FONLL (red) [13]. Right: Summed NRQCD and FONLL predictions compared to the same data.

and iii) in red to a non-prompt J/ψ Fixed-Order Next-to-Leading Logarithm (FONLL) calculation by Cacciari *et al.* [13]. The summed NRQCD and FONLL calculations agree with ALICE data over the full p_T range and five orders of magnitude in σ.

4 J/ψ production in p–Pb collisions at $\sqrt{s_{NN}}$ = 8.16 TeV

In 2016, ALICE took data in two beam configurations corresponding to the proton momentum (p-Pb collisions) or to the lead-nucleus momentum (Pb-p collisions) beam oriented towards the muon spectrometer. Measurements were obtained of the nuclear modification factor R_{pPb} (Fig. 4) of J/ψ at forward (2.03< y_{cms} <3.53) and backward (-4.46< y_{cms} <-2.96) rapidity corresponding to an x region of $1.1\times10^{-5} < x < 5\times10^{-5}$ (p-Pb) and $7.3\times10^{-3} < x < 3.3\times10^{-2}$ (Pb-p).

The corresponding data samples for the reported results corresponds to an integrated luminosity of $L_{int}{}^{pPb} = 7.3$ nb^{-1} for p–Pb collisions and $L_{int}{}^{Pbp} = 10.2$ nb^{-1} for Pb–p collisions.

The results reported in Fig. 4 (top) extend the p_T range from 7 to 20 GeV/c from the previous lower energy result at $\sqrt{s_{NN}}$ = 5.02 TeV [14]. These new measurements are compared to theoretical calculations which include effects due to PDF modifications, color glass condensate, energy loss and final state effects. All models describe the data within the theoretical uncertainties [14].

The new preliminary results at $\sqrt{s_{NN}}$ = 8.16 TeV confirm with an increased precision the features already observed at $\sqrt{s_{NN}}$ = 5.02 TeV (Fig. 4 bottom). Mainly that the nuclear modification factor as function of centrality is smaller than one for forward rapidity and near unity for backward rapidity.

5 Summary

The inclusive J/ψ and ψ(2S) differential cross sections as well as ψ(2S)-to-J/ψ cross section ratios as a function of p_T and y measurements have been presented in pp collisions at $\sqrt{s}$ = 13 TeV with

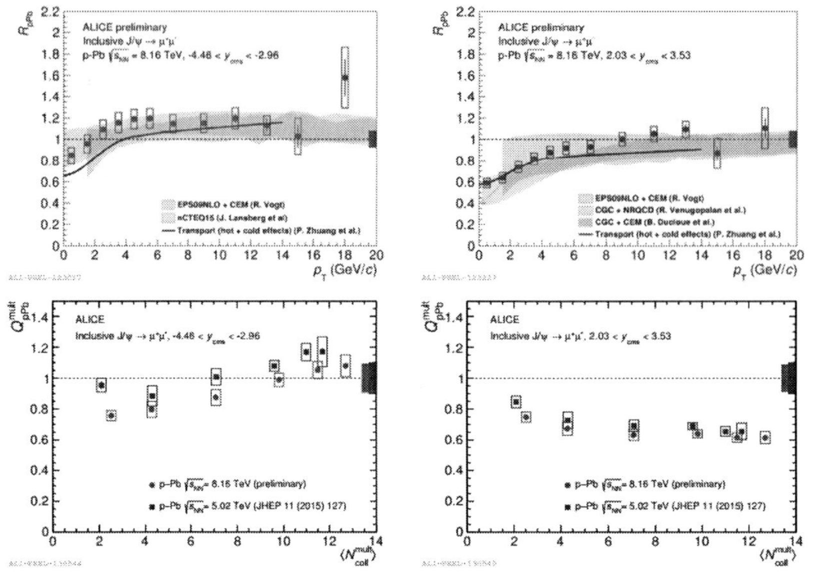

Figure 4. Top: J/ψ's R_{pPb} as a function of p_{T} at backward (left) and forward rapidity (right). Bottom: same measurements as a function of centrality and compared between two center-of-mass energies $\sqrt{s_{\mathrm{NN}}}$ = 8.16 and 5.02 TeV. See text for details.

the ALICE detector. Comparisons of these measurements to NLO NRQCD and LO NRQCD+CGC prompt-charmonium calculations have been presented. Agreement is observed between data and theory for both particles and as a function of both p_{T} and y provided that the non-prompt contribution to the inclusive cross section is included using FONLL. We have presented the measurement of the centrality dependence of inclusive J/ψ production in p–Pb collisions at $\sqrt{s_{\mathrm{NN}}}$ = 8.16 TeV. At forward rapidity, preliminary results show a suppression which slightly increases towards central collisions. At backward rapidity a different trend is observed, with the nuclear modification factor increasing from peripheral to central collisions. Results obtained at this new $\sqrt{s_{\mathrm{NN}}}$ are compatible with those previously measured by ALICE in p–Pb collisions at $\sqrt{s_{\mathrm{NN}}}$ = 5.02 TeV.

References

[1] N. Brambilla et al., Eur. Phys. J. C **71**, 1534 (2011), 1010.5827
[2] A. Andronic et al., Eur. Phys. J. **C76**, 107 (2016), 1506.03981
[3] K. Aamodt et al. (ALICE Collaboration), JINST **3**, S08002 (2008)
[4] B. Abelev et al. (ALICE Collaboration), Int. J. Mod. Phys. **A29**, 1430044 (2014), 1402.4476
[5] J. Adam et al. (ALICE Collaboration), Phys. Rev. **C91**, 064905 (2015)
[6] R. Aaij et al. (LHCb Collaboration), JHEP **10**, 172 (2015), [Erratum: JHEP05,063(2017)], 1509.00771
[7] B. Abelev et al. (ALICE Collaboration), Phys. Lett. **B718**, 295 (2012), [Erratum: Phys. Lett.B748,472(2015)], 1203.3641
[8] J. Adam et al. (ALICE Collaboration), Phys. Lett. **B766**, 212 (2017), 1606.08197
[9] B.B. Abelev et al. (ALICE Collaboration), Eur. Phys. J. **C74**, 2974 (2014), 1403.3648
[10] J. Adam et al. (ALICE Collaboration), Eur. Phys. J. **C76**, 184 (2016), 1509.08258
[11] Y.Q. Ma, K. Wang, K.T. Chao, Phys. Rev. Lett. **106**, 042002 (2011), 1009.3655
[12] Y.Q. Ma, R. Venugopalan, Phys. Rev. Lett. **113**, 192301 (2014), 1408.4075
[13] M. Cacciari et al., JHEP **10**, 137 (2012), 1205.6344
[14] ALICE-PUBLIC-2017-007 (2017), https://cds.cern.ch/record/2272151

Exploration of particle production mechanisms via angular correlations of π, K, p, Λ with ALICE in pp collisions at $\sqrt{s} = 7$ TeV

Małgorzata Anna Janik[1,*]

[1] *Faculty of Physics, Warsaw University of Technology, Poland*

Abstract. Two-particle correlations as a function of $\Delta\eta$ and $\Delta\varphi$ are used in many colliding systems to study a wide range of physical phenomena. Examples include the collective behavior of the quark-gluon plasma medium, jets, quantum statistics or Coulomb effects, conservation laws, and resonance decays. In this work, measurements of the correlations of identified particles and their antiparticles (for π, K, p, Λ) are reported in pp collisions at $\sqrt{s} = 7$ TeV at low transverse momenta. The analysis reveals differences in particle production between baryons and mesons. The correlation functions for mesons exhibit the expected peak dominated by the effects of mini-jet fragmentation and are reproduced well by general purpose Monte Carlo generators. For baryon pairs where both particles have the same baryon number, an anti-correlation structure is observed instead of a peak centered at $(\Delta\eta, \Delta\varphi) = (0, 0)$; an observation which presents a challenge to models typically used to describe pp data (PYTHIA, PHOJET). This baryon anti-correlation is further interpreted in the context of baryon production mechanisms in the fragmentation processes.

1 Introduction

Studies of particle production mechanisms in elementary collisions date back to the times of R. Feynman and R. Field, who proposed in 1977 a simple mechanism describing the principles of creation of the so-called "jets", which are collimated streams of particles [1]. They proposed rules for how the particles are created and the energy is distributed, and also considered limitations connected to the conservation laws. Elements of the proposed scheme are used even today in the most popular fragmentation models (such as the "Lund model" employed in the PYTHIA generator). However, the implementation details have to be compatible with the experimental data. It is then the task for the experiment to provide basic information: How strong should the correlations be between created hadrons? How does this correlation change, when two or more baryons or strange particles are created? E.g. answers to the questions about the details of strangeness and baryon number conservation have been searched so far only in e^+e^- collisions, at much lower energies than those achieved by contemporary high energy physics experiments and on substantially smaller data samples [2, 3].

We propose to study particle production mechanisms using two-particle angular correlations. This tool allows for a broad exploration of the underlying physics phenomena of particle production in collisions of both protons and heavy ions by measuring angular distributions in $\Delta\eta\Delta\varphi$ space (where $\Delta\eta$

*e-mail: majanik@if.pw.edu.pl

is the pseudorapidity difference and $\Delta\varphi$ is the azimuthal angle difference between two particles). The analysis is performed for identified particles, that is pions, kaons, protons, and lambda particles, produced in pp collisions at $\sqrt{s} = 7$ TeV recorded by the ALICE detector in 2010 [4]. The measured correlations should be sensitive to conservation laws as well as details of particle production mechanisms, including the parton fragmentation. In order to interpret the data in this context, dedicated Monte Carlo simulations were performed.

2 Analysis

The studies were done separately for particle–particle and anti-particle–anti-particle pairs, and for four particle species (π, K, p, Λ) [4]. All particles used for the analysis were measured at low transverse momenta, up to 2.5 GeV/c, within the pseudorapidity range $|\eta| < 0.8$. The particle identification of pions, kaons, and protons was performed on a track-by-track basis using information from the Time Projection Chamber and Time-Of-Flight detectors[5]. The applied procedure resulted in a purity above 99% for pions and protons and above 96% for kaons. The lambda baryons were reconstructed using their distinctive decay topology in the channel $\Lambda(\overline{\Lambda}) \to \mathrm{p}\pi^-(\overline{\mathrm{p}}\pi^+)$. The Λ purity was found to be above 95% [4].

The reported experimental correlation function is constructed as

$$C(\Delta\eta, \Delta\varphi) = \frac{S(\Delta\eta, \Delta\varphi)}{B(\Delta\eta, \Delta\varphi)}, \tag{1}$$

where $\Delta\eta = \eta_1 - \eta_2$ is the difference in pseudorapidity, $\Delta\varphi = \varphi_1 - \varphi_2$ is the difference in azimuthal angle. $S(\Delta\eta, \Delta\varphi)$ is the distribution of correlated pairs and $B(\Delta\eta, \Delta\varphi)$ is the reference distribution, calculated using the mixed-events technique, reflecting the single-particle acceptance. Both the S and B distributions are normalized by the respective number of pairs, therefore, the reported distribution is a ratio of probabilities.[1] All details of the analysis can be found in [4].

3 Results

In Fig. 1, correlation functions are presented for like-sign pairs for (a) pions, (b) kaons, (c) protons, and (d) lambdas.[2] Results show significantly different behavior of identical meson and baryon pairs: mesons (pions and kaons) exhibit a near-side peak for $\Delta\varphi = 0$, while for baryons (p and Λ) a depression is observed in this region.

The correlation functions are compared to predictions of Monte Carlo (MC) models. The following MC event generators were used: PYTHIA6.4 tunes Perugia-0 and Perugia-2011 [6, 7], PYTHIA8 Monash tune [8, 9], and PHOJET version 1.12 [10]. The correlations between mesons are qualitatively reflected by the models (the difference visible for pions comes from Bose–Einstein correlations which are absent in the studied MC samples). However, for baryon pairs (Fig. 1-bottom row) significant differences can be seen, both quantitative and qualitative. Models fail to reproduce the depression visible for like-sign baryons. All studied generators frequently produce two baryons close in phase-space (e.g. within the mini-jet peak), which is not reflected in the experimental data. Further studies performed with EPOS-LHC [11] and HERWIG [12] models (not shown on the plot) do not differ qualitatively from PYTHIA and PHOJET. In case of baryon-antibaryon pairs (results for $p\overline{p}$ can be seen in Fig. 2-left) we observe a significant correlation, as expected from contributions from minijets and baryon number conservation. However, MC models still fail to reproduce the magnitude of the correlation.

[1] A conditional probability to observe a particle with azimuthal angle φ_1 and pseudorapidity η_1 if a particle with azimuthal angle φ_2 and pseudrapidity η_2 is observed as well. In the absence of correlations, the ratio should equal unity.

[2] For the full collection of results including two-dimensional $\Delta\eta\Delta\varphi$ correlations see [4].

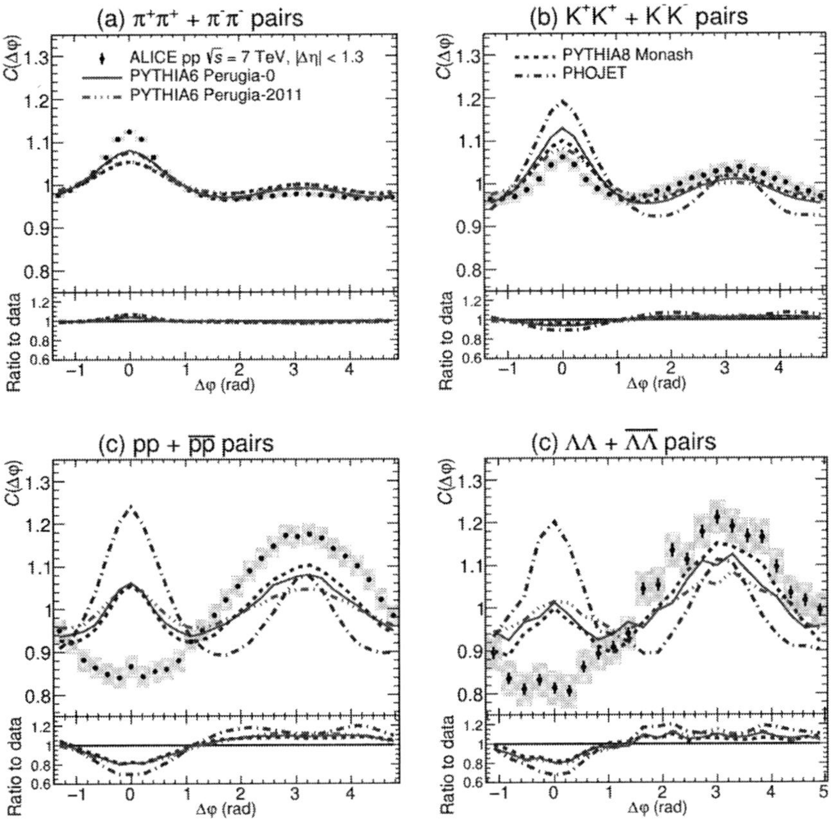

Figure 1. Projections of correlation functions integrated over $\Delta\eta$ for combined pairs of (a) $\pi^+\pi^+ + \pi^-\pi^-$, (b) $K^+K^+ + K^-K^-$, (c) $pp + \overline{pp}$ and (d) $\Lambda\Lambda + \overline{\Lambda\Lambda}$, obtained from ALICE data and four Monte Carlo models (PYTHIA6 Perugia-0, PYTHIA6 Perugia-2011, PYTHIA8 Monash, PHOJET). Bottom panels show ratios of ALICE data to MC models. Statistical (bars) and systematic (boxes) uncertainties are plotted. Plot from [4].

As shown in Fig. 2-right, multiple baryon pair combinations were studied to determine which physics effects (not included in the models) can influence the shape of baryon-baryon correlation functions.

- The Coulomb effect plays a marginal role: the shape of the correlation function for all studied baryon–baryon (and baryon–anti-baryon) pairs is similar, regardless of the electric charge of the particles.
- Fermi-Dirac quantum statistics is not the cause of the observed depression: the same magnitude of depression is observed for pp, $\Lambda\Lambda$ (identical particles), and $p\Lambda$ (non-identical particles).

Moreover, local baryon number conservation is not the only source of the depression. All studied models include this mechanism, but still are not able to reproduce the experimental data.

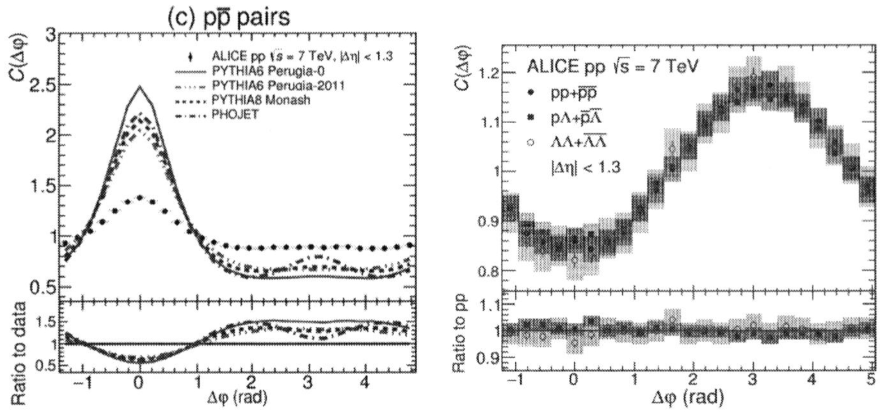

Figure 2. Projections of correlation functions integrated over $\Delta\eta$ for (left) $p\overline{p}$ obtained from ALICE data and four Monte Carlo models, (right) combined pairs of $pp + \overline{pp}$, $p\Lambda + \overline{p\Lambda}$, and $\Lambda\Lambda + \overline{\Lambda\Lambda}$. Statistical (bars) and systematic (boxes) uncertainties are plotted. Plots from [4].

4 Summary

Angular correlations of identified particles (π, K, p, Λ) were analyzed in pp collisions at $\sqrt{s} = 7$ TeV recorded with the ALICE experiment. A significant depression at $(\Delta\eta, \Delta\varphi) \approx (0, 0)$ is observed for the baryon–baryon and anti-baryon–anti-baryon pairs, which is not seen for mesons nor for baryon–anti-baryon pairs. This depression is not reproduced by Monte Carlo models. This suggests that (a) the jet fragmentation is not the dominant mechanism involved in the production of baryons in the studied p_{T} range as models indicate, or (b) the fragmentation mechanisms employed in PYTHIA and PHOJET are incomplete. The latter scenario would further suggest that some additional, not identified mechanism must exist. Such mechanism would suppress the production of more than one baryon–anti-baryon pair during a single fragmentation. Therefore, the presented results may suggest the need to modify particle production mechanisms as well as the modification of fragmentation functions in models.

Acknowledgements

This work was supported by the Polish National Science Centre under decisions no. 2014/13/B/ST2/04054, no. 2015/19/D/ST2/01600 and no. 2016/22/M/ST2/00176.

References

[1] R.D. Field, R.P. Feynman, Nucl. Phys. **B136**, 1 (1978)
[2] H. Aihara et al. (TPC/Two Gamma), Phys. Rev. Lett. **57**, 3140 (1986)
[3] P.D. Acton et al. (OPAL), Phys. Lett. **B305**, 415 (1993)
[4] J. Adam et al. (ALICE), Eur. Phys. J. **C77**, 569 (2017), `nucl-ex/1612.08975`
[5] K. Aamodt et al. (ALICE), JINST **3**, S08002 (2008)
[6] T. Sjostrand, S. Mrenna, P.Z. Skands, JHEP **0605**, 026 (2006), `hep-ph/0603175`
[7] P.Z. Skands, Phys.Rev. **D82**, 074018 (2010), `hep-ph/1005.3457`
[8] T. Sjostrand et al., Comput. Phys. Commun. **178**, 852 (2008), `hep-ph/0710.3820`
[9] P. Skands, S. Carrazza, J. Rojo, Eur. Phys. J. **C74**, 3024 (2014), `hep-ph/1404.5630`
[10] R. Engel, Z. Phys. **C66**, 203 (1995)
[11] T. Pierog et al., Phys. Rev. **C92**, 034906 (2015), `hep-ph/1306.0121`
[12] G. Corcella et al., JHEP **01**, 010 (2001), `hep-ph/0011363`

Energy and multiplicity dependence of strange and non-strange particle production in proton-proton collisions at the LHC with ALICE

Fiorella Maria Celeste Fionda[1], on behalf of the ALICE Collaboration*

[1] *University of Bergen (Norway)*

Abstract. The study of energy and multiplicity dependence of hadron production in proton-proton collisions provides a powerful tool to understand similarities and differences between small and large colliding systems. In this work we present mid-rapidity measurements of the $p_{\rm T}$ spectra and yields of identified hadrons, namely pions, kaons, protons, K^0_S, Λ, Ξ and Ω in pp collisions at $\sqrt{s}$ = 7 and 13 TeV. The comparison of results at $\sqrt{s}$ = 13 TeV to earlier results at 7 TeV provides insights about the energy dependence of the strangeness enhancement. Comparisons between data and expectations from commonly-used Monte Carlo event generators will be presented.

The primary goal of the ALICE experiment is the study of the strongly-interacting Quark-Gluon Plasma (QGP), which is expected to be produced in ultra-relativistic heavy-ion collisions. In order to interpret correctly heavy-ion results proton-proton (pp) and proton-nucleus (p-A) collisions ("small" systems) are used as a reference and control experiments. However, in recent measurements from the highest LHC energies, it was observed that small systems showed striking commonalities with heavy-ion collisions when multiplicity dependent studies are performed. Several effects, like near-side long-range correlations and mass-dependent hardening of $p_{\rm T}$ distributions, which in nuclear collisions are typically attributed to collectivity, have been observed in high-multiplicity pp and p-A collisions at the LHC (see for example [1, 2]). These commonalities raise questions about the possibility to employ a description based on the assumption of thermal and chemical equilibrium, commonly used in the heavy-ion context, also in small systems. The study of identified hadron production, both integrated yields and differential $p_{\rm T}$-spectra, as a function of the charged particle multiplicity represents a powerful tool to shed light on these open points. In these proceedings a selection of multiplicity dependent results obtained by ALICE at mid-rapidity will be discussed.

A detailed description of the ALICE apparatus and of its performance can be found in [3, 4]. The main subdetectors used to measure identified hadrons at mid-rapidity are: *(i)* the Inner Tracking System (ITS) which is composed of six cylindrical layers of silicon detectors employing three different technologies; *(ii)* the Time Projection Chamber (TPC) which provides track reconstruction with up to 159 three-dimensional space points per track, as well as particle identification (PID) via the measurement of the specific energy deposit dE/dx in the gas; *(iii)* the Time-Of-Flight detector (TOF), a cylindrical detector equipped with Multi-gap Resistive Plate Chambers (MRPCs) dedicated to PID. The V0 detector consists of two arrays of 32 scintillators, each placed on both sides of the interaction

*e-mail: fiorella.fionda@cern.ch

region and covering the pseudo-rapidity intervals of -3.7 < η < -1.7 and 2.8 < η < 5.1. It is used for triggering as well as to define multiplicity event classes based on the total charge deposited in the scintillators. For each event class the average pseudorapidity density of primary charged particles $\langle dN_{ch}/d\eta \rangle_{|\eta|<0.5}$ is measured at mid-rapidity ($|\eta| < 0.5$).

The transverse momentum distributions for several particle species measured at mid-rapidity ($|y| < 0.5$) in pp collisions at $\sqrt{s}$ = 7 TeV are shown in Fig. 1, for different event multiplicity selections, with class I denoting the highest multiplicity and class X the lowest. Transverse momentum spectra are observed to become harder as charged particle multiplicity increases, with changes of the spectra happening at low-p_T and being more pronounced for heavier particles. In heavy-ion collisions this behaviour is interpreted in terms of collective expansion (radial flow) which may require the system to be in local thermodynamical equilibrium. It is possible to further investigate

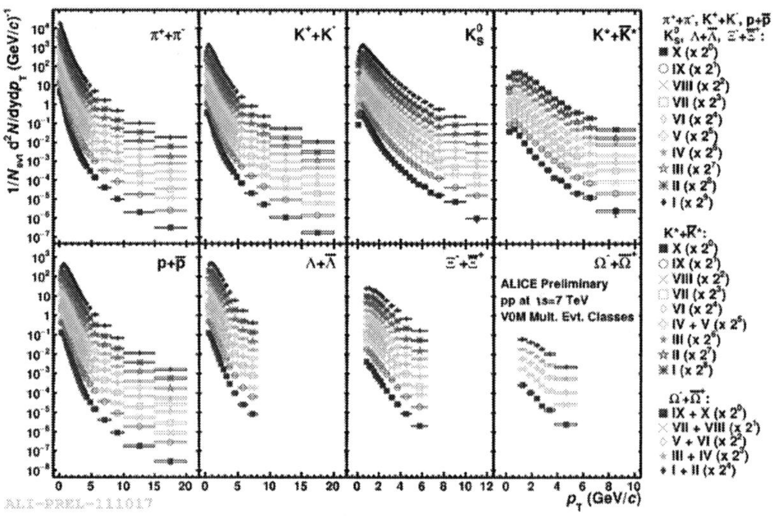

Figure 1. p_T-differential spectra for several particle species plotted for different event multiplicity classes (indicated by different symbols and colours). The different spectra from the various event multiplicity selection are scaled by a of factor 2^n in order to improve visibility.

the multiplicity dependence of p_T-spectra and the hypothesis of the collective expansion by studying the baryon-to-meson ratio as a function of the transverse momentum. The p/π ratio as a function of p_T is shown in Fig. 2 for pp ($\sqrt{s}$ = 7 TeV), p-Pb ($\sqrt{s_{NN}}$ = 5.02 TeV) and Pb-Pb ($\sqrt{s_{NN}}$ = 2.76 TeV) collisions at both low and high multiplicity. The ratio at high multiplicity in pp collisions exhibits a characteristic depletion at low-p_T (~ 0.7 GeV/c), an enhancement at intermediate-p_T (~ 3 GeV/c) and no modification at high-p_T if compared to the corresponding results at low multiplicity. This behaviour is qualitatively similar to the one observed in p-Pb and Pb-Pb collisions. From a quantitative point of view the observed changes in these particle ratios are very different in the various colliding systems. However it is important to take into account that also the corresponding average charged particle multiplicity densities at mid-rapidity differ significantly. Studying carefully the measured ratios as a function of $\langle dN_{ch}/d\eta \rangle_{|\eta|<0.5}$ in specific p_T intervals, a remarkable smooth trend across all colliding systems is observed. This is shown in Fig. 3.

Concerning the hadrochemistry one of the most intriguing results recently reported by the ALICE

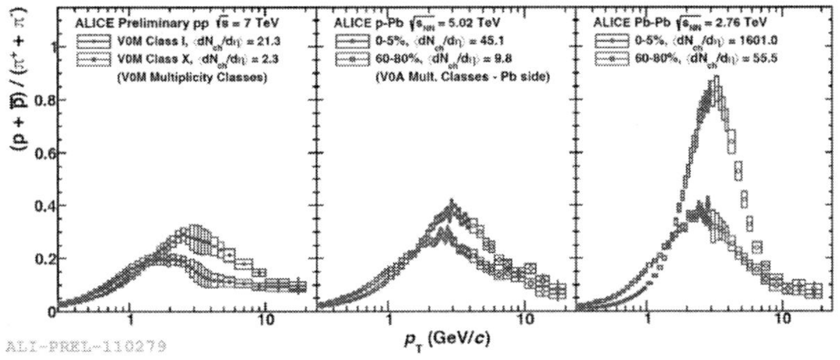

Figure 2. p/π ratio as a function of p_T in pp (left), p-Pb (middle), Pb-Pb (right) collisions for high and low multiplicity intervals.

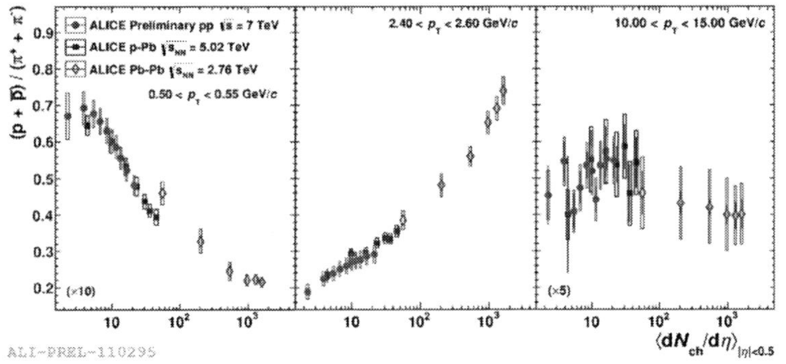

Figure 3. p/π ratio in low, mid and high-p_T intervals in pp, p-Pb, Pb-Pb collisions as a function of $\langle dN_{ch}/d\eta\rangle_{|\eta|<0.5}$.

collaboration is the enhancement observed in the production of (multi-)strange hadrons relative to pions as a function of multiplicity in pp collisions at $\sqrt{s}$ = 7 TeV [5]. These are surprising observations, because strangeness enhancement was considered historically one of the signatures for the deconfinement in heavy-ion collisions [6], and also because none of the commonly-used pp Monte Carlo models reproduces quantitatively the existing data. In particular, QCD-inspired models such as PYTHIA [7, 8] predict a flat dependence of strange-to-pion ratios as a function of multiplicity. Models that include effects from densely packed strings as implemented in DIPSY [9], core-corona mechanism as in EPOS [10] or statistical-thermal models implementing strangeness canonical suppression [11] can reproduce qualitatively the observed trends.

The energy dependence of the strange hadron yields as a function of $\langle dN_{ch}/d\eta\rangle_{|\eta|<0.5}$ is reported in Fig. 4, where multiplicity dependent results in pp collisions at $\sqrt{s}$ = 7 and 13 TeV are compared. The yields of strange hadrons increase linearly with the charged particle multiplicity, and the production rate is the same at the two energies for a given multiplicity. It should also be noted that the yields of particles with larger strange quark content increase faster as a function of multiplicity as already reported in [5]. Data are compared to several Monte Carlo models. The comparison shows clearly that both PYTHIA 6 and PYTHIA 8 are not able to describe the multiplicity dependence of strange particle yields, and the disagreement increases with the strangeness content of the considered hadron. EPOS predicts the strangeness enhancement but it overestimates the increase of multi-strange hadrons

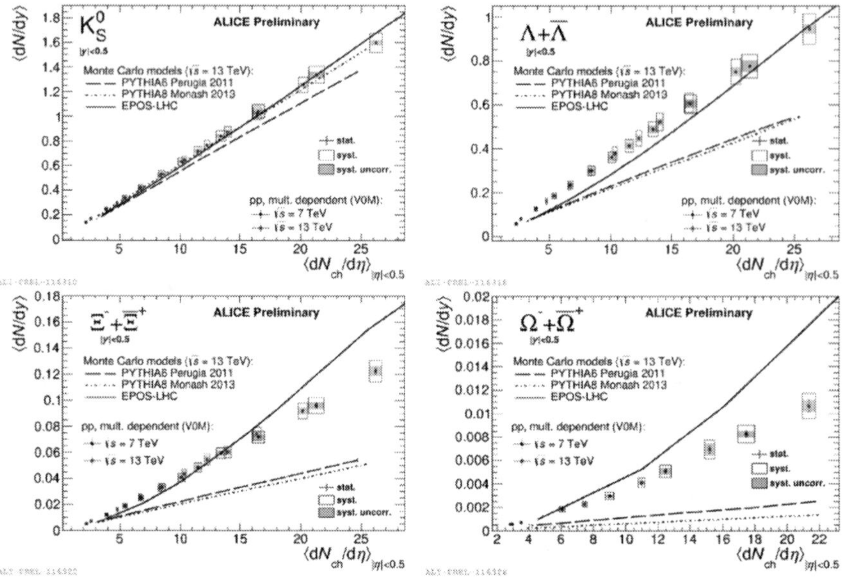

Figure 4. Integrated yields of K^0_S, Λ, Ξ and Ω as a function of $\langle dN_{ch}/d\eta\rangle_{|\eta|<0.5}$ in pp collisions at $\sqrt{s}$ = 7 and 13 TeV. The results are compared to predictions from several Monte Carlo models.

at high multiplicity.

In summary, ALICE provides a comprehensive set of results on the production of identified hadrons as a function of charged particle multiplicity in pp collisions. These results show striking commonalities with heavy-ion collisions, and none of the tested Monte Carlo models succeeds to quantitatively reproduce the data. These results provide an important input to the theory community for further tuning or implementation of new model ingredients, in order to improve significantly our knowledge about the microscopic mechanisms that are responsible of these intriguing features observed in small systems.

References

[1] CMS Collaboration, V. Khachatryan *et al.*, JHEP **09** 091 (2010).
[2] ALICE Collaboration, B. Abelev *et al.*, Phys. Lett. B **728** 25-38 (2014).
[3] ALICE Collaboration, K. Aamodt *et al.*, JINST **3** S08002 (2008).
[4] ALICE Collaboration, B. Abelev *et al.*, Int. J. Mod. Phys. A **29** 1430044 (2014).
[5] ALICE Collaboration, J. Adam *et al.*, Nature Phys. **13** 535–539 (2017).
[6] J. Rafelski and B. Müller, Phys. Rev. Lett. **48** 1066 (1982).
[7] T. Sjostrand, S. Mrenna, and P. Z. Skands, JHEP **05** 490 (2006).
[8] P. Skands, S. Carrazza, and J. Rojo, Eur. Phys. J. C **74** 3024 (2014).
[9] C. Bierlich, G. Gustafson, L. Lnnblad, and A. Tarasov, JHEP **03** 148 (2015).
[10] T. Pierog, I. Karpenko, J. Katzy, E. Yatsenko, and K. Werner, Phys. Rev. C **92** 034906 (2013).
[11] V. Vislavicius and A. Kalweit, arXiv:1610.03001 [nucl-ex].

Heavy quark flow as better probes of QGP properties

Zi-Wei Lin[1,2,⋆], *Hanlin* Li[3,4,⋆⋆], and *Fuqiang* Wang[4,5,⋆⋆⋆]

[1] *Key Laboratory of Quarks and Lepton Physics (MOE) and Institute of Particle Physics, Central China Normal University, Wuhan 430079, China*
[2] *Department of Physics, East Carolina University, Greenville, North Carolina 27858, USA*
[3] *College of Science, Wuhan University of Science and Technology, Wuhan, Hubei 430065, China*
[4] *Department of Physics and Astronomy, Purdue University, West Lafayette, IN 47907, USA*
[5] *School of Science, Huzhou University, Huzhou, Zhejiang 313000, China*

Abstract. In earlier studies we have proposed that most parton v_2 comes from the anisotropic escape of partons, not from the hydrodynamic flow, even for semi-central Au+Au collisions at $\sqrt{s_{NN}} = 200$ GeV. Here we study the flavor dependence of this escape mechanism with a multi-phase transport model. In contrast to naive expectations, we find that the charm v_2 is much more sensitive to the hydrodynamic flow than the lighter quark v_2, and the fraction of v_2 from the escape mechanism decreases strongly with the quark mass for large collision systems. We also find that the light quark collective flow is essential for the charm quark v_2. Our finding thus suggests that heavy quark flows are better probes of the quark-gluon-plasma properties than light quark flows.

1 Introduction

Azimuthal anisotropies in heavy ion collisions, such as the elliptic flow v_2, are important tools for the study of the properties of the quark-gluon plasma (QGP). Recent studies with parton transport models suggest [1, 2] that most parton v_2 comes from the anisotropic escape of partons, not from the hydrodynamic flow, even for semi-central Au+Au collisions at $\sqrt{s_{NN}} = 200$ GeV. This escape mechanism converts the spatial anisotropy in the overlap volume very efficiently into azimuthal anisotropies in final state particles, even though the parton cross section and average number of collisions of each parton are small [1–4]. Thus it naturally explains the similar azimuthal anisotropies observed in small and large collision systems. However, it poses a challenge to the current perfect-fluid paradigm for heavy ion collisions, at least when the collision system and/or energy are not very large.

Our earlier studies [1, 2] looked at all quarks (regardless of their flavors) and only investigated d+Au and Au+Au systems at 200*A* GeV. Here we study the flavor dependence of the parton escape mechanism, especially the charm quarks [5, 6]. As in the earlier studies, we use the string melting version of a multi-phase transport (AMPT) model [7] with the same parameters , which can reasonably describe the experimental data for the bulk matter for high energy heavy ion collisions [8]. We follow the entire evolution history of quarks of different flavors in AMPT and then analyze the developments

⋆e-mail: linz@ecu.edu
⋆⋆e-mail: lihl@wust.edu.cn
⋆⋆⋆e-mail: fqwang@purdue.edu

of light (u and d quarks), strange, and charm v_2 in three systems: p+Pb collisions at 5*A* TeV and impact parameter $b=0$ fm, Au+Au collisions at 200*A* GeV and $b \in (6.6, 8.1)$ fm, and Pb+Pb collisions at 2.76*A* TeV and $b=8$ fm. Note that only the results for the Pb+Pb collisions are shown in the figures here, and the results of quarks and antiquarks of the same flavor have been combined.

2 Results

To follow the parton collision history in the AMPT model, we define N_{coll} as the number of collisions suffered by a parton. At any given N_{coll} value, we study three groups of quarks of a given flavor: freezeout partons (partons that freeze out after exactly N_{coll} collisions), non-freezeout partons (partons with more than N_{coll} collisions), and all (active) partons (sum of the previous two groups).

Figure 1a shows the v_2 of light (black), strange (blue), and charm (red) quarks within $|\eta| < 1$ as functions of the number of collisions N_{coll} for 2.76*A* TeV Pb+Pb collisions at $b = 8$ fm from AMPT simulations. We see a clear mass ordering in that $v_{2,u/d} > v_{2,s} > v_{2,c}$ at low N_{coll} but the opposite at high N_{coll}, indicating that charm quarks need more scatterings to generate their v_2. The normalized N_{coll} distribution and the average number of collisions of each quark flavor (of all pseudorapidities) are shown in Fig. 1b, where we see that charm quarks typically have more collisions than lighter quarks. This should be related to the initial spatial and momentum distributions, which are different for each quark flavor. For example, we find that, in comparison with light quarks, a bigger fraction of charm quarks is produced in the inner region of the overlap volume, consistent with the hard production nature of charm quarks and their scaling with binary collisions. In addition, we find that the above features in Fig.1 are true for all three collision systems in our study.

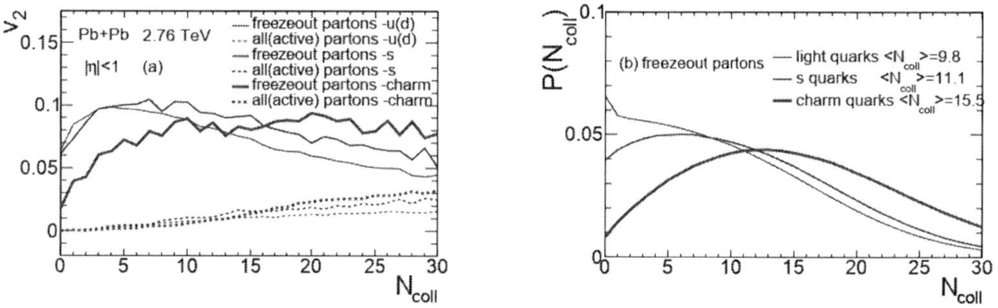

Figure 1. AMPT simulations of Pb+Pb collisions at 2.76A TeV and $b=8$ fm: (a) v_2 of light, strange, and charm quarks within $|\eta| < 1$ as functions of N_{coll}, where solid curves represent freezeout parton and dashed curves represent all (active) partons; (b) normalized N_{coll}-distributions of different quark flavors.

For partons that freeze out (i.e. hadronize) at $N_{coll} = 0$, their finite positive v_2, as shown in Fig. 1a, is due to the fact that it is easier to avoid collisions along the impact parameter direction in the transverse plane. As these partons have not been affected by the collective flow, their v_2 comes purely from the anisotropic escape probability, an interaction-induced response to the anisotropic geometrical shape that we named the escape mechanism. For partons that freeze out at $N_{coll} \neq 0$, however, their v_2 comes partly from the interaction-induced response to geometry and partly from the collective flow that is also anisotropic. In order to identify the contribution from the escape mechanism, we have designed the azimuth-randomized test to remove the collective flow [1, 2], where we randomize the azimuth angle of each of the two final state partons after every parton scattering. As a result, anisotropic flows in the azimuth-randomized simulations are generated only by the escape mechanism.

We show in Fig. 2 the v_2 as functions of N_{coll} for light, strange, and charm quarks within $|\eta| < 1$ from both normal AMPT simulations (solid curves) and azimuth-randomized AMPT simulations (dashed curves) of Pb+Pb collisions. Note that the average parton v_2 is the freezeout v_2 shown here summed with the weight given by the N_{coll}-distribution shown in Fig. 1b. First we see that the v_2 results from azimuth-randomized calculations are finite, although they are mostly lower than that from normal calculations due to the lack of the anisotropic collective flow. We also see that the reduction of v_2 going from normal to azimuth-randomized calculations is more obvious for heavier quarks, indicating that a smaller fraction of v_2 comes from the escape mechanism for heavier quarks.

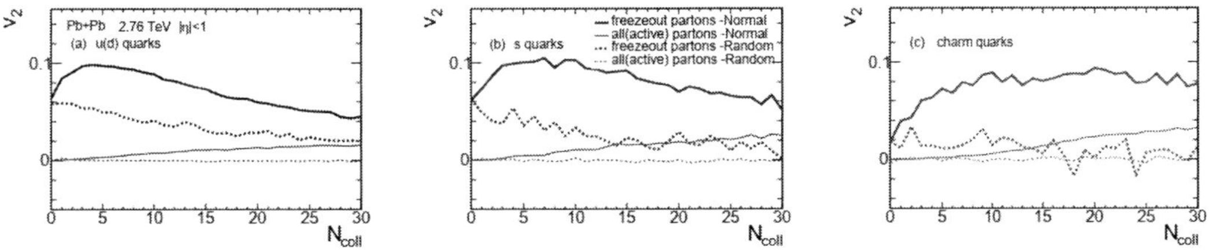

Figure 2. The v_2 of (a) light, (b) strange, and (c) charm quarks within $|\eta| < 1$ as functions of N_{coll} for Pb+Pb collisions at 2.76A TeV and $b = 8$ fm.

Another interesting test is to do azimuth-randomized calculations only for u/d/s quarks and antiquarks but not on charm, i.e., only charm quarks and antiquarks are allowed to keep their collective flow developed in the parton cascade. Results from this test are shown in Fig. 3a for charm quarks within $|\eta| < 1$ in Pb+Pb collisions, where we see that the freezeout charm v_2 is much reduced when compared to the normal charm v_2 (thick solid curve in Fig 2c) and is similar to the freezeout charm v_2 when all quark flavors are azimuth-randomized (thick dashed curve in Fig 2c). This means that charm quarks cannot develop a significant v_2 without the light quark collective flow.

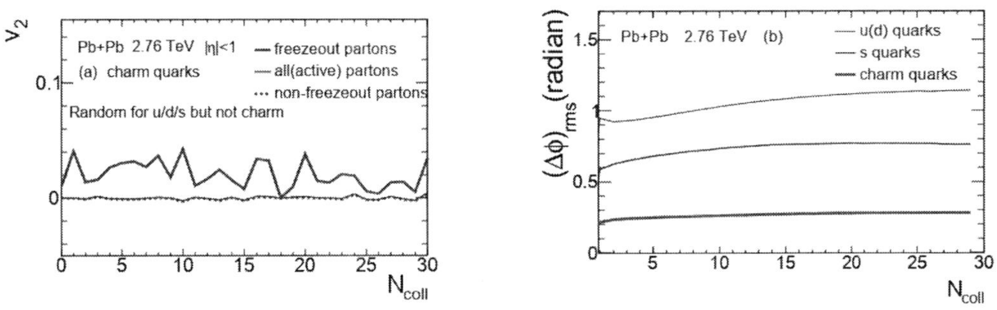

Figure 3. AMPT results of 2.76A TeV Pb+Pb collisions at $b = 8$ fm: (a) v_2 of three charm quark groups within $|\eta| < 1$ as functions of N_{coll} in simulations where u/d/s (but not charm) (anti)quarks are azimuth-randomized; (b) the rms change of azimuth due to the $N_{coll}-th$ collision for different quark flavors in normal simulations.

To further understand the flavor dependence, we have also looked at the effect of each parton collision on the azimuth of different flavors. Figure 3b shows the root-mean-square (rms) change of the azimuth angle, in radian, as a function of N_{coll} for different quark flavors (of all pseudorapidities) in Pb+Pb collisions. There is a clear mass ordering in that the average azimuth change is much smaller for heavier quarks, consistent with the expectation that it is more difficult to deflect a heavier quark in the parton cascade. Note that AMPT uses the same light quark mass as PYTHIA [9]: $m_u = 5.6, m_d = 9.9$, $m_s = 199$ MeV/c^2, and we use $m_c = 1.2$ GeV/c^2 for charm in this study.

Table 1 lists the ratios of the $\langle v_2 \rangle$ from azimuth-randomized AMPT over that from normal AMPT, where $\langle v_2 \rangle$ represents the v_2 averaged over all partons within $|\eta| < 1$, in four different collision systems at RHIC and LHC energies. This ratio represents the fraction of v_2 that comes from the escape

Table 1. Ratio of the averaged v_2 from azimuth-randomized simulations over that from normal simulations, which represents the fraction from the escape mechanism.

	d+Au at 200 GeV (b=0 fm)	p+Pb at 5 TeV (b=0 fm)	Au+Au at 200 GeV (b=6.6 – 8.1 fm)	Pb+Pb at 2.76 TeV (b=8 fm)
u/d	93% (all flavors)	72.9%	65.6%	42.5%
s		59.1%	47.4%	26.5%
c		56.8%	21.8%	8.5%

mechanism, i.e., from the interaction-induced response to the anisotropic geometry. Note that the d+Au result comes from our earlier study on quarks of all flavors [1]. We can see that the escape contribution decreases with the quark mass, more strongly for larger systems. It also decreases with the collision energy and/or system size, which is expected. Therefore these results show that the hydrodynamical collective flow contributes more to the v_2 of heavier quarks, especially in large systems at high energies, suggesting that heavy quark flows can better reflect the properties of the quark-gluon plasma. We note that similar claims have been made in studies from other points of view [10, 11].

3 Summary

We have followed the complete parton collision history to study the v_2 of light, strange and charm quarks in small and large collision systems at both RHIC and LHC energies using the string melting version of the AMPT model. We find that the fraction from the interaction-induced response to the anisotropic spatial geometry (the escape mechanism) decreases not only with the system size and collision energy but also with the quark mass (especially for large systems at high energies). The escape mechanism is no longer dominant for the light quark average v_2 in semi-central Pb+Pb collisions at 2.76 TeV but its contribution is stil significant. On the other hand, most of the charm quark v_2 comes from the hydrodynamical collective flow for the large systems covered in this study. We also find that the collective flow of light quarks is essential for the generation of charm quark v_2. These results indicate that heavy quark flows are better probes of QGP properties than light quark flows.

This work is supported in part by the NSFC of China under Grants Nos. 11628508, 11647306, and US Department of Energy Grant No. DE-SC0012910. HL acknowledges the financial support from the China Scholarship Council.

References

[1] L. He, T. Edmonds, Z. W. Lin, F. Liu, D. Molnar and F. Wang, Phys. Lett. B **753**, 506 (2016).
[2] Z. W. Lin, L. He, T. Edmonds, F. Liu, D. Molnar and F. Wang, Nucl. Phys. A **956**, 316 (2016).
[3] A. Adare *et al.* [PHENIX Collaboration], Phys. Rev. Lett. **115**, 142301 (2015).
[4] J. D. Orjuela Koop, R. Belmont, P. Yin and J. L. Nagle, Phys. Rev. C **93**, 044910 (2016).
[5] H. Li, Z. W. Lin and F. Wang, J. Phys. Conf. Ser. **779**, 012063 (2017).
[6] H. L. Li, Z. W. Lin, and F. Wang, in preparation.
[7] Z. W. Lin, C. M. Ko, B. A. Li, B. Zhang and S. Pal, Phys. Rev. C **72**, 064901 (2005).
[8] Z. W. Lin, Phys. Rev. C **90**, 014904 (2014).
[9] T. Sjostrand, Comput. Phys. Commun. **82**, 74 (1994).
[10] R. Esha, M. Nasim and H. Z. Huang, J. Phys. G **44**, 045109 (2017).
[11] V. Greco, Nucl. Phys. A **967**, 200 (2017).

Subthreshold Ξ^- production in proton-nucleus collisions in a BUU model

Miklós Zétényi[1,2,⋆] and *György* Wolf[1]

[1] *Wigner Research Center for Physics, Konkoly-Thege Miklós út 29-33, 1121 Budapest, Hungary*
[2] *ExtreMe Matter Institute EMMI, GSI Helmholtzzentrum für Schwerionenforschung, Planckstrasse 1, 64291 Darmstadt, Germany*

Abstract. We study the production of the doubly strange Ξ baryon in subthreshold $p + A$ collisions using a BUU type transport model. We propose a new mechanism for Ξ creation in a two-step process via hyperon-nucleon collisions. We study the influence of the anisotropy of hyperon production in $N + N$ collisions on the Ξ multiplicity. Applying reasonable assumptions on the unknown elementary cross sections, we are able to reconstruct the Ξ yield observed by the HADES collaboration (GSI, Darmstadt) in subthreshold p+Nb collisions.

1 Introduction

Particle production in nuclear collisions near or below the kinematical threshold of elementary $N + N$ reactions is sensitive to the details of the reaction dynamics, e.g. Fermi motion, collision of secondaries or in-medium modification of particle properties. Due to strangeness conservation in the strong interaction, strange particles are created in pairs, which leads to higher production thresholds. Thus subthreshold strangeness production is a suitable tool to study the reaction dynamics at higher energies where higher baryon densities are reached.

Proton-nucleus ($p + A$) collisions are an intermediate step between $N + N$ and $A + A$ collisions. Here properties of hadrons at normal nuclear matter density are studied. In $p + A$ collisions fewer elementary production channels of a specific hadron species are possible, since e.g. the collision of two secondaries is rather unlikely.

The HADES collaboration has studied the production of the doubly strange Ξ^- baryon in both p+Nb reactions at the subthreshold energy of $\sqrt{s_{NN}} = 3.2$ GeV [1], and in Ar+KCl collisions at the deeply subthreshold energy of $\sqrt{s_{NN}} = 2.61$ GeV [2]. In both reactions a surprisingly high Ξ^- multiplicity has been found. The THERMUS statistical model gives a good description of most strange and non-strange hadron yields in both reactions. The only exception is the Ξ^- yield which the model underestimates by a factor 15 in the case of the reaction Ar+KCl and by a factor 8 in the case of the reaction p+Nb [3].

In earlier transport model calculations, Ξ production in strangeness exchange reactions has been considered, in particular in antikaon-hyperon collisions, $\bar{K}+Y \to \pi+\Xi$ ($Y = \Lambda$ or Σ), hyperon-hyperon collisions, $Y + Y \to \Xi + N$ and the reaction $\eta + \Lambda \to \Xi + K$. All these reaction channels involve the

⋆e-mail: zetenyi.miklos@wigner.mta.hu

collision of two secondaries, therefore they cannot account for the high Ξ multiplicity in subthreshold $p + A$ collisions.

At present only the UrQMD transport model is able to reproduce the Ξ multiplicities found by HADES in subthreshold p+Nb and Ar+KCl reactions [4]. In that study it was assumed that the high mass tail of some higher nucleon resonances decays to the $\Xi + K + K$ final state with a branching ratio of 10 %.

In the present contribution we propose a new production mechanism of the Ξ baryon in nuclear collisions via the collision of a secondary hyperon and a target nucleon. In Sec. 2 we discuss this new production mechanism, give some estimates of the elementary cross sections, and discuss the relevance of the angular distribution of hyperon production to the Ξ multiplicity. BUU transport calculations are presented in Sec. 3, where our results are compared to the HADES data.

2 Ξ production via hyperon-nucleon collisions

We propose the following production mechanism for Ξ production in p+A collisions: first, the projectile proton collides with a nucleon in the target and they create a hyperon via the reaction $p + N \to N + K + Y$, ($Y = \Lambda$ or Σ). This hyperon then collides with a second target nucleon and they create a Ξ baryon via $Y + N \to N + K + \Xi$.

The threshold energy for $\Lambda(\Sigma)$ production in a $p + N$ collision is $\sqrt{s_{NN,\text{thr}}} = 2.55$ GeV (2.62 GeV), which is lower than the CM energy $\sqrt{s_{NN}} = 3.2$ GeV available in the first collision of the projectile proton and a target nucleon in the p+Nb collision studied by HADES. Furthermore, in the ideal case of a hyperon emitted in the forward direction, the maximum available CM energy in the collision of the Λ hyperon with a second target nucleon is $\sqrt{s_{\Lambda N,\text{max}}} = 3.05$ GeV, which is above the threshold of $\sqrt{s_{\Lambda N,\text{thr}}} = 2.75$ GeV for Ξ production. The energetical situation with a Σ hyperon instead of a Λ is very similar. This means that both steps of the above production mechanism are kinematically allowed in the p+Nb experiment of HADES. We note that, according to the proposed mechanism, not only the energy but also the two units of strangeness are accumulated in two steps.

Estimate of the elementary cross section

We assume that Ξ production in $Y + N$ collisions proceeds via an intermediate Λ^* or Σ^* resonance, as shown in Fig. 1. In order to obtain a rough estimate of the cross section of this process, we first assume

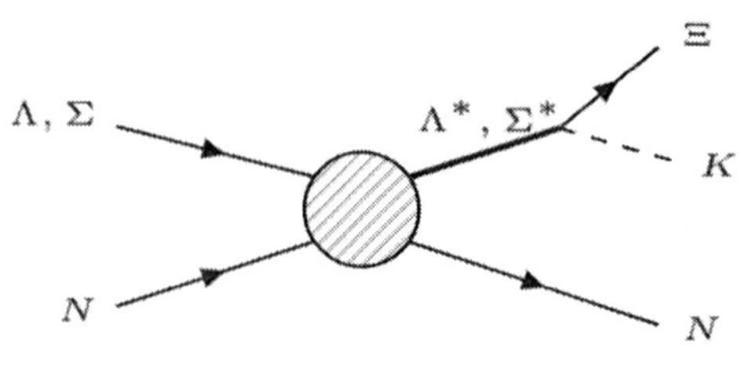

Figure 1. Schematic diagram for Ξ production in a hyperon-nucleon collision.

that the total $Y + N$ cross section is of the same order of magnitude as the proton-proton cross section at the same excess energy $\sqrt{s} - \sqrt{s_{\text{thr}}} \approx 0.3$ GeV, that is, a few times 10 mb. If resonance dominance is assumed, i.e. all final states in $Y + N$ reactions are reached via intermediate (strange or non-strange) resonances, then the total $Y + N$ cross section equals the total resonance production cross section in the same reaction. The cross section for Λ^* and Σ^* creation, which is a considerable part of the above, should be of the order of 10 mb at the maximum energy available in the HADES experiment. The Ξ production cross section is obtained via multiplication by the $\Lambda^*(\Sigma^*) \rightarrow \Xi + K$ branching ratio. The only available information on this branching ratio is for the $\Lambda(2100)$ resonance, for which an upper limit of 3 % is given by the Particle Data Group [5]. Using 10 mb for the $\Lambda^*(\Sigma^*)$ creation cross section and 3 % for the ΞK branching ratio, we get an estimate of 0.3 mb for the $Y + N \rightarrow N + \Xi + K$ cross section, which should be valid at an excess energy of about 0.3 GeV.

In order to model the energy dependence of the elementary cross section of Ξ production, we study the phase space effects of the process depicted on Fig. 1. We calculate the production cross section of Λ^* and Σ^* resonances assuming a constant matrix element of the processes $\Lambda(\Sigma) + N \rightarrow \Lambda^*(\Sigma^*) + N$ and we consider the mass dependence of their decay widths. We include the 3 and 4 star Λ^* and Σ^* states in the relevant mass range and choose the constant matrix element in such a way that the Ξ production cross section of 0.3 mb at 0.3 GeV excess energy, the estimate obtained in the previous paragraph, is roughly reproduced. The resulting energy dependence of the $\Lambda + N \rightarrow N + \Xi + K$ cross section is shown in Fig. 2. For the corresponding cross section with the Σ hyperon in the initial state we obtained a very similar energy dependence.

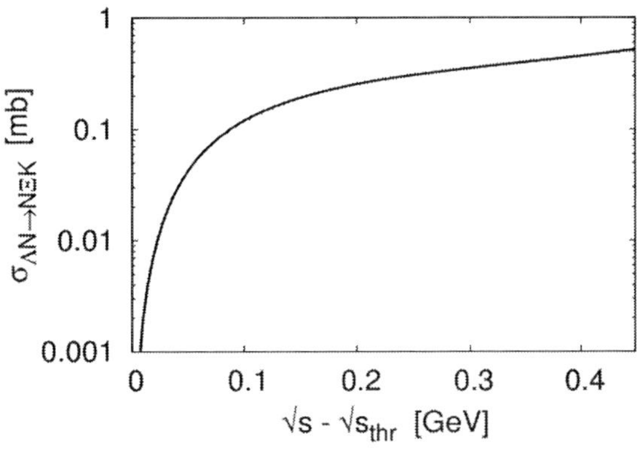

Figure 2. Cross section of the process $\Lambda + N \rightarrow N + \Xi + K$ as a function of the excess energy.

Relevance of the angular distribution of hyperon production

Λ and Σ hyperons emitted in the forward direction in the first $p + N$ collision have more energy available in their collision with a second target nucleon than those emitted in a different direction. Thus, the angular distribution of Λ and Σ creation in $N + N$ collisions is expected to influence the Ξ multiplicity in subthreshold $p + A$ collisions. This angular distribution has been studied by the COSY-TOF collaboration and they have found an increased emission probability of hyperons in the forward direction [6], which can result in an enhancement of Ξ production in $p + A$ collisions.

3 BUU calculations

We use the BUU code developed in a Budapest-Rossendorf collaboration, that has been successfully applied to strangeness production near and below the production threshold [7]. The elementary cross sections of Λ and Σ production in $N + N$ collisions are implemented in the code based on an effective Lagrangian model study, which assumes the dominance of resonance contributions [8]. For the $Y + N \rightarrow N + \Xi + K$ cross section we use our estimate shown in Fig. 2.

Assuming an isotropic distribution of the produced Λ and Σ hyperons in the elementary $N + N$ reaction in the center of momentum frame of the colliding nucleons, we obtained a Ξ^- multiplicity of $P_{\Xi^-} = 1.4 \times 10^{-4}$ and a Ξ to hyperon ratio of $P_{\Xi^-}/P_{\Lambda+\Sigma^0} = 0.68 \times 10^{-2}$. Implementing the anisotropy of hyperon production found in [6], we get the increased values of $P_{\Xi^-} = 2.16 \times 10^{-4}$ and $P_{\Xi^-}/P_{\Lambda+\Sigma^0} = 1.14 \times 10^{-2}$. These have to be compared to the experimental values by the HADES collaboration, $(P_{\Xi^-})_{\mathrm{exp}} = (2.0 \pm 0.4 \pm 0.3) \times 10^{-4}$ and $(P_{\Xi^-}/P_{\Lambda+\Sigma^0})_{\mathrm{exp}} = (1.2 \pm 0.3 \pm 0.4) \times 10^{-2}$.

4 Conclusions

We have proposed a new mechanism for the production of the doubly strange Ξ baryon in subthreshold proton-nucleus collisions via a reaction of a secondary Λ or Σ hyperon with a target nucleon. We implemented this mechanism in our BUU transport code and studied the reaction p+Nb at $\sqrt{s_{NN}} =$ 3.2 GeV energy, which has been measured by the HADES collaboration. In the new mechanism, both the energy needed to create the Ξ and the associated kaons, as well as the two units of strangeness are accumulated in two steps. The angular distribution of the first step process (hyperon creation) has an influence on the energy available in the second step process (Ξ creation in hyperon-nucleon collision). We have found that the anisotropy of hyperon production in $N + N$ collisions enhances the Ξ^- multiplicity by about 50 % in p+Nb collisions at the studied subthreshold energy.

Applying reasonable assumptions on the cross section of the experimentally unknown process of Ξ production in hyperon-nucleon collisions, we were able to reproduce the high Ξ^- multiplicity found in p+Nb collisions in the HADES experiment. The new mechanism utilizes the first two energetic elementary collisions in the $p + A$ reaction. It is clear that in a thermalised medium formed in the same reaction, the energy needed for Ξ creation would not be available in a hyperon-nucleon collision. This is consistent with the finding that a thermal model calculation strongly underestimates the experimentally observed Ξ^- multiplicity.

Acknowledgements

This research was supported by the Hungarian OTKA Fund No. K109462 and by the ExtreMe Matter Institute EMMI at the GSI Helmholtzzentrum für Schwerionenforschung, Darmstadt, Germany.

References

[1] HADES Collaboration (G. Agakishiev et al.), Phys. Rev. Lett. **114**, 212301 (2015).
[2] HADES Collaboration (G. Agakishiev et al.), Phys. Rev. Lett. **103**, 132301 (2009).
[3] HADES Collaboration (G. Agakishiev et al.), Eur. Phys. J. A **52**, 178 (2016).
[4] J. Steinheimer and M. Bleicher, J. Phys. G **43**, 015104 (2016).
[5] C. Patrignani et al. (Particle Data Group), Chin. Phys. C **40**, 100001 (2016).
[6] COSY-TOF Collaboration (M. Abdel-Bary et al.), Eur. Phys. J. A **46**, 27–44 (2010).
[7] H.W. Barz, M. Zétényi, Gy. Wolf, B. Kämpfer, Nucl. Phys. A **705** 223 (2002); H.W. Barz, M. Zétényi, Phys. Rev. C **69**, 024605 (2004); H. Schade, Gy. Wolf, B. Kämpfer, Phys. Rev. C **81** 034902 (2010).
[8] K. Tsushima, A. Sibirtsev, A.W. Thomas, and G. Q. Li, Phys. Rev. C **61**, 029903 (2000).

Particle Production and Collectivity in High Multiplicity pp and pPb at the LHC

George S. F. Stephans[1,*] (For the CMS Collaboration)

[1]*Massachusetts Institute of Technology, Cambridge, MA 02139 USA*

Abstract. Recent unexpected evidence for collectivity in high multiplicity pp and pPb collisions at LHC energies has challenged the notion that such small systems do not exhibit any of the properties that have been used to study the quark gluon plasma in heavy ion collisions. An overview of recent results concerning particle production and collectivity in such collisions using the CMS detector at the LHC is presented.

1 Introduction

For roughly the first decade of heavy ion physics at RHIC and the first few years of similar studies at the LHC, heavy ion physicists assumed that no interesting physics (in the quark gluon plasma, QGP, sense) happened in pp collisions and almost no interesting physics happened in pA (or dA) collisions. These systems were viewed as too small for any possibility of forming a system of quarks and gluons that could be considered a medium. Data from pp collisions were used almost exclusively as references to determine what similar heavy ion data (for example jet spectra) would look like in the absence of the QGP. Results from pA were studied to look for cold nuclear matter effects in order to verify that any interesting observations seen in AA were not due to initial state effects. More recently, some pA data were found to be useful to constrain nuclear PDFs, but this usage served more as a fine tuning of assumptions about the initial state, as opposed to being directly related to QGP physics.

These attitudes towards these small systems have been challenged by recent observations of evidence for collectivity, first in very high multiplicity pp collisions [1] and later in high multiplicity pA collisions [2]. Studies of two-particle correlations of charged particles found a clear "ridge", a peak at small relative azimuthal angle, $\Delta\phi$, that extended over the entire accessible range of relative pseudorapidity, $\Delta\eta$. Collecting sufficient quantities of these very rare events is only possible through the use of a dedicated online high multiplicity selection implemented in the CMS trigger system.

This paper will summarize what has been learned recently about the properties of pp and pA collisions using the CMS detector at the LHC. For the sake of brevity, only the major conclusions will be described, additional information can be found in the references. Details of all heavy ion results from CMS can be found on the heavy ion physics website [3].

2 Properties of the Pb nucleus

Recent pPb data from CMS at $\sqrt{s_{NN}}$ = 8.16 TeV has pushed the study of the properties of charged particle multiplicity in collisions involving projectiles larger than a proton to the highest nucleon-nucleon

*e-mail: gsfs@mit.edu

center-of-mass energy ever [4]. Combined with data at lower energies, these results can constrain the physics processes in various particle production models and thereby improve our understanding of using similar models to describe the soft background in AA. As just one example, it is found that models without shadowing significantly overpredict the observed multiplicities [4].

In addition, various observables from pA collisions can be used to constrain the nuclear PDFs used in modeling the initial state in AA interactions. The differences between parton properties in nuclei and in protons can modify, in particular, the production of heavy particles and therefore must be taken into account in interpreting the nuclear modification factor, R_{AA}, for example for J/Ψ or Z bosons. One observable that has been found to be particularly sensitive is the differences between dijet pseudorapidity distributions in various p_T ranges for pp and pPb [5]. While calculations including the EPS09 nPDFs give a better agreement than other options, some discrepancies remain, suggesting that some additional tuning is needed. The asymmetry of these collisions has also been exploited by looking at the forward-backward asymmetry as a function of the rapidity in the nucleon-nucleon center-of-mass frame, $|y_{cm}|$, where again the data are better matched by EPS09 [6]. Similar studies have been done using the rapidity dependence of the yield of prompt J/Ψ [7], although in this case the effect of comovers might also be important.

3 High p_T suppression (or not) in pPb

Initially, the study of R_{AA} in pA was motivated by the need to determine the influence of initial state effects on the large suppressions observed in AA. In all cases, it was found that such effects were small, and in many cases went in the wrong direction. For example, the R_{AA} of charged particles in pPb shows an *enhancement* of about 5-15% over a broad range of p_T for $|\eta| < 1$ [8]. For jets, small effects are seen as a function of η and p_T and it is found that next-to-leading order calculations including EPS09 are able to match the data [9]. Expanding these studies to rarer objects such as jets tagged as containing a c quark have shown R_{AA} values consistent with 1.0 [10], although the statistical and systematic uncertainties are still large.

In light of the evidence for collective effects in small systems at high multiplicities (see next section), renewed interest has been paid to looking in more detail at the possibility of quenching in pA. However, three caveats need to be kept in mind. First, all studies to date have used inclusive pA collisions containing the high p_T particle of interest. Extending these studies to higher multiplicity events remains to be done. Furthermore, essentially all models of quenching assume some length dependence. It is quite possible that even if a QGP-like medium is formed in some high multiplicity pA events, the length traversed by emerging partons will be too small to generate a measurable energy loss. Finally, when comparing particle production in high multiplicity pA to either scaled pp or to AA, the question of the correct reference data arises. High multiplicity AA events typically result from a large number of incoherent NN interactions, each of which produces multiplicities similar to inclusive pp. However, similar multiplicity events in pA must arise from a small number of highly unusual NN collisions that produce many more particles than a typical pp interaction. It's not at all clear whether comparing particle production in such events to that in scaled inclusive pp is meaningful.

It should be stressed, however, that these recent discoveries do not shed doubt on the previous uses of pp and pA collisions as reference systems to normalize or interpret the results from AA data. As mentioned previously, even when quite peripheral, a typical heavy ion collision consists of a very large number of independent nucleon-nucleon interactions. Therefore, taking minimum bias pp or pA data as a reference system for comparing to data from AA collisions remains appropriate.

4 Recent results in small system collectivity

After the early discoveries of evidence for collectivity in high multiplicity pp and pPb, more recent work has focussed on exploring other aspects of these events. Particular emphasis has been put on observables that are expected to be sensitive to the presence of a collective evolution of the system.

The earliest example from CMS work on pPb data is the finding that triangular flow, v_3, has essentially identical magnitudes in pPb and PbPb events having the same charged particle multiplicity [11]. Unlike elliptic flow, v_2, which in AA is sensitive to the lenticular overlap region for the colliding nuclei, v_3 results from fluctuations in the initial energy density. The fact that this is a collective effect relating the entire system, and not some unusual feature of correlations between a few particles is established by using cumulants and Lee-Yang Zeroes to extract v_2 for correlations of larger groups of particles. The results show that when correlating 4 or more particles, the elliptic flow magnitudes do not change [12]. Finally, as expected for pressure-driven hydrodynamic flow, v_2 as a function of p_T shows a clear mass ordering when comparing charged particles to K^0_S and Λ [13].

Recent CMS work finds that these collective features seen in pPb, including both mass ordering and higher-order cumulants for v_2, are also present in high multiplicity pp data [14]. Another observation expected for a pressure-driven expansion is a mass ordering of the yields as a function of p_T with heavier particles being enhanced at higher transverse momenta. Again using K^0_S and Λ, the expected pattern is found in both pp and pPb, with the ratios of the yields of the two particles for a given p_T and event multiplicity being very similar for the these two systems, as well as for PbPb [15].

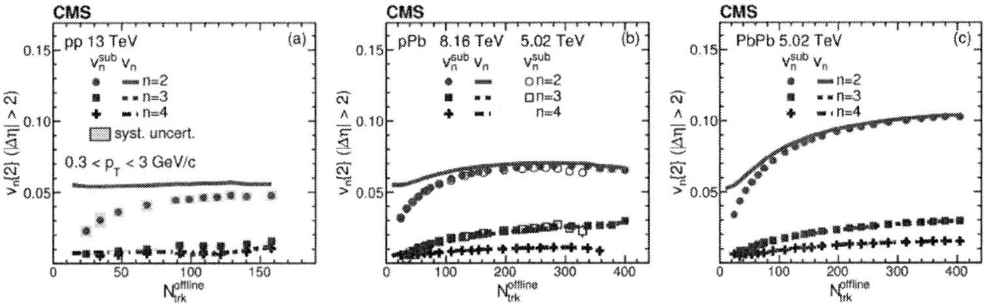

Figure 1. The v_2, v_3, and v_4 flow coefficients found using long-range ($|\Delta\eta|$ >2.0) two-particle correlations for charged particles with $0.3 < p_T < 3$ GeV/c as a function of total event particle multiplicity, $N_{trk}^{offline}$, in collisions of pp at $\sqrt{s}$ = 13 TeV (a), pPb at $\sqrt{s_{NN}}$ = 5.02 and 8.16 TeV (b), and PbPb at $\sqrt{s_{NN}}$ = 5.02 (c). The results with (without) correction by subtraction of correlations in low-multiplicity events are denoted as v_n^{sub} (lines v_n). The gray boxes represent systematic uncertainties [16]. Note that the abscissa range is expanded for the pp data.

Comparisons of similarities and differences of Fourier coefficients of various orders for pp, pPb, and PbPb over a very broad ranges of multiplicity have continued. Extending the previous comparisons of v_2 and v_3 for pPb and PbPb, very similar values of both v_3 and v_4 are found for these two systems, as well as for pp, when studied at the same multiplicity (see Fig. 1) [16]. As these analyses are extended to somewhat lower multiplicities, and especially in the case of v_2 for pp, it is critical to perform a careful subtraction of the contributions of other sources of particle correlations such as jets. In the CMS work, these effects are accounted for using the lowest multiplicity events studied. As expected due to contributions from the shape of the initial nuclear overlap region, v_2 in PbPb shows increasing deviations from the other two systems at the higher multiplicities studied. To investigate what is happening in even more detail, event-by-event correlations between the different orders have been extracted. When normalized by the product of the average values, the correlations between v_2

and v_3 do not show a strong system dependence (although the statistical precision of the pp data is still poor), while the v_2/v_4 correlation, which can be affected by initial shape effects, does show differences [16]. Finally, studying properties of the particles at freeze-out, system sizes found using HBT correlations of identified identical particle pairs, for both pions and kaons, are in many cases quite similar for all three systems at lower multiplicites [17]. Small differences for the radii found at higher multiplicities may be related to differences in temperature or lifetime of the various systems. Even in these cases, the dependence of the radii on the pair average transverse momentum, another effect of radial expansion, is independent of multiplicity and does not depend on the system studied.

Many current interpretations of the collective signals in pp and pA data begin with assuming a non-uniform energy density formed at the earliest stages of the interaction. That non-uniformity must arise almost exclusively from fluctuations in the initial state of the proton in the case of pp and such fluctuations also contribute for pA (see [18] for a recent review). Investigating the properties of these exotic events opens a new window in the study of QCD matter under extreme conditions.

5 Summary

In conclusion, it is now clear that studies of both pp and pA collisions can serve many uses beyond simple reference systems for the "more interesting" AA data. Detailed studies of production of various particles in pA can help constrain the nuclear PDFs which are important input to interpretation of AA results. When explored in the realm of very high particle multiplicity, both pp and pA exhibit evidence for some unexpected collective phenomena. In addition to expanding the scope of studies of the formation and evolution of QGP (or "QGP-like") media, these data can provide insight on additional interesting physics, for example fluctuations in the initial state of the proton. Finally, comparing and contrasting results from the various systems provides the opportunity to explore the onset of collectivity in small systems over a broad range of conditions, which may provide insight into similar transitions in systems other than QCD matter.

References

[1] CMS Collaboration, JHEP **09**, 091 (2010)
[2] CMS Collaboration, Phys. Lett. B **718**, 795 (2013)
[3] CMS heavy ion physics results, https://twiki.cern.ch/twiki/bin/view/CMSPublic/PhysicsResultsHIN
[4] CMS Collaboration, CMS-PAS-HIN-16-021, http://cds.cern.ch/record/2264548 (2017)
[5] CMS Collaboration, CMS-PAS-HIN-16-003, http://cds.cern.ch/record/2201538 (2016)
[6] CMS Collaboration, Phys. Lett. B **759**, 36 (2016)
[7] CMS Collaboration, Eur. Phys. J. C **77**, 26 (2017)
[8] CMS Collaboration, JHEP **04**, 039 (2017)
[9] CMS Collaboration, Eur. Phys. J. C **76**, 372 (2016)
[10] CMS Collaboration, submitted to Phys. Lett. B, https://arxiv.org/abs/1702.00630 (2017)
[11] CMS Collaboration, Phys. Lett. B **724**, 213 (2013)
[12] CMS Collaboration, Phys. Rev. Lett. **115**, 012301 (2015)
[13] CMS Collaboration, Phys. Lett. B **742**, 200 (2015)
[14] CMS Collaboration, Phys. Lett. B **765**, 193 (2017)
[15] CMS Collaboration, Phys. Lett. B **768**, 103 (2017)
[16] CMS Collaboration, Submitted to Phys. Rev. Lett., https://arxiv.org/abs/1709.09189 (2017)
[17] CMS Collaboration, PAS-HIN-14-013, https://cds.cern.ch/record/1703272 (2015)
[18] W. Li, Nucl. Phys. **A967**, 59 (2017) 59, arXiv:1704.03576

https://doi.org/10.1051/epjconf/201817120001

Thermo-magnetic dependent couplings for quark matter

V. S. Timóteo[1,⋆] , *R. L. S. Farias*[2,⋆⋆] , *S. S. Avancini*[3,⋆⋆⋆] , *M. B. Pinto*[3,⋆⋆⋆⋆] and *W. R. Tavares*[3,†]

[1]*Faculdade de Tecnologia, Universidade Estadual de Campinas - UNICAMP, Limeira, SP, Brasil*
[2]*Departamento de Física, Universidade Federal de Santa Maria - UFSM, Santa Maria, RS, Brasil*
[3]*Departamento de Física, Universidade Federal de Santa Catarina - UFSC, Florianpolis, SC, Brasil*

Abstract. In this work, we improve our model by adding the strange quark and the Polyakov loop. We then build a thermo-magnetic dependence for the G and K couplings of the SU(3) PNJL model by fitting lattice QCD calculations for the average and the difference of u and d quark condensates under a strong magnetic field.

1 Introduction

NJL based interactions have been used as effective models for QCD for quite some time due to their simplicity and clear description of fundamental properties like chiral symmetry breaking and light hadron spectrum [1–4]. With development of lattice techniques, quantities like quark condensates are now calculated directly from QCD so that effective models have now a new reference to constraint their parametrization. If we take a standard parametrization of the PNJL model, with no magnetic field, to calculate the quark condensates and compare to lattice QCD simulations, we can see in Fig. 1 that not only the position of the phase transition is wrong but also its sharpness. This suggests that the effective model needs to be modified in order to describe the behavior of the condensates revealed by lattice QCD calculations. In the following we describe how we construct a thermo-magnetic coupling based on lattice QCD results.

2 Thermo-magnetic couplings

In recent letter [5], we presented a an SU(2) NJL model with a coupling which depends on the magnetic field in order to calculate the neutral pion mass at T = 0. We also considered the finite temperature case and studied thermodynamic quantities for magnetized quark matter [6] based on a thermo-magnetic dependent coupling for the SU(2) NJL model. Here we extend the same scheme to the SU(3) PNJL model in order to obtain temperature and magnetic field dependent couplings $G(eB,T)$ and $K(eB,T)$.

The first attempt to modify the NJL coupling in order to improve the description of lattice QCD calculations was made by Farias et al. [7]. They guessed an analytical formula for the coupling

⋆e-mail: varese@ft.unicamp.br (Grupo de Óptica e Modelagem Numérica - GOMNI)
⋆⋆e-mail: ricardo.farias@ufsm.edu.br
⋆⋆⋆e-mail: sidney.avancini@fsc.ufsc.br
⋆⋆⋆⋆e-mail: marcus.benghi@fsc.ufsc.br
†e-mail: william.tavares@posgrad.ufsc.br

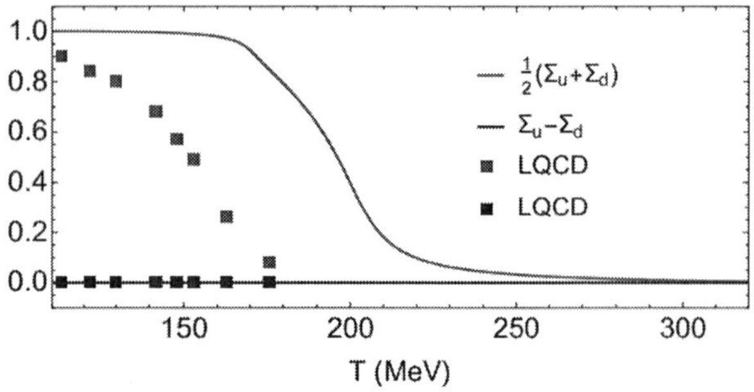

Figure 1. Average and difference of light quark condensates with no magnetic field, $B = 0$, from the PNJL model (solid lines) compared to lattice QCD results (squares).

inspired by confinement and showed an improvement in the condensates when compared to lattice results. Later we refined the idea by performing a fit for the NJL coupling to lattice calculations for different temperatures and magnetic fields [6].

For selected values of the magnetic field and temperature given by the lattice QCD collaboration [8–10], we determine the value of couplings G and K that makes the PNJL model to match lattice QCD results for the light quark condensates. Then we make a secondary fit in order to have the couplings at any temperature for any magnetic field.

3 Numerical results

We start with the $B = 0$ case: in the upper panel of Fig. 2 we show the values of the couplings G (red squares) and K (blue squares) obtained by fitting the condensates shown in the lower panel. It is remarkable that a small linear decrease in both couplings is enough to provide a perfect match between PNJL model and lattice QCD calculations.

Next we look at the case with a constant magnetic background where the PNJL model also fails to describe the behavior of the condensates. In Fig. 3 we show the results of our fit for $eB = 0.2$ GeV^2. Again the fit suggests linear behavior for the couplings but with different slope signs before and after the phase transition for both G and K.

Finally we apply the thermo-magnetic couplings to the PNJL model in the calculation of thermal susceptibilities and compare to the results obtained with the standard fixed couplings. This is shown in Fig. 4, where the thermal susceptibilities for $B = 0$ (left panels) and $eB = 0.2$ GeV^2 (right panels) have been computed with the fixed set of couplings (upper panels) and with the modified couplings (lower panels).

As the new couplings are tied to lattice QCD condensates, the PNJL description of the chiral transition becomes perfect and the thermal susceptibilities enhance, sharpen and their peaks shift from $T \sim 200$ MeV to $T \sim 160$ MeV, for both $B = 0$ and $eB = 0.2$ GeV^2 cases.

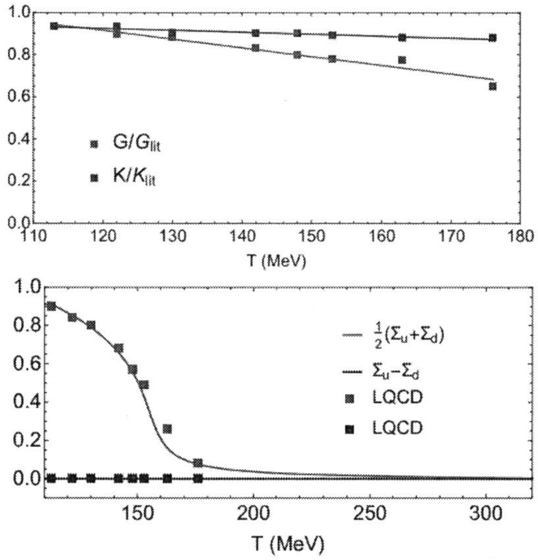

Figure 2. Temperature dependence of PNJL couplings G and K for $B = 0$ normalized by the literature values (upper panel) and the quark condensates resulting from the thermo-magnetic couplings (lower panel).

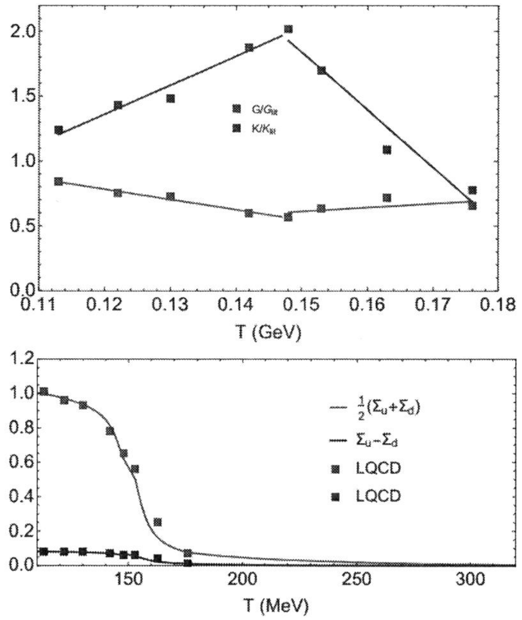

Figure 3. Temperature dependence of PNJL couplings G and K for $eB = 0.2$ GeV^2 normalized by the literature values (upper panel) and the quark condensates resulting from the thermo-magnetic couplings (lower panel).

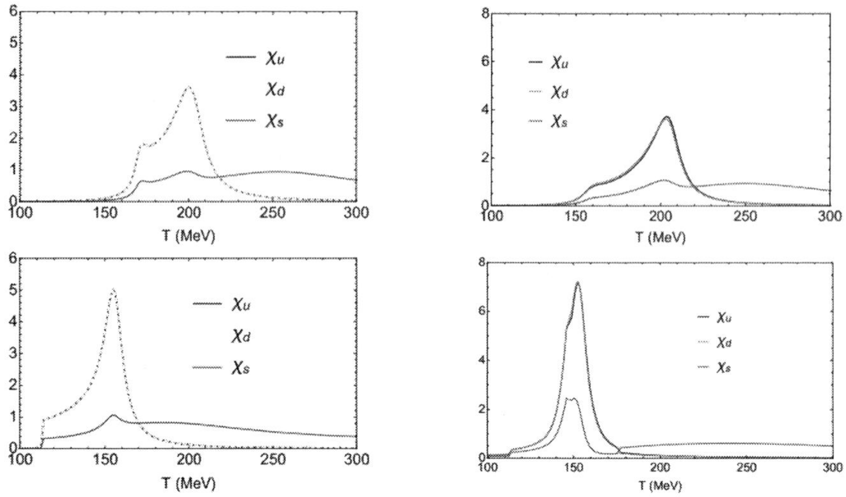

Figure 4. Thermal susceptibility χ as a function of temperature for $B = 0$ (left) and $B = 0.2$ GeV^2 (right) with constant couplings (upper panels) and thermo-magnetic couplings (lower panels).

4 Final remarks

We propose a set of thermo-magnetic couplings for the PNJL model obtained directly by fitting lattice QCD computations of the quark condensates. For $eB = 0$, we find a smooth linear decrease of both G and K while for a finite magnetic background of $eB = 0.2$ GeV^2 the couplings also behave linearly, but present change in the slope sign.

Acknowledgements

We are grateful to FAPESP, CNPq and FAEPEX for financial support.

References

[1] Y. Nambu, G. Jona-Lasinio, Phys. Rev. **122**, 345 (1961).
[2] U. Vogl, W. Weise, Prog. Part. Nucl. Phys. **27**, 195 (1991).
[3] S. P. Klevansky, Rev. Mod. Phys. **64**, 649 (1992).
[4] M. Buballa, Phys. Rept. **407**, 205 (2005).
[5] S. S. Avancini, R. L. S. Farias, M. B. Pinto, W. R. Tavares, V. S. Timóteo, Physics Letters B **767**, 247-252 (2017).
[6] R. L. S. Farias, V. S. Timóteo, S. S. Avancini, M. B. Pinto, G. Krein, Eur. Phys. J. A **53**, 101 (2017).
[7] R. L. S. Farias, K. P. Gomes, G. Krein, M. B. Pinto, Phys. Rev. C **90**, 025203 (2014).
[8] G. S. Bali, F. Bruckmann, G. Endrödi, Z. Fodor, S. D. Katz, A. Schäfer, Phys. Rev. D **86**, 071502(R) (2012).
[9] G. S. Bali, F. Bruckmann, G. Endrödi, Z. Fodor, S. D. Katz, S. Krieg, A. Schäfer, K. K. Szabó, JHEP **1202**, 044 (2012).
[10] G. S. Bali, F. Bruckmann, G. Endrödi, S. D. Katz, A. Schäfer, JHEP **1408**, 177 (2014).

Strange matter prospects within the string-flip model

Niels-Uwe F. Bastian[1,⋆], *David B.* Blaschke[1,2,3], *Mateusz* Cierniak[1], *Tobias* Fischer[1], *Mark A. R.* Kaltenborn[4,5], *Michał* Marczenko[1], and *Stefan* Typel[6,7]

[1] *Institute of Theoretical Physics, University of Wroclaw, Wroclaw, Poland*
[2] *Bogoliubov Laboratory for Theoretical Physics, Joint Institute for Nuclear Research, Dubna, Russia*
[3] *National Research Nuclear University (MEPhI), Moscow, Russia*
[4] *Department of Physics, The George Washington University, Washington DC, USA*
[5] *Los Alamos National Laboratory, Los Alamos, USA*
[6] *Institut für Kernphysik, Technische Universität Darmstadt, Darmstadt, Germany*
[7] *GSI Helmholtzzentrum für Schwerionenforschung, Darmstadt, Germany*

Abstract. In this contribution we extend the recently developed two-flavor quark-matter string-flip model by including strange quarks. We discuss implications for compact stars.

1 Introduction

The state of matter at extreme densities encountered, e.g., in the interior of neutron stars is highly unknown. In particular, the appearance of additional particle degrees of freedom, such as hyperons or the transition to the quark-gluon plasma, are challenged by the observations of pulsars with masses on the order of 2 $M_\odot$ [1–3], as they tend to generally soften the equation of state (EOS). Despite the yet unresolved hyperon problem, it has been demonstrated recently that vector repulsion between quarks in the deconfined phase provides the necessary stiffness to yield massive neutron stars of 2 $M_\odot$ or more [4–6]. Solutions of Quantum Chromodynamics (QCD) – the theory of strong interactions among quarks and gluons – at vanishing chemical potential and finite temperature predict a smooth crossover transition at $T \simeq 154 \pm 9$ MeV [7] from hadrons to quarks and gluons. For astrophysical studies at low temperatures and large densities, only phenomenological approaches are available, such as, the thermodynamic bag model [8] and models based on the Nambu-Jona-Lasinio (NJL) approach [9–11]. These models feature a first-order hadron-quark phase transition for a given hadronic EOS and take (de)confinement into account effectively. Note that perturbative QCD is valid only at extremely high densities and temperatures [12], when quarks and gluons are no longer strongly interacting. This goes far beyond the conditions encountered in astrophysical applications, e.g., neutron stars and supernovae. Recently, a new approach has been developed, which starts from an effective relativistic density-functional [13] that implements quark confinement based on the string-flip model (SFM) [14, 15]. According to the SFM, the color interactions between quarks are saturated among nearest neighbors. In hadronic matter they are confined to the same hadron, while in deconfined matter the situation is described by a string-length distribution and an effective string tension as sketched in Fig. 1.

⋆e-mail: niels-uwe@bastian.science

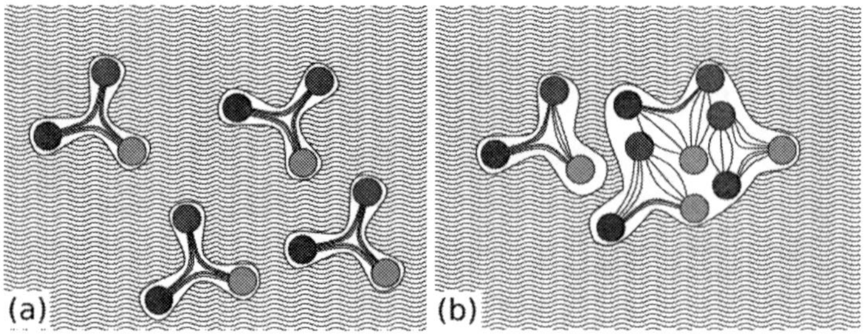

Figure 1. Illustration of the effective reduction of the string tension (density of color field lines) at high densities. At low densities (a) the field lines are compressed to thin flux tubes by the dual Meissner effect while at high densities (b) this pressure is reduced and consequently the effective string tension is lowered. Figure from [13].

2 String-flip model for two-flavor quark matter

In the mean-field approximation, the correlation energy can be obtained by folding the string-length distribution function for a given density with some interaction potential [15–17]. Moreover, the average string length between quarks in uniform matter is related to the scalar density, n_s, being proportional to $n_s^{-1/3}$. Hence, one obtains a contribution to the energy density functional of quark matter correspondingly [18]. In analogy to the Walecka model of nuclear matter [19], the relativistic density-functional approach to interacting quark matter can be obtained from the path integral approach based on the partition function [13],

$$\mathcal{Z} = \int \mathcal{D}\bar{q}\mathcal{D}q \exp\left\{\int_0^\beta d\tau \int_V d^3x \left[\mathcal{L}_{\rm eff} + \bar{q}\gamma_0\hat{\mu}q\right]\right\}, \quad \text{with } \mathrm{q} = \begin{pmatrix} q_u \\ q_d \end{pmatrix}, \tag{1}$$

with effective Lagrangian density $\mathcal{L}_{\rm eff} = \mathcal{L}_{\rm free} - U(\bar{q}q, \bar{q}\gamma_0 q)$. The interaction is given by the potential $U(\bar{q}q, \bar{q}\gamma_0 q)$, which is a nonlinear functional of the scalar and vector quark field-currents. In the mean-field approximation, this potential can be expanded around the expectation values of the field currents, n_s and n_v respectively,

$$U(\bar{q}q, \bar{q}\gamma_0 q) = U(n_s, n_v) + (\bar{q}q - n_s)\Sigma_s + (\bar{q}\gamma_0 q - n_v)\Sigma_v + \dots, \tag{2}$$

with scalar and vector self-energies, Σ_s and Σ_v. Moreover, the following density functional of the interaction is adopted,

$$U(n_s, n_v) = D(n_v)n_s^{2/3} + an_v^2 + \frac{bn_v^4}{1 + cn_v^2}. \tag{3}$$

The first term captures aspects of (quark) confinement through the density dependent scalar self-energy, $\Sigma_s = \frac{2}{3}D(n_v)n_s^{-1/3}$, defining the effective quark mass $M = m + \Sigma_s$ (see Fig. 2). We also employ higher-order quark interactions [4], by inclusion of the third term in Eq.(3), for the description of hybrid stars (neutron stars with a quark matter core) in order to obey the observational constraint of 2 $M_\odot$. To this end, the denominator in the last term of Eq.(3) guarantees that for the appropriate choice of the parameters b and c, causality is not violated (i.e., the speed of sound $c_s = \sqrt{\partial P/\partial \varepsilon}$ does not exceed the speed of light). All terms in Eq.(3) that contain the vector density contribute to the shift defining the effective chemical potentials $\mu^* = \mu - \Sigma_V$, $\Sigma_v = 2an_v + \frac{4bn_v^3}{1+cn_v^2} - \frac{2bcn_v^5}{(1+cn_v^2)^2} + \frac{\partial D(n_v)}{\partial n_v} n_s^{2/3}$. The SFM modification takes into account the effective reduction of the in-medium string tension, $D(n_v) = D_0\Phi(n_v)$. It is understood as a consequence of the modification of the pressure on the color field lines by the dual Meissner effect, since the reduction of the available volume corresponds to the reduction of the non-perturbative dual superconductor QCD vacuum that determines the strength of the confining potential between the quarks. The reduction of the string tension is modeled via a Gaussian functional, $\Phi(n_v) = e^{-\alpha n_v^2}$, where α is a parameter for the available volume fraction. A detailed discussion of the role of the parameters a, b, c and α is given in Ref. [13].

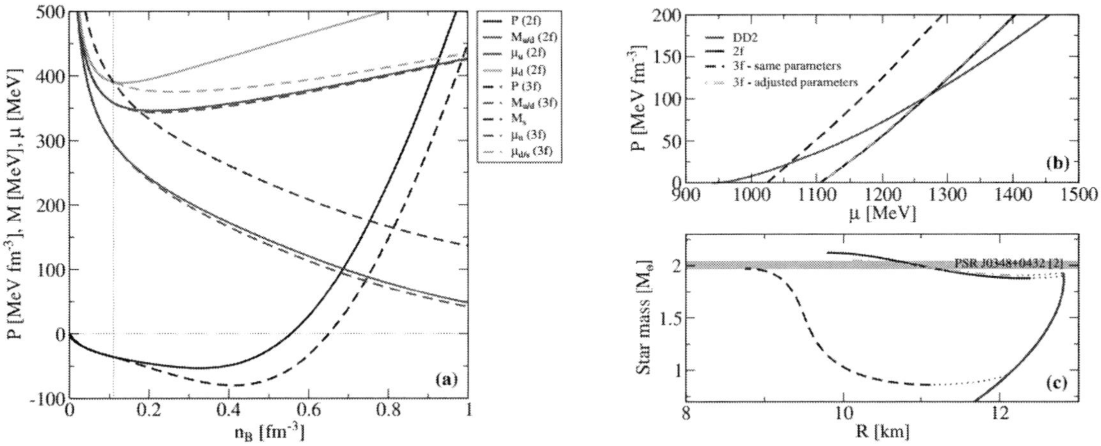

Figure 2. β-equilibrium SFM EOS for two- and three-flavor quark matter, as well as corresponding hybrid-star configurations, for the parameters $\alpha = 0.2$ fm^6, $\sqrt{D_0} = 240$ MeV, $a = -2.0$ MeV fm^3, $b = 2.0$ MeV fm^9, and $c = 0.036$ fm^6. (a) Effective quark masses, effective chemical potentials and pressure. The vertical dotted line shows the onset of strange quarks. (b) Maxwell construction of the phase transition between the DD2 hadronic and three quark matter EOS. (c) Mass-radius relation for the three resulting hybrid EOS: dotted lines indicate instability of the configuration against gravitational collapse.

3 Extension to strange matter – implications for compact stars

In order to extend the SFM to three-flavors, we extend quark spinor field, $(q_u, q_d) \longrightarrow (q_u, q_d, q_s)$. Note that flavor-dependent interactions are neglected at this stage, and, hence, all quark flavors couple and contribute to the mean fields equally. Consequently, the total scalar and vector densities of quarks are given by the following expressions, $n_v = n_v^u + n_v^d + n_v^s$ and $n_s = n_s^u + n_s^d + n_s^s$, respectively. We are interested in applications of the SFM to compact stars, where matter is in the state of β-equilibrium at zero temperature via weak processes, resulting in the relation between the chemical potentials for electrons (μ_e), up (μ_u), down (μ_d) and strange (μ_s) quarks, so that $\mu_s = \mu_d = \mu_e + \mu_u$. However, strange quarks have a significantly higher bare mass compared to the light up and down quarks. Here we assume $m_s = 100$ MeV and $m_u = m_d = 5$ MeV. The density dependence of the effective chemical potentials and effective masses is illustrated in Fig. 2 (a). This aspect is similar to chiral quark models like the NJL. The difference is in the behaviour at the onset. Up to the density where $\mu_s = m_s$, two- and three-flavor quark matter are indistinguishable. Beyond this point (vertical dashed line in Fig. 2 (a)) up- and down-quark masses drop significantly faster with density. Note further that since strange and down quarks carry the same negative charge, their chemical potentials are somewhat lower in the three-flavor case than the down quark chemical potential in the two-flavor case.

Since we work in the mean-field approximation, the SFM cannot provide the EOS of the confined phase. For the hadronic EOS we select the relativistic mean-field EOS with density-dependent couplings (DD2) from Ref. [20] (Fig. 2 (b)). The hadron-quark phase transition is obtained by a Maxwell construction. SFM parameters (given in the caption of Fig. 2) are selected such that massive hybrid star configurations are obtained with maximum masses $\gtrsim 2$ M$_\odot$ (see Fig. 2 (c)), featuring here the 'twin-phenomenon' [21, 22] where stable hadronic and hybrid star branches are disconnected by an unstable branch [23] (thin black dotted lines in Fig 2 (c)). It has recently been shown [24] that this feature can also be obtained within the multi-polytrope approach [25]. The appearance of strange quarks softens the EOS significantly, which results in the shift of the phase-transition onset to somewhat lower densities for the same SFM parameters (see Fig 2 (b)) and a lower onset mass (see Fig 2 (c)). This is compensated when re-adjusting the parameter set for the three-flavor case accordingly ($\alpha = 0.14$ fm^6, $\sqrt{D_0} = 240$ MeV, $a = -1.0$ MeV fm^3, $b = 4.0$ MeV fm^9, $c = 0.036$ fm^6).

4 Summary

Consequences of strangeness in quark matter for compact stars are studied within the SFM, a new phenomenological approach to quark-matter with an effective confining mean field. We observe an onset of strange quarks at much lower densities than the deconfinement phase transition. This leads to the simultaneous appearance of all flavours, in contrast to the commonly employed NJL-type models with sequential deconfinement [5, 26]. We have found that the inclusion of the strange quark flavor lowers the onset of the deconfinement transition, but SFM parameters can be selected such that hybrid star sequences in the mass-radius diagram of two- and three-flavor quark matter are nearly indistinguishable.

Acknowledgments This work was supported by the Polish National Science Center (NCN) under grant numbers UMO-2014/13/B/ST9/02621 (NUFB, MC, MM and MARK), UMO-2016/23/B/ST2/00720 (TF) and UMO-2011/02/A/ST2/00306 (DBB). DBB acknowledges support by the MEPhI Academic Excellence Project under contract No. 02.a03.21.0005. ST was partially supported by the DFG through grant SFB1245. This work was supported in part by the COST Actions MP1304 "NewCompStar", CA15213 "THOR" and CA16117 "ChETEC".

References

[1] P. Demorest, T. Pennucci, S. Ransom, M. Roberts and J. Hessels, Nature **467**, 1081 (2010).
[2] J. Antoniadis *et al.*, Science **340**, 6131 (2013).
[3] E. Fonseca *et al.*, Astrophys. J. **832**, no. 2, 167 (2016).
[4] T. Klähn, *et.al.*, Phys. Lett. B **654**, 170 (2007).
[5] T. Klähn and T. Fischer, Astrophs. J. **810**, 134 (2015).
[6] S. Benic, D. Blaschke, D. E. Alvarez-Castillo, T. Fischer and S. Typel, Astron. Astrophys. **577**, A40 (2015).
[7] S. Borsányi, Z. Fodor, C. Hoelbling, S. D. Katz, S. Krieg, and K. K. Szabó, Phys. Lett. B **730**, 99 (2014).
[8] E. Farhi and R. Jaffe, Phys. Rev. D **30**, 2379 (1984).
[9] Y. Nambu anf G. Jona-Lasinio, Phys. Rev. **122**, 345 (1961).
[10] S. Klevansky, Rev. Mod. Phys. **64**, 649 (1992).
[11] M. Buballa, Phys. Rept. **407**, 205 (2005).
[12] A. Kurkela, E. S. Fraga, J. Schaffner-Bielich, A. Vuorinen, Astrophys. J. **789**, 127 (2014).
[13] M. A. R. Kaltenborn, N. U. F. Bastian and D. B. Blaschke, Phys. Rev. D **96**, 056024 (2017).
[14] C. J. Horowitz, E. J. Moniz and J. W. Negele, Phys. Rev. D **31**, 1689 (1985).
[15] G. Röpke, D. Blaschke and H. Schulz, Phys. Rev. D **34**, 3499 (1986).
[16] C. J. Horowitz and J. Piekarewicz, Nucl. Phys. A **536**, 669 (1992).
[17] C. J. Horowitz and J. Piekarewicz, Phys. Rev. C **44**, 2753 (1991).
[18] A. S. Khvorostukin, V. V. Skokov, V. D. Toneev and K. Redlich, Eur. Phys. J. C **48**, 531 (2006).
[19] J. I. Kapusta, "Finite Temperature Field Theory", Cambridge University Press, Cambridge (1989).
[20] S. Typel, G. Röpke, T. Klähn, D. Blaschke and H. H. Wolter, Phys. Rev. C **81**, 015803 (2010).
[21] N. K. Glendenning and C. Kettner, Astron. Astrophys. **353**, L9 (2000).
[22] A. Bhattacharyya, S. K. Ghosh, M. Hanauske and S. Raha, Astron. Astrophys. **418**, 795 (2004).
[23] M. G. Alford, S. Han and M. Prakash, Phys. Rev. D **88**, no. 8, 083013 (2013).
[24] D. E. Alvarez-Castillo and D. B. Blaschke, Phys. Rev. C 96, 045809 (2017)
[25] K. Hebeler, J. M. Lattimer, C. J. Pethick and A. Schwenk, Astrophys. J. **773**, 11 (2013).
[26] D. Blaschke, F. Sandin, T. Klahn and J. Berdermann, Phys. Rev. C **80** (2009) 065807

Muon bundles from the Universe

P. Kankiewicz[1], M. Rybczyński[1,*], Z. Włodarczyk[1], and G. Wilk[2]

[1] Institute of Physics, Jan Kochanowski University, 25-406 Kielce, Poland
[2] National Centre for Nuclear Research, Department of Fundamental Research, 00-681 Warsaw, Poland

Abstract. Recently the CERN ALICE experiment, in its dedicated cosmic ray run, observed muon bundles of very high multiplicities, thereby confirming similar findings from the LEP era at CERN (in the CosmoLEP project). Significant evidence for anisotropy of arrival directions of the observed high multiplicity muonic bundles is found. Estimated directionality suggests their possible extragalactic provenance. We argue that muonic bundles of highest multiplicity are produced by strangelets, hypothetical stable lumps of strange quark matter infiltrating our Universe.

1 Introduction

This talk is based on [1], where the relevant details can be found. Cosmic ray physics is our unique source of information on events in the energy range which will never be accessible in Earth-bound experiments [2, 3]. This is why one of the most important aspects of their investigation is the understanding of the primary cosmic ray (CR) flux and its composition. In this respect the recent measurement performed by the ALICE experiment at CERN LHC in its dedicated cosmic ray run [4, 5] is of great importance. A number of events with muon bundles of high multiplicity was registered in the so called Extensive Air Showers (EAS) produced by cosmic ray interactions in the upper atmosphere [4]. A special emphasis has been given to the study of high multiplicity events containing more than 100 reconstructed muons. Similar events have already been studied in the previous LEP experiments at CERN (in the so called CosmoLEP program) such as ALEPH, DELPHI and L3.

In this talk we shall concentrate mainly on the first, high multiplicity events. To describe them we propose seriously to consider, for a moment, the possibility of the existence in the flux of incoming CR a component with very high atomic number, of the order of $A \sim 10^3$. In fact we propose to return to our old idea that muon bundles of extremely high multiplicities could be produced by *strangelets*, hypothetical stable lumps of strange quark matter (SQM) infiltrating our Universe [6]. Strangelets with such masses, much larger than the masses of ordinary nuclei, could easily produce extremely large groups of muons in collisions with the atmosphere.

2 Searches for Strange Quark Matter in cosmic rays

The idea of SQM originated some time ago in Ref. [7], but it remains alive, cf. [8]. In short, the basic, commonly accepted view is that that SQM (understood as a combination of roughly an equal number

*e-mail: maciej.rybczynski@ujk.edu.pl

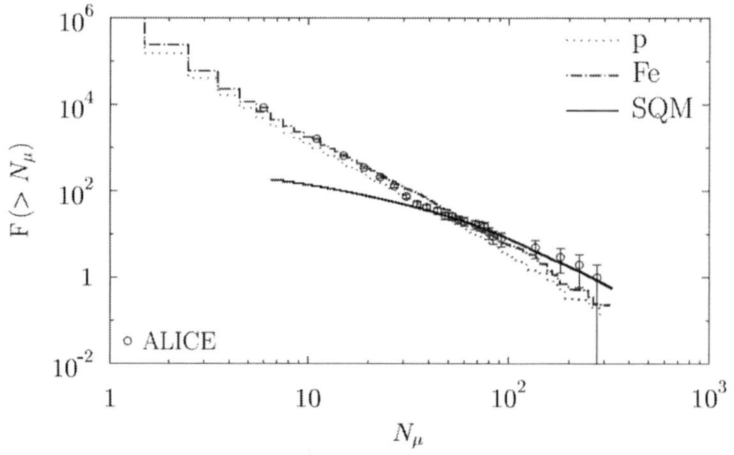

Figure 1. Integral multiplicity distribution of muons for the ALICE data (circles) published in [4]. Monte Carlo simulations for primary protons (dotted line); iron nuclei (dashed dot line) and primary strangelets with mass A taken from the $A^{-7.5}$ distribution (full line) with abundance of the order of $2 \cdot 10^{-5}$ of the total primary flux.

of up, down and strange quarks) might be the true ground state of quantum chromodynamics (QCD). It is therefore reasonable to expect that it exists in some form in the Universe and can be detected. This supposition resulted in a number of searches for strange stars and quark stars, in which such a form of matter would be dominant and which could therefore be a possible source of strangelets penetrating outer space [9]. The experimental data mentioned above lead to a flux of strangelets which follows the $A^{-7.5}$ behaviour, which in turn coincides with the behaviour of the abundance of normal nuclei in the Universe. Strange quark matter fills a vast gap in the distribution of all known forms of stable matter, cf. chart of nuclides presented in Ref. [10]. It places itself exactly between the heaviest atomic elements and neutron stars. The estimated flux of strangelets is also consistent with the astrophysical limits and with the upper limits given experimentally [11].

3 High multiplicity muon bundles from SQM

We start with a short reminder of our proposed possible scenario of propagation of strangelets through the atmosphere [12]. The apparent contradiction between the large initial size of the incoming strangelets and their required strong penetrability in the atmosphere can be resolved by assuming that strangelets reaching deeply into the atmosphere are formed from the original large strangelet which loses mass in many successive interactions with air nuclei when penetrating the atmosphere. To provide numerical estimate we limit ourselves to the two most extreme pictures of the collision of a strangelet of mass number A with an air nucleus target of mass number A_t: (*i*) All quarks of A_t which are located in the geometrical intersection of the two colliding projectiles are involved and one assumes that each quark from the target interacts with only one quark from the strangelet; (*ii*) All quarks from both nuclei which are in their geometrical intersection region participate in the collision. In the first case, during the interaction up to $(3A_t)$ quarks from the strangelet could be used up and its mass could drop to a value of $A - A_t$, at most. The second case is an analogue of the so called tube model used occasionally in nuclear collisions. After the collision the atomic mass of the strangelet diminishes to a value equal to $A - A^{1/3} \cdot A_t^{2/3}$. While this is a rather extreme variant, it is still useful in providing an estimate of the maximum possible destruction of the quarks in the strangelet.

In this study we have used suitably modified SHOWERSIM [13] modular software. We performed Monte Carlo simulations for primary nuclei composed of protons and iron nuclei and for primary strangelets with mass A taken from the $A^{-7.5}$ distribution. The results of our simulations are shown in Fig. 1. Note that whereas the lower and medium multiplicities can be reproduced by the ordinary nuclei, the extremely large groups of muons can be described only by allowing (a relatively minute, of the order of 10^{-5} of the total primary flux) admixture of SQM of the same total energy.

4 Anisotropy of arrival directions of strangelets

The ALICE data also turns out to be very valuable for another reason. Namely, they show the angular distribution of the muon events in the spherical reference frame with zenith angle (θ) and azimuth angle (Φ).

Taking into account the orientation of the axes, these coordinates can be transformed into a horizontal reference frame: $(\Phi, \theta) \rightarrow (Az, h)$, where Az is the Azimuth used in navigation/astronomy (in this particular case: $Az = \Phi + \alpha_0$), and h is the elevation above the horizon (where $h = 90° - \theta$). Finally, we can transform the coordinates from the horizontal to the equatorial reference frame: $(Az, h) \rightarrow (\alpha, \delta)$. Such a system of spherical coordinates is often used in astronomy and is expressed by right ascension (α) and declination (δ).

Using the usual spherical triangle formulas or rotation matrices one can convert between the horizontal and equatorial coordinates. For the five high-multiplicity muon events we obtained the celestial equatorial coordinates $(\alpha_{2000}, \delta_{2000})$ and the estimated limits of their positional errors are $\sim 10°$ [1]. The centroid (average location) of the five considered events is marked by a star in Fig. 2. The coordinates of this centroid are: $\alpha_{centr} = 11\ h\ 15\ m$, $\delta_{centr} = 39°\ 29'$.

We can conclude that the distribution of the directions of these five events suggests their possible extragalactic provenance. Most of them are located in the vicinity of the galactic north pole, far from the galactic plane. If we temporarily assume that all these events have the same source, the most probable source would be the blazar Mrk 421 which is located close to the centroid of the five events. It is one of the brightest blazars known, with major outbursts, active at all wavebands [14]. However, there are also many other known extragalactic high energy sources located close to the coordinates of our events [14], [15] and, at the moment, they cannot be excluded as potential sources.

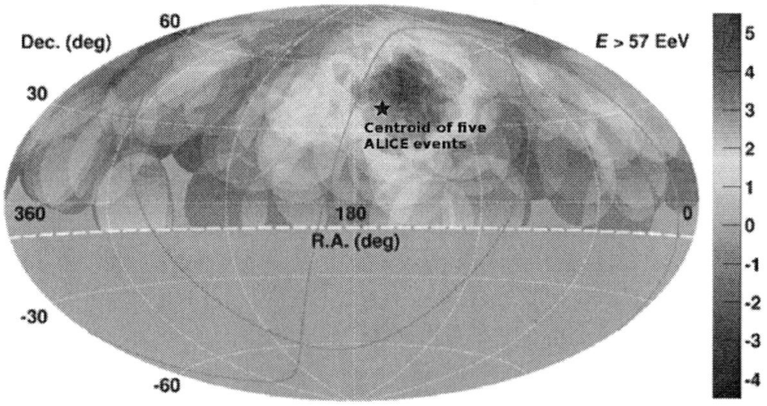

Figure 2. Aitoff projection of the UHECR map in equatorial coordinates taken from Telescope Array Collaboration data [16]. Location of the centroid of five ALICE events is marked with a star.

5 Conclusions

There is an ongoing discussion concerning the possibility of the existence in the Universe of some stable forms of SQM and the feasibility of its detection. Whereas until now we do not have fully convincing observations of SQM, nevertheless there are numerous indications of the possible existence of strange stars, or even of the existence of quark stars. Strangelets could originate, for example, from violent collisions between such objects. The recent results of ALICE reporting the existence of bundles of high-multiplicity muons could serve, notwithstanding the weakness of this signal, as a direct (almost) signal of stable lumps of SQM called strangelets arriving at the Earth from outer space. As a result, we have found that whereas the low multiplicities of muon groups measured by the CERN ALICE experiment favour light nuclei as primaries and the medium multiplicities show behaviour specific for heavier primaries, the muon groups (bundles) of really high multiplicities (of the order of ~ 100) apparently cannot be described by the common interaction models. We have shown here that the situation can be rescued by allowing for a relatively small (of the order of 10^{-5} of the total primary flux) admixture of strangelets of the same total energy. Our estimation of their flux does not contradict the results obtained recently by the SLIM Collaboration [11]. The ALICE data allowed us to estimate the directionality of the five events of the highest muon multiplicities, and to identify their most probable extragalactic source(s).

Acknowledgments

This research was supported by the Polish National Science Centre, grant 2016/23/B/ST2/00692 (MR) and contract 2016/22/M/ST2/00176 (GW).

References

[1] P. Kankiewicz, M. Rybczynski, Z. Włodarczyk and G. Wilk, Astrophys. J. **839**, no. 1, 31 (2017).
[2] A. Dar and A. De R ujula, Phys. Rep. **466** 179 (2008).
[3] A. Letessier-Selvon and T. Stanev, Rev. Mod. Phys. **83** 907 (2011).
[4] J. Adam *et al.* [ALICE Collaboration], J. Cosm. Astrop. Phys. **01** 032 (2016).
[5] J. Adam *et al.* [ALICE Collaboration], ALICE-PUBLIC-2016-003.
[6] M. Rybczyński, Z. Włodarczyk, G. Wilk, Nucl. Phys. B (Proc. Suppl.) **97**, 85 (2001).
[7] E. Witten, Phys. Rev. D **30** 272 (1984).
[8] M. G. Alford, Nucl. Phys. A **830** 385c (2009).
[9] K. S. Cheng, Z. G. Dai, D. M. Wei and T. Lu, Science **280** 407 (1998).
[10] H. J. Crawford and C. H. Greiner, Sci. Am. **270** 58 (1994).
[11] Z. Sahnoun *et al.*, Radiat. Meas. **44** 894(2009).
[12] G. Wilk and Z. Wlodarczyk, J. Phys. G **23** 2057 (1997).
[13] A. Wrotniak, SHOWERSIM/84, University of Maryland Preprint, 85 (1984).
[14] D. Horan and T. C. Weekes, New Astr. Rev. **48** 527 (2004).
[15] C. F. Turley *et al.* [Astrophysical Multimessenger Observatory Network) Collaboration], Astrophys. J. **833**, no. 1, 117 (2016).
[16] R. U. Abbasi *et al.* [Telescope Array Collaboration], Astrophys. J. **790**, L21 (2014).

Gravitational waves from binary compact star mergers in the context of strange matter

Matthias Hanauske[1,2,⋆], *Zekiye Simay* Yilmaz[1], *Christina* Mitropoulos[1], *Luciano* Rezzolla[1,2], and *Horst* Stöcker[1,2,3]

[1] *Institut für Theoretische Physik, Max-von-Laue-Straße 1, 60438 Frankfurt, Germany*
[2] *Frankfurt Institute for Advanced Studies, Ruth-Moufang-Straße 1, 60438 Frankfurt, Germany*
[3] *GSI Helmholtzzentrum für Schwerionenforschung GmbH, 64291 Darmstadt, Germany*

Abstract. In this article we will focus on the appearance of the hadron-quark phase transition and the formation of strange matter in the interior region of the hypermassive neutron star and its conjunction with the spectral properties of the emitted gravitational waves (GWs). A strong hadron-quark phase transition might give rise to a mass-radius relation with a twin star shape and we will show in this article that a twin star collapse followed by a twin star oscillation is feasible. If such a twin star collapse would happen during the postmerger phase it will be imprinted in the GW-signal.

1 Introduction

The four confirmed detections of the gravitational waves (GWs) emanated from the inward spiral and merger of pairs of black holes marked the beginning of a new era in observational astrophysics. The new field of gravitational-wave astronomy will uncover violent, highly energetic astrophysical events that could not be explored before by humankind. Without doubt, GWs from a binary compact star merger will soon be announced by the LIGO-VIRGO collaboration including an electromagnetic counterpart.[1] By analysing the power spectral density profile of the post-merger emission, the GW signal can set tight constraints on the high density regime of the equation of state (EOS) of elementary matter. The modification of the EOS due to a potential influence of a hadron-quark phase transition (HQPT) and the impact of strange quark matter on the EOS, which is currently solely probed in relativistic heavy ion collisions, might be imprinted in the post-merger phase of the emitted GW of a merging compact star binary. Hybrid star mergers represent optimal astrophysical laboratories to investigate the QCD phase structure and in addition with the observations from heavy ion collisions will possibly provide a conclusive picture on the QCD phase structure at high density and temperature [1].

2 Gravitational waves from binary compact star mergers

Numerical simulations of merging neutron star binaries in full general relativity show that the maximum (central) value of the rest-mass density $\rho_{\max}$ of the produced hypermassive neutron star (HMNS)

⋆e-mail: hanauske@th.physik.uni-frankfurt.de
[1]The article was written before the announcement of *GW170817*.

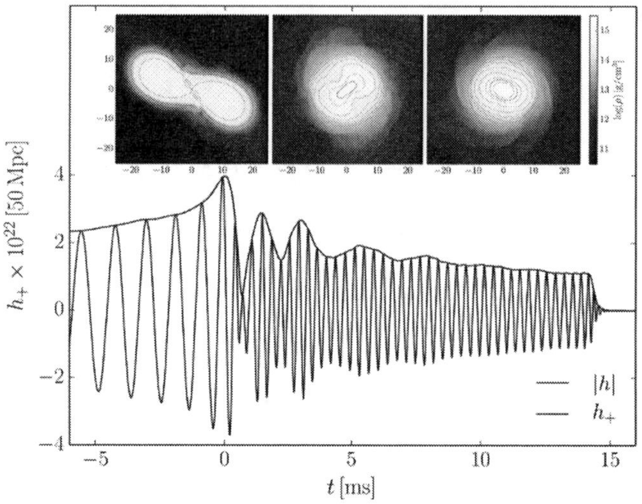

Figure 1. Logarithm of the rest-mass density profile (time snap-shots at $t = -0.17, 4.05, 13.16$ ms) and gravitational wave amplitude $|h|$ and h_+ at a distance of 50 Mpc for the ALF2-EOS with $M_{tot} = 2.7 M_\odot$. Soon after the merger ($t := 0$) the density reaches values above $\rho_{trans} = 3\rho_{nuc}$, forming a mixed phase inner region of deconfined quark matter. At $t_{BH} = 14.16$ ms the HMNS collapses and the free quark matter will be macroscopically deconfined by the event horizon of the formed rotating black hole.

increases with time until it either collapses to a black hole or reaches a quasi-stationary hydrostatic equilibrium [2]. The emitted GWs of the merger and post-merger phases are strongly determined by the high density region of the EOS reaching values $\rho_{max}/\rho_{nuc} \approx 2 - 6$ ($\rho_{nuc} := 2.705 \times 10^{14}$ g/cm^3). Fig.1 shows the GW-amplitude of an equal-mass neutron star binary merger simulation, wherein the used EOS (ALF2) has incorporated a slight phase transition to color-flavor-locked quark matter [2]. The upper panel of Fig.1 depicts the logarithm of the rest mass density profiles and the boundary of the HQPT is marked with a red curve. Although the ALF2-EOS comprises a HQPT, the properties of the resulting hybrid stars are hardly distinguishable from purely hadronic stars as the transition to the deconfined phase is very weak. So far, no simulation of a binary compact star merger containing a strong HQPT has been performed. However, the effects of a strong HQPT have been investigated in the context of static [4] and uniformly rotating hybrid stars [5] and the results show that tremendous changes in the star properties might occur including the existence of a third family of compact stars - the so called "twin stars" [6].

3 The Twin Star Collapse

The possibility of neutron star twins has been discussed in the context of different types of phase transitions including pion condensation, HQPTs and phase transitions to hypermatter [3]. A twin star behaviour is present, if the third stable sequence of compact stars is separated from the second one by an unstable region. Fig. 2 shows a typical twin star behaviour for static hybrid stars calculated using the Tolman-Oppenheimer-Volkoff equation. For the hadronic part of the EOS a DD2-RMF parametrization has been used, while the HQPT was modelled using a soft piecewise polytrope mixed phase region ($\Gamma_{MP} = 1.07 \in \rho/\rho_{nuc} = [3, 4.5]$) followed by a stiff deconfined quark phase ($\Gamma_{QP} = 5.7 \in \rho/\rho_{nuc} =]4.5, \infty[$). In [4] it was argued that the unstable region (dotted curves in Fig. 2) opens the possibility of a catastrophic rearrangement of the twin star from one configuration to the other with a prompt burst of neutrinos (with energies of about 100 MeV) followed by a gamma ray burst (with photon energies of about 1 MeV) and a total release of energy of about 10^{52} erg. However, dynamical simulations of such a twin star collapse have not yet been performed. In the following we will present for the first time the results of numerical simulations of a twin star collapse performed

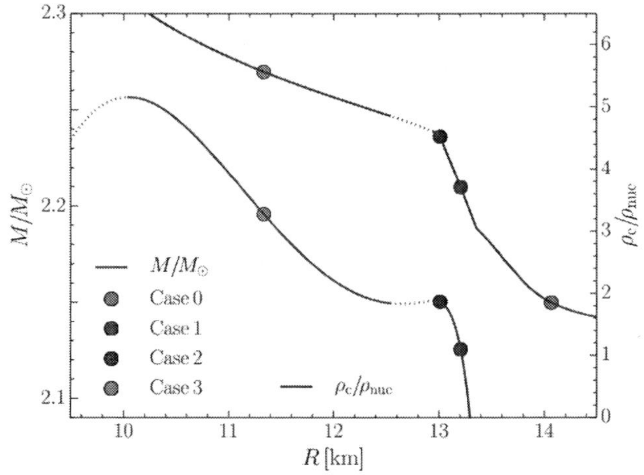

Figure 2. Mass-radius and (central rest-mass)-radius relations for a hybrid star. The underlying EOS has a strong hadron-quark phase transition implemented, where the mixed-matter phase exists in the rest-mass density range $3\,\rho_{nuc} \leq \rho \leq 4.5\,\rho_{nuc}$). The two stable "twin star" solutions are separated by a small unstable region which is visualized using dotted curves. The coloured circles display the initial configurations of the four different stars used in the underlying dynamical simulations (see Fig. 3).

in full general relativity using the `Einstein Toolkit` and the `WhiskyTHC` code. The four different initial configurations of the compact stars (coloured circles in Fig. 2) were perturbed by a radially inward directed initial velocity kick. Fig. 3 shows the time dependence of the maximum value of the rest-mass density and the minimum of the lapse function. With the exception of *Case 0* the unstable twin star region is reached during the evolution and the compact objects oscillate between their two twin star configurations. To illustrate these twin star oscillations, the evolution of the rest-mass density profile for several time snap-shots is displayed in Fig. 4. The black curve shows the profile of the initial star of *Case 2*, which is placed in the stable part of the mass-radius relation in the vicinity of the maximum mass star of the second sequence (see Fig.2). This initial star is radially perturbed using an inward directed velocity kick and as a result its density increases and the star shrinks (see solid blue curve in Fig. 4). During the collapse the density in the core reaches values $\rho/\rho_{nuc} > 4.5$ where

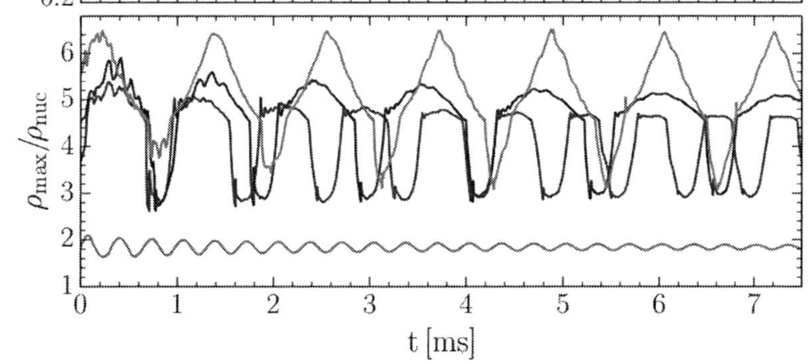

Figure 3. Minimum value of the lapse function α_{min} (upper panel) and maximum of the rest-mass density ρ_{max} (lower panel) versus time in milliseconds. The initial stars configuration has been radially perturbed in the same way for all of the different cases. Due to the low value of the initial rest-mass in *Case 0* (green curve), the interior composition of neutron star never reaches the values where the HQPT starts and as a consequence the oscillations show a regular property.

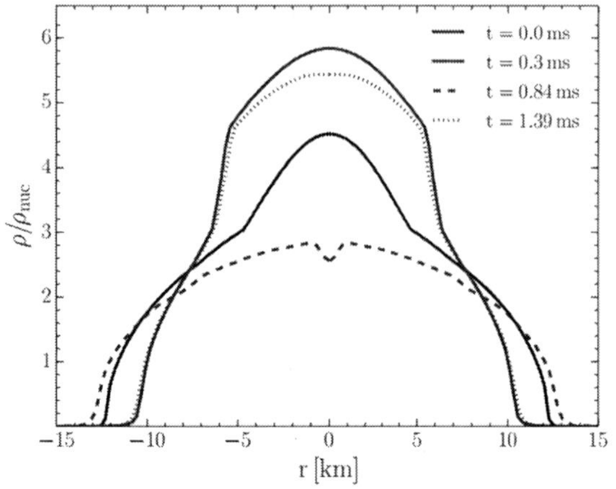

Figure 4. Rest-mass density profiles $\rho(r)/\rho_{\rm nuc}$ for four different time snap-shots calculated using the initial condition of *Case 2*. The black curve shows the initial profile which contains a mixed phase region but no pure quark phase in its inner core ($r \le 4.5$ km). The solid blue curve depicts the profile of the first maximum of the twin star oscillation and shows the structure of a hybrid star with a large pure quark phase. The dashed curve depicts the profile near the first minimum of the twin star oscillation, while the dotted curve shows the profile in the second maximum.

the stiff part of the pure quark phase generates a high pressure which counteracts against the strong gravitational attraction. As a result, the large deconfined pure quark core pushes the fluid outward, causing the first twin oscillation. The dashed curve depicts the end point of the first oscillation at $t \approx 0.84$ ms. The shape of the profile clearly shows that this star is solely composed of hadronic matter. It should be mentioned that the expected emission of neutrinos and high energetic photons or viscosity effects, which are not included in the simulations, will damp the oscillations.

4 Summary

Neutron star mergers represent optimal astrophysical laboratories to investigate the QCD phase structure using a spectrogram of the post-merger phase of the emitted gravitational waves. As gravitational waves emitted from merging neutron star binaries are on the verge of their first detection, it is important to understand the main characteristics of the underlying merging system in order to predict the expected GW signal. Numerical-relativity simulations of merging neutron star binaries show that the emitted GW and the interior structure of the generated hypermassive neutron stars depends strongly on the equation of state. The appearance of the hadron-quark phase transition in the interior region of the HMNS will change the spectral properties of the emitted GW if it is strong enough. If the unstable twin star region is reached during the "post-transient" phase, the f_2-frequency peak of the GW signal will change rapidly due to the sudden speed up of the differentially rotating HMNS.

References

[1] M. Hanauske, J. Steinheimer, L. Bovard, A. Mukherjee, S. Schramm, K. Takami, J. Papenfort, N. Wechselberger, L. Rezzolla, and H. Stöcker, Journal of Physics: Conference Series, **878**, p. 012031 (2017)

[2] M. Hanauske, K. Takami, L. Bovard, L. Rezzolla, J.A. Font, and H. Stöcker, Physical Review D, **96**, p. 043004 (2017)

[3] J. Schaffner-Bielich, M. Hanauske, H. Stöcker and W. Greiner, Phys. Rev. Lett., **89**, 171101 (2002)

[4] I.N. Mishustin, M. Hanauske, A. Bhattacharyya, L.M. Satarov, H. Stöcker and W. Greiner, Phys. Lett. B, **552**, p. 1-8 (2003)

[5] S. Banik, M. Hanauske, D. Bandyopadhyay, and W. Greiner, Phys. Rev. D, **70**, p. 123004 (2004)

[6] N.K. Glendenning, and C. Kettner, Astronomy Astrophysics, **353**, p. L9-L12 (2000)

The Small Acceptance Vertex Detector of NA61/SHINE

M. Deveaux[1,⋆], *A.* Aduszkiewicz[2], *Y.* Ali[3], *M.* Baszczyk[4], *W.* Brylinski[5], *P.* Dorosz[4], *S.* Di Luise[6], *G.* Feofilov[7], *M.* Gazdzicki[1], *S.* Igolkin[7], *M.* Jablonski[3], *V.* Kovalenko[7], *M.* Koziel[1], *W.* Kucewicz[4], *D.* Larsen[3], *T.* Lazareva[7], *P.* Martinengo[8], *A.* Merzlaya[3,7], *L.* Mik[4], *R.* Planeta[3], *A.* Snoch[1], *V.* Vechernin[7], *D.* Tefelski[5], *M.* Suljic[9], *P.* Staszel[3], and *the NA61/SHINE collaboration*

[1] *Goethe University Frankfurt, Germany*
[2] *University of Warsaw, Warsaw, Poland*
[3] *Jagiellonian University, Cracow, Poland*
[4] *AGH University of Science and Technology, Cracow, Poland*
[5] *Warsaw University of Technology, Warsaw, Poland*
[6] *Eidgenössische Technische Hochschule Zürich, Zürich, Switzerland*
[7] *St. Petersburg State University, St. Petersburg, Russia*
[8] *European Organization for Nuclear Research (CERN), Geneva, Switzerland*
[9] *University & INFN Trieste, Italy*

Abstract. Charmonium production in heavy ion collisions is considered as an important diagnostic probe for studying the phase diagram of strongly interacting matter for potential phase transitions. The interpretation of existing data from the CERN SPS is hampered by a lack of knowledge on the properties of open charm particle production in the fireball. Moreover, open charm production in heavy ion collisions by itself is poorly understood. To overcome this obstacle, the NA61/SHINE was equipped with a Small Acceptance Vertex Detector (SAVD), which is predicted to make the experiment sensitive to open charm mesons produced in A-A collisions at the SPS top energy. This paper will introduce the concept and the hardware of the SAVD. Moreover, first running experience as obtained in a commissioning run with a 150 *A*GeV/*c* Pb+Pb collision system will be reported.

1 Introduction

The NA61/SHINE [1] experiment is a fixed target hadron spectrometer located at the CERN SPS. The experiment can be operated with hadron (π, K, p) and heavy ion beams (Be, Ar, Xe, Pb). The beam energy amounts 13 – 158 *A*GeV/*c* for ions and up to 400 GeV/*c* for protons. Since 2009, NA61/SHINE is undertaking a systematic scan of the nuclear phase diagram, which relies on measurements with different beam energies and system sizes. This phase scan aims for obtaining an understanding of a potential first order phase transition between hadronic matter and QGP and the identification of the predicted critical endpoint of this phase transition.

J/ψ-melting has been proposed as a suited diagnostic probe for identifying this phase transition [2] and was soon experimentally observed at the CERN-SPS. However, a precise and quantitative measurement of the yield and properties of open charm particles is required to discriminate between

⋆e-mail: deveaux@physik.uni.frankfurt.de

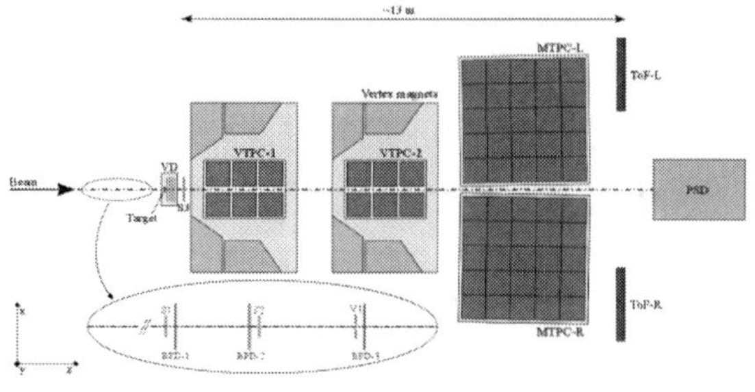

Figure 1. Sketch of the NA61/SHINE detector setup used during the Pb+Pb 2016 data taking. The position of the SAVD is shown.

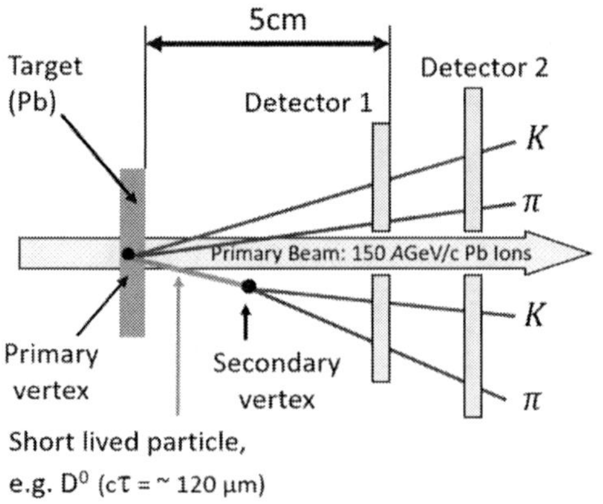

Figure 2. Reconstruction strategy for open charm (not to scale). For details, see section 2.

the onset of phase transition and alternative explanations of the findings [3]. The Small Acceptance Vertex Detector (SAVD) of the NA61/SHINE experiment was build to provide the necessary data by means of doing direct measurements of open charm mesons via their hadronic decay channel (e.g. $D^0 \rightarrow K^- + \pi^+$) at the SPS top energies.

2 Particle reconstruction strategy and detector requirements

Open charm mesons show a typical life-time of $c\tau \approx 100$ μm. Their daughter particles, namely K and π, are produced in large numbers by other sources in heavy ion collisions. To distinguish the daughter particles of open charm particles from those primary hadrons, one aims to accept only hadron pairs generated in a secondary decay vertex (see Figure 2). This requires to separate the latter from the primary vertex of the heavy ion collision, which comes with the need to measure the position of both vertices with excellent precision. This measurement is done by extrapolating the trajectories back to the production target and identifying intersection points. The primary vertex will typically appear as intersection point of multiple tracks while the number of tracks leaving a decay vertex remains small and well-defined.

The uncertainties of a vertex measurement are dominated by the multiple coulomb scattering of the particle in the detector material and by the limited position accuracy of the detector. To minimize both uncertainties, the detector has to provide an excellent spatial resolution. Moreover, it must be build from a minimal amount of material and located the closest possible to the primary vertex.

3 The SAVD

3.1 Detector concept

The SAVD of NA61/SHINE was designed to perform a vertex reconstruction with ~70 μm accuracy. It consists of two spectrometer arms composed each from four detector stations, which are located 5, 10, 15 and 20 cm downstream the target in the stray field of the vertex magnet of NA61/SHINE (see Figure 1). The number of stations was motivated by the wish for a limited stand-alone tracking capability, which eases to match the SAVD tracks with the tracks found in the TPCs. The position of the first station forms a compromise between the need to handle the high radiation load and track

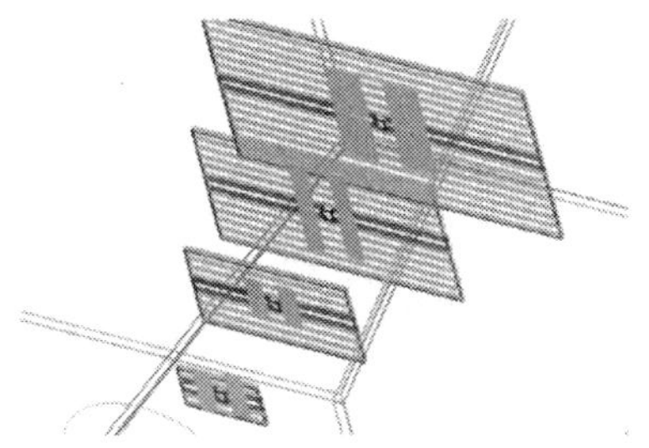

Figure 3. Active surface of the SAVD (orange) vs. the nominal geometrical acceptance of NA61/SHINE in the vertex detector region (green).

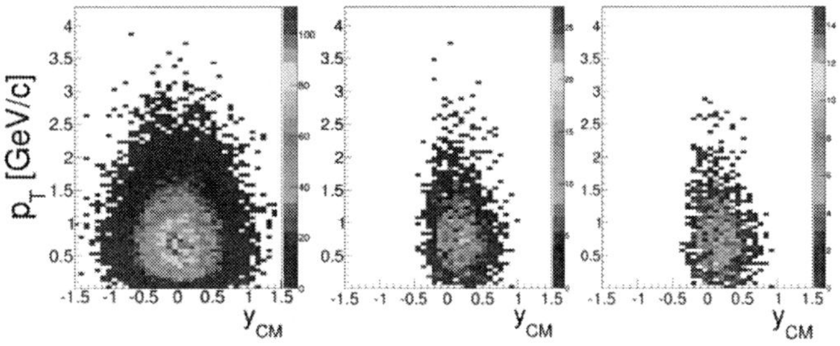

Figure 4. Simulated acceptance of the SAVD for $D^0 \rightarrow K^- + \pi^+$ for 150 AGeV/c Pb+Pb collision systems. Left: Anticipated 4π-emission, middle: full NA61/SHINE acceptance, right: SAVD acceptance.

density found nearby the collision point (which is eased by increasing the distance) and the need to measure the vertex position with good precision (which calls to reduce the distance).

The stations are equipped with the ultra-light, only 50 μm thin, MIMOSA-26AHR CMOS sensors [4, 5] developed by the IPHC Strasbourg. Those silicon sensors provide a very fine pixel pitch of 18.4×18.4 μm^2 resulting in an outstanding spatial resolution of 3.5 μm. The 1.06×2.13 cm^2 active surface of the sensor is covered by 1156 columns made each from 576 pixels. The readout of the pixel is done by means of a rolling shutter and the readout time is 115.2 μs. MIMOSA-26AHR tolerates an integrated non-ionizing dose of $> 10^{13}$ n_{eq}/cm^2 [5] and an integrated ionizing dose of > 150 krad [6].

The SAVD had to be realized with limited resources, which required to minimize rigorously its size and complexity. As illustrated in Figure 3, each of the first two layers of the SAVD hold only one sensor per arm. The third station and the last station hold two and four sensors per arm respectively. This turns into a total of 16 sensors with a surface of 32 cm^2 and 10^7 pixels, which covers only a small fraction of the geometrical acceptance of NA61/SHINE. However, thanks to the forward boost of the center-of-mass at 150 AGeV/c, it is sufficient to cover mid rapidity and a substantial part of the forward rapidity (see Figure 4). Simulations [7] suggest that the SAVD accepts geometrically 35% of all D^0 emitted into the full NA61/SHINE acceptance. During the simulations, we assumed the mean multiplicity of $D^0 + \bar{D}^0$ mesons per central (20%) 150 AGeV/c Pb+Pb collision to be 0.1. Then, the number of reconstructed $D^0 + \bar{D}^0$ mesons from 4×10^6 events amounts 1500 [8]. The S/B was estimated to amount 5 (0.5) assuming perfect (no) particle identification respectively.

3.2 Running conditions

To reduce the beam-gas interactions, the SAVD is operated in a detector vessel flooded with helium, which also contains the target. The magnetic field in the detector region is not homogeneous and amounts $B < 0.25$ T. The SAVD is exposed to radiation from three major sources, which are hadrons emitted from the primary collisions, delta-electrons kicked out of the target by the primary beam and direct beam ions. It was estimated that the two first radiation sources generate a radiation dose of $< 10^{12}$ n_{eq}/cm^2 and 25 krad during a run of 40 days. The radiation damage caused by the direct beam ions depends decisively on the beam tuning. The radiation load was estimated based on historic data from the NA61/SHINE beam counters. It was found that the most exposed point of the detector might receive $\sim 10^9$ ions per cm^2 and run. According to the Bethe-Bloch equation, this corresponds to about 230 krad. This slightly exceeds the guaranteed capabilities of the sensor, but was considered as acceptable because only few mm^2 detector surface are affected. Moreover, radiation tolerance studies

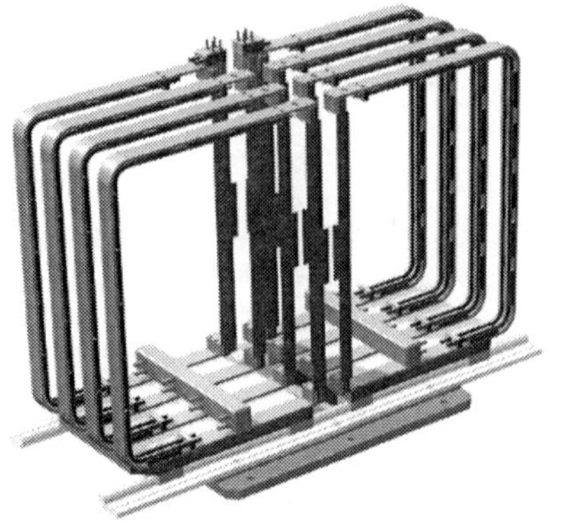

Figure 5. C-frame structure of the SAVD. The ladders, extension plates and cooling pipes are visible.

Figure 6. Photograph of the integrated detector. The sensors, extension plates, ladders and cables can be seen.

presented in [9] suggest that the reaction of MIMOSA-26AHR to modest local overdoses restricts itself likely to local losses of efficiency. The integrated non-ionizing dose was estimated to amount 4×10^{11} n_{eq}/cm^2 by extrapolating theoretical results found in [10].

3.3 Detector implementation

The construction of the detector became possible thanks to the availability of high performance building blocks, which were kindly made available by the PICSEL group of IPHC Strasbourg and the ALICE- and CBM-collaboration. The MIMOSA-26AHR sensors were integrated on 30 cm long, ultra-light, actively cooled carbon fibre ladders developed for the upgrade of the ALICE Inner Barrel [11]. The width of the ladders in the sensor region was extended by means of a thin carbon fibre plate. The sensors and the dedicated, 20 cm long single layer flex print cables used for steering them were first glued to this plate and wedge bonded hereafter. In a next step, the integrated plate was glued to ladder. This integration step relied on tools and experiences obtained from the prototype of the CBM Micro Vertex Detector (MVD) [12].

The design of the flex print cables between the sensors and the related FEE-boards was optimized to minimize the material budget of the SAVD within the acceptance of NA61/SHINE. This was done to preserve the sensitivity of the experiment outside of the SAVD acceptance. The decision was taken knowing that, unlike to newer sensors of the MIMOSA-series, MIMOSA-26AHR is somewhat sensitive to pick-up noise injected by long cables. External filter capacitors were added to alleviate this effect. The front end boards biasing the sensors, and the TRB-data processing boards [13] providing their fast and slow control, were copied from the updated [14] CBM MVD prototype. The related firmware and software [15] was also used but had to be adapted to the needs of NA61/SHINE. Among others, the intrinsically free streaming local DAQ of the MVD prototype had to be interfaced with the triggered DAQ of NA61/SHINE.

The ladders are held by dedicated C-frames made from aluminium. During the December 2016 Pb+Pb run reported in this work, they were complemented by one native ALICE-ITS ladder equipped with ALPIDE sensors. This was done in order to test the sensors to the high particle fluxes found in fixed target experiments. The C-frames of each arm are mounted on a magnet field tolerant moving table, which allows to move the sensors into a safe position during beam tuning or in case of a beam loss. This aims to avoid a direct impact of the beam into the sensors. Moreover, the moving engines allow to extend the only 6 mm wide beam gap in case necessary beam focusing is not reached. We found that the tables available on the market are too slow to react in a reasonable way to an unexpected beam displacement. Fortunately, we also observed that the sensors tolerate direct beam impacts without significant loss of performance (see next section). The target holder was mounted on a separate moving table.

A light and gas tight detector box was realized with metallized Plexiglas with beam entrance and exit windows made from aluminium foil. The cooling system relied on an industrial chiller using demineralized water, which was kept at modest underpressue. This underpressure and the reduced conductivity of the coolant was to reduce the consequences of a potential leakage in the cooling pipes on the nearby electronics.

4 First operation experience

4.1 Immunity to heavy ions

In 2015, the SAVD and its components were tested in beam times with a 30 *A*GeV/*c* Pb beam. The first beam time focused on testing the immunity of the MIMOSA-26AHR sensors to direct beam hits. This included irradiating unpowered MIMOSA-34 test sensors with up to 1.2×10^{10} 30 *A*GeV/*c* Pb-ions per cm^2 and testing them hereafter. As discussed in more detail in [16], the test sensors survived the illumination without significant radiation damage, which is is compatible with the extrapolated predictions of [10].

Independently, an individual thick MIMOSA-26 sensor was operated directly in the heavy ion beam. This test aimed to observe the response of the sensor to heavy ions and to spot the probability of latch-ups and destructive radiation damage on the sensor. Due to missing reference detectors, the experiment was not suited to observe more subtle radiation effects like moderate losses of detection efficiency. It was found that heavy ion impacts generate clusters with about 200 fired pixels. A ring shaped after-glow is sometimes observed. As the sensor was found to be encouragingly immune to ions, it was moved to the beam spot and operated at an ion rate of about 1 kHz (during spill) for 24 hours. We observed no hard error (e.g. stop of operation, reset, data corruption) of the sensor. Three latch-up events causing a modest over-current below the threshold of the related protection system were seen. They were recovered by manual power cycling.

4.2 First observations from the commissioning run

A first commissioning beam time for the SAVD with 150 *A*GeV/*c* Pb+Pb collision system was carried out in December 2016. During this beam time, the SAVD was operated together with the full NA61/SHINE detector. The trigger of NA61/SHINE was tuned to record 20% central collisions, which was not optimal for open charm search but allowed for carrying out other detector tests in parallel. A total of 1.3×10^5 central collisions (after quality cuts, preliminary) were collected.

15 of the nominal 16 sensors were found operational and the detector operated without failure or significant damage during the full beam time. One detector arm generated about 10% of corrupted events, which was associated to a version conflict within the readout and steering software.

The data obtained from the beam test is being analysed. A track finding based on a Hough transform [17] is being used for the stand alone tracking in the SAVD. Identified tracks are extrapolated to a matching plane and merged with TPC-tracks based on position and inclination. Preliminary results show that the data obtained from the SAVD and the remaining NA61/SHINE-detector can be synchronized and that track matching is possible. After a preliminary alignment of the SAVD, a spatial resolution of ~ 6 μm (hit to track residuals) was observed. This very satisfactory result allowed to measure the primary vertex with an about 30 μm precision along beam direction, which matches once more the expectations.

5 Summary and outlook

The SAVD of NA61/SHINE is to reconstruct open charm particles generated in Pb+Pb collisions at highest SPS energies. Its for stations integrate MIMOSA-26AHR CMOS sensors as provided by the IPHC Strasbourg and the ultra-light carbon fibre ladders developed for the ALICE ITS. The DAQ components and integration techniques were developed for the prototype of the CBM-detector. Despite its small active surface of 32 cm^2, the SAVD covers the forward rapidity of the fireball at the nominal beam energy of 150 *A*GeV/*c*. The detector was constructed, and tested among others in a beam time with 150 *A*GeV/*c* Pb+Pb collision system. The results confirm the feasibility of using CMOS sensor based vertex detectors in a fixed target heavy ion environment and can also be considered as a successful test of the building blocks used. Among the positive findings is the unexpectedly high immunity of the sensors to direct heavy ion hits. Moreover, a preliminary analysis of the data confirms the anticipated good spatial and vertexing resolution of the device, which provides an optimistic prospective on the success of future open charm measurements within a production run. Remaining issues like the e.g. the above mentioned software version issue are being eliminated. The improved SAVD will be used and re-tested in a Xe-La beam time end 2017 before performing a high statistics measurement in a 150 *A*GeV Pb+Pb collision system (scheduled for end 2018). Hereafter, it is planned to decommission the SAVD and to replace it by a full-size vertex detector (Large Acceptance Vertex Detector, LAVD) during the long shut down two of the CERN SPS.

Acknowledgements

This work became possible thanks to know-how and support from the ALICE-collaboration, the CBM-collaboration, the TRB-collaboration, and the PICSEL group of IPHC Strasbourg. We wish to thank P. Klaus, J. Michel and M. Wiebusch for valuable help and training on the CBM - MVD prototype DAQ. This work was supported by the Polish National Center for Science (2014/15/B/ST2/02537). M. Deveaux is supported by HIC for FAIR.

References

[1] N. Abgrall et al. (NA61/SHINE), JINST **9**, P06005 (2014)
[2] H. Satz, Nucl.Phys. **B326**, 613 (1989)
[3] H. Satz, Adv. High Energy Phys. **2013**, 242918 (2013)
[4] A. Dorokhov et al., Nucl. Instr. and Meth. **A650**, 174 (2011)
[5] M. Deveaux et al., Journal of Instrumentation **6**, C0 (2011)
[6] M. Koziel et al., Nucl. Instr. and Meth. **A624**, 437 (2010)
[7] Y. Ali, P. Staszel et al., Journal of Physics: Conference Series **509**, 012083 (2014)
[8] NA61/SHINE, CERN-SPSC-2015-038 / SPSC-P-330-ADD-8 (2015)
[9] D. Deoring, Ph.D. thesis, Goethe University Frankfurt (2014)
[10] M. Xapsos et al., IEEE TNS **51**, 3250 (2004)
[11] B. Abelev et al. (ALICE), J. Phys. **G41**, 087002 (2014)
[12] M. Koziel et al., Nucl. Instr. and Meth. **732**, 515 (2013)
[13] J. Michel et al., IEEE TNS **58**, 1745 (2011)
[14] P. Klaus et al., JINST **11**, C03046 (2016)
[15] B. Milanovic et al., PoS **TIPP2014**, 404 (2014)
[16] NA61/SHINE, CERN-SPSC-2016-038 / SPSC-SR-197 (2016)
[17] A. Merzlaya et al., Journal of Physics: Conf. Series **798**, 012072 (2017)

Heavy Ion Acceleration at J-PARC

Susumu SATO[1,2,] for the J-PARC-HI (J-PARC Heavy Ion) Collaboration*

[1]Advanced Science Research Center, Japan Atomic Energy Agency, Tokai-mura, Ibaraki, 319-1195 Japan
[2]J-PARC center, Tokai-mura, Ibaraki, 319-1195 Japan

Abstract. J-PARC, the Japan Proton Accelerator Research Complex, is an accelerator, which provides a high-intensity proton beam. Recently as a very attractive project, the acceleration of heavy ions produced by supplementary ion sources, called *J-PARC-HI*, is seriously contemplated by domestic as well as international communities. The planned facility would accelerate heavy ions up to U^{92+} with a beam energy 20 AGeV ($\sqrt{S_{NN}}$ of 6.2 AGeV). The highlight of the J-PARC-HI project is its very high beam rate up to ~10^{11} Hz, which will enable the study of very rare events. Taking advantage of this high intensity, J-PARC-HI will carry out frontier studies of new and rare observables in this energy region: (i) nuclear medium modification of chiral property of vector mesons through low-mass di-lepton signal, (ii) QCD critical point characterization through event-by-event fluctuation signals of particle production, (iii) systematic measurements related to the equation of state through collective flow signal or two-particle momentum correlation signal, or (iv) the search of hyper nuclei with multi strangeness including or exceeding $S = 3$. The current plan of J-PARC-HI aims to carrying out the first experimental measurements in 2025.

1 J-PARC as of today

At J-PARC, which provides one of the most intense proton beams in the world, a number of experimental programs are ongoing. J-PARC consists of a LINAC, Rapid Cycle Synchrotron (RCS), and Main Ring synchrotron (MR). In J-PARC, three experimental facilities are in operation: materials and life-science facility (MLF), neutrino facility (NU), and hadron experimental hall (HD).

#	Title of the Physics Experiments	PAC stage
E07	Systematic Study of Double Strangeness System with an	2 (data taking)
E10	Production of Neutron-Rich Lambda-Hypernuclei with the	2 (finished [4])
E13	Gamma-ray spectroscopy of light Hypernuclei	2 (finished [4])
E15	A Search for deeply-bound Kaonic Nuclear states by in-flight	2 (data taking
E19	High-resolution Search for Θ^+ Penta quark in $\pi^- p \rightarrow K^- X$	2 (finished [4])
E27	Search for a nuclear K-bar bound state $K^- pp$ in the	2 (data taking)
E31	Spectroscopic study of hyperon resonances below KN	2 (data taking)

Table 1. Highlighted PAC-approved experiment at J-PARC HD, as of January 2017 (full list [2]).

Beam commissioning started in November 2006, when all the accelerator and monitoring devices [1] of the LINAC were ready. In January 2007, the proton beam was successfully accelerated to 181 MeV, and the commissioning of RCS and MR followed. Then the production of the first beam of neutrons, muons, kaons, and neutrinos was achieved in May, September 2008, February, and April 2009, respectively. After the summer shutdown in 2010, a slow-extraction experiment (beginning with the E19 experiment) in the HD facility was aiming for operation with more than 5 kW power and more than 15% duty factor, and the fast-extraction experiment (namely T2K experiment) in the NU facility was running at a power of over 100kW, which would increase toward the design power. As of January 2017, the J-PARC physics advisory committee (PAC) approved 14 experiments at the Stage-1 (which is a partial approval to admit scientific interests), and 17 experiments at Stage-2 (which is the final approval after feasibility review and safety reviews). The latter experiments include 4 experiments, which have completed data taking and 5 experiments whose data taking is on-going. The PAC-approved proton-beam experiments can be found in Ref. [2]. Among the variety of experimental programs at the J-PARC accelerator complex [3], several highlighted experiments at the hadron experimental hall are listed in Table 1, and also described in the following subsections.

*Corresponding author: susumu.sato@j-parc.jp

1.1 Experiments at Hadron Experimental Hall with Proton Beam

• *E07 experiment* (*S* = *-2 hypernuclei*): The purpose of the experiment is a systematic study of double-strangeness nuclei with 10 times higher statistics than the previous experiment (KEK-E373). The key reaction is: $K^- + "p" \rightarrow K^+ + X^-$, where "$p$" indicates proton in the nucleus used in the emulsion target, which is also used as precise 3-dimensional tracking detector for event decay topology determination. In total during experimental periods in 2016 and 2017, some 1×10^9 K^- [per emulsions-set] × 118 [emulsions-sets] were irradiated. The data are now being analyzed.
• *E13 experiment* (*γ-ray hypernuclei*): Charge symmetry is studied by comparing the corresponding energy spacing in the mirror hypernuclei, $^4_\Lambda He$ and $^4_\Lambda H$. The key reaction for $^4_\Lambda He$ is: $K^- + {}^4He \rightarrow \pi^- + {}^4_\Lambda He$. The energy spacing between the spin-doublet bound state of $^4_\Lambda He(1^+, 0^+)$ was determined to be 1.406 ± 0.002 (stat.) ± 0.002 (syst.) MeV, by measuring γ-rays for the $1^+ \rightarrow 0^+$ transition with a high efficiency germanium detector array. By comparing with existing data for the same baryon number but with one nucleon difference (a neutron instead of a proton (i.e. $^4_\Lambda H$) which is 1.09 ± 0.02 MeV, the result provides a measurement of the charge symmetry breaking (CSB) in ΛN interaction [4].

cycle 5520msec

Beam area	LI RCS MLF MR HD NU
HD status	Beam Inhibit
Run number	75
Shot number	370313
Last shot time	17/06/28 05:32:43
MR Power	37.46 kW
MR Intensity (P3)	4.3e+13 PPP
SX Duty	39.45 %
SX spill length	2.04 sec
SX extraction efficiency	99.28 %
HD Power	36.29 kW
HD Intensity	4.2e+13 PPP

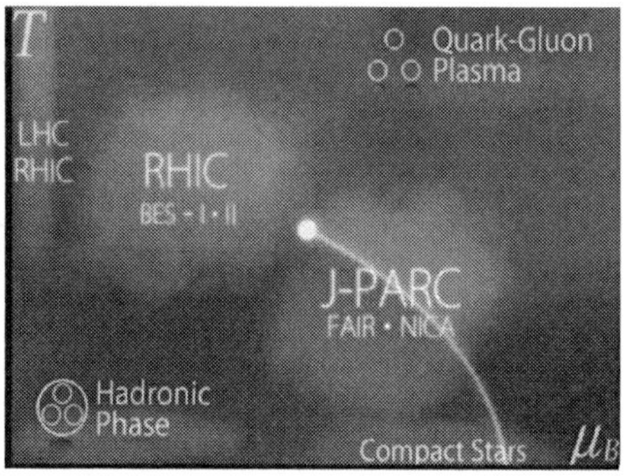

Fig. 1. (left) Example of performance achieved in slow-extraction mode with the MR synchrotron, during the stable physics running period in June 2017. **(Right)** Schematic view of the phase diagram [15] in temperature (*T*) and in baryon chemical potential (μ_B), for QCD/hadronic matter.

1.2 Accelerator Performance with Proton Beam

J-PARC is an accelerator complex currently utilizing a high intensity proton beam. In June 2017, during which stable running for more than one month were performed, 4.3×10^{13} protons per pulse for slow extraction (SX) mode were successfully provided to the hadron experimental hall. In 2016, an intensity as high as 5.9×10^{13} proton per pulse was achieved. Figure 1 shows the screenshot of the accelerator performance during the running period of data taking for the E07 experiment in 2017. In SX mode, the extraction efficiency was 99.28%, and in 2016 as high as 99.54%. Secondary Kaon beam of 2.8×10^5 [per pulse] were provided. (Also in 2017, as high as 3.1×10^5 Kaons per pulse were achieved.) In fast extraction (FX) mode, 2.44×10^{14} protons per pulse (~470 kW, and recorded in Feb. 2017) were provided.

2 Heavy Ion Physics in J-PARC

In order to survey the phase diagram of QCD in temperature and in baryon chemical potential (μ_B), a beam energy scan including lower energy is needed [5] (see Figure 1, right). By performing a beam-energy scan using heavy-ion projectiles, changes are experimentally produced in (i) initial extreme condition, (ii) development of the hydro-dynamical system (with a specific viscosity and entropy), (iii) hadronization (chemical-freeze-out) point, and (iv) kinetic freeze-out point. These data are systematically integrated to evaluate phase boundaries and a location (if any) of critical point. Also, to extract the equation of state for QCD/hadronic matter is one of the ultimate purposes of heavy-ion collision experiments including J-PARC-HI. In heavy-ion collisions at J-PARC energies (see section 2.2), the initial momenta of nucleons from nuclei are sufficiently decreased by the interaction as to overlap with each other at the colliding point ("baryon stopping"). Therefore, a medium at high μ_B can be created. The μ_B of the medium created in the collisions changes depending on the collision energy. This $\sqrt{s_{NN}}$ dependence of the collision dynamics is clearly observed in the rapidity distribution of net-proton numbers. The chemical freeze-out picture for hadron yields suggests that the maximum baryon density at the freeze-out is realized at $\sqrt{s_{NN}} = 3 - 5$ GeV [8]. This energy range is covered by J-PARC-HI. Analyses based on dynamical models (e.g. JAM model [5]) indicate that the highest baryon density of $\rho/\rho_0 > 5$ at the equilibration time is achieved at these collision energies.

2.1 Higher Interaction Rate

There are several projects, which plan to utilize heavy-ion beams in the energy region up to a few tens of GeV in $\sqrt{s_{NN}}$. At J-PARC, interaction rate of approximately 10^7 to 10^8 Hz, by using higher beam rate up to 10^{11} Hz (with typical target of 0.1% interaction length) is aiming to be utilized [6]. From those interaction rates, the observation of rare events [7] and of event-by-event fluctuation in temperature and in density [8], is expected. Furthermore, the production of rare particles exhibiting strangeness of 3 or larger is expected. By measuring those quantities, a variety of observables can be examined.

• *Chiral Property*: At higher density, a change of chiral properties occurs, and the mass modification and the yield modification are key observables to study this phenomenon (preceding experiment, KEK-E325 [4]). The invariant-mass spectrum of dileptons is directly connected to the spectral function in the vector channel. One of the most important physics goals, which can be addressed, is the in-medium modification of vector mesons.

• *Particle Production Rate*: By changing the beam energy, a variety of transition phenomena of particle-production have been measured in this energy region [9]. At the AGS, only particles/nuclei with a multiplicity (times branching ratio) larger than 10^{-2} were measured, down to antiprotons in charged hadrons, and only ${}^{3}_{\Lambda}H$ in hypernucleus. Neither dileptons nor charmed mesons were measured [7]. Assuming a beam rate of 10^{11} Hz at J-PARC and a target with thickness of 0.1% interaction length, the calculated interaction rate is 10^8 Hz. When we select 0.1% most central events, the trigger rate is 100 kHz. In one month of experiment, we expected to measure 10^7–10^9 dielectron decays of ρ, ω, ϕ, 10^5–10^6 D and J/Ψ at the beam energy of 20 AGeV, and 10^5 –10^{10} hypernuclei. Also the ratios K^+/π^+ and Λ/π^- have a peak called the "horn" at $\sqrt{s_{NN}} = 5 - 10$ GeV. There are many discussions, which relate these structures to the onset of the deconfinement phase transition.

• *Exotic Particle Production*: There are expectations for the exotic particle production with strangeness of 3 or more. In heavy-ion collisions, hypernuclei are produced both in the mid-rapidity and the target and the beam rapidity regions. In the mid-rapidity region, a few nucleons and hyperons are coalesced to form mainly a small hypernucleus. The yields in the mid-rapidity region become maximal around the J-PARC energies. In the target- and the beam-rapidity regions, fragments are embedded with hyperons produced in nucleon-nucleon collisions to form larger hypernuclei. Therefore, in heavy-ion collisions, a variety of hypernuclei may be produced with a different size, with $S = -1, -2, -3$ and so on with different proton/neutron constitutions.

2.2 Plan up to U^{92+} with 20 AGeV ($\sqrt{s_{NN}}$ = 6.2 AGeV) into the operating J-PARC

There are several merits for J-PARC (proton accelerator) to upgrade into a heavy-ion accelerator: (1) the 2 Synchrotrons (RCS and MR) are already available and functioning (2) only an injector (= "HI-Linac"+ "HI-Booster") is needed (3) the base line data for *p+p* (and *p+A*) are ready to measure (4) a variety of particle species (from lighter HI to ${}^{238}U$) can be accelerated. The plan is accelerating U^{92+} with a beam energy of 20 AGeV ($\sqrt{s_{NN}}$= 6.2 AGeV). Beam loss in RCS synchrotron, is simulated for U^{86+} ion of 4×10^{11}/cycle. The beam loss is less than 0.05% [10], and no any unexpected beam losses. And moreover, the beam loss is very localized at the ring collimator.

2.3 Large Acceptance Spectrometer and A Spectrometer for Hypernuclei Search

For systematic measurements of charged particles, a large acceptance spectrometer called JHITS with general tracking (for higher order of flow measurements, etc.) and with di-muon identification is planned [11]. It is also worth noting that a complementary program using existing proton beam on *A*-target is also possible, which is J-PARC E16 (the 2nd stage PAC- approved) experiment ($\phi \rightarrow e^+e^-$). In order to cover large acceptance both for hadrons and leptons, and to make a magnetic field free volume from the target to RICH, the spectrometer is being designed based on a toroidal magnet. In the toroidal spectrometer, there are silicon vertex trackers (SVTs) around the target, a RICH, toroidal coils, a Time-of-Flight counter (MRPC-TOF), an electro-magnetic calorimeter (PbWO4), and a muon tracker system (steel absorbers + GEM trackers). Also there are many GEM trackers between each detector. Also, a dedicated spectrometer with closed geometry is being designed for a systematic search of hypernuclei [12]. A spectrometer for hyper-nuclear studies is designed at the beam rapidity. It consists of a first dipole magnet, which sweeps out most of produced hadrons that stop at the tungsten collimator. The second dipole contains a TPC, which measures hyper-nuclear decays.

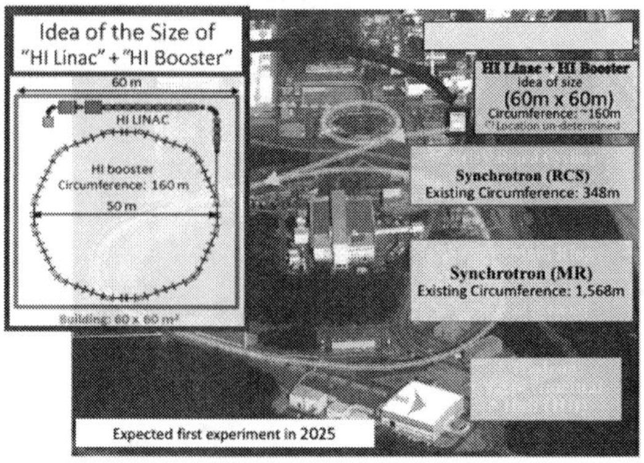

Fig. 2. An idea of size and possible location of the heavy-ion injector (= "HI-Linac" + "HI-Booster"). Note that: neither the size nor the location is determined. The size is based on [11].

2.4 Idea of the Size of "HI-Linac" + "HI-Booster"

An idea of the size and possible location of the heavy ion injectors is schematically shown in Figure 2. And please note that: neither the size nor the location are yet determined. The letter of intent was submitted to the J-PARC PAC (the 22nd PAC, in 2016 June), as "Letter of Intent for J-PARC Heavy-Ion Program [11]". The full list of international collaborations can be found in the LOI. Now, the efforts have been made toward the J-PARC's staging review, and toward the administrational approvals, for experiments starting in 2025.

3 Summary

J-PARC is currently running with the most intense proton beam at 30 GeV. HI acceleration at J-PARC is motivated partially from phase diagram study, at lower beam energy, where high-baryon density is expected. The idea to utilize the existing 2 Synchrotrons (RCS, and MR) at J-PARC, with a new compact-sized injector (HI-Linac and HI-Booster), is the one of the main merits of accelerating heavy-ion at J-PARC.

References

1. S. Sato et al., *Proc. of the 10th European Particle Acc. Conf. (EPAC)*, 1151-1153 (2006)
2. https://j-parc.jp/researcher/Hadron/en/ PAC_for_NuclPart_e.html
3. S. Sato, Nucl. Phys. A862-863 (2011) 238
4. H. Sugimura, et al., Phys. Lett. B 729 (2014) 39; T. O. Yamamoto et al., Phys. Rev. Lett. 115 (2015) 222501; K. Shirotori et al., Phys. Rev. L 109 (2012) 132002, M. Moritsu et al., Nucl. Phys. A914 (2013) 91; Y. Sada PTEP 2016 p051D01-1 (2016); R. Muto et al., Phys. Rev. Lett. 98 (2007) 042501
5. L. Kumar Nucl. Phys. A904 (2013) 256c; J. Randrup et al., Phys. Rev. C74 (2006) 047901; Y. Nara, A. Ohnishi et al., Phys. Rev. C61 (2000) 024901
6. V. D. Kekelidze et al., Prospects for the dense baryonic matter research at NICA. QM2015 (for J-PARC, assuming w/o experimental device limit)
7. FAIR Tech. Rep. (2006), A. Andronic, Phys. Lett. B697 (2011) 203 and P. B. Munzinger, J. Phys. G 21 (1995) L17
8. A. Ohnishi, 2002 Proc. of the 2nd theory W.S. on JHF *KEK PROC* 2002-13
9. A. Andronic, D. Blaschke, P. B. Munzinger, K. Fukushima et al., Nucl. Phys. A 837 (2010) 65
10. P. K. Saha et al., Proc. of 13th int'l conf. on Heavy Ion Acc. Technology, 127-129 (2015)
11. LOI for J-PARC-HI Program, 2016, http://j-parc.jp/researcher/Hadron/en/Proposal_e.html
12. H. Tamura, Reimei Workshop, 10 Aug. 2016, Tokai, Japan

AUTHOR INDEX

AUTHOR INDEX

AUTHOR INDEX